THE LIFE OF

VERTEBRATES

THE LIFE OF VERTEBRATES

BY

J. Z. YOUNG
M.A., F.R.S.
PROFESSOR OF ANATOMY AT
UNIVERSITY COLLEGE, LONDON

SECOND EDITION

OXFORD
AT THE CLARENDON PRESS
1962

Oxford University Press, Amen House, London E.C.4
GLASGOW NEW YORK TORONTO MELBOURNE WELLINGTON
BOMBAY CALCUTTA MADRAS KARACHI LAHORE DACCA
CAPE TOWN SALISBURY NAIROBI IBADAN ACCRA
KUALA LUMPUR HONG KONG

FIRST EDITION 1950
SECOND EDITION 1962

PRINTED IN GREAT BRITAIN
AT THE UNIVERSITY PRESS, OXFORD
BY VIVIAN RIDLER
PRINTER TO THE UNIVERSITY

PREFACE TO THE SECOND EDITION

FOR this edition every part of the book has been revised and corrected, but the basic plan and balance of interests have not been altered. Changes of arrangement and emphasis might have suited some types of reader but I have thought it better that the book should continue to show the idiosyncracies and interests of the author. One of the dangers of a textbook is, surely, that unsophisticated readers may suppose that they are getting the authentic and complete treatment of the subject. Some obvious imbalances may, therefore, even be an advantage as reminders of the relativity of all statements.

Nevertheless, I have attempted to make the treatment rather more complete and systematic than before. For example, the descriptions of the parts of the body are now arranged more nearly similarly for all groups. With the help of many friends mistakes have been removed and accounts of recent work added. The anatomy of Mammals is not dealt with in the same detail as that of other groups, being covered separately in *The Life of Mammals* (Clarendon Press, 1957), where also there is a fuller account of the comparative embryology of vertebrates.

During the revision I have become even more conscious of the defects of the work, both in general form and detailed treatment. It is still not possible to see more than the vaguest outlines of a proper science of comparative biology. We are faced with a great series of wonderful systems, differing slightly from each other and maintaining themselves in slightly different surroundings. But we have no proper scientific words with which to talk about them. For example, it is absurd that this book contains so little reference to genetics, biochemistry, or control theory. No doubt this is partly my fault, but the fact is that these more exact sciences have yet to show us how to treat the organization of a whole creature.

Fortunately, the animals remain as fascinating as ever, indeed the search for exact ways of describing makes them even more so. Those of us who have revised the book will be well rewarded for our trouble if others are helped to look and think for themselves. If they do they will find a really astonishing array of experiments made by natural selection with every part of the vertebrate organization. To take one example, we are offered the opportunity to learn how the endocrine control system works by examining hundreds of different variants of it. The more one thinks of it the more surprising it is that biology has made so little use of the experiments that have been done for us by nature. Surely soon someone will come along with sufficient knowledge and logical and mathematical ingenuity to show us how to study vertebrate organization.

Besides those mentioned below who have assisted in the revision of particular sections, I should like to thank the many people, including teachers and students, who have written about particular points, and especially Professor J. Lever of Amsterdam for his many detailed comments. My grateful thanks are also due to Mrs. J. Astafiev, who has redrawn many of the figures, Mr. C. Marmoy for assistance with the Bibliography, Mr. P. N. Dilly, who has helped throughout, also my secretaries and especially Miss S. Thistleton and Miss J. Everard, for continuous help with the manuscript. It is also a pleasure to thank the members of the Clarendon Press and in particular Miss M. Gregory for the help with the revision.

J. Z. Y.

February 1962

PREFACE TO THE FIRST EDITION

THE history of textbooks is often dismissed by the contemptuous assertion that they all copy each other—and especially each other's mistakes. Inspection of this book will quickly confirm that this is true, but there is nevertheless an interest to be obtained from such a study, because textbooks embody an attitude of mind; they show what sort of knowledge the writer thinks can be conveyed about the subject-matter. It may be that they are more important than at first appears in furthering or preventing the change of ideas on any theme.

The results of the studies of scholars on the subject of vertebrates have been summarized in a series of comprehensive textbooks during the past hundred years. Most of these works are planned on the lines laid down by the books of Gegenbaur (1859), Owen (1866), and Wiedersheim (1883), lines that derive from a pre-evolutionary tradition. This partly explains the curiosity that in spite of the great importance of evolutionary doctrine for vertebrate studies, and vice versa, vertebrate textbooks often do not deal directly with evolution. They derive their order from something even more fundamental than the evolutionary principle. The essential of any good textbook is that it should be both accurate and general. As Owen puts it in his Preface: 'In the choice of facts I have been guided by their authenticity and their applicability to general principles.' The chief of the principles he adopted was 'to guide or help in the power of apprehending the unity which underlies the diversity of animal structures, to show in these structures the evidence of a predetermining Will, producing them in reference to a final purpose, and to indicate the direction and degrees in which organisation, in subserving such Will, rises from the general to the particular'. He confessed 'ignorance of the mode of operation of the natural law of their succession on the earth. But that it is an "orderly succession"—and also "progressive"—is evident from actual knowledge of extinct species.'

These principles were essentially sound, and Owen's treatment was to a large extent the basis of the work that appeared after the Darwinian revolution. In English, following the translation of Wiedersheim's book by W. N. Parker (1886) we have H. J. Parker and Haswell's work, now in its 6th edition. The books of Kingsley and Neal and Rand are in essentially the same tradition, though they incorporate much new work, especially from the neurological studies of Johnston and Herrick. Further exact studies on these same general morphological lines made possible the books of Goodrich (1930) and de Beer (1935), which have provided the morphological background for the present work. Throughout these works on Comparative Anatomy the emphasis is on the evolution of the form of each organ system rather than on the change of the organization of the life of the animal as a whole.

Meanwhile many other treatises appeared dealing with the life and habits of the animals, rather than with morphological principles. Among these we may mention Bronn's *Tierreich* (1859 onwards), the *Cambridge Natural History*, and many works dealing with particular groups of vertebrates. The palaeontologists produced their own series of textbooks, mainly descriptive, such as those of Zittel and Smith Woodward, culminating in Romer's admirably detailed and concise book, to which the present work owes very much. The results of embryological work have been summarized by Graham Kerr (1919), Korscheldt and Heider (1931), Brachet (1935), Huxley and de Beer (1934), and Weiss (1939), among others. Unfortunately there has been little summarizing of what is commonly called the comparative physiology of vertebrates. Winterstein's great *Handbuch der vergleichenden Physiologie* (1912) covers much detailed evidence, but comes no nearer than do the comparative anatomists to giving us a picture of the evolution of the life of the whole organism.

All of these books deal in some way with the evolution of vertebrates, and yet curiously enough they speak of it very little. It is hardly an exaggeration to say that they leave the student to decide for himself what has been demonstrated by their studies. Huxley's *Anatomy of Vertebrated Animals* (1871) is an exception in that it deals with the animals rather than their parts, and at a more popular level. Brehm's *Thierleben* (1876) gives a picture of the life of the animals, though in this case not of their underlying organization. Kukenthal's great *Handbuch der Zoologie* has the aim of synthesizing a variety of knowledge about each animal-group, and some of the volumes dealing with vertebrates make fascinating reading—notably that of Streseman on birds. But the size of the work and the multiplicity of authors make it impossible for any general picture of vertebrate life to appear from the mass of details.

The position is, then, that we have good descriptions of the structure, physiology, and development of vertebrates, of the discoveries of the palaeontologists and accounts of vertebrate natural history, but that there is no work that attempts to define the organization of the whole life and its evolution in all its aspects. Indeed, none of these works defines what is being studied or tries to alter the direction of investigation—all authors seem prepared to agree that biological study is adequately expressed through the familiar disciplines of anatomy, physiology, palaeontology, embryology, or natural history. In passing, we may note the extraordinary fact that there are no detailed works on the comparative histology or biochemistry of vertebrates—surely most fascinating fields for the future, as is, indeed, hinted by the attempts that have been made in older works, such as that of Ranvier (1878), and the newer ones of Baldwin (1937 and 1945).

The present book has gradually grown into an attempt to define what is meant by the life of vertebrates and by the evolution of that life. Put in a more old-fashioned way, this represents an attempt to give a combined account of the embryology, anatomy, physiology, biochemistry, palaeontology, and ecology of all vertebrates. One of the results of the work has been to convince me more than ever that these divisions are not acceptable. All of their separate studies are concerned with the central fact of biology, that life goes on, and I have tried to combine their results into a single work on the way in which this continuity is maintained.

A glance through the book will show that I have not been successful in producing anything very novel—others will certainly be able to go much farther, and in particular to introduce to a greater extent facts about the evolution of the chemical and energy interchanges of vertebrates, here almost omitted! However, I have very much enjoyed the attempt, which has provided the stimulus to try to find out many things that I have always wanted to know.

For any one person to cover such a wide field is bound to lead to inexactness and error in many places. I have tried to verify from nature as often as possible, but a large amount has been copied, no doubt often wrongly. Throughout, the aim has been to provide wherever possible an idea of the actual observations that have been made, as well as the interpretations placed upon them. A proper appraisal of general theories can only be reached if there is first a knowledge of the actual materials, which is the characteristic feature of scientific observation. A book such as the present has value only in so far as it leads the reader to make his own observations and helps him to know the world for himself.

Mammalian organization requires more detailed treatment than that of other groups, and in providing this the work grew to beyond the length of a single book. Mammalian structure, function, and development will therefore be dealt with in a separate volume, which will also include a survey of comparative embryology.

The original plan was that the palaeontological parts of the book would be written by J. A. Moy-Thomas. Had he lived this aspect of the work would have been very much better, and his common sense and laughter would have lightened the whole. I have tried to give some compensation at least by the speculation that is possible from a single point of view. To protect the reader against the limitations of my ignorance I have consulted specialists on every part of the work, and my deepest thanks are due to those who have helped in this way. They have done wonders in correcting mistakes, but, of course, are not responsible for any that remain, or for views expressed. Among those who have helped in this way with particular parts are Professor G. R. de Beer, Mr. R. B. Freeman, the late Professor W. Garstang, Dr. A. Graham, Professor J. B. S. Haldane, Professor W. Hollingworth, Dr. W. Holmes, Dr. J. S. Huxley, Dr. D. Lack, Mr. Maynard Smith, Dr. F. S. Russell, Dr. Tyndell Hopwood, Mr. H. G. Vevers, Professor D. M. S. Watson, and Professor S. Westoll. They have been patient and severe critics, and the reader and I owe them very much.

One of the main problems of such a work is its illustration, and here I have been extraordinarily fortunate in having the help of Miss E. R. Turlington, who has not only provided brilliantly clear and beautiful pictures, but has taken extremes of care to ensure their accuracy by drawing from live animals, from dissections, and from skeletons, as well as by research into the illustrations of others. Miss J. de Vere has also given much help with drawing. We have borrowed good pictures unhesitatingly and should like to thank those who have given permission for their reproduction.

I should also like to thank particularly my secretary, Miss P. Codlin, who has played a large part in making the book possible, and my daughter Cordelia for help with the index.

Finally, I have to thank the Secretary and members of Oxford University Press for the care with which the book has been produced, and for their friendly co-operation, which has made the work a pleasure.

J. Z. Y.

1950

CONTENTS

ACKNOWLEDGEMENTS

My grateful thanks are due to the following, who spent much time reading and criticizing various parts of the manuscripts and proofs:

E. H. Ashton, A. d'A. Bellairs, Q. Bone, B. B. Boycott, P. M. Butler, S. Crowell, D. H. Cushing, F. C. Fraser, R. B. Freeman, H. Greenwood, I. Griffiths, R. J. Harrison, R. A. Hinde, K. A. Kermack, J. Lever, N. B. Marshall, D. R. Newth, C. Nicol, F. R. Parrington, P. Robinson, A. J. Sutcliffe, H. G. Vevers, E. I. White, M. Whitear.

Thanks for permission for reproductions of illustrations are due to:

G. C. Aymar, E. J. W. Barrington, J. Berrill, T. H. Bullock, A. J. E. Cave, W. E. Le Gros Clark, E. Crosby, F. G. Evans, Helen Goodrich, A. Gorbman, J. Gray, W. K. Gregory, W. J. Hamilton, J. E. Harris, L. Hogben, A. Holmes, E. Hoskings, W. W. Howells, J. S. Huxley, F. Knowles, D. Lack, N. A. Mackintosh, G. H. Parker, A. T. Phillipson, R. J. Pumphrey, E. C. R. Reeve, A. S. Romer, E. S. Russell, F. K. Sanders, A. H. Schultz, G. G. Simpson, N. Tinbergen, G. L. Walls, L. Waring, T. S. Westoll, E. I. White, F. F. Zeuner,

and to the following publishers and other bodies:

The American Museum of Natural History; Baillière, Tindall and Cox; E. Benn, Ltd.; *Biological Reviews*; the *British Medical Journal*; the British Museum; the Cambridge University Press; the Company of Biologists, Ltd.; Dodd, Mead & Co.; Editors of the *Ibis*; Longmans, Green & Co., Ltd.; the Physiological Society; Putnam, Ltd.; the *Scientific Monthly*; the Wilson *Bulletin*; the Wistar Institute; Zoological Society of London.

I

EVOLUTION OF LIFE IN RELATION TO CLIMATIC AND GEOLOGICAL CHANGE

1. The need for generality in zoology

THE aim of any zoological study is to know about the life of the animals concerned. Our object in this book is, therefore, to help the reader to learn as much as possible about all the vertebrate animal life that has ever been. Thinking of the great numbers of types that have existed since the first fishes swam in the Palaeozoic seas, one might well be appalled by such a task: to describe all these populations in detail would indeed demand a huge treatise. However, in a well-developed science it should be possible to reduce the varied subject-matter to order, to show that all differences can be understood to have arisen by the influence of specified factors operating to modify an original scheme. Animal and plant life is so varied that it has not yet proved possible to systematize our knowledge of it as thoroughly as we should wish. Thinking, again, of the variety of vertebrate lives, it may seem impossible to imagine any general scheme and simple set of factors that would include so many special circumstances. Yet nothing less should be the aim of a true science of zoology. Too often in the past we have been content to accumulate unrelated facts. It is splendid to be aware of many details, but only by the synthesis of these can we obtain either adequate means for handling so many data or knowledge of the natures we are studying. In order to know life—what it is, what it has been, and what it will be—we must look beyond the details of individual lives and try to find rules governing all. Perhaps we may find the task less difficult than expected. Even an elementary anatomical and physiological study shows that all vertebrates are built upon a common plan and have certain similarities of behaviour. Our object will be to come to know the nature of this plan of life, of structure, and action, to show how it is modified in special cases and how each special case is also an example of a general type of modification.

Since the problem arises from the variety of animals that have lived and live today, our central task is obviously to inquire into the reason for the existence of so much difference. If vertebrate life began as one single fish-like type, why has it not continued as such until now? Why, instead of numerous identical fishes, are there countless different kinds, while descendants of most unfish-like form are found living out of the water and even in the air and under the ground?

To put it in a way more familiar, though perhaps less clear: what are the forces that have produced the changes of animal form? Knowing these forces, and the original type, it would be possible to construct a truly general science of zoology, with sure premisses and deductions. Even if we cannot reach this end, we should at least try, hoping that after investigation of the biology of vertebrates it will be possible to retain something more than a mass of detailed information. At the end of such a study, if we deal with the subject right, we should surely be better able to answer some of the fundamental biological questions. We should be able to say something about the nature of evolution and of the differences between types, to know whether there have been rhythms of change at work to produce these differences, and also—the acid test of any true science—to forecast how these changes are likely to proceed in the future.

2. What do we mean by the life of an animal?

In biology we make much use of analogies, attempting to grasp the nature of the processes at work by comparison with man-made machines. We have a science of anatomy, which we are told is concerned with the 'structure' of animals, and we feel that we understand what 'structure' means. Physiology is the study of 'function', and this, too, we seem to understand. We take the analogies from our machines, which have what we call 'structure' and 'function'. However, the difficulty at once arises that the living things make and control themselves. The whole scheme fails us when we ask what is it, then, that we call the 'life' of the animal, and what is it that is passed on from generation to generation, and that changes through the ages by the process we call evolution? It has gradually become apparent that the body is not a fixed, definite 'structure' as it appears to casual observation or when dissected. In life there is ceaseless activity and change going on within the apparently constant framework of the body. The movement of the blood is one sign of this activity, and since Harvey's discovery of the circulation (1628) we have learnt of innumerable others. Everyone knows that the skin is continually being renewed by growth from below, and many other types of cell are similarly replaced; for instance, red blood-cells last only for a few weeks in man. Even in the cells that are not completely destroyed and replaced, such as the nerve-cells, there is continual change of the molecules that make up their substance. The full extent of this exchange has been shown by using isotopes to discover for how long individual atoms remain in the body; the work of Schoenheimer (1942), which by this means first clearly established the rapidity of the turnover, is a classic of modern biology.

There are no man-made machines that replace themselves in this way, but in recent years there has been much study of machines that control their own operations. Such work provides us with new analogies and new mathematical techniques with which we can analyse the control of living systems (see Yockey and Quastler, 1958). As yet we have no means of grasping the enormously complicated network of activities that constitutes a single life. Throughout this book, however, an attempt will be made to approach that end by use of certain clues to help us to concentrate on significant features, to see the rhythms or patterns common to the lives of the animals, and thus to carry in mind many details. It is possible in this way to bring together information collected by morphologists, geneticists, embryologists, physiologists, biophysicists, and biochemists to give a single view of the life of the organisms concerned. The task is admittedly a hard one and the success achieved only partial. Continually one slips into the discussion of particular structures, substances, or processes, forgetting the whole life. A detail of form or of chemical composition attracts, and thus distracts, attention; perhaps it can hardly be otherwise if we are to describe exactly. But it is surprising how practice improves the powers of selecting and emphasizing those patterns or details of knowledge that are significant for the study of each life as a whole.

The first difficulty is to force oneself to remember all the time that a living animal or plant system is in a continual state of change. When making any observations, whether by dissection or with the microscope, with a test-tube, microelectrode, or respirometer, it is necessary continually to think back to the time when the tissue was active in the living body, and to frame the observation so that it shall reveal something significant of that activity. This means that every biologist must know as much as possible of the life of the whole organism with which he deals; indeed, something of the whole population from which the specimen was drawn.

3. Living things tend to preserve themselves

The clue by which we recognize significant features during any biological study is that living activity tends to ensure the continuance of its own pattern. The processes of life draw materials into the system, organize them there, and then send them out again, all in such a way that the arrangement or pattern of the processes remains almost unchanged as the molecules pass through it. We see analogies in the way that a waterfall or a human institution such as the Catholic Church remains the same, though its components change. Our business is to try to describe this arrangement or pattern of processes that is preserved. It is this pattern that we call the life of the species. The activities that go to make up one sort of life are not necessarily all to be found in any one individual, still less in any part of an individual. The pattern is not to be seen in any single creature or part. Though we speak of 'individuals' they are no more the final units than are the cells, the heart, or the brain, the bones, hair, or nails. A whole interbreeding population is the unit of life that tends to preserve the type, assisted, in social species, by individuals that play a part in the life without participation in reproduction, such as worker-bees.

A wide range of activities, therefore, goes to make up any one type of life, and we shall only appreciate these activities properly if we study that whole life as it is normally lived in its proper environment. The way to study animals or men is, first and foremost, to examine them whole, to see how their actions serve to meet the conditions of the environment and to allow preservation of the life of the individual and the race. Then, with this knowledge of how the animal 'uses' its parts we may be able to make more detailed studies, down to the molecular level, and show how together the activities form a single scheme of action.

A living animal is continually doing things. Even when it is asleep it is breathing, its heart beating and brain pulsing, while countless chemical changes go on throughout its tissues. The waking life, of course, shows this restless action even more clearly. Animals may indeed sometimes be still, but they are never wholly inactive. It is not difficult to see startling glimpses of this activity if we watch animals alive, especially when they are in groups. A hawk wheeling, a pond full of tadpoles, or a crowd of people moving on a city street will remind us that if we are to see the interesting side of life we have to study activity and not, as is more easy to do and so often done, to spend all our time examining the 'structure' or 'chemistry' of the dead.

The peculiarity of this activity of animals is that much, perhaps most, of it has the effect of maintaining the integrity of the body and, indeed, even of increasing the bulk, or of reproducing more bodies like the first (homeostasis). The search for food provides raw materials giving to the muscles energy for further search. If the situation demands still greater efforts these efforts will themselves lead to 'hypertrophy', or increase in the muscle substance and power. Similarly, the muscular movements of respiration provide the oxygen by which these same movements and others are made possible.

We could go on indefinitely describing how the activity of each part of the body tends, with some exceptions, to ensure the continuation of the whole. The mere statement of the existence of this tendency to self-maintenance does not, perhaps, sufficiently emphasize the power that it represents. It is one of the great 'forces' that control the matter of the earth. It causes huge masses of material to be moved annually to the tops of high trees and millions of wonderfully built animals to roam daily to find and consume uncounted tons of food or, not finding it, to search on and maintain their activity while any calorie remains available. The power of life is sufficient to bring about the incorporation of an appreciable part of the matter of the earth's surface into living things. Within the appropriate range of conditions, found chiefly near the surface of the sea and on the damper parts of the earth, life dominates the lifeless and provides a main influence on the matter present.

Animals and plants are able to take these actions that tend to their own preservation because they contain stores of information about the conditions that are likely to be met with and the means by which adverse changes may be prevented. A fish is born with a body so shaped that it may swim, a gull can soar on air currents, and a monkey leap from branch to branch. Every type may thus be said to *represent* the environment in which it lives, that is to say, it has a hereditary store of information about it. Moreover, this hereditary store provides it with receptor organs and brain with which it can acquire further information during its lifetime. The study by engineers of the means by which information may be coded, transmitted, and stored has provided biologists with further means for study of the living memory stores, which are comparable in some ways with those of machines.

4. What do we mean by awareness of life?

A man states that he is aware that he is alive. He says that he knows his needs and that he feels satisfaction when they are fulfilled. One of the most difficult problems of biology is to decide how to relate such statements about 'subjective experience' to what may be called the 'objective' descriptions of science. This is clearly a philosophical problem too large and important to be discussed properly here but it must be approached. Perhaps it begins to find a solution when we remember that in speaking of all these matters we are using the words of a conventional code, trying with them to convey information to our fellows or somehow to influence them. Then we shall stop asking such questions as 'what is consciousness?' substituting 'what sort of information does he transmit when he says "I am conscious"?'

This will help with the particular aspect of the problem that concerns us here. In trying to define what we mean by the 'life' of an animal should we assume that, in addition to the actions of its body, which we describe, there are also actions of some other entity, its 'mind'? It is true that we should feel that any description of our own lives that left out 'awareness' was ludicrously incomplete. Since we have evolved from animals, so the argument runs, why should we deny that they have some form of 'consciousness'? This seems logical but overlooks that the essential feature of statements such as 'I am aware that I am alive' or 'I feel pain' is that they are part of the means by which man, the communicating animal, controls and influences his environment. Statements are part of human life, just as swinging from branches is a feature of the life of monkeys and flying of birds. It will at once be objected that these animals also communicate, but the point is that communication must be considered as a part of the life system of each animal, like respiration, locomotion, or reproduction.

This leaves us with the baffling problem of finding words with which to describe the describing system itself. Where indeed can we find a sure basis from which to start? Here, I think, we can only proceed by humbly admitting both ignorance and inadequacy. In our language we have a communication system with which we can convey to each other incomparably more information than passes between other animals. Our system is improving every year, but it is still grossly inadequate to describe the more subtle features of the world, and especially of living things. We may show the greatest respect for the depth of these mysteries by recognizing that they are still too great for us to describe in our simple language. To provide a good description of all the marvellous features of the life of a man or an animal requires a complicated and subtle terminology, for which we are striving. In pre-scientific language all such problems are simplified by supposing that the actions of any system are produced by some agent rather like a human being that resides within it. Thus a child says that the clouds move 'because they want to'. So we are accustomed to say that the body moves because it is guided by 'the mind'. This may indeed be the best way of speaking for some occasions, with our imperfect language, but it is a feeble descriptive technique. I do not believe that it is satisfactory for biology and especially not for zoology. By the life of an animal we mean all those activities that make a certain pattern and serve to maintain that pattern. In so far as we can describe this as a whole it is by comparing it with other self-maintaining systems and particularly with those self-controlling machines that we have made for ourselves. Biology today has a great opportunity to explore the means by which animals remain alive, using many sorts of descriptive technique, chemical, electrical and, not least, the means by which mathematicians and engineers describe whole complicated self-maintaining systems. It is in such language that a fuller and richer account of living things can be given. It is curious that objections to the use of scientific terminology often claim that it somehow 'reduces' or 'restricts' our view of life. Exactly the reverse is the case. Explaining human or animal life in terms of 'spirits', good or bad, is only describing them by comparison with themselves. Scientific description allows us to break out of our narrow prison and to show how each of the many aspects of life can be measured and compared with the forces that can be detected throughout the universe.

5. The influence of environment on life

Growth is the addition of material to that which is already organized into a living pattern. But the pattern is not fixed and invariate, even throughout any one life. Each individual changes through its lifetime, develops, as we say, and moreover is modified by the action upon it of its surroundings. Those parts that are exercised by the interaction of the animal's tendencies and the surrounding circumstances increase in amount (hypertrophy), while any disused parts undergo atrophy or reduction. The pattern is thus able to conform to a considerable extent to the exigencies of change in the external world. It could be imagined that a sufficiently plastic animal organization would be able in this way, if its tendencies to survival were strong enough, to mould itself to all the changes of climate through the millennia, so that a great variety of animal types would arise by use and disuse alone. Only a limited degree of change is possible in this way, however, and it is not such changes either of development or by the direct influence of environment that we call 'evolution'. There is abundant evidence that the result of such interaction between organism and environment is not handed on in the genetic code. Acquired characters are not inherited.

6. What is it that heredity transmit?

What is passed on is a coded pattern or plan controlling the organization of the life processes of the next generation. The plan takes the physical form of a series of molecules of deoxyribose nucleotides (DNA) in the chromosomes. These by the specific arrangements of the four types of base that they contain somehow organize the proper linear sequences of the twenty or so amino-acids that make up the proteins of each species. By the emergence at the proper time during development of the appropriate proteins, enzyme systems are produced that ensure the development and functioning of the embryo and later the adult. We cannot fully understand how all these processes are regulated but we see in outline how it all follows if the DNA molecules provide a code from which natural selection has chosen in the past those items that are suitable to provide viable organisms for a particular environment.

The organization of life is very rarely identical in any two individuals; there is, therefore, a considerable range of potential patterns resident in all those animals of a population that are capable of mating together. The sum of those variants of the hereditary materials constitutes the pattern or mould, as it were, of the life of the whole species. Evolution consists in a change in this hereditary genotype, producing, of course, a new set of adults. The genotype probably rarely stays for long quite the same. Even in species that do not seem to be changing rapidly there are continual adjustments, for example in the power to produce antibodies or to manufacture enzymes. Evolution, proceeding by mutation, recombination, and selection, is not some remote or rare thing occurring only sporadically. It is a 'physiological' process as much as is a change in respiration rate or in number of red cells, but it has a longer time course than these. Evolution is the process by which the whole population adjusts its control system to meet changing needs. Over long periods of years these adjustments produce the new forms of life that appear as, say, the first fishes, or land animals or mammals. Our aim is to try to discover the conditions under which each new main group of vertebrates arose and so to understand the processes that have been at work, modifying the basic organization.

We must therefore direct our studies continually to populations, rather than to single individuals, thinking of all the creatures of a kind, spread out wherever a suitable habitat for them occurs. They will not all be alike genetically, and the circumstances of the lives of some members of the group may become sufficiently dissimilar to produce further divergences by use and disuse. Limitations of intermating may occur on account of limitation of movement, accentuated by partial and, perhaps, eventually complete geographical barriers. Such variations in external circumstance become matched by divergences in type, until two new races are produced, at first relatively and then absolutely infertile, so that there are then two separate populations or species instead of one.

7. The increasing complexity of life

The acquisition of new matter, and hence growth and reproduction, occurred in the earliest animals by relatively simple means, as it still does today in the bacteria, lower plants, and some protozoa. It is not easy to provide rigid criteria for the definition of 'simple'; perhaps some of the chemical changes involved may be quite complex, but the whole system can, with meaning, be said to be simple. The number of parts that it contains is relatively small and the number of 'adaptive' actions that it can take is limited. A population of bacteria in a suitable culture medium obtains its raw materials by diffusion; the chief device that it uses to secure these materials is to provide a large number of spores, so that some may come to rest in suitable surroundings. Such a life can be said to be more simple than that of a vertebrate, whose system includes many special devices for obtaining access to the raw materials that it needs. We can say that a species of bacteria transmits less information than a vertebrate. Unfortunately there are no satisfactory counts of the number of genes available; the amount of DNA in bacteria is said to be about 0·05 mgm per gm and in rat liver 2 mgm per gm. Bacteria of any one species are able to alter their enzymes to suit the substrates available, but their life does not depend upon the differentiation into numerous cell types each with its special functions. The variety of information available in the 'higher' genotypes enables them to take actions that ensure survival under conditions where the 'lower' organisms would die. Of course, each type has its own special 'niche' and the comparison of higher and lower is only possible if we can show exact quantitative differences.

8. The progression of life from the water to more difficult environments

In general, the new environments colonized have involved ever wider departures from that watery one in which life first arose. This is shown most strikingly if we contrast the simple way in which the means of life are obtained by a marine bacterium with the complicated activities that go to maintain a man alive in a city. Yet all living systems, even those that have changed most markedly since their first origin, are still watery, and must have salt and nitrogenous compounds with which to make proteins and so on. Perhaps, indeed, the basic plan of the living activity differs less in the various types than one might suppose. 'Protoplasm' is certainly not identical in all creatures, but it may be that it differs less than do the outward forms that support it.

In order to provide the conditions necessary for the maintenance of such a watery system, in very different environments, many auxiliary activities have been developed. It is these that give added complexity to the higher animals and plants, enabling them to undertake what can be called more difficult ways of life. In order to do this their activity must also be physically greater than is necessary in more lowly types. It may be presumed that more energy is transferred to maintain a given mass of living matter in the less 'easy' environments, and in this sense the higher animals are less efficient than the lower, by a very crude criterion of efficiency.

According to this conception, then, evolution has involved a change in the relationship between organism and environment. Life has come to occupy places in which it did not exist before. Perhaps the total mass of living matter has thus been greatly increased. It must not, of course, be supposed that every evolutionary change has produced an increase in complexity in this way; examples of 'degeneration' are too well known to need quoting. We have, however, a clear impression that through the years there has been, in general, some change in animals and plants and that in a sense some of the later organisms are 'higher' than the earlier. It is hardly possible to deny that there is some meaning in the assertion that man is a higher animal than amoeba. Our thesis attempts to specify more clearly what we can know about this evolutionary change, by saying that it consists of a colonization by life of environments more and more different from that in which life arose. This colonization was made possible by the gradual acquisition of a store of instructions enabling adjustments to be made by which life could be maintained in conditions not tolerable before.

It is not easy to enumerate the complexity of any animal or to define quantitatively the nature of its relations with its environment, and for this reason it is difficult to prove our thesis rigorously. This book nevertheless makes an attempt to show how the organization of vertebrate life has become more complex since it first appeared, and that the increasing complexity is related to the adoption of modes of life continually more remote from the simple diffusion of substances from the sea. Of course, even the earliest vertebrates had already departed a long way from the first conditions of life and were quite complex organisms. However, in the history of their life through nearly 500 million years since the Ordovician period we can trace considerable further changes in complexity. During this time vertebrate life has left the sea to live in fresh water, on swampy land, and finally on dry land and in the air. It has produced special types able to support life by such an astonishing variety of devices that we cannot possibly specify them all. We shall only direct attention to a few, and thus attempt to obtain an impression of the scheme of life of the vast hordes of vertebrate animals, which, in one shape or another, have swarmed and still swarm in the waters and over the earth. We shall try to discern whether there is reason to suppose that all this variety is related in some way to changes in the surrounding world and we may therefore finish this introduction by a brief survey of the evidences for climatic and geographical changes such as may have been responsible for the changes in organic life.

9. Changes of climate and geological periods

9.1. *Changes of level of the continents*

Changes of geography are mostly so slow that they cannot in themselves influence individual lives. On the other hand, nearly all living things must be suited to daily and annual cyclical changes, unless they live where no light enters. There is indirect evidence of further changes in climate and geography, occurring with such long periods that they are without appreciable effect on individual organisms, but may greatly affect the history of the race.

The idea of geographical change is made familiar by the fact that coast-lines and river-courses have changed appreciably in historical times. We are familiar with stories of destruction of some houses or of a village by the sea, though it may come as a shock to learn that the sea-level has changed so much that England and France were connected by land 8,000 years ago, and that man-made instruments fished up from the Dogger Bank show that it was an inhabited peat bog 6,000 years B.C. These changes in height of the land are signs of the 'diastrophic movements', which are major features of long-period geological evolution. The earth forces that produce these movements are still obscure but they lead to repeated elevation and sinking of the land masses. The action of frost, wind, and rain continually breaks up

Acc. #84-97

and carries away the surface of the land, at a rate of the order of 1 ft per 4,000 years, the processes known as weathering and denudation. The material carried away is deposited in the river-beds and in the lakes and shallow seas around the river mouths (sedimentation) (Fig. 1). Here it builds the sedimentary rocks, which may be many thousands of feet in thickness, the whole continental platform continuing to sink for long periods, perhaps with intervals during which it becomes raised above the water. Fossil remains are usually the result of

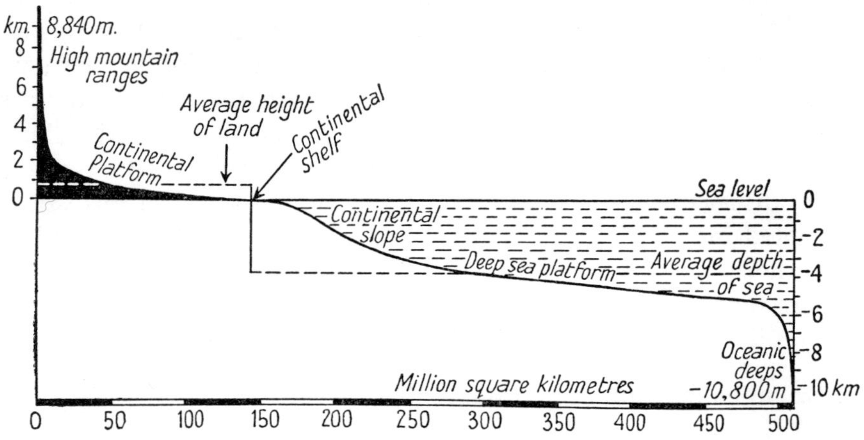

FIG. 1. Curve showing the areas of the earth's solid surface in relation to the sea level. (From Holmes.)

the preservation of the harder parts of animals in sedimentary deposits, and the most complete series of fossils are likely to be those of animals living in the seas.

The surface crust of the earth is not a layer of uniform thickness and density but consists of irregular masses of lighter material, rich in silicon and aluminium (sial), forming the continents, and heavier material, rich in magnesium (sima), under the ocean beds. The reason for this non-uniform distribution is obscure, but it has the effect of making the continents stand higher, floating on the plastic denser medium beneath the crust. When material is removed from the continents by denudation they rise; conversely the addition of millions of tons of ice will depress them. The continents are thus said to rest in isostatic equilibrium, and following the small changes in level the sea leaves more or less of the continental shelf uncovered. Such upward and downward movements profoundly influence the climate. Oceanic climatic influences tend to produce a damp, equable climate, with large areas of marsh and forest. When the land stands higher extremes of climate develop, some parts being cold, others forming large, dry interior plains.

Besides changes in the balance produced by denudation and the advance of ice-caps there are also from time to time marked movements of uplifting or lowering of the land, which may be called independent earth movements. Such vertical movements of the continental masses are produced by internal forces of unknown origin. They are doubtless related to a second series of major movements of crustal deformation that are due to tangential forces and lead to the formation of new mountain ranges (orogenesis) by compression, or to fracturing by tension. The upwelling of lava from the inside of the earth at these times makes the igneous rocks, usually devoid of fossils.

Changes in geography are, then, mainly changes in the height of the land and the amount of it that is above water. Where the continent is surrounded by a rather shallow continental shelf, this leads to considerable changes in appearance of the land-masses. The general opinion is that the main outline of the continental masses has remained much as at present, at least since Cambrian times. However there have probably been considerable movements of the land-masses in relation to each other. Some hold that the continents of lighter material are continually expanding, at least in certain directions, having grown from small centres to their present size. According to the hypothesis of Wegener, the continents have all been formed by the splitting up of one or a few land-masses. There is indeed evidence from both geophysics and biology that the continents have been drifting apart (Bullard, 1959). The direction of magnetization of rocks, which is determined at the time of their formation, shows that the land-masses must have changed their positions greatly. For example, such data show that during the Triassic period the British Isles lay in the tropics and in confirmation of this we find that many salt deposits (formed only in very warm climates) lie in the Triassic formation (Droitwich, Bath, Nantwich, &c.).

9.2. *Changes of climate*

Evidence of marked changes of climate is the finding in England and other regions now temperate of animal and plant remains appropriate to warmer or colder conditions (corals and woolly rhinoceros, for instance). There is thus every reason to think that there have been great changes from hot to cold and wet to dry conditions, in conjunction with the changes in latitude and in level of the land.

These fluctuations in geography and climate are obviously of great importance to the biologist. We can hardly expect to treat animals and plants as stable systems if the environment around them is changing. In order to be able to assess the influence of such changes on life we must know more about the rates at which they occur, and careful study shows that some of the climatic changes are rhythmic. Rhythmic changes of climate are, of course, very familiar to us in the cycles of days, months, and years, and the immense importance of these short-period changes for animal and plant life must not be forgotten.

Here we are more concerned with changes of longer periodicity, of which the best known are fluctuations of the amount of solar radiation received at any given part of the earth's surface. These are likely to be especially important since plants, and hence ultimately animals, depend for their energy on sunlight. The cycle of number of sunspots (11·4 years) involves a change in amount of radiation, and this is associated with some biological cycles, for instance in the distribution of the rings of growth made by trees. Longer-period fluctuations in the amount of radiation received on any part of the earth's surface depend on the perturbations of the earth's orbit, particularly on changes in the obliquity of the ecliptic. The effect of these perturbations can be calculated, and the results show that at any one place there are rhythmical variations in the amount of radiation received, and in its seasonal distribution. The periodicity of these calculated changes is about 40,000 years, with considerable irregularities and variations in the sizes of the maxima (Fig. 2).

During the last million years (the Pleistocene epoch) there has been a series of waves of glaciation (ice ages); the ice-caps have several times advanced towards the equator and then retreated again. These changes are usually classified into four periods of glaciation, separated by interglacial periods. However, the last (fourth Pleistocene) glaciation, of which we know the most, certainly had three separate climaxes of cold. The correspondence of these with especially marked minima in the curve of solar radiation is not perfect (Fig. 2), but it suggests that the basic periodicity may have been something like 40,000 years, and that the division of the whole Pleistocene period into four periods of glaciation obscures a change with much shorter periodicity. From about 120,000 to 180,000 years B.P. (Before Present) there were no marked minima in the solar radiation curve, and this agrees with other evidence of a long interglacial period (third Pleistocene interglacial). Two marked minima agree with the other signs of a penultimate (third Pleistocene) glaciation, and this was preceded by a very long warmer period, the second inter-glacial. As we go farther back the study becomes more and more difficult, but the available evidence suggests that fluctuations of climate considerable enough to alter the entire fauna and flora may have taken place at a periodicity of something over 40,000 years. It is a measure of the difficulty of geological science that we cannot yet give a systematic account of the chronology or climatic changes even of the relatively recent Pleistocene period (variously estimated at 600,000 to 1,800,000 years) during which these glaciations occurred.

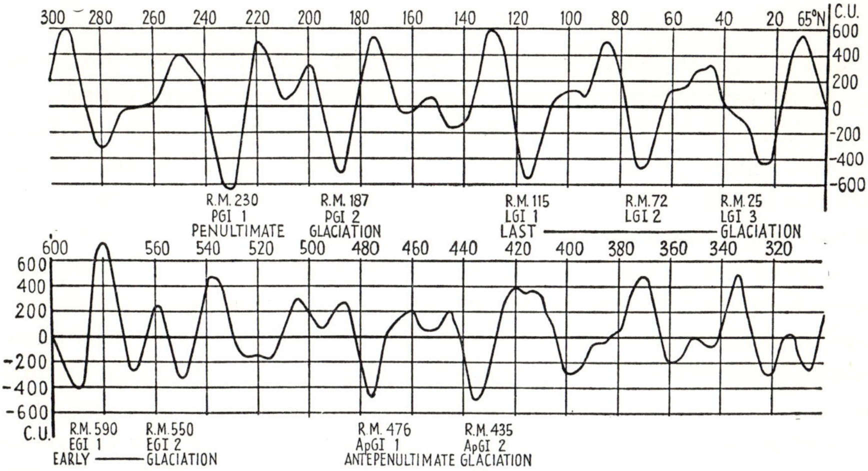

FIG. 2. Curve of solar radiation received at 65° N. lat. in the summer. The radiation is expressed in 'canonic units' (related to the solar constant in calories). Time in thousands of years. R.M. 25, &c., indicate the radiation minima. (From Zeuner, based on the tables of Milankovitch.)

As we proceed to study times still more remote our vision becomes increasingly blurred. We can now only rarely distinguish periodicities as short as 40,000 years, though there is evidence that they existed, for instance from varved Cretaceous sediments. All we can see in the study of geological deposits are the very marked changes produced by the major movements of orogenesis and by the isostatic readjustments. The surprising thing is that these immensely slow changes have been sufficiently regular to leave layered deposits, allowing the development of a system of geological classification. The process of sedimentation was interrupted by periods when the continental shelf on which the rocks rest was raised above the water surface and underwent denudation for a while, before being again lowered below the sea and covered with a new deposit. During the interval, while the shelf was raised above the water, the animals and plants in the sea became changed; thus rather sharp breaks appear in the series of fossils. The occurrence of these breaks has been used by geologists to define the major geological periods, which thus correspond to cycles of elevation and depression of the continents. By comparing the fossils contained in the rocks major geological periods have been recognized in various parts of the world. The times of submergence and emergence differ from region to region, however, and no very close detailed comparison is possible. It is easy to forget that climates and land levels do not always change in the same direction in different parts of the world.

9.3. *Geological time*

Until recently most geologists assumed that there was a regular cycle of raising and lowering (diastrophism) and that comparable periods could be recognized everywhere. It is now widely doubted whether there has been any such 'pulse of the earth'. The rock series are not the same in all the continents. For example, in South Africa three long series, known as Cape, Karoo, and Cretaceous formations, occupy the time covered in Europe by the many elevations and depressions between Silurian and Cretaceous times. Probably the conditions under which rocks were formed have remained about the same throughout geological time but have been interfered with by periods of elevation, depression, and folding that are peculiar to each region.

The study of fossils often establishes the order in which the rocks were laid down, but other methods have to be used to discover the period of time covered by each stage. This is especially important to the biologist, who wants to know the rate at which animals or plants have evolved. Reliable knowledge of the ages of the rocks has only begun to accumulate since the discovery of radioactivity. Uranium and thorium disintegrate, producing lead, at rates that are unaffected by any known conditions. The age of any rock since its deposition can therefore be calculated if we can estimate the amount of breakdown products of these elements present in it. The lead present in a rock is often not all derived from the uranium and thorium there, but separation of the lead isotopes enables those of radioactive origin to be estimated, and the age of the deposit can then be determined, assuming that the breakdown of uranium to lead began when the rock was crystallized in its present position. Other methods of estimating the ages of rocks from isotope ratios have been developed. Especially promising is the determination of the ages of the deposition of sedimentary rocks from the ratio of A^{40}/K^{40} and Sr^{87}/Rb^{87} in deposits formed by erosion of micas or granites.

The time at which the crust of the earth assumed its present form is now thought to have been 4,500 million years ago (Holmes, 1959) but the rocks laid down during the greater part of this long period contain no undoubted animal or plant remains. Cambrian rocks, when fossils become readily discernible, were laid down about 600 million years ago.

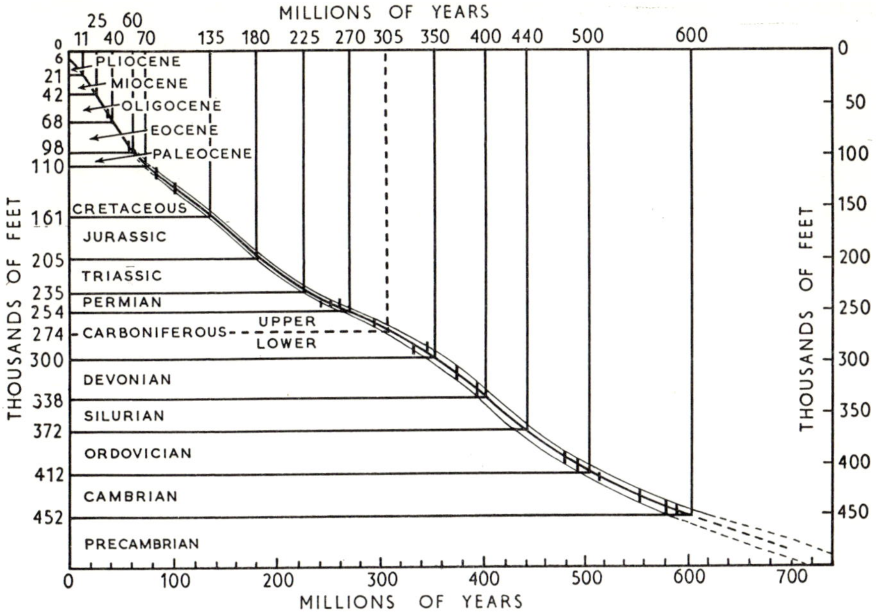

FIG. 3 shows the maximum thickness of sediment in each period plotted against estimates of the absolute date. The error attached to these determinations is shown by the marginal lines. Apparently the rate of sedimentation has not been constant (modified after Holmes).

Classical geology is based mainly on studies in Europe and North America. Although a terminology based on absolute time is beginning to emerge, it is still necessary to use that based mainly on stratigraphic studies, begun by William Smith in the British Isles early in the nineteenth century. In this system, the time since the Cambrian is divided into eleven major periods, but several of these were double or triple periods of advance and retreat of the sea. Even the most carefully compiled radioactivity data are not yet adequate to provide us with definite estimates of the durations of the periods, though there is agreement on a total period of about 600 million years since the Cambrian. Fig. 3 shows the maximum thickness of sediment in each period plotted against estimates of the absolute dates. The error attached to these determinations is shown. Apparently the rate of sedimentation has not been constant.

It is conventional to postulate a series of crustal revolutions. The extent of the movements has not been equal throughout and some of them, more marked than others, were times of building of great mountain chains such as the Alps or Andes (Fig. 4). There were also many lesser rises and falls and changes of climate with shorter periods,

TABLE I

MAXIMUM THICKNESSES AND REVISED TIME-SCALE
(ACCORDING TO HOLMES)

THICKNESSES IN THOUSANDS OF FEET.			TIME SCALE IN MILLIONS OF YEARS	
WORLD MAXIMA	CUMULATIVE MAXIMA	PERIODS	SINCE BEGINNING OF PERIOD	DURATION OF PERIOD
6	6	PLEISTOCENE	1	1
15	21	PLIOCENE	11	10
21	42	MIOCENE	25	14
26	68	OLIGOCENE	40	15
30	98	EOCENE	60	20
12	110	PALEOCENE	70 ± 2	10
51	161	CRETACEOUS	35 ± 5	65
44	205	JURASSIC	80 ± 5	45
30	235	TRIASSIC	25 ± 5	45
19	254	PERMIAN	70 ± 5	45
20 } 46 26 }	300	CARBONIFEROUS (UPPER / LOWER)	350 ± 10	80
38	338	DEVONIAN	400 ± 10	50
34	372	SILURIAN	450 ± 10	40
40	412	ORDOVICIAN	500 ± 15	60
40	452	CAMBRIAN	600 ± 20	100

such as those of about 40,000 years that we can detect in the later part of the Pleistocene. Many modern geologists are sceptical about the existence of any regularities or rhythms in these changes (see Herbert, 1952, and Gilluly, 1949). It is useful when trying to adjust the mind to periods of 30 million years to remember the frequent changes of level and climate that have occurred in the last 100,000 years. In spite of all that we know about the history of the earth's surface, it is necessary every time that we make statements about the influence of presumed climatic changes on organic evolution to remember how scanty our knowledge is.

9.4. *Classification of geological history*

The period isolated as 'Cambrian' by geologists lasted 100 million years and almost certainly included several inundations, perhaps three. The Ordovician lasted for 60 million years and included three floods in North America. There were powerful earth movements at the end of this period, at any rate in North America, known as the Taconian revolution. The Silurian, lasting for 40 million years, apparently included a single main cycle of inundation, ending in an elevation of the land, which though slight in America, was marked in Europe as the Caledonian revolution, producing the range of mountains stretching across Scandinavia to Scotland and Ireland.

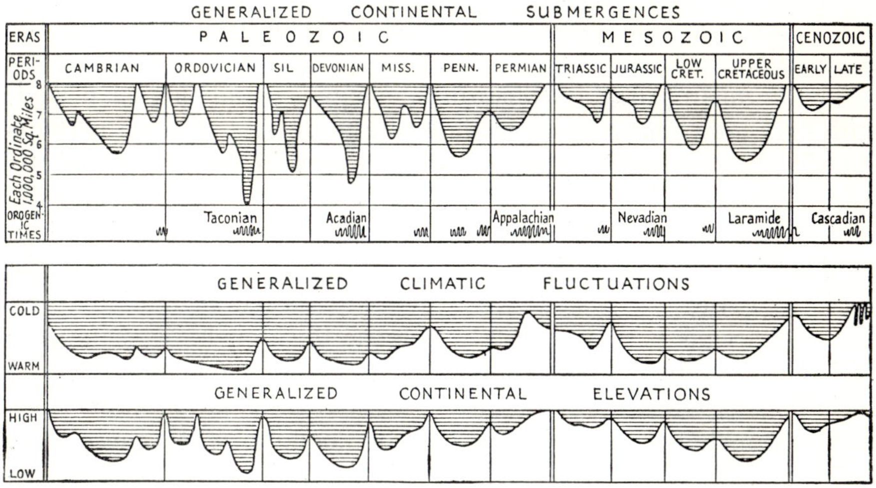

FIG. 4. Diagrams of main changes of areas of land and water and in climatic conditions since the Cambrian. The chief times of mountain-building (orogenesis) in America are also shown. (Redrawn by permission from *Textbook of General Zoology*, 2nd ed. by W. C. Curtis and M. J. Guthrie, John Wiley & Sons, Inc., 1933.)

Throughout these early Palaeozoic periods the fossils are entirely those of aquatic animals, except for some traces of land plants and arthropods at the end of the Silurian. The oldest remains of vertebrates are fish-scales from the Ordovician (p. 125). Details of the Palaeozoic climatic changes are not clear, but the fact that corals, which can now live only in warm water, were alive over a considerable part of the earth's surface suggests that conditions were warmer than at present at least at some early Palaeozoic times.

The Devonian is considered by some to include a single main period, about 50 million years long, with one flood at the middle and more arid conditions at the end, but other authorities divide it into several periods. The first forests appeared at this time, and here, also, are found the first signs of vertebrate terrestrial life, in the form of fossil lung-fishes and amphibians (p. 296). The period recognized as Carboniferous in Europe includes two major periods of about 40 million years each in America, the Mississippian and Pennsylvanian. Throughout this long time conditions varied widely in different parts of the world. In the early Mississippian there were many swamps in North America. In the northern hemisphere the Pennsylvanian was probably a time of warm, moist conditions, with no cold winters, but there are signs that for part of this time India and Africa were covered with an ice-sheet. The coal measures show us the remains of the forests of spore- and seed-bearing plants that were then produced, and the land conditions evidently favoured the life of the Amphibia.

The Permian probably constitutes a single 45-million-year period, with very active orogenesis, leading to a more arid climate, perhaps showing large seasonal changes, with deserts in some parts of the world and glaciation in others. These conditions continued into the Triassic, when the continents lay high. The reptiles, first found in the Permian, developed throughout the Triassic and flourished in the succeeding Jurassic period, which probably lasted 45 million years. The Cretaceous period, during which the thick chalk deposits were laid down, probably lasted for rather more than 60 million years, including two major cycles of inundation. The lower Cretaceous certainly included extensive periods of flooding, when there were large shallow seas. Then later, towards the end of the upper Cretaceous, there were extensive orogenic movements, the Laramide revolution, producing the Rockies and the Andes. The temperature was warm until near the end of the Cretaceous, and we do not know what condition led to the break that is found between the animals of the Cretaceous and Eocene. Some groups of dinosaurian reptiles seem to have died out suddenly, but it is important to notice that not all disappeared at the same time, for instance, the stegosaurs and pterodactyls (p. 569) disappeared well before the end of the Cretaceous. However, it is probable that great changes went on at the end of this period, and we may guess that a factor leading to the development of the birds and mammals was the great rise of the continents, perhaps accompanied by a fall in temperature over wide areas that had enjoyed warmer weather. As always, when we look closely at such problems, we are appalled by the vast lengths of time involved and the scanty nature of our clues about them. The land lay very high at this time, and the apparent abruptness of the break between Cretaceous and Eocene fauna may be an artifact due to the scarcity of fossils. In North America there is evidence from terrestrial deposits of a long Paleocene period between the Cretaceous and Eocene.

It is usual to divide the last main geological period, the Tertiary,[1] into epochs, Paleocene, Eocene, Oligocene, Miocene, Pliocene, and Pleistocene, the names originally referring to the percentage of fossil genera surviving to the present day (see p. 571). Probably the whole time since the end of the Cretaceous has been about 70 million years. During the early part of the Tertiary period the climate was cold, but as erosion of the mountains that had been produced at the end of the Cretaceous proceeded the conditions became warmer, and throughout the Eocene and Oligocene there were large forests and humid conditions. Then during the Miocene there were marked earth movements, leading to elevation of the land and accompanied by more arid conditions, with wide areas of prairie and the widespread appearance of important new food plants—the grasses. The weather probably became gradually colder through the Pliocene, no doubt with many fluctuations, culminating in the ice ages of the Pleistocene. Here we come back to the period of which we have more detailed knowledge, and are reminded that the ice age was not continuous, but interrupted by many warmer periods.

This very brief survey of geological history in the northern hemisphere can hardly do more than remind us of the depths of our ignorance. We see enough to be sure that climatic conditions have varied throughout the millions of years, but we cannot yet see sufficient details to allow us to discover whether there is any rhythm of major cycles. It is easy to talk glibly of 'Carboniferous forests' or 'arid conditions of the Permian', forgetting that these periods lasted for a time that we can only roughly record in numbers and not properly imagine in terms of our experience, although we are among the longest lived of animals. The evidence suggests that conditions did not remain stable for such a vast length of time as a whole geological period, but fluctuated markedly, either irregularly or with complicated rhythms of greater and lesser magnitude. We must not forget that very profound 'climatic' changes occur every day, others every year, and some every eleven years. It is not impossible that these shorter-period changes, necessitating continual readjustment of animal and plant life, have been as important as the slower changes in producing evolution.

10. Summary

To reduce to order our knowledge of vertebrate life we shall try to discover its general organization and then examine the factors that

[1] This word is a survival from an old-fashioned classification of rocks, the Tertiary being the period since the Cretaceous.

have produced all the varied types. The pattern of organization we have to study is that of the animal as an active system maintaining itself in its environment. This tendency to maintenance and growth is the central 'force' that produces the variety of life. The opportunity for change is provided by the fact that reproduction seldom produces an exact copy of the parent, and thus a range of types is provided. The tendencies to grow and to vary lead animals to colonize new environments and produce the variety of life. As evolution has proceeded animals have come to occupy environments differing ever more widely from the sea in which life probably arose. Life in these more difficult environments is made possible by the development of special devices, making the later animals more complex than the earlier and in this sense 'higher'. It remains uncertain what influences have been responsible for producing the changes in organic form. Geological evidence shows that there have been many changes in climate and geography, some of them proceeding at very slow rates in comparison with the rhythms of individual animal lives. It is uncertain whether evolutionary changes follow these slow geological changes, or are a result of the instability imposed on living things by climatic rhythms with shorter periods, such as those of days, years, and the sunspot cycles.

II

THE GENERAL PLAN OF CHORDATE ORGANIZATION: AMPHIOXUS

1. The variety of chordate life

THE Chordata occupy a greater variety of habitats and show more complicated mechanisms of self-maintenance than any other group in the whole animal kingdom. They and the arthropods and the pulmonate molluscs have fully solved the problem of life on the land—which they now dominate. This domination is achieved by most delicate mechanisms for resisting desiccation, for providing support, and for conducting many operations that are harder in the air than in water. By even more wonderful devices the body temperature is raised and kept uniform and thus all reactions accelerated. Finally, use is made of this high rate of living for the development of the nervous system into a most delicate instrument, allowing the animal not only to change its response to a given stimulus from moment to moment, but also to store up and act upon the fruits of past experience.

Besides these more developed types of chordate that dominate the land and air there are also great numbers of extremely successful aquatic and amphibious types. The frog is often referred to as a somewhat lowly and unsuccessful animal, but frogs and toads are found all over the world. The sharks and bony fishes share with the squids and whales the culminating ecological position in the food chains of the sea, while the bony fishes are the only animals that have achieved considerable size and variety in fresh water. Among the still more lowly chordates the sea-squirts take a very important, though not dominant, position among the animal and plant communities that occupy the sea bottom, but they have never entered fresh water.

One could continue indefinitely with particulars of the amazing types produced by this most adaptable phylum. Yet through all their variety of structure the chordates show a considerable uniformity of general plan, and there can be no doubt that they have all evolved from a common ancestor of what might be called a 'fish-like' habit. In the very earliest stages only the larva was fish-like, and the life-history probably also included a sessile adult stage, such as the tunicates still show today (p. 66). This bottom-living phase was then eliminated by paedomorphosis, the larvae becoming the adults. Therefore the essential organization of a chordate is that of a long-bodied, free-swimming creature. All the other types can be derived from such an ancestor, though in some cases only by what is often called 'degeneration'.

2. Classification of chordates

We may conveniently divide the Phylum Chordata into four subphyla:

Subphylum I. Hemichordata
Balanoglossus; Cephalodiscus; Rhabdopleura
Subphylum 2. Cephalochordata (= Acrania)
Branchiostoma
Subphylum 3. Tunicata
Ciona, Sea-squirts
Subphylum 4. Vertebrata

The Vertebrata, the largest of these groups, may be subdivided:
Subphylum Vertebrata
Superclass I. AGNATHA
Class I. Cyclostomata. Lampreys and hag-fishes
Class 2. *Cephalaspidomorphi. **Cephalaspis*
Class 3. *Pteraspidomorphi. **Pteraspis*
Class 4. *Anaspida. **Birkenia*, **Jamoytius*

Superclass 2. GNATHOSTOMATA
Class I. *Placodermi. **Acanthodes*
Class 2. Elasmobranchii. Dogfishes, skates, and rays
Class 3. Actinopterygii. Bony fishes
Class 4. Crossopterygii. Lung-fishes
Class 5. Amphibia
Class 6. Reptilia
Class 7. Aves
Class 8. Mammalia.

3. Amphioxus, a generalized chordate

It has long been realized that through their great variety all these types show certain common features, often referred to as the *typical chordate characters*. It is better to regard these not as a list of isolated 'characters' but as the signs of a certain pattern of organization that is characteristic of the group. There is much reason to suppose that this basic chordate organization was that of a free-swimming marine animal, probably feeding by the collection of minute particles. We are fortunate in having still alive a little animal, amphioxus, the lancelet, which shows nearly all of these features in diagrammatic form. Study of amphioxus will go a long way to show the basic plan on which all later chordates are built, and, indeed, gives us a strong indication of what the early chordates must have been like.

Though it can swim freely through the water, amphioxus is essentially a burrowing animal, and many of its special features are connected with this habitat. It lives in the sand, at small depths, and has been found all round the oceans of the world. Evidently, in spite of its simplicity, it is a successful type. It is found on British coasts and, indeed, the first individual described was sent (preserved) from Cornwall to the German zoologist Pallas, who supposed it to be a slug and called it *Limax lanceolatus* (1774). It was first figured and given the name *Amphioxus lanceolatus* by Yarrell in 1836. However, the name *Branchiostoma* had been given in 1834 by Costa and by the rules of priority this is the official name of the genus. We may keep amphioxus as a common name. Some eight species of *Branchiostoma* are recognized, and in addition there is a group of six species referred to the genus *Asymmetron*. These resemble *Branchiostoma* in general organization, but they have gonads only on the right side.

The adult *Branchiostoma lanceolatum* is rather less than 2 in. long and has the typical fish-like organization, whose main external features are related to the methods of locomotion and feeding (Fig. 5). The body is elongated, and flattened from side to side. The skin has no pigment, and the muscles can be easily seen as a series of blocks, the myotomes, serving to bend the body into folds. As the name implies, the body is pointed at both ends; there is no recognizable head separated from the body. Indeed, there are no separate eyes, nose, or ears, and no jaws, so that the fundamental plan of chordate organization appears in almost its fullest simplicity from one end of the body to the other. The front end is, however, marked by a series of buccal cirri, which form a sieve around the opening of the oral hood and are provided with receptor cells.

Although the animal is provided with a large number of gill-slits these do not appear externally, being covered by lateral folds of the body, which enclose a ventral space, the atrium, opening posteriorly by an atriopore. The outside edges of the atrium project as a pair of metapleural folds, giving the body a triangular shape in transverse section. The alimentary canal opens posteriorly by an anus, in front of the hind end of the body, thus leaving a definite tail—a region of the body not containing any part of the alimentary canal.

The general arrangement of the organs can best be understood by considering the body as consisting of two tubes, the outer skin (ectoderm) and the inner alimentary canal (endoderm), with a space between (the coelom) lined by a third layer (the mesoderm). This arrangement is actually found during the course of the development (Fig. 18). The mesoderm at first forms thin layers, the somatopleure applied to the outer body wall and the splanchnopleure to the gut. Very soon the inner layer becomes much thickened where it is applied to the nerve-cord and notochord, and here it forms the myotomes, or muscle-blocks. In this dorsal part of the mesoderm the coelom, known here as the myocoele, soon becomes obliterated, leaving the ventral splanchnocoele around the gut. Besides the muscle that forms in the myotomes, non-myotomal muscles develop in the somatopleure and splanchnopleure. These are not divided into segments and are innervated by the dorsal nerve-roots, the ventral roots supplying only the myotomes.

4. Movement of amphioxus

The adult myotomes are blocks of striated muscle-fibres, running along the body, separated by sheets of connective tissue, the myocommas. This repetition or segmentation is characteristic of the organization of all chordates. The myocommas do not run straight down the body from dorsal to ventral side but are V-shaped (Fig. 5). However, each muscle-fibre runs straight from before backwards, and the contraction of the whole myotome therefore bends the body. A full discussion of the means by which forward motion is achieved by such a system will be given later (p. 133). Essentially, contraction of the myotomes results in transverse motion of the body inclined at varying angles in such a way as to result in forward propagation. Each myotome must therefore contract after that in front of it—the effect being to produce an S-bend that moves backwards through the water as the fish moves forward.

For our present purpose the point is that the contraction is serial, that is to say, it depends on the breaking up of the longitudinal muscle into blocks. It was probably the need for division of the musculature that led to the development of the segmentation, and this, affecting primarily the muscles, has come to influence a great part of chordate organization.

Contraction of the longitudinally arranged muscle-fibres will only produce a sharp bending of the body if there is no possibility of shortening of the whole. To prevent telescoping, an incompressible and elastic rod, the notochord, runs down the centre of the body. It is usually stated that this is a 'supporting structure', but, of course, an animal such as a fish in water needs no 'support'. Nor is the notochord a lever to which muscles are attached, as they are to the bones of many higher forms. No muscles pull on it directly, though the myocommas are attached to its sheath. Its function is to prevent the shortening of the body that would otherwise be the result of contraction of longitudinal muscles. In fact, it serves to make that contraction efficient in bending the body; its elasticity may also play an important part.

The notochord is composed of a series of flattened plates surrounded by a fibrous sheath. The plates are arranged in a regular manner with their flat surfaces in the transverse plane of the body. They are of two sorts, fibrous and homogeneous, which alternate with each other. Each plate develops as a highly vacuolated cell, the nuclei being later pushed aside to the dorsal or ventral edge. This structure is well suited by the turgidity of its cells enclosed in the sheath to resist forces tending to shorten the body. The cord of amphioxus is peculiar in that it extends from the very tip of the head to the end of the tail, projecting, that is to say, beyond the level of the myotomes, a condition presumably associated with the burrowing habit.

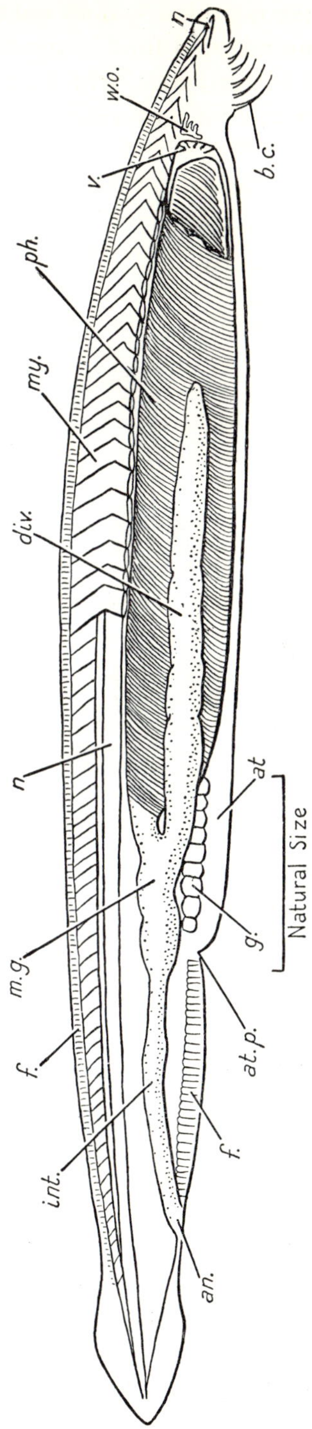

FIG. 5. Amphioxus, dissected.

The body wall and atrial wall (*at.*) have been removed on the right side, showing the pharynx (*ph.*), mid-gut (*m.g.*) with its diverticulum (*div.*), and intestine (*int.*). The oral hood has been cut away on the right, leaving the left buccal cirri (*b.c.*), wheel organ (*w.o.*), and velum (*v.*). *an.* anus; *at p.* atriopore; *f.* fin-ray boxes; *g.* gonads; *my.* myotomes; *n.* notochord.

Amphioxus probably does not often swim free in the water and the body is not adapted for fast movements. It has no elaborate fins such as those of later fishes, which ensure static stability like the feathers on an arrow, or are movable, to allow active control of the direction

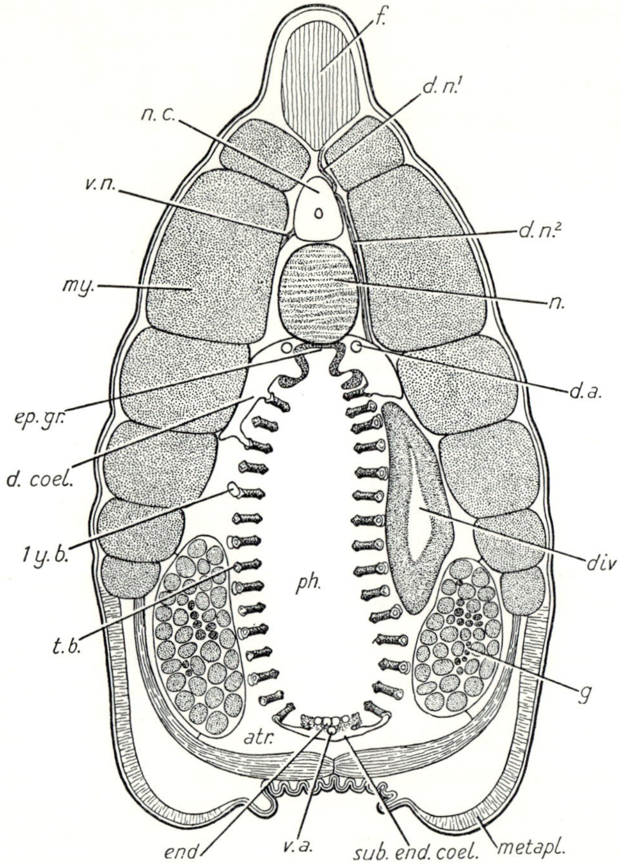

FIG. 6. Transverse section through amphioxus in the region of the pharynx.
atr. atrium; *d.a.* dorsal aorta; *d. coel.* dorsal portion of coelom; *div.* intestinal diverticulum; *d.n.*[1] and *d.n.*[2], branches of the dorsal nerve-root; *end.* endostyle; *ep. gr.* epipharyngeal groove; *f.* fin-ray box; *g.* gonad; *1 y.b.* primary gill bar containing coelom; *my.* myotome; *metapl.* metapleural fold; *n.* notochord; *n.c.* nerve-cord; *ph.* pharynx; *sub.end.coel.* subendostylar coelom; *t.b.* tongue bar; *v.a.* ventral aorta; *v.n.* ventral nerve-root. (After Krause.)

of swimming (p. 136). There is a low dorsal ridge, which continues behind as a small caudal fin. There are no definite paired fins, but the metapleural folds might perhaps be considered comparable to the lateral fin folds from which all vertebrate limbs are probably derived. They are distended with coelomic fluid and, with the dorsal ridge, probably serve to protect the body during the rapid dives by means of which the creature enters the sand. The habit of swimming with the front end downwards suggests the presence of a gravitational receptor mechanism. The larvae of lampreys swim in a similar way (p. 114).

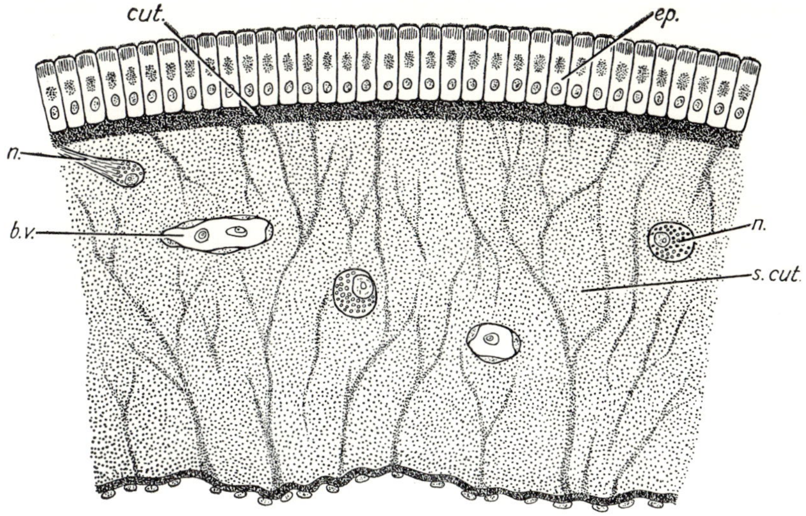

FIG. 7. Section of the skin of amphioxus.
b.v. blood-vessel; *cut.* cutis; *ep.* epidermis; *n.* nerves; *s.cut.* sub-cutis. (After Krause.)

5. Skeletal structures of amphioxus

Around the notochordal sheath is a further layer of gelatinous material containing fibres. There are no cells within this material but it is secreted by cells around the outside, which retain the epithelial arrangement of the mesoderm from which they were derived. This connective tissue continues as a sheath around the nerve-cord and above this into a series of structures known as fin-ray boxes, which support the median ridge. These are more numerous than the segments and each contains a more rigid material referred to as 'cartilage'. The relationship of these structures to the fin supports of vertebrates is obscure. Other skeletal rods occur in the cirri around the mouth and in the gill bars.

6. Skin of amphioxus

The epidermis differs from that of vertebrates in being very thin, composed of a single layer of cells, ciliated in the young, and with the outer border slightly cuticularized in the adult (Fig. 7). It is not known whether this cuticle contains a substance similar to the keratin produced by the many-layered skin of later forms. There are receptor cells but no glands or chromatophores in the skin.

Below the epidermis is a fibrous cutis, and below this again a gelatinous material containing fibres, the sub-cutis. Both these layers are secreted by scattered cells having some similarity to the fibroblasts

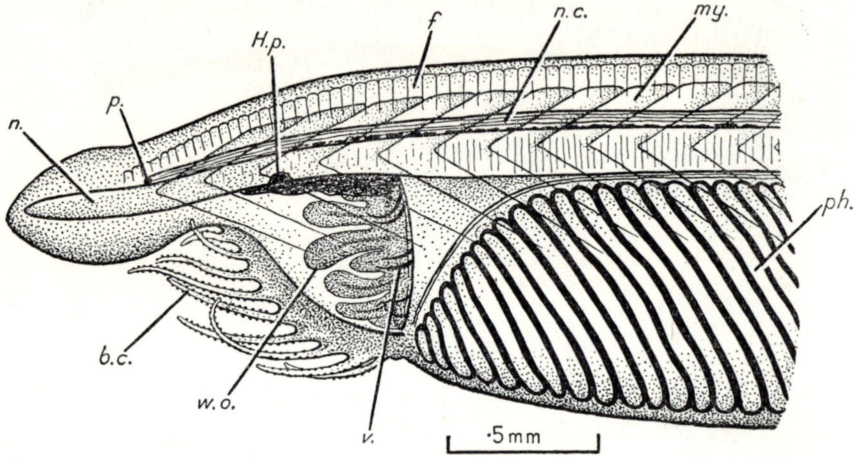

FIG. 8. Anterior end of amphioxus, from a stained and cleared preparation of a young animal.

b.c. buccal cirri; *f.* fin-ray box; *H.p.* Hatschek's pit; *my.* myotome; *n.* notochord; *n.c.* nerve-cord; *p.* pigment spot; *ph.* pharynx; *v.* velar tentacles; *w.o.* wheel organ.

of higher forms. They contain a system of cutaneous canals, with endothelial lining (Fig. 14).

7. Mouth and pharynx and the control of feeding

Amphioxus obtains its food by extracting small particles from a stream of water, which it draws in by means of cilia. In all animals that use cilia for this purpose a very large surface is provided (e.g. lamellibranchs, ascidians), and the pharynx and gill bars of amphioxus occupy more than one-half of the whole surface area of the body. Special arrangements are made for the support and protection of this ciliated surface, the wall of the pharynx being so greatly subdivided that it needs the protection of an outer layer, the atrium.

The mouth lies covered by an oral hood whose edges are drawn out into buccal cirri, provided with sense-cells (Fig. 8). When feeding the cirri are curved to form a funnel-like sieve preventing the entry of large particles. Around the mouth itself there is a further ring of sensory tentacles, the velum. The oral hood contains a complex set of ciliated tracts, the 'wheel organ' of Müller, and this plays a part in sweeping the food particles into the mouth (Figs. 8 and 9). Near its centre is a groove, Hatschek's pit, formed as an opening of the left first coelomic sac to the exterior (p. 44).

The main operation of food collection is performed by the pharynx, a large tube, flattened from side to side, whose walls are perforated by nearly 200 oblique vertical slits, the number increasing as the animal gets older. The slits are separated by bars containing skeletal rods and further subdivision is provided by cross-bars (synapticulae). Since the bars slope diagonally many of them are cut in a single transverse section, but it must be remembered that they are essentially the vertical portions of the main walls of body and pharynx, where these have not been perforated by a gill-slit. Such a portion of the body wall must contain a coelomic space and this can in fact be seen in the original or primary gill bars. However, an increase of the ciliary surface is produced by downgrowth of secondary or tongue bars from the upper margin, dividing each primary slit; these secondary bars contain no coelom. The coelomic spaces in the primary bars, of course, communicate above and below with continuous longitudinal coelomic cavities (Fig. 6).

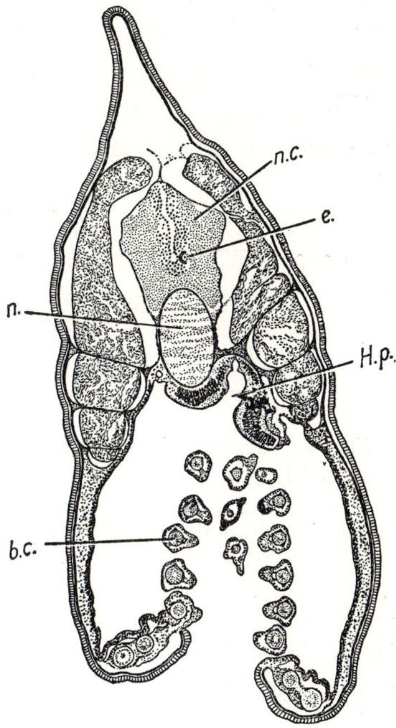

FIG. 9. Transverse section through front end of amphioxus.

b.c. buccal cirri; *e.* eyespot; *H.p.* Hatschek's pit; *n.* notochord; *n.c.* nerve-cord.

There are cilia on the sides and inner surfaces of the gill bars, the lateral ones being mainly responsible for driving the water outwards through the atrium and thereby drawing the feeding current of water in at the mouth. In the floor of the pharynx lies the endostyle, containing columns of ciliated cells, alternating with mucus-secreting cells, which produce sticky threads in which food particles become entangled. Various currents then draw the sticky material along until it reaches the mid-gut. The frontal cilia of the gill bars produce an upward current, driving the mucus from the endostyle into a median dorsal epipharyngeal groove, in which it is conducted backwards. The cilia of the endostyle also move mucus along the peripharyngeal ciliated tracts, behind the velum, to join the epipharyngeal groove. Radioactive iodine is concentrated by one of the columns of the endostyle and secreted with the mucus. Barrington (1958) suggests that these may be regarded as the precursors of the thyroid cells, serving to produce iodinated mucoproteins, which are then absorbed farther down the gut (see p. 119).

The pharynx narrows at its hind end to open dorsally into a region best known as the mid-gut, the name stomach being inappropriate.

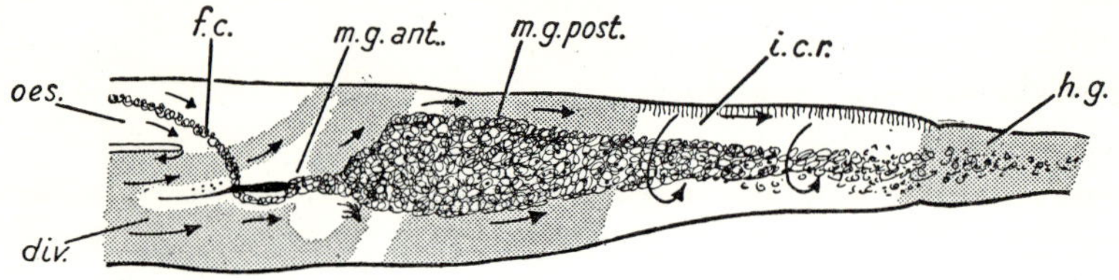

FIG. 10. Currents in the mid-gut of amphioxus, showing the appearance when an animal is placed in a medium containing carmine particles. Arrows show the chief ciliary currents.

div. diverticulum; *f.c.* food cord; *h.g.* hind-gut; *i.c.r.* ileo-colon ring; *m.g.ant.* and *m.g.post.* anterior and posterior parts of mid-gut; *oes.* oesophagus. (After Barrington.)

A large mid-gut diverticulum reaches forward from this region on the right-hand side of the pharynx. From its position this organ is often called the liver, but Barrington has given reasons for supposing that it is the seat of the production of digestive enzymes. Zymogen cells, similar to those of the mid-gut, are found in its walls. Its strong dorsal and ventral ciliation maintains in it a circulation of food materials and secretion, and its cells are capable of phagocytosis as well as secretory activity. Amphioxus thus combines intracellular with extracellular digestion, doubtless in connexion with its microphagous habit. Particles placed in the diverticulum are swept backwards and join the main food cord that passes through the mid-gut (Fig. 10).

The hind end of the mid-gut is marked by a specially ciliated region, the ileo-colon ring, whose cilia rotate the cord of mucus and food. The movement is transmitted to the portion of the food cord in the mid-gut and presumably assists in the taking up of the enzymes that emerge from the diverticulum. Extracellular digestion takes place in the mid-gut and the enzymes responsible have been studied by Barrington. The pH of the contents varies from 6·7 to 7·1. An amylase is present in extracts of the diverticulum, mid-gut, and hind-gut, but not in those of the pharynx. Lipase and protease are present in the same regions, the latter having an optimum action at about pH 8·0, being, that is to say, a tryptic type of enzyme. There is no sign of any protease with an acid optimum, similar to the pepsin of higher forms.

Behind the ileo-colon ring the intestine runs as a straight hind-gut to the anus. Absorption of food takes place here, and perhaps also in the mid-gut, apparently partly by intracellular digestion, since ingested carmine particles are taken into the cells.

The feeding current is regulated by the rate of beat of the cilia and the degree of contraction of the inhalent and exhalent apertures. The walls of the atrium contain an elaborate system of afferent and efferent nerve-fibres. The receptors include a set of large peripheral nerve-cell bodies, lying beneath the atrial epithelium and sending axons in by way of the dorsal roots. The motor fibres also pass through the dorsal roots and run without synapse to the cross-striated fibres of the pterygial muscle, which forms the floor of the atrium. The stream flowing into the pharynx is tested by the receptors of the velum and atrium, and if noxious material is present, the water is expelled by closing the atriopore and contracting the pterygial muscle, producing a 'cough'. The system can distinguish between suspensions of food material and inorganic particles. When sufficient food has been taken, collection is suspended until it has been digested (Bone, 1960).

The atrial nervous system probably regulates spawning as well as feeding. It has often been compared with the sympathetic system of craniates but there are almost no close similarities. The nerve cells in it are receptors and there is no sign of the peripheral synapse on the efferent pathway that is so characteristic of the true autonomic system. The atrial system is developed in relation to filter feeding and has perhaps been completely lost in higher forms that feed by other methods and have developed new methods to control them (p. 117).

8. Circulation

The blood-vessels of amphioxus show in diagrammatic form the fundamental plan on which the circulation of all chordates is based (Fig. 11). Slow waves of contraction occur in various separate parts in such a way as to drive the blood forwards in the ventral vessels, backwards in the dorsal ones. Below the hind end of the pharynx there is a large sac, the sinus venosus, into which blood from all parts of the body is collected. From this there proceeds forwards a large endostylar artery (truncus arteriosus or ventral aorta) from which spring vessels carrying blood up the branchial arches. At the base of each primary bar there is a little bulb, functioning as a branchial heart. From the gill bars blood is collected into paired dorsal aortae, which join behind the pharynx. From the paired and median aortae blood is carried to the system of lacunae that supplies the tissues. There are no true capillaries. From the lacunae blood is collected into veins, the most important of which are the caudals, cardinals, and a plexus on the gut. The cardinals are a pair of vessels in the dorsal wall of the coelom, and they collect blood from the muscles and body wall. They lead to the sinus venosus by a pair of vessels, ductus Cuvieri, which pass ventrally and across the coelom to join the sinus venosus on the floor of the gut. The caudal veins join the plexus on the gut, from which blood is collected by a large subintestinal vein running on to the liver; from here another plexus leads to the sinus venosus.

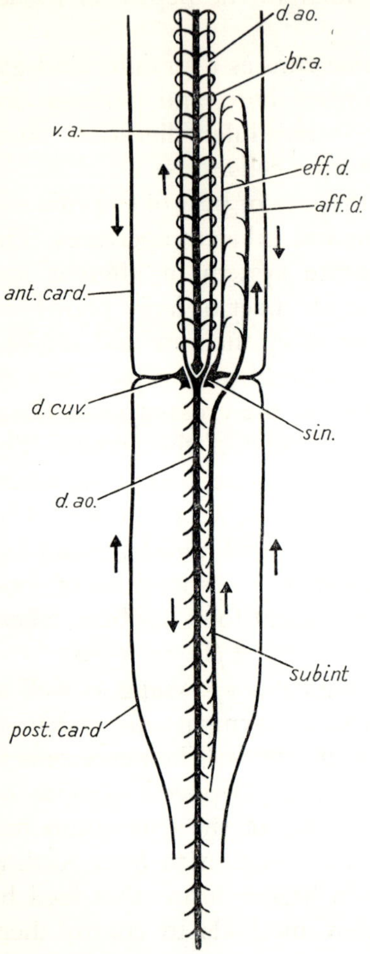

FIG. 11. Diagram of the circulation of amphioxus.

aff.d. afferent vessel of diverticulum; *ant. card.* anterior cardinal vein; *br.a.* branchial arch; *d.ao.* dorsal aorta; *d.cuv.* ductus Cuvieri; *eff.d.* efferent vessel of diverticulum; *post.card.* posterior cardinal vein; *sin.* sinus venosus; *subint.* subintestinal vein; *v.a.* ventral aorta. (After Grobben and Zarnik.)

Contractions arise independently in the sinus venosus, branchial bulbs, subintestinal vein, and elsewhere. The rhythms are very slow (once in two minutes), irregular, and apparently not coordinated by any control system.

The blood is colourless and is not known to contain any respiratory pigment. It contains no cells. Presumably the tension of dissolved oxygen acquired by simple solution is sufficient for the small energy needs of the animal, which spends most of its life at rest. It is by no means certain that any oxygenation of the blood takes place in the gills. Orton has suggested that since these, through their cilia, do much of the work of the body, the blood actually leaves the gills less rich in oxygen that when it enters them. Oxygenation probably takes place chiefly in the lacunae close to the skin, perhaps especially those of the metapleural folds.

9. Excretory system of amphioxus

One of the most mysterious features about the organization of amphioxus is that there are flame-cells, comparable with those found

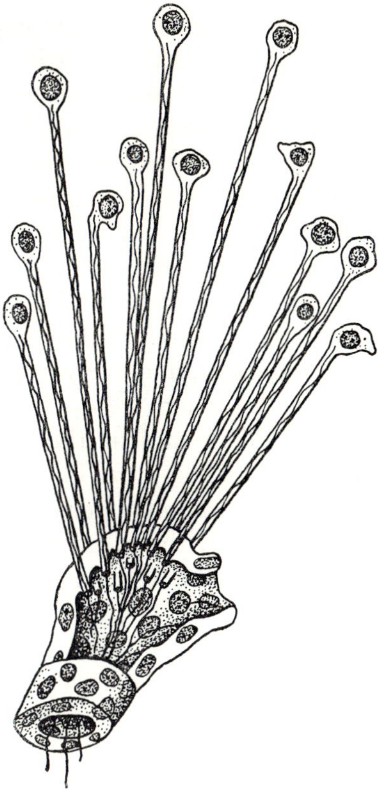

FIG. 12. Solenocytes of amphioxus, showing the nuclei, long flagella, and the openings into the main excretory canal, which leads to the atrium. (After Goodrich.)

in platyhelmia, molluscs, and annelids. The excretory organs, therefore, do not conform to the basic chordate plan, and are in fact very different from those not only of all other chordates but also from any found in the remote invertebrate allies of the chordates which, as we shall presently see, include the echinoderms, brachiopods, and polyzoa.

The nephridia lie above the pharynx. To each primary gill bar there corresponds a sac, opening by a pore to the atrium and studded with numerous elongated flame-cells (solenocytes) (Fig. 12). These flame-cells do not open internally, but are in close contact with special blood-vessels (glomeruli) whose walls separate the flame-cells from the coelomic epithelium. Assuming that there are 200 of these nephridia, each with 500 solenocytes 50 μ long, Goodrich, who has provided the most accurate information about these organs, shows that the total length available for excretion is no less than 5 metres. It is assumed that excretion takes place by diffusion through the flame-cell wall, the liquid being driven down the tube by cilia. Coloured particles injected into the blood-stream are not excreted by the nephridia.

In development these remarkable organs arise from groups of cells close to the meeting-place of ectoderm and endoderm; almost certainly they are derived from the former. They have no relation whatever to the mesoderm and this fact alone sufficiently indicates that they are in no way comparable to the pronephros of vertebrates, as is sometimes stated. There is no organ in vertebrates with which they can be compared, nor is there any trace in amphioxus of organs comparable to the vertebrate kidney system. In fact we have here a remarkable case of an isolated feature; evidently separate items of the genotype may vary independently, and the whole bodily organization does not necessarily change together.

The brown funnels are blind sacs at the front of the atrium, invaginating into the epibranchial coelom. They are probably receptor organs. Some parts of the atrial wall may perform excretory functions. Masses of cells in the atrial floor, the atrial glands, contain granules that may be excretory but may have been taken up from the food current.

In the gonads, especially the testes, there are large yellow masses, containing uric acid, which are extruded with the gametes.

10. Nervous system

Amphioxus possesses a hollow dorsal nerve-cord similar to that of vertebrates. Though this is somewhat modified at the front end, it is not there enlarged into an elaborate brain. The nervous system is connected with the periphery by a remarkably simple set of nerve-roots, a dorsal and a ventral on each side in each segment. The roots do not join (Fig. 13): the ventral roots lie opposite the myotomes, to which they carry motor-fibres, and these end on the muscle-fibres with motor end-plates. The dorsal root runs out between the myotomes and carries all the afferent fibres of the segment and motor-fibres for the non-myotomal muscles of the ventral part of the body. This is the fundamental pattern of the roots in all vertebrates.

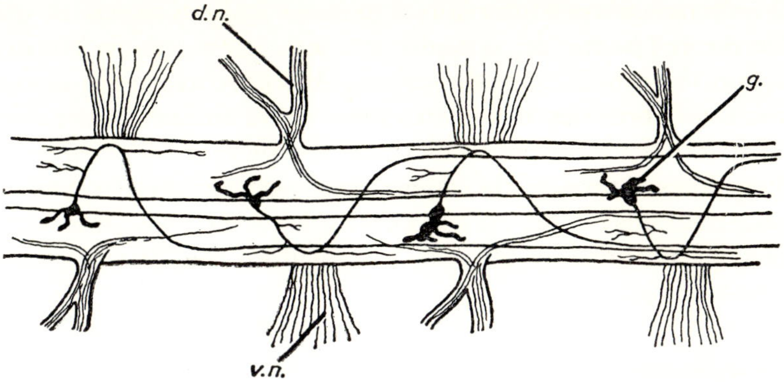

FIG. 13*a*. Nerve-cord of amphioxus.
d.n. dorsal nerve-root; *g.* 'giant' nerve-fibres; *v.n.* ventral nerve-root. (After Retzius.)

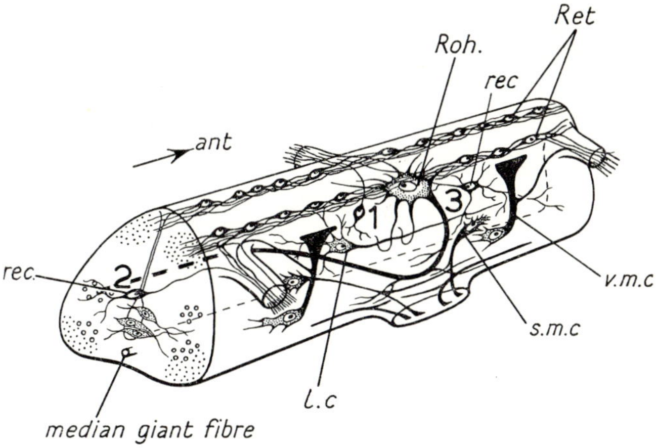

FIG. 13*b*. Stereogram illustrating the structure of the spinal cord in an adult amphioxus.
The receptor system is made up of a more or less continuous column of bipolar cells of Retzius (*Ret.*), together with smaller cells of various types (*rec.*). According to Johnston these receptor cells (1, 2 and 3) can be regarded as equivalent to the dorsal root ganglion cells of vertebrates. The other type of receptor cell is the giant Rohde cell (*Roh.*), which has a large axon and elaborate dendritic system. It is probable that at least some of these cells possess a peripheral axon running in the dorsal root. *l.c.* longitudinal connective cell.
The visceral motor cells (*v.m.c.*) are arranged segmentally, one per segment.
The somatic motor cells (*s.m.c.*) lie at a different level in the cord from the ventral roots.
Other cells in the cord are internuncials of various types. (After Bone.)

The fibres of the peripheral nerves differ from those of vertebrates in that they have no thick myelin sheath that will blacken with osmium tetroxide. The nerve trunks are surrounded by an epineurium with connective tissue cells but there seem to be no Schwann cells accompanying the nerve-fibres (Bone, 1958).

The afferent fibres of the dorsal roots are unique among chordates in that the cell bodies are not collected into spinal ganglia but mostly lie within the central nervous system. At least three types of central neuron send fibres that terminate as free nerve endings in the skin. In addition, on the head and tail there are peripheral receptor cells, sending fibres centrally, also complicated encapsulated organs in the metapleural folds (Bone, 1960). There are numerous large multipolar nerve-cells, presumably afferent, just beneath the atrial epithelium.

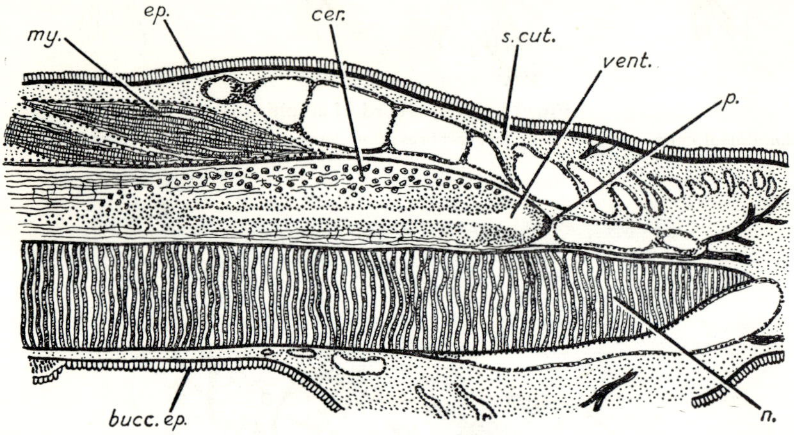

FIG. 14. Sagittal section through the front end of amphioxus.
bucc.ep. buccal epithelium; *cer.* cerebral vesicle, with large nerve-cells; *ep.* epidermis; *my.* myotome; *n.* notochord; *p.* pigment spot; *s.cut.* subcutis; *vent.* ventricle of cerebral vesicle. (After Krause.)

These cells have many branched dendrites and an axon that runs through a dorsal root to the spinal cord. Their status is discussed on p. 33.

The spinal cord has only a narrow lumen and its elements are arranged as in vertebrates, namely, ependyma close to the canal, cell layer ('grey matter'), and outer fibrous layer ('white matter'). The cells are not arranged clearly in horns as they are in vertebrates. The most conspicuous cells are the giant cells, which lie dorsally in the anterior and posterior parts but are absent from about the 13th to 39th segments. Each of these cells has many dendrites, branching in the region of entry of the dorsal root fibres, and a single axon, which runs backwards in the front part of the body, forwards in the hind, passing in each case for the whole length of the cord. A median giant fibre, which runs ventrally for the length of the cord, lies close to the viscero-motor cells that probably produce the 'coughing' movements of the atrium (p. 33).

Ten Cate has investigated the movements of amphioxus and found that it responds to all stimuli by movements of 'flight'. There are no isolated or local movements; the effect of any stimulus such as touch on the side of the body is to produce waves of myotomal contraction. These may, however, vary from strong waves going the whole length

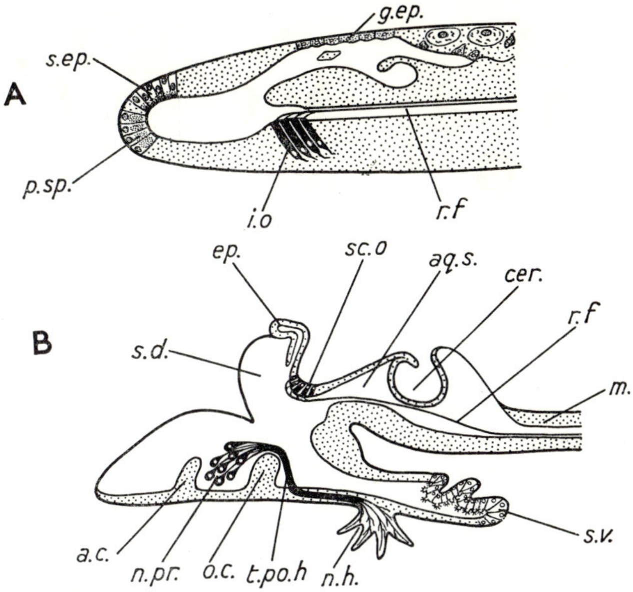

FIG. 15. Diagram of (A) the anterior end of the nervous system of amphioxus and (B) the brain of a fish (*Polypterus*).

A. Amphioxus. *g.ep.* granulated ependyma in the wall of the 'dorsal central canal'; *i.o.* infundibular organ; *p.sp.* pigment spot; *r.f.* Reissner's fibre in the central canal; *s.ep.* sensory epithelium. B. *Polypterus, a.c.* anterior commissure; *aq.s.* aqueductus Sylvii; *cer.* cerebellum; *ep.* epiphysis; *m.* medulla spinalis; *n.h.* neurohypophysis; *n.pr.* nucleus praeopticus; *o.c.* optic chiasma; *r.f.* Reissner's fibre in fourth ventricle; *s.c.o.* subcommissural organ; *s.d.* saccus dorsalis; *s.v.* saccus vasculosus with primary sense cells; *t.po.h.* tractus praeoptico-hypophyseus. (After Olson and Wingstrand.)

of the body to single rapid twitches. The giant cells participate in the spread of these waves. It seems likely that the arrangement ensures that touch on the anterior part of the body, normally exposed when feeding, produces backward movement (i.e. withdrawal into the sand) but touch on the hind part the reverse movement of emergence and escape.

At the front end the central canal is enlarged to form a cerebral vesicle (Fig. 14). The whole neural tube is hardly wider here than in the region of the spinal cord and there is no thickening of the walls, which are indeed mostly formed of a single layer of ciliated epithelial cells (Fig. 15). This is a striking indication of the lack of cephalization of the animal. From the region of the cerebral vesicle spring the first two dorsal roots, to which there are no corresponding ventrals. These roots carry impulses from the receptors of the oral hood and its tentacles.

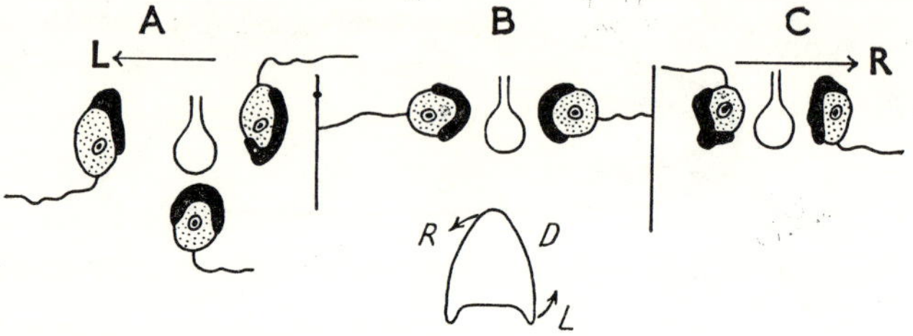

FIG. 16. Diagram to show the direction of the eye-spots of amphioxus. A, anterior, B, middle, and C, posterior regions of the body. The eyes are shown as if seen from behind. D shows the direction of spiralling of the animal when swimming—as seen from in front. (After Franz.)

The infundibular organ (Fig. 15) is composed of tall cells with long cilia, which beat in the opposite direction to those of the rest of the vesicle. From them fibres run backwards down the cord. The organ is also the site of origin of Reissner's fibre (Fig. 15). This is a thread of non-cellular material, present in all vertebrates at the centre of the neural canal. It is secreted at the front end and then passed backwards and is often collected and absorbed in a sac at the hind end of the spinal cord. In vertebrates it arises from secretory ependymal cells of the subcommissural organ, lying dorsally in the diencephalon (Fig. 15). The infundibular organ of amphioxus is clearly not exactly similar, yet the Reissner's fibres are clearly comparable; an interesting problem in homology.

A further complication is that the cells of the infundibular organ contain material that stains with the Gomori method, and is similar to the neurosecretory material found in the fibres of the hypophysial tract (Fig. 15). The organ thus seems to occupy a central position in the control system as a receptor, originator of nerve-fibres, and of two sorts of secretion. There is clearly much to be learned from this about the origin and significance of the control systems of the diencephalon.

In young stages the cerebral vesicle opens by an anterior neuropore, and at the point where the closure takes place there develops a depression of the skin, lined by special epithelium, and known as Kölliker's pit. It is said to receive no special innervation. The cells at the front end of the cerebral vesicle contain pigment and there have been attempts to show that this represents an eye. More probably it serves to prevent rather than to receive photic stimulation; there are other cells lying in the spinal cord that are clearly photoreceptors (Fig. 9). In the front part of the body these are unprotected by pigment, whereas more posteriorly they are so pigmented as to be protected asymmetrically from the light (Fig. 16). This asymmetry may be connected with the fact that when swimming free in the water amphioxus moves spirally about its axis, turning clockwise as seen from behind. It was established by Parker that a small beam of light produces movements of amphioxus only when it is directed on to the region of the body or tail, not when it shines on the head. Since the animal normally lies with the head protruding we may suppose that the pigment spot serves to prevent light that strikes down vertically from stimulating the photoreceptors in the cord.

Amphioxus is therefore provided with receptor and motor systems that serve to keep it in its sedentary position, able to collect food from the current that it makes by the cilia (p. 33). There are mechanisms that help it to make appropriate movements of escape when it is touched or when the body (but not head) is illuminated. The touch receptors of the buccal cirri produce rejection of large particles and those of the velum are chemo-receptors. The infundibular organ may be some form of gravity or pressure receptor. By means of these receptor organs and its simple movements of swimming, burrowing, and closing the oral hood, the animal is maintained, probably mainly by trial and error (phobotactic) behaviour, in an environment suitable for its life. There are none of those elaborate mechanisms that we find in higher chordates for 'seeking' special environments or for so 'handling' or managing them that they may prove habitable by the animal. Amphioxus must take and leave the world very much as it finds it. The 'correct' environment is chosen for it by the selective settling of the larvae.

11. Gonads and development of amphioxus

The gonads of amphioxus are hollow segmental sacs with no common duct. Each sac develops from mesoderm cells, perhaps originally from a single cell, at the base of the myotomes in the branchial region, the genital cells themselves developing on the walls (Fig. 6). The sexes are separate and the genital products are shed by dehiscence into the atrium, the aperture by which they escape closing and the gonad developing afresh.

Extrusion of the gametes occurs in spring, on warm evenings

following stormy weather. Fertilization is external and development then occurs free in the water. Numerous eggs are produced and they are small but yolky. Complex flowing movements take place in them after fertilization, and cleavage is then rapid and complete, producing a blastula composed of a dome of somewhat smaller and a floor of rather larger cells (Fig. 17). These latter then invaginate to make the archenteron, opening by a wide blastopore, which later becomes the anus. At about this stage the gastrula becomes covered with flagella, by which it rotates within the egg case.

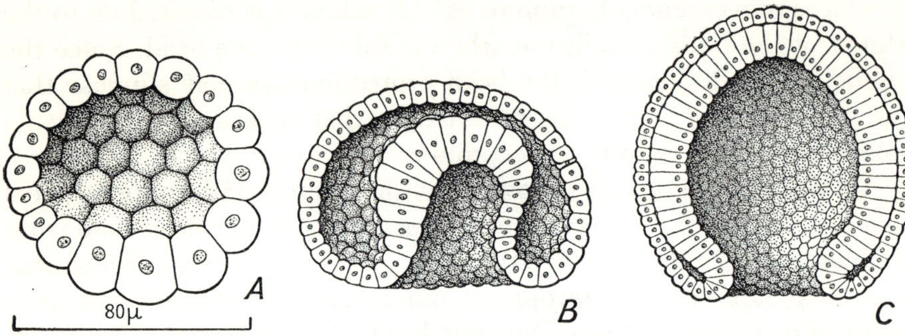

FIG. 17. Three stages in the development of amphioxus as seen in stained preparations.

A, the blastula; B, early, and C, later gastrula.

The creature now elongates and its dorsal side flattens and eventually sinks in to form the neural tube (Fig. 18 A). At about this time the dorsal side of the inner layer begins to fold near to the front end, in such a way as to make a pair of lateral pouches. The walls of these pouches are the future mesoderm and the cavity is the coelom. As in other early chordates, therefore, the coelom is continuous at first with the archenteron. The roof of the archenteron also arches up dorsally and forms the notochord, the gut wall being completed by the approximation of the edges of the remaining portion of the inner layer, which is now the definitive gut wall or endoderm.

The analysis of the processes of development now enables us to say something of the forces by which these formative foldings and cell movements are produced. The formation of the neural tube, mesoderm and notochord and the completion of the gut roof all involve an upward movement of cells towards the mid-dorsal line. This process of 'convergence' is a very marked feature of the development of all chordates (Young, 1957, p. 609).

As the animal elongates, further mesodermal pouches are produced, each separating completely from the endoderm and from its neighbours. The cells of each pouch push down ventrally on either side of the gut, the outer ones applying themselves to the body wall to form the somatopleure, the inner to the gut wall as splanchnopleure (Fig. 18 D). The inner wall of the mesoderm on either side of the nerve-cord thickens to form the myotome, and a tongue of cells growing up between this and the nerve-cord forms the sheaths of the latter and

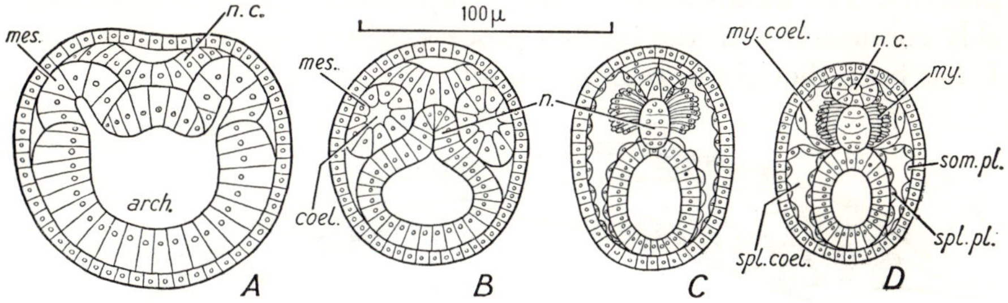

FIG. 18. Further stages in the development of amphioxus as seen in transverse sections.

A, stage of three somites; B, six somites; C, nine somites; D, eleven somites. *arch.* archenteron; *coel.* coelom; *mes.* mesoderm; *my.* myotome; *my.coel.* myocoele; *n.* notochord; *n.c.* nerve-cord; *som.pl.* somatopleure; *spl.pl.* splanchnopleure; *spl.coel.* splanchnic coelom. (After Hatschek.)

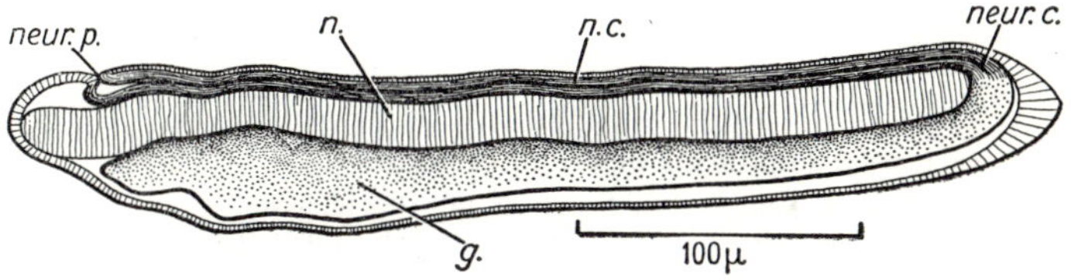

FIG. 19. Young amphioxus, soon after hatching.

g. gut; *n.* notochord; *n.c.* nerve-cord; *neur.c.* neurenteric canal; *neur.p.* neuropore.

probably also the fin-ray boxes and other 'mesenchymal' tissues. The upper part of the coelomic cavity, the myocoele, becomes separated from the ventral splanchnocoele. Whereas the former becomes almost completely obliterated, the latter expands to form the adult coelom, the cavities between the adjacent sacs breaking down.

While this differentiation of the mesoderm has been proceeding the animal has elongated into a definitely fish-like form. The neural tube is a small dorsal canal, opening by an anterior neuropore and continuous behind through a neurenteric canal with the gut (Fig. 19). The larva hatches when only two segments have been formed and swims at the sea surface by means of its ciliated epidermis, turning on its axis from right to left as it proceeds with the front end forwards.

The mouth now appears as a circular opening and then moves over to the left side and becomes very large. From this time onward the whole development is markedly asymmetrical, presumably in connexion with the spiral movement and method of feeding. The first gill-slit also forms near the midline but moves up on to the right side (Fig. 20). At about the same time the right side of the pharyngeal wall develops into a V-shaped thickening, the endostyle. Behind this there forms a tube, the club-shaped gland, joining the pharynx to the outside and formed by the closure of a groove in the side of the pharynx. The significance of this organ is still obscure; it is presumably connected with the feeding process, which begins at this stage. It has been thought to represent a gill-slit.

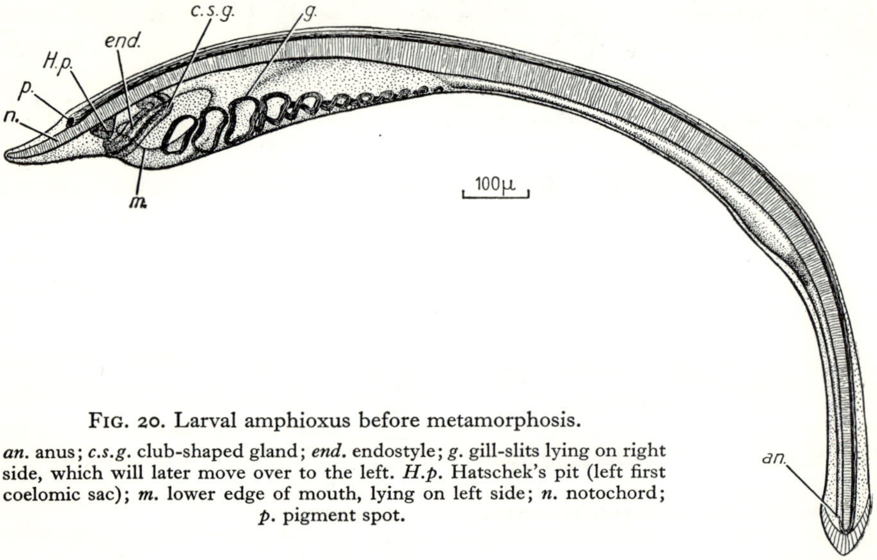

FIG. 20. Larval amphioxus before metamorphosis.

an. anus; *c.s.g.* club-shaped gland; *end.* endostyle; *g.* gill-slits lying on right side, which will later move over to the left. *H.p.* Hatschek's pit (left first coelomic sac); *m.* lower edge of mouth, lying on left side; *n.* notochord; *p.* pigment spot.

The first two coelomic pouches differentiate, asymmetrically, at this time. That on the right becomes the coelomic cavity of the head region, while the left one acquires an opening to the exterior and a heavily ciliated surface. This is perhaps also connected with the feeding-systems and becomes developed into Hatschek's pit of the adult. Its interest to the morphologist lies in the fact that the first coelomic cavity opens to the exterior in other early chordates and in some vertebrates (p. 206). The pit has thus some claim to be considered the equivalent of the hypophyseal portion of the pituitary gland.

Further gill-slits develop in the mid-ventral line and move over on to the right side until fourteen have been so formed. Meanwhile, a further row of eight slits appears *above* that already formed. These are the definitive slits of the right side and presently the larva proceeds to become symmetrical by movement of eight of the first row of slits over to the left side, the remainder disappearing. At this 'critical stage' with eight pairs of slits the larva pauses for some time before further changes. It is interesting that this is the time at which it most nearly represents what might have been an ancestral craniate, with eight branchial arches (p. 145). Further slits are then gradually added in pairs on both sides. Each slit becomes subdivided, soon after its formation, by the downgrowth of a tongue bar. The atrium is absent from the early larva. Metapleural folds then appear on either side and are united from behind forwards to form a tube below the pharynx. During the later stages of development the larva sinks and finally rests on the bottom while undergoing the migration of gill-slits that constitutes its metamorphosis. In other species the larva remains longer in the plankton, becoming large and even showing quite large gonad rudiments. These were at first thought to be adults of a new genus (*Amphioxides*).

The development of amphioxus, like its adult organization, shows us many features of the plan that is typical of all chordates and was presumably present in the earliest of them. Thus the cleavage, invagination, and mesoderm formation recall those of echinoderms and other forms similar to the ancestors of the chordates, and also show a pattern from which all later chordate development can be derived. Unfortunately we cannot pursue this study as far as we should like because of the difficulty of investigating the development of amphioxus. Modern embryologists aim at tracing the morphogenetic movements by which the organism is built, and ultimately at discovering the forces responsible for these processes. We still remain ignorant of the details of these morphogenetic movements, and can only guess that the system of cell activities by which an amphioxus is built represents quite closely the original set of morphogenetic processes of vertebrates (Young, 1957, p. 633).

There are, of course, some special features connected with the method of life of the larva, and especially with its asymmetry. The strange sequence of gill formation, the immense left-sided larval mouth, perhaps the club-shaped gland, and Müller's organ, may show considerable modifications of relatively recent date. However, the earliest chordates probably fed by means of cilia and were planktonic, so we must not too hastily assume that even these asymmetrical features are novelties.

The division of the mesoderm of amphioxus into a series of sacs presents an interesting problem. The segmentation of the mesoderm of vertebrates is restricted to the dorsal region. In the lowest chordates (see p. 51), as in their pre-chordate ancestors, there are three coelomic cavities, but it is probable that the many segments of vertebrates arose in order to provide a set of muscles able to contract in a serial manner for the purpose of swimming. Their segmentation would thus be a relatively late development, not related to the segmentation of annelids, which divides the whole body into rings. Accordingly the ventral part of the vertebrate coelom usually remains unsegmented. But in amphioxus (and in the lamprey) it is subdivided from its first appearance and only becomes continuous later. The best interpretation of this condition is to suppose that in order to provide a series of myotomes a rhythmic process subdividing the mesoderm was adopted. In its earliest stages this affected the whole mesoderm, ventral as well as dorsal, but later became restricted to the dorsal region. New morphogenetic processes may often pass through stages of refinement and simplification in such ways.

12. Amphioxus as a generalized chordate

Amphioxus provides us, then, with a valuable example of a chordate that retains the habit of ciliary feeding, which was probably that of the earliest ancestors of our phylum. No doubt in connexion with this, and the bottom-living habit, there are many specializations; the enormously developed pharynx with its atrium, the asymmetry, and so on; but the general arrangement of the body is almost diagrammatically simple, and it may well be that amphioxus shows us a stage very like that through which the ancestors of the craniates evolved. Perhaps next the larva remained longer in the plankton and became mature there. The *Amphioxides* larvae show signs of such a change.

This might give rise to a suspicion that amphioxus is not an ancestral type but a simplified derivative of the vertebrates, perhaps a paedomorphic form. It possesses, however, sufficient peculiar features to make this view unlikely. Neoteny might explain the regular segmentation, separate dorsal and ventral roots, and other features, but can hardly account for the method of obtaining food, for the condition of the skin, or for the presence of nephridia. It may be, therefore, that amphioxus shows us approximately the condition of the early fish-like chordates, living in the Silurian some 400 million years ago, and that it has undergone relatively little change in all the time since.

III

THE ORIGIN OF CHORDATES FROM FILTER FEEDING ANIMALS

1. Invertebrate relatives of the chordates

WE have seen in the organization of amphioxus the plan of chordate structure as it may have existed in Palaeozoic times. Before proceeding to discuss the later forms that evolved from animals of this sort we may first look yet farther backwards to discuss the origin of the whole chordate phylum from still earlier ancestors. The great difficulty of such an inquiry is itself a stimulus and a challenge. Typical fish-like chordates were undoubtedly established by the Ordovician period, but we have no good fossil record of their earliest form and this must therefore be deduced from study of amphioxus and later animals. No fossils that suggest chordate affinities have been found in the still earlier rocks. There are, however, certain strange animals alive today which, though not of fish-like type, show undoubted relationship with our group. These might, of course, be degenerate offshoots from later periods, but careful comparison suggests that they have been separated for a very long time and provide us with relics of some of the early stages of our history.

The first step in our inquiry, however, before discussing these forms, should be to find out, if possible, which of the main lines of invertebrate animals shows the closest affinity with the chordates. Almost every phylum in the animal kingdom has been suggested, including the nemertines. Many still suppose that the annelids and arthropods, because of their metameric segmentation, are related to the chordates, but closer examination shows that the similarities are superficial. The segmentation of these annulate animals is an almost complete division of the whole body into rings, and all the organ systems are affected by it to some extent. In chordates only the dorsal myotomal region is segmented; even the mesoderm is not divided in its ventral region in most animals. Moreover, the whole orientation of the body differs in the two groups. The vertebrate nerve-cord is dorsal to the gut, in annulates the nerve-cord is below and the 'brain' above. The blood circulates in opposite directions, the limbs are based on quite different plans, and so on. Attempts have been made to get over these difficulties by turning the invertebrate upside down! Patten and Gaskell carried such theories to extremes and tried to show a relationship of chordates with the eurypterids, heavily armoured arachnids of the Cambrian and Silurian. These animals show a certain superficial resemblance to some early fossil fishes, the cephalaspids of the Devonian (Fig. 83), and these workers, with great ingenuity, claimed to find in them evidence of the presence of many chordate organs.

The safest evidence of affinity is a similarity of developmental processes: animals that develop very differently are unlikely to be closely related. The development of modern annulates is utterly different from that of chordates. The cleavage by which the fertilized egg is divided into blastomeres follows in annulates a 'spiral' plan, in which every blastomere arises in a regular way and the future fate of each can be exactly stated. In later annulates, such as the arthropods, this plan is complicated by the presence of much yolk, but even in these animals the cleavage does not resemble that of chordates, which is radial or 'irregular', the cells not forming any special pattern. This characteristic has been used to divide the whole animal kingdom into two major groups, Spiralia and Irregularia.

The next stage of development, gastrulation, by which the ball of cells is converted into a two-layered creature, also occurs very differently in the two groups. Our knowledge of the mechanics of the processes by which this change is produced is still imperfect, in spite of recent advances, but in lower chordates it occurs by invagination, the folding in of one side of the ball of cells to form an archenteric cavity communicating with the exterior. In annulates this is never seen; the cells that will go to form the gut migrate inwards either at one pole or all round the sphere and only later form themselves into a tube, which comes to open secondarily to the outside. It is probable that when we know more of the forces by which the gastrulation is produced the difference will appear even more marked than it does from this crude and formal statement that gastrulation in chordates is by invagination, in annulates by immigration.

The same applies to the method by which the mesoderm and coelom are formed. In lower chordates the third layer is produced by separation from the endoderm, so that the coelom is continuous with the archenteron and is said to be an enterocoele. In annulates cells separate in various ways to form the mesoderm and a coelom then arises within this solid mass as a schizocoele. It is true that in some, indeed many, of the higher chordates the coelom is never continuous with the archenteron, but its method of development shows it to be a modified enterocoele.

In all these points of development the chordates differ from the annulates, but resemble the echinoderms and their allies. Further features support this latter relationship. One of the most important of these is that the echinoderm-like animals, and some of the early chordates, have a larva with longitudinal ciliated bands, very different from the trochophore larva, in which the bands run transversely round the body, which is found in the other line of animals. The nervous system of annulates consists of a set of ganglionated cords, whereas in echinoderm-like animals it is a diffuse sheet of cells and fibres below the epidermis. The nerve-cord of the chordates can be derived from the latter but not easily from the former condition. Many further points could be cited, for instance, the presence of a mesodermal skeleton in both chordates and echinoderms, but not in annulates. It may be that there are also fundamental biochemical differences. Most of the spirally cleaving types of animal conduct their energy transfers with arginine phosphate, whereas vertebrates, amphioxus, ascidians, and ophiuroid echinoderms use creatine phosphate. *Balanoglossus* and echinoids have both.

In the study of evolution it is not sufficient merely to make formal comparisons, we must try to find out and compare the plan of development and structure common to all members of two groups, a technique often requiring great knowledge and good sense. When this is done in the present case it will be found that the essential plan of development of annulates involves spiral cleavage, gastrulation by immigration, and a coelom formed as a schizocoele, a trochophore-like larva, and full segmentation of the mesoderm. It is exceedingly unlikely that such animals have given rise to chordates with their very different development, which we may crudely define as showing radial cleavage, gastrulation by invagination, and larva of echinoderm type.

Extending this method we may divide the whole world of Metazoa by similar criteria into Spiralia or Polymera and Irregularia or Oligomera. The former include besides the annulates the molluscs and platyhelmia, whereas the latter group contains, in addition to the chordates, the echinoderms, brachiopods, polyzoa (ectoprocta), graptolites, pogonophora, and *Phoronis*. The animals in this latter group seem at first sight to be very different from the chordates in outward form, but the farther we look into their fundamental organization, the more we become convinced that the ancestors of the fish-like animals are to be found here. By study of the relics of the early chordates it is possible to trace the history of this strange change with some plausibility, though its full details will probably never be known.

2. Subphylum Hemichordata (= Stomochordata)

Class 1. Enteropneusta

Balanoglossus; Glossobalanus; Ptychodera; Saccoglossus

Class 2. Pterobranchia

Cephalodiscus; Rhabdopleura

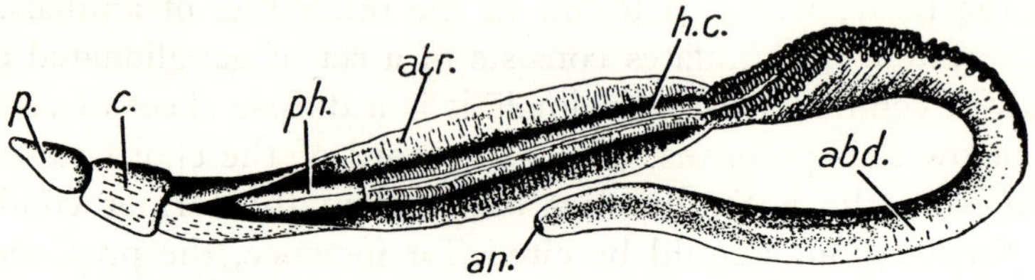

FIG. 21. *Balanoglossus*, removed from its tube and seen from the dorsal side. *abd.* abdomen; *atr.* atrium; *an.* anus; *c.* collar; *h.c.* hepatic caeca; *p.* proboscis; *ph.* pharynx. (From van der Horst.)

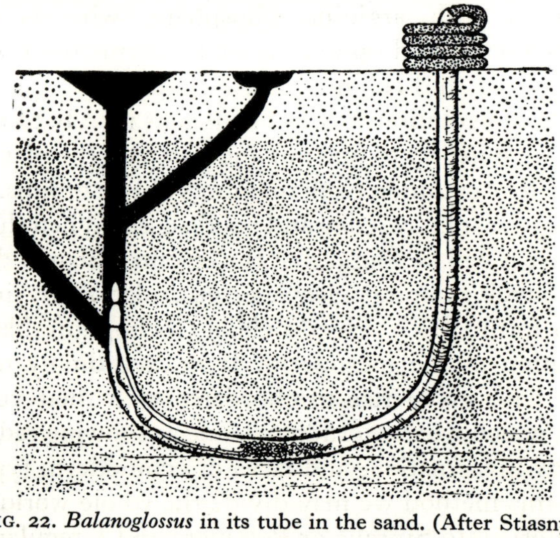

FIG. 22. *Balanoglossus* in its tube in the sand. (After Stiasny.)

In the Hemichordata are placed animals of two types, the worm-like *Balanoglossus* and its allies (Enteropneusta) and two sedentary animals, *Cephalodiscus* and *Rhabdopleura* (Pterobranchia). The Enteropneusta are mostly burrowing animals (Figs. 21 and 22) varying in different species from 2 cm to over 2 metres long. Several genera are recognized (e.g. *Balanoglossus*, *Saccoglossus*, *Ptychodera*) and they occur in all seas. *Saccoglossus* occurs around the British coast. The body is soft, without rigid skeletal structures, and divided into proboscis, collar, and trunk. The animals are very fragile and it is difficult to collect specimens in which the hind part of the trunk ('abdomen') is intact. The proboscis, collar, and trunk each contain a coelomic cavity, and the coeloms of the proboscis and collar are distensible by intake of water through a single proboscis pore and paired collar pores. The skin is richly ciliated all over the body. The outer epithelium is thus unlike the squamous, layered skin of higher forms (Fig. 23). It contains numerous gland-cells, whose secretion is very copious, so that the animals are always covered with slime.

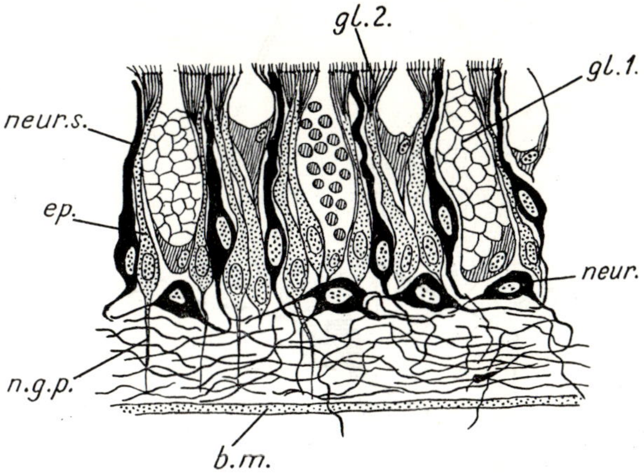

FIG. 23. Section of the epidermis of an enteropneust.
b.m. basement membrane; *ep.* epidermal cell; *gl.* 1 and 2, different types of gland cell; *neur.* neuron; *neur.s.* neuro-sensory cell; *n.g.p.* process of epidermal cell acting as neuroglia in the nerve net. (After Bullock, v. der Horst and Grassé.)

A characteristic feature is an unpleasant smell, resembling that of iodoform, which possibly serves, like the mucus, as a protection.

Below the skin is a nerve plexus receiving the inner processes of receptor cells and containing ganglion cells (Fig. 23). Deep to this are muscles running in various directions. It is said that the animal moves by first pushing the proboscis and collar forward through the sand and then drawing the body after it. Protrusion of the proboscis cannot, however, be very vigorous. It may perhaps be produced by ciliary action distending the coelom as is usually stated—more probably by circular muscles, but these are weak. Numerous longitudinal muscles are present, however, in the proboscis and trunk and are partly attached to a plate of skeletal tissue in the collar. This tissue is attached to the ventral side of a forwardly directed diverticulum of the pharynx. The wall of this is thick, composed of vacuolated cells, and bears a certain resemblance to a notochord (Fig. 24). A notochord extending throughout the length of the body would clearly be disadvantageous for an animal whose main movements are lengthening and shortening. It is possible that the diverticulum and plate found in the collar represent the remains of a notochord, serving as a fixed point by which the body is drawn forward on to the proboscis. However, many prefer to call it a 'stomochord' to avoid too close a

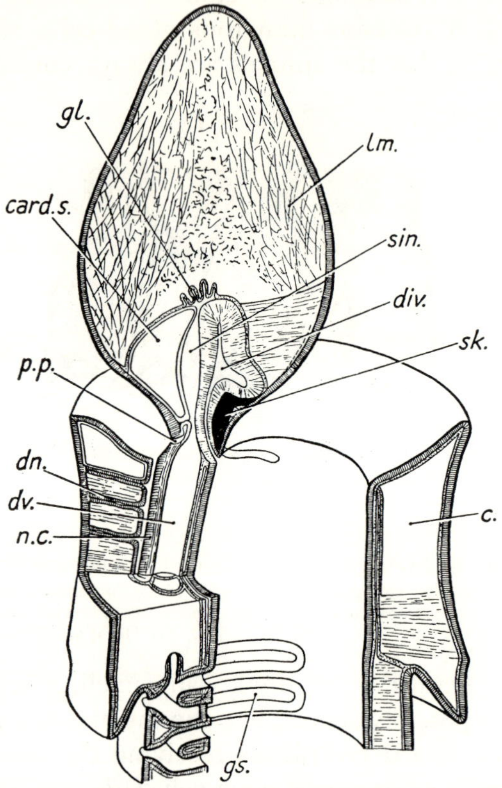

FIG. 24. Diagrammatic section of front end of *Balanoglossus*.

c. collar coelom; *card.s.* sac around heart; *div.* pharyngeal diverticulum ('stomochord'); *dn* dorsal nerve-root; *dv.* dorsal vessel; *gl.* glomerulus; *gs.* gill-slit; *lm.* longitudinal muscles of proboscis; *n.c.* nerve-cord; *p.p.* proboscis pore; *sk.* skeletal plate. (Modified after Spengel.)

comparison with the notochord. The external cilia probably play a considerable part in locomotion; possibly they are the chief burrowing organs, the muscles serving mainly to perform escape movements.

The mouth lies in a groove between the proboscis and collar (Fig. 25). The proboscis contains many mucus-secreting cells and the food particles are captured on its surface and conveyed by ciliary currents to the mouth. In the anterior part of the trunk there is a wide pharynx, opening by a series of gill-slits (Figs. 24, 26). These resemble the gills of amphioxus in the presence of a supporting skeleton in the gill bars; there are also tongue bars dividing the slits from above, and

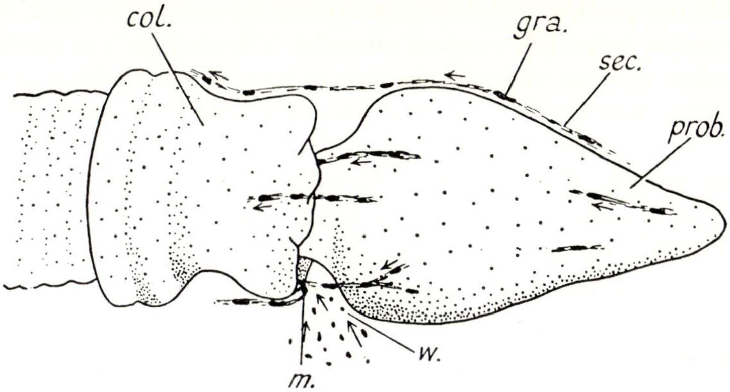

FIG. 25. Feeding-currents on proboscis of *Glossobalanus*, shown by placing the animal in water containing carmine particles. The particles (*gra.*) are either taken directly into the mouth (*m.*) as at W., or are caught up in strands of mucus (*sec.*) and passed backwards. (From Barrington, *Quart. J. Micr. Sci.* **82,** by permission.)

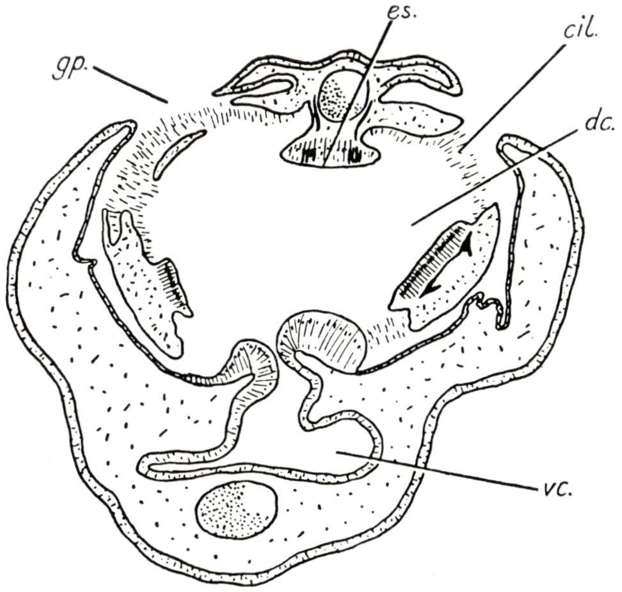

FIG. 26. Transverse section of the pharynx of *Glossobalanus*.
cil. cilia of the gill bars; *dc.* dorsal chamber of pharynx; *es.* epibranchial strip; *gp.* gill pore; *vc.* ventral chamber of pharynx. (From Barrington. With permission as for Fig. 25.)

horizontal synapticulae strengthening the gill arches. The slits open in some species into an atrium formed by lateral folds, usually turned upwards to leave a long mid-dorsal opening. In some species each slit opens to a gill pouch. The whole branchial apparatus perhaps assists in the process of feeding, probably by serving to filter off the excess water from the material already collected on the proboscis, which often consists of large amounts of sand or mud. Relative to the size of the animal the pharynx is less extensive than in amphioxus, presumably because ciliary surfaces are provided on the outside and also large masses of sand are forced into the mouth during locomotion.

There is no endostylar apparatus, but the ventral part of the pharynx is often partly separated from the rest (Fig. 26). Along this groove the matter ingested is passed to a straight oesophagus and intestine opening by a terminal anus. There is no true tail in the adult but a post-anal region is present in some species during development. Numerous

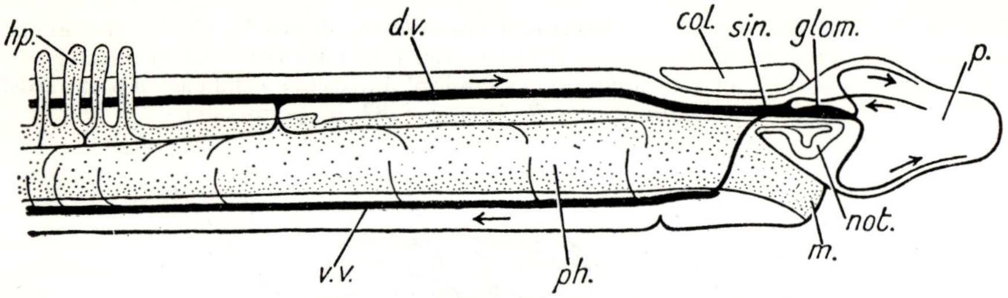

FIG. 27. Diagram of the blood system of *Balanoglossus*.
col. collar; *d.v.* dorsal vessel; *glom.* glomerulus; *hp.* hepatic caeca; *m.* mouth; *not.* 'notochord'; *p.* proboscis; *ph.* pharynx; *v.v.* ventral vessel. (After Bronn.)

hepatic caeca in the anterior part of the intestine can be seen from the outside as folds of the body wall, often highly coloured.

The blood system consists of a complex set of haemocoelic spaces, communicating with large dorsal and ventral vessels (Fig. 27). The former enlarges into a sinus anteriorly and this is partly surrounded by the wall of a pericardial cavity, which contains muscles and may be said to be the heart, though clearly lying in a very different position from that of other chordates. From the sinus, vessels proceed to the proboscis and round the pharynx to the ventral vessel. The blood is said to move forwards in the dorsal and backwards in the ventral vessels. The front of the sinus forms a series of glomeruli, covered by a region of the proboscis coelom specialized to form excretory cells, the nephrocytes, some of which drop off into the coelom. The blood is red in some species but usually colourless. It contains a few amoebocytes.

The nervous system is one of the most interesting features of Enteropneusta. It resembles that of echinoderms in consisting of a sheet of nerve-fibres and cells lying beneath the epidermis all over the body (Fig. 23). This sheet is thick in the mid-dorsal and mid-ventral lines, and in the dorsal part of the collar region it is rolled up as a hollow neural tube, open at both ends (Fig. 24). These unmistakable resemblances not only to the uncentralized sub-epithelial plexus of echinoderms but also to the hollow dorsal nerve-cord of vertebrates are most instructive, showing the affinity of the groups and the origin of the general plan of the vertebrate nervous system. There are no organs of special sense, unless this is the function of a patch of special ciliated cells on the collar. Receptor cells all over the body send their processes into the nerve plexus (Fig. 23), on the primitive plan of neurosensory cells found elsewhere in vertebrates only in the olfactory epithelium and the retina. The plexus is remarkable in receiving fibres from the outer ciliated epithelial cells, which thus represent the ependyma, the earliest form of neuroglia (Fig. 23). Nothing is known of the organization of pathways or of the connexions with the muscles. The collar nerve-cord contains giant nerve-cells whose axons proceed backwards to the trunk and forward to the proboscis (Fig. 28). They are probably responsible for rapid contractions (Knight-Jones, 1951).

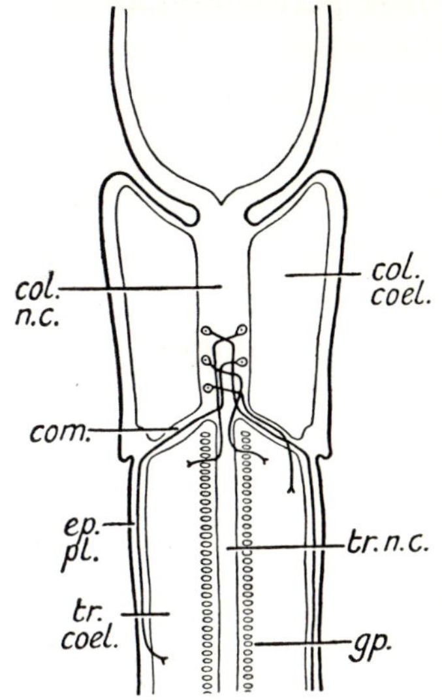

FIG. 28. Diagram of certain tracts in the nervous system of *Balanoglossus*. *com.* circular connective; *col.coel.* collar coelom; *col.n.c.* collar nerve-cord; *ep.pl.* nerve plexus in epidermis of trunk; *gp.* gill pore; *tr. coel.* trunk coelom; *tr.n.c.* trunk nerve-cord. (From Bullock, *J. Comp. Neurol.*, vol. **80**, by permission.)

Bullock has investigated the behaviour of the animals and found only one clear-cut reflex, namely, a contraction of the longitudinal muscles in response to tactile stimulation. Isolated pieces of the body are able to show reflex responses, moving away from light or tactile stimuli. Such local actions are an interesting sign of the uncentralized nature of the nervous system, and similar actions are found in echinoderms. A further sign of lack of special conducting pathways is that stimulation of flaps of body wall partly severed from the rest produces generalized contraction, proving that conduction can occur in all directions. The dorsal and ventral nerve-cords do, however, act as quick conduction pathways, and contraction of the trunk following stimulation of the proboscis is delayed or absent if one, and especially if both, cords have been cut.

Perhaps the most interesting behaviour observed was the activity shown by an isolated proboscis, collar, trunk, or portion of trunk. These organs may move around vigorously in an exploratory manner; evidently the main nerve-cords are not necessary for the initiation of action, as is the central nervous system of higher chordates.

There are nerve-fibres in the walls of the pharynx and oesophagus, where peristaltic movements have been observed. Their relationship

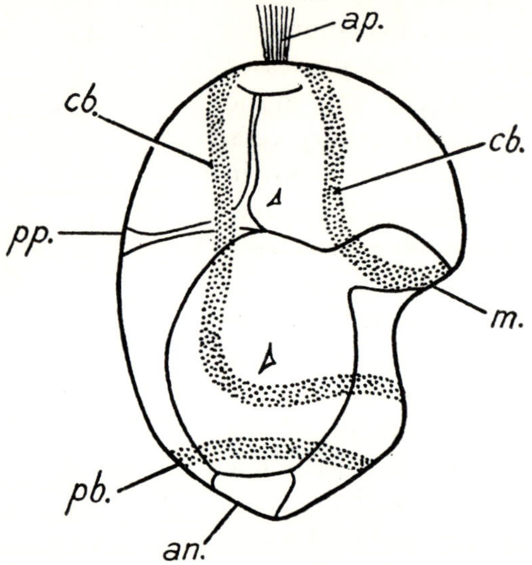

FIG. 29. Young tornaria larva, seen from the side.
an. anus; *ap.* apical organ; *cb.* longitudinal ciliated band; *m.* mouth; *pb.* posterior ciliated band; *pp.* proboscis pore. (After Stiasny.)

to the rest of the nervous system is unknown. They may represent the beginnings of an autonomic nervous system.

The sexes are separate in enteropneusts and the gonads resemble those of amphioxus in being a series of sacs developing from cells just outside the coelom. These proliferate and bulge into the coelom, covered by the somatopleure. They acquire a cavity and each opens by a narrow duct to the exterior, fertilization being external. The development is remarkably like that of echinoderms. Cleavage is holoblastic and resembles that of amphioxus and ascidians, gastrulation is by invagination, and the coelom is formed as an enterocoele, later becoming subdivided into proboscis, collar, and trunk coeloms. Hatching occurs to produce a pelagic tornaria larva, with a ciliated band that has exactly the relations found in the dipleurula larva of echinoderms. The band passes in front of the mouth, down the sides of the body, and in front of the anus (Fig. 29). It then divides into more dorsal and ventral sections, exactly as in the production of the bipinnaria larva of a starfish. This arrangement differs essentially from the rings of cilia that pass round the body in the trochophore larva found in the annelids and other spirally cleaving forms. In later tornaria larvae there is, however, in addition to the longitudinal bands always a posterior ring of stout cilia (telotroch), and in large oceanic forms (which may reach 8 mm in length) the longitudinal band itself is prolonged into prominent tentacle-like loops (Fig. 30). The cilia of the posterior ring are purely locomotive, while those of the band set up feeding-currents converging to the mouth. As the larva becomes larger the ciliary surface needed for locomotion and feeding has to increase relatively faster than the increasing mass of the body, the latter following the cube but the former only the square of the linear dimensions. Accordingly the cilia of the locomotive ring become broadened and flame-like, while the convolutions of the longitudinal (feeding) band reach fantastic proportions. In some types, however (*Saccoglossus*), the pelagic phase is brief and the telotroch alone is formed.

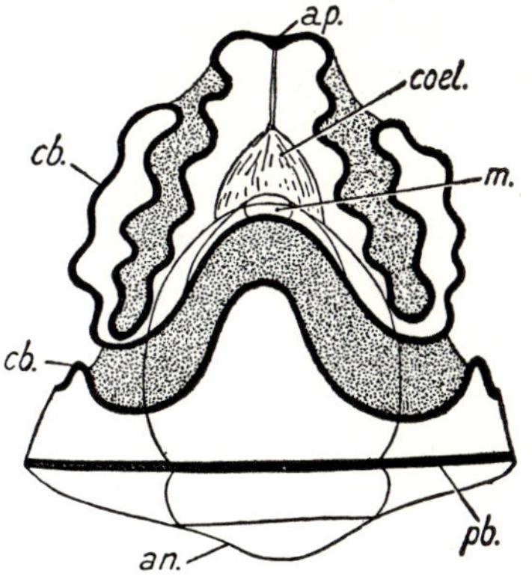

FIG. 30. Older tornaria larva seen from ventral surface.

Letters as Fig. 29; *coel.* proboscis coelom. (After Stiasny.)

Finally the larva sinks, becomes constricted into three parts, and undergoes metamorphosis into the worm-like adult. This development is so like that of an echinoderm that it would be necessary to consider the enteropneusts to be related to that group even if no other clues existed. Such close similarity in the fundamentals of development cannot be due to chance.

These animals thus provide a very remarkable and sure demonstration that the chordates are related to the echinoderms and similar groups. The general arrangement of the nervous system as a sub-epithelial plexus, as well as the whole course of the development, show the affinity with the invertebrate groups, whereas the hollow dorsal nerve-cord and the tongue-barred gill-slits are by themselves sufficient to show affinity with the chordates, this affinity being also perhaps suggested by other features, such as the 'notochord'. As we have seen already, affinities are not to be determined by single 'characters' but by the general pattern of organization of animals and especially that of their development. The organization of the enteropneusts is certainly highly specialized for their burrowing life, but showing through the special features we can clearly see a plan that has similarity with both the echinoderms and the chordates. The special value of study of these animals is that it proves decisively that an affinity between these groups exists. Exactly how they are all related is a more speculative matter, which we shall deal with later (see p. 74).

3. Class Pterobranchia

These are small, colonial, marine, sedentary animals, which show some signs of the general echinoderm-chordate plan of organization we have been discussing. *Cephalodiscus* (Fig. 31) has been found on the sea bottom at various depths, mainly in the southern hemisphere: there are several species. The colony consists of a number of zooids held together in a many-chambered gelatinous house. The zooids are formed by a process of budding, but do not maintain continuity with each other. Each zooid has a proboscis, collar, and trunk; there are coeloms in each of these parts, and proboscis and collar pores. The collar is prolonged into a number of ciliated arms, the lophophore, by means of which the animal feeds. There is a large pharynx, opening by a single pair of gill-slits, which serve as an outlet for the water drawn in by the cilia of the tentacles for the purpose of bringing food. The intestine is turned upon itself, so that the anus opens near the mouth. A thickening in the roof of the pharynx corresponds exactly in position with the stomochord and contains vacuolated cells. The blood system consists of a series of spaces arranged on a plan similar to that in *Balanoglossus*. There is a dorsal ganglion in the collar, but this is not hollow. The gonads are simple sacs and development takes place in the spaces of the gelatinous house. Gastrulation is by invagination at least in some species and the coelom is formed as an enterocoele. The larva somewhat resembles that of ectoproctous polyzoa, which is not closely similar to the echinoderm larvae, but could be derived from the same plan.

Rhabdopleura occurs in various parts of the world, including the North Atlantic and northern part of the North Sea. The zooids are connected together and have proboscis, collar, and trunk, ciliated arms, coelomic spaces with pores (not 'nephridia' as is sometimes stated) and stomochord, but no gill-slit. The development is not known.

The Pterobranchia thus show undoubted signs of the enteropneust-chordate plan of organization and provide also an interesting suggestion of possible affinities with Polyzoa, Brachiopoda, and *Phoronis*.

Like the Pterobranchia the Polyzoa Ectoprocta are sessile, with mouth and anus pointing upwards. They feed by means of the cilia borne on a horseshoe-ring of tentacles (the lophophore); but there is no division

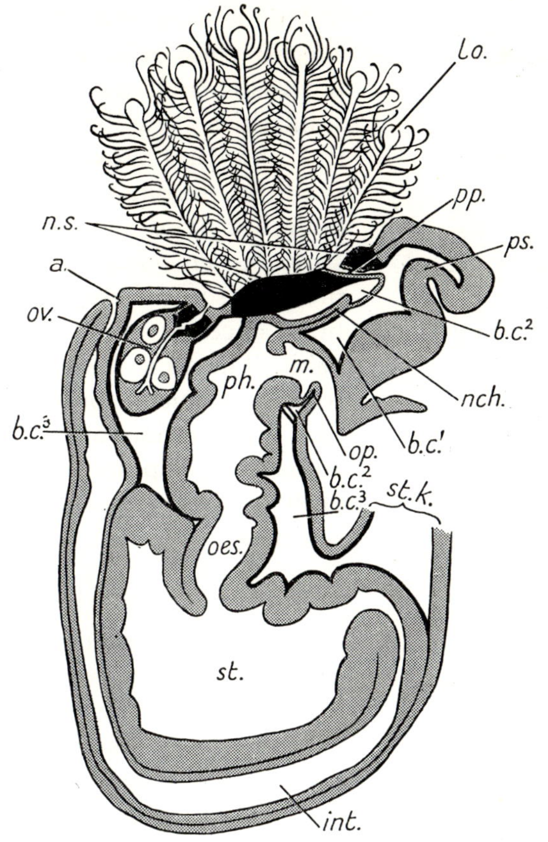

FIG. 31. Longitudinal median section of *Cephalodiscus*.
a. anus; *b.c.* 1, 2, and 3 body cavities; *int.* intestine; *lo.* lophophore; *m.* mouth; *nch.* 'notochord'; *n.s.* nervous system; *oes.* oesophagus; *op.* operculum (collar); *ov.* ovary; *ph.* pharynx; *pp.* proboscis pore; *ps.* proboscis; *st.* stomach; *st.k.* stalk.
(Modified after Harmer, *Cambridge Natural History*, Macmillan.)

into proboscis, collar, and trunk, and no tripartite coelom. The nervous system is in the condition of a sub-epithelial plexus, which is folded, around the base of the lophophore, to form a hollow tube—a remarkable point of similarity to the chordates. Even though it is difficult to compare this tube exactly with the nerve-cord of chordates, it is at least evidence of the organization of the nervous system on a plan that allows of such folding. It is probable that the modern pterobranchs are the surviving members of the ancient group of graptolites, but these are known only from the skeleton. The Pogonophora may also be distantly related, their larva can be regarded as of tornaria type, the coelom develops as in enteropneusts and the larval body shows three parts, as does that of the adult in some species.

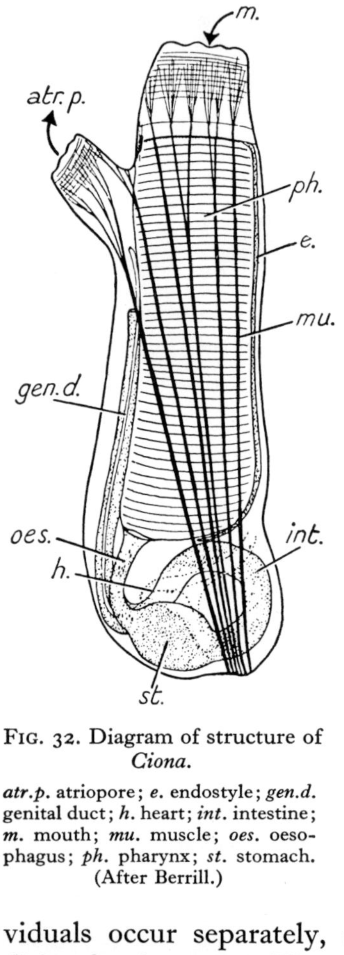

FIG. 32. Diagram of structure of *Ciona*.

atr.p. atriopore; *e.* endostyle; *gen.d.* genital duct; *h.* heart; *int.* intestine; *m.* mouth; *mu.* muscle; *oes.* oesophagus; *ph.* pharynx; *st.* stomach. (After Berrill.)

Although it would be unwise to suggest close relationship between the polyzoans and the pterobranchs, the similarities are sufficient to suggest that the chordates arose from sedentary creatures, feeding by means of ciliated tentacles. The evidence is sufficiently strong to encourage us to look for the presence somewhere in the line of vertebrate ancestry of an animal with this habit. The difficulties of this view arise when we come to consider how the fish-like organization of a free-swimming animal first appeared, a question better dealt with after consideration of the tunicates.

4. Subphylum Tunicata. Sea squirts

In the adult ascidians or sea squirts there is no obvious trace of the fish-like form at all. The majority of these animals are sac-like creatures living on the sea floor and obtaining their food by ciliary action. Often the separate individuals are grouped together to form large colonies, but in *Ciona intestinalis*, common in British waters, the individuals occur separately, and this is possibly the primitive condition for the group. The whole of the outside of the body is covered by a tunic, in which there are only two openings, a terminal mouth and a more or less dorsal atriopore, both carried upon siphons (Fig. 32). The tunic is made mainly of a carbohydrate, tunicin, closely related to cellulose, with which is combined about 20 per cent. of glycoprotein. It is secreted by the epidermis but contains special cells that have arrived there by migration from the mesoderm. In some tunicates calcareous secretions of various shapes are found in the tunic. The mantle that lines the tunic is covered by a single-layered epidermis.

Ascidians are often brightly coloured, the pigment being either in the tunic or the underlying body, which shows through the transparent tunic. The colour can change, at least over a period of some days. Little is known about the origin of the pigment, but it is sometimes derived from the blood-pigment and may lie in pigment cells.

The mantle is provided with muscle-fibres running in various directions but mainly longitudinally, and serving to draw the animal together, with the production of the jet of water from which the animals derive their common English name.

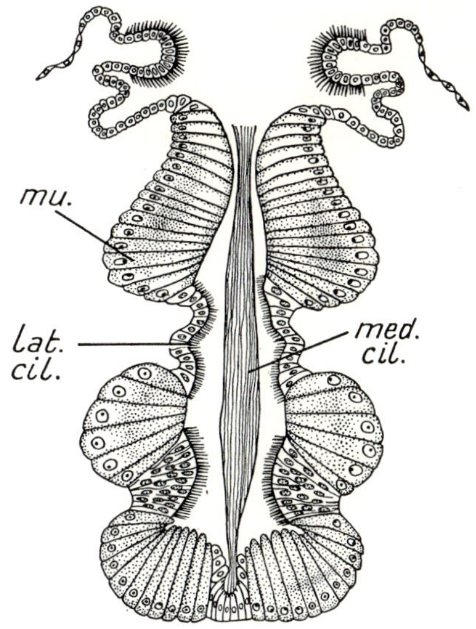

FIG. 33. Transverse section of the endostyle of *Ciona*.

lat. cil. lateral cilia; *med. cil.* long median cilia; *mu.* mucous cell. (After Sokoloska.)

The greater part of the body is made up of an immense pharynx, beginning below the mouth and forming a sac reaching nearly to the base (Fig. 32). The sac is attached to the mantle along one side (ventral) and is surrounded dorsally and laterally by a cavity—the atrium. This pharynx is, of course, the food-collecting apparatus; its walls are pierced by rows of stigmata (gill-slits) whose cilia set up a food current entering at the mouth and leaving from the atriopore. The entrance to the pharynx is guarded by a ring of tentacles, which may be compared with the velum of amphioxus. The stigmata are very numerous vertical cracks, all formed by sub-division of three original gill-slits. Tongue bars grow down to divide each slit and then from each tongue bar grow horizontal synapticulae. This arrangement has clear resemblance to that of amphioxus and results in the production of a pharyngeal wall pierced by numerous holes. Immediately within the stigmata there is a series of papillae, provided with muscles and cilia. There is an endostyle, which has three rows of mucus cells on each side, separated by rows of ciliated cells and with a single median set of cells with very long cilia (Fig. 33). The mucus secreted in the endostyle is caught up on the papillae, whose muscles move them rhythmically, spreading a curtain of mucus over the inside of the pharynx. Food particles are caught in the mucus, which moves upwards and is then passed back to the oesophagus by the cilia of a dorsal lamina or of a series of hook-like 'languets'. Autoradiographs made from tunicates that have been provided with isotopes of iodine show that iodination occurs in certain cells lying above the glandular tracts of the endostyle. Iodine is also abundant in the tunic, as it is in the exoskeletal structures of molluscs and insects. When it became of metabolic value its production may have become concentrated in the pharynx (see p. 118).

The extensive ciliated surface of the pharyngeal wall ensures the passage of large volumes of water inwards at the mouth and out at the atriopore. Rapid change of the water is also produced by periodic muscular contractions (p. 65). The pressure of the exhalant current is sufficient to drive the water that has been used well away from the animal.

The oesophagus leads to a large 'stomach' with a folded wall containing gland-cells, which produce digestive enzymes. These include much amylase, invertase, small amounts of lipase, and a protease of the tryptic type. The organ is therefore not to be compared with the stomach of vertebrates. A branching 'pyloric gland' opens into the lower end of the stomach. From the stomach a rather short intestine leads upwards to open inside the atriopore; this is apparently the absorptive region of the gut.

The heart lies below the pharynx and is a sac, surrounded by a pericardium (see p. 63) and communicating with a system of blood spaces derived from the blastocoele. The larger of these spaces have an endothelial lining; the biggest is a hypobranchial vessel below the endostyle, from which branches pass to the pharynx. From the opposite end of the heart springs a large visceral vessel and others pass to the dorsal side of the pharynx, tunic, body wall, &c. The heart is peculiar in that the beat can proceed in either direction. After passing blood into the hypobranchial vessel and gills for a few beats, its direction reverses, passing the blood to the viscera. This reversal is produced by the presence of two pacemaker centres, each capable of initiating rhythmical contractions, one at either end of the heart. Stimulation of these by warming and cooling allows control of the reversal of the beat. There are no capillaries and the blood system is a haemocoele. The blood-plasma is colourless but contains corpuscles, some of which are phagocytes, while others contain orange, green, or blue pigment (in different species). The green and other pigments are remarkable in that they contain vanadium. In some ascidians (*Molgula*) some individuals contain vanadium, others niobium (Carlisle, 1958).

The vanadocytes contain much sulphuric acid and the metal is associated with a chain of pyrrol rings. This haemovanadin is able to reduce cytochrome but it remains uncertain what part the pigment plays in respiration. The blood turns blue in air but cannot take up more oxygen than can sea water.

The blood is isotonic with sea water, and ascidians appear to have little or no power of regulating their osmotic pressure; none of them is found in fresh water. They are not even able to colonize brackish waters or those of low salinity. For example, they are rare in the Baltic Sea, from which only six species have been reported. Only one species, *Molgula tubifera*, has been reported from the Zuider Zee (salinity 8·4 per mille).

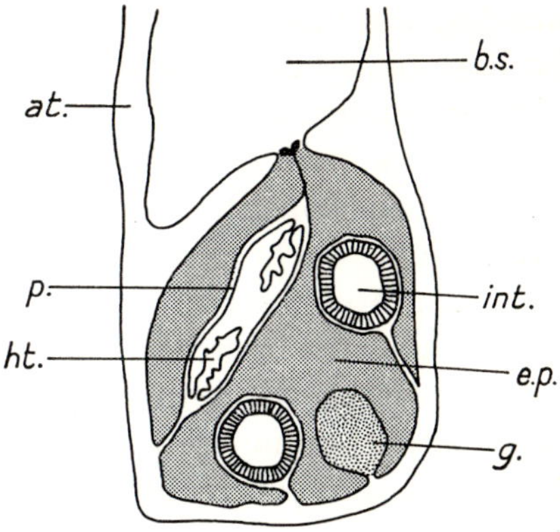

FIG. 34. Section through base of *Ciona*, showing heart, *ht.*, in pericardium, *p.*, and the epicardia, *e.p.*, opening into the pharynx, *b.s.*

at. atrium; *g.* gonad; *int.* intestine.

A possible reason for this inability to regulate the internal composition is perhaps the need to expose a large surface to the water. There are no tubular excretory organs such as could be used to maintain an osmotic gradient. Ninety-five per cent of the nitrogen is excreted as ammonia. Cells known as nephrocytes found in the blood and elsewhere contain concretions within the cytoplasm and these may in some cases be stored in an excretory sac until the animal dies.

There has been much debate as to whether the tunicates possess a coelomic cavity. The heart develops from a plate of cells arising early from the mesoderm and lying between ectoderm and endoderm. This becomes grooved and folded to make the heart itself and the pericardium. The irregular system of haemocoelomic spaces around the pharynx and elsewhere is usually said to consist of 'mesenchyme' and to be derived from the blastocoel and therefore not coelomic, but its walls are mesodermal. The situation is complicated by the presence of a pair of outpushings from the pharynx, the epicardia, or perivisceral sacs, which end blindly on either side of the heart (Fig. 34). Berrill and others have suggested that these epicardia may be compared with coelomic cavities. Their function in the open condition in which they are found in *Ciona* is perhaps to allow sea water to circulate about the heart and hence to help excretion (and respiration ?). In other ascidians the epicardium loses its connexion with the pharynx. The closed sac functions in some cases as an excretory organ, containing concretions of uric acid, whereas in other animals it becomes the main source of the cells that make the asexual buds.

The central nervous system consists of a round, solid ganglion (Fig. 36), lying above the front end of the pharynx. The ganglion has a layer of cells around the outside and a central mass of neuropil and is therefore quite unlike the nerve-cord of a vertebrate. From the ganglion nerves proceed to the siphons, other parts of the mantle, muscles, and viscera. Receptor cells with nerve-fibres ending around the base have been described, especially in the siphons. The gut is said to contain a plexus of cells and fibres, whose relation to the autonomic system of higher forms remains uncertain.

Movement consists mainly of contraction and closure of the apertures. Light touching of either siphon causes closure proportional to the strength of the stimulus. Stronger stimuli cause closure of both siphons and if very strong there is contraction of the whole body and ejection of the water in the pharynx and atrium. Stimulation just inside either siphon produces closure of the other one and also, if strong enough, contraction of the body, ensuring that a jet of water sweeps out the aperture that received the stimulus. These crossed reflexes depend upon the integrity of the ganglion.

The surface of the body is sensitive to changes in light intensity, and these are followed by local or total contractions, according to their extent. After removal of the ganglion the wider reflexes can no longer be obtained but local responses continue, suggesting the presence of nerve-cells in the body wall. Electrical stimulation also provides evidence of this. One shock may produce only a small response but if a second shock follows shortly afterwards there is marked facilitation and a large contraction occurs. These responses are also seen after removal of the ganglion. The various parts of the body are not all equally sensitive to light, the highest sensitivity being in the region of the ganglion. The 'ocelli' are cup-like collections of orange-pigmented cells around the siphons; according to Hecht they are not photoreceptors.

The neuromuscular system thus appears to function mainly as a reflex apparatus for producing protective movements in response to certain stimuli. This is the role that might be expected of it in an animal that remains fixed in one place. The 'initiative' for food-gathering activities comes from the continuous action of the cilia of the pharynx. The nervous system shows little sign of those continuous activities that produce the varied and 'spontaneous' acts of behaviour in higher forms. Nevertheless, it would be unwise to suppose that the nerves are only activated by external stimuli. There are some suggestions that even in these simple animals rhythmical activities are initiated from within. The food-collecting operations of the pharyngeal wall involve rhythmical movement of the papillae by their muscles. Further, in many species of ascidians there are regular contractions of the siphons and body musculature in rotation, with

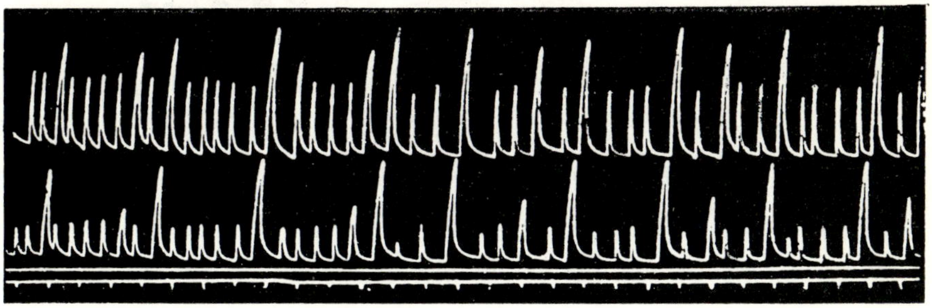

FIG. 35. Rhythmical 'spontaneous' contractions of *Styela* shown by attaching levers to the two siphons. Branchial siphon above, atrial siphon below. The time-marker shows intervals of 5 minutes. (From Yamaguchi.)

a frequency of 8–27 per hour (Fig. 35). These contractions are especially marked when the animal is in filtered water and they may be some form of 'hunger' contraction, directed towards the obtaining of food. More water is moved by these contractions than by the ciliary current. Their presence is a striking warning of the dangers of assuming that even the simplest nervous system operates only when stimulated from outside.

The neural gland is a sac lying beneath the ganglion and opening by a ciliated funnel on the roof of the pharynx. It arises mainly from the ectoderm of the larval nervous system, in part from the pharynx. This double embryological origin, and its position, suggest that the neural gland may be compared with the infundibulum and hypophysis of vertebrates. There is an obvious similarity with Hatschek's pit of amphioxus. Both seem to be receptor organs, testing the water stream and also producing mucus. The subneural gland has also been held to have a similarity to the pituitary in that it controls the release of gametes. When eggs or sperms of the same species are present in the water, signals from the neural gland apparently produce discharge from the gonad. The pathway of the signals is said to be partly hormonal, partly nervous. Discharge is produced by injection of extract of neural gland or of mammalian gonadotropin, but these act through the ganglion, since they produce no effect if the nerves leading from this (and the dorsal strand) are cut.

Further similarities with the pituitary have been claimed, such as the presence of vasopressor and oxytocic substances in the subneural gland. However, oxytocin is present elsewhere in the tunicate and in

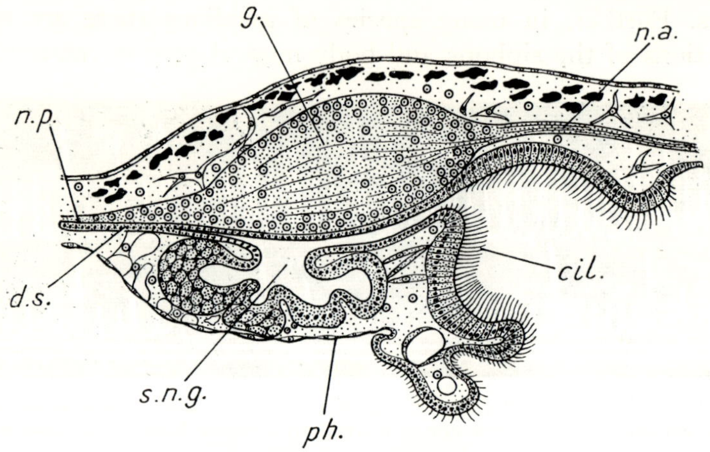

FIG. 36. Longitudinal section of the ganglion (*g.*) and subneural gland (*s.n.g.*) of an ascidian.

cil. ciliated funnel; *d.s.* dorsal strand; *n.a.* and *n.p.* anterior and posterior nerves; *ph.* wall of pharynx. (After L. Bertin from Grassé.)

any case differs from that of vertebrates. It cannot be claimed that the relationship with the pituitary is clear, but it seems likely that there is some. As in the thyroid, a pharyngeal mucus-secreting organ stimulated by the environment has evolved into a glycoprotein-secreting endocrine organ, controlled by substances reaching it in the blood. (Barrington, 1959, in Gorbman, *Symposium on Comparative Endocrinology*.)

5. Development of ascidians

Tunicates are hermaphrodite, the ovary and testis being sacs lying close to the intestine and opening by ducts near the atriopore. Fertilization is external in the solitary forms but internal in those that form colonies, the development in the latter taking place within the parent. The details of cleavage and gastrulation show a remarkable general similarity to those of amphioxus. Indeed, the whole development is so strikingly like that of chordates that it establishes the affinities of the tunicates far more clearly than the vague indications

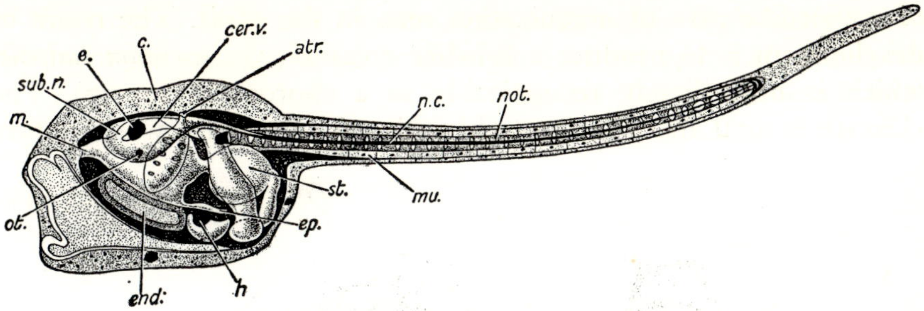

FIG. 37. Ascidian tadpole of *Clavelina*.

atr. atriopore; ***c.*** mantle; ***cer.v.*** cerebral vesicle; ***e.*** eye-spot; ***end.*** endostyle; ***ep.*** epicardium; ***h.*** heart; ***m.*** mouth; ***mu.*** muscle-cells; ***n.c.*** nerve-cord; ***not.*** notochord; ***ot.*** otocyst; ***st.*** stomach; ***sub.n.*** subneural gland.

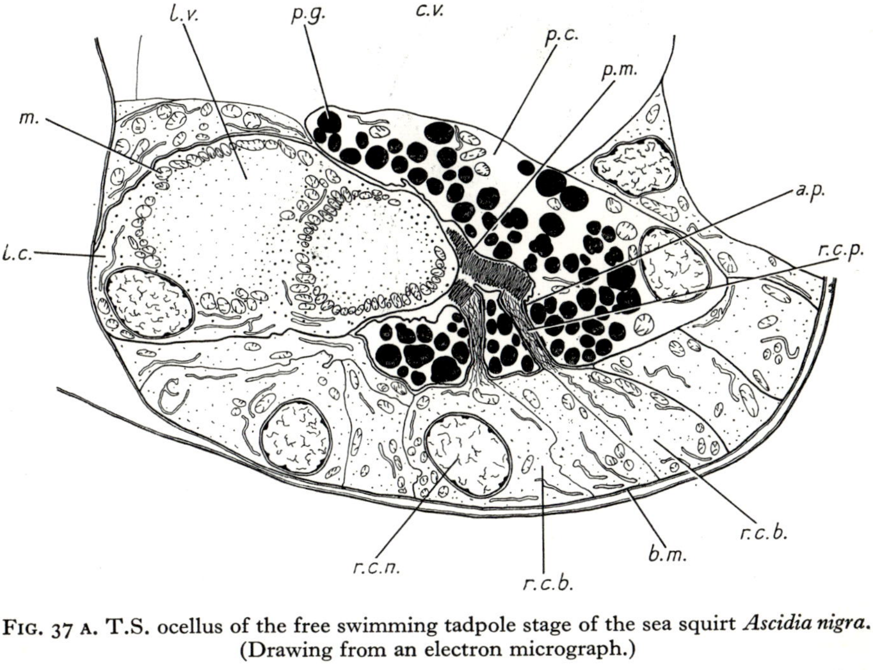

FIG. 37 A. T.S. ocellus of the free swimming tadpole stage of the sea squirt *Ascidia nigra*. (Drawing from an electron micrograph.)

The ocellus is situated in the posterior wall of the cerebral vesicle. It consists of three parts, a lens cell, a pigment cell, and a retina. The lens cell usually contains three lens vesicles, which are spheres of cytoplasm bounded by mitochondria. The pigment cell contains granules of melanin, which protect the photoreceptor from stray light. The retinal cells have processes that penetrate the pigment cell. They are similar to vertebral rods, composed of a pile of membranes, closely applied to the inner edge of the lens cell.

a.p. attachment plaque, a membrane specialization thought to function as an anchor of the retinal cell process to the pigment cell membrane; ***b.m.*** basement membrane, the outer limit of the cerebral vesicle; ***c.v.*** cavity of the cerebral vesicle; ***l.c.*** lens cell; ***l.v.*** lens vesicle; ***m.*** mitochondrion; ***p.c.*** pigment cell; ***p.g.*** pigment granule; ***p.m.*** piled menbrane of photoreceptor part of the retinal cell; ***r.c.b.*** retinal cell body; ***r.c.n.*** retinal cell nucleus; ***r.c.p.*** retinal cell process.
(From a preparation by N. Dilly.)

of a chordate plan of organization seen in the adult. The result of development is to produce a fish-like creature, the *ascidian tadpole*, which is immediately recognizable as a chordate (Fig. 37). The cleavage is total and produces a blastula with few cells, whose future

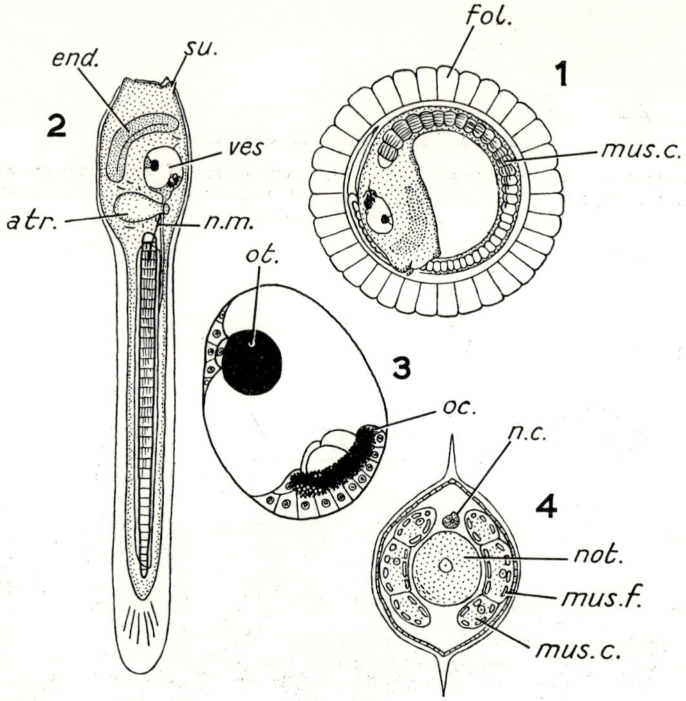

FIG. 38. The ascidian tadpole (*Ascidia* or *Ciona* type). 1. Tadpole ready to hatch. 2. Tadpole. 3. Sensory vesicle. 4. Cross section of tail.

atr. atrium; *end.* endostyle; *fol.* follicle cells; *mus.c.* muscle cells; *mus.f.* muscle fibrils; *n.c.* nerve-cord; *n.m.* nerve to tail muscles; *not.* notochord; *oc.* ocellus; *ot.* otolith; *su.* sticking gland; *ves.* sensory vesicle. (After Berrill.)

potentialities are already determined. Gastrulation by invagination follows and the creature then proceeds to elongate into the fish-like larva. This possesses an oval 'head' and long tail, the latter supported by a notochord formed by cells derived from the archenteric wall. Forty of these cells make up the entire rod, becoming vacuolated and elongated by swelling.

On either side of this notochord run three rows of muscle-cells, eighteen on each side, derived from mesoderm that arises from yellow-pigmented material already visible in the egg and later forming part of the wall of the archenteron. Other cells of this tissue migrate ventrally to make the pericardium, heart, and mesenchyme. The muscle-cells contain cross-striated myo-fibrils at the periphery, these being continuous from cell to cell.

The nervous system is formed by folds essentially similar to those of vertebrates, making a hollow, dorsal tube, extending into the tail and enlarged in front into a cerebral vesicle, within which is an ocellus and also a unicellular otolith (Fig. 37 A). Nerve-fibres proceed only to the front end of the rows of muscles and the rest of the cord contains no nerve-cells or fibres (Fig. 38).

The larva takes no food and the gut is not well developed. There is a pharynx with usually a single pair of gill-slits opening into an atrium, which develops as an ectodermal inpushing. Below or around the mouth various forms of sucker are formed.

The whole process of development occupies only one or two days, and the larva, in the species in which it is set free, is positively phototropic and negatively geotropic and so proceeds to the sea surface. But its life here is also limited. Within a day or two, depending on the conditions, its tropisms reverse so that it passes to the bottom, turns to any dark place and thus finds a suitable surface. It attaches by the suckers, loses its tail, develops a large pharynx, and grows into an adult ascidian. Presumably its short life in the chordate stage is sufficient to ensure distribution, and the simple nervous system serves to find a place in which to live.

In addition to the sexual reproduction, tunicates have great powers of regeneration and also often multiply by budding. The bud consists of an outer epicardial, mesenchymal, pharyngeal or atrial tissue. The epidermis develops only more tissue like itself and all the other tissues are formed from the inner mass. This occurs by a process of folding to make a central cavity; the nervous system, intestine, and pericardium are then formed by further foldings. The bud thus begins in a condition comparable to a gastrula but develops directly into an adult, without passing through the tadpole stages. The fact that a complete new animal is thus formed from one or two layers shows that the separation into three layers during development does not involve any fundamental loss of potentialities, as would be required if the 'germ layer' theory held rigorously. The germinal tissue of the bud is not necessarily derived from that of the parent.

6. Various forms of tunicate

Besides some 2,000 species of sessile tunicates, about 100 species have become secondarily modified for a pelagic life. These pelagic animals are perhaps all related, but the whole subphylum is conveniently subdivided into three classes.

Class 1. Ascidiacea.

Typical bottom-living forms such as *Ciona* (solitary), *Botryllus* (colonial).

Class 2. Thaliacea.

Pelagic forms, simple or colonial, swimming by means of circular muscle bands. *Salpa*, *Doliolum*, *Pyrosoma*.

Class 3. Larvacea.

Pelagic tunicata without metamorphosis; the adult has a tail and resembles the tadpole of the other groups. *Oikopleura*.

7. Class Ascidiacea

The typical sessile ascidians are found in all seas. They may be divided into those that live as single individuals (Ascidiae simplices) and those forming colonies (Ascidiae compositae). Both types include many different forms, however, and the division is not along phylogenetic lines. The colonial forms produced by budding may consist simply of a number of neighbouring individuals (*Clavelina*) or of a common gelatinous test in which the individuals are embedded (*Botryllus*, *Amaroucium*). The form of the body is related to the type of bottom upon which they are found; there has thus been an adaptive radiation within the group; a great variety of habitats is available for bottom living creatures, and the animals become adapted accordingly.

Most of the species live in the littoral zone, but a few deep-sea forms are known, such as *Hypobythius calycodes*, found below 5,000 metres.

Many ascidians probably live only for a short time, becoming mature in their first year and dying thereafter. In some species the animals live over a second winter, during which they become reduced in size, growing and budding again in the following spring (*Clavelina*).

8. Class Thaliacea

These are pelagic tunicates living in warm water. They have circular bands of muscle, enabling the animal to shoot through the water by jet propulsion. In *Doliolum* and its allies the muscle-bands pass right round the body (Cyclomyaria), whereas in *Salpa* the rings are incomplete (Hemimyaria). The mouth and atriopore are at opposite ends of the body. The tunic is thin and, like the rest of the body, transparent.

The life-history of these forms involves a remarkable alternation of

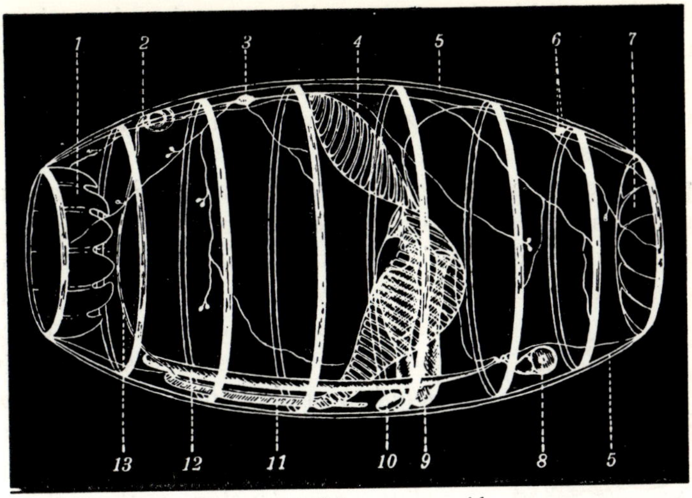

FIG. 39. *Doliolum*, gonozooid.
1, inhalent aperture; 2, ciliated pit; 3, ganglion and nerves; 4, pharynx; 5, mantle; 6, sense-cells, 7, exhalant aperture; 8, ovary; 9, intestine; 10, heart; 11, endostyle; 12, testis; 13, ciliated groove. (After Neumann.)

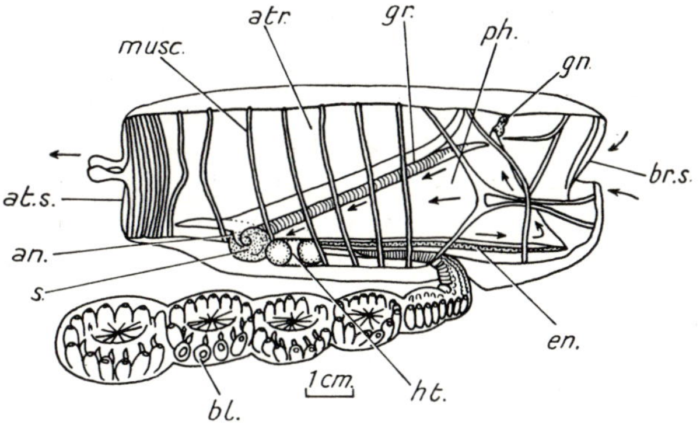

FIG. 40. *Cyclosalpa affinis*, oozooid with chain of five wheels of blastozooids.
an. anus; ***atr.*** atrium; ***at.s.*** atrial siphon; ***bl.*** blastozooid with egg; ***br.s.*** branchial siphon; ***en.*** endostyle; ***gn.*** ganglion; ***gr.*** gill ridge; ***ht.*** heart; ***musc.*** muscle ring; ***ph.*** pharynx; ***s.*** stomach. ($\times \frac{1}{2}$ modified. After Ritter and Johnson and Berrill.)

generations. In *Doliolum* the ascidian tadpole develops into a mother or nurse zooid (oozooid). This by budding gives rise to a string of daughter zooids, which it propels along by its muscles. The daughter zooids are of three types: (1) sterile, nutritive, and respiratory individuals, the trophozooids, permanently sessile on the parent; (2) sterile nurse forms, which are eventually set free (phorozooids); (3) sexual forms (gonozooids, Fig. 39), nursed and carried by the phorozooids until sexually mature, when they also break loose.

In *Salpa* the sexual form (blastozooid), produces only a single egg, which develops within the mother without passing through a tadpole stage, nourished by a diffusion placenta, whose cells also migrate into the developing embryo. This becomes the asexual oozoid and produces a long chain of blastozooids, which it tows about until these break away by sections (Fig. 40).

The pelagic colonial *Pyrosoma* of warm seas consists of a number of individuals associated to form an elongated barrel-shaped colony. The mouths open outwards and the atria inwards into a single cavity with a terminal outlet from which a continuous jet emerges. The mode of budding from the epicardium and other features suggest an affinity with *Doliolum* and *Salpa*, but *Pyrosoma* also resembles the ascidians in that its zooids are all sexual and capable of budding. The yolky eggs develop within the parent, without forming a larva. The outstanding characteristic of the creatures is the powerful light that they shine. This is produced in photogenic organs on each side of the pharynx. The photogenic cells contain curved inclusions about 2 μ in diameter (Fig. 41). These are considered by some to be symbiotic luminescent bacteria, but this is doubtful. The light is so powerful that when large masses of *Pyrosoma* occur together the whole sea is illuminated sufficiently to allow of reading a book. A remarkable feature of the phenomenon is that the light is not produced continuously but only when the animal is stimulated, as by the waves of a rough sea. If one individual is stimulated others throughout the colony may show their lights, but the mechanism of this effect is not known and the groups of cells that form the luminescent organs receive no nerves. Other types of animal with luminescent bacteria emit light continuously. The sudden flashes of light probably serve as a dymantic reaction (p. 302), giving protection against enemies by producing a flight-reaction in the same way as do sudden manifestations of colour or black spots by other animals. It has been observed in the laboratory that colonies of *Pyrosoma* that are dying and do not light up may be eaten by fishes, whereas any that light up when seized may then be dropped.

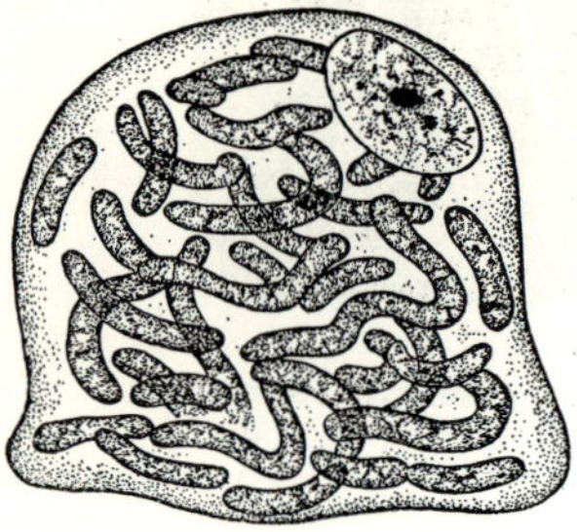

FIG. 41. Photogenic cell of *Pyrosoma*. (After Kukenthal.)

9. Class Larvacea

The (Appendicularia) Larvacea (Figs. 42 and 43) are minute neotenous tunicates that live in the plankton. Instead of the test, each

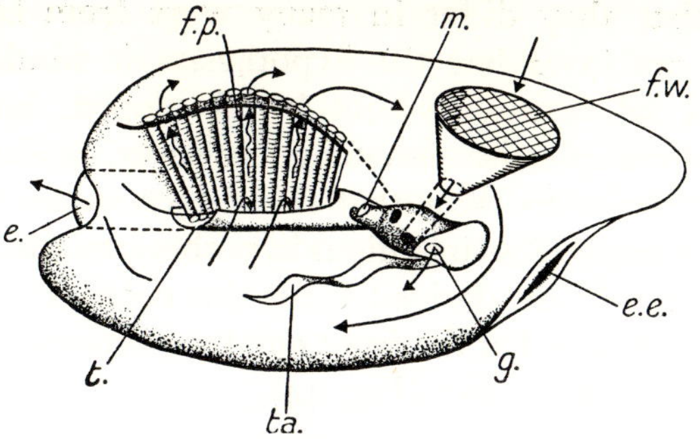

FIG. 42. *Oikopleura*, one of the Larvacea, in its house, showing the feeding-currents.

e. exhalant aperture; *e.e.* 'emergency exit'; *f.p.* filter pipes; *f.w.* filter window; *g.* gill-slit; *m.* mouth; *t.* trough; *ta.* tail. (After Garstang; this and Figs. 44 and 45 by permission of the Editors of the *Quarterly Journal of Microscopical Science.*)

individual builds a 'house', by secretion from a special part of the skin, the 'oikoplastic epithelium'. The tail is a broad structure held at an angle to the rest of the body; its movement produces a current in which the food is carried and caught by a most elaborate filter arrangement in the house (Fig. 42). Water enters the house by a pair of posterior 'filtering windows' and is passed through a system of filter pipes in the part of the house in front of the mouth. The very minute flagellates of the nanonplankton are stopped by these pipes and sucked back to the mouth. The pharynx has two gill-slits, also an endostyle and peripharyngeal bands. The general organization is that of a typical ascidian tadpole, and there can be no doubt that these forms have arisen from tunicates by the acceleration of the rate of development of the alimentary organs and gonads so that the metamorphosis and normal adult stage are eliminated. This may, of course, have happened long ago, so that the modern Larvacea are not closely related to any living forms,

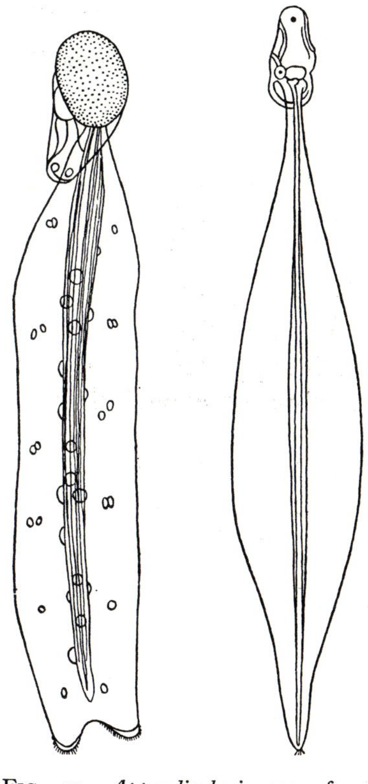

FIG. 43. *Appendicularia* seen from the side and from below. (After Lehmann.)

but the fact that they differ in many ways from known ascidian tadpoles does not invalidate the hypothesis; it would be expected that many special features would be developed during evolution after the paedomorphosis. Garstang, however, believed that there is sufficient evidence to show that the Larvacea are related to the Doliolidae and suggested an ingenious hypothesis by which the appendicularian home could be derived from the doliolid test, the animal itself remaining attached at the front end by gelatinous threads, which came to make the filter tubes (Fig. 44).

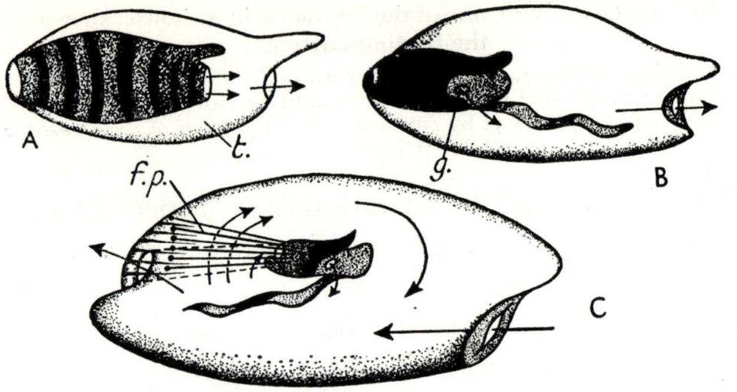

FIG. 44. Sequence of stages by which the Larvacea may have been evolved from a doliolid type.

A, Thaliacean type of individual in its test (*t*). B, Paedomorphosis has occurred so that a tailed creature is found in the test; *g*. gill-slit. C, The tadpole has moved away from the inhalant aperture, leaving a series of threads that become the filter pipes (*f.p.*), the inhalant aperture becoming exhalant and vice versa. (After Garstang.)

The tail is a highly developed organ, serving for locomotion, nutrition, and in the building of the house. It has a wide, continuous fin and is supported by a notochord of 20 cells. Bands of 10 large striped muscle-cells extend down each side, giving an appearance that has been compared with metameric segmentation. The small number of the cells makes any such comparison very difficult. Moreover, the muscles are not developed from anything resembling myotomes. The nerve-cord is a hollow tube with ganglionic thickenings, each containing one to four nerve-cells. From these cells fibres proceed to the muscles and to the skin in a series of roots that usually remain separate, the motor being more dorsal.

10. The formation of the chordates

We can now recapitulate the points that we have established about the origin of the chordates and attempt to piece together the evidence to show the sequence of events that led to the production of a free-swimming, fish-like animal. The chordates are related to the echinoderms and their allies. This is established by the similarities in early development (cleavage, gastrulation, mesoderm formation); by the presence in early members of both groups of three separate coelomic cavities, some with pores; by the similarity of the larva of enteropneusts to the dipleurula, and by other points of general morphological and biochemical similarity between early chordates and echinoderms, especially the arrangement of the nervous system and presence of a mesodermal skeleton.

The echinoderms we have to consider are not the modern starfishes and sea-urchins, which are relatively active animals, but their sessile Palaeozoic ancestors. These were sedentary, often stalked animals, the cystoids, blastoids, and crinoids, feeding by ciliary action. Surviving animals of related phyla, such as Polyzoa Ectoprocta and *Phoronis* suggest that the ancestor for which we are looking may have possessed a ciliated lophophore for food-collecting. For purposes of dispersal its life-history presumably included a larval stage with a longitudinal ciliated band, similar in plan to that of the auricularia.

One might well ask how such an animal could possibly become converted into a motile, metameric fish, feeding with its pharynx. Yet the evidence of the lower chordates is sufficient to establish that this change has occurred, and even provides us with an outline of the main stages in the process of the change. *Cephalodiscus*, which is in some ways the most primitive of surviving chordates, with its lophophore also possesses gill-slits. This suggests that the pharyngeal mechanism was substituted for the lophophore as a means of feeding in the adult stage. There are other possible interpretations. It has been suggested that *Cephalodiscus* was derived from a larval enteropneust (Burden-Jones). However, it is possible that ciliary mechanisms developed in the pharynx first to deal with food collected outside by tentacles or proboscis. Later the pharynx became developed into a self-contained feeding mechanism, making unnecessary the tentacles, which provide a tempting morsel for predators. The adoral band of cilia of the auricularia probably serves to carry food into the mouth, and for this purpose it is actually turned in to the floor of the pharynx. Garstang suggests that the endostyle has been derived from this loop of the adoral band.

The pharyngeal method of food-collecting thus replaced the tentacles in the adult and the whole apparatus of an endostyle and an atrium to protect the gills became developed. We may notice here the remarkable similarity of this arrangement of the pharynx in tunicates, amphioxus, and cyclostome larvae, and the partial similarity in *Balanoglossus*.

The tunicates show us a stage in which branchial feeding has fully replaced tentacle feeding in a sessile adult. But they have a larva that is beyond all question a fish-like chordate. If the adult tunicate has evolved from a modified lophophore-feeding creature, how has the

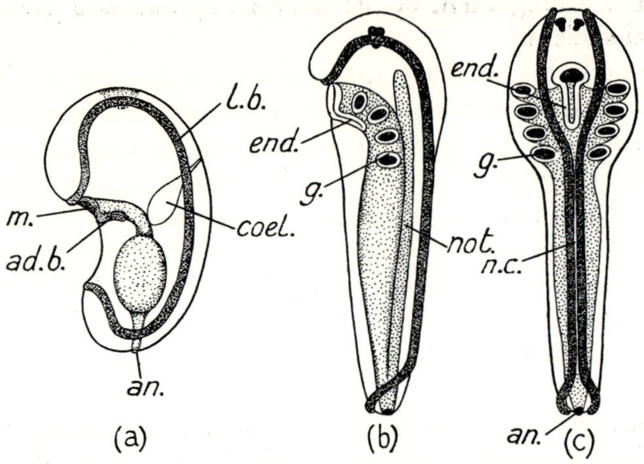

FIG. 45. To show the method by which a protochordate animal might have been derived from an echinoderm larva such as the auricularia.

a. Auricularia in side view; *b.* protochordate in side view; *c.* same, dorsal view. *ad.b.* adoral band; *an.* anus; *coel.* coelom; *end.* endostyle; *g.* gill-slit; *l.b.* longitudinal ciliated band; *m.* mouth; *n.c.* nerve-cord; *not.* notochord. (After Garstang.)

ascidian tadpole arisen from the auricularia larva? Garstang's auricularia theory, first propounded in 1894, provides a possible answer. As a ciliated larva grows its means of locomotion becomes inadequate because the ciliated surface increases only as the square of the linear dimensions, the weight as the cube. Muscular locomotion is not subject to this difficulty, and some of the starfish larvae actually show flapping of the elongated processes, movements that presumably assist them to remain at the surface. Garstang suggests that the fish-like form arose by development of muscles along the sides of the elongated body, the ciliated bands being pushed upwards and eventually rolled up with their underlying sheets of nerve plexus to form the neural tube. The adoral ciliated band might then well be the endostyle (Fig. 45).

This theory may seem at first sight fantastic. It is necessarily speculative, but it has certain strong marks of inherent probability. It violates no established morphological principles and certainly enables us to see how a ciliated auricularia-like larva could be converted by progressive stages into a fish-like creature with muscular locomotion, while the adults, at first sedentary, substituted gill-slits and endostyle for the original lophophore. The alternative is to suppose that the ascidian tadpole arose as a purely tunicate development, providing sufficient receptor and muscular organs to allow for the finding of suitable sites on the bottom (Berrill, 1955).

We may plausibly regard the adult tunicate organization as directly derived from that of sessile lophophore-feeding creatures, and the larval organization as descended from an echinoderm-like larva. There is no need, on this view, to regard the sessile adult tunicate as a 'degenerate' chordate. The problem that remains is in fact not 'How have sea-squirts been formed from vertebrates?' but 'How have vertebrates eliminated the sea-squirt stage from their life-history?' It is wholly reasonable to consider that this has been accomplished by paedomorphosis. Advance of the time of development of the gonads relative to that of the soma is well known to occur in certain special cases such as the axolotl. The example of the Appendicularia shows that a similar process can happen among tunicates! Various workers have stressed the differences between the ascidian tadpole and the adult appendicularian, in attempts to show that the two are not comparable. But the differences, though considerable, are superficial: the similarity of organization is profound. Any sensible biologist with an understanding of the way in which the characteristic forms of animals arise by change in the rate and degree of development of features can see how the Appendicularia may represent modified ascidian larvae.

The appendicularians do, indeed, carry certain characters of the 'adult' sea-squirt, in particular they have gill-slits, though of simple form. Nothing is more likely, however, than that some features of the sessile adult would be adumbrated in its larva and capable of fuller development therein if advantageous. Larva and adult, it must be remembered, possess the same genotype; the remarkable feature in all animals with metamorphosis is the difference between the two stages, not the similarity. Any characteristic may appear at either larval or adult stage or be transferred by evolutionary selection from one to the other. There is no serious objection to the view that the early adult free-swimming chordates arose by paedomorphosis of some tunicate-like metamorphosing form. If the creatures abandoned the habit of fixation it would be possible for characters previously present separately in larva and adult to become combined in a single stage. This is indeed what has happened in the Appendicularia.

Strangely enough, one of the chief difficulties of this theory is to find the position of the enteropneusts. Since the larva is still in the ciliated-band stage there should be no sign of organs characteristic of the muscle-swimming, fish-like pro-chordate. Yet such signs are present in the adult *Balanoglossus*; there is a hollow nerve-cord and some sign of a notochord. These features almost compel us to suppose that the group has at one time possessed a free-swimming, fish-like stage. The only escape from this conclusion would be by supposing the hollow nerve-tube to be a case of convergence, for which a parallel might be cited in the hollow nervous system of Polyzoa. But there is no clear reason why the nerve-cord should become rolled up in the collar, and it is easier to suppose it a vestige. This imposes two further hypotheses on us. First that a fish-like stage once *followed* an advanced ciliated-band stage in ontogeny, and secondly that this fish-like stage later became adapted to a burrowing life, in fact that *Balanoglossus* is a 'degenerate' chordate. Neither of these propositions is impossible, but it must be admitted that the position of the enteropneusts is not clear. Showing a combination of ciliated larva and chordate characters they provide a valuable proof of the affinity of chordates and echinoderm-like creatures, but these very chordate characters become an embarrassment when we try to explain in detail how they have arisen!

There is strong reason to suppose that what we may call the Bateson–Garstang theory of the origin of chordates is correct. There is little doubt that chordates are related to the sessile lophophore-feeding type of creature rather than to any annulate, and we can reconstruct the course of events by which the lophophore-feeder may have come to have a pharynx with gill-slits and its larva to have muscles, a notochord, and a nerve-tube. Then by paedomorphosis the sessile stage disappeared and the free chordates began their course of evolution. There are some reasons for supposing that a type such as amphioxus could have been derived from a creature not distantly related to the simpler Appendicularia and this in turn from a neotenous doliolid or some similar ancestral type.

We need not, however, follow the theory into its details, which are speculative. The whole treatment provides a conspicuous example of close morphological reasoning, allied with proper consideration of general biological principles, and establishes with some probability the main outlines of the origin of our great phylum of active creatures from such humble sedentary beginnings.

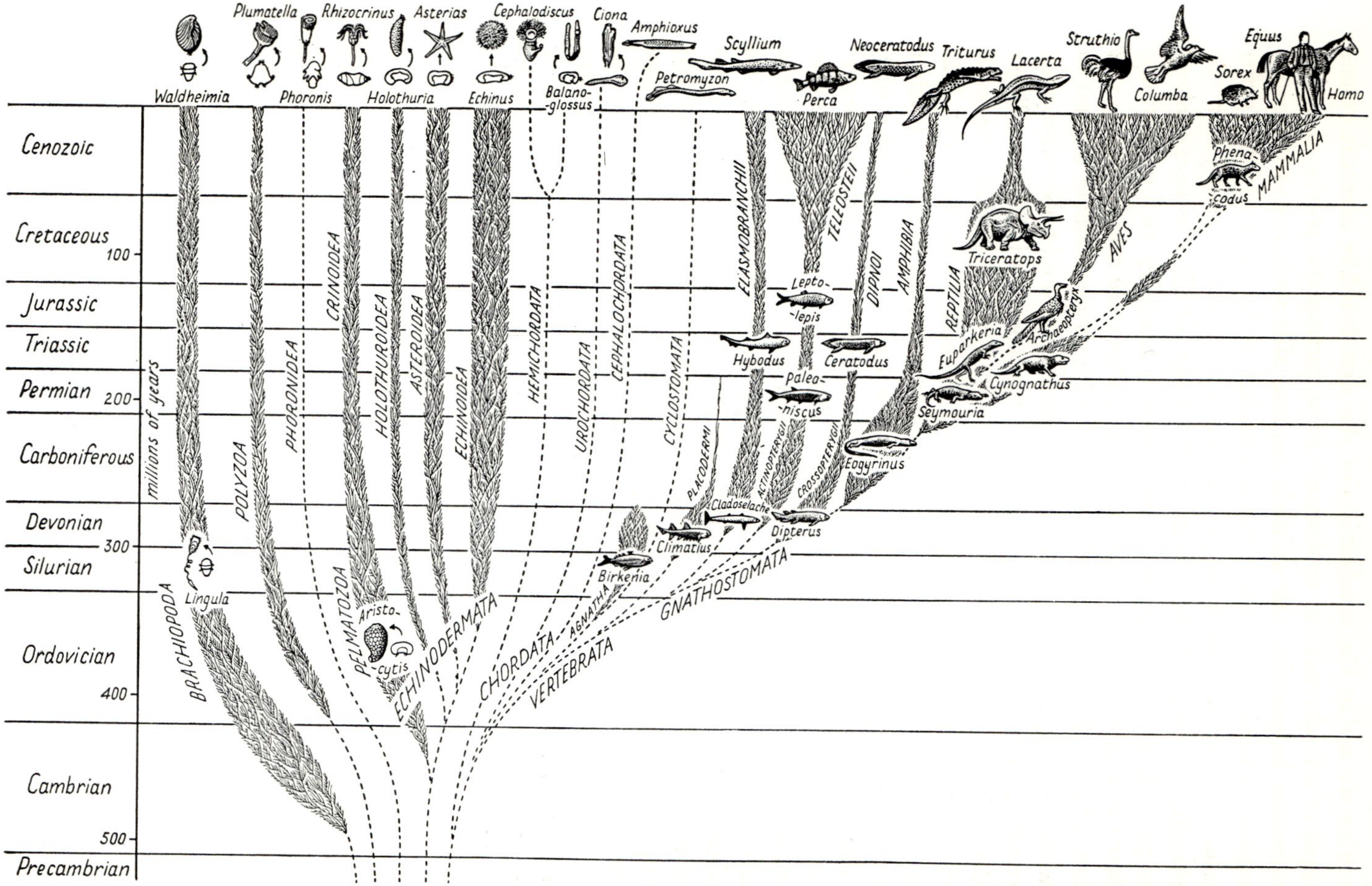

FIG. 46. Table to show the probable times of origin and affinities of the chordates and related groups of animal.

Can we see in the production of the first fish-like creatures clear signs of an 'advance' in evolution? In acquiring the power of active muscular locomotion the animals became able to live and feed in a variety of habitats, either at the sea surface or on the bottom. Forms with a sedentary adult stage are limited by the necessity for the presence of a sea bottom of suitable character. The larvae were evolved to provide the information to make sure of reaching such conditions. But whereas suitable situations on the bottom are not common, and are liable to change, the sea surface provides a generalized habitat in which there is always abundant food, though no doubt also strenuous competition for it. Paedomorphosis in this case, as in others, allows the race to eliminate from its life-history the stage passed in a 'special' environment, which is difficult to find. Although the fish-form that was thus produced proved to have great possibilities for further evolution, the change was not at first a strikingly progressive one. The surface of the sea is perhaps the most general of all environments; possibly it was the seat of the origin of life. Races that have devised means of living on the sea bottom may therefore be said to have advanced, because they have invaded a more difficult habitat. To abandon the sedentary life might in this sense be regarded as a retrograde step. The peculiar feature of the early fishes, however, was that they developed powers of active movement in a relatively large organism provided with efficient receptors, and by making use of the feeding mechanism developed at first by the bottom-living adult were able to live successfully at the sea surface. They acquired their dominance at this stage not by invading new habitats but by developing effective means of living in the richly populated plankton.

IV

THE VERTEBRATES WITHOUT JAWS. LAMPREYS

1. Classification

Phylum Chordata
Subphylum 4. Vertebrata (= Craniata)
  Superclass 1. Agnatha
    Class 1. Cyclostomata
      Order 1. Petromyzontia
        Petromyzon; Lampetra; Entosphenus; Geotria; Mordacia
      Order 2. Myxinoidea
        Myxine; Bdellostoma
    Class 2. *Osteostraci. Silurian–Devonian
      **Cephalaspis;* **Tremataspis*
    Class 3. *Anaspida. Silurian–Devonian
      **Birkenia;* **Jamoytius*
    Class 4. *Heterostraci. Ordovician–Devonian
      **Astraspis;* **Pteraspis;* **Drepanaspis*
    Class 5. *Coelolepida. Silurian–Devonian.
      **Thelodus;* **Lanarkia*
  Superclass 2. Gnathostomata

2. General features of vertebrates

All the remaining chordates are alike in possessing some form of cranium and some trace of vertebrae; they make up the great subphylum Vertebrata, also called Craniata. The organization of a vertebrate is similar to that of amphioxus, but with the addition of certain special features. A few of these novelties may now be surveyed, with emphasis on those that provide the basis for the capacity to live in difficult environments that is so characteristic of the vertebrates. Firstly the front end of the nervous system is differentiated into an elaborate brain, associated with special receptors, the nose, eye, and ear. Through these receptors the vertebrates are able to respond to more varied aspects of the environment than are any other animals. Some of them have the ability to discriminate between visual shapes and colours, and in the auditory field between patterns of tones, also between a host of chemical substances. The motor organization allows the performance of delicate movements to suit the situations that the receptors reveal. The swimming process, by the passage of waves down the body, is itself perfected by improvements in the shape of the fish, allowing rapid movements and turns. Besides the median fins there develop lateral paired ones, serving at first a stabilizing and steering function and then converted, when the land animals arose, into organs of locomotion on the ground or in the air and finally, in the shape of the hands, into a means of altering the environment to suit the individual.

The brain itself, at first mostly devoted to the details of sensory and motor function, comes increasingly to preside, as it were, over all the bodily functions, and to give to the vertebrates the 'drive' that is one of their most characteristic features. The skull is developed as a skeletal thickening around the brain, probably at first mainly for protection, but later serving for the attachment of elaborate muscle systems. The study of vertebrates is especially identified with study of the skull, because in so many fossils this is the only organ preserved.

The food of the earliest vertebrates was collected by ciliary action, but this habit has long been abandoned and only in rare cases today does the food consist of minute organisms. The pharynx of most vertebrates is small, there are relatively few gill-slits and these are respiratory. In all except the most ancient forms the more anterior of the arches between the gills became modified to form jaws, serving not only to seize and hold the food but also to 'manipulate' the environment.

The blood system shows two of the most characteristic vertebrate features, namely, the presence of a heart that has at least three chambers and thus provides a rapid circulation, and of haemoglobin within corpuscles, serving to carry large amounts of oxygen to the tissues. The efficiency of this system must have been a major factor in producing the dominance of the vertebrate animals. In the air-breathing forms, and especially the warm-blooded birds and mammals, the respiratory and circulatory systems allow the expenditure of great amounts of energy per unit mass of animal, so that quite extravagant devices can be used, allowing survival under conditions that would otherwise not support life.

The excretory system is based on a plan quite different from that of amphioxus. It consists of mesodermal funnels, leading primarily from the coelom to the exterior. It may be that this type of kidney arose in connexion with the abandoning of the sea for fresh water. Probably all but the earliest vertebrates have passed through a fresh-water stage, and it is significant that all except *Myxine* have less salt in their blood than there is in sea water. Elaborate devices for regulation of osmotic pressure have been developed, and the mesodermal kidneys play a large part in this regulation.

This outline only gives a few suggestive features of vertebrate organization. The details differ bewilderingly in the different types and it is our business now to survey them. In the earliest forms the more special mechanisms are absent or at least function only crudely, and passing through the vertebrate series we find more and more devices adopted, along with more and more delicate co-ordination between the various parts, culminating in the extremely highly centralized control of almost every aspect of life that is exercised by the mammalian cerebral cortex.

3. Agnatha

The earliest vertebrates, while showing most of the characteristic features of the group, differ from the rest in the absence of jaws and are therefore grouped together in a superclass Agnatha, distinguished from the remaining vertebrates, which have jaws, and are therefore called Gnathostomata. The only living agnathous animals are the Cyclostomata, lampreys and hag-fishes, but the first vertebrates to appear in the fossil series, mostly heavily armoured and hence known as 'ostracoderms', found in Silurian and Devonian strata, also show the agnathous condition, and have some other features in common with the Cyclostomata. This group of agnathous vertebrates shows some interesting experimentation in methods of feeding, before the jaw-method became adopted. The modern cyclostomes are parasites or scavengers, in the adult state, but as larvae the lampreys still feed on microscopic material, using an endostyle resembling that of amphioxus in many ways, but making use of muscular contraction rather than ciliary action to produce a feeding current. The methods of feeding of the Devonian forms are not known for certain, but probably included shovelling detritus from the bottom.

The Cyclostomata are therefore worth special study as likely to show us some of the characteristics possessed by the earliest vertebrate populations.

4. Lampreys

The most familiar cyclostomes are the lampreys, of which there are various sorts found in the temperate zones of both hemispheres. All lampreys have a life-history that includes two distinct stages: the ammocoete larva lives in fresh water, buried in the mud, and is microphagous: the adult lamprey has a sucking mouth, and usually lives in the sea, where it feeds on other fishes. *Lampetra* the lamprey (Fig. 47), is a typical example, common in Great Britain. The adult is an eel-like animal about 30 cm long, black on the back, and white below. The surface is smooth, with no scales. The skin is many-layered (Fig. 48). The outermost cells have a striated cuticular border. Mixed with these epithelial cells the lamprey, like most aquatic vertebrates, has many gland-cells for producing slime. Below the epidermis lies the dermis, a layer of bundles of collagen and elastin

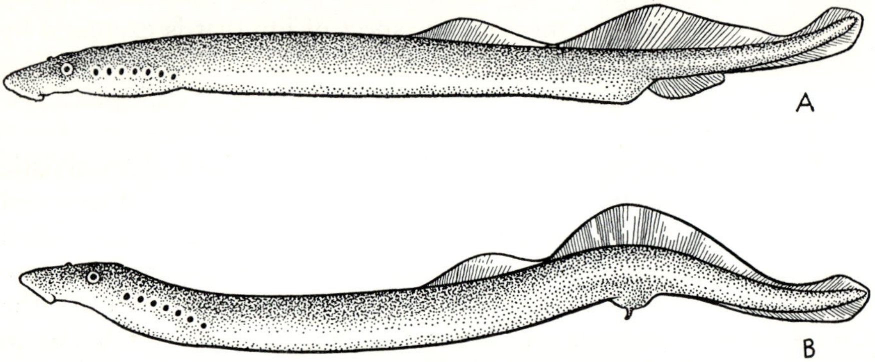

FIG. 47. Brook lampreys, *Lampetra planeri*.
A, ripe female, with anal fin; B, ripe male; note shape of dorsal fin and presence of copulatory papilla. (Curves due to fixation.)

fibres, running mostly in a circular direction. This tissue is sharply marked off from a layer of subcutaneous tissue containing blood-vessels and fat, as well as connective tissue. There are pigment cells in the dermis and a thick layer of them at the boundary of dermis and subcutaneous tissue. The chromatophores are star-shaped cells whose pigment is able to migrate, making the animal dark or pale. This change is especially marked in the larva and is produced by variation in the amount of a pituitary secretion (p. 107).

The head of the lamprey bears a pair of eyes and a conspicuous round sucker. On the dorsal side is a single nasal opening, and behind this there is a gap in the pigment layers of the skin through which the third or pineal eye can be seen as a yellow spot. There are seven pairs of round gill openings, which, with the true eyes (and some miscounting or perhaps inclusion of the nasal papilla), are responsible for the familiar name 'nine eyes'. There is no trace of any paired fins, but the tail bears a median fin, which is expanded in front as a dorsal fin. There are sex differences in the shape of the dorsal fins of mature individuals and the female has a considerable anal fin (Fig. 47).

The lamprey swims with an eel-like motion, using its myotomes in the serial manner that has been mentioned in amphioxus and will be discussed later (p. 133). The waves that pass down the body are of short period relative to the length, so that the swimming is mechanically inefficient; lampreys show great activity, but their progress is not rapid. The animal often comes to rest, attaching itself with the sucker to stones (hence the name, 'suck-stone') or to its prey. In this position water cannot of course pass in through the mouth, but both

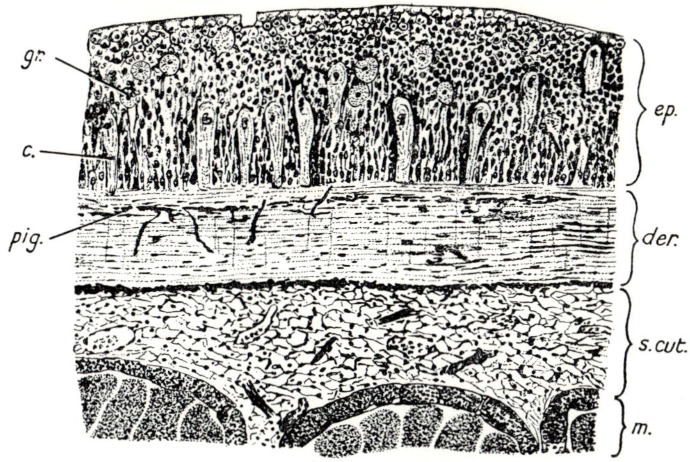

FIG. 48. Section of skin of lamprey.

c. club cells; *der.* dermis; *ep.* epidermis; *gr.* granular gland-cells; *m.* myotomal muscle; *pig.* pigment cells; *s.cut.* subcutaneous connective tissue. (After Krause.)

enters and leaves by the gill openings. When swimming the backward jet of water may assist in locomotion.

The trunk musculature consists of a series of myotomes separated by myocommas. Each myotome has a W-shape, instead of the simple V of amphioxus. The muscle-fibres run longitudinally and they are striped, but of a somewhat peculiar fenestrated type.

5. Skeleton of lampreys

The skeleton of lampreys consists of the notochord and various collections of cartilage. This latter is partly of the typical vertebrate type, that is to say, consists of large cells in groups, separated by a matrix of the protein chondrin, which they secrete. In other regions a tissue containing more cells and less matrix is found, the so-called fibro-cartilage, and this more nearly resembles fibrous connective tissue and serves to emphasize that no sharp line can be drawn between these tissues. There is also, in the larva, a tissue known as muco-cartilage, which is an elastic material serving more as an antagonist to the muscles than for their attachment.

The notochord remains well developed throughout life as a rod below the nerve-cord. It consists of a mass of large vacuolated cells,

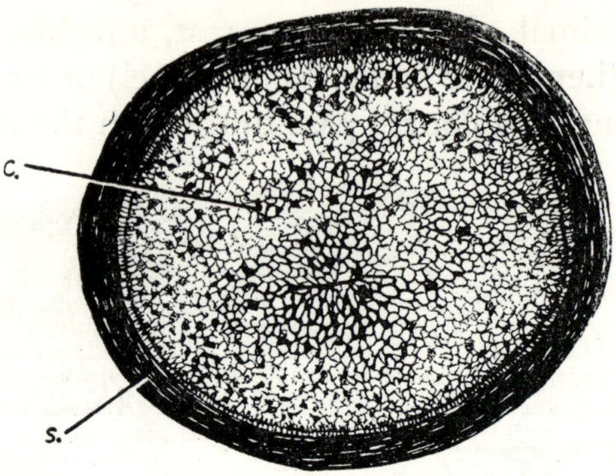

FIG. 49. Transverse section through notochord of lamprey. *c.* cells; *s.* sheath. (After Krause.)

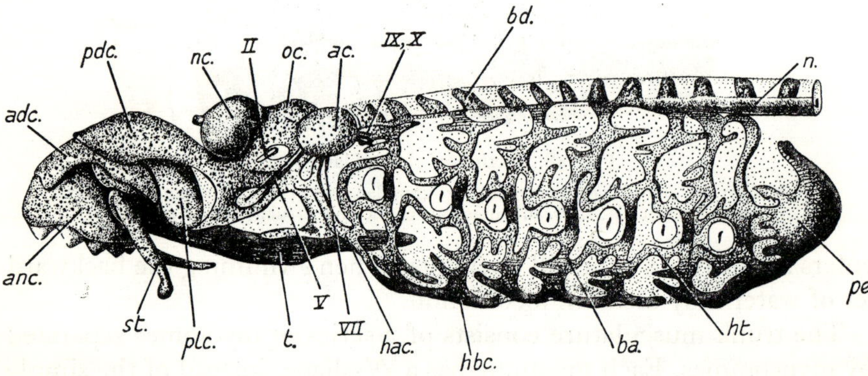

FIG. 50. Lateral view of skeleton of head and branchial arches of *Petromyzon*. *ac.* auditory capsule; *adc.* antero-dorsal cartilage; *anc.* annular cartilage; *ba.* branchial arch; *bd.* basidorsal; *hac.* hyoid cartilage; *hbc.* hypo-branchial rod; *ht.* horizontal bar; *n.* notochord; *nc.* nasal capsule; *oc.* orbital cartilage; *per.* pericardial cartilage; *pdc.* postero-dorsal cartilage; *plc.* posterolateral cartilage; *st.* styliform cartilage; *t.* tendon of tongue; *II–X*, cranial nerves. (After Parker.)

enclosed in a thick fibrous sheath (Fig. 49). The rigidity of the whole rod depends on the turgor of the cells and it often collapses completely in fixed and dehydrated material (Fig. 59). No doubt in life it serves, like the notochord of amphioxus, to prevent shortening of the body when the myotomes contract.

The notochordal sheath is continuous with a layer of connective tissue, which also surrounds the spinal cord and joins the myocommas and thus eventually the subcutaneous connective tissue. Within this connective tissue there develop certain irregular cartilaginous thickenings that are of special interest because they may be compared with vertebrae, perhaps with the basi-dorsal element (p. 132). They lie on either side of the spinal cord (Fig. 50), that is to say, above the notochord, and consist either of one nodule on each side of the segment, through the middle of which the ventral nerve-root emerges, or of two separate nodules, with the nerve between them. Rods of cartilage extend dorsally and ventrally into the fins, but are not attached to the 'vertebrae'.

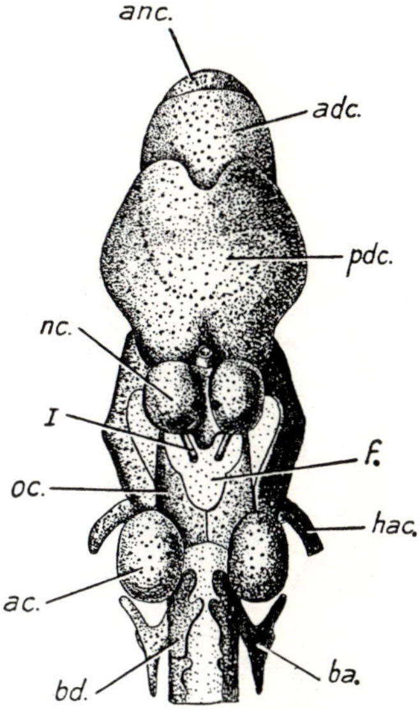

FIG. 51. Dorsal view of skull of *Petromyzon*.
Lettering as Fig. 50. *I*, olfactory nerve; *f*. hole in roof of cranium. (After Parker.)

The lamprey skull shows even in the adult the basic arrangement found only in the embryo of higher vertebrates. The floor is formed of paired parachordals on either side of the notochord and in front of this paired trabeculae. Attached to this base is a series of incomplete cartilaginous boxes surrounding the brain and organs of special sense (Fig. 51). To this skull is attached the skeleton that supports the sucker and gills. The arrangement of the skull differs considerably from that of later vertebrates. The cranium has a floor around the end of the notochord, and in front of this there is a hole containing the pituitary gland. The side walls are strong but the roof is composed only of a tough membranous fibro-cartilage. The auditory capsules are compact boxes surrounding the auditory organs at the sides. The olfactory capsule, imperfectly paired, is also almost detached from the cranium. Other ridges of cartilage lie below the eyes and there is a complex support for the sucker.

The skeleton of the branchial region consists of a system of vertical plates between the gill-slits, joined by horizontal bars above and below them. This cartilage lies outside the muscles and nerves and is therefore difficult to compare with the branchial skeleton of higher fishes, which lies in the wall of the pharynx. The elastic action of the cartilages produces the movement of inspiration. A backward extension of the branchial basket forms a box surrounding the heart.

6. Alimentary canal of lampreys

The sucker is bounded at the edges by a series of lips, which besides being sensory serve also to make a tight attachment when the lamprey sucks (Fig. 52). In the sucker are numerous teeth, whose arrangement varies in the different types of lamprey. These teeth are horny epidermal thickenings, supported by cartilaginous pads, and are therefore not comparable with the teeth of vertebrates, which are derived

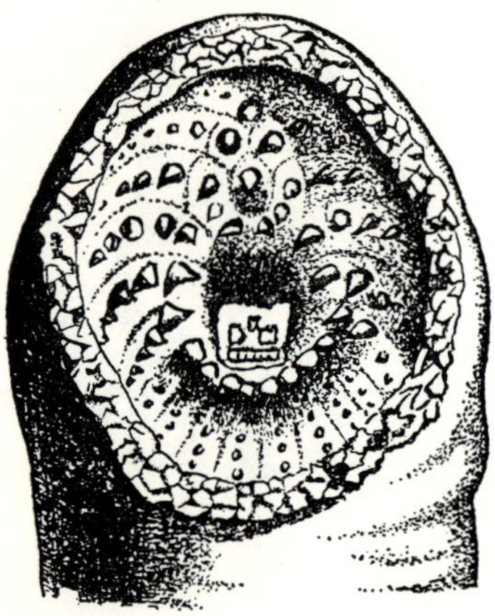

FIG. 52. Sucker of *Petromyzon* showing outer circular lip, teeth, and tongue, with special teeth, at the centre. (After Parker.)

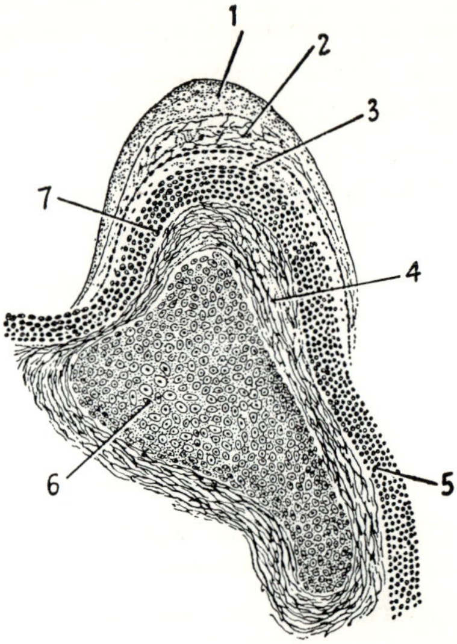

FIG. 53. Section through tooth of lamprey. 1, horny cap; 2, stellate tissue; 3, cap to replace 1; 4, connective tissue; 5, epidermis of mouth; 6, cartilage; 7, proliferative layers of epidermis that produce the horny cells. (After Hansen, from Kukenthal.)

mainly from mesodermal tissues (Fig. 53). The sharper and larger teeth are borne on a movable tongue, which is used as a rasp (Fig. 54).

An annular muscle runs round just above the lips of the sucker and presumably serves to narrow the margin and hence to release the fish. The remaining muscles are mostly attached to the tongue and base of the sucker. The largest of these muscles, the m. cardioapicalis, is attached posteriorly to the cartilage surrounding the heart and in front is prolonged into a conspicuous lingual tendon, which is attached to the tongue and serves to pull it backwards. Presumably the action of this muscle deepens the oral cavity and is thus the main agent securing attachment of the sucker. There is a collar of circular fibres around the front end of the cardio-apical muscle, serving to lock the tendon and maintain the suction. Dorsal and ventral to the main tendon are groups of muscles that rock the tongue up and down to produce a rasping action. The muscles of the sucker are all derived from the lateral plate and are innervated from the trigeminal nerve; their fibres are striated.

The mouth is a small opening above the tongue and leads into a large buccal cavity. At the hind end this divides into a dorsal passage, the oesophagus, for the food, and a ventral respiratory tube, which leads to the gill pouches but is closed behind. At the mouth of the

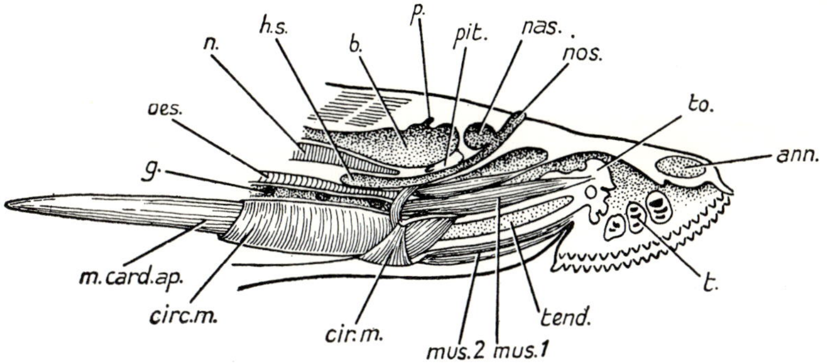

FIG. 54. Longitudinal section through head of lamprey.

ann. annular muscle of sucker; *b.* brain; *circ.m.* circular fibres of tongue-muscles; *oes.* oesophagus; *g.* gill aperture; *h.s.* hypophysial sac; *m.card.ap.* cardio-apical muscle; *mus. 1* and *2*, muscles that rock the tongue; *n.* notochord; *nas.* nasal sac; *nos.* nostril; *p.* pineal; *pit.* pituitary gland; *t.* tooth; *tend.* tendon of tongue, pulled back by *m.card.ap.*; *to.* tongue. (Partly after Tretjakoff.)

respiratory tube is a series of velar tentacles, corresponding exactly in position to those of amphioxus, and serving to separate the mouth and oesophagus from the respiratory tube while the lamprey is feeding. The seven branchial sacs are lined by a folded respiratory epithelium and surrounded by muscles, and these, together with the elastic cartilages and appropriate valves, ensure the pumping of the water tidally, in and out of the external openings. In front of the first sac is the remains of an eighth pouch, whose surface is not respiratory.

The 'salivary' glands are curious organs of which little is known. They are a pair of pigmented sacs, embedded in the hypobranchial muscles. Each has a folded wall, from which a duct proceeds forward to open below the tongue. The salivary glands produce a secretion that prevents coagulation of the blood of the fishes on which the lamprey feeds. The nature of this secretion is not known, but it rapidly turns black on exposure to the air and the glands for this reason appear to be pigmented. It has been observed that in lampreys taken from fishes the intestine is filled with red corpuscles, and there is therefore no doubt that they feed mainly on the blood of their prey.

Little is known of the habits of lampreys in the sea, but in North America there are races of lampreys that are land-locked and feed on the fishes in the lakes, where they have recently become a most serious pest (Fig. 55).

The oesophagus (fore-gut) leads directly into a straight intestine (mid-gut); there is no true stomach in lampreys (Fig. 56). The surface of the intestine is increased by a typhlosole, running a somewhat spiral course. There is a liver, gall-bladder, and bile-duct of typical vertebrate plan, but no separate pancreas. However, in the wall of the

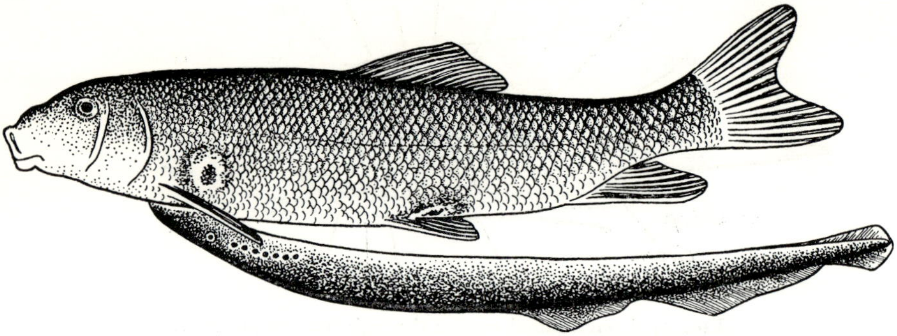

FIG. 55. Lake lamprey attached to a bony fish, which also shows the scars of the attacks of other lampreys. (After Gage.)

anterior part of the intestine there are large patches of cells that resemble those of the acini of the pancreas of higher forms and contain secretory granules. Barrington has shown that extracts of this region have a high proteolytic power, the enzyme being of the tryptic type, with its optimum between pH 7·5 and 7·8. Some of this tissue is collected in the walls of short diverticula, reaching forwards. The situation is therefore essentially similar to that found in amphioxus, and we may regard these patches of zymogen cells, or the diverticula, as the forerunners of the exocrine portions of the pancreas. In the lampreys the endocrine portion, not yet identified in amphioxus, also appears. Around the junction of the fore-gut and intestine are groups of follicles that do not communicate with the lumen of the intestine. These 'follicles of Langerhans' were, appropriately enough, first seen by the discoverer of the islets in higher forms, and Barrington has now shown that following destruction of this tissue by cautery there is a rise in blood-sugar. Moreover, after injection of glucose, vacuolation of the cells occurs. We may safely conclude that these cells are involved in carbohydrate metabolism, but only one type of cell is present.

7. Blood system of lampreys

The blood vascular system is arranged on the same general plan as in amphioxus but there is a well-developed heart. This lies behind the gills and can be considered as a portion of the sub-intestinal vessel, folded into an S-shape and divided into three chambers. The heart is suspended in a special portion of the coelom, the pericardium, whose walls are supported by cartilage. In the larva the heart first appears as a straight tube and owing to an abnormality of development it sometimes fails to develop its S-shape. Contractions can neverthe-

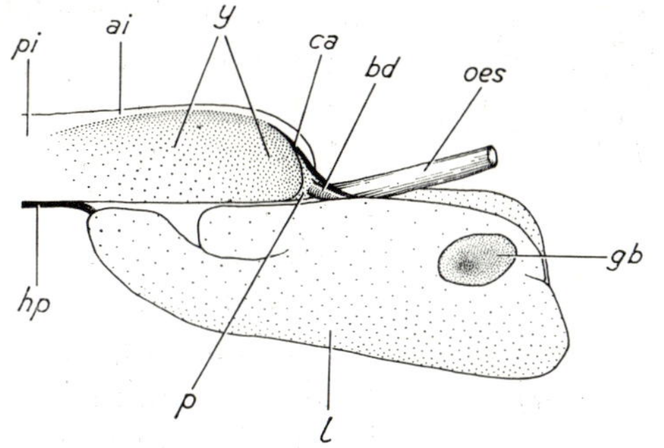

FIG. 56. Mid-gut of larval lamprey.

ai. anterior region of intestine; *bd.* bile-duct; *ca.* coeliac artery; *gb.* gall-bladder; *hp.* hepatic portal vein; *l.* liver; *oes.* oesophagus; *p.* position of 'pancreas', containing islet tissue; *pi.* posterior intestine; *y.* yellow area where wall of intestine contains zymogen cells. (From Barrington.)

less be seen in these abnormal hearts, passing from behind forwards along the straight tube. Similarly in the normal heart contraction proceeds in the chambers from behind forwards. The most posterior chamber is a thin-walled sinus venosus, into which the veins pour blood. This leads to an auricle (atrium), also thin-walled, lying above the sinus. The atrium passes blood into the ventricle below it, a thick-walled chamber, providing the main force for sending the blood round the body.

The heart receives nerve-fibres from the vagus nerve and contains nerve-cells, some of which give a chromaffin reaction suggesting the presence of adrenalin-like substances. Stimulation of the vagus nerve produces acceleration of the heart-beat, followed by slowing. Acetyl choline also accelerates the heart. In *Myxine* there are no nerves to the heart or nerve-cells in it and acetyl choline has no effect. Both hearts contain much adrenaline and similar substances but show little change when adrenaline is added to a perfusate.

Blood leaves the ventricle by a large ventral aorta, running forwards between the gill pouches, to which it sends a series of eight afferent branchial arteries. These break up into capillaries in the gills, and efferent branchial arteries collect to a pair of dorsal aortae, running backwards, which join and form the main dorsal aorta. This passes down the trunk and carries blood to all the parts of the body by means of series of segmental arteries and special vessels to the gut, gonads, and excretory organs. A curious feature is that many of these arteries are provided with valves at the point at which they leave the main trunks (Fig. 57). It may be significant that such valves are not found where the efferent branchials join the dorsal aorta, nor at the points of exit of the renal arteries, so that perhaps the valves serve to reduce the pressure in the majority of the arteries, while leaving it high in those to the kidneys. The removal of large quantities of water is an important problem in all freshwater animals and is facilitated by a high pressure in the kidneys. This must be difficult to maintain in an animal with a branchial circulation and hence a double set of capillaries.

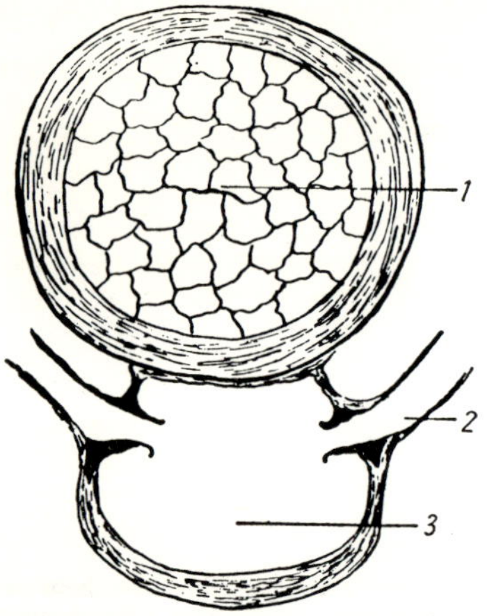

FIG. 57. Valves at the origin of segmental arteries of a lamprey.
1, notochord; 2, segmental artery; 3, aorta.
(From Kukenthal, after Keibal.)

The venous system consists of a network of sinuses, with contractile venous hearts in various places. There is a large caudal vein, dividing where it enters the abdomen into two posterior cardinals. These run forward in the dorsal wall of the coelom, collecting blood from the kidneys, gonads, &c., and opening into the heart by a single ductus Cuvieri on the right-hand side, this being the remains of a pair found in the larva. Anterior cardinals collect blood from the front part of the body, and there is also a conspicuous ventral jugular vein draining venous blood from the muscles of the sucker and gill pouches. Besides the veins proper there is a large system of venous sinuses, especially in the head. Blood from the gut passes by a hepatic portal vein through a contractile portal heart to the liver, from which hepatic veins proceed to the heart.

The blood of lampreys, like that of all vertebrates, contains the respiratory pigment haemoglobin, enclosed in corpuscles, here nucleated. This arrangement immensely increases the oxygen-carrying power of the blood. Haemopoietic tissue occurs in the intestinal wall of the larva and this has been regarded by some as representing the spleen. In the adult the blood-forming tissue lies below the spinal cord and in the kidney. White corpuscles resembling lymphocytes and polymorphonuclear cells occur, produced by lymphoid tissue in the kidneys and elsewhere. However, there is no distinct system of lymphatic channels.

8. Urinogenital system of lampreys

The excretory and genital systems of vertebrates consist of a series of tubes opening from the coelom to the exterior and serving to carry away both excretory and genital products. This plan of organization is

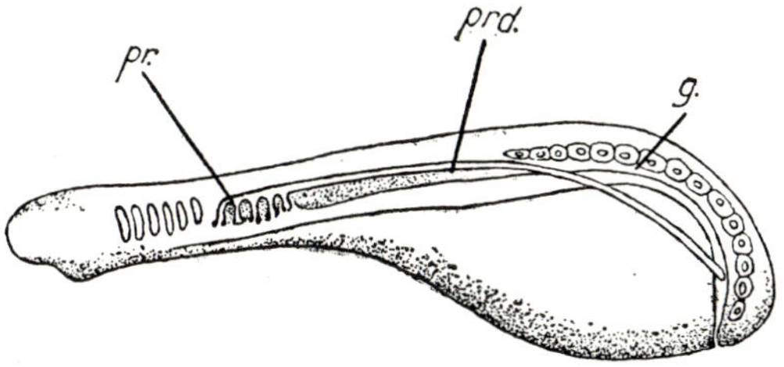

FIG. 58. Diagram to show arrangement of the pronephros in a freshly hatched lamprey.
g. gonad; *pr.* pronephros; *prd.* pronephric duct. (After Wheeler.)

quite different from that found in amphioxus and represents a new acquisition by the vertebrates. It is not clear whether the excretory or genital component of the complex is the primary one, nor indeed why they are associated. The gonads develop from the walls of the coelom in all animals possessing that cavity; some hold that the coelom represents an enlargement of a sac that at first served purely as a gonad. Genital ducts leading from the coelom to the exterior are common in invertebrates, and we may guess that at their first appearance the urinogenital tubules of vertebrates served only for genital products.

The conversion of these tubules to excretory purposes may have been a result of the adoption of the freshwater habit. The blood of lampreys, when in fresh water, contains a higher concentration of salts than the surrounding water. Little is known about the condition in sea lampreys, where blood is probably hypotonic to the sea. When in the river the animals must deal with the tendency for water to flow in. This water must be removed without losing salt; accordingly in most freshwater animals, including vertebrates, we find some system by which the separation can be achieved.

The region that gives rise to the kidney during development lies between the dorsal scleromyotome and the more ventral lateral plate

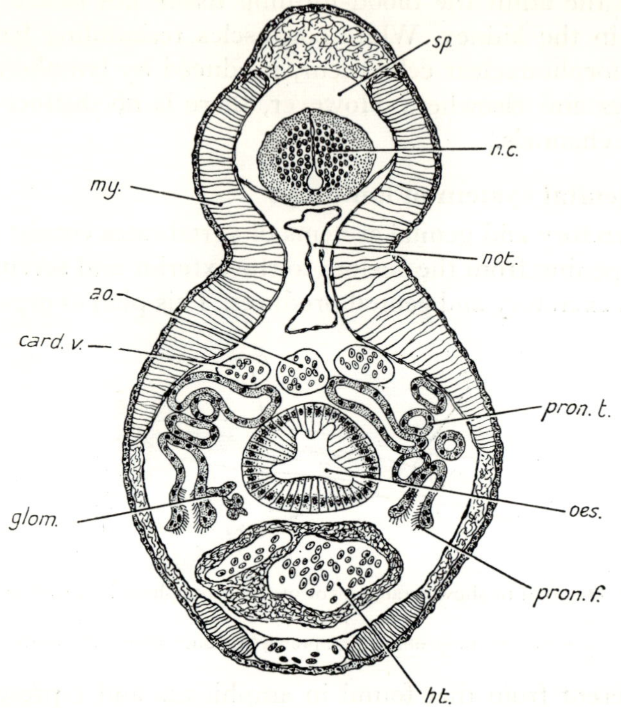

FIG. 59. Section through newly hatched larvae of *Lampetra* behind the pharynx. *ao.* aorta; *card.v.* cardinal vein; *glom.* glomerulus; *ht.* heart; *my.* myotome; *n.c.* nerve-cord; *not.* notochord, which has collapsed because of lack of turgidity after fixation; *oes.* oesophagus; *pron.f.* ciliated funnel of pronephros; *pron.t.* twisted pronephric tubule; *sp.* space around nerve-cord.

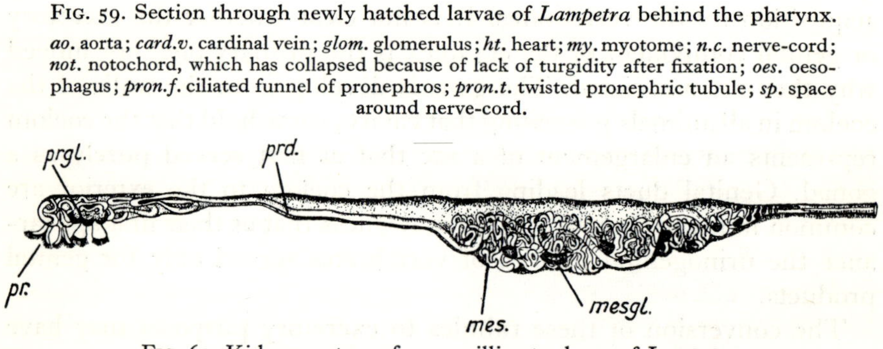

FIG. 60. Kidney system of a 22-millimetre larva of *Lampetra*. *mes.* mesonephric tubules; *mesgl.* mesonephric glomeruli; *pr.* pronephric funnels; *prd.* pronephric duct; *prgl.* pronephric glomeruli. (After Wheeler.)

mesoderm; it is known as the nephrotome. This tissue differentiates during development from in front backwards, making a series of segmental funnels, opening into a common archinephric duct (Fig. 58). The most anterior funnels open into the pericardium; usually there are four of these in a freshly hatched larva, opening into a single duct, which reaches back to an aperture near the anus. Close to each funnel there develops a tangle of blood-vessels, the glomerulus (Figs. 59 and 60). Presumably the osmotic flow of water into the body is relieved by the pressure of the heart-beat forcing water out from the glomeruli into the coelomic fluid, whence it is removed by the funnels, with the aid of their cilia. The tubules become longer and twisted after hatching and may perhaps serve for salt-reabsorption.

These anterior funnels constitute the pronephros. As the animal grows they are replaced by a more posteror set, the mesonephros. There is, however, a gap of several segments in which no tubules appear (Fig. 60), a strange and unexplained discontinuity, common to all vertebrates. The pronephric tubules gradually disappear and finally in the adult all that remains of the organ is a mass of lymphoid tissue. Meanwhile the mesonephros develops as a much larger fold, hanging into the coelom and containing very extensive winding tubules. These do not open to the coelom (at least in the adult) but each to a small sac, the Malpighian corpuscle, which contains a portion of the coelom and the glomerulus.

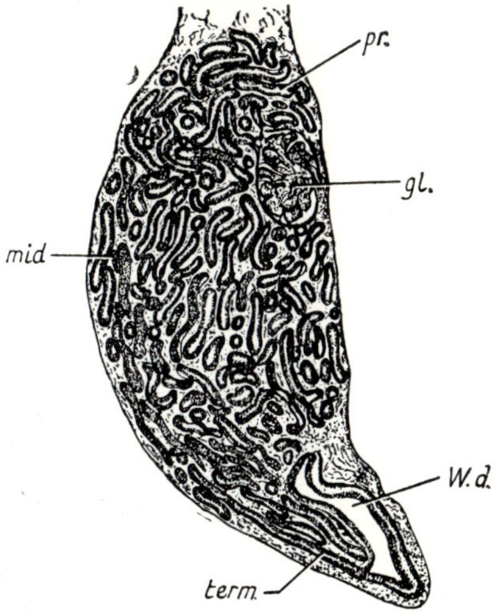

FIG. 61. Transverse section of kidney of *Lampetra*.

gl. glomerulus; *mid.* middle section of tubule; *pr.* proximal region of tubule; *term.* terminal region, opening into *W.d.* Wolffian duct.
(After Krause.)

This is obviously a more efficient method for allowing the heart to pump excess water out of the blood and down the tubules. The latter themselves have become greatly elongated and make up the main bulk of the organ (Fig. 61). The segmental arrangement is therefore much obscured and as extra glomeruli are added it disappears completely. The mesonephros extends at its hind end as the animal grows, until it forms the adult kidney, a continuous ridge of tissue reaching back to the hind end of the coelom. Besides the excretory apparatus the kidney also contains much lymphoid tissue and fat, and it probably plays a part in the formation and destruction of red and white corpuscles.

The gonads are unpaired ridges medial to the mesonephros. Primordial germ-cells, set aside very early in development, migrate into these ridges and develop into eggs or sperms. The differentiation of the gonad occurs relatively late in lampreys, so that in young ammocoetes the organ is 'hermaphrodite', containing developing oocytes and spermatocytes together. The ripe ovary consists of ova each surrounded by single-layered follicular epithelium, which finally ruptures and liberates the egg into the coelom, whence it escapes by pores to be described presently. The testis consists of a number of follicles containing sperms; it is unique among vertebrates in that the follicles

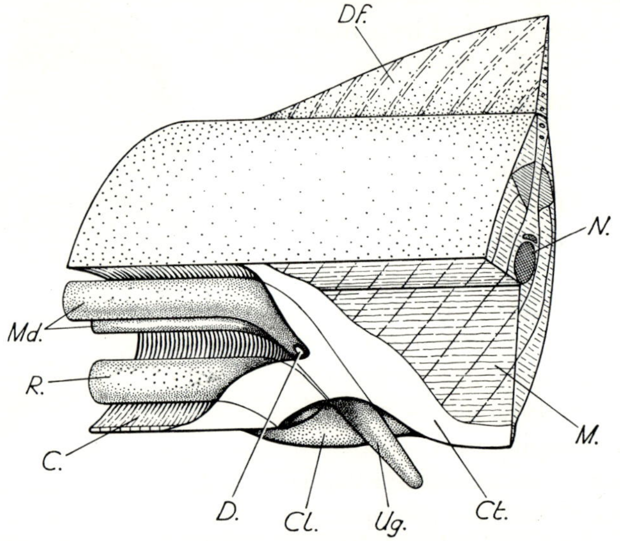

FIG. 62. Cloacal region of fully adult *Lampetra*.

C. coelom; *Cl.* lips of cloaca; *Ct.* connective tissue; *D.* duct leading from coelom to the mesonephric duct; *Df.* dorsal fin; *M.* muscle; *Md.* mesonephric ducts; *N.* notochord; *R.* rectum; *Ug.* urinogenital papilla. (After Knowles.)

have no ducts; when ripe they rupture into the coelom, which becomes filled with spermatozoa and these escape, like the ova, by pores.

These apertures by which the gametes escape are similar in the two sexes and consist of short channels, one on each side, leading from the coelom to the lower end of the kidney duct (Fig. 62). They normally become open only a few weeks before spawning, but Knowles has shown that injections of oestrone or anterior pituitary extract will cause perforations of the ducts in young lampreys, indeed even in the ammocoete larve.

Fertilization is external, but there are modifications of the cloaca in the two sexes to assist in ensuring fertilization and proper placing of the eggs in the 'nest' (p. 113). The lips of the cloaca of the ripe male are united to form a narrow penis-like tube. The cloacal lips of the female are enlarged and often red; in addition she has an anal fin, probably used, as in salmon and trout, to make a nest. These sex differences, which develop shortly before spawning, can also be initiated by injection of anterior pituitary extracts (p. 107).

9. Nervous system of lampreys

The nervous system of the cyclostomes is very much better developed than that of amphioxus and shows the characteristic plan that is present in all vertebrates. The essence of the vertebrate nervous organization may be said to be that it consists of large amounts of tissue and is highly centralized. The brains of vertebrates contain much larger aggregates of nervous tissue than are to be found in any other animals, and this tissue produces by its actions the most characteristic features of vertebrate life. Vertebrates are active, exploratory creatures, and their behaviour is much influenced by past experience.

We shall return later to detailed discussion of the organization of the central nervous system; now we may look briefly at the plan found in the lamprey, as an introduction to that of other vertebrates. As compared with amphioxus there has been a very high degree of cephalization. The front end of the spinal cord is enlarged into a complicated brain, and the nerves connected with a number of the more anterior segments have become modified to form special cranial nerves.

The spinal nerves, however, still show the plan found in amphioxus in that the dorsal and ventral roots do not join. In amphioxus the ventral roots contain motor-fibres for the myotomes and some proprioceptive fibres, while the dorsal roots contain sensory fibres and motor-fibres for the lateral plate musculature (p. 36). The details of the composition of the nerves of lampreys are still unknown, but there are hints of considerable deviations from this plan. The ventral roots contain many motor fibres passing to the myotomes. The dorsal roots consist largely of sensory fibres with bipolar cell bodies collected into dorsal root ganglia including proprioceptor fibres from the myotomes: it is not known whether the dorsal roots also contain any efferent fibres. In the young larva many of the afferent fibres are the processes of cells lying in the spinal cord (Rohon-Beard cells), which are typical of the early stage of many chordates. There are few types of cells in the cord at this time, allowing for only the simplest reflex arcs.

The autonomic nervous system shows some generalized and some special features. The gut is mainly innervated by the vagus, which extends far back along the intestine. There is little contribution of fibres from the spinal nerves to the alimentary canal, since this has no mesentery, being attaached only at its cranial and caudal ends. There are, however, numerous fibres from the spinal nerves to the rectum, ureters, and cloacal region, and numerous postganglionic neurons are found here. Nerve-cells are also found in the intestinal plexuses.

The sympathetic system consists of isolated fibres running in both dorsal and ventral roots. Many of these run directly to their endings, for instance in the arteries, without interpolation of neurons. A few postganglionic cells are present, however, but they are seldom collected into ganglia. The system is therefore even more scattered than in elasmobranchs (p. 173). The 'adrenal' system is also diffuse. There are scattered masses of interrenal (cortical) tissue and large groups of suprarenal (medullary) cells, especially in the walls of the veins and the heart. The suprarenal tissue receives 'preganglionic' fibres from the spinal nerves. Its cells sometimes seem to be connected with each other by fibres like those of neurons and they may operate a form of control intermediate between nervous and hormonal (Johnels, 1956).

The nerve-fibres in the nervous system of cyclostomes are not provided with myelin sheaths; in this they resemble the nerves of amphioxus. Conduction is slow in such non-medullated fibres, the only case actually investigated in cyclostomes being the lateral line nerve of *Bdellostoma*, found by Carlson to conduct at the low rate of 5 metres a second (frog about 50 m/sec., mammals up to 100 m/sec.).

The spinal cord is of a uniform transparent grey colour and is flattened dorso-ventrally, apparently to allow access of oxygen, and metabolites, no blood-vessels being present within the cord. However, vessels are present in *Myxine* in which the cord is also flat. The nerve-cell bodies lie, as in higher vertebrates, towards the centre, but the synaptic contacts are not made in this 'grey' matter but at the periphery, in what would correspond to the white matter of higher forms. The outer part of the cord is thus made up of a neuropil or nerve feltwork, formed of the terminations of the incoming sensory fibres and the dendrites of the motor-cells. These cells (Fig. 63) lie in the ventral part of the cord, their axons running out to make the large fibres of the ventral roots and their dendrites passing to all parts of the peripheral regions of both the same and the opposite sides of the cord. They are thus presumably able to be stimulated directly by impulses in the processes of the afferent fibres that end in these regions.

Direct control of the spinal cord from the brain is obtained through

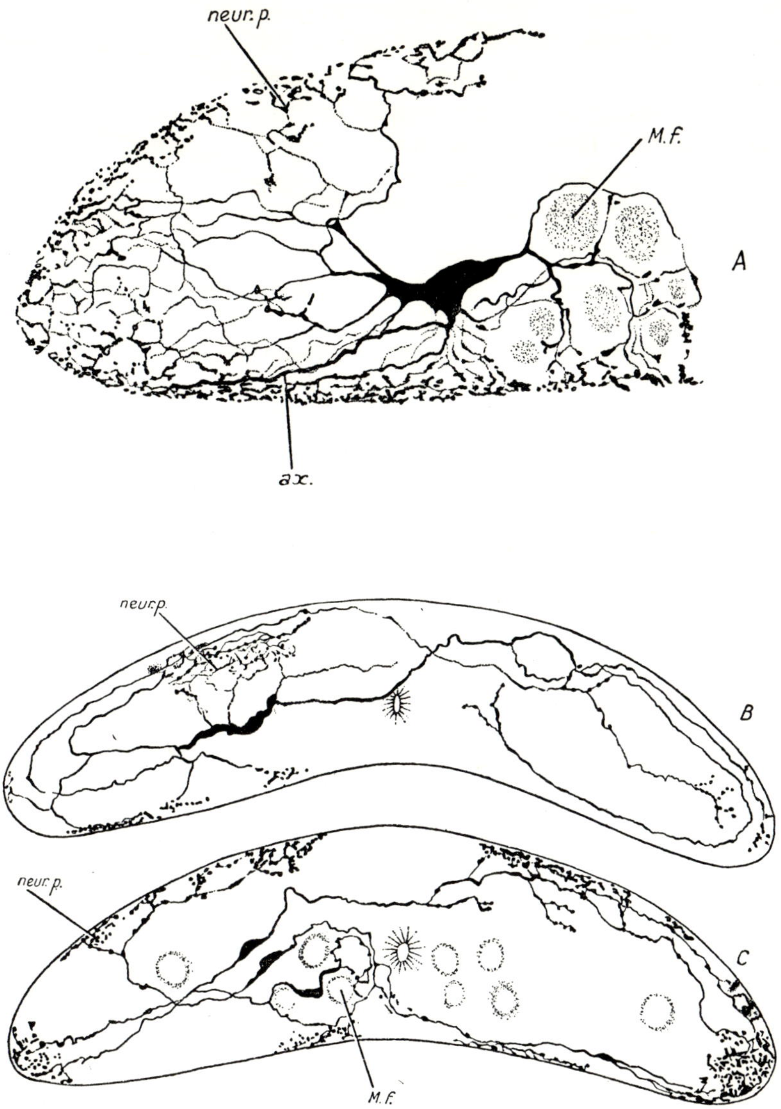

FIG. 63. Cells of the spinal cord of the larva of *Lampetra*.

***A* and *B*, large** motor-cells, with dendrites reaching to the opposite side; *C*, small cells with **widespread dendrites,** but no axon; *ax.* axon; *M.f.* Müller's fibres; *neur.p.* neuropil at periphery of spinal cord. (After Tretjakoff.)

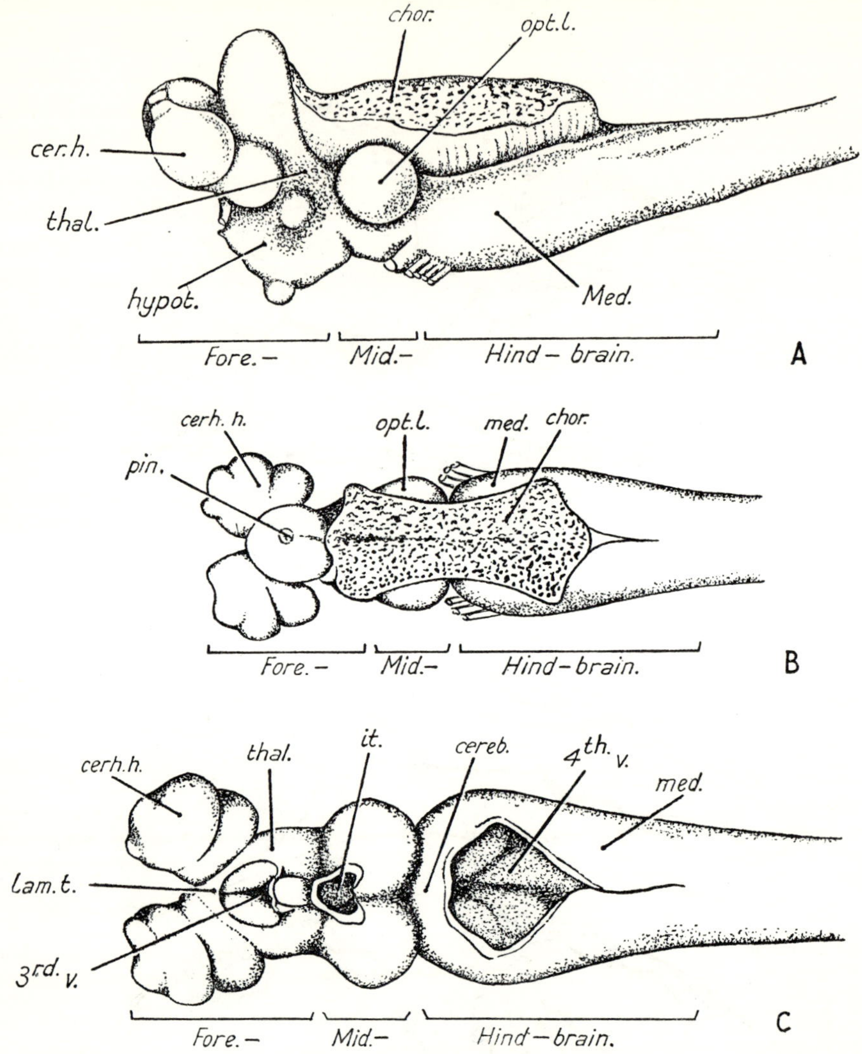

FIG. 64. Brain of the lamprey.

A, side view; B, dorsal view with choroid plexus intact; C, after removal of choroid. *cereb.* cerebellum; *cer.h.* cerebral hemisphere; *chor.* choroid plexus; *hypot.* hypothalamus; *it.* iter between third and fourth ventricles; *lam.t.* lamina terminalis (thickened anterior wall of third ventricle); *med.* medulla oblongata; *opt.l.* optic lobe; *pin.* pineal eye; *thal.* thalamus; *3rd v.*, *4th v.*, third and fourth ventricles. (After Sterzi.)

a number of very large Müller's fibres, originating from giant cells in the reticular formation of the brain, whose large dendrites (Fig. 65) receive fibres from several higher centres, providing an uncrossed final common pathway to the spinal cord. There is some difference of opinion as to whether any branches of these large fibres proceed directly into the ventral roots; probably they do not do so but the dendrites of the motor-cells branch around them and thus receive stimulation (Fig. 63). In the earliest larva co-ordination is by a pair of giant Mauthner cells, with dendrites among the entering fibres of the eighth nerve and an axon descending on the opposite side. Such cells are present in the earliest stages of nearly all fishes and amphibians.

Other nerve-cells in the more dorsal parts of the cord have no long axons and apparently serve to connect the neuropil of the various regions. The afferent fibres reaching the cord in the dorsal roots give off branches that ascend for a short distance and descend for long distances. The pathways to the brain thus pass through multiple relays.

The brain itself (Fig. 64) is built on the typical vertebrate plan, as an enlargement of the front end of the spinal cord, with thickenings and evaginations corresponding to the various organs of special sense. Although we know little of its internal functional organization in lampreys, it is probably not far wrong to regard it as chiefly consisting of a series of hypertrophied special sensory centres; thus the forebrain is connected with smell, midbrain with sight, hind-brain with acoustico-lateral and taste-bud systems. The forebrain and olfactory sense are moderately well developed in adult lampreys, as is the visual sense, with its chief centre in the midbrain. The auditory and acoustico-lateral systems are not very well marked, and the cerebellum is small. Taste is also much less developed than in the higher fishes (p. 220).

Parts of the brain

Forebrain (prosencephalon)	Cerebral hemispheres (telencephalon)
	Between-brain (diencephalon)
Midbrain (mesencephalon)	Optic lobes
Hind-brain (rhombencephalon)	Cerebellum (metencephalon)
	Medulla oblongata (myelencephalon)

The upper surface of the brain is covered by an extensive vascular pad, the choroid plexus or tela choroidea (Fig. 64). This extends into the ventricles of the brain at three points—into the third ventricle of the diencephalon, into the iter (duct) leading through the midbrain from third to fourth ventricles, and into the fourth ventricle itself. The roof of the brain is thus non-nervous in these regions. In later vertebrates the choroid extends only into the third and fourth ventricles. Presumably the vascular membranes of the brain are highly developed in lampreys because of the absence of cerebral blood vessels.

From the lower part of the mid- and hind-brain arise all the cranial nerves except the olfactory and optic. These nerves follow the same plan as those of gnathostomes but they are difficult to make out by dissection in the lamprey and will be left for consideration in connexion with the dogfish, in which they can easily be dissected. The

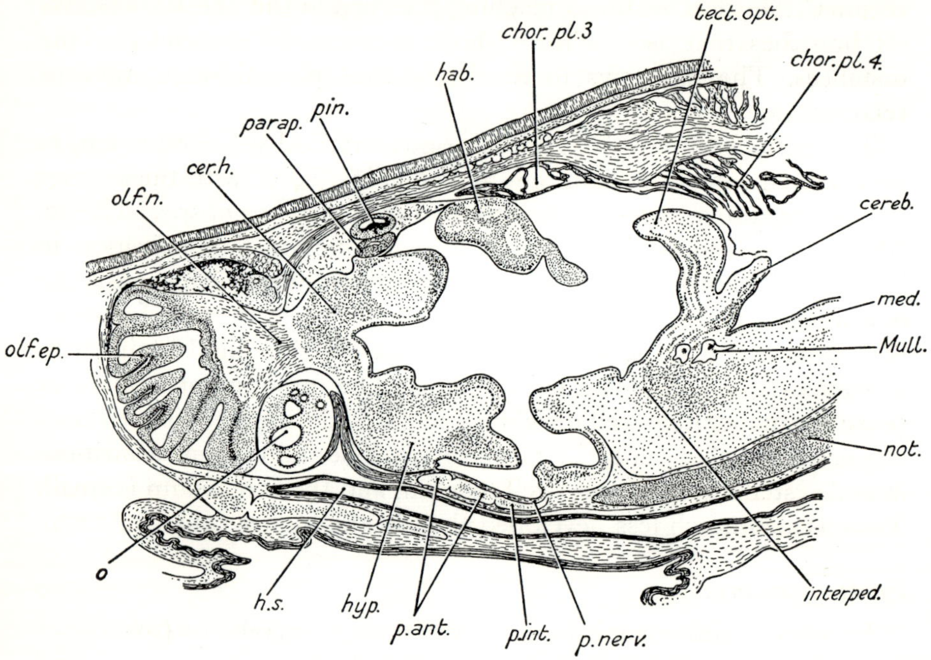

FIG. 65. Sagittal section through head of lamprey.

cereb. cerebellum; *cer.h.* cerebral hemisphere; *chor.pl. 3 & 4,* choroid plexuses of the 3rd and 4th ventricle, extending also into the midbrain; *h.s.* naso-hypophysial tube; *hab.* habenular region; *hyp.* hypothalamus; *interped.* interpeduncular region; *med.* medulla oblongata; *Mull.* Müller's cell; *not.* notochord; *o.* glandular organ of nasal sac; *olf.ep.* olfactory epithelium; *olf.n.* olfactory nerve; *p.ant.*, *p.int.*, and *p.nerv.* partes anterior, intermedia, and nervosa of the pituitary gland; *parap.* parapineal; *pin.* pineal; *tect.opt.* tectum opticum.

cranial nerves represent nerves similar to the dorsal and ventral nerve-roots of the trunk, much modified as a result of the special development of the head (p. 148). They carry afferent fibres from the skin of the head and gills and motor-fibres for moving the eyes, sucker, and branchial apparatus.

From the relative sizes of the parts of the brain it can be seen that the various special sensory centres are still small. The largest part of the brain is the medulla oblongata, which is well developed because of the extensive sucking apparatus, innervated from the trigeminal nerve.

The forebrain consists of a pair of large cerebral hemispheres and these open by the foramina of Munro into a median third ventricle, whose walls constitute the diencephalon or between-brain (Fig. 65). This diencephalon, besides connecting the forebrain with the mid-brain, includes the thalamus and serves important functions of its own. Its ventral part, the hypothalamus, is well developed in all vertebrates as a central organ controlling visceral activities and the internal life of the organism. Nerve-fibres from the supraoptic nucleus of the hypothalamus proceed to the pars nervosa of the pituitary and, as in other vertebrates, are filled with granules of neurosecretory material, which presumably controls pituitary action. A simple portal system of blood-vessels connects the hypothalamus with the pituitary.

10. The pineal eyes

The diencephalon is also the region of the brain from which the eyes are formed. In lampreys, besides the usual pair of eyes, there is also, attached to the roof of the between-brain, the so-called third, epiphysial, or median eye, better developed in these animals than in any other living vertebrate except perhaps certain reptiles.

This organ is actually not median but consists of an unequally developed pair of sacs, that on the right, the pineal, being larger and placed dorsal to the morphologically left parapineal (Fig. 66). The sacs form by evagination from the brain and remain connected with the dorsal epithalamic or habenular region of the between-brain by two stalks. The two organs are similar in structure, consisting of irregular flattened sacs with a narrow lumen. Both upper and lower walls of each organ contain receptor cells, with processes that project into the lumen and nerve-fibres directed outwards. These fibres apparently mostly end within the organ, in contact with ganglion cells whose axons run to unequal right and left habenular ganglia. In addition there are supporting and pigment cells in the retinas. Knowles has shown that the retinal cells of the pineal make movements, being arranged differently under conditions of illumination and darkness. The significance of these photomechanical changes is unknown but they demonstrate that the pineal cells are sensitive to light.

The structure of these pineal organs shows that they consist of portions of the diencephalic wall where the ciliated cells of the ependyma are specialized as photoreceptors. They show the same general plan as the paired eyes, but with no differentiated dioptric apparatus.

It has been possible to find out something of the part that these organs play in the life of the lamprey. When a bright spot of light is directed upon the pineal region of a stationary ammocoete larva movement is usually initiated, but only after illumination for many seconds.

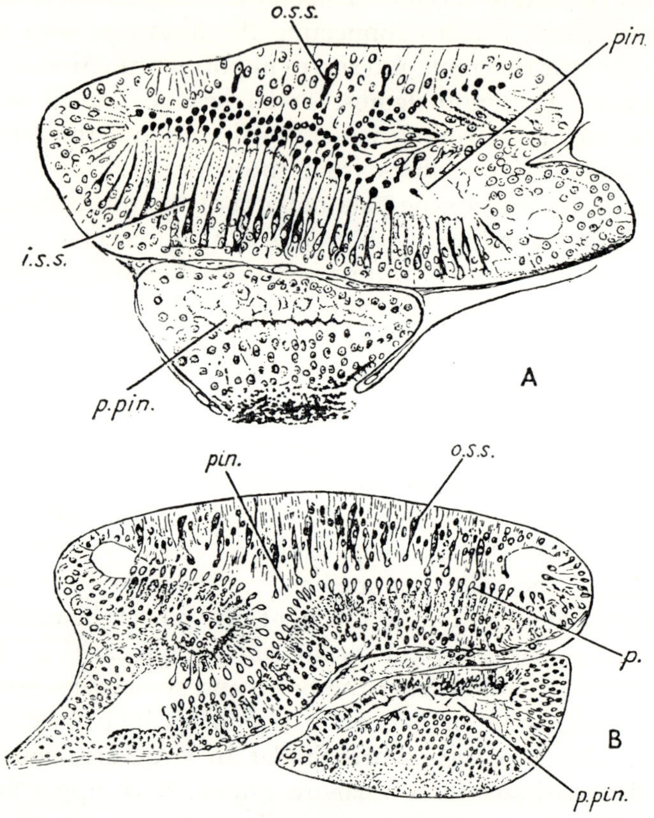

FIG. 66. Pineal and parapineal organs of adult *Lampetra fluviatilis* *A.* larva, *B.* adult. Sagittal section.

i.s.s., *o.s.s.*, inner and outer sensory cells; *p.* process; *pin.* pineal; *p.pin.* parapineal. (After Tretjakoff.)

Moreover, these movements can be elicited even after the pineal organs have been removed! In the larval lamprey the paired eyes are deeply buried below pigmented skin, so the movement is not likely to be due to them; indeed it continues when they too have been taken out! Evidently there must be still other receptors, able to respond to changes of light intensity in the wall of the diencephalon. This recalls the fact that photoreceptors are found within the substance of the nervous system of amphioxus. This power of response to changes of illumination has been retained in the vertebrates, and persists in some as yet unknown cells in the brain, even after the paired and pineal eyes have become specialized for light reception. The whole study is of special interest as showing the stages by which the eyes may have been evolved. Higher fishes also show the power of responding to

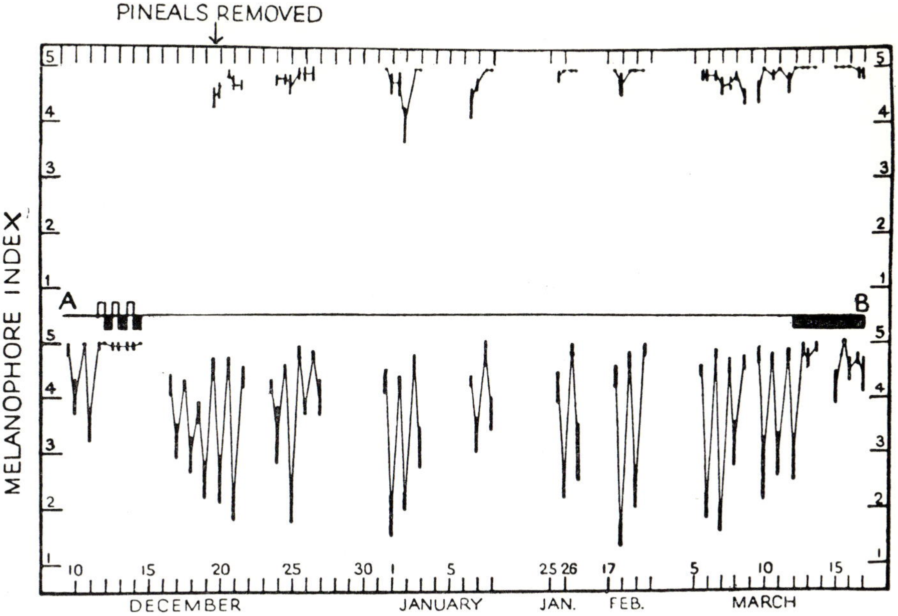

FIG. 67. Colour-changes of larval lampreys, measured by the melanophore index (see p. 300). Animals kept out of doors except as shown along the line *AB*, where rectangles above the line show illumination with electric light and below the line total darkness. Normal animals show a regular daily rhythm, becoming pale at night. Reversal of normal day and night illumination stops the change. On 19 December the pineal eyes were removed from five out of the ten individuals and these thereafter remained dark (upper chart); the other five continued to show the normal rhythm, until placed in total darkness. (After Young.)

changes of illumination after the paired eyes and epiphysis have been removed (p. 210).

If the pineal eyes are not essential for the initiation of movement, what is then their function? In the ammocoete larva there is a daily rhythm of change of colour, the animals becoming dark in the daytime and pale at night. After removal of the pineal eyes this change no longer occurs: the animals remain continually dark (Fig. 67). This effect on the colour is produced by the action of influences from the pineal, passing to the pituitary gland (see p. 103). It seems that the pineal apparatus is an organ concerned with adjustment of the internal activities of the animal to correspond to the changing conditions of illumination. The control may be effected by impulses carried in the large tract that proceeds from the habenular ganglion to the hypothalamus, in the floor of the diencephalon. The latter is known to be concerned, throughout the vertebrate series, with the integration of the internal activities of the animal.

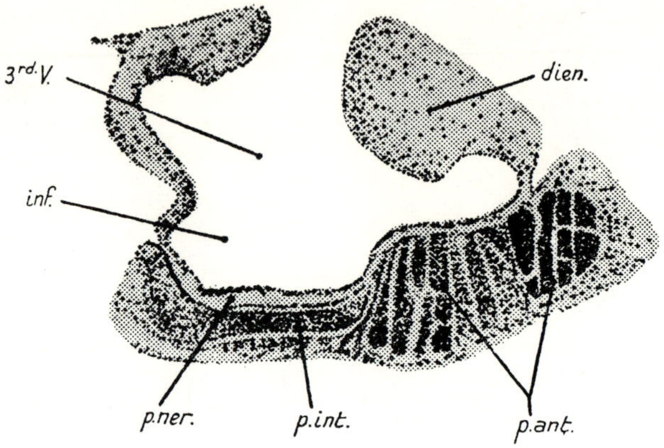

FIG. 68. Sagittal section of the pituitary gland of lamprey. *dien.* diencephalon; *inf.* infundibulum; *p.ant.*, *p.int.*, and *p.ner.* partes anterior, intermedia, and nervosa, there are two types of cell in the pars anterior; *3rd V.* third ventricle. (After Stendell.)

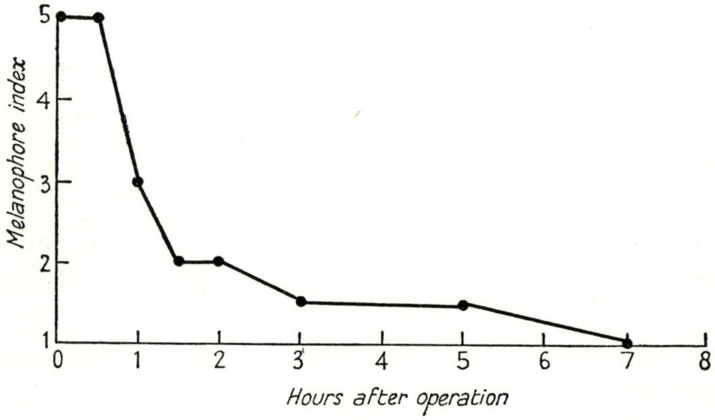

FIG. 69. Onset of pallor in a larval lamprey after removal of the pituitary gland, as shown by the decline in the melanophore index. (From Young.)

11. Pituitary body and hypophysial sac

The lower portion of the diencephalon, the hypothalamus, forms a prominent pair of sacs, the lobi inferiores, which contain a partly separated diverticulum of the third ventricle and end below in the infundibulum (Fig. 65). The pituitary gland (hypophysis) is pressed against the underside of the hypothalamus (Fig. 68). The lower wall of the brain in this region consists not of nerve-cells but of a single epithelial layer, corresponding to the pars nervosa of the pituitary of higher forms. The major portion of the pituitary gland is a mass of secreting cells in which two parts can be recognized, the partes anterior and intermedia. After experimental removal of the intermediate portion of the pituitary lampreys become permanently pale in colour (Fig. 69), showing that, as in other vertebrates (p. 299), a melanophore-expanding substance is liberated into the blood by this gland, the secretion being presumably under the control of the pineal eyes (p. 105). The lamprey pituitary has been shown to contain oxytocic and 'water balance' hormones as well as one producing melanophore expansion. Moreover injections of mammalian anterior pituitary extracts induce appearance of the secondary sexual characters of lampreys. Evidently the functions of the pituitary have remained essentially the same through the whole chordate series.

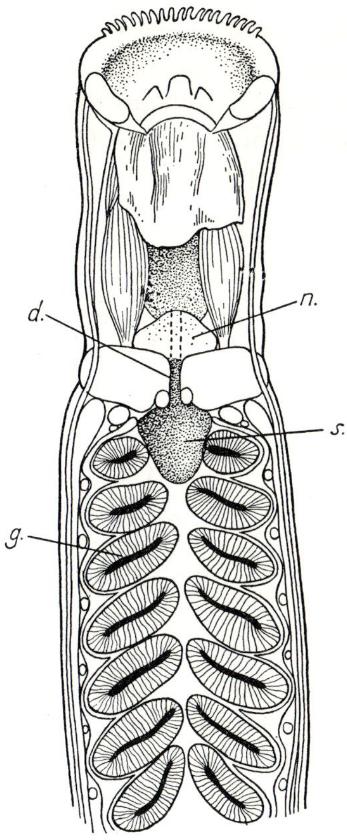

FIG. 70. Dissection of lamprey from the ventral surface after injection of coloured gelatine to show the outline of the naso-hypophysial sac (*s*) and its duct (*d*), which is shown dotted where it runs upwards between the nasal sacs (*n*). *g*, gill pouches. Contraction of the branchial apparatus squeezes the sac *s*, so that water is drawn in at each relaxation.

The pituitary of lampreys is peculiar because of great development of the naso-hypophysial sac (Fig. 70). Characteristically in vertebrates the pituitary body develops by the formation of a pocket of buccal ectoderm, whose walls then become folded, so that the part in front of the lumen becomes the pars anterior, that behind the pars intermedia. In nearly all vertebrates the lumen then loses its connexion with the exterior. In lampreys the hypophysial rudiment is continuous with that of the olfactory epithelium. The latter then moves dorsally and the two remain connected throughout larval life by a strand of cells. At metamorphosis this acquires a lumen and forms a tube extending from the nostril below the pituitary and brain. Because of its development this is sometimes called the naso-hypophysial tube but others doubt that it represents the cavity of the hypophysis and prefer the name nasopalatine canal.

Inside the single nostril, guarded by a valve, are openings into the nasal sacs, which are cavities with folded walls. Some of the cells of these walls are the olfactory receptors and give off the axons that make up the olfactory nerves, entering the olfactory bulbs on the anterior end of the hemisphere (Fig. 65). Behind the nasal sacs lie numerous glandular follicles opening into the sac in the larva, but completely closed in the adult (Fig. 65). They may be comparable to Jacobson's organ (p. 405).

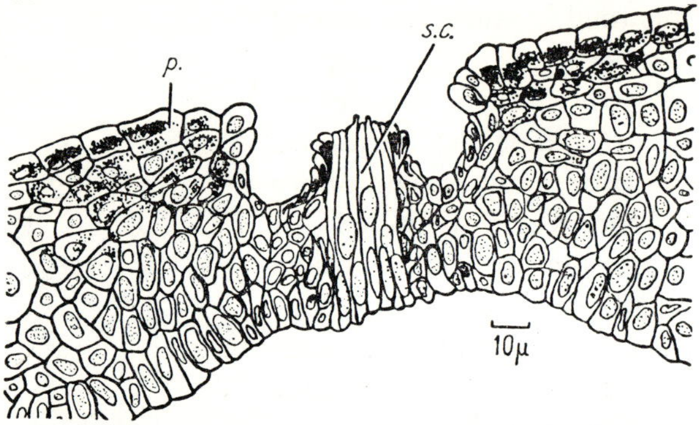

FIG. 71. Section of lateral line organ of tail of adult *Lampetra*. *p.* pigmented cells around the pit; *s.c.* receptor cells (not showing long hairs). (After Young.)

The naso-hypophysial tube proceeds back behind the pituitary to a closed sac lying between the first pair of gill pouches (Fig. 70). During the movements of respiration this sac is squeezed and water is expelled with some force through the nostril. When the gills relax water flows in at the nostril, and in this way the olfactory organ is provided with samples. If the naso-hypophysial opening is closed with a plug of plasticine the lamprey no longer reacts to solutions, for instance of alcohol, to which it normally responds by freeing its sucker and swimming away.

12. Lateral line organs of lampreys

The lateral line receptors, peculiar to fish-like vertebrates, are little patches of sensory cells found along certain lines on the head and trunk. They are all innervated by cranial nerves, those on the body and tail being served by a special backward branch of the vagus nerve. The receptor cells carry long hairs and are thus able to detect either movement of the water relative to the fish or of the fish itself. Objects moving nearby set up disturbances that may also be detected (p. 218).

In the lamprey the lateral line organs are very simple (Fig. 71), being open to the exterior and not sunk in a canal as in higher forms. The rows are somewhat irregular, especially those on the body.

13. Vestibular organs of lampreys

The labyrinth may be considered as a specialized portion of the lateral line system, concerned with recording the position of the head

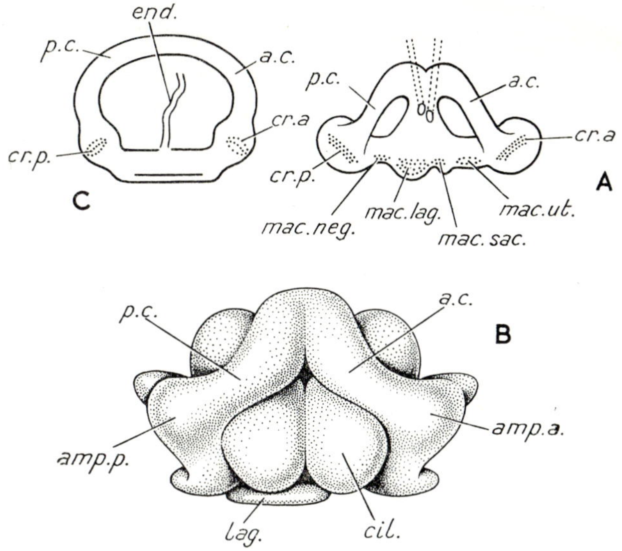

FIG. 72. Labyrinth of right side, seen in lateral view. A and B, *Lampetra*, C, *Myxine*.

a.c. anterior canal; *amp.a.* and *p.* anterior and posterior ampullae; *cil.* ciliated chamber; *cr.a.* and *p.* cristae; *end.* endolymphatic duct; *lag.* lagena; *mac. lag.*, *neg.*, *sacc.*, *ut.* maculae of the lagena, neglecta, saccule, and utricle; *p.c.* posterior canal. (After de Burlet.)

and angular accelerations. There is no evidence to decide whether lampreys can respond to sound. The labyrinth develops by an in-pushing of the wall of the head, and this then becomes closed off from the exterior. Internal foldings divide up the sac into a number of chambers, which differ considerably from those of gnathostomes. There is a large central vestibule, into which open below several partially separate sacs, provided with patches of sensory hairs. These correspond, from in front backwards, to the maculae of the utricle, saccule, and lagena of higher forms (Fig. 72). The hairs of the maculae are loaded with otoliths. There are only two broad semicircular canals, corresponding to the anterior and posterior vertical canals of other vertebrates, each with an ampulla, containing a receptor ridge, the crista. Also opening to the vestibule are two large sacs, covered with cilia (Fig. 72), whose beat produces complicated counter currents in the dorso-ventral plane. It has been suggested that these function as a gyroscope, compensating for the absence of a horizontal canal.

In *Myxine* the condition is even simpler, there being only a single vertical semicircular canal (Fig. 72). However, it is claimed that this

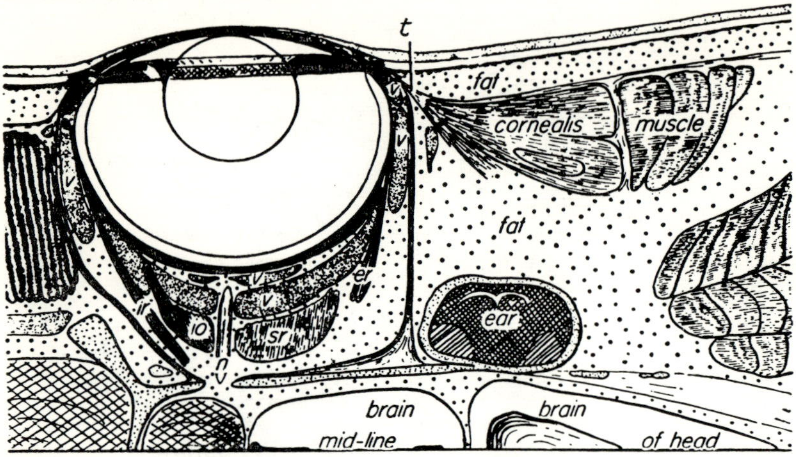

FIG. 73. Horizontal section of the eye of a lamprey.
er., *ir.*, *sr.*, external, internal, and superior rectus muscles; *v.v.* venous sinuses which cushion the eye. (From Walls, *The Vertebrate Eye*, Cranbrook Institute of Science.)

has cristae at both ends. The macular system also does not show the characteristic subdivisions but is a single *macula communis*.

14. Paired eyes of lampreys

The structure of the paired eyes is similar to that in other vertebrates. They are formed, like the pineal eyes, by evaginations of the wall of the diencephalon; the so-called optic nerve is therefore not really a peripheral nerve but a portion of the brain; it should strictly be called the optic tract. The eyes are moved by extrinsic muscles arranged in a somewhat unusual manner. Accommodation is effected by a process found in no other vertebrates. The cornea consists of two distinct layers, separated by a gelatinous substance. Attached to the outer (or dermoid) cornea is a cornealis muscle, apparently of myotomal origin, which flattens the cornea and pushes the lens closer to the retina (Fig. 73). There is an iris, outlining a round pupil, which changes little, if at all, in diameter under different illuminations. Most species of lampreys are diurnal animals. They are said to move towards white objects and probably use both the eyes and the nose to find their prey. In the ammocoete larva the paired eyes are buried below the pigmented skin and the animal makes no movements when light is shone on to this region.

The optic tracts of adult lampreys end in the roof of the midbrain (tectum opticum) which is a highly differentiated, stratified region. Besides the optic fibres it receives also impulses from fibres ascending from the spinal cord and others from the auditory and lateral line

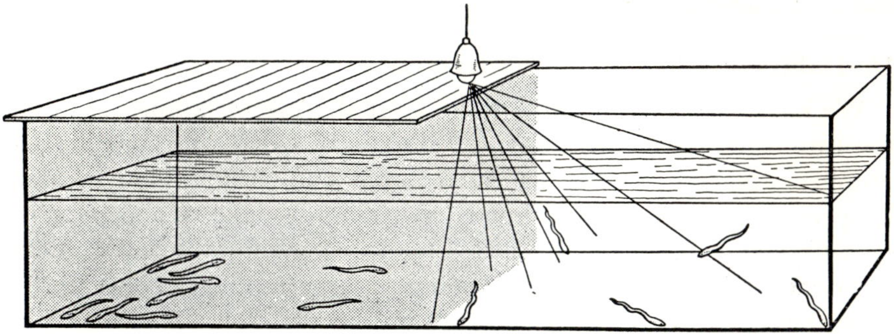

FIG. 74. Experiment to show behaviour of larval lampreys when illuminated. The tank is left in total darkness and the larvae settle in all parts. When the light is switched on those in the illuminated part begin to swim and continue to do so until by chance they arrive in the darkened part, where they settle down. (From Young.)

centres. The midbrain is therefore undoubtedly one of the most important parts of the brain in lampreys, though nothing is known in detail of its functions. Its cells control movements of the animal, by means of fibres that run to make connexion with the dendrites of the large Müller's cells, whose axons pass down the spinal cord; other fibres from the tectum opticum reach to various parts of the brain, and it is probable that its activities are closely correlated with those of many other regions.

15. Skin photoreceptors

Like many lower vertebrates the lamprey has light-sensitive cells in the skin, as well as those in the eyes. These receptors are abundant in the tail and if a light is shone on to this region the animal rapidly moves away (Figs. 74 and 76). If the spinal cord is cut just behind the head and a light then shone on to the tail, the *head* will be seen to move. This suggests that the impulses are carried forwards by means of the lateral line nerves, which is confirmed by the fact that if these latter are sectioned, leaving the spinal cord intact, then no movements follow when the tail is illuminated. This sensitivity of the lateral line organs to light is not found in other fish-like vertebrates. Indeed the receptors are not strictly lateral line organs but pigmented epidermal cells. The sensitivity curve shows a sharp peak at 530 mμ, this being the region of the spectrum at which light penetrates farthest into sea water. The pigment is probably a porphyropsin (Steven, 1950).

In hag-fishes (*Myxine*) the head and cloacal regions are more sensitive to light than is the rest of the body. The impulses from the skin are conducted through the spinal nerves in these animals, not the lateral line nerves.

16. Habits and life-history of lampreys

We have very little information about the life of lampreys during the time that they are in the sea. They are caught in considerable number attached to other fishes. It is not known how many years a lamprey spends in the sea, but it returns only once to the river for spawning and dies after this act. The up-river migration of *L. fluviatilis* occurs in the autumn, for instance large numbers come up the River Severn and are caught in traps on the way, for use as food. The spawning migrations of lampreys may take them for hundreds of miles, for example, those of the eastern Pacific ascend to the headwaters of the Columbia River. They are said to perform remarkable feats of climbing, leaping from stone to stone and hanging on by their suckers. During this period of migration some lampreys assume brilliant orange and black colour patterns. On the other hand, lampreys land-locked in the lakes of New York (*Petromyzon marinus unicolor*) feed in fresh water and ascend only a few miles up streams to breed.

Once in the river the lampreys do not feed again but live over the winter on the reserves accumulated in the form of fat, especially under the skin and in the muscles. During the winter the gonads ripen progressively and the secondary sexual characters begin to become apparent only in February. The females then develop a large anal fin, while in the male a penis-like organ appears (Fig. 47) and the base of the dorsal fin becomes thickened.

Spawning occurs in the spring and is preceded by a form of nest-building. Numerous lampreys collect together, usually at a place below a weir where the water is shallow and rather swift, and the bottom both stony and sandy. Stones are then dragged by the mouth in such a way as to make a small depression. Fertilization is secured by a process of copulation in which the male fixes by the sucker on to the fore-part of the female and the two then become intertwined and undergo rapid contortions, the eggs being squeezed into the water, while sperms are ejected through the 'penis' (Fig. 75). Fertilization is therefore external, but the sperms must be placed very close to the

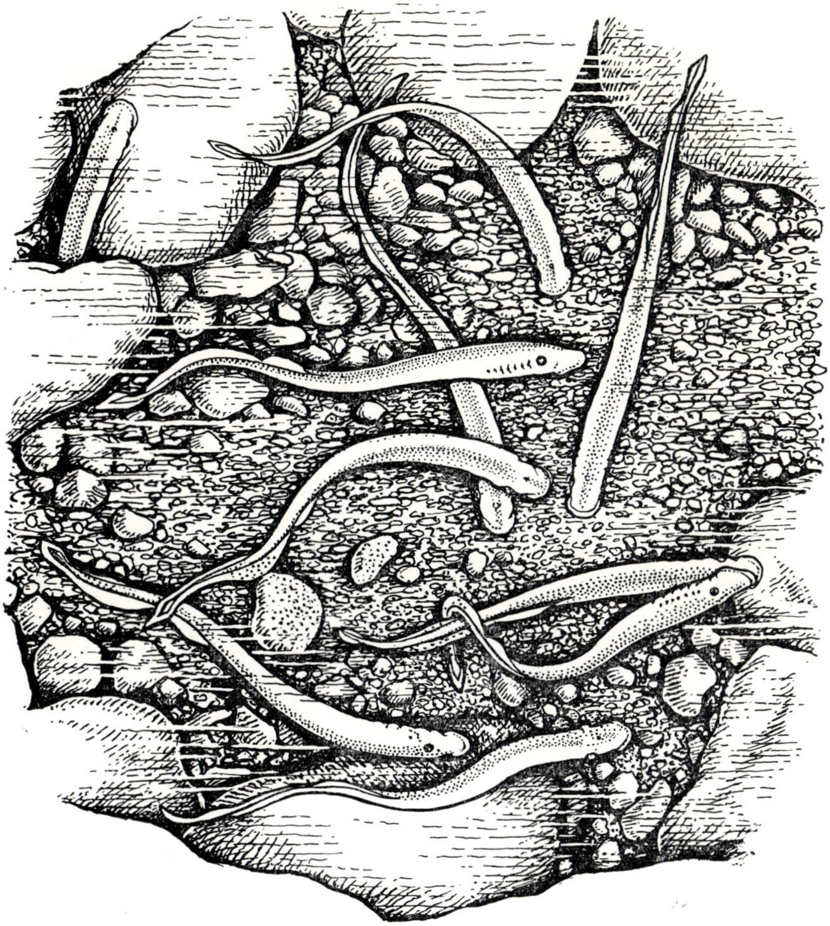

FIG. 75. Spawning lampreys seen in their nest. (After Gage.)

eggs, for they remain active only for about one minute after entering the fresh water, which provides the stimulus that activates them. The eggs and sperms are not all laid at once; mating is repeated several times until all the products have been shed, after which the animals are exhausted and soon die. The movements of the animals stir up the sand in the nest (this is probably the function of the anal fin of the female) ensuring that the eggs are covered up as they are carried away by the current.

17. The ammocoete larva

The eggs contain a considerable quantity of yolk, but their cleavage is total and proceeds in a manner not unlike that of the frog. After about three weeks the young hatches as the ammocoete larva, about 7 mm long. At first this is a tiny transparent creature, but its larval life lasts for a long time, during which it grows into an opaque eel-like fish, up to 170 mm long (Fig. 76).

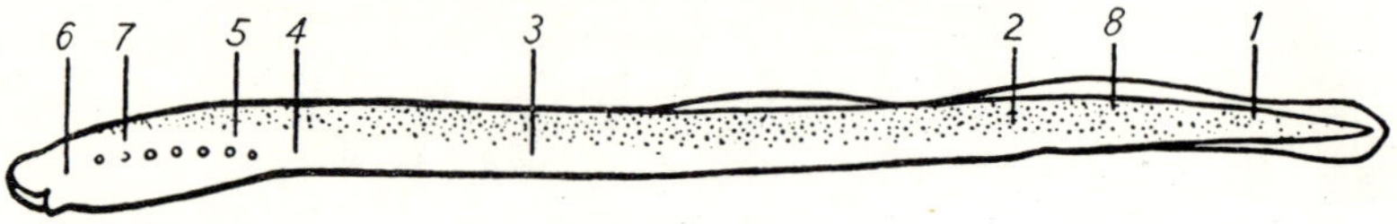

FIG. 76. Ammocoete larva of *Lampetra planeri*, showing the effect of shining a narrow beam of light on to various parts of the side of the body. Illumination of 1, 2, or 8 is followed by movement after a few seconds, but no movement follows illumination at points 3–7. (From Young.)

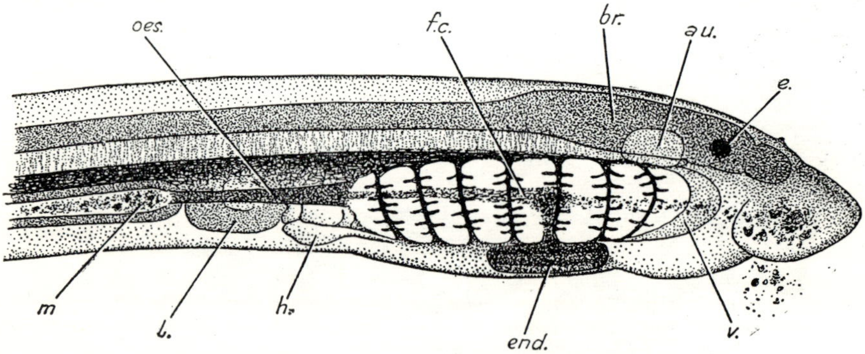

FIG. 77. Young ammocoete larva of lamprey fixed while feeding on green flagellates and detritus and then stained and cleared.

au. auditory sac; *br.* brain (covered by meninges); *e.* eye; *end.* endostyle; *f.c.* food cord in pharynx; *h.* heart; *l.* liver; *m.* mid-gut; *oes.* oesophagus; *v.* velar fold.

This portion of life is spent buried in the mud, the animals emerging only occasionally to change their feeding-ground, presumably if the mud is not sufficiently nutritious. There is no sucker, the mouth being surrounded by an oral hood rather like that of amphioxus (Fig. 77). The paired eyes are covered by muscles and skin. The head at this stage is little sensitive to light, but the animal quickly begins to swim if the tail is illuminated. We have seen already (p. 108) that in lampreys there are photoreceptors in the tail, connected with the lateral line nerves. In the larva these are the main photoreceptors, and they ensure that the animal lies completely buried.

If a number of larvae are left in a vessel with a layer of mud on the

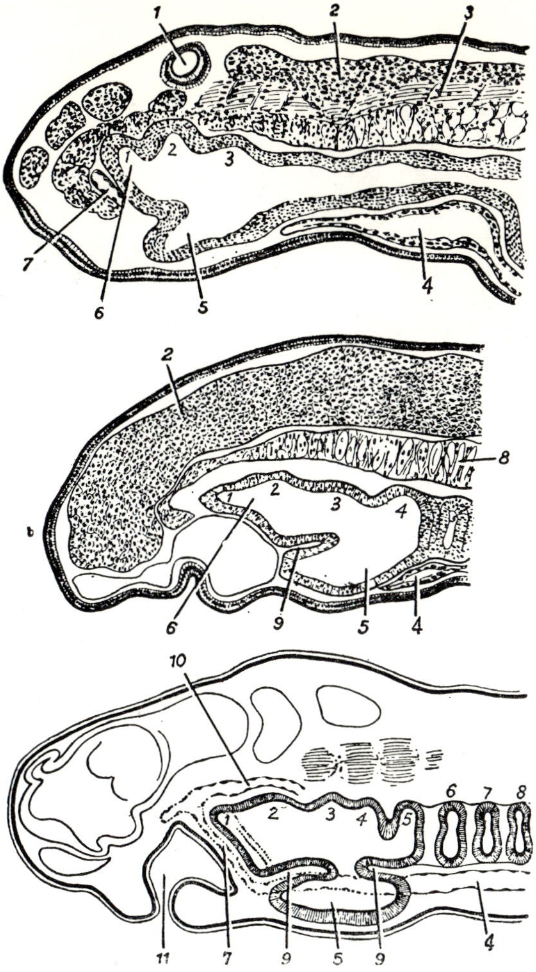

FIG. 78. Development of the endostyle of the lamprey. Sagittal sections through the head at three stages.

1, auditory sac; 2, medullary tube; 3, myotome; 4, conus arteriosus; 5, endostyle; 6, first gill-slit; 7, first arterial arch; 8, notochord; 9, inpushings which cut off the endostyle from the pharynx; 10, aorta; 11, stomodaeum. (After Dohrn, from Kukenthal.)

bottom they rapidly disappear and remain hidden indefinitely, the heads perhaps just visible in small depressions made by the rhythmic respiratory movements. When disturbed they always swim with the head downwards and in contact when possible with the ground. This habit leads them to burrow rapidly. It is not known whether they have other receptors to guide them to mud rich in possible food organisms. The nasal and hypophysial sacs are poorly developed in the larva, and the sense of smell can hardly serve this purpose.

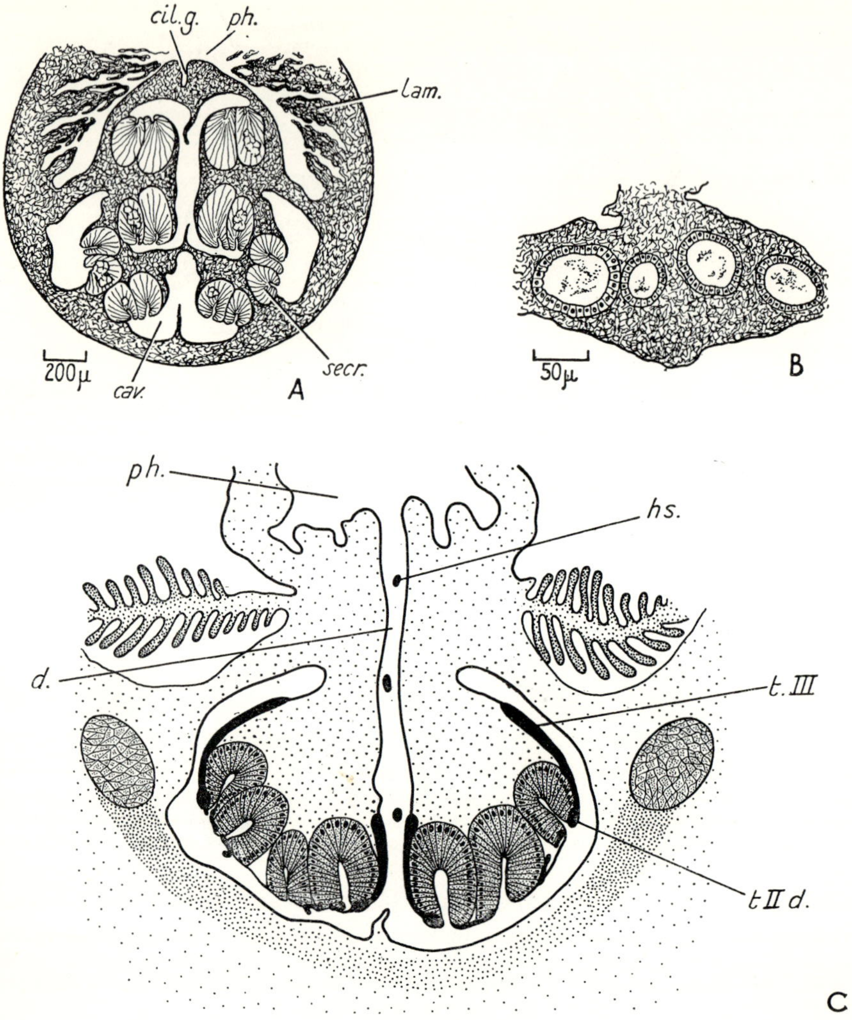

FIG. 79. A, Transverse section of endostyle of ammocoete larva.

cav. cavity of the gland; *cil.g.* ciliated groove in floor of pharynx; *lam.* lamellae of gills; *ph.* cavity of pharynx; *secr.* mucus-secreting cells of the gland.

B, Transverse section of thyroid follicles of adult. (After Young and Bellerby.)

C, Cross section of endostyle of ammocoetes larva of *Petromyzon marinus* at level where it is connected by a duct to the pharynx. Autoradiograph showing distribution of protein-bound I^{131}. The radioactive clumps of cellular debris in the glandular lumen and in the duct suggest that the material represents a holocrine secretion, which will probably be absorbed in the intestine.

d. duct; *hs.* cellular debris; *ph.* pharynx; *t II d.* type II dorsal cells; *t III* epithelial cells.

Feeding takes place by the intake of water through the mouth and the separation of small food particles from it in the pharynx (Fig. 77). For this purpose there is used a great quantity of mucus, which is secreted by the endostyle and gathered into a strand by the cilia of the pharynx. This endostyle is a most remarkable organ, forming early in development as a sac below the pharynx (Fig. 78). It consists of a pair of tubes, on the floor of which there are four rows of secretory cells (Fig. 79). There is a single opening to the pharynx, by a slit at about the middle of the length. As development proceeds the inner rows of cells at the hind part of the organ become coiled upwards, and at the end of larval life the endostyle therefore forms a very large mass below the pharynx, composed of tubes lined partly by secretory and partly by ciliated cells. Probably no enzymes are secreted by the endostyle, its function being to produce mucus in which the food particles become entangled. Although it resembles the endostyle of amphioxus in the arrangement of the secretory columns, there is a difference in that the organ in the ammocoete larva is not an open groove. There is, however, a ciliated groove in the floor of the pharynx, that is to say, on the roof of the endostyle (Fig. 79).

The details of the feeding-currents of the ammocoete larva are not understood. An important difference from the arrangement in amphioxus is that the current is produced by muscular rather than ciliary action. The velum, a pair of muscular flaps, provides the main current when the animal is at rest. The branchial basket can also be expanded and contracted by an elaborate system of muscles. It is not easy to observe how the food particles are taken up from the current, but apparently a strand of mucus shoots from the endostyle and occupies the whole of the centre of the pharynx (Fig. 77). This strand probably rotates and as it passes backwards into the eosophagus it catches the particles.

Evidently the system enables the animals to feed efficiently on the small unicellular algae and bacteria of the mud. In amphioxus the ciliated pharynx, occupying a considerable proportion of the whole surface, is only able to support a tiny creature, but the muscular feeding-system of the ammocoete allows a relatively small pharynx to feed a fish 170 mm long and weighing up to 10 grams. This use of muscles for moving the gills was evidently an important step in chordate evolution. It allowed the animals to escape from the limitation of size imposed by the ciliary method of feeding. After the development of jaws to form a still more efficient feeding mechanism the rhythmic movement of the branchial apparatus persisted for the purpose of respiration. We cannot be certain about changes which occurred so long ago, but it seems likely that the respiratory movements of a fish were first introduced to provide food rather than oxygen.

The endostyle therefore shows the survival of the primitive feeding-methods of chordates, but it also undergoes at metamorphosis an astonishing change into a thyroid gland. The mucus-secreting columns shrink and the whole organ becomes reduced to a row of closed sacs, lying below the pharynx (Fig. 79 B). Each of these sacs is lined by an epithelium, contains a structureless 'colloid' substance, and is therefore closely similar to a thyroid vesicle. Moreover, experiments have shown that extracts of this organ contain iodine and exert an accelerating effect on the metamorphosis of frog tadpoles. Although nothing is known of the part played by the secretion of this gland in the life of the adult lamprey, we may safely conclude that we have here the conversion of an externally secreting feeding-organ into a gland, of internal secretion. The actual mucus-secreting cells are not transformed into those of the thyroid follicles, these latter are derived from epithelial cells in the wall of the larval organ. One cannot avoid speculating on this extraordinary change of function. It may perhaps be significant that the endocrine gland that regulates basal metabolism (the thyroid) is derived from the part of the feeding-system that in the earliest chordates was responsible for providing the raw materials of metabolism. Experiments with radioactive iodine show that this element is concentrated in certain cells of the larval endostyle (Fig. 79 C). Moreover, after addition of the anti-thyroid substance thiourea to the water there are changes in the endostyle. Thyroxine has been extracted from the gland and it probably has an endocrine function as well as secreting mucus, though no one has ever produced any changes in larval lampreys by administering thyroid hormones. Lampreys thus show, as larvae, a stage in which the accumulation of iodoproteins, previously widespread, becomes concentrated in the pharynx. Perhaps at this site there were already cells specialized for halide transport (cf. the chloride-secreting cells of teleosts, used for osmogulation, p. 203). In adult lampreys and all higher chordates the iodoprotein is secreted into the blood under the control of blood-borne signals (Fig. 80). The change may well be related to developments in the regulation of metabolism, which, in the animals with a fully endocrine thyroid becomes more nearly independent of variations in the external supply of iodine.

The great change in the endostyle is only part of the complete metamorphosis by which the ammocoete larva changes into an adult lamprey. The mouth becomes rounded and its teeth, tongue, and complex musculature develop. The paired eyes (previously buried) appear; the olfactory organ becomes internally folded, and the olfactory nerve and tracts much enlarged. The naso-hypophysial sac grows backwards to the gills. In the pharynx the gills develop into sacs opening to the branchial chamber. Changes also take place in the

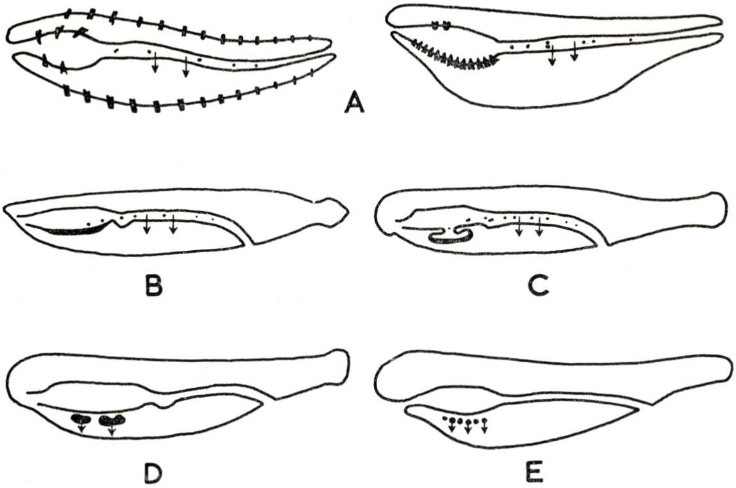

FIG. 80. Diagram to show distribution of iodoproteins, at first in exoskeletal structures, as in many invertebrates and in tunicates (A). Some of this material is concentrated in the pharynx. This tendency is exaggerated in amphioxus and the ammocoete larva, and in the adult lamprey and later animals this pharyngeal material forms the thyroid. A. Many invertebrates and tunicates; B. Amphioxus; C. Ammocoetes; D. Metamorphosis of ammocoetes; E. General vertebrate type. (After Gorbmann, A., in *Comparative Endocrinology*. Wiley, New York.)

intestine. The yellow-brown colour of the larva gives place to the black with silver underside of the adult. The animal more and more frequently leaves the mud and finally migrates to the sea to begin its parasitic life.

18. Races of lampreys, a problem in systematics

Besides the river lampreys, such as *L. fluviatilis* (Linn.), which show this characteristic migratory life-history, there are also in various parts of the northern hemisphere small brook lampreys ('prides'), such as *L. planeri* Bloch, which remain throughout their life in fresh water. These prides are very abundant in many English rivers and streams, but since the greater part of their life is passed in the ammocoete stage they are not often seen. The larvae remain in the mud probably for three years and undergo metamorphosis in late summer and autumn. The characteristic of this type of lamprey is that the adults never migrate and never feed. The gonads are already well developed at metamorphosis and ripen during the winter. Spawning takes place in March or April and the animals then die.

There has been much dispute about the status of these freshwater races. In structure the adult *L. planeri* is nearly if not quite identical with an adult *L. fluviatilis*, except that the latter is much the larger and has sharper teeth. Crossing of the two sorts could presumably never take place in nature, on account of the size difference, but by artificial stripping of the adults cross-fertilization in both directions can easily be achieved. Unfortunately the hybrid larvae have never been reared to maturity; we cannot therefore say whether the small size and failure to migrate of the *planeri* forms are inherited characters or are produced by the influence of the environment. The effect of the non-migratory condition is to enable the lampreys to colonize very fully rivers that, because of effluents, they would be unable to occupy if a migration to the sea was necessary. By this process of acceleration of the development of the gonads a dangerous stage in the life-history has been avoided.

Similar pairs of migratory and non-migratory forms of lamprey are found in Japan and in North America. Indeed, the condition appears to be developing independently in several river systems in the United States. Since it may be difficult for the brook lampreys to spread from one river system to another it is possible that many of the *planeri* forms have evolved separately, perhaps quite recently. If so, this is a remarkable example of a similar response produced in different parts of a population by a similar environmental stimulus, in this case the effluents. This process of alteration in the relative times of metamorphosis and sexual maturity (paedomorphosis) has occurred also in certain amphibians (the axolotl) and in tunicates (Larvacea). Similar changes in rates of development may have been essential factors in the development of the whole chordate phylum (p. 77).

In one race, found in Italy, ammocoetes with mature gonads have been reported. However, in most of these lampreys the paedomorphosis is only partial: metamorphosis does take place, but is immediately followed by maturity. Since in mammals injections of anterior pituitary extracts accelerate development of the gonads, it was thought possible that complete neoteny might be produced by making such injections into larvae of *L. planeri*. No completely sexually mature ammocoetes have yet been produced by this method, but following the injections the larvae assumed the secondary sexual characters, which are normally shown only at maturity, namely, swelling of the cloaca, opening of the pore from coelom to exterior, and the changes in body form. No signs of metamorphosis were produced by these injections and we are left without information as to the cause of that change in the lamprey. In Amphibia even very young larvae undergo metamorphosis when treated with thyroid extracts, but similar treatment of ammocoete larvae has failed to produce any change. Further investigation of the problem should be very interesting, since it seems likely that the differences between the *fluviatilis* and *planeri* forms are the result of an endocrine factor accelerating the onset of sexual maturity in the latter. The fact that the change is occurring in various parts of the world adds further interest to this example of evolution in progress.

Besides all these relatively small lampreys, there is a much larger form, the sea lamprey, *Petromyzon marinus* Linn., reaching to over a metre in length. This animal differs from *Lampetra* in body form, structure of sucker, and other features, as well as in size. Like most other groups of animals lampreys therefore present several problems of nomenclature. Linnaeus included the three types that occur in Europe in the one genus *Petromyzon*; since they are all rather alike in shape this is in some ways a reasonable procedure. But are we then also to include in the same genus forms that differ more widely, such as those occurring in the southern hemisphere? As so often happens, systematists have chosen the course of splitting up the Linnaean genus, even though several of the resulting genera have only one species. Thus Gray suggested the genus *Lampetra* for the brook and river lampreys, keeping *Petromyzon* for the larger species of sea lamprey. Other genera have been added, such as *Entosphenus* Gill for some of the North American forms and *Mordacia* Gray and *Geotria* Gray for the forms from the southern hemisphere (Chile, Australia, and New Zealand). Such distinctions, though they may seem irritating at first sight, are an advantage in that they call attention to the differences which exist. For instance, it is a striking fact that lampreys are found in temperate waters of both hemispheres, but not in the tropics, and it is interesting to learn that the forms from New Zealand, Australia, and South America (there are none in South Africa) show distinct peculiarities. Thus *Geotria* possesses a large sac behind the sucker.

A special problem of nomenclature arises from the fact that the river and brook lampreys are almost identical in structure and differ mainly in size, time of sexual maturity, and habits. A further complication is that the germ-cells of the two races allow cross-fertilization, although this probably never occurs in nature! We may take Dobhzansky's definition of species as 'groups of populations which are reproductively isolated to the extent that the exchange of genes between them is absent or so slow that the genetic differences are not

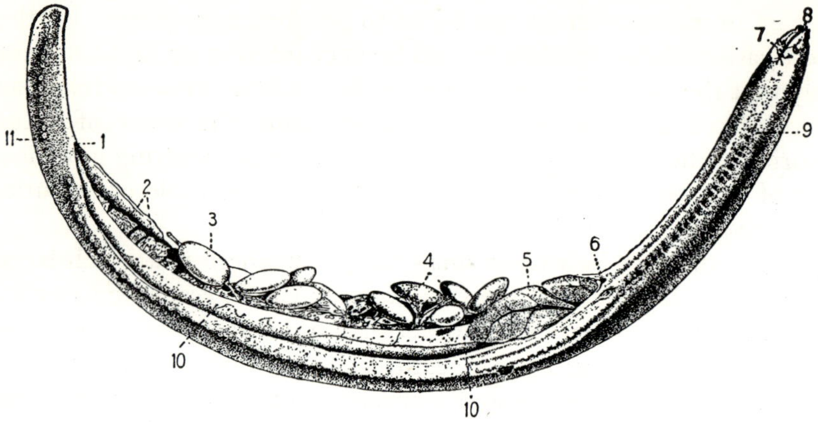

FIG. 81. *Myxine*, partly dissected.

1, cloaca; 2, testis; 3 and 4, ovary with eggs; 5, liver; 6, branchial opening; 7, mouth; 8, nostril; 9 and 11, slime glands; 10, intestine. (After Retzius, from Kukenthal.)

diminished or swamped', and in this sense we may retain the specific names *L. fluviatilis* and *L. planeri* for the two populations.

19. Hag-fishes, order Myxinoidea

The hag-fishes, *Myxine* and *Bdellostoma* (Fig. 81), are animals highly modified for sucking. They live buried in mud or sand and probably eat polychaetes and other invertebrates, as well as scavenging dead fishes. The eyes are functionless rudiments, though the animals are sensitive to changes of illumination, through skin receptors. There are sensory tentacles around the mouth, and in both hag-fishes the teeth and sucking apparatus are well developed. They burrow into the bodies of dead or dying fishes. As many as 123 *Myxine* have been taken from a single fish. Since the introduction of trawling they have become less common in the North Sea, where they used to be a serious source of loss to fishermen by their attacks on fishes caught in drift nets or on lines. They seem to find fish when they are dying or just dead, and entering by the mouth of their prey eat out the whole contents of the body, leaving a sack of skin and bones. When they are themselves caught on lines (for instance, with a salted herring bait) the hook is swallowed so deeply that it may be found near the anus!

The gills are modified into pouches (6–14 in *Bdellostoma*, 6 in

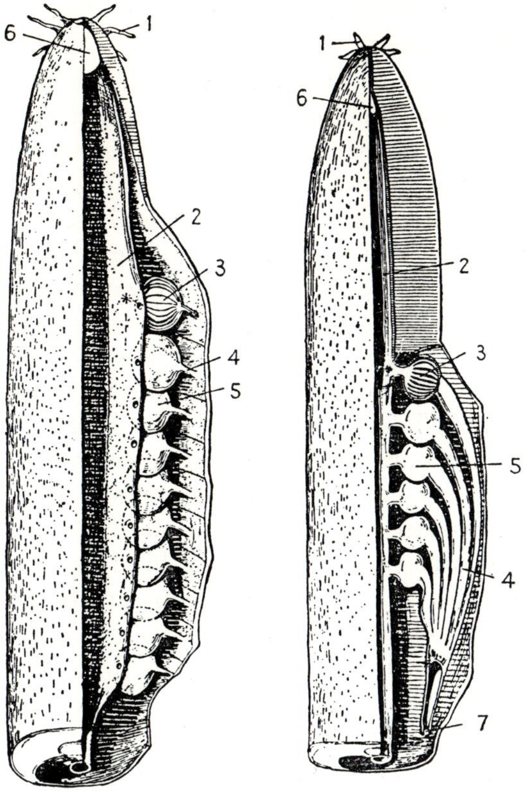

FIG. 82. Arrangement of gills in *Bdellostoma* and *Myxine*.

1, tentacles; 2, wall of pharynx; 3, branchial sac opened to show gill lamellae; 4, branchial duct; 5, branchial sac; 6, mouth; 7, common branchial aperture in *Myxine*. (From Kukenthal after Dean.)

Myxine), opening by tubes into the pharynx, and to the exterior (Fig. 82). In *Myxine* all the tubes are joined and open by a single posterior aperture on each side. Water enters at the nostril and is pumped backwards by a muscular velum through the gill chambers and out behind. There is also a single posterior oesophago-cutaneous duct on the left side, which is probably closed during normal respiration but is opened to allow expulsion of large particles. If the nostril is closed experimentally with a plug no water enters by the mouth or posterior apertures but the fish survives well, presumably respiring through the skin.

The thyroid gland consists of a long series of sacs formed by evagination from the floor of the pharynx.

Down the sides of the body are pairs of slime glands, able to secrete large amounts of mucus, which may be protective and is said also to be produced under the operculum to hasten the end of a dying fish that the hag has attacked.

A curious difference from the nervous system of lampreys is that the dorsal and ventral roots join, though the details suggest that the union is not similar to that found in gnathostome vertebrates. The brain shows several features of reduction and simplification and no pineal eyes are present. There is only one semicircular canal in the ear (p. 109). The kidneys show a more generalized condition than in any other vertebrate in that the pronephros persists in the adult and is hardly marked off from the mesonephros, so that an almost continuous series of funnels and glomeruli can be recognized. Moreover, there is a regular series of mesonephric glomeruli, a pair in each segment.

The development is known only in *Bdellostoma*, where the egg is yolky and cleavage partial, leading to the formation of an embryo perched on a mass of yolk. It is often stated that *Myxine* is a protandric hermaphrodite, because individuals are found in which the front end of the gonad contains eggs, whereas the hind part is testis-like (Fig. 81). No ripe sperms have ever been found in this region, however, and, moreover, individuals with fully testicular gonads do occur. Since it is known that in other vertebrates (including the lampreys) the gonads go through a hermaphrodite stage during development it seems likely that *Myxine* is not a functional hermaphrodite but that the double-sexed gonad shows a rather late persistence of the indeterminate stage.

The hag-fishes all live in the sea and their blood differs from that of other chordates in that it is isosmotic with sea water. However, the individual ions are regulated; sodium and phosphate exceed their values in sea water, and the other ions are present in lower concentration. It is usually assumed that fishes, with their glomerular kidneys, evolved in fresh water. However, the very earliest fragments of armoured agnathans are from Ordovician deposits that may be littoral or marine and it might be that the condition of the blood and kidney of *Myxine* is that of the earliest agnathans and that the glomerulus was not evolved as an adaptation to freshwater life, as is often supposed (Robertson, 1954).

The organization of the lampreys and hag-fishes shows that they preserve many characteristics from a very early stage of chordate evolution, probably that of about the Silurian period. Their special interest for us is in giving an insight into the organization possessed by the vertebrates before jaws were evolved. However, no doubt many changes have gone on during cyclostome evolution and we must not suppose that all Silurian vertebrates were like lampreys. Indeed, we may now complete our picture of this stage of evolution by examining the fossil fishes known to have existed at that period. We shall find them superficially so different from modern cyclostomes that only careful morphological comparison reveals the similarities. The inquiry will show us once again how a common plan of organization can be found in animals of very different superficial form and habits.

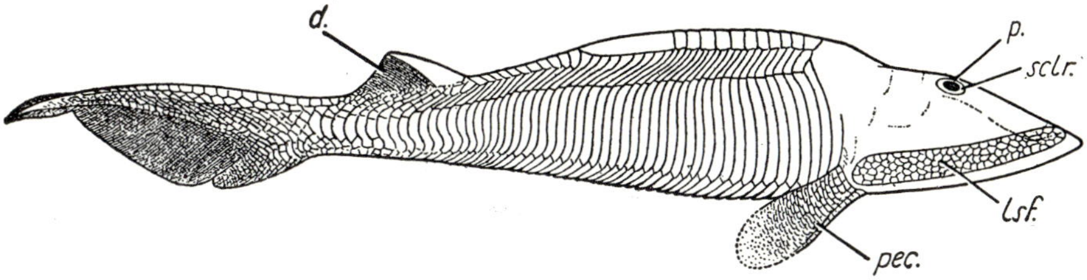

FIG. 83. A cephalaspid restored (*Hemicyclaspis*).
d. dorsal fin; *lsf.* lateral field; *pec.* pectoral fin; *p.* pineal; *sclr.* sclerotic ring.
(From Stensiö.)

20. Fossil Agnatha, the earliest-known vertebrates

The ostracoderms are fossil forms from freshwater Silurian and Devonian deposits. They are therefore the oldest fossil vertebrates known to us (except for a few Ordovician fragments), and this makes it specially interesting that they show affinity with the cyclostomes. These are fossils that are rarely found complete, particularly the pteraspids, but a quarry in Herefordshire yielded numerous whole specimens of *Cephalaspis* and *Pteraspis* of Old Red Sandstone age, probably all from a single dried-up pool.

In the cephalaspids (Osteostraci) the head was flattened and composed largely of a shield. The rest of the body was fish-like, with an upturned tail (heterocercal, see p. 136) covered with heavy bony scales (Fig. 83). A pair of flaps behind the gills may have functioned like pectoral fins.

On the dorsal surface of the shield are two median holes, one behind the other, which served a naso-hypophysial opening and a pineal eye. The whole outline of the cranial cavity is preserved and shows a brain remarkably like that of a lamprey, with a naso-hypophysial canal below it (Fig. 84). There were paired eyes and only two

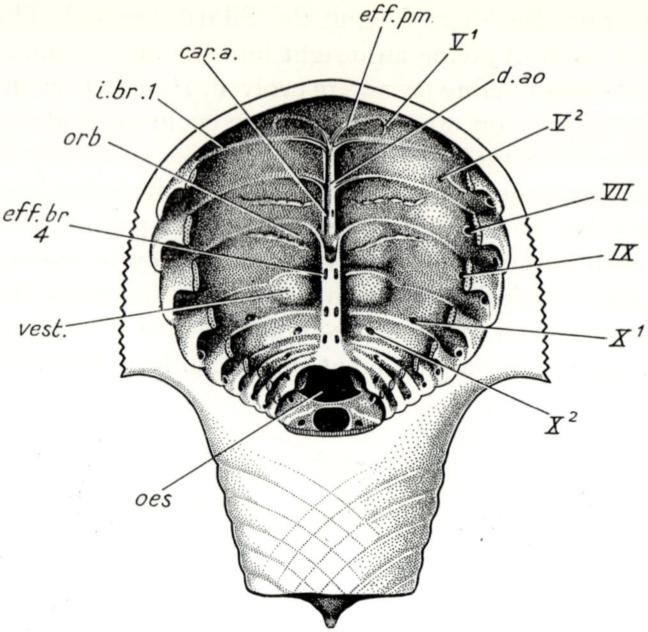

FIG. 84. Head shield of the cephalaspid *Kiaeraspis*, see from below. 5 × natural size. ***car.a.*** internal carotid; ***d.ao.*** dorsal aorta; ***eff.br.4.*** 4th efferent branchial; ***eff.pm.*** efferent branchial of 1st arch; ***i.br.1.*** 1st interbranchial ridge; ***oes.*** oesophagus; ***orb.*** depression made by orbit; ***vest.*** depression made by vestibular apparatus. *V1–X2*. Cranial nerves. (After Stensiö.)

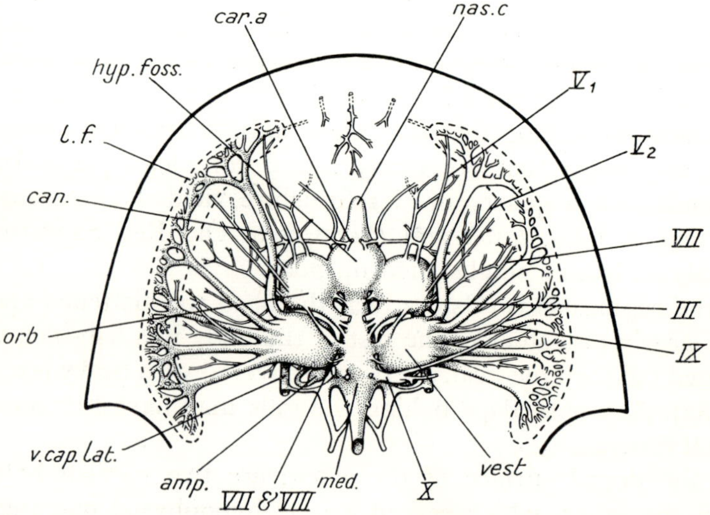

FIG. 85. Cast of the endocranium and system of canals in the head shield of *Kiaeraspis*. ***amp.*** ampulla of posterior semicirculr canal; ***can.*** canal leading to field; ***car.a.*** carotid artery; ***hyp.foss.*** hypophysial fossa; ***l.f.*** lateral 'electric' fields; ***med.*** medulla oblongata; ***nas.c.*** naso-hypophysial canal; ***orb.*** orbit; ***v.cap.lat.*** vena capitis lateralis; ***vest.*** vestibular apparatus; ***III–X*** cranial nerves. (After Stensiö.)

semicircular canals. Long tubes leading through the shield contained the cranial nerves, which can be reconstructed in detail (Fig. 85). On the under side of the shield is a series of ridges, which outline a set of ten pairs of branchial pouches. The first of these lies far forward at the sides of the mouth and the ridge in front of it is probably the premandibular arch; it carries the profundus nerve (p. 152), which was large. The ventral surface of the head was flat and covered with small scales. Probably the gills were pouches, as in lampreys. The canals of the aorta, epibranchial arteries, and some features of the veins and heart have been preserved.

The mouth was a slit at the extreme front end with which the animals may have scooped decaying matter from the lake floor. On the dorsal surface there are sunken areas, covered by small scales, known as the median and lateral fields, and supposed by some to have contained electric organs. They were apparently served by a very rich blood-supply and a system of wide canals leads to the vestibular region. These canals might have contained nerves, but Watson makes the far more likely suggestion that they housed tubular extensions of the labyrinth and served to carry pressure waves to the ear, perhaps providing a substitute reinforcement for the defective lateral line system.

We therefore know in some respects as much about these fossils as of many living fishes. They show in the complete segmentation of the head the most primitive condition known among craniates. Many of their features are very like those of modern lampreys and there can be little doubt that, as Stensiö suggests, the latter represent their surviving descendants, which have lost the bony shield.

The Anaspida (mostly Silurian) are placed by Stensiö near the Cephalaspids but they are less well known. They were small fishes (up to 7 in. in length) covered with rows of bony scales (Fig. 87). The tail shows a lower lobe larger than the upper ('hypocercal'). This would presumably serve to drive the head end upwards perhaps to compensate for the weight of its armour. The opposite ('heterocercal') condition, found in cephalaspids and many modern fishes (for instance, the dogfish), produces a tendency to negative pitch and is associated with the presence of pectoral fins (p. 136). The anaspids possessed a curious ventral or ventro-lateral fin fold (Fig. 87) or perhaps a series of them. There were large paired eyes, median holes presumed to be nasal and pineal and a series of up to fifteen small round gill openings.

We may consider here the fossil *Jamoytius* from the Silurian. The notochord was persistent and there was no calcified endoskeleton. There were long continuous lateral fin folds and a hypocercal tail. A series of transverse structures were at first interpreted as myotomes

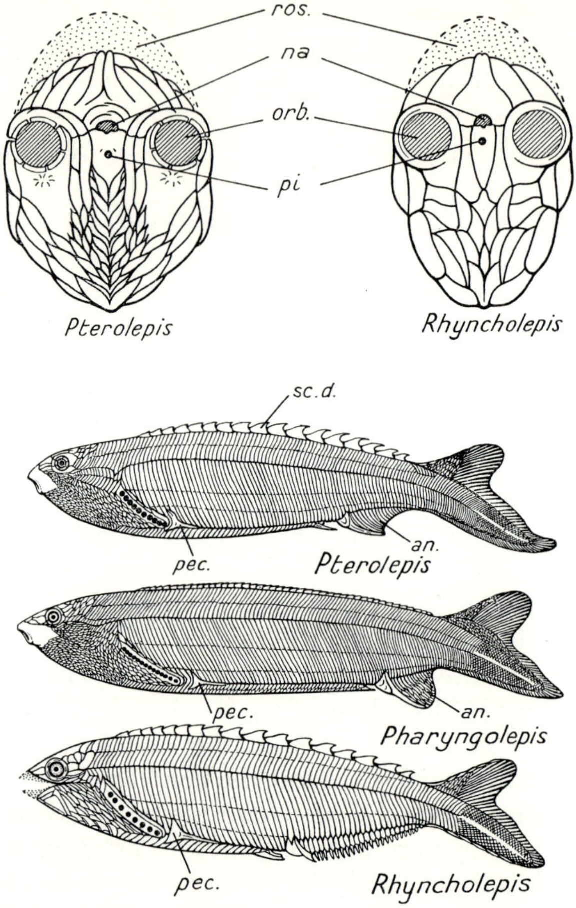

FIGS. 86 and 87. Anaspids seen in dorsal and lateral views.
an. anal fin; *na.* nasal aperture; *orb.* orbit; *pec.* pectoral spine; *pi.* pineal foramen; *ros.* rostrum; *sc.d.* dorsal scales. (After Stensiö and Kiaer and Grassé.)

but Stensiö and Ritchie (1960) consider these to be scales and place *Jamoytius* with the Anaspida. In either case, the form is of the greatest interest, and represents as White says 'the most primitive of the "vertebrate" series of which we have knowledge'. It is suggested that it might be the ammocoete larva of an ostracoderm (Newth).

The Heterostraci are actually the oldest known craniates, since their scales occur in the Ordovician. They were common in the Silurian and lower Devonian. There were ventral as well as dorsal shields (Fig. 88), and a long series of gill pouches, but only a single pair of exhalent branchial apertures, suggesting to Watson respiration by a moving flap (velum). The shields were of cell-less bone (isopedin) covered with dentine. The body was covered with scales of similar material. The tail was hypocercal and there were lateral horizontal keels but no fins. There were paired eyes, two semicircular canals and clearly

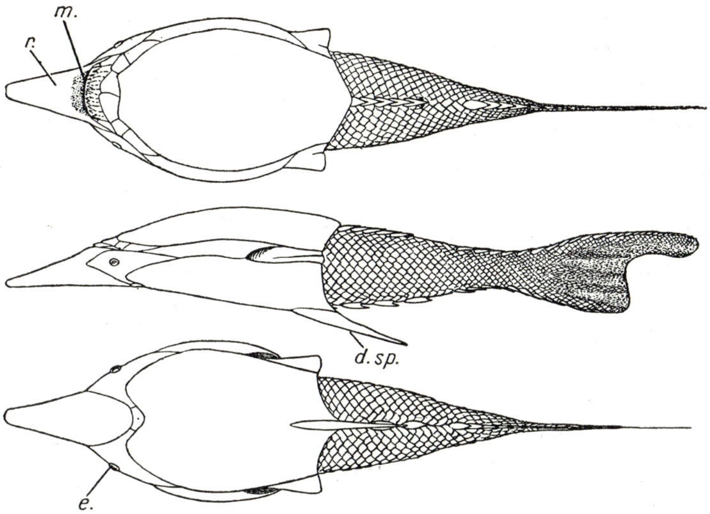

FIG. 88. Three views of a restoration of *Pteraspis*.
d.sp. dorsal spine; *e.* eye; *m.* mouth; *r.* rostrum. (From White.)

marked lateral line canals. There was a pineal opening, closed in the adult, but no sign of the nostril, which may have opened into the mouth. The latter was surrounded by long plates, suggesting that it formed a protrusible apparatus, which could be pushed out to form 'a kind of scoop or shovel (Fig. 88) whereby mud and decaying refuse could be taken off the bottom, for it seems likely that such were their food and habit' (White).

The coelolepids or thelodonts are the least known group of agnathans. The outer surface was covered with fine, placoid-like scales or hollow spines, which in isolation are often found in late Silurian and Early Devonian rocks. The anterior end was usually flattened and wide but the body behind was narrow, with a forked, probably hypocercal tail. Structures that are probably eye-spots occurred widely separated near the front margin. The mouth was ventral and traces of seven branchial arches have been found. There were flap-like extensions on each side of the head but no paired fins. The only median fin was the anal.

The affinities of these ostracoderm fossils with each other and with the cyclostomes have been much disputed. Lankester claimed that pteraspids were related to cephalaspids 'because they are found in the same beds, because they have a large head shield and because there is nothing else with which to associate them'. At the other extreme Stensiö holds that we have sufficient evidence to assert that the pteraspids have given rise to the myxinoids, and the cephalaspids to the lampreys. Except for the absence of jaws there is indeed little in common among the fossil forms. The differences in the shape of the tail are especially baffling. As White points out, an animal with a heterocercal tail and pectoral fins can hardly have lost either of these organs independently. He suggests that the earliest vertebrates possessed straight ('diphycercal') tails and that from these were evolved on the one hand the pteraspids with hypocercal tails and on the other the cephalaspids with upturned heterocercal tails. The modern cyclostomes are perhaps derived from the latter, but which, if either, group gave rise to the earliest gnathostomes is unknown.

The Agnatha were the first animals of the chordate type to become large, and they apparently all did so by feeding on the detritus at the bottom of rivers and lakes. They evolved into various types, mostly rather heavily armoured and perhaps slow-moving forms. The lampreys and hag-fishes have been derived from early Agnatha by the evolution of a sucking mouth, perhaps with loss of the bony skeleton and paired limbs. However, it was the unknown forms that evolved a biting mouth that made the next great advance in vertebrate evolution.

V

THE APPEARANCE OF JAWS. THE ORGANIZATION OF THE HEAD

1. The elasmobranchs: introduction

IN all parts of the sea there are to be found members of the class of the elasmobranchs (literally 'plate-gilled' fishes), including sharks ranging from monsters of 50 ft long to the common dogfish *Scyliorhinus caniculus* of 1–2 ft. Nearly all the fishes in the group are carnivorous or scavengers: the skates and rays are bottom-living relatives, feeding mostly on invertebrates. Although they are not quite so fully masters of the water as are the bony fishes, they are yet well enough suited to that element to survive in great numbers in all oceans. Perhaps the skill and cunnning of a shark is exaggerated by the frightened boatman or bather, who is apt to mistake a keen nose and the persistence of hunger for intelligence, especially when he is faced at intervals with a well-armed mouth; but the sharks have a large brain and their active, predacious habits enable many of them to live by eating the more elaborately organized bony fishes.

Evidently such active creatures have changed considerably if they have been evolved from the heavily armoured and probably slow-moving agnathous vertebrates that shovelled up food from the bottom of Palaeozoic seas. It used to be supposed that these elasmobranch or cartilage fishes represent a very primitive stock, but we now realize that there have been great changes since the biting mouth was first evolved; we cannot be sure that any features we find in the elasmobranchs were possessed by the earliest gnathostomes.

The typical shark is a long-bodied fish, swimming by the passage of waves of contraction along the metamerically arranged muscles. As in the lampreys and eels, the wave that passes down the body is of short period, relative to the length of the fish, and is therefore evident as it travels along. This is probably a less efficient system than is provided by the longer period waves of the most highly developed bony fishes; the sharks are good swimmers, but except for the mackerel sharks (Isuridae) not among the swiftest. Stability and control of direction are ensured by the upturned tail and the fins. The tail, with its dorsal lobe larger than the ventral, is called heterocercal, and tends to drive the head downwards. This is corrected by the flattened shape of the head itself and by the pectoral fins, which act as 'aerofoils', allowing steering in the horizontal plane (p. 140). There are two dorsal fins, which secure stability against rolling, and also assist in making possible the vertical turning movements.

The muscles for the production of these movements are a serial metameric set, with longitudinal fibres, essentially like those of the lamprey or amphioxus. The central axis is no longer simply a rod; the notochord has become surrounded and partly replaced by a series

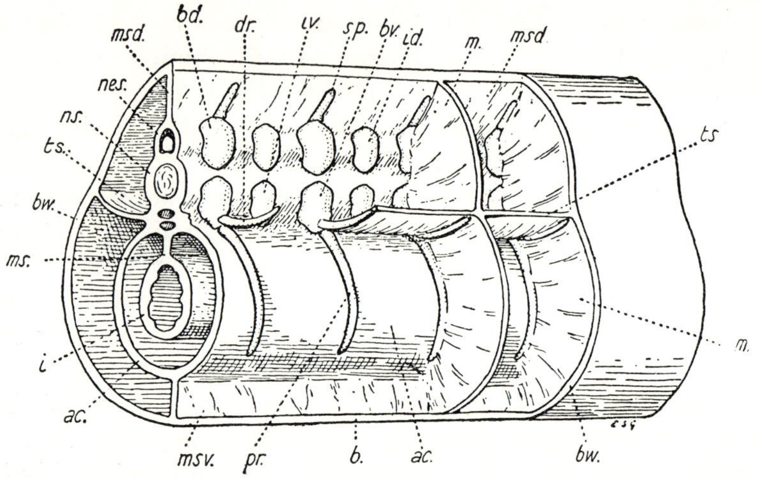

FIG. 89. Diagram of the organization of a vertebrate.

ac. wall of abdominal coelom; *b.* body wall; *bd.* basidorsal; *bv.* basiventral; *bw.* body wall; *dr.* dorsal rib; *i.* intestine; *iv.* interventral; *m.* myocomma; *ms.* mesentery; *msd.* median dorsal septum; *msv.* ventral mesentery; *nes.* neural tube; *ns.* notochordal sheath; *pr.* ventral (pleural) rib; *sp.* neural spine; *ts.* horizontal septum. (From Goodrich.)

of vertebrae (Fig. 89). These develop as two pairs of cartilaginous nodules in each segment, the basidorsals and basiventrals behind, and smaller elements, the interdorsals and interventrals, in the front. The basiventrals, lying on either side of the notochord, form the centrum of each vertebra, invading and almost interrupting the notochord, which widens again, however, between the vertebrae. The vertebrae are held together by ligaments, but are not articulated by complex facets as they are in land animals. The basidorsals form neural arches above the nerve-cord, and the interdorsals make intercalary arches. The interventrals partly separate the centra. Attached to each basiventral is a pair of transverse processes, which in the anterior region bear short ribs and in the tail are fused in the midline to make the haemal arches.

The median and paired fins are supported by cartilaginous rods, the radials, and their edges are further strengthened by special horny rays, the ceratotrichia. The radials of the paired fins form a series attached to larger rods at the base. These more basal rods are attached to a 'girdle' of cartilage embedded in the body wall. The pectoral

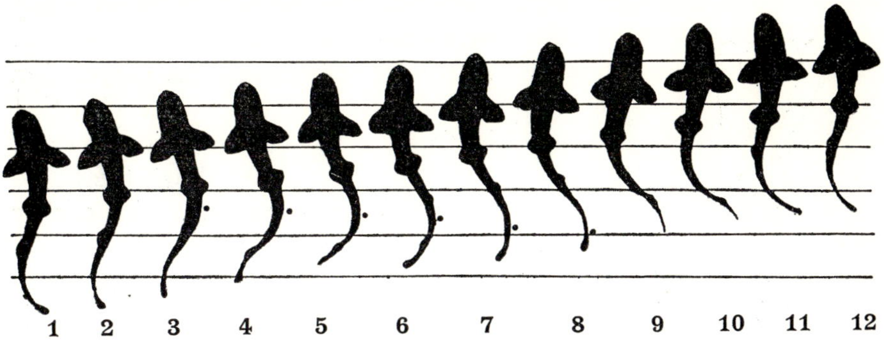

FIG. 90. Successive positions of a swimming dogfish at intervals of 0·1 sec. The lines are 3 in. apart. The passage of a wave is marked by dots. (After Gray.)

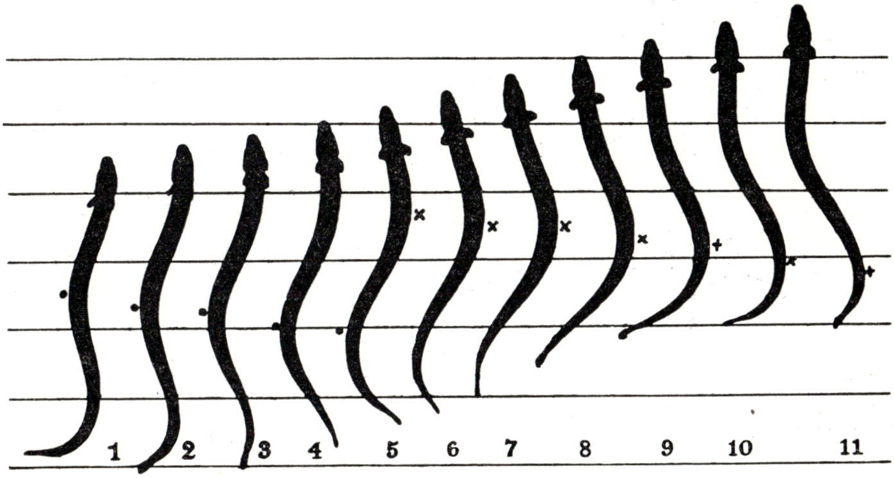

FIG. 91. Successive positions of a swimming eel at intervals of 0·05 sec. Scale 3 in. The wave-crests are marked. (After Gray.)

girdle is a hoop extending some way round the body, but the pelvic girdle is simply a transverse rod in the abdominal wall. The origin of these girdles and of the fins will be discussed later (p. 136).

2. The swimming of fishes

The propulsive forces that move a fish through the water are usually produced by the longitudinal muscle-fibres of the myotomes, but in some forms the propulsion is produced by movement of the fins, whose function is usually rather to give the fish its stability, enabling it to keep on a constant course, and also to change its course.

The myotomes consist of blocks of longitudinal muscle-fibres, placed on either side of an incompressible central axis, the notochord or vertebral column. The effect of contraction of the muscle-fibres in any myotome is therefore to bend the body. In forward swimming the contraction of each myotome takes place after that in front of it. In this way waves of curvature are passed down the body, alternately on each side. This can be illustrated by a series of photographs of a fish such as the dogfish or eel in which the amplitude of the waves is large (Figs. 90 and 91).

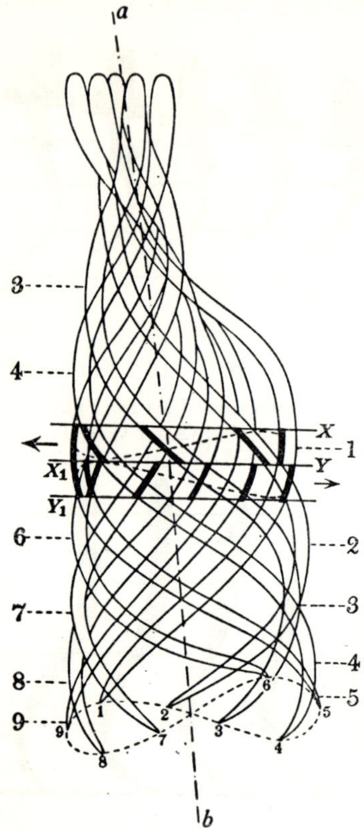

FIG. 92. Enlarged drawings of successive photographs of a young eel superimposed on each other so that the tips of the head are on the same transverse axis and the longitudinal axes of motion (*ab*) are made to coincide. As the wave passes the section XY it first moves to the left and is directed backwards and to the left, whereas $X_1 Y_1$ moves in the opposite direction. The tip of the tail follows a figure of 8. (From Gray.)

In other fishes the waves are not so immediately obvious, but serial photographs show that even in such forms as the mackerel and whiting there is a backward movement of waves. The number of waves per minute in steady swimming varies from 54 in the dogfish to 170 in the mackerel, the corresponding velocities of the waves being 55 and 77 and of the whole fish 29 and 42·5 cm/sec.

Gray has shown how the muscle contractions produce movements of the parts of the body, related to one another in such a way as to transmit a backward momentum to the water. Fig. 92 shows superposed drawings of an eel, made from successive photographs. The region marked XY is moving from right to left and that X_1Y_1 from left to right and evidently, as Gray puts it, 'all parts of the fish's body which are in transverse motion have their leading surfaces directed backwards and towards the direction of transverse movement, but the angle of inclination is most pronounced when the segment is crossing the axis of longitudinal motion, and at this point the segment of the body is travelling at its maximum speed. Each point of the body is travelling along a figure 8 curve relative to a transverse line which is moving forward at the average forward velocity of the whole fish. The track of any point on the body (relative to the earth) is a sinusoidal curve whose pitch or wave length is less than that of a curve which defines the body of the fish. There is therefore a definite angle between the surface of the fish and its path of motion.'

Each portion of the side of the fish can thus be considered as moving like the blade of an oar used for sculling at the back of a boat. The principle used, that of an inclined plane, is the same as in screw propulsion, the essential feature being that the moving surface is inclined at an angle to its line of motion. The effect of the movement is greatly increased by the fact that the amplitude of the oscillations grows passing backwards, as is necessary to produce additive effects in any coupled system of screws or turbines. The whole fish thus operates as a single self-propelling system.

The magnitude of the forward thrust thus generated depends among other things on (*a*) the angle that the surface of the fish makes with its own path of motion, (*b*) the angle between the surface of the fish and the axis of forward movement of the whole fish, and (*c*) the velocity of transverse movement of the body (Gray). These are evidently factors that will vary with the shape of the body and the action of its muscles. The body form of the faster-moving types of bony fishes provides substantial advantages for swimming over that of the more elongated types. The essential differences are that the bony fishes have (1) large caudal fins, (2) a much smaller length of the body relative to its depth, (3) less flexibility.

The role of the large caudal fin is to resist transverse movements; its effect is, again quoting Gray, 'to keep the leading surface of the body directed obliquely backwards during both phases of its transverse movements and thereby to exert a steady pressure on the water'. Since, however, the tail does execute transverse movements, and at the same time is being rotated towards and away from the axis of motion, it exerts a very large propulsive effect, probably as much as 40 per cent. of the total thrust.

The effect of the caudal fin, combined with the shortness of body and reduced flexibility, is that the front part of a bony fish makes only small transverse movements; the track of the head is therefore nearly straight and the whole front of the body presents a streamlined surface with little resistance. Further, the muscles just in front of the tail exert their tension with very little change in length.

No doubt the shape of the body also has an important influence on the effect of the fish on the water and hence on the turbulence in the flow of water and the resistance that must be overcome. Gray has shown that in a dolphin the resistance cannot be that of a rigid model towed at the speed at which the animal moves, since this would require that the muscles generate energy at a rate at least seven times greater than is known in the muscles of other mammals. By watching the flow of particles past the body of fish-like models he showed that movements such as those produced in swimming accelerate the water in the direction of the posterior end, and this would greatly reduce the turbulence.

Something is known of the nervous mechanism responsible for the production of the swimming waves. An eel can swim if its whole skin has been removed. If a region of the body is immobilized by a clamp, swimming waves can pass along. Therefore the rhythm is determined by some intrinsic activity of the spinal cord and not by any mechanism such as proprioceptor impulses arising in active muscles and causing others to contract.

Experiments in which the spinal cord was cut across show that in the eel the rhythm is only initiated when suitable impulses reach the cord either from spinal afferents or from the brain. Thus the spinal eel can be made to swim either by fixing a clip on to its caudal fin or by electrical stimulation of the cut end of the spinal cord. Though the cord requires such afferent stimuli for its functioning, they do not determine the frequency of the rhythm, which bears no relationship to that of the applied stimuli.

In the dogfish the isolated spinal cord is able to initiate rhythmic swimming. After transection behind the brain the posterior portion of the fish exhibits continuous swimming movements for many days. Light touch on the sides of the body inhibits these movements, but some sensory impulses are necessary for their initiation; after complete de-afferentation, by section of all the dorsal roots, the movements cease.

The information available does not yet enable us to understand fully how the swimming rhythm is initiated and maintained, nor how it is influenced by the brain. It would be very interesting to have further knowledge on these topics, especially because the locomotor rhythms of land animals are probably based on the serial contractions of their fish ancestors.

3. Equilibrium of fishes in water; the functions of the fins

Making use of the methods of investigation of aeronautical engineers, studies have been made of the forces that operate to keep a fish stable as it moves through the water, or allow it to become temporarily unstable and hence to change direction. Instead of attempting to study a living or dead fish moving in water, Harris made models and supported them in a wind-tunnel in an apparatus suitable for measuring the forces at work in the various directions. Such a method, in which no compensating movements of the fins are allowed, makes it possible to investigate the so-called 'static stability' of the fish, that is to say,

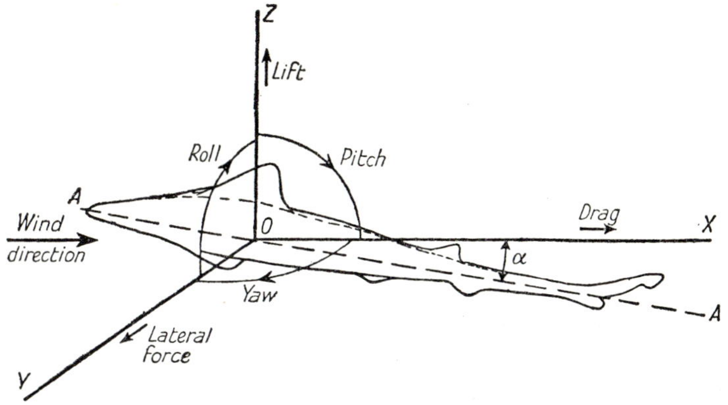

FIG. 93. Diagram of model of the dogfish *Mustelus*, showing the conventional terms for describing deviations of motion. The longitudinal axis X is that of the wind tunnel and Y (horizontal) and Z (vertical) are at right angles to it. The arrows show the directions known as positive rolling, pitching, and yawing, which occur about the X, Y, and Z axes respectively. α is the angle of attack between the axis of the model and the X axis. (From Harris, *J. exp. Biol.* **13**.)

to see whether the body and fins are so shaped as to provide forces that tend to bring the fish back into its previous line of movement after it has deviated in any direction. Any body such as a fish or aeroplane is said to be in stable motion if when it veers slightly from its line of progress the new forces produced upon its planes tend to restore the original direction of motion.

The forces acting on the fish are measured along three primary axes, longitudinal, horizontal, and vertical. Deviation from the line of motion about the longitudinal axis is known as rolling, about the transverse axis as pitching, and about the vertical axis as yawing (Fig. 93). The forces along these three axes are known as drag, lateral force, and lift.

In order to discover the effect of the median fins and tail on the stability, these fins were removed, the heterocercal tail being replaced by a cone having the same taper as the actual caudal fin. The model was then placed in the wind-tunnel with a wind at 40 m.p.h., which corresponds to a motion of 3 m.p.h. in water. The lateral force was measured when the body was made to yaw at various angles. The

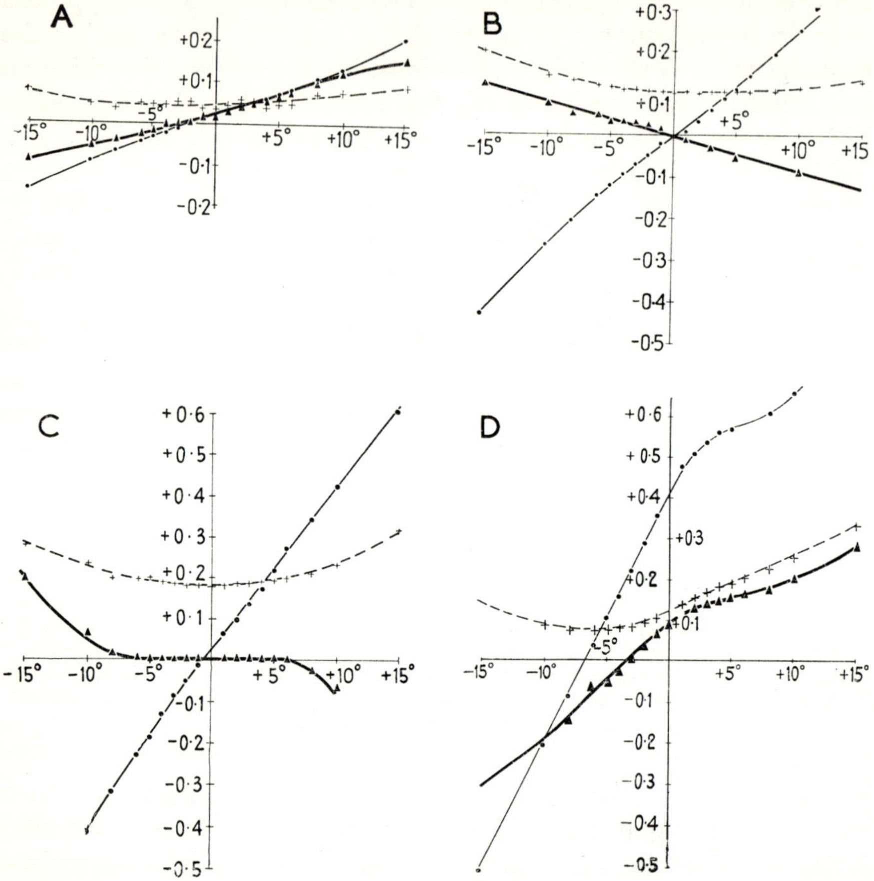

FIG. 94. A. Results of yawing test on model of *Mustelus* without fins. The lateral force is plotted as a light full line, drag force as a light broken line; yawing moment about centre of gravity as full heavy line. Abscissae show the angle of attack in degrees, ordinates the lateral force and drag in pounds weight, yawing moment in in.-lb. $\times \frac{1}{10}$.
B. Yawing test similar to (A) but with the fins behind the centre of gravity in place.
C. Yawing test with all median fins in place.
D. Pitching test on model of *Mustelus* with all fins intact and pectoral fins set at an angle of incidence of 8°. Lift force is shown as a light full line, drag force as a light broken line, pitching moment about the centre of gravity as a heavy full line.
(From Harris.)

results showed that the equilibrium in this plane is quite unstable; a slight turn off the direct course would produce a turning moment tending to increase still further the deflection (Fig. 94). This is a well-known property of all airship hulls, and is known as the 'unstable moment' of the hull. It is corrected in the airship by the addition of suitable horizontal and vertical fin surfaces at the rear end, when the airship becomes in effect a feathered arrow. The forces operating on the fins tend to bring the body back into the original line of motion.

The fins of the fish operate in a similar manner. If the experiment is performed with a model to which all the fins behind the centre of gravity have been added, namely, the caudal, anal, and second dorsal fins, it is found that the curve for the yawing moment now has a steep negative slope (Fig. 94 B), that is to say, every deviation produces forces that tend to give directional stability. With the first dorsal fin also in position the model possesses a remarkable neutral equilibrium (Fig. 94 C). Deviations by as much as 10° produce no resultant yawing moment about the centre of gravity. The form of the dorsal fins is therefore definitely such as to maintain stable swimming and prevent yawing.

Turning of a fish is produced either by the propagation of a wave down one side only of the body or by asymmetrical braking with the pectoral fins (see below). The former type of turn has been investigated by Gray in the whiting, where there is a large caudal fin. This gives great lateral resistance, so that the first part of the turn is executed by bending the front part of the fish on the tail as a fulcrum. This enables the animal to turn through 180° within a circle of the diameter of its own length. After removal of the caudal fin the turns are much less effective.

In both elasmobranchs and teleosts the dorsal fins are well developed in the active swimmers. In most elasmobranchs they are fixed, but in many teleosts the dorsal fin can be folded up and down, and it is observed that the fin is raised during turning. This would have the effect of increasing the yawing moment produced by asymmetrical action of the body muscles or by unilateral braking with the pectoral fins.

Since the body is so markedly flexible in the lateral plane and there are powerful muscles available for turning it in this direction, the part played by the fins in determining the stability is important mainly when the body is held straight. The fish thus has the double advantage of great stability (by keeping the body straight) and great controllability (by bending it). In a body unable to change its shape in this way, stability and controllability would be inversely related. This is the case for the stability of the fish in the vertical plane, in which the body is little flexible. Fig. 94 D shows the positive slope of the curve for the pitching moment and clearly the equilibrium in this plane is quite unstable. The pectoral fins contribute more than any others to movement in this plane, and since they lie in front of the centre of gravity they greatly increase the instability. The fish must be able to alter direction in the vertical plane, and it has apparently sacrificed static stability for controllability. The equilibrium in this plane is a dynamic one, controlled by the movable pectoral fins, and it is so unstable that only a small movement of these fins is necessary to produce a deflecting force that restores the original direction of motion.

The pectoral fins, lying in front of the centre of gravity, tend to produce a movement of positive pitch, that is to say, they force the head upwards. This effect is normally compensated by a component produced by the heterocercal tail. The upper lobe of this is rigid and the lower more flexible, therefore the lateral motion given by the swimming movements of the body produces a vertical lift force on the tail, giving, of course, negative pitch. After amputation of the hypocaudal lobe and anal fin a dogfish swims continually along the bottom of the tank: in order to compensate for the absence of negative pitch the pectoral fins are held horizontally and hence there is no moment to counteract the weight of the fish. If the pectoral fins are then also removed the anterior end of the body is pointed upwards, often so much so as to cause the fish to swim with its head out of the water. This is the result of an over-strenuous attempt to compensate, by raising the head, for the negative pitch produced by the tail. The system is no longer suitable for making the continuous adjustments necessary to ensure stability.

This analysis makes it clear why a heterocercal tail is found in almost all the primitive swimming chordates; it is almost a necessity for an animal with a specific gravity in excess of the medium and little flexibility in the vertical plane. The component of positive pitch could be provided by the flattened head or by continuous lateral fin folds, such as may have been present in early fishes, and adjusted by the limited flexibility possible in the fin. The development of movable pectoral fins confers much greater control. Since the useful portions of a fin fold for this purpose would be those well in front of and behind the centre of gravity, we can perhaps see the reason why the intervening portion has become lost. In the modern sharks the pelvic fins have little influence on the stability and are perhaps retained only for their modification as claspers.

It is not surprising that races of fishes with stability ensured by systems of this sort should tend to adopt a bottom-living habit, with dorso-ventral flattening, such as is found in the skates and rays.

Expansion of the front end is developed at first to compensate the effect of the tail, but the pectoral fin becomes expanded to allow vertical adjustments and then reduction of the hypocaudal lobe of the tail accompanies the adoption of life on the sea bottom. Eventually all

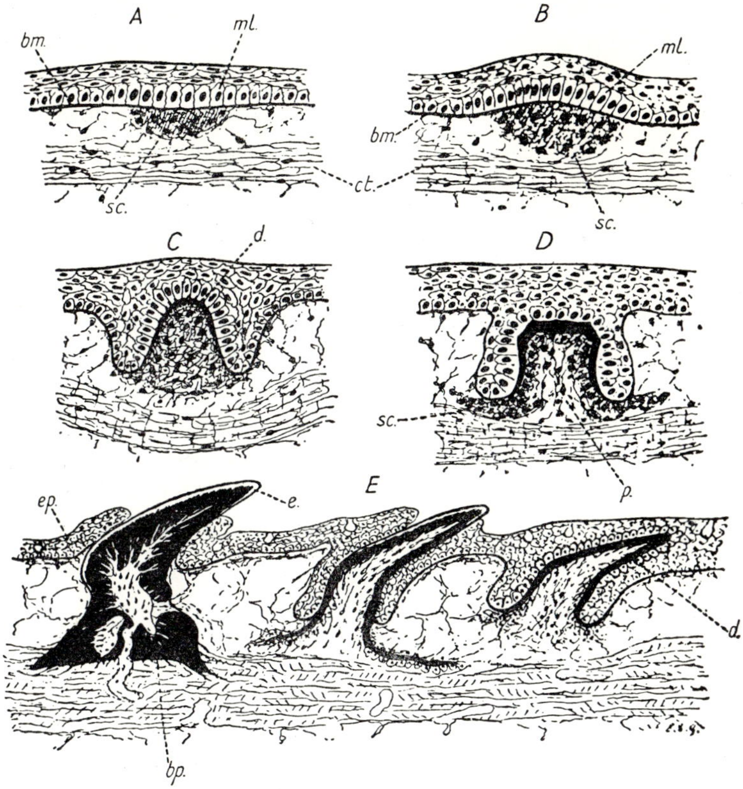

FIG. 95. Development of denticles in the dogfish.

A and B, first gathering of odontoblasts (*sc.*) below the basement membrane (*bm.*); *ml.* are the epidermal cells that will become modified. C, first deposition of dentine (*d.*). In D there is more dentine and a pulp cavity (*p.*) is seen. In E are shown stages in the formation of enamel (*e.*) and of the basal plate (*bp.*) while the denticle cuts the epidermis (*ep.*). (From Goodrich, *Vertebrata*, A. & C. Black, Ltd.)

locomotion is produced by undulatory movements of the fins, which were at first used only to raise the fish off the bottom.

4. Skin of elasmobranchs

Being swift and predatory animals, more attackers than attacked, the sharks do not possess a very heavy external armament. The skin itself is tough, being covered by layers of epidermis. Beneath this is a thick dermis of connective tissue with fibres arranged at right angles as in a carpet, giving a tissue of great strength and flexibility, able to maintain the shape of the body. Scattered over the skin are the characteristic denticles or placoid scales (Fig. 95). Each of these consists of a pulp cavity, around the edge of which lies a layer of odontoblasts secreting the calcareous matter of the scale, known as dentine. This has a characteristic structure resulting from the fact that the odontoblasts send fine processes throughout its substance. The outside of the dentine is covered by a layer of enamel, secreted by the overlying ectoderm. Usually the denticles pierce through the ectoderm, after which no further enamel can be added to their surface. Obviously the scales are similar to teeth, which are indeed to be considered as specialized denticles developed on the skin of the jaws. It has often been supposed that the denticle is the primitive type of fish scale, from which others have been derived, but it now seems more likely that the earliest covering was a continuous layer, later broken into large scales, from which the denticle was ultimately derived (p. 269).

The skin also gives protection to the fish by its colour, produced by a layer of chromatophores beneath the epidermis. Many sharks have a spotted or wavy pattern, which breaks up their visible outline as they move in the water, especially near the surface. They are able to change their colour, though only slowly, becoming darker on a dark background (see p. 164).

5. The skull and branchial arches

In general organization a dogfish follows closely the fish plan, which we have already considered. Most of its special new features are in the head, and we may now turn to a consideration of the organization of the head and jaws of a gnathostome vertebrate. The jawless vertebrates of the Silurian and Devonian included freshwater animals of various sorts, but the vertebrate type began to flourish and increase more abundantly with the appearance of creatures with jaws in the late Silurian. From this stage onwards we have to follow the parallel history of numerous orders and families, as the vertebrate plan of structure became adapted for various habitats. It seems likely that the development of a biting mouth greatly increased the range of possibilities of vertebrate life. The most obvious use of a mouth is for attacking other animals, but it may also have been used to collect plant food from all sorts of situations where it would not be available to the microphagous or shovelling Agnatha. Probably the mouth was also early used for defence, and in this way influenced the whole bodily organization, making unnecessary the heavy armature that is so characteristic of many early vertebrates. Modern research has shown that the armour has become progressively reduced along various lines of fish evolution. Older ideas of comparative anatomy regarded the 'cartilage fishes' as showing a primitive stage, preceding the appearance of bone. We now realize that this is the opposite of the truth and that the dogfish and its relatives represent a higher type,

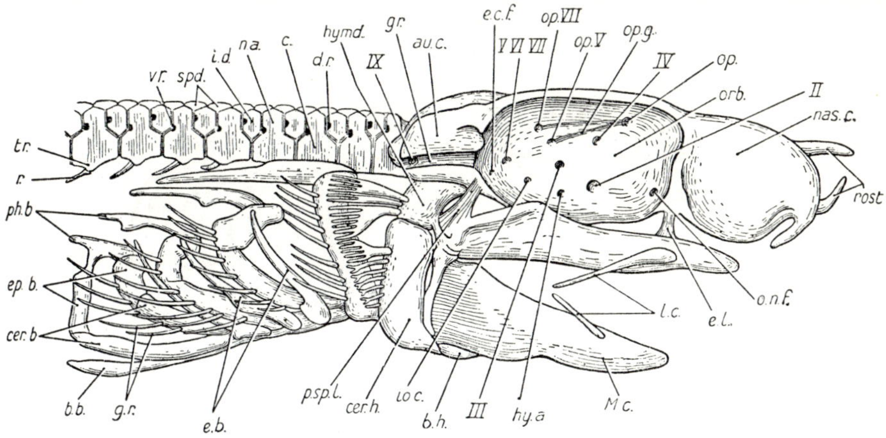

FIG. 96. Skull and branchial arches of the dogfish (*Scyliorhinus*).

au.c. auditory capsule; ***b.b.*** basibranchial; ***b.h.*** basihyal; ***c.*** centrum; ***cer.b.*** ceratobranchials; ***cer.h.*** ceratohyal; ***d.r.*** foramen for dorsal root; ***e.b.*** extrabranchials; ***e.c.f.*** external carotid foramen; ***e.l.*** ethmoid ligament; ***ep.b.*** epibranchials; ***gr.*** groove for anterior cardinal sinus; ***g.r.*** gill rays; ***hy.a.*** foramen for hyoid artery; ***hymd.*** hyomandibula; ***i.d.*** interdorsal; ***io.c.*** interorbital canal; ***l.c.*** labial cartilages; ***M.c.*** Meckel's cartilage; ***na.*** neural arch; ***nas.c.*** nasal capsule; ***o.n.f.*** orbito-nasal foramen; ***op.*** foramen for ophthalmic nerve; ***op.g.*** groove for ***op.V***; ***op.V***, ***op.VII*** ophthalmic branches of *V* and *VII*; ***orb.*** orbit; ***ph.b.*** pharyngobranchials; ***p.sp.l.*** prespiracular ligament; ***r.*** rib; ***rost.*** rostral cartilages; ***spd.*** supradorsals; ***tr.*** transverse process; ***vr.*** foramen for ventral root; *II–IX*, foramina for cranial nerves. (After Borradaile.)

able to defend themselves by mobility, by biting, and by efficient sensory and nervous organization. Heavy defensive armour is a primitive form of protection for animals, as for man.

Besides its use in feeding and defence, the mouth can also be used as a means of 'handling' the environment, for instance in the nest-building activities of many fishes. Indeed, it is difficult for us to realize the utility of the jaws for an animal not provided with any other means of seizing hold of objects.

The development of the mouth to a point at which it could be used in these varied ways was, therefore, a very important stage in evolution. Recognition of the Gnathostomata as a separate group of animals is far more than a matter of classificatory convenience, it marks the achievement of the possibility of life in a greatly increased range of environments.

Morphological analysis enables us to see how this biting mouth was produced, by modification of one or more of the gill-slits. The main differences that separate the gnathostome from cyclostome vertebrates are therefore in the head and its skeleton. Although the modern elasmobranchs show the skull and jaws in a modified and reduced condition, they provide by their simplicity a good starting-point for discussion. The 'skull' of a dogfish consists of a series of

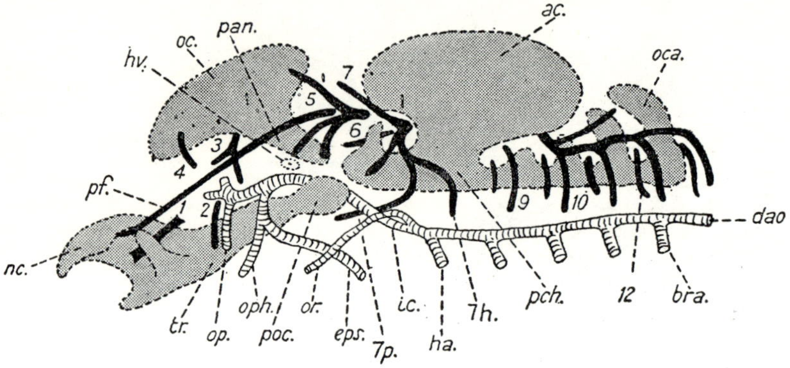

FIG. 97. Diagram of skull of selachian embryo before fusion of the main cartilages; cranial nerves black, numbered; arteries cross-lined.

ac. auditory capsule; *bra.* epibranchial; *dao.* dorsal aorta; *eps.* efferent pseudobranchial; *ha.* efferent hyoid; *hv.* hypophysial vein; *i.c.* internal carotid; *nc.* nasal cartilage; *oc.* orbital cartilage; *oca.* occipital arch; *op.* optic; *oph.* ophthalmic; *or.* orbital; *pan.* pila antotica; *pf.* profundus nerve; *pch.* parachordal; *poc.* polar cartilage; *tr.* trabecula. (From Goodrich.)

cartilaginous boxes surrounding the brain and receptor organs (Fig. 96). The nasal capsules, orbital ridges, and auditory capsules are largely fused with the main cranium, producing a single continuous structure, the chondrocranium. It is interesting to consider how this structure has arisen during the process of cephalization. Presumably parts of it represent the modified sclerotomes of trunk regions. We shall see presently that there is strong evidence that the head has arisen by modification of a segmental arrangement such as is seen in the trunk; the morphogenetic processes that build the skull must therefore be related in some way to those of the vertebrae. The first rudiment of the skull in the embryo consists of two pairs of cartilaginous rods, the parachordals and trabeculae (Fig. 97). The former lie on either side of the notochord, the trabeculae in front of the notochord. These first rods fuse up to make a continuous plate; from this grow sides and roof, completing the cartilaginous neuro-cranium around the brain. Meanwhile cartilaginous capsules form around the nose, eyes, and ears, and become joined to the neuro-cranium. Posteriorly, behind the auditory capsules, the cranium is completed by the addition of a number of segmented elements, evidently modified vertebrae.

The problem is, therefore, to determine the nature of the pro-otic part of the skull. Before we can settle this we must consider the visceral or branchial arches. These are pairs of rods of cartilage developed in the walls of the mouth and pharynx, between the gill-slits. In the dogfish each typical branchial arch (Fig. 96) consists of a series of four pieces, the pharyngo-, epi-, cerato-, and hypo-branchials. Ventrally some of the arches join a median basibranchial plate. These rods lie in the pharynx wall and on their outer sides carry a series of projecting rods, the branchial rays and extrabranchial cartilages, whose function is to support the lamellae of the gills.

There are five such branchial arches, differing only slightly from each other. In front of these lie two arches, the hyoid and mandibular, which, though modified, are obviously of the same series. The hyoid the more nearly resembles a typical branchial arch. Its most dorsal element, the hyomandibular cartilage, is a thick rod attached dorsally to the skull by ligaments and at its lower end forming the support for the hind end of the jaw. It apparently corresponds to the epibranchials. The more ventral elements, cerato- and basihyal, resemble the corresponding members of more posterior arches. The jaws themselves (mandibular arches) depart more widely from the form of a typical branchial arch, but the two thick rods of which each is composed, the upper palato-pterygo-quadrate bar and the lower Meckel's cartilage, are recognizably members of the branchial series. Looking at the whole apparatus with a thought to the embryological processes that have produced it, with as it were a manufacturer's eye, we can see at once that the jaws and hyoid arch have been produced by a modification of the processes that make the branchial arches.

6. The jaws

Study of the serial relationship of the jaws and branchial arches gives us an understanding of the course of evolution of the mouth. We may suppose that the ancestors of the gnathostomes possessed a nearly terminal mouth, either on the front end of the body or on the ventral surface. The pharynx was pierced by a series of gill pouches, beginning shortly behind the mouth and separated by arches, each containing a set of cartilaginous bars (Fig. 96). There is some evidence that this condition persisted in the cephalaspids (p. 125), where there is found to be a series of ten pairs of gill-slits, beginning far forward on either side of the mouth. The muscles moving the more anterior parts of the pharynx wall and the anterior arches could be called into play to help in the collection of food. In this way the mouth came to be used for prehension, and the grasping jaws of the gnathostomes appeared as the more anterior arches became modified to allow more efficient seizing, and the skin over them was modified to form the teeth. The mouth probably shifted backwards during this process and its lateral edges joined the first gill-slit. The rods supporting the posterior wall of that slit thus became bent over into the characteristic position of the vertebrate jaws.

There is some uncertainty as to the means of support of the jaws in the earlier stages of their evolution. The front end of the palato-pterygo-quadrate bar is attached to the cranium in the dogfish by the ethmo-palatine 'ligament'. In most elasmobranchs the hind end of the upper jaw is not fixed to the cranium but is slung from the latter by the hyomandibula and by a prespiracular ligament. This means of support, known as hyostylic, was for long supposed to have been the original one. But the earliest gnathostomes (the acanthodians) do not have this arrangement (p. 187), indeed, their hyoid arch is an almost typical branchial arch, not modified to support the jaw. In the primitive condition one would not expect the hyoid arch to have any connexion with the mandibular. In the acanthodians the jaw is supported by direct attachments to the cranium at its hind as well as front end, a condition known as autodiastylic.

The early elasmobranchs themselves do not have a hyostylic jaw support, but an arrangement in which the upper jaw is both attached to the cranium and also supported by the hyomandibula. This amphistylic condition persists to-day in the primitive shark *Hexanchus*. Apparently the jaws, which at first swung from the skull, later became fixed at the hind end to the hyoid, and this finally became the only means of support posteriorly. The advantage of this last arrangement is presumably that it allows a wide gape for swallowing the prey whole. As the sharks sought to eat larger and larger fishes, those in which the hind end of the upper jaw was less firmly fixed to the skull were the more successful and so the hyostylic condition was achieved.

If this theory of the origin of the jaws is correct we may expect to find some trace of a cartilaginous support for the side wall of the pharynx in front of the original first gill-slit, a premandibular visceral arch. Many sharks have two pairs of labial cartilages in this position, which have been held to represent arches. However, there are strong grounds for believing that this is represented by the trabeculae cranii, the rods lying on each side in front of the parachordals and contributing to the floor of the skull (Fig. 98). Many points indicate that these rods are not part of the axial skeleton. The main axis of the body presumably ends at the front end of the notochord, that is to say, at the level of the front ends of the parachordals. Indeed there is much confirmatory evidence to show that this level represents the end of the

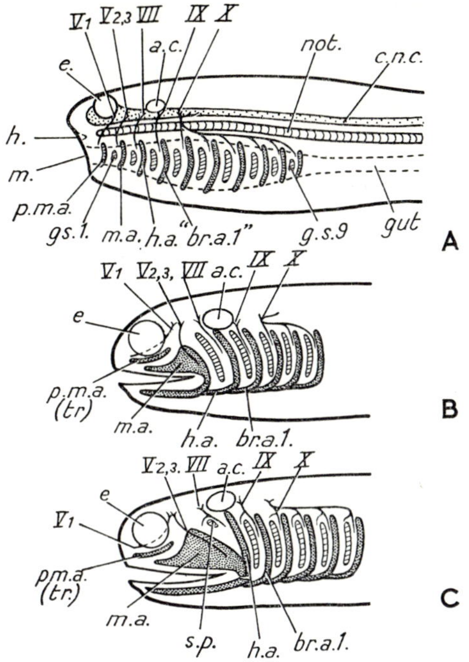

FIG. 98. Diagrams to show the condition of the visceral arches and jaws in early vertebrates.

A. cephalaspid; *B.* acanthodian; *C.* elasmobranch. ***a.c.*** auditory capsule; "***br. a.*** 1" first branchial arch; ***c.n.c.*** nerve-cord; ***e.*** eye; ***gs. 1.*** first gill-slit; ***h.*** hypophysis. ***h.a.*** hyomandibular arch; ***m.a.*** mandibular arch; ***m.*** mouth; ***not.*** notochord; ***p.m.a.*** premandibular arch (trabecula); ***sp.*** spiracle; *V1*, profundus nerve; *V2, 3,* trigeminal nerve; *VII*, facial; *IX*, glossopharyngeal; *X*, vagus. (Modified after Westoll.)

segmented part of the body, everything in front of this level being as it were pushed forward from above or below. The trabeculae have exactly the relations to the most anterior nerves and blood-vessels that would be expected of visceral arches. Confirmation of the theory comes from the discovery that the cartilage of the front part of the trabeculae, like that of the visceral arches, is formed by material streaming down from the neural crest, that is to say, from ectoderm.

The branchial arches, hyoid, jaws, and trabeculae thus all constitute a single series, the result of the working of a repetitive or rhythmic process, appropriately modified at each level.

7. Segmentation of the vertebrate head

The rhythmicity or metamerism seen in the cartilages can be traced throughout the structure of the head. Although in higher vertebrates the head appears as a distinct structure, separated from the body by a neck, yet there is every reason to think that it has arrived at that state by gradual modification of the anterior members of an originally complete metameric series. The jaws, the receptor-organs, and the brain have become developed at the front end of the body, producing what zoologists conveniently if pretentiously call cephalization.

The fundamental segmentation of the head is not very easily apparent to superficial observation; the working out of its details is an excellent exercise in morphological understanding. Recognition of the segmental value of the various structures also makes them the more easily remembered. For instance, the nerves found in the head have been named and numbered for centuries by anatomists in an arbitrary series:

I. Olfactorius
II. Opticus
III. Oculomotorius
IV. Trochlearis (patheticus)
V. Trigeminus
VI. Abducens
VII. Facialis
VIII. Acousticus
IX. Glossopharyngeus
X. Vagus
XI. Accessorius
XII. Hypoglossus

Morphological study has shown that these nerves are not isolated structures, each developed independently, but that they represent a regular series of segmental dorsal and ventral roots of the head somites. The satisfaction and simplification given by this generalization is one of the clearest advantages of morphological insight. More important still, such understanding of the morphology of a structure shows us how to look for the morphogenetic processes that produce it; such knowledge of how organs are made is an essential step in mending or remaking them.

The idea of the essential similarity of structure of the head and trunk was early developed by Goethe, who tried to show that the mammalian skull is a series of modified vertebrae. Unfortunately this view cannot be maintained in detail and the theory was brought to ridicule by T. H. Huxley and others. The segmental value of the skull floor and sides is not at all easy to determine; the parachordals arise as a pair of unsegmented rods on either side of the notochord.

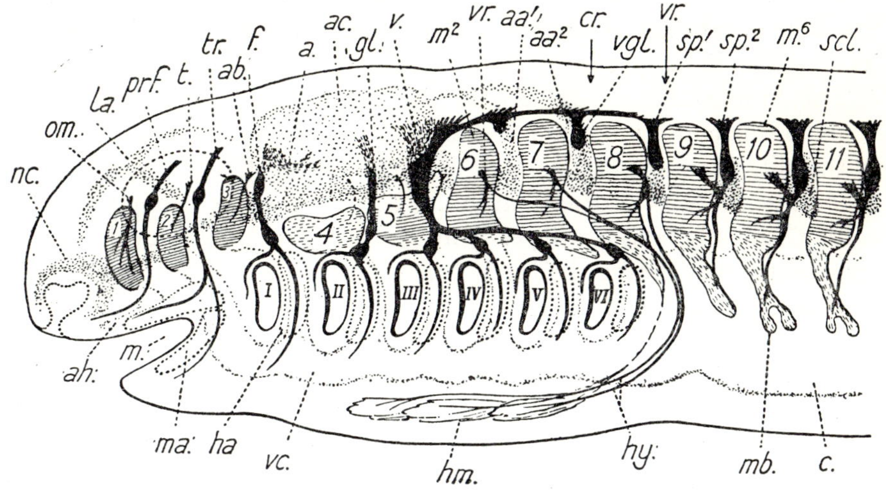

FIG. 99. Diagram of the segmentation of the head of a dogfish.

cr. limit of neurocranium; *vr.* limit of visceral arch skeleton; *a.* auditory nerve; *aa.1*, pre-occipital arch; *aa.2*, occipital arch; *ab.* abducens nerve; *ac.* auditory capsule; *ah.* anterior head cavity; *c.* coelom; *f.* facial nerve; *gl.* glossopharyngeal nerve; *ha.* hyoid arch; *hm.* hypoglossal muscles; *hy.* hypoglossal nerve; *la.* pila antotica; *m.* mouth; *m.2–6*, myomere 2–6; *ma.* mandibular arch; *mb.* muscle-bud; *nc.* nasal capsule; *om.* oculomotor nerve; *prf.* profundus nerve; *scl.* sclerotome of segment 10; *sp.1–2*, ganglion of spiral nerve 1–2; *t.* trochlear nerve; *tr.* trigeminal nerve; *v.* vagus nerve; *vgl.* vestigial ganglion of segment 7; *vc.* ventral coelom; *vr.* ventral root of segment 6. (From Goodrich.)

8. The pro-otic somites and eye-muscles

Ideas about the segmentation of the head were first correctly formulated by F. Balfour. In his studies of the development of elasmobranchs (1875) he showed that three myotomes, the pro-otic somites, can be recognized during development in front of the auditory capsule (Fig. 99). The auditory sac, pushing inwards and becoming surrounded by cartilage, then breaks the series of myotomes, so that several are missing in the adult, though the series is complete in the embryo.

If this analysis is correct we should be able to recognize that the nerves of the head belong to a series of dorsal and ventral roots, similar to that in the trunk, the ventral roots being those for the myotomes and the dorsal roots, running between the myotomes, carrying sensory fibres for the segment and motor-fibres for any non-myotomal musculature present (p. 36). In the spinal region the dorsal and ventral roots join, but this is not the primitive condition (witness amphioxus and the lampreys), and in the head region the earlier state of affairs is retained, the dorsal and ventral roots remain separate. Presumably the arrangement we find in the head today was laid down in very early times, in the Silurian period or earlier, when the dorsal and ventral roots were still separate. The head, in spite of its specializations, preserves for us a relic of that ancient condition.

The branchial nerves, such as the glossopharyngeal, show clear signs of this condition. Each has a small pre-trematic branch in front of the slit, a larger post-trematic branch behind it, and a pharyngeal branch to the wall of the pharynx. The pre-trematic branch usually contains mostly sensory fibres from the skin, the pharyngeal branch visceral sensory fibres, including those from taste buds. The post-trematic branch contains both motor and sensory fibres. In addition to these more ventral branches the branchial nerves also usually provide dorsal rami to the skin of the back.

The three pro-otic somites become completely taken up in the formation of the six extrinsic muscles of the eye, arranged similarly in all gnathostome vertebrates. The four recti roll the eye straight upwards, downwards, forwards, or backwards, and the two obliques, lying farther forward, turn it, as their name suggests, upward or downward and forward (Fig. 100). Of these muscles the superior, anterior, and inferior rectus and inferior oblique are all derived from the first myotome and are innervated by the oculomotor (third cranial) nerve. The superior oblique, innervated by the trochlear nerve (fourth cranial), is the derivative of the second and the posterior rectus (external rectus of man), innervated by the abducens (sixth cranial), of the third somite. These three nerves are evidently the ventral roots of the three pro-otic somites. At some early stage of vertebrate evolution all the myotomal musculature of the front part of the head became devoted to the movement of the eyes. The muscles originally forming part of the swimming series became attached to a cup-like outgrowth from the brain.

Most of the rest of the musculature of the head, including that of jaws and branchial arches, is derived from the somatopleure wall of the coelom and is therefore lateral plate or visceral musculature. This lateral plate muscle is indeed better developed in the head than in the trunk, where all the muscles, even of the more ventral parts of the body, are formed by downward tongues from the myotomes. The lateral plate origin of the jaw-muscles at once gives us the clue to the nature of some more of the cranial nerves, the fifth, seventh, ninth, and tenth. These nerves all carry ganglia containing the cell bodies of sensory fibres and these are comparable to the spinal dorsal root

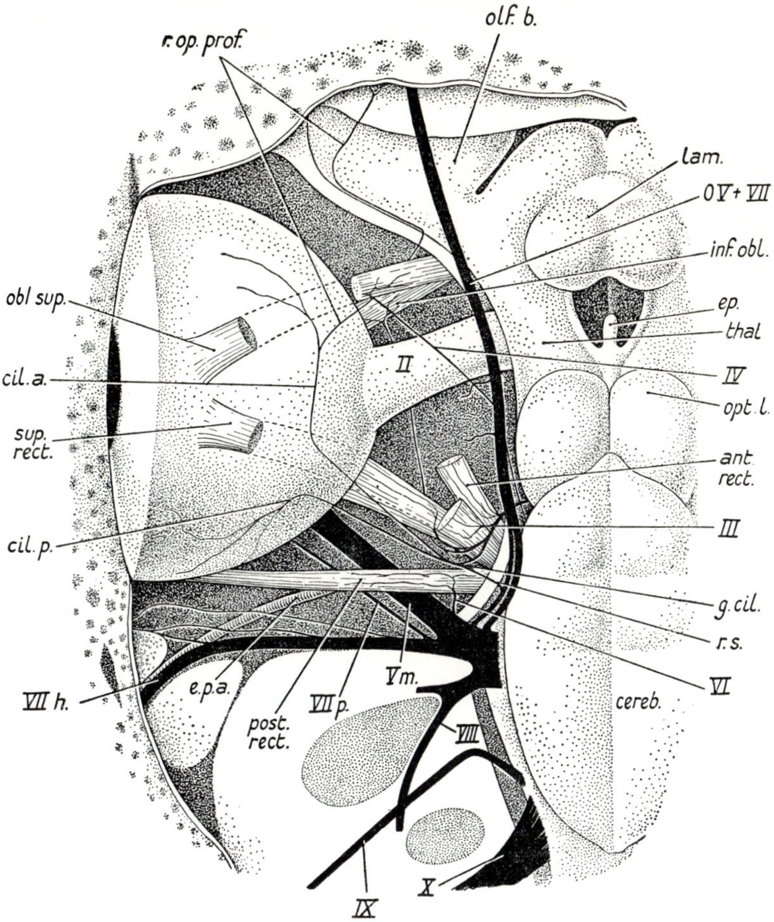

FIG. 100. Orbit of the dogfish.

ant.rect. anterior rectus; *cereb.* cerebellum; *cil.a.*, *cil.p.* anterior and posterior ciliary nerves; *epa.* anterior carotid artery; *ep.* epiphysis; *g.cil.* ciliary ganglion; *inf.obl.* inferior oblique; *lam.* lamina terminalis of cerebrum; *O.V.* and *VII*, superficial ophthalmic branch of trigeminal and facial; *obl.sup.* superior oblique; *olf.b.* olfactory bulb; *opt.l.* optic lobe; *post.rect.* posterior rectus; *r.op.prof.* ramus ophthalmicus profundus of trigeminal; *rs.* sensory root of ciliary ganglion; *sup.rect.* superior rectus; *thal.* thalamus; *II* to *X*, cranial nerves. (After Young, *Quart. J. Micr. Sci.* **75**.)

ganglia. But the nerves also transmit motor-fibres to the muscles of the jaws and branchial arches. They are in fact mixed roots, just as we have seen that the primitive dorsal roots should be, carrying the sensory fibres for the segment and motor-fibres for the non-myotomal muscles (p. 36).

9. The cranial nerves of elasmobranchs

These nerves are more easily studied in elasmobranchs than in any other vertebrates, because of the relatively soft and transparent cartilage through which they run. We may therefore take this opportunity to examine the whole series of cranial nerves in some detail in the dogfish, beginning with the oculomotor nerve, the first ventral root. Examination after removal of the brain will show clearly in any vertebrate, including man, that this nerve arises from the ventral surface, at the level of the hind end of the midbrain (optic lobes). This is not true of the trochlear or pathetic nerve, which emerges from the dorso-lateral surface of the brain but nevertheless is the ventral root of the second segment. Its cells of origin lie close behind those of the oculomotor nerve, in the ventral part of the brain. The reason for the dorsal emergence is that the muscle lies dorsally and the nerve has been modified so as to reach its muscle by running partly within the tissues of the brain. The third ventral root (abducens), which is very short, is clearly ventral.

In looking for the dorsal roots that correspond to these three segments we have to examine the trigeminal, facial, and auditory nerves. The trigeminal of the dogfish, like that of man, has ophthalmic, maxillary, and mandibular branches (Fig. 100), but can be shown to represent the dorsal roots of the two first segments. The ophthalmic branch is a sensory nerve carrying fibres for skin sensation from the snout. The maxillary branch supplies sensory fibres to the upper jaw, whereas the mandibular is a mixed nerve to the skin and muscles of the lower jaw. Besides these main branches there is also a small but important sensory branch from the trigeminal to the eyeball (Fig. 100). This joins a motor root from the oculomotor nerve where the latter swells slightly to form a ciliary ganglion. Two ciliary nerves then carry motor and sensory fibres to the eyeball. In some specimens a branch of the more anterior ciliary nerve leaves the eyeball anteriorly, runs between the oblique muscles, and out of the orbit again to end in the skin of the snout (Fig. 100). Though this branch is small and inconstant in the dogfish, its course corresponds exactly with that of a much larger nerve in the related shark *Mustelus* and in skates and rays. In these animals there are two ophthalmic branches of the trigeminal nerve; one, having a course similar to that of the main nerve in the dogfish, is the ramus ophthalmicus superficialis; the second, the ramus ophthalmicus profundus, runs across within the orbit, gives off the long ciliary nerve to the eyeball, passes between the oblique muscles, and leaves the orbit for the skin of the snout. In higher vertebrates the nasociliary nerve and the long ciliary, innervating the eyeball, represent the profundus, while the rest of the ophthalmic nerve of mammals corresponds to the superficial ophthalmic of elasmobranchs.

The relations of these nerves to other structures shows that the so-called trigeminal nerve really includes the dorsal roots of two segments combined. The profundus can be traced back in development to a nerve that is obviously the dorsal root of the first somite, of which the oculomotor nerve is the ventral root. Indeed it may be noticed that the profundus and oculomotor partly join, at the ciliary ganglion. The ramus ophthalmicus profundus and the oculomotor nerve thus constitute the dorsal and ventral roots of the first or pre-mandibular somite, whose corresponding branchial arch is presumably the trabecula cranii (p. 147). The dorsal root does not show the full structure of a branchial nerve, presumably because there is no gill-slit. The profundus represents only the dorsal branch of a typical branchial nerve, innervating the skin.

The ramus ophthalmicus superficialis, and the maxillary and mandibular branches together constitute the dorsal root of the second pro-otic somite, whose ventral root is the trochlear nerve. The corresponding gill arch is the mandibular (palato-pterygo-quadrate bar and Meckel's cartilage), whose gill-slit we have suggested has been incorporated with the edge of the mouth. The trigeminal nerve shows considerable similarity to a branchial nerve, its maxillary branch represents the pre-trematic and the mandibular the post-trematic ramus, while the ophthalmicus superficialis is the dorsal branch to the skin. There is no pharyngeal branch. An anomalous feature of the trigeminal is that it contains sensory fibres whose cells of origin lie within the brain (mesencephalic root). These fibres are probably proprioceptors from the masticatory muscles and eye-muscles. The latter run from the eye-muscle nerves to join the trigeminal.

The dorsal root of the third segment, whose ventral root is the abducens, includes the whole of the facial and also the auditory nerve. The facial is a large mixed nerve in the dogfish. Its ophthalmic branch runs to the snout, carrying mainly fibres for the organs of the lateral line system that lie there. A large buccal branch supplies sensory fibres to the mouth and a palatine branch joins the trigeminal. A small prespiracular branch carries sensory fibres from the skin in front of the spiracle, and the main portion of the nerve continues behind the spiracle as the hyomandibular nerve, dividing up into motor branches for muscles of the hyoid arch and sensory ones for the skin of that region.

This nerve is obviously the branchial nerve to the spiracle; we can safely say that the facial and abducens are the dorsal and ventral roots of the third or hyoid segment. The auditory nerve is included as part of the dorsal root of the third somite because the auditory sac is formed by sinking in of a portion of the ectoderm within the territory of the facial nerve. The labyrinth still communicates with the surface of the head in the adult dogfish by a canal, the aquaeductus vestibuli. The nerve that innervates the auditory sac, whatever complexities it may acquire, is to be regarded morphologically as a portion of the dorsal root of the hyoid segment.

The segmental nature of the structures in the pro-otic region can therefore be made out without serious difficulty. The disturbance introduced by the auditory capsule makes the segmental arrangement of the more posterior region of the head somewhat confused. The series of dorsal roots is uninterrupted; the ninth (glossopharyngeal) nerve is the dorsal root of the fourth segment of the series and runs out through the cartilage of the auditory capsule. The dorsal roots of the succeeding segments are then fused to form that very puzzling nerve the vagus. The branches it sends to the gills are clearly typical branchial nerves, but why should they all come off together from the medulla oblongata, and if there is any advantage in this union, why is the ninth nerve not also so incorporated? Above all, why does the vagus send two branches far outside the segments of its origin, the lateral line branch carrying fibres to the organs right to the tip of the tail and the visceral branch fibres to the heart, stomach, and probably small intestine?

Evidently these 'wanderings', from which the vagus gets its name, began very long ago. The nerve reaches as far back in cyclostomes as in any other vertebrates. It is easy to understand that if visceral functions are to be directed from the medulla oblongata there is an advantage in having sensory impulses sent direct to that region of the brain and motor impulses sent out direct to the viscera. It may be that these advantages allowed the centralization of these visceral functions, while the need for serial contraction of the swimming muscles led to the retention of the segmental arrangement of the spinal cord. It is an interesting thought that but for the swimming habits of our ancestors our nervous system might by now consist of a central ganglion with nerves passing from it direct to all the organs. Indeed we are tending in that direction, as the spinal cord shortens and becomes more and more nearly a simple pathway between the brain and the periphery.

However this may be, the vagus is certainly a nerve compounded of the dorsal roots of several segments and it is a mixed nerve, containing both receptor and motor fibres. Some of the more posterior rootlets of this series are separated off in higher animals (not the dog-fish) to form the eleventh cranial nerve, the accessorius or spinal accessory, which in mammals sends motor-fibres to certain muscles of the neck, the sternomastoid, and part of the trapezius. Its motor nature has led some to suppose that this nerve is a ventral root, but these muscles are derived from lateral plate musculature and the accessorius represents the motor portion of the hinder dorsal roots of the vagus series.

The ventral roots of this post-otic region have become much reduced. Several myotomes are always missing completely, so that there are no ventral roots corresponding to the glossopharyngeal and first three or four vagal segments. The more anterior of the surviving post-otic somites are to be found not in the dorsal region but ventrally, as the hypoglossal musculature of the tongue. The muscle-buds have grown round into this portion behind the gill-slits, and the nerve (hypoglossal) that innervates them represents the ventral roots of the more posterior segments of the vagus-accessorius series (Fig. 104). The origin of this nerve from the floor of the medulla is a clear sign that it is a ventral root.

Thus the entire series of cranial nerves is:

Segment	*Arch*	*Dorsal root*	*Ventral root*
Pre-mandibular	Trabecula	R. op. profundus V	Oculomotorius III
Mandibular	Palato-pterygo-quadrate bar and Meckel's cartilage	Rr. op. superficialis, maxillaris, and mandibularis V	Trochlearis IV
Hyoid	Hyoid	Facialis VII Acousticus VIII	Abducens VI
1st Branchial	1st Branchial	Glossopharyngeus IX	(absent)
2nd Branchial	2nd Branchial	Vagus X	
3rd Branchial	3rd Branchial	+	Hypoglossus XII
4th Branchial	4th Branchial	Accessorius XI	
5th Branchial	5th Branchial		

Two cranial nerves have not yet been considered, the first, olfactory, and second, optic. Our thesis is that all connexions between centre and periphery are made by means of a segmental series of dorsal and ventral roots and therefore these nerves, too, should be fitted into the series. No embryological or other studies have enabled this to be done and the reason in the case of the optic nerve is quite clear. It is not morphologically a peripheral nerve at all. The eye is formed as a vesicle attached to the brain; the optic 'nerve' therefore develops as a bundle of fibres joining two portions of the central nervous system; in fact it is now usually called the optic tract, not the optic nerve.

This reasoning will not apply to that very peculiar and interesting structure the olfactory nerve. This is unique among all craniate nerves in consisting of bundles of fibres whose cell bodies lie *at the periphery*. The cells of the olfactory epithelium, like the sensory cells in invertebrates and some of those of amphioxus, are neurosensory cells, that is to say, their inner ends are prolonged to make the actual nerve-fibres that pass into the brain. This fact does not by itself solve the problem of fitting the nerve into the series of dorsal and ventral roots, but it reminds us that the nerve is very ancient, and suggests that it does not fall into the rhythm of the rest of the series because it precedes the other cranial nerves either in time or space, or perhaps even both. The olfactory nerve may have existed before any segmental structure appeared, possibly as the nerve of sense-organs on the front end of the ciliated larva which we suppose gave rise to our stock (p. 76). Alternatively we can say that the olfactory nerve is as it is because it lies in front of the region over which the segmenting process operates; it is, as it were, 'prostomial'. If we wish we can hold both these views together.

There are one or two other exceptions to the rhythmic arrangement of nerves, perhaps more difficult to account for than the first and second cranial nerves. If all connexions between centre and periphery are made by dorsal and ventral roots what is the status of the fibres that run down the infundibular stalk to reach the cells of the pituitary body? This glandular tissue, derived from the epithelium of the hypophysial folding of the roof of the mouth, is undoubtedly a peripheral organ. Does it receive its nerve-fibres direct from the brain? If so presumably we must say that the pituitary, like the nose, is prostomial, lying in front of the segmental region, and this is reasonable enough from its position. There is good reason to believe that it is an extremely ancient organ, already present in the earliest chordates.

A still more puzzling exception is the nervus terminalis. This is a small bundle leaving the brain ventrally behind and below the olfactory nerves and running to the olfactory mucosa or to the accessory olfactory organ of Jacobson, where this is present (p. 350). In some vertebrates it carries a small ganglion. The fibres are probably afferents and they run backwards through the brain tissue to the pre-optic nucleus of the hypothalamus. A possible clue to its origin is that this is the region of the brain where the morphologically ventral region of the neuraxis ends (p. 147). The nervus terminalis may represent the ventral olfactory nerve, the much larger main nerve being morphologically dorsal.

A further puzzle of some importance which may be mentioned here is the course of the proprioceptor fibres for those muscles that are supplied purely by ventral roots. The eye-muscles contain proprioceptor organs and Sherrington and others have shown that the afferent fibres connected with these run to the brain through the third, fourth, and sixth nerves, that is to say, through ventral roots. Similarly, it has been shown that there are afferent fibres in the hypoglossal nerves in mammals. Conversely it is now known that there are efferent fibres running from the brain to many receptor organs. For example, such fibres run in the auditory nerve. To pursue these questions farther would lead us into discussion of the factors that control the making of connexions within the nervous system Here we are concerned only with analysis of the plan that produces the main outlines of the structures in the head, a plan which, with all its modifications, is essentially segmental.

10. Respiration

The function of the branchial arches is not merely to support the gills but to allow the movements of the pharynx wall by which the respiratory current of water is produced. It is for this reason that the jointed system of rods is present. The respiratory movements consist in a lowering of the floor of the mouth by means of the hypoglossal muscles, with at the same time an expansion of the walls of the pharynx. This causes an inrush of water through the mouth, which is then closed and the floor raised, forcing the water out through the gill-slits. The whole movement is worked by the 'visceral' (lateral plate) muscles of the pharynx wall, innervated by the trigeminal, facial, glossopharyngeal, and vagus nerves, in co-operation with the myotomal hypoglossal muscles, innervated by the hypoglossal nerve.

The gill filaments bear lamellae that meet at the tips, leaving minute channels for the water. The blood flows through the lamellae in the opposite direction to the water so that just before leaving the gills the blood meets the highest concentration of oxygen and lowest of carbonic acid.

11. The gut of elasmobranchs

The digestive system of sharks shows several changes from the plan found in lampreys, especially the presence of a true stomach, characteristic of all gnathostomes. Apparently little or no digestion goes on in the mouth and pharynx. The teeth consist of rows of backwardly directed denticles (Fig. 101). They are carried on special folds of skin lining the jaws and are continually replaced as they are worn away on

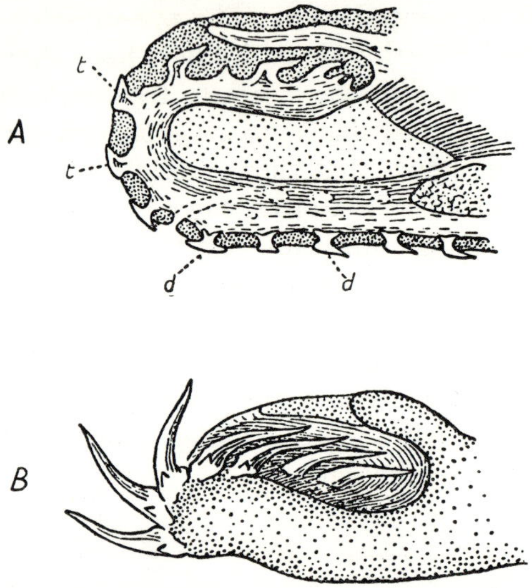

FIG. 101. Sections through the jaws of A, dogfish (*Scyliorhinus*), and B, sand-shark (*Odontaspis*), showing the transitions between dermal denticles (*d*) and teeth (*t*). (From Norman, partly after Gegenbaur.)

the edge. The replacement of milk by permanent teeth in mammals is a relic of such serial replacement in a fish. The 'gill rakers' are rods attached to the branchial cartilages and serving to prevent the escape of prey. The basihyal supports a short non-protrusible tongue.

The wall of the pharynx is lined by a stratified epithelium on to which open numerous mucous glands, sometimes complex. The mucus serves to assist the passage of the food, but probably has no strictly digestive function, though the salivary glands of higher vertebrates no doubt originate from a modification of these mucous glands.

The pharynx narrows to an oesophagus with thick muscular walls, leading without sharp transition to the stomach. We have seen that in cyclostomes the oesophagus opens directly into the region of gut that receives the bile and pancreatic secretion. The stomach, which we now meet for the first time, has probably been formed as a special portion of the oesophagus. Barrington suggests that it evolved with the jaws, serving originally as a receptacle for the large pieces of food, or even whole fishes, which could now be swallowed. The mucous glands became modified to produce acid, since this prevents bacterial decay. Finally, an enzyme, pepsin, was evolved able to digest proteins in acid solution. In the dogfish this condition has been fully established and the stomach has essentially the structure and functions found in all higher vertebrates. However, in the gastric glands only one type of cell is recognizable, there are no separate pepsin-secreting and acid-producing cells. Nevertheless, there is a pepsin-like enzyme present and the contents are acid. The stomach is divided into two parts, a descending cardiac and ascending pyloric limb, the significance of the divisions being unknown.

The region where the stomach joins the intestine is guarded by a powerful pyloric sphincter, immediately beyond which open the bile and pancreatic ducts. The liver is a large two-lobed organ, receiving the hepatic portal blood from the gut. It serves as a storage organ containing much glycogen and fat and sometimes the hydrocarbon squalene. It probably also plays a part in the destruction of red blood corpuscles. Bile is carried away to a gall-bladder, from which a bile-duct leads to open at the front end of the spiral intestine.

The pancreas, hardly recognizable as a distinct organ in the lamprey, forms in the dogfish an elongated body between the stomach and intestine. It contains both exocrine and endocrine cells and its duct enters the intestine shortly below the pylorus. The 'small' intestine of elasmobranchs is of a peculiar form, being short but with its surface greatly increased by the presence of a spiral ridge or 'valve'. The intestinal contents are alkaline and contain trypsin, amylase, and lipase. There is no constant fauna of commensal bacteria. Absorption presumably takes place wholly in this organ, for the remaining length of gut consists only of a short rectum, to which is attached an organ of unknown function, the rectal gland, containing branched glands and much lymphoid tissue.

12. The circulatory system

The heart develops as a specialization of the subintestinal vessel between the place where it receives the veins from the liver and the body wall and the gills, which are to be supplied under high pressure. It consists of a single series of three main chambers, sinus venosus, atrium, and ventricle, all of which are muscular, and there is also a muscular base to the ventral aorta, the conus arteriosus, provided with valves (Fig. 102).

The five afferent branchial arteries carry blood to the gill lamellae, whence it is collected by a system of four efferents and connecting vessels into a median dorsal aorta, carrying blood to all parts of the body. Oxygenated blood is supplied to the head from three sources. (1) From the top of the first gill a carotid artery leaves the efferent branchial and runs forwards and towards the midline: it then divides

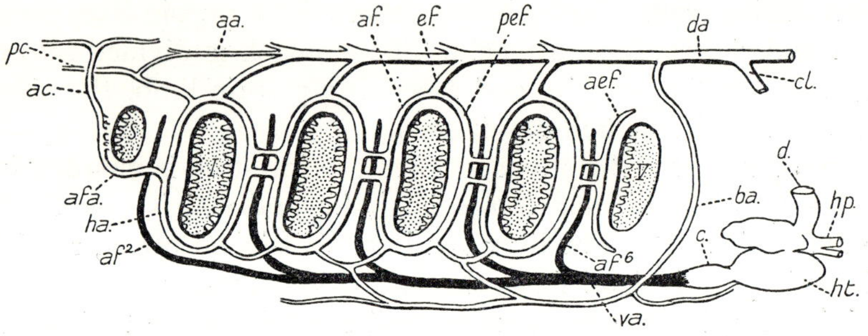

FIG. 102. Diagram of the branchial circulation of an elasmobranch fish.

aa. median anterior prolongation of aorta; *ac.* anterior carotid; *afa.* afferent vessel of spiracular gill; *aef.* afferent vessel from last hemibranch; *af.* anterior efferent vessel; *af.*$^{2-6}$, five afferent vessels from ventral aorta; *afa.* afferent artery of spiracular gill; *c.* conus leading to ventral aorta; *cl.* coeliac artery; *d.* ductus Cuvieri; *da.* dorsal aorta; *ef.* epibranchial artery; *h.a.* hyoid afferent vessel; *hp.* hepatic veins; *ht.* heart; *pc.* posterior carotid; *pef.* posterior efferent vessel; *s.* spiracle; *va.* ventral artery; I–V, branchial slits. (From Goodrich, *Vertebrata*, A. & C. Black, Ltd., after Parker.)

into an external carotid to the upper jaw and internal carotid to the brain. (2) The dorsal aorta divides at its front end into branches, which join the carotids before their division. (3) From the vessel that collects blood from the first gill arises a hyoidean artery, carrying oxygenated blood to the spiracle. From here the hyoidean artery runs on as the anterior carotid (Fig. 102) across the floor of the orbit to join the internal carotid within the brain-case.

The heart is supplied by a cardiac artery arising from the dorsal aorta behind the gills. The blood-pressure in the ventral aorta is 30–40 mm Hg and there is a drop of 10–20 mm Hg across the gills. The circulation is slow, with a mean circulation time as low as 2 minutes. The venous return of the fishes is ensured by a system of very large sinuses. The pericardium is almost completely enclosed in a cartilaginous framework by the basibranchial plate above and pectoral girdle below it. It may be that this produces a negative pressure in the veins. There is a passage of unknown function, the pericardio-peritoneal canal, leading from the pericardium to the abdominal coelom and the hinder end of this is very narrow.

A caudal sinus from the tail opens into a renal portal system above the kidneys. From the latter, and from the muscles of the back, blood is collected into the pair of very large posterior cardinal sinuses, lying on the dorsal wall of the coelom. Above the heart these receive the openings of other large sinuses, such as the anterior cardinal sinus, running above the gills and collecting blood from the head by way of an orbital sinus, and the jugular, lateral cardinal, subclavian, and other sinuses from the body wall. Blood then passes round the oesophagus in the two ductus Cuvieri into the sinus venosus, where hepatic sinuses also open.

The resistance offered by a vessel to flow within it decreases with approximately the fourth power of the diameter, therefore the large size of these vessels substantially assists in allowing return to the heart. The heart-muscles, like any others, require antagonists; they can contract in one direction only, and each chamber therefore needs to be actively dilated. It will be noted that the fish heart consists of a series of three muscular chambers, presumably because the low venous pressure is able to dilate only a chamber with very thin walls, such as the sinus venosus. Contraction of the sinus then inflates the auricle, and the auricle inflates the ventricle, which thus constitutes the third step in this serial pressure-raising system. In the land animals, where most of the blood only passes through a single set of capillaries, a two-step system (auricle and ventricle) is sufficient for each part of the circulation.

Little is known of the control of the circulation but it is probably less effective than in higher animals. There is a cardiac branch from the vagus ending in an elaborate plexus in the sinus venosus (Fig. 104). Stimulation of this nerve slows the heart. There is no anatomical or physiological evidence of a sympathetic nerve to the heart, but abundant sympathetic fibres run to the arteries. Small doses of adrenaline cause prolonged rise of blood-pressure.

There are receptors in the efferent branchial vessels and in the post-branchial plexus above the cardinal veins (p. 175). Nerve impulses from these receptors can be recorded in the vagus at each systole and are increased by raising the blood-pressure. Their reflex effects are to slow the heart and respiration and decrease the blood-pressure, perhaps for protection of the gill capillaries. These reflexes are presumably the ancestors of the carotid sinus and similar reflexes of land vertebrates.

13. Urinogenital system

The blood of the elasmobranchs differs from that of all other vertebrates in its very high content of urea. As measured by the depression of the freezing-point the blood is isotonic with the surrounding sea water (say, 3·5 per cent. NaCl); it may even be slightly hypertonic. But there is far less salt in the blood than in the sea, in fact only about 1·7 per cent. NaCl. Although the blood is nearly isotonic with the sea its composition is therefore regulated (homeosmotic). This arrangement is apparently a legacy of the fact that the ancestors of the elasmobranchs were originally fresh-water animals (p. 187). The return passage to the sea has been accomplished by the elasmobranchs through the device of urea retention. The gill surfaces, in which alone the blood comes into close contact with sea water, are not permeable to urea, but this substance penetrates freely into the tissues, as it does in other animals. Elasmobranch tissues if placed in sea water are therefore in contact with a strongly hypertonic medium. They are so habituated to the presence of urea that they are unable to function unless it is present in a concentration that would be toxic to most animals.

This arrangement has presumably been responsible for the fact that few of the elasmobranchs have returned to fresh water. In the case of the saw-fish *Pristis*, which lives some hundreds of miles up the Mississippi and various rivers in China, Smith found that a considerable concentration of urea is still maintained in the blood, thus further increasing the work that these fishes, like any fresh-water animal, must do in order to maintain an osmotic concentration above that of the surrounding water. One shark (*Carcharhinus nicaraguensis*) and some rays, *Trygon*, also live in fresh water.

In the ordinary marine elasmobranchs the high urea concentration is maintained by the presence of a special urea-absorbing section of the kidney tubules. The urinary apparatus is a mesonephros and these fishes show a considerable specialization in that the urinary functions of this organ are separated from its generative ones in the male. The hinder part of the kidney (sometimes called opisthonephros, the term metanephros should be used only for the definitive kidney of amniotes, which has a different method of development) consists of a mass of tubules ending in very large glomeruli, and a section of each tubule has the function of urea absorption. All the tubules join to form a series of five urinary ducts and these enter a urinary sinus, opening to the cloaca. The sinus can be compared functionally with a bladder, but it is a mesodermal structure, derived from the main kidney duct, and not strictly comparable to the endodermal bladder of tetrapods. The urinary sinus is a small organ and the volume of liquid excreted is small.

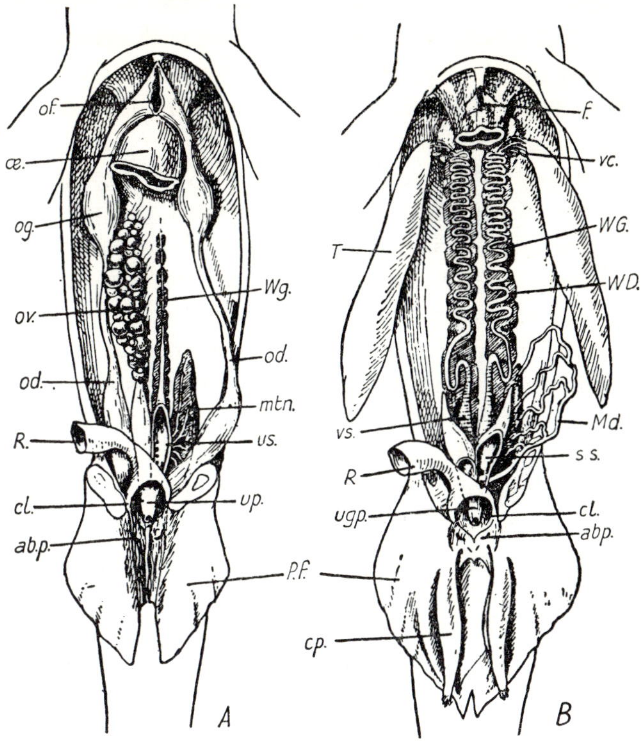

FIG. 103. A. Urinogenital system of the female, B, of the male dogfish. *ab.p.* abdominal pores; *cl.* cloaca; *cp.* claspers of the male; *F.* rudiment of the oviducal opening in the male; *Md.* urinary ducts; *mtn.* hinder (excretory) part of mesonephros; *od.* oviduct; *oe.* cut end of oesophagus; *of.* oviducal funnel; *og.* oviducal gland; *ov.* ovary; *P.f.* pelvic fins; *R.* rectum; *s.s.* sperm-sacs; *T.* testis; *up.* urinary papilla in the female; *ugp.* urogenital papilla in the male; *us.* urinary sinus; *vc.* vasa efferentia; *vs.* vesicula seminalis; *WD.* Wolffian duct; *Wg.* Wolffian gland or mesonephros. (After Bourne, from Goodrich, *Vertebrata*, A. & C. Black, Ltd.)

The genital system is highly specialized to allow internal fertilization and the production of a few very yolky and well-protected eggs. There is a single large ovary, from which the eggs are carried by the cilia of the peritoneum to a pair of funnels lying on either side of the liver behind the heart (Fig. 103). These are apparently formed from pronephric funnels and the Müllerian duct (oviduct) separates from the original nephric duct. In the adult it becomes a thick-walled muscular tube, bearing a swelling, the nidamental gland, the upper part of which produces albumen, the lower the horny egg case.

The testes are paired and sperms are collected at their front ends by vasa efferentia leading into the anterior or reproductive portion of the mesonephros. This consists of a much coiled, thick-walled, vas deferens, whose glands produce material that aggregates the sperms into spermatophores. The vas expands into a broader ampulla (seminal vesicle), which at its lower end gives off a forwardly directed blind diverticulum, the sperm sac, developmentally the lower end of the Müllerian duct, reduced of course in the male; small funnels are still visible at the upper end.

Transmission of the sperms is produced by a large and complicated pair of claspers. These are modified parts of the pelvic fins of the male, developed into scroll-like organs and containing a pumping mechanism and erectile tissue; they are inserted into the female cloaca. The mechanism of erection is operated by nerves and may involve the liberation of adrenaline; experimental injection of that substance will produce erection, and it is perhaps significant that the male possesses a reserve of adrenaline-producing tissue (see p. 167).

Development of elasmobranchs is by partial cleavage, producing a blastoderm, perched on the top of a large mass of yolk. The egg is protected by an elaborate egg-case, the 'mermaid's purse', within which development proceeds until the yolk has been used up.

In several elasmobranchs development is viviparous, the oviduct forming a 'uterus'. In *Mustelus* there is a yolk-sac placenta, but in *Trygon* 'uterine milk' is secreted into the embryo by villi (trophonemata) inserted through the spiracles.

14. Endocrine glands of elasmobranchs

Elasmobranch fishes possess the full complement of endocrine glands, but these show some interesting differences from those of higher vertebrates. The pituitary body lies in the usual place below the diencephalon and anterior, intermediate, tuberal, and neural divisions can be recognized. Little is known of the functions of the gland. The gonads of the dogfish retrogress after removal of the pituitary. Little or no vasopressin or oxytocin is present. The neuro-intermediate lobe contains a substance that produces the expansion of melanophores. Hogben and Waring have also produced evidence that the pars anterior produces a substance causing contraction of the melanophores, but this has not yet been isolated and the evidence for its existence is indirect.

The thyroid is formed by a downgrowth from the floor of the pharynx, to which it often remains attached by a narrow stalk con-

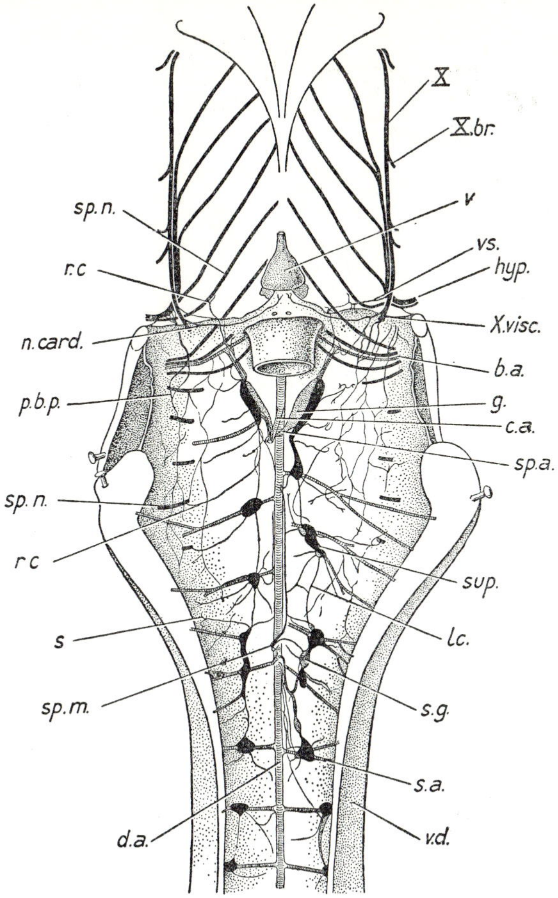

FIG. 104. Dissection of suprarenal bodies and sympathetic nervous system of the dogfish.

b.a. brachial artery; *c.a.* coeliac artery; *d.a.* dorsal aorta; *g.* first sympathetic ganglion; *hyp.* hypoglossal nerve; *l.c.* longitudinal sympathetic 'chain'; *n.card.* cardiac branch of vagus; *p.b.p.* post-branchial plexus; *r.c.* ramus communicans; *s.* sensory fibre; *s.a.* segmental artery; *sg.* sympathetic ganglion; *sp.a.* anterior splanchnic nerve; *sp.m.* middle splanchnic nerve; *spn.* spinal nerve; *sup.* suprarenal body; *v.* ventricle; *v.d.* vas deferens; *vs.* vago-sympathetic anastomosis; *X.* vagus; *X.br.* branchial branch of vagus; *X.visc.* visceral branch of vagus. (After Young, *Quart. J. Micr. Sci.* **75**.)

taining a small ciliated pit, a reminder that the organ was once a ciliated mucus-secreting gland.

The adrenal tissue is especially interesting because the two parts, so closely associated in mammals, are here found widely separated. A segmental series of glands, the suprarenals, are rich in noradrenaline. They project into the dorsal wall of the posterior cardinal sinus and can be seen when it is opened (Fig. 104). The more anterior ones are fused to form an elongated structure on either side of the oesophagus.

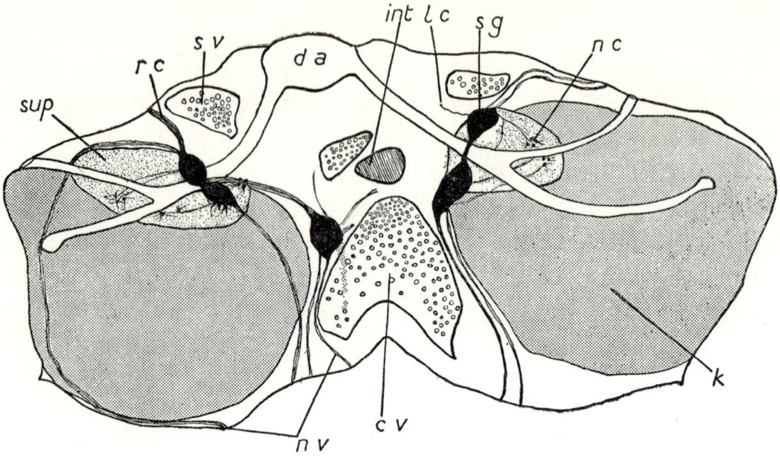

FIG. 105. Diagram of transverse section through hind region of mesonephros of dogfish.

cv. cardinal vein; *da.* dorsal aorta; *int.* interrenal body; *k.* mesonephros; *l c.* longitudinal sympathetic connective; *n c.* sympathetic nerve-cells scattered in suprarenal body; *nv.* sympathetic nerves; *rc.* ramus communicans; *sg.* sympathetic ganglion; *sv.* subcardinal vein; *sup.* suprarenal body. (From Young, *Quart. J. Micr. Sci.* **75**.)

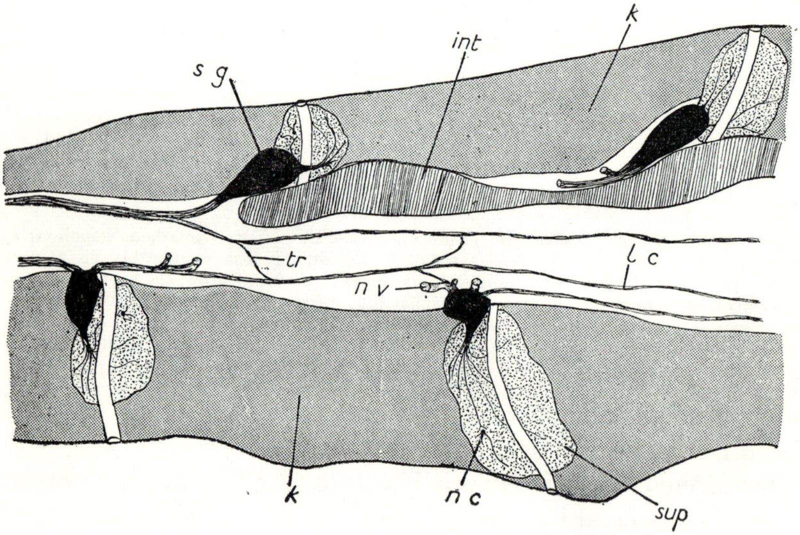

FIG. 106. Diagram of arrangement in hinder mesonephric region of dogfish.

Lettering as Fig. 105. *tr.* transverse sympathetic nerve. (From Young, *Quart. J. Micr. Sci.* **75**.)

The sympathetic ganglia are closely associated with these suprarenal bodies, as would be expected from their common origin from cells of the neural crest. The segmental series continues along the whole length of the abdomen, the more posterior members being embedded in the kidney tissue (Fig. 105). These posterior suprarenal bodies are larger in the male than in the female, but only the central part of the male glands shows the reaction with chrome salts that indicates the presence of adrenaline. The peripheral portion of each gland appears to consist of non-functioning cells, possibly a reserve used only during reproduction (see p. 164).

The part of the adrenal corresponding to the cortex of mammals is represented in elasmobranchs by the interrenal bodies, lying medially in some species, paired in others, in the kidney region (Figs. 105 and 106). The cells of these organs resemble cortical adrenal cells. Since they are not in contact with the suprarenals at any point, it would seem that the association of the two parts is not necessary for their functioning, at least in these animals. Removal of the interrenal is always fatal. The gland is stimulated by 'stress' or by mammalian ACTH. Extracts of it prolong the life of adrenalectomized rats. There is evidence that it influences carbohydrate metabolism and activity of the gonads but not electrolyte balance.

The islets of Langerhans contain two cell types as in mammals. The pineal body is small and without any trace of eye-like structure.

The gonads contain endocrine organs, producing steroid hormones. These are formed by interstitial cells in the testes. Oestrogens probably come from the outer (theca) cells of the follicles that surround the eggs. The inner (granulosa) cells of the capsule assist in yolk production but may also produce progesterone and in viviparous species they develop into a distinct corpus luteum after ovulation.

15. Nervous system

The brain is large and well developed in elasmobranchs, having a structure characteristically different from that of both the cyclostomes and bony fishes (Fig. 100). The forebrain is large and has cerebral hemispheres thickened both in floor and roof, whereas in teleosts the roof is thin. The hemispheres are wide relative to their length and the end of the unpaired portion of the forebrain between the hemispheres, the lamina terminalis, is also much thickened. Attached to the ends of the cerebral hemispheres are large olfactory bulbs and there are also large nasal sacs. Evidently the olfactory sense is well developed in these animals and they depend greatly on it for hunting.

All parts of the cerbral hemispheres receive fibres from the olfactory bulbs and the forebrain serves mainly for analysing the olfactory impulses. However, it is stated that there are fibres reaching forward to one area at the back of the roof of the hemispheres from other centres. Johnston therefore called this region the 'general somatic area' and suggested that it represents the beginnings of that development so characteristic of mammals by which all the senses are centred on the cerebral hemispheres. Further work is needed to confirm the

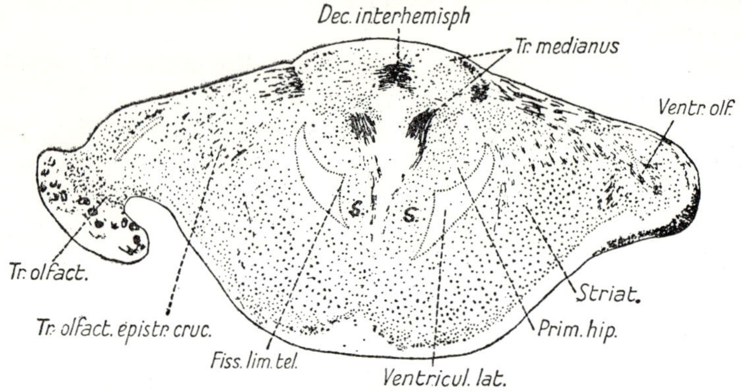

FIG. 107. A cross-section through the forebrain of a shark.

Dec.interhemisph. decussatio interhemispherica; *Fiss.lim.tel.* fissura limitans telencephali; *Prim.hip.* primordium hippocampi; *S.* septum; *Striat.* striatum; *Tr.medianus.* tractus medianus; *Tr.olfact.* tractus olfactorius; *Tr.olfact.epistr.cruc.* tractus olfacto-epistriaticus cruciatus; *Ventricul.lat.* ventriculus lateralis; *Ventr.o.f.* ventriculus olfactorius. (From Kappers, Huber, and Crosby.)

existence of this pathway, and even if present its significance must not be exaggerated. There is of course no cortical arrangement of tissue in the hemispheres. The cells form thick masses around the ventricle (Fig. 107). The roof is quite thick and contains decussating fibres in the midline. The sides and floor make up the main bulk of the organ, the lateral wall being known as the striatum, its upper part the epistriatum. The medial wall is known as the septum and its upper portion is often referred to as the primordium hippocampi, having a position similar to that of the hippocampus of mammals. The main efferent pathways are tracts leading to the hypothalamus and to the optic lobes. After removal of the forebrain the sense of smell is lost but the fish shows no obvious disturbance of posture, locomotion, or behaviour.

The diencephalon is a narrow band of tissue, there are no extensive tracts leading forward through it, and the optic and other pathways do not end here as they do in higher animals. The lower part of the between-brain, the hypothalamus, is, however, as well developed (relatively) in these animals as in mammals. Its hind part (inferior lobes) receives olfactory impulses via the forebrain (the 'fornix' of higher vertebrates) and gustatory pathways from the medulla. Its efferent fibres run to reticular centres. The more anterior part of the hypothalamus lies above the pituitary and contains the supraoptic nucleus, whose axons form the hypophysial tract, ending in the intermediate lobe. The supraoptic cells of all vertebrates are large and contain granules of neurosecretory material that is probably passed down the axons and liberated in the pituitary. The anterior hypothalamus is a higher centre for visceral control, regulating, for example, circulation, respiration, and many metabolic activities. Attached to the hind end of the hypothalamus of fishes is a peculiar organ, the saccus vasculosus, with folded, pigmented walls. It has been suggested that this acts as a pressure receptor, since it is well developed in deep-sea fishes. It is one of the characteristic features that the sharks and bony fishes have in common.

The midbrain, as in cyclostomes and teleosteans, is very large and is perhaps the dominant centre of the brain. The optic tracts end in its roof (tectum opticum) after complete decussation below the brain. The cells of the tectum are arranged in a complicated pattern of layers. Other sensory centres that send tracts to the optic lobes are the olfactory (cerebral hemispheres), acustico-lateral, cerebellar, gustatory, and probably also the general cutaneous centres of the spinal cord. Efferent tracts leave the midbrain roof to the base of the midbrain and extend backwards into the medulla, perhaps into the spinal cord. The efferent midbrain fibres have direct influence on the spinal cord, and electrical stimulation of points on the tectum opticum produces various movements of the fins, suggesting a system of control similar to that exercised over spinal centres by the cerebral cortex of mammals through the pyramidal tract. Various forced movements follow injury to the midbrain.

The cerebellum is a very large organ in elasmobranchs, as in all animals that move freely in space. Its main source of sensory fibres is from the ear and from the organs of the lateral line system, whose afferent fibres enter through the seventh, ninth, and tenth cranial nerves. The internal structure of the cerebellum is very uniform and essentially similar in all vertebrates. Removal of portions of it from dogfishes produces aberrations of swimming.

The medulla oblongata is the region from which most of the cranial nerves spring and especially those that regulate the respiration and visceral functions. In mammals this control is indirect, but in fishes the nerves that spring from the medulla directly innervate the respiratory muscles of the gills and floor of the mouth. It is no doubt for this reason that the centre for the initiation of the respiratory rhythm developed in the medulla.

16. Receptor-organs of elasmobranchs

The paired nasal sacs have much-folded walls. Water enters by a single opening but this may be partly divided by a fold, making a groove, which may open to the mouth. There are taste-buds scattered over the wall of the pharynx. It has been shown experimentally that, as in higher animals, these are receptors for sampling the food after it has been brought close to the animal, whereas the nose acts as a distance receptor. Smell and taste are therefore different senses for a dogfish, as for us. By training fishes to discriminate between various substances it can be shown that those that we should smell are detected by the nose in the dogfish, but its organs of taste, like ours, can discriminate only between a few qualities, including salt, sour, and bitter.

The eyes are well developed in sharks and no doubt serve as an important means of finding the prey and avoiding enemies. However, the retina usually contains only rods, and visual discrimination is probably poor, but there are cones in *Mustelus* and *Myliobatis*. Unfortunately details as to the functional performance of the eyes, ability to discriminate shapes, &c., are scanty. Behind the retina there is often a reflecting layer, the tapetum lucidum. This may be provided with pigment cells, which expand in the light but contract in darkness, allowing the underlying guanophores to reflect, thus increasing sensitivity. The lens is spherical and very hard, as in all fishes, since it must perform the whole work of refraction. It is provided with a protractor-lentis muscle, presumed to produce active accommodation for near vision by swinging the lens forward. The iris is peculiar in those elasmobranchs that hunt by day; when it narrows it divides the pupil into two slits by the descent of an upper flap or operculum. The muscles of the iris are better developed in elasmobranchs than in most bony fishes and the pupil makes wide excursions. The sphincter iridis muscle, which narrows the pupil, works as an independent effector. It is stimulated to contract by light, but its movements are not controlled by any nervous mechanism. The radial dilatator fibres, which open the pupil, receive motor-fibres from the oculomotor nerve. The closure of the iris when illuminated is relatively slow. If the whole eye is cut out from the head, in the dark, the sphincter, being an independent effector, still closes when illuminated. The muscle, being without nerves, is not affected by any of the usual drugs that mimic action of the autonomic nervous system, though some of these affect the innervated dilatator muscle. We have therefore the curious situation that no 'autonomic' drugs applied to the isolated dark adapted eye cause closure of the pupil; this can only be produced by illumination (Fig. 108).

The ear of elasmobranchs contains receptors concerned (1) with maintenance of muscle tone, (2) with angular accelerations, (3) with

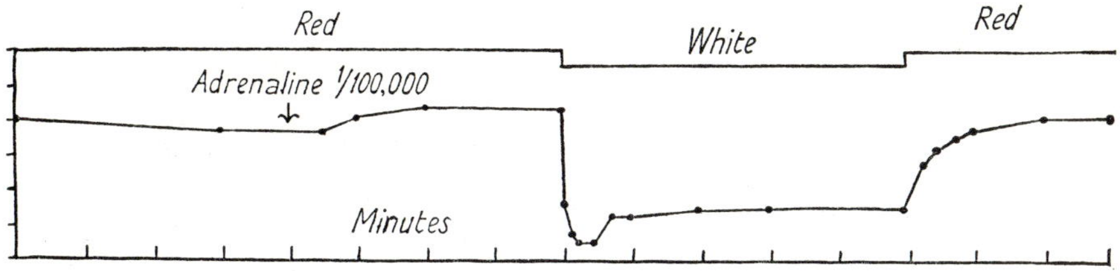

FIG. 108. Movements of margin of pupil of an isolated iris of the shark *Mustelus*, followed by plotting with a camera lucida and here shown magnified 53 ×. Addition of adrenaline causes slight dilation of the already dilated pupil and illumination then causes closure. Acetyl choline even in concentrations of 1 in 10,000 has a similar *dilatory* effect. (From Young, *Proc. Roy. Soc.* B. **112**.)

gravity, (4) perhaps with hearing. There are three pairs of semicircular canals, each with an ampulla containing receptor cells, whose hairs are embedded in a gelatinous cupula. This behaves as a highly damped torsion pendulum, swinging with movement of the fluid. These receptors discharge impulses continuously and during angular rotations the frequency is either increased or decreased in the appropriate ampullae, initiating compensatory movements of the eyes and fins.

The otolith organs include three patches of receptor cells in partially distinct sacs, the utricle, saccule, and lagena. The endolymphatic duct is an open canal and in some species serves to admit sand grains, which are attached to the maculae as gravity receptors. The utricle seems to be the main receptor producing appropriate postures in relation to gravity. The lagena shows a maximum discharge rate near the normal position of the head and thus serves as an 'into level' receptor. The areas of these maculae that carry otoliths do not respond to vibrational stimuli but carry only gravitational receptors. Vibration responses in the form of nerve impulses have been seen in rays but only up to 120 c/sec, although vestibular microphonics up to 750 c/sec occur. At high intensity there is much synchronization of units. These results suggest that the ear may function as a vibration receptor, but there are no conditioning experiments to show whether these fishes can hear.

There is a well-developed system of lateral line organs, whose function is considered later (p. 218). The organs of this system on the head are highly modified in elasmobranchs to form the ampullae of Lorenzini, long canals filled with mucus. Sand showed that these

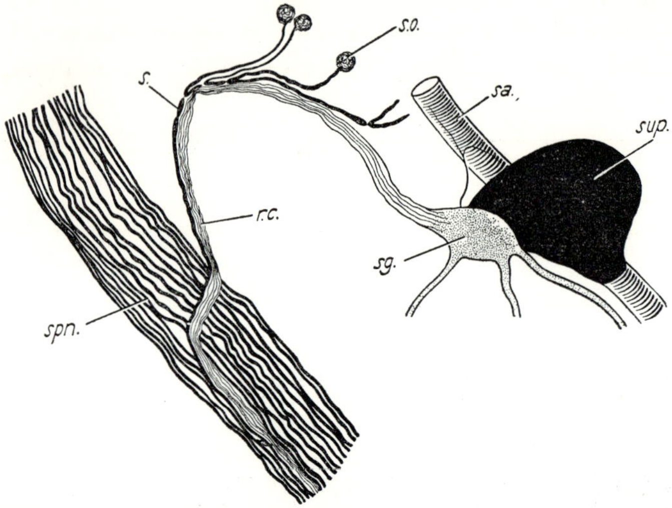

FIG. 109. Drawing of a sympathetic ganglion and related structures in a dogfish. Lettering as in Figs. 104 and 105. *s.o.* sense-organs. (After Young, *Quart. J. Micr. Sci.* **75**.)

organs increase their discharge of nerve impulses with very slight falls of temperature, and he suggested that their function is to detect such changes. They are also sensitive to weak tactile stimulation and to small voltage gradients in the water. Their function therefore remains uncertain. It may be related to determining changes of hydrodynamic pressure distribution over the surface of the aerofoil-like body, especially in skates and rays. They may thus act as mechano-receptors detecting local changes of pressure near the body surface.

No doubt elasmobranchs, like other animals, have many senses referred to the skin, such as we call touch, pain, and the like, but few studies of these exist. Sand has shown the presence of volleys of impulses in the nerves connected with muscles when the latter are stretched. Proprioceptors have been demonstrated histologically in the muscles of *Raja*. This agrees with the fact that after severance of the spinal cord the swimming rhythm only continues if some afferent nerves are intact.

17. Autonomic nervous system

The sympathetic system of elasmobranchs consists of an irregular series of ganglia, approximately segmental, lying dorsal to the posterior cardinal sinus and extending back above the kidneys. These ganglia contain motor nerve-cells (post-ganglionic cells) whose ascons end in the smooth muscles either of the arterial walls or of the viscera. The cells themselves are controlled by pre-ganglionic nerve-fibres whose cell bodies lie in the spinal cord and whose processes run out in the ventral spinal roots and rami communicantes (Fig. 109). In higher animals the sympathetic ganglia send post-ganglionic fibres back to the spinal nerves for distribution to the skin ('grey rami communicantes') but these are absent in elasmobranchs and correspondingly there is no evidence of sympathetic control of skin functions (e.g. chromatophores); a very different condition is found in bony fishes (p. 222). Another peculiarity of the sympathetic system of elasmobranchs is that it does not extend into the head. This condition is unique among vertebrates, but it is not clear whether it is primary or the result of a secondary loss.

In mammals it is usual to recognize a parasympathetic system acting in antagonism to

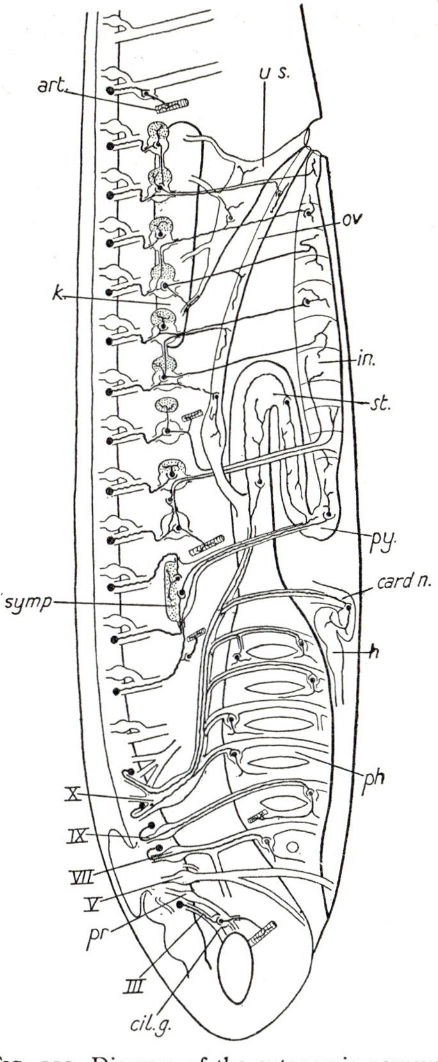

FIG. 110. Diagram of the autonomic nervous system of the dogfish.

art. artery; *card.n.* cardiac nerve; *cil.g.* ciliary ganglion; *h.* heart; *in.* intestine; *k.* mesonephros; *ov.* oviduct; *ph.* pharynx; *pr.* profundus nerve; *py.* *pylorus*; *st.* stomach; *symp.* sympathetic ganglion (with suprarenal near it); *u.s.* urinogenital sinus; *III, V, VII, IX, X,* cranial nerves. (From Young, *Quart. J. Micr. Sci.* **75**.)

the sympathetic, but this is not easy to define in the elasmobranchs (Fig. 110). The vagus, it is true, is well developed, with branches to the heart and gut, but little is known of autonomic fibres in the other cranial nerves, or of a special 'sacral' parasympathetic system. Stimulation of either the vagus or the sympathetic nerves causes contraction of the stomach. A ciliary ganglion connected with the oculomotor nerve is present as in other animals, but there is no sense in which it can be called antagonistic to the sympathetic system, since the latter does not extend into the head. The post-branchial plexus is a network of fibres and cells connected with the vagus but stretching back above the posterior cardinal sinus (Fig. 104). Receptors in this plexus and in the afferent branchials (Fig. 109) may be concerned with vascular reflexes (p. 161).

VI

EVOLUTION AND ADAPTIVE RADIATION OF ELASMOBRANCHS

1. Characteristics of elasmobranchs

THE organization of a shark used to be considered to show the earlier stages of fish evolution, but we have seen evidence that this is a mistake (p. 131). The sharks and skates and rays are highly developed creatures; in particular, the absence of bone is a secondary feature; they have been able to give up their defensive armour because of the development of other means of protection, swift swimming, good sense-organs and brain, and powerful jaws. We can now examine the history of these changes and study the varied creatures that can be classified as elasmobranchs. As usual in examining such histories we must try to discover evidence about the forces that have operated to produce the changes of type, and look for signs of any consistent trends, persisting for long periods of years.

2. Classification

Superclass Gnathostomata
Class Elasmobranchii (= Chondrichthyes)
Subclass 1. Selachii
*Order 1. Cladoselachii. Devonian–Permian
**Cladoselache; *Goodrichia*
*Order 2. Pleuracanthodii. Devonian–Trias
**Pleuracanthus*
Order 3. Protoselachii. Devonian–Recent
**Hybodus; Heterodontus*
Order 4. Euselachii. Jurassic–Recent
Suborder 1. Pleurotremata. Jurassic–Recent
Division 1. Notidanoidea. Jurassic–Recent
Hexanchus; *Chlamydoselache*
Division 2. Galeoidea. Jurassic–Recent
Scyliorhinus; Mustelus; Cetorhinus; Carcharodon
Division 3. Squaloidea. Jurassic–Recent
Squalus; Squatina; Pristiophorus; Alopias
Suborder 2. Hypotremata. Jurassic–Recent
Raja; Rhinobatis; Pristis; Torpedo; Trygon

Superclass Gnathostomata (*cont.*)

Subclass 2. Bradyodonti. Devonian–Recent

*Order 1. Eubradyodonti. Devonian–Permian

**Helodus*

Order 2. Holocephali. Jurassic–Recent

Chimaera

The elasmobranchs form a very compact group of fishes, nearly always marine and of predaceous habit, having a great quantity of urea in the blood, with no bone in the skeleton, no operculum over the gills, and no air-bladder. The tail is usually heterocercal. The pectoral fin is anterior to the pelvic and the latter is usually provided with claspers, fertilization being internal. The body is more or less completely covered with placoid scales (denticles) and these are specialized in the mouth to form rows of teeth. The intestine is short and provided with a spiral valve. The typical cartilage-fishes with these characters may be placed in the subclass Selachii, to distinguish them from an early aberrant offshoot the Bradyodonti, represented today by the peculiar creature *Chimaera* (p. 184).

3. Palaeozoic elasmobranchs

The selachians are among the most numerous of the various predatory animals in the sea. There have, however, been many side-branches of the main shark line and we may now survey the history of the group from its first appearance. The characters we have used in our definition mark the elasmobranchs off from the earliest-known gnathostomes, the acanthodians and other placoderm types (Fig. 111), which we shall consider later (p. 186). Presumably the elasmobranchs were derived from some placoderm, but the earliest evidence of the existence of true sharks is in the form of isolated teeth and scales from Middle Devonian deposits, and the earliest type about which full information exists is **Diademodus* from the Upper Devonian, 'an early and not distant offshoot from the primitive Chondrichthyan stock, the main line of which led through **Ctenacanthus* and the hybodonts to the modern elasmobranchs; **Cladoselache* is a specialized side-line of this main stock and is not an appropriate ancestral type for the Chondrichthyes' (Harris). The teeth of **Diademodus* are many-cusped and resemble the scales more closely in sculpturing than in other primitive sharks. The jaw suspension was amphistylic and the notochord unconstricted. The pectoral fin was continuous posteriorly with the body wall and there was no well-developed

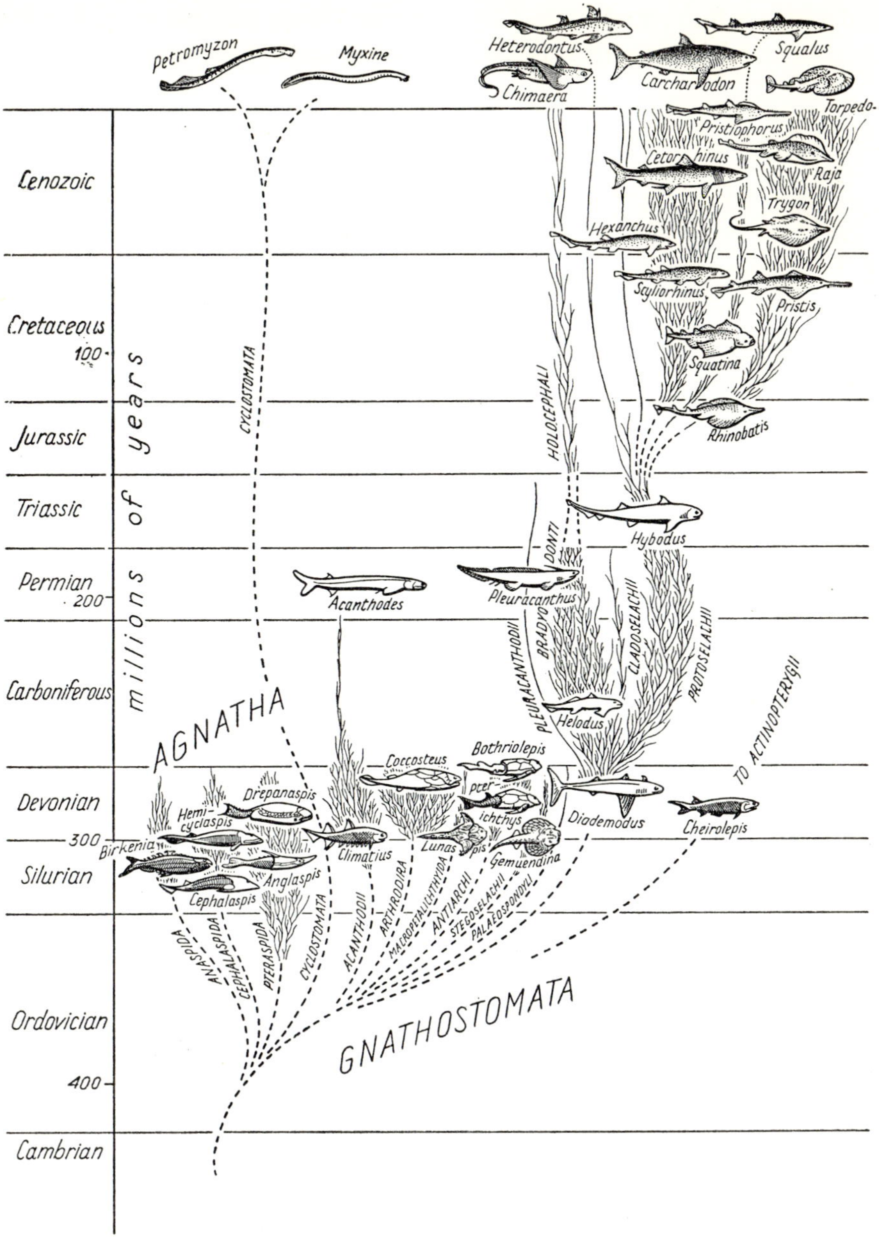

FIG. 111. The early evolution of vertebrates.

pectoral girdle. The tail was heterocercal and there are no signs of skeletal support for lateral keels. All of these Harris regards as primitive features; **Diademodus* was specialized in having no spines in front of the dorsal fin and no clasper on the head. Both of these features are frequent in hybodonts and in **Cladoselache* there is a large spine

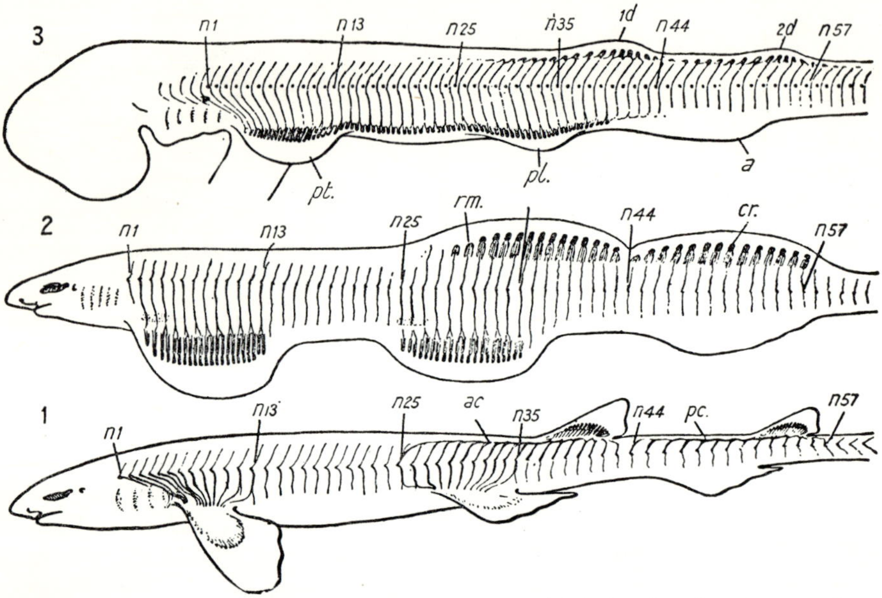

FIG. 112. Development of the fins of the dogfish. 1, Adult showing the nerve-supply of the fins; 2, adult with the fins shown expanded and their nerves and muscles shown as if concentration had not taken place; 3, a 19-mm. embryo, showing the actual condition.

a. anal fin; *ac.* anterior collector nerve of first dorsal fin; *cr.* (black) cartilaginous radial partially hidden by the radial muscle; *n.* 1–57, spinal nerves and ganglia; *pc.* collector nerve of second dorsal fin; *pl.* pelvic fin; *pt.* pectoral fin; *rm.* radial muscle; *1d.* and *2d.* first and second dorsal fins. (From Goodrich, *Vertebrata*, A. & C. Black, Ltd.)

in front of the first dorsal fin. These fishes were thus like modern sharks in their general form, but the fins were remarkable in having a broad base, not sharply marked off from the body-wall. It has been suggested by Goodrich and others that this was the earliest condition of the pectoral fin, perhaps showing its derivation from a continuous or extended fin-fold (Fig. 113). This theory has the advantage that it agrees with the embryological development of the fin by concentration of a series of segments (Fig. 112). It also seems likely that anterior and posterior fins expanded in the horizontal plane would be necessary for stabilization (p. 136). Moreover, this theory of the origin of paired fins has the great advantage that it compares them with the median fins, which are also continuous folds. It has been argued, however, that the cladoselachians are very far from the earliest known fishes and that in both ostracoderms (p. 125) and placoderms (p. 186) fins are known that have a narrowly constricted base. We cannot yet say for certain what has been the course of evolution of the paired fins, but the fin-fold theory has much plausibility, in spite of the difficulties raised by palaeontologists.

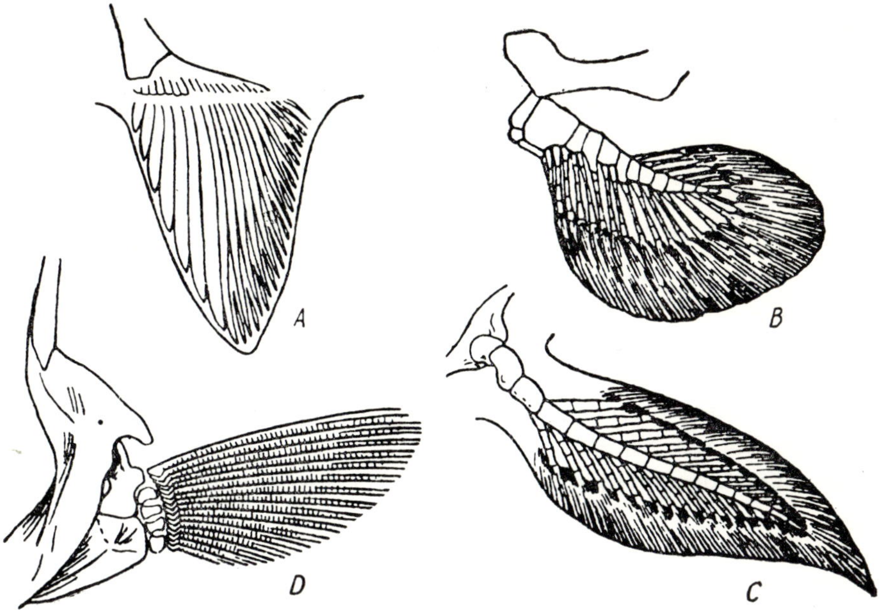

FIG. 113. Pectoral fins of various fishes.
A, **Cladoselache*; B, **Pleuracanthus*; C, *Neoceratodus*; D, *Gadus*. (From Norman.)

The cladoselachians represent the ancestral Devonian sharks, from which all later forms have been derived. Animals of similar type were fairly common in late Devonian and Carboniferous seas. The ctenacanths, such as **Goodrichia*, reached a length of 8 ft. Later radiation of the selachians took place along three different lines, represented by the three remaining orders shown in the classification. The pleuracanthodians (**Pleuracanthus*) were a specialized group of freshwater carnivores. The tail was straight (diphycercal) and the paired fins had become modified accordingly (see p. 137). The axis was completely freed from the body wall to give a paddle-like fin, with pre- and post-axial rays, a type known as archipterygial (Fig. 113), because it was once supposed to be ancestral to all others. A large spine on the head gives the group its name. Claspers were present. These animals were common in the Carboniferous and Lower

Permian, but in subsequent times they disappeared without leaving descendants.

4. Mesozoic sharks

After flourishing in Palaeozoic seas the shark line seems to have become nearly extinct during the Permian and Trias. During this period there was probably little fish life in the sea and the stock seems only to have survived by adopting a varied diet, including invertebrate food. The protoselachian or heterodont sharks of this period had two types of tooth, pointed ones in front and flattened ones, for crushing molluscs, behind. *Heterodontus*, the Port Jackson shark of the Pacific, is a surviving form having a dentition of this type.

There is total cleavage of the yolk of the egg. The meroblastic form typical of modern elasmobranchs and teleosts was therefore a relatively late development and other survivors of the mesozoic period besides *Heterodontus* also show holoblastic cleavage (pp. 184–236). In later Triassic times sharks again became more abundant, and this agrees with the presence of numerous bony fish types, on which they presumably fed. Some of the Triassic sharks still possessed a heterodont dentition (**Hybodus*), though otherwise much like the modern forms.

In Jurassic times or earlier, however, the sharks divided into the main lines that exist today. In the suborder Pleurotremata or true sharks the teeth all became sharp and the animals swift swimmers. In the suborder Hypotremata, on the other hand, the teeth remained flattened and sometimes became highly specialized for a mollusc-eating diet (Fig. 114), producing the flattened bottom-living creatures, the skates and rays. The stages of this transition can be followed, and some of the intermediate types still exist. Thus in *Rhinobatis*, the banjo-ray (Fig. 111), the pectoral fins are enlarged but still distinct from the body. Almost identical creatures have been found in Jurassic rocks. It is probable that several separate lines showed this flattening of the body.

5. Modern sharks

The Pleurotremata may be divided into three divisions all dating from the Jurassic. The Notidanoidea show many primitive features, such as an amphistylic jaw, the presence of six or seven gill-slits, and an unconstricted notochord. *Hexanchus* and *Heptranchias*, are long-bodied, slow-moving sharks from warm waters. They are viviparous but without placentae. *Chlamydoselache*, the frilled shark, lives in deep water and feeds on cephalopods. The division Galeoidea is much larger and includes the sharks with two dorsal fins, not supported by spines. Here belong the dogfishes *Scyliorhinus* and *Mustelus*, both mainly bottom-living animals feeding on a mixed diet, including

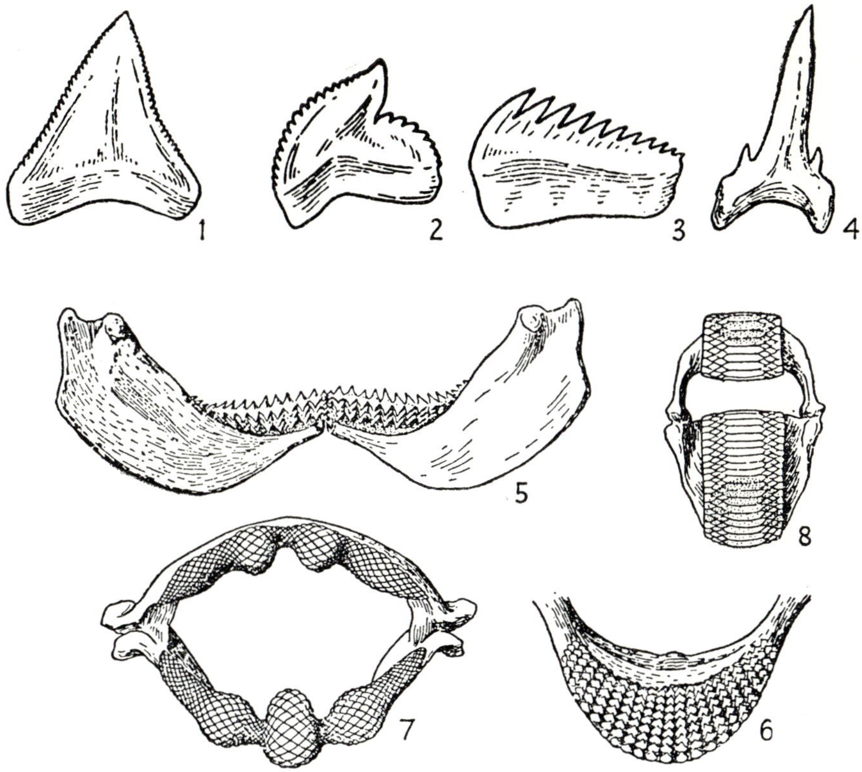

FIG. 114. Teeth of various elasmobranch fishes.

1, Man-eater (*Carcharodon*); 2, tiger shark (*Galaeocerdo*); 3, comb-toothed shark (*Hexanchus*); 4, sand-shark (*Odontaspis*); 5, blue shark (*Carcharinus*); 6, nurse shark (*Ginglymostoma*); 7, guitar fish (*Rhina*), 8, eagle-ray (*Myliobatis*), (After Norman.)

crustaceans and molluscs. In *Cetorhinus*, the basking shark, the predaceous habit of the group has been abandoned in favour of straining small food directly from the plankton by means of special combs on the gills (gill rakers), an arrangement recalling that of the whalebone whales. The great effectiveness of this method of feeding may be seen in the length of 35 ft or more attained by some of these sharks. Basking sharks produce very numerous small eggs, which develop within an 'uterus', but without placentae. *Rhineodon*, the whale shark, is also a plankton feeder and becomes very large. It is not closely related to the basking sharks. It moves up and down vertically, the mouth open, sucking in plankton. In this group there are also many of the fiercest man-eating sharks, such as *Carcharodon*, often 30 ft long, found in many seas. Some fossil forms of this genus are estimated to have reached a much greater length, possibly of 90 ft.

The division Squaloidea includes those sharks in which there is a spine in front of each dorsal fin. They are not, however, otherwise different in habits from the other sharks. The spiny dogfish (*Squalus*) is a well-known type and here belong also the saw-sharks (*Pristiophorus*) and a group of bottom-living forms, the angel-fishes or monks (*Squatina*), which acquire a superficial similarity to the skates and rays. *Alopias*, the thresher, is said to differ from most sharks in that instead of seizing the prey as it is presented, it hunts systematically, several sharks working together and using their whip-like tails to drive smaller fishes such as mackerel into shoals, where they are then seized.

6. Skates and rays

The Hypotremata, skates and rays, have become specialized for life on the bottom of the ocean in shallow waters, feeding mainly on invertebrates, and usually having blunt teeth (Fig. 114). Locomotion is no longer by transverse movements of the body but by waves that pass backwards along the fins. In the earlier stages, such as *Rhinobatis*, the banjo-ray, which has existed from the Jurassic period to the present, the edges of the fins are still free and the tail is well developed. In *Pristis*, another saw-fish type, outwardly similar to *Pristiophorus* and known since the Cretaceous, the head is drawn out into a long rostrum armed with denticles. Its use is uncertain but the head strikes from side to side among shoals of fishes. There are species in India, China, and the Gulf of Mexico that live in fresh water. In *Raja*, first found in the Cretaceous, the pectoral fins are attached to the sides of the body and the median fins are very small, whereas in the more recent *Trygon* and other sting-rays the tail is reduced to a defensive lash, the dorsal fin persisting as a poison spine. In the eagle-rays (*Myliobatis*) the teeth are flattened to form a mill able to grind mollusc shells (Fig. 114). The sea-devils (*Mobula*) have expansions of the fins at the front of the head, which they use to chase fishes to the mouth, hunting in packs. In *Torpedo*, the electric ray, the fins extend so far forward that the front of the animal presents a rounded outline. The animal is protected by a powerful electric organ, formed by modified

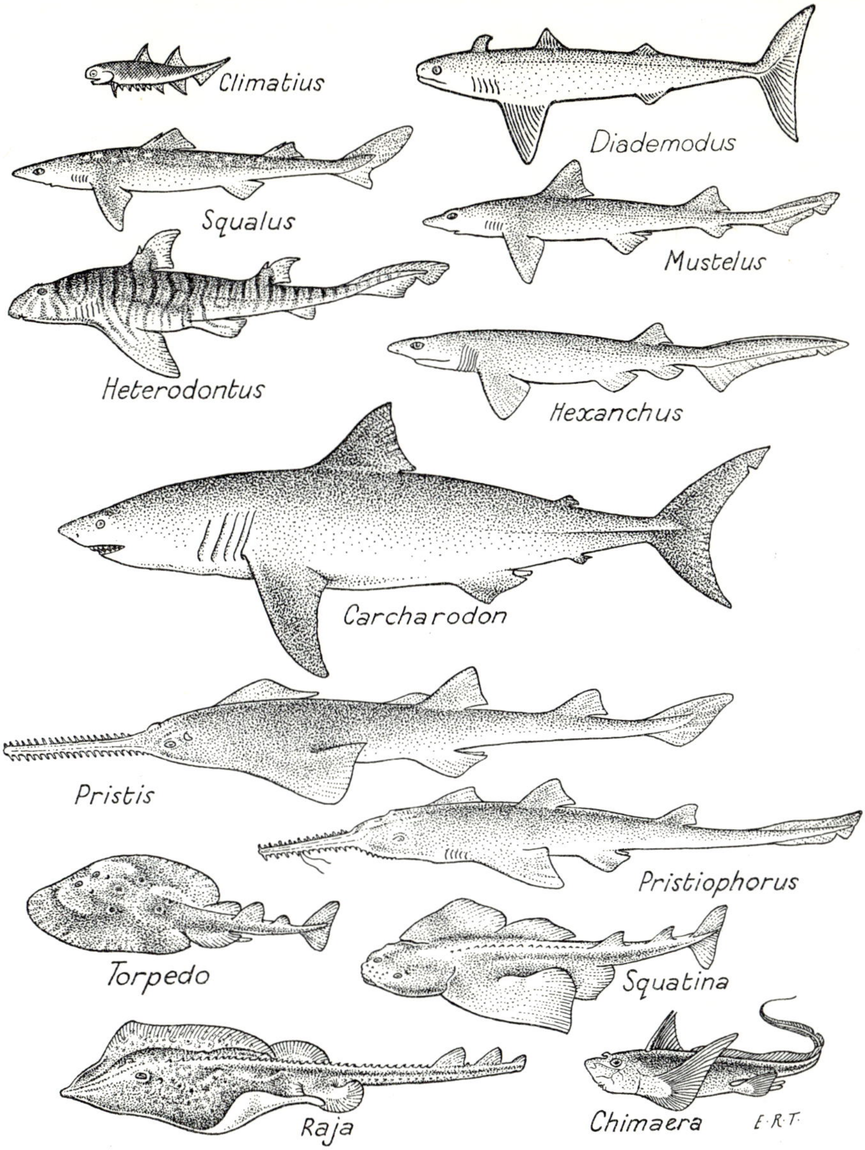

FIG. 115. Various elasmobranch fishes.

lateral plate muscle, innervated by cranial nerves. Several species of *Raja* have weak electric organs perhaps used for guidauce (p. 253).

Life on the bottom has produced many further modifications in the skates and rays. In those that live in shallow and hence well-illuminated waters the colour of the upper surface is often elaborate, the under side being white. In certain species of *Raja*, for example, there is a pattern of black and white marks, which probably serves to break up the outline of the fish.

The eyes of the skates and rays have moved on to the upper surface of the head and are protected by well-developed lids. In most forms the pupil is able to vary widely in diameter and often has an operculum by which the aperture can be reduced to two small slits.

There is a special modification of the respiratory system so that water is drawn in not through the mouth but by the spiracle, which is provided with a special valve that shuts at expiration, as the water is forced out over the gills. The Hypotremata have therefore developed many special features for their bottom-living habits and have diverged among themselves into many varied lines. They have been very successful and are among the commonest fishes in the sea.

7. Chimaera and the bradyodonts

Finally we must consider an aberrant group, the bradyodonts, which diverged from the main stock at least as early as the Carboniferous and preserves for us today some features of elasmobranch life at that time as the strange *Chimaera*, the rat-fish of deep seas (Fig. 115). Instead of the usual large, toothed mouth these Holocephali have a small aperture surrounded by lips, giving the head a parrot-like appearance. The teeth are large plates firmly attached to jaws, and the upper jaw is remarkable in being fused to the skull ('holostylic'), the hyoid arch being free. There is no stomach or spiral intestine. These peculiarities are probably associated with a capacity to eat small pieces of animal food. The Holocephali differ further from the Selachii in the presence of an opercular flap attached to the hyoid arch. There are also extra claspers in front of the usual pelvic ones and an organ on the head of the male known as the cephalic clasper, whose function is obscure. The notochord is unconstricted and the vertebrae reduced to separate nodules. The cleavage is holoblastic, as in other fishes with features of mesozoic type (p. 180).

Many of the internal features resemble those of selachians, for instance the conus arteriosus, and urinogenitals in which there are separate urinary and spermatic ducts. The brain has a peculiar shape on account of the large size of the eyes, which almost meet above the brain, so that the diencephalon is long and thin.

These strange creatures appear in the Jurassic, apparently descended from the somewhat similar bradyodonts (such as **Helodus*), which were common in the Carboniferous and Permian. They preserve some primitive features (vertebrae, jaw support, open lateral line canals, cleavage) but have developed many specializations in the teeth, operculum, fins, and brain, probably in connexion with life on the bottom of deep seas.

8. Tendencies in elasmobranch evolution

The elasmobranchs have been in existence ever since the Devonian, and for much of this long period of nearly 400 million years we can follow their changes with some accuracy. This type of fish was first formed by loss of the heavy bony armour of the earliest gnathostomes, associated with the adoption of a rapidly moving and carnivorous habit. The resulting shark-like form has remained with relatively little change through the whole history of the group; cladoselachians from the Devonian are remarkably like modern sharks, and it would be difficult to assert that the latter show clear signs of being in any way of a 'higher' type. Both are in fact suited to the same mode of life.

If our interpretation of the evidence is right, however, the modern shark type has been evolved from the Devonian type through a heterodont stage. During the late Permian and Trias there was little fish food for the sharks and they appear to have taken to living on invertebrates. Eating this diet was presumably easier for animals possessing the two types of teeth described on p. 180, and the animals also became rather flattened with their life on the bottom. On the reappearance of numerous fishes in the sea, in the Jurassic, some of these heterodonts resumed the shark-like habit, lost the crushing teeth, and developed into the varied fish-eating types alive today. Others of the heterodonts, however, became still more specialized for bottom life, as the modern skates and rays.

It is difficult to see any persistent tendency in all this, except to eat other animals of some sort. When fishes are available sharks will eat them, and the bodily organization for doing so seems to have been evolved at least twice. Similarly other members of the same stock ate molluscs and crustacea and became modified for this. The tendency is for survival or continuance of the animals and this leads them to adopt whatever habits are possible given their surroundings. In meeting the circumstances certain types will be suitable at one time, others at another. We know that genetic variations will produce fluctuations of type—at a time when circumstances force the animals to strive in one direction those with a particular bodily type, say, broad, 'heterodont' teeth will be selected. When fish food again becomes available those animals born with quicker habits and sharper teeth will be able to eat the fish and the shark type returns.

The method of ensuring stability in the pitching plane adopted by elasmobranchs (p. 136) necessitates a certain flattening of the front end of the animal. It is not therefore surprising that this tendency is often exaggerated and has several times produced flattened bottom-living creatures, such as the skates and rays. The Actinopterygii show the opposite tendency, to lateral flattening (p. 248). We might imagine that most of the modern skates and rays had become so modified in structure that only life on the bottom is possible for them and that there could be no return to a free-swimming, fish-eating habit, but it would not be true to say that this is certain or that the past history of the group shows undoubted evidence of such irreversible specialization.

The only general conclusion from our study of elasmobranchs since the Devonian, then, is that they have tended to keep alive by eating fish or invertebrates, that some have changed little during this time but that, judging especially from the modern forms, the group tends to produce varied types at any one time, each able to find its food in a special manner. It is not clear that the group has advanced, in any sense, since the Devonian. The type has always been a successful one, able to produce specialized carnivores. We do not know enough to be sure whether the number of creatures with this organization has changed greatly, but it seems that, except for a reduction in numbers in the Triassic, they have always been moderately abundant and are perhaps at present on the increase.

9. The earliest Gnathostomes, Placoderms

*Class Placodermi (= Aphetohyoidea)

- *Order 1. Acanthodii. Silurian–Permian (**Climatius*)
- *Order 2. Arthrodira. Silurian–Devonian (**Coccosteus*)
- *Order 3. Macropetalichthyida. Devonian (**Lunaspis*)
- *Order 4. Antiarchi (= Pterichthyomorphi). Devonian (**Bothriolepis*)
- *Order 5. Stegoselachii. Devonian–Carboniferous (**Gemundina*)
- *Order 6. Palaeospondyli. Devonian (**Palaeospondylus*)

It has already been mentioned that the earliest gnathostome vertebrates found in the rocks do not have the shark-like form, and present so many peculiarities that they are placed in a distinct class. In the past the fossils included here have been referred to various groups, usually either to the agnatha or the elasmobranchs, and there is still some doubt as to their position. In many respects they are highly specialized, but they all have one feature that may be presumed to have existed in the ancestral gnathostome, namely, that the hyoid arch played no part in the support of the jaws and the spiracle was therefore a typical gill-slit. For this reason they are often given the name Aphetohyoidea, but we shall prefer to call them Placodermi, to emphasize that they all have a heavy armour of bone-like material. The class contains several orders, not obviously very closely related to each other; all are fossil forms, none of which is known to have survived the Permian.

The best-known, earliest, and perhaps most interesting group is the acanthodians, found in freshwater deposits extending from the Upper Silurian to the Permian but chiefly in the Devonian. These were small fishes with a fusiform body (Fig. 115), with heterocercal tail and two, or later one, dorsal fins. The lateral fins consisted of a series of pairs, often as many as seven in all, down the sides of the body. The effect of these in stabilizing the fish would presumably be different from that of a continuous fold, and the problem of the form and function of the earliest paired fins remains obscure. The fins were all supported by the large spines from which the group derives its name.

The whole surface of the body was covered with a layer of small rhomboidal scales, composed of layers of material ressembling bone, covered with a shiny material similar to the ganoin of early Actinopterygii. On the head these scales were enlarged to make a definite pattern of dermal bones, numerous at first but fewer in the later forms. The pattern of the bones has no close similarity to that of later fishes. The reduced bones of the later acanthodians are related to the lateral line canals, which have an arrangement similar to that in other fishes, but run between and not through the scales and bones of the head. The teeth are formed as a series of modified scales. The skull is partly ossified—important evidence that the boneless condition of elasmobranchs was not typical of all early gnathostomes.

The jaws of acanthodians were attached by their own processes to the skull (autodiastyly) and are remarkable in that four separate ossifications take place in them (two in the upper and two in the lower jaw), making a series of elements similar to that found in the typical branchial arches. The hyoid was an unmodified branchial arch. At first the mandibular, hyoid, and each of the branchial arches were provided with small flap-like opercula, but in later forms the mandibular operculum became especially developed and covered all the gills.

These animals might well represent the ancestors of many if not all other groups of gnathostomes. They have not the peculiar features that we characterize as shark-like, and though they may well have been carnivorous they are not very highly specialized for that mode of life. Whether or not the known acanthodians represent the actual ancestors of the other gnasthostome groups, it is clear that knowledge of their anatomy forces us to discard two conclusions which have often been accepted in the past, namely, that lack of bone and an amphistylic jaw support are primitive gnathostome features. Here already in the Silurian we find animals that possessed both endochondral bone and scales composed of bony substance. Moreover, some of them have no trace of denticles and we must therefore regard with suspicion any theory that considers the placoid scale as the original type of all scales. It is at least as likely that scales composed of simple layers of bone in the dermis were the ancestral type and that placoid forms with a pulp cavity were a later specialization.

Several other types of placoderm fish are known, mostly from the Devonian strata. The Arthrodira, Macropetalichthyida, and Antiarchi (Fig. 111) were mostly heavily armoured fishes with dermal bones on the head and often a large shield over the body. There was usually a heterocercal tail and a covering of scales. The earlier fishes were mostly from fresh water, the later from the sea. Many were rather flattened, probably bottom-living and invertebrate-eating forms. The bony plates on the head were often arranged in characteristic patterns, none of which, however, shows close similarity to the pattern of bones on the head of bony fishes or tetrapods. Lateral line canals of typical arrangement were present and the 'bones' follow these to some extent.

**Gemundina* was a flattened animal, superficially similar to a skate, from marine Lower Devonian deposits. The skin was covered with denticles, but under these were large plates, apparently of bone. This fish is placed in a special order Stegoselachii and its affinities are unknown, but it shows again that the tendency to develop a flattened form has been present from the earliest appearance of fishes. **Palaeospondylus* from the Devonian is another isolated form, in the past often classed with the cyclostomes. Moy-Thomas showed, however, that jaws were present and that probably the hyomandibula was not suspensory. He therefore classed the fish with the placoderms, in spite of the absence of any dermal skeleton. So far as can be discovered, all these placoderm fishes except the acanthodians were specialized types and have not left any later descendants. Indeed it may well be that they have been preserved only because of the great extent of their armour; less heavily protected relatives may have existed but have not survived as fossils. The remains that are known are sufficient to establish the fact that there were, in the Devonian period, numerous types of fish possessing a bony skeleton.

VII

THE MASTERY OF THE WATER. BONY FISHES

1. Introduction: the success of the bony fishes

THE acanthodians and some other of the late Silurian and Devonian gnathostome fishes possessed bony skeletons; from these, or some placoderm animals like them, may have been derived not only the elasmobranchs but also the bony fishes and the lung-fishes, which gave rise to the land animals. These presumed descendants of the placoderms can be divided into three groups: first the elasmobranchs, secondly, the crossopterygians, the lobed-fin or lung-fishes, including the Devonian forms that led to the amphibia, and thirdly, the actinopterygian or rayed-fin fishes, culminating in the modern bony fishes. In Devonian times the Crossopterygii and Actinopterygii were very alike and both, like the placoderms, contained bone. The term bony fishes or Osteichthyes is often applied to these two groups together, since they have some features in common and distinct from the elasmobranchs.

The great group of Actinopterygii, which, for all the importance of the elasmobranchs, must be reckoned as the dominant fish type at the present time, includes most of our familiar fishes, perch, pike, trout, herring, and many other types of 'modern' fish. In addition there are placed here some surviving relics of the stages that have been passed before reaching this condition, such as the bichir, sturgeons, bow-fin, as well as related fossil forms.

Many groups of animals have been successful in the water; crustacea, for instance, are very numerous and so are cephalopod molluscs and echinoderms, but the success of the bony fishes surpasses that of all others. From a roach or perch in a stream, to a huge tunny or a vast shoal of herrings in the sea, they all have the marks of mastery of the water. They can stay almost still, as if suspended, dart suddenly at their prey or away from danger. They can avoid their enemies by quick and subtle changes of colour. Elaborate eyes, ears, and chemical receptors give news of the surrounding world and complex behaviour has been evolved to meet many emergencies. Reproductive mechanisms may be very complex, involving elaborate nest-building and care of the young; social behaviour is shown in swarming movements, which may be accompanied by interchange of sounds (p. 217).

Bony fishes abound not only in the sea but also in fresh water, which has never been effectively colonized by cephalopods or elasmobranchs. They can exist under all sorts of unfavourable or foul-water conditions and a considerable number of them breathe air and live for a time on land. Perhaps the majority are carnivorous, but others feed on every type of food, from plankton to seaweeds.

To whatever feature of fish life we turn we find that the bony fishes excel in it in several different ways in different species. It is small wonder that with all these advantages they are excessively numerous. There are some 3,000 species of living elasmobranchs, but more than 20,000 species of bony fish have been described.

The number of individuals of some of the species must be really astronomical. For instance, at least 3,000 million herrings are caught in the Atlantic Ocean each year, so that the whole population there can hardly be less than a million million. Again, it is estimated that a thousand million blue-fish collect every summer off the Atlantic coast of the United States and, being very voracious carnivores, they consume at least a thousand million million of other fishes during the season of four months. This gives some idea of the tremendous productivity of the sea, and of the way the bony fishes have made use of it. Needless to say, man has also made considerable use of the bony fishes, which indeed provide, with the elasmobranchs, a not inconsiderable portion of the total of human food.

2. The trout

Salmo trutta, the brown trout, may be taken as an example of a bony fish; we shall also refer at intervals to conditions in other common freshwater fish such as the dace, *Leuciscus*, and perch (*Perca fluviatilis*). There is considerable confusion about the various types of trout and their close relatives the salmon. The brown trout is abundant in rivers and streams throughout Europe and is commonly about 20 cm long at maturity, though it may grow larger. It is grey above and yellowish below, with a number of dark spots scattered down the sides of the body (Fig. 116).

The body form is typical of that of teleostean fishes in being short, narrow in the lateral plane but deep dorso-ventrally, in fact more obviously streamlined than the shape of elasmobranchs. The movements of a trout do not at first sight obviously involve the bending of the body into an S; nevertheless, the method of swimming is essentially by the propagation of waves along the body by the serial contraction of the longitudinally directed fibres of the myotomes (p. 133).

The tail differs from that of elasmobranchs in being outwardly symmetrical, though internally there are still traces of the upturned tip of the vertebral column (Fig. 118). Besides the typical caudal 'fish-tail', supported by bony rays, there are two dorsal fins and a ventral fin, but the hinder dorsal fin differs from the others in having no rays to support it and is called an adipose fin, because of its flabby structure. The paired fins are rather small and it is from their structure that the whole group derives the name Actinopterygii or rayed-fin fishes. There is no lobe projecting from the body and containing

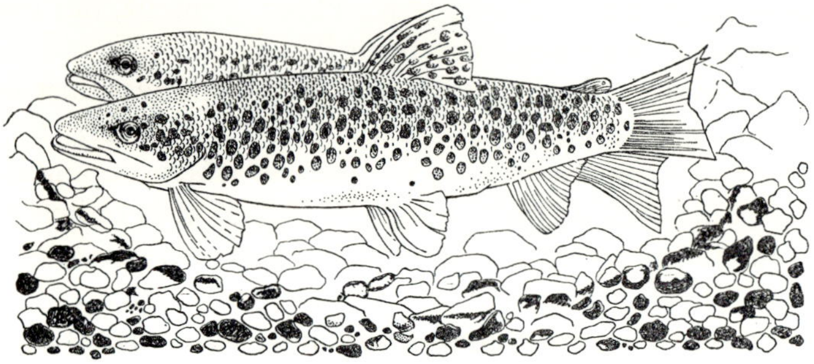

FIG. 116. Male and female brown trout (*Salmo trutta*) spawning. The male is quivering—a short sequence of rapid shudders of whole body which excites the female. (After J. W. Jones, *The Salmon*.)

basal fin supports, as there is in the fin of lung-fishes. All the basal apparatus of the fin is contained within the body wall and only the fin rays project outwards, as a fan. The pelvic fin of bony fishes often lies relatively far forward; in the trout, however, it is unusually far back, just in front of the anus; in other types it may be level with the pectoral fin, or even anterior to it (Fig. 118). The significance of the shape of the body and fins in swimming will be discussed later (p. 244).

The skin consists of a thin epidermis and thicker dermis, the former has stratified squamous layers but contains no keratin (Burgess, 1958). It contains mucous glands. The mucus of some eels and other fishes has remarkable powers of precipitating mud from turbid water. The mesodermal dermis provides an elaborate web of connective tissue fibres. It also contains smooth muscle, nerves, chromatophores, and scales. The latter are thin overlapping bony plates, covered by skin, that is to say, they do not 'cut the gum' as do placoid denticles. The exposed part of each scale bears the pigment cells, which control the colour of the animal, in a manner presently to be described. The bone of the scales is absorbed at intervals by scleroclasts, making a series of rings, which, like the growth-rings on a tree, are due to the fact that growth is not constant but occurs fast in the spring and summer and hardly at all in the winter. The age of the fish can therefore be determined from these rings (Fig. 117), or from the similar markings on the ear stones (p. 216). While an adult salmon is in fresh water no growth occurs, leaving a spawning mark on the scale.

The head of the trout shows some of the most specialized and typical teleostean features (Fig. 119). There are two nostrils on each side, but no external sign of ears. The mouth is very large and its edges are supported by movable bones, to be described below. The maxillary and mandibular valves are folds of the buccal mucosa, serving to prevent the exit of water during respiration. The tongue, as in Selachians, has no muscles, but may carry teeth and taste-buds. Behind the edge of the jaw is the operculum, a flap covering the gills and also supported by bony plates. In connexion with these special developments of jaws and gills the skull has become much modified and has developed complex and characteristic features (Fig. 118).

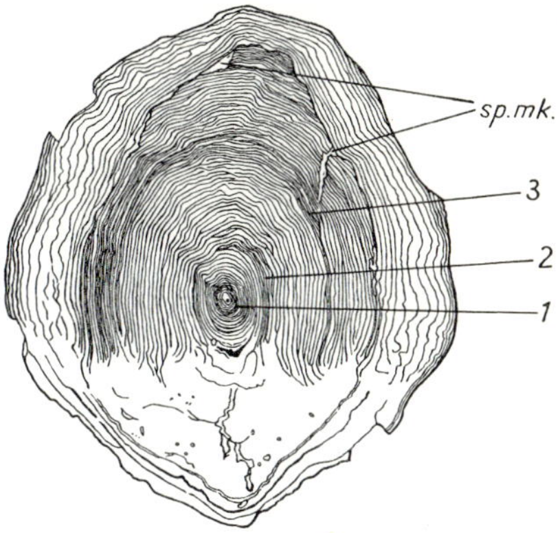

FIG. 117. Spawning mark (*sp. mk.*), the result of erosion or absorption of the scale margin due to a calcium deficiency fasting period. (After J. W. Jones, *The Salmon*.)

1. 1st river winter. 2. 2nd river winter. 3. 1st sea winter.

3. The skull of bony fishes

The main basis for the skull is a chondrocranium and set of branchial arches, exactly comparable to those of the elasmobranchs. In the early stages of development there is a set of cartilaginous boxes around the nasal and auditory capsules, brain and eyes, and a series of cartilaginous rods in the gill arches. Bones are then added in two ways: either (1) as cartilage bones (endochondral bones) by the replacement of some parts of the original chondrocranium, or (2) as membrane or dermal bones, laid down as more superficial coverings and considered to be derived from a layer of scales in the skin. This outer position of the bones can be clearly seen in many cases by the readiness with which the membrane bones can be pulled away from the rest of the skull.

The skull bones are arranged in a regular pattern, whose broad outlines can be seen in all fishes and in their tetrapod descendants. However, there are many confusing variations and the naming of

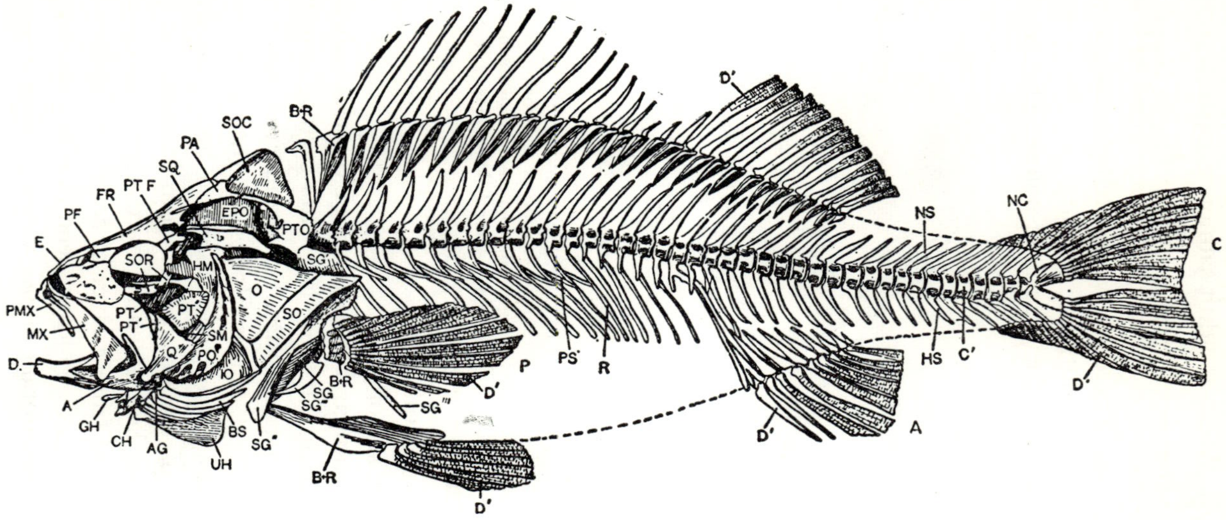

FIG. 118. Skeleton of the perch.

A. anal fin; AG. angular; B+R. basal and radial cartilages; BS. branchiostegal rays; C. caudal fin; CH. ceratohyal; D. dentary; D′. dermal rays of fins; E. ethmoid; EPO. epiotic; FR. frontal; GH. glossohyal (hypobranchial); HM. hyomandibula; HS. haemal spine; IO. interoperculum; MX. maxillary; NC. notochord; NS. neural spine; O. operculum; P. pectoral fin; PA. parietal; PF. prefrontal; PMX. premaxillary; PO. preoperculum; PS. supporting processes of ribs; PT, PT′, and PT″. ecto-, ento-, metapterygoid; PTF. post-frontal ; PTO. pterotic; Q. quadrate; R. ribs; SG. shoulder-girdle; SG′. dorsal process of SG. SG″; outer rim of SG; SG‴. posterior processes of SG.; SM. symplectic; SO. suboperculum; SOC. crest of supra-occipital; SOR. suborbital ring; SQ. squamosal, UH. urohyal. (After Zittel, from Dean, *Fishes, Living & Fossil*, The Macmillan Co.)

bones has given much controversy. No generally acceptable theory has yet appeared, perhaps because we know little of the factors that cause separate bony elements to develop. There is some evidence of relations between bones and teeth and bones and lateral line organs and the pattern of the latter may play a large part in determining the plan of the skull (p. 325). Provisionally we may recognize four classes of dermal bones (1) canal bones, (2) tooth bones, (3) 'ordinary' bones, whose determination is unknown, (4) extra bones, filling special areas (Wormian bones).

The arrangement of the numerous bones is made less difficult to understand and remember if they are considered in the following order. First the endochondral ossifications in the original neurocranium, then the dermal bones that cover this above and below; next the endochondral bones formed within the original cartilaginous jaws, then the dermal bones that cover the edges of the jaws, and finally the ossifications in the branchial arches and pectoral girdle, which latter is in bony fishes attached to the skull.

The endochondral ossifications may be considered by beginning at the hind end of the skull: here the floor ossifies as the basioccipital, the sides as the exoccipitals, and the roof, over the spinal cord, as the supra-occipital bone; these posterior bones are not well marked off from each other in the adult skull. In the auditory capsule are five separate otic bones, of these the epiotic and pterotic can be seen externally (Fig. 118).

The floor in front of the basioccipital is occupied by a basisphenoid bone and the walls above this by alisphenoids. The eyes nearly meet in the midline and the orbits are here separated only by a thin orbitosphenoid. The only more anterior part of the chondrocranium to ossify is the region between the nasal capsule and the orbit, forming the ectethmoid.

The dermal bones that cover this partly ossified neurocranium may be identified as on top and in front a pair of frontals and a median supraethmoid, behind which are large paired parietals and small paired post-parietals. These names have been inferred from study of crossopterygians and early amphibians (p. 325), which showed that the homologies earlier accepted were wrong. Fig. 118 carries the old nomenclature in which the large paired bones were called frontals. Around the eyes is a ring of circum-orbitals, and on the floor of the skull two median bones, the parasphenoid and vomer.

The jaw bones are numerous, including both endochondral and dermal elements, and the relation of the method of support to that found in other animals is not clear. In the embryo palato-pterygo-quadrate bars and Meckel's cartilages are seen. The upper jaw bears inward projections, which extend towards the chondro-cranium and probably represent the traces of an autostylic means of support (see p. 187). But the effective support in the adult is achieved by the ossified hyomandibular cartilage. The palato-pterygo-quadrate bar ossifies in several parts and palatine, pterygoid, mesopterygoid, metapterygoid, and quadrate bones appear, some of them partly formed in membrane. The only part of Meckel's cartilage to ossify is the articular bone, at the hind end. The actual edges of the jaws are supported by membrane bones, the premaxilla, maxilla, and jugal, covering the upper jaw. The dentary covers most of the lower jaw, except for a small bone, the angular, that lies on the inner side at the posterior end.

The hyomandibular bone runs from an articulation with the otic capsule to the upper end of the quadrate. The symplectic is a small separate ossification at the lower end of the hyomandibula. The rest of the hyoid arch is present as epi-, cerato-, and hypohyals, which support a large toothed tongue. Bony fishes only rarely possess an open spiracle and immediately behind the hyoid are attached the bones supporting the operculum that covers the gills. The branchial arches are formed of several pieces, as in elasmobranchs, each being ossified separately.

The effect of this complicated set of bones is to provide an efficient apparatus for the protection of the brain and sense-organs, support of the jaws and teeth and of the respiratory apparatus. Teeth are found on the vomers, palatines, premaxillae, maxillae, dentary, and on the tongue. Covering the typical dentine (orthodentine) is a layer of harder vitrodentine, poor in organic matter and perhaps derived partly from ectodermal ameloblasts. The teeth are usually spikes pointing in a backwards direction, used to prevent the escape of the food and not usually for biting or crushing. They may, however, form plates or be firmly attached to the bones. Folds of the mucous membrane, supported by cartilage and carrying gill-rakers, are found in species that feed on small prey.

4. Respiration

Limitations are imposed on the respiration of fishes by the facts that water is 800 times more dense than air and the dissolved oxygen is 30 times more dilute. The cost of respiration is therefore high. There is a 70 per cent. increase in metabolism when a trout increases its ventilation volume four times in water poor in oxygen.

Respiration is produced by a current passing in a single direction, as in elasmobranchs, namely, in at the mouth and out over the gill lamellae, but the mechanism by which the current is produced is somewhat different. The pumping action is produced by a buccal pressure pump and opercular suction pumps, resulting from sideways move-

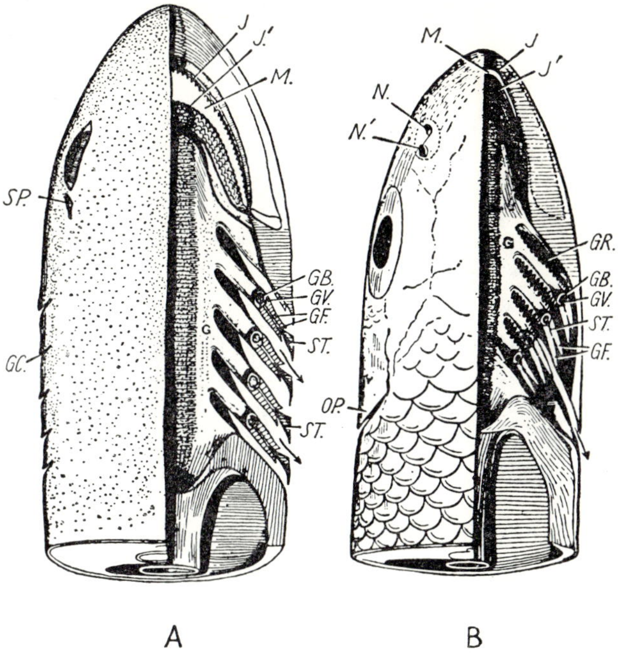

FIG. 119. Arrangement of organs of head and gills in A, a shark, and, B, a teleostean fish.

GB. gill bar; GC. outer opening of gill cleft; GF. gill filament; GR. gill-rakers; GV. gill vessels; J, J'. upper and lower jaw; M. mouth; N, N'. openings of nasal chamber; OP. operculum; SP. spiracle; ST. septum between gill filaments. (From Dean, *Fishes, Living & Fossil*, The Macmillan Co.)

ments of the operculum, enlarging the branchial cavity. The branchiostegal folds, below the operculum, prevent inflow of water from behind. When the operculum moves inward dorsal and ventral flaps in the throat prevent the exit of water forwards.

The gill lamellae differ from those of elasmobranchs in the great reduction of the septum between the respiratory surfaces (Fig. 119). This has the effect of leaving the lamellae as free flaps, increasing the surface available for respiration.

The area of the gills varies greatly, being relatively larger in more active species. The rate of respiration is controlled by a medullary centre but the most active rate of respiratory exchange is only some four times the standard rate (as against twenty times in man). The

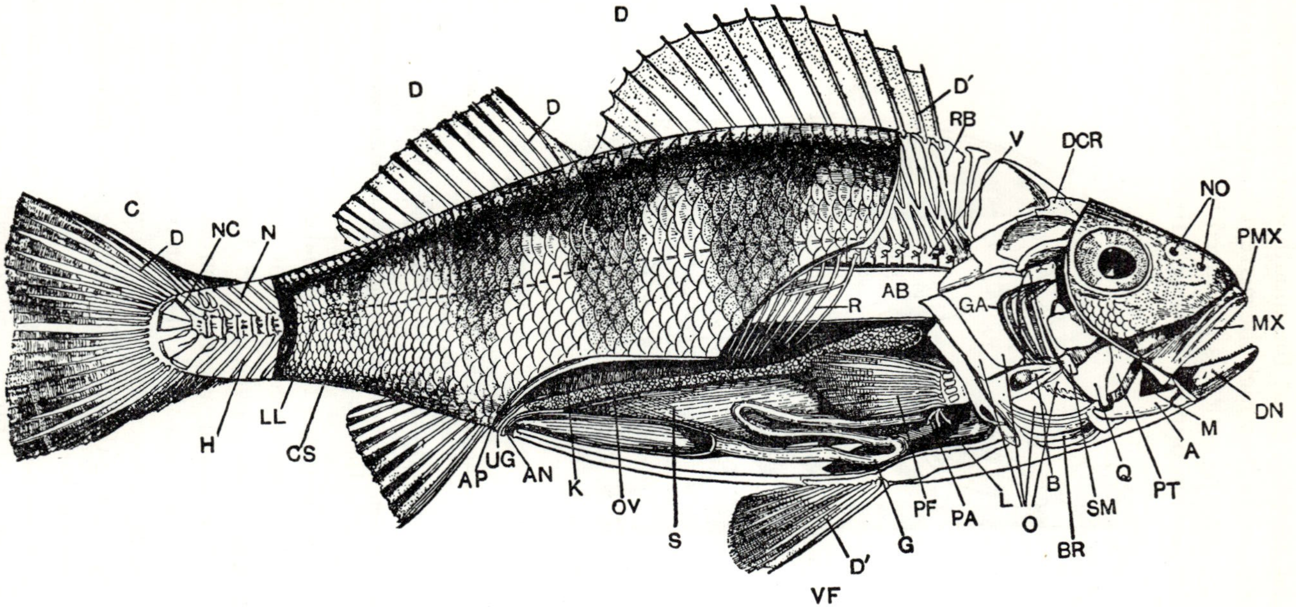

FIG. 120. The general anatomy of a teleostean fish.

A, angular; AB. air-bladder; AN. anus; AP. abdominal pore; B. bulbus arteriosus; BR. branchiostegal rays; C. caudal fin; CS. cycloidal scales; D. dorsal fin; D'. dermal supports of fins; DN. dentary bone; DCR. dermal bones roofing cranium; G. intestine; GA. gill arches; H. haemal arch and spine; K. kidney (mesonephros); L. liver; LL. lateral line; N. neural arch and spine; NO. anterior and posterior nares; O. opercula; OV. ovary; PA. pyloric appendices; PF. pectoral fin; PMX. premaxillae; PT. pterygoid; Q. quadrate; R. ribs, showing accessory supports; RB. radial and basal fin supports; S. stomach; UG. urinogenital opening; V. vertebra (centrum); VF. ventral fin. (From Dean, *Fishes, Living & Fossil*, The Macmillan Co.)

area of the respiratory surface is thus an important limiting factor in the movement and growth of fishes. During activity of a fish lactic acid accumulates in the blood and the pH falls. The fish is thus able to display a considerable burst of activity and then to repay the oxygen debt over a long subsequent period.

5. Vertebral column and fins of bony fishes

The vertebral column of bony fishes performs the same function as in other fishes, namely, to prevent shortening of the body when the longitudinal muscles contract. It has, however, become very complicated and with the ribs and neural and haemal arches forms an elaborate system serving to maintain the body form under the stresses of fast swimming. Like other parts of the skeleton it is extensively ossified, and the necessary lateral flexion is obtained by division of the column into a series of sections joined together. Typically there is one such section (vertebra) corresponding to each segment, but in the tail region of *Amia* there are twice as many vertebrae as segments.

Each vertebra consists of a centrum, neural arch and neural spine, and in the tail region, in addition, haemal arch and haemal spine. These parts are formed partly by ossification of cartilaginous masses, the basidorsal and basiventral, interdorsal and interventral, such as we saw in elasmobranchs, and partly by extra ossification in the sclerogenous tissue around the notochord and nerve-cord and between the muscles. The vertebrae are inter-segmental, the middle of each lying opposite the myocomma that separates two muscle segments.

The centra are concave both in front and behind (amphicoelous), and in the hollows between them are pads made of the remains of the notochord, an arrangement that allows the column to resist longitudinal compression and yet remain flexible; similar flat or concave articulations of the centra are found in other aquatic vertebrates from the elasmobranchs to the whales. Extra processes on the front and back of the vertebrae ensure the articulation and are comparable to the zygapophyses found in tetrapods. The ribs, which are so prominent in the backbone of many fishes, are of two sorts; pleural ribs between the muscles and the lining of the abdominal cavity, and more dorsal intramuscular ribs. Both sorts are attached to the centrum. The bony rods attached above the neural and below the haemal arches are often called neural and haemal spines, though it is doubtful whether they correspond to the neural spines of land vertebrates. They form the supporting rods or radials of the median fins and are usually divided into two or three separate bones in each segment. In addition to these radials the fins are also supported by a more superficial set of bony rods, the dermal fin rays (dermotrichia or lepidotrichia), which may be considered as modified scales and accordingly lie superficial to the radials. These dermal fin rays are usually forked at their tips. They make an extra support for the fin margin and to them are attached the muscles that serve to throw the fin into folds.

In the tail region the internal skeleton is not quite symmetrical and shows signs of origin from an animal with a heterocercal tail. The

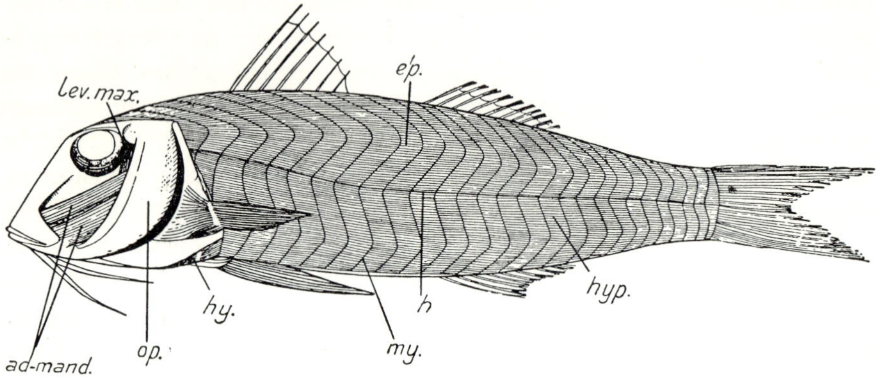

FIG. 121. Muscles of a teleostean fish, mainly based on *Mullus*. *ad-mand.* adductor mandibulae; *ep.* epaxonic muscles; *h.* horizontal myoseptum; *hy.* hypobranchial muscles; *hyp.* hypaxonic muscles; *lev. max.* levator maxillae; *my.* myocomma; *op.* operculum. (From Ihle.)

notochord turns up sharply at the tip, so that the neural spines are very much shorter than the haemal spines, known here as hippural bones. The final portion of the notochord is often surrounded by a single ossification, the urostyle, and the whole makes a rigid support for the dermotrichia of the tail. Such a tail with internal asymmetry but external symmetry is said to be homocercal.

The myotomes are arranged in a complicated pattern having the effect that contraction of each affects a considerable section of the body (Fig. 121); in fast swimmers such as the tunny each myotome may overlap as many as nineteen vertebrae. Between the lateral and ventral muscle masses there is in many fishes a layer of red muscle and this is especially well developed in the tunnies and bonitos.

The paired fins are similarly supported by ossified radials, covered by dermal fin rays. At the base the radials are connected with 'girdles' lying in the body wall. The pectoral girdle (Fig. 118) consists of a cartilaginous endo-skeletal portion in which ossify the scapula, coracoid, and sometimes mesocoracoid, while dermal bones, large cleithrum, and one or more small clavicles, become attached superficially.

Above these a further series of dermal bones, the supra-clavicle and post-temporal, attach the pectoral girdle to the otic region of the skull.

The pelvic girdle is very simple, consisting only of a single bone, the basipterygium.

6. Alimentary canal

The food of the trout consists mainly of small invertebrates such as *Gammarus*, *Cyclops*, and other crustaceans, and aquatic insects and their larvae, together with the fry of other fishes and perhaps sometimes larger pieces of 'meat'. The food is mostly swallowed whole, being helped down the pharynx by the mucous secretions, but these, as in elasmobranchs, contain no enzymes. The entrance to the stomach is guarded by a powerful oesophageal sphincter, no doubt serving to prevent the entry of the water of the respiratory stream. The stomach is divided into cardiac and pyloric portions, though the distinction is less clear than in elasmobranchs. The duodenum is beset by a number of wide-mouthed pyloric caeca, serving to increase the intestinal surface (Fig. 120). The intestine and caeca are lined throughout by a simple columnar epithelium and there are no specialized multicellular glands such as the Brunners glands or crypts of Lieberkühn of mammals. The exocrine pancreas consists of numerous diffuse glands in the mesentery. The endocrine portion, however, forms a compact mass of tissue. This is very rich in insulin and after its removal a fish shows hyperglycemia and glycosuria. The intestine is relatively longer than in elasmobranchs and often coiled; its internal surface may be increased by folds, but there is no true spiral valve, though this was present in the ancestors of the Teleostei (p. 233). There is no gland attached to the rectum.

7. Air-bladder

Dorsal to the gut is a very large sac with shiny, whitish walls, the air-bladder, filled with oxygen. A narrow pneumatic duct connects this with the pharynx in the more primitive forms. The origin and functions of the air-bladder will be discussed below (p. 261); it serves as a hydrostatic organ, enabling the animal to remain suspended in the water at any depth.

8. Circulatory system

The general plan of the circulation is similar to that of an elasmobranch (Fig. 122), that is to say, there is a single circuit and all the blood passes through at least two sets of capillaries. The heart contains a series of three chambers, sinus, auricle, and ventricle, but the muscular conus arteriosus is absent, there being only a thin-walled bulbus arteriosus at the base of the ventral aorta. The walls of the bulbus are elastic but not muscular, and study of its action by means of X-rays shows that it is dilated by the ventricular beat and then contracts, thus maintaining the pressure against the capillaries of the gills. The ventral aorta is short, but the arrangement of the afferent and efferent branchial vessels is essentially as in elasmobranchs.

The blood-pressure in the ventral aorta is less than 40 mm Hg in most fishes at rest, and in the dorsal aorta about half this. The venous pressures are around zero, the pericardium being fibrous but not rigid as it is in elasmobranchs (p. 160). There is no communication between the pericardial and peritoneal chambers. There is a vagal cardiac depressor nerve, but no sympathetic nerve to the heart.

There is a well-developed lymphatic system beneath the skin and in the muscles and viscera. Lymphoid tissue is abundant in various organs but there are no lymph-nodes along the vessels. There is a large spleen concerned with haemopoiesis, which also proceeds in the kidneys. The red cells are smaller in bony fishes (8–10 μ) than in elasmobranchs (up to 20 μ). A continuous series of white cells is present and acidic and basic granules may occur in the same cell.

9. Urinogenital system and osmoregulation

The kidneys are mesonephric in the adult and consist of an elongated brown mass above the air-bladder. The ducts of the two kidneys join posteriorly and are swollen to form a bladder which, being mesodermal, must be distinguished from the endodermal cloacal bladder of tetrapods. The urinary duct opens separately behind the anus, there being no common cloaca.

Nitrogenous elimination is a function mainly of the gills, which excrete as ammonia and urea more than six times as much nitrogen as the kidneys. The latter excrete creatine, uric acid, and the weak base trimethylamine oxide, which is present in large amounts in the blood of marine teleosts.

One of the most striking features of the life of bony fishes is that they occur both in fresh water and the sea, and many, such as the trout itself, can move from one to the other. It is supposed by some that the earliest gnathostomes were freshwater animals (p. 187), and the bony fishes might be said to show evidence of this in that the concentration of salt in their blood is always less than in the sea, in the neighbourhood of 1·4 per cent. NaCl against 3·5 per cent. outside.

In fishes in fresh water the blood is more dilute, about 0·6 per cent. NaCl, but is, of course, more concentrated than the surrounding medium, which contains only traces of inorganic ions. Freshwater fishes are able to take up salts from the water through the gill surfaces.

The kidney apparatus, with its filtration system of glomeruli and tubules for salt reabsorption, was probably developed for life in fresh water and still serves in this way in the freshwater forms. Various special devices are adopted in fresh water for minimizing the tendency

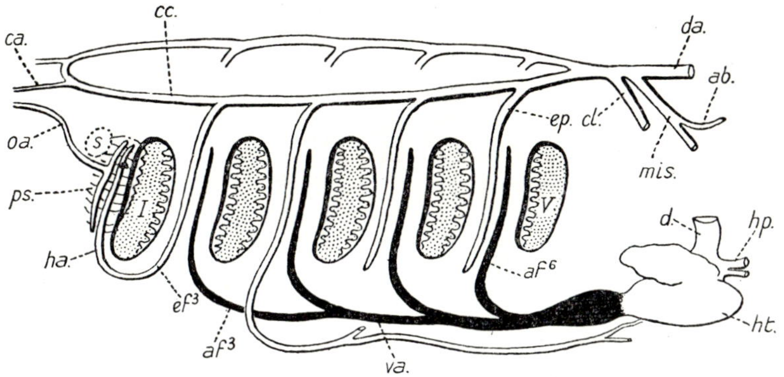

FIG. 122. Diagram of the branchial circulation of a teleostean fish.

ab. artery to air-bladder; *af*$^{3-6}$. four afferent vessels from ventral aorta; *ca.* carotid artery; *cc.* circulus cephalicus; *cl.* coeliac artery; *d.* ductus Cuvieri; *da.* dorsal aorta; *ef*^{3}. efferent vessel of first branchial arch; *ep.* epibranchial artery; *ha.* hyoidean artery (afferent vessel of pseudobranch); *hp.* hepatic vein; *ht.* heart; *mis.* mesenteric artery; *oa.* ophthalmic artery (efferent vessel of pseudobranch); *ps.* pseudobranch (hyoidean gill, possibly with spiracular gill); *s.* position of spiracle (closed); *va.* ventral aorta; *I–V.* five branchial slits. (From Goodrich, *Vertebrata*, A. & C. Black, Ltd., after Parker.)

to gain water and lose salt. The skin is little vascularized and probably makes an almost waterproof layer. The production of mucus assists in this waterproofing, and abundant mucus is secreted when an eel is transferred from salt to fresh water: the full change cannot be made suddenly without killing the fish.

In marine teleosts the problem is the opposite one of keeping water in, or keeping out salt. The usual kidney mechanism is clearly ill suited for this and it is found that the glomeruli are few, or often completely absent from the kidneys. This no doubt reduces the loss of water, but is not enough by itself to solve the problem, which is met by taking in water and salts and excreting the salts. For this purpose special chloride-secreting cells are present in the gills and it has been shown that the amount of oxygen they use, and hence the work they do in diluting the blood, is proportional to the difference of concentration between the inside and the outside. A marine fish is able to drink and absorb sea water in spite of the fact that this is more concentrated than its blood. The chloride-secreting cells dispose of the excess salts. It remains to explain the means by which a solution passes against the osmotic gradient from the cavity of the gut to the blood; the membranes here must have some special properties of which we are ignorant. Sodium and chloride enter with the water but magnesium is excluded and may be precipitated in the intestine.

The genital system is nearly completely separated from the excretory in both sexes. The testes (soft roes) are a large pair of sacs opening into the base of the urinary ducts. The ovaries (hard roes) are also elongated in the trout and the eggs are shed free into the coelom (Fig. 123) and passed to the exterior by abdominal pores. This condition is unusual among teleosts, in most the ovaries are closed sacs, continuous with the oviducts.

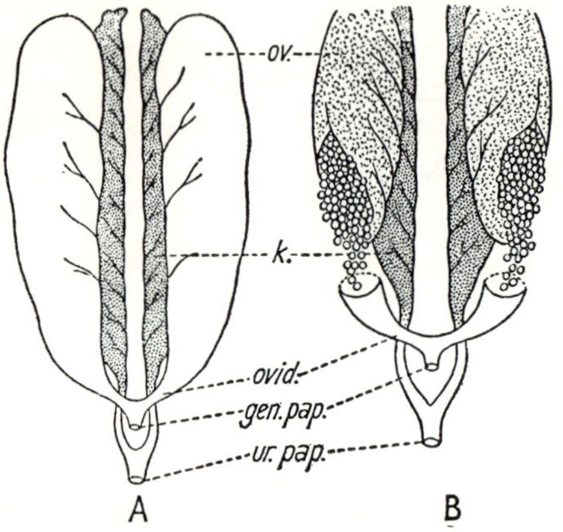

FIG. 123. Ovaries and kidneys of A, typical teleostean fish, B, trout and some other fishes where the eggs are shed into the coelom. (From Norman, after Rey.)

Fertilization is external and the eggs of the trout are shed in small pits or depressions in the sand; being sticky, they become attached to small stones. The eggs are very yolky and cleavage is therefore only partial, forming a cap of cells, the blastoderm, which eventually differentiates into the embryo. After hatching, the young fish may still carry the yolk sac and obtain food from it for some time, while beginning to eat the small crustaceans and other animals that are its first food.

10. Races of trout and salmon and their breeding habits

There is considerable confusion about the various races of trout and their allies the salmon. In both trout and salmon the adult originally spent the great part of its life in the sea but returned to the rivers to breed. Trout and salmon that do this are still abundant on the West Atlantic coasts and ascend all suitable rivers to breed, the process being known as the 'run'. But the trout has produced many races of purely fluviatile animals, living either in lakes (where they often become very large) or rivers, and never returning to the sea to breed. These freshwater races differ in small points from each other and are given various common names (phinock, severn, Loch Leven trout, brook trout, &c.). There can be no doubt that interesting genetic differences between these forms exist, but they have not yet been fully studied. The salmon are much less prone to form purely fresh-water races, though such are known.

During the breeding-season characteristic changes take place in the fishes and differences between the sexes appear. In the salmon the jaws become long, thin, and hooked, especially in the male. The animals make pairs and the males fight with others that approach the female. As the gonads ripen, the other parts of the fish, which were well supplied with fat at the beginning of the run, become progressively more watery. Finally, spawning takes place, the female laying the eggs in a shallow trough (redd), which she has 'cut' in the gravel by movements of her tail, while the male sheds sperms over them. She then covers the eggs with gravel by further cutting movements. The young male salmon (parr), which have not yet been to the sea, may become sexually mature. They accompany the fully grown fish, hanging around the cloacal region and shedding their sperms at the same time as the large male. It is possible that this development of a kind of third sex serves to increase the variability of the population. The spent parr eat some of the eggs and they then proceed to grow, migrate to the sea, and return later.

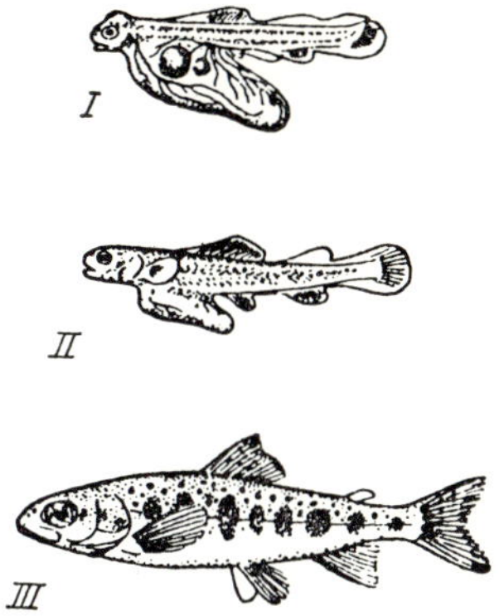

FIG. 124. Three stages in the development of the salmon. I and II are alevins; III, parr. (From Norman.)

Male trout will follow a spawning salmon and fertilize her eggs if her own male is not looking. Hybrids formed in this way can develop, but are said to be less fertile than the normal types; indeed, the males are wholly sterile.

After fertilization the salmon are very exhausted (known as kelts); the males seldom return to the sea. The females, however, may recover and after a period in the sea return to breed again, and this process may be repeated several times.

Very young trout or salmon are known as alevins or fry and remain mostly among the stones (Fig. 124). When they emerge they are called parr and have a number of characteristic parr-marks along their sides. After two to four years spent as parr in fresh water salmon acquire a silver colour and pass to the sea as smolts. Young salmon returning for the first time to breed are called maidens. If they have spent only one and a half years in the sea they are called grilse and may then return to the sea as kelts. Others ascend for the first time after three years or more at sea.

It is well established that salmon nearly always return to breed in the river in which they were born, and it is certain that they may journey for considerable distances in the sea. The mechanisms by which these migrations are initiated and guided are only partly known.

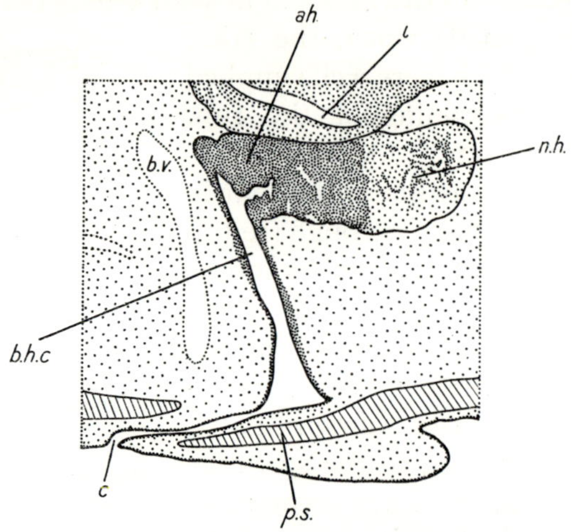

FIG. 125. Pituitary gland of the primitive teleost *Elops*, showing the persistent Rathke's pouch in the form of a hollow bucco-hypophysial canal, piercing the parasphenoid bone.

ah. adenohypophysis; *b.h.c.* bucco-hypophysial canal; *b.v.* blood-vessel; *c.* continuation with pharynx; *i.* infundibulum; *n.h.* neurohypophysis; *p.s.* parasphenoid. (After Olsson.)

They probably involve endocrine changes, for example the thyroid is very active in the smolt as they begin to migrate. The return to the home river may be a result of olfactory conditioning (see p. 221).

11. Endocrine glands of bony fishes

The pituitary gland occupies the same central part in the endocrine signalling system that it has in mammals. Neural and glandular regions are present and the adenohypophysis has three parts, the two more posterior corresponding to the mammalian intermediate and anterior lobes. The most anterior glandular region may be comparable to the pars tuberalis. Experiments by removal and injection have shown that the middle portion produces hormones that stimulate growth, the gonads, the thyroid, adrenal, and probably the pancreas. The posterior lobe produces a melanophore-dispersing hormone

(p. 260) and there may be a melanophore-concentrating one in the anterior lobe. Oxytocin and vasopressin are present but there is no evidence that excretion is controlled by the pituitary.

The thyroid tissue is not aggregated into a compact gland but forms scattered masses along the ventral aorta. Its hormones appear to be

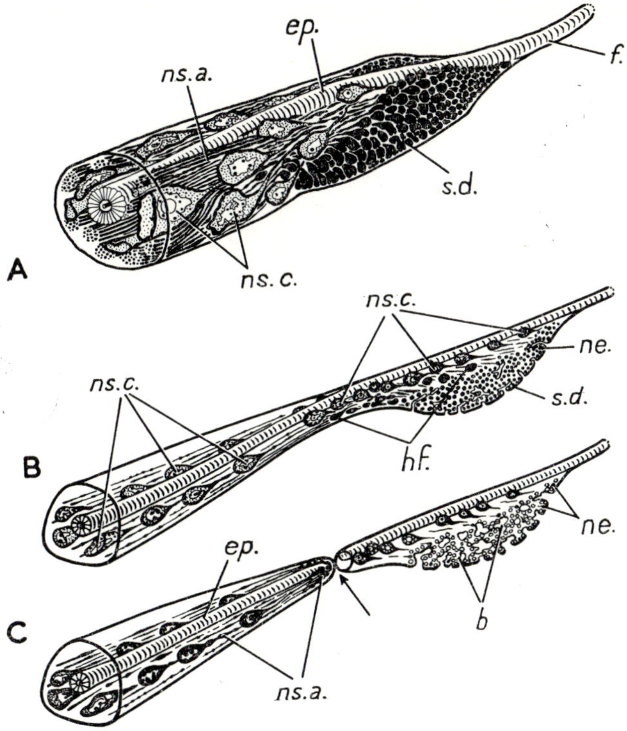

FIG. 126. Urohypophysis of A, eel; B, loach (*Misgurnus*); C, the same in a loach after sectioning the spinal cord and injecting hypertonic saline.
b. blood-vessels; *ep.* ependyma; *ft.* filum terminale; *hf.* lumps of secretion (?'Herring bodies'); *ne.* nerve endings; *ns.a.* neurosecretory axon; *ns.c.* neurosecretory cells; *s.d.* storage depot of neurosecretion. (After Enami, N., in *Symposium on Comparative Endocrinology*. Wiley, New York.)

identical with those of mammals, including mono- and di-iodotyrosine and thyroxin. Thyroid follicles are often found in the kidneys, heart, eye, and elsewhere in the body of fishes, especially those deprived of iodine.

The suprarenal and interrenal tissues are partly associated in masses around the thickened walls of the posterior cardinal veins. Because of the difficulty of isolating these tissues there is little information as to their function. The corpuscles of Stannius are groups of gland cells dorsal to the kidneys, they have been held to be related to the adrenals, but their nature is still uncertain.

The ultimobranchial gland is a mass of cells developed from the last branchial pouch and perhaps related to the parathyroids.

The hormone rennin, which raises the blood pressure, is said to be present in freshwater teleosts, with their high glomerular filtration, but not in marine ones.

The gonads produce steroid hormones as in elasmobranchs (p. 167). The secondary sex characters depend upon their presence, thus

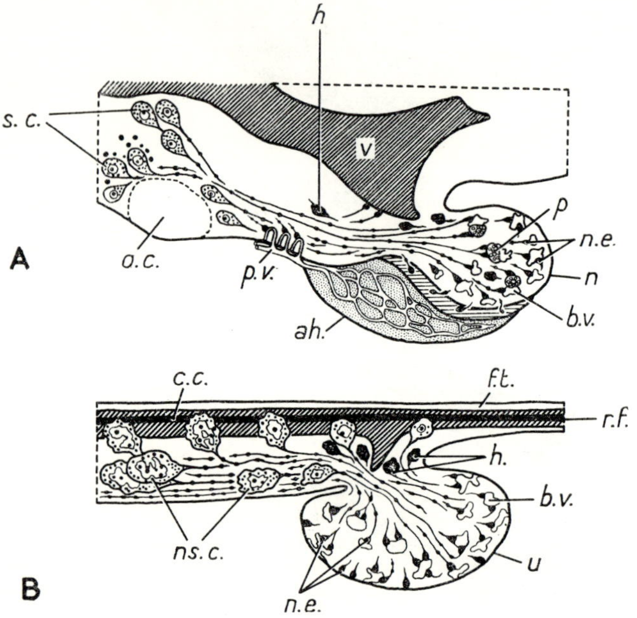

FIG. 127. Comparison of pattern of organization of the caudal neurosecretory system (B) with the hypothalamo-hypophysial system (A).

ah. adenohypophysis; *b.v.* blood-vessels; *c.c.* central canal; *f.t.* filum terminale; *h.* 'Herring bodies', neurosecretory products; *n.* neurohypophysis; *n.e.* nerve endings; *ns.c.* neuro-secretory cells; *o.c.* optic chiasma; *p.* pituicytes; *p.v.* hypophysial portal vessels; *r.f.* Reissner's fibre; *u.* urohypophysis; *v.* ventricle. (After Enami, N., in *Symposium on Comparative Endocrinology.* Wiley, New York.)

the gonopodium of the male of viviparous fishes (p. 267) is developed if sex hormone is added to the water.

At the hind end of the spinal cord of fishes is a small lump consisting of masses of secretion produced by neurosecretory cells of the spinal cord and hence called the urohypophysis (Figs. 126 and 127). In function it appears to be connected with salt regulation; injection of hypertonic NaCl produces hypersecretion, the products accumulating at the cut surface if the cord has been severed. Injection of extracts produces changes in the sodium content of the fish and also changes in buoyancy, perhaps due to an influence on the carbonic anhydrase of the gas bladder.

12. Brain of bony fishes

The brain of bony fishes is built on the same general functional and structural plan as that of elasmobranchs, namely, the development of a number of separate centres, one concerned with each of the main receptor systems. The forebrain is often large, but it is characterized

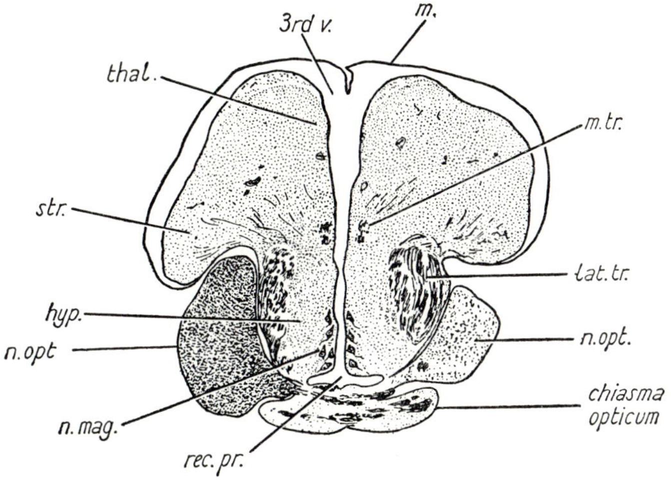

FIG. 127 (*a*). Cross-section of the forebrain of the cod.
lat.tr. and *m.tr.* lateral and medial tracts between olfactory region and hypothalamus; *hyp.* hypothalamus; *m.* membranous roof of forebrain; *n.mag.* nucleus magnocellularis preopticus; *n.opt.* optic tracts (that on the right has atrophied in this specimen); *rec. pr.* preoptic recess; *str.* hind end of striatum; *thal.* thalamus; *3rd v.* third ventricle. (From Kappers, Huber, and Crosby.)

by great development of its ventral regions (the 'corpus striatum'), the roof being wholly membranous (Figs. 127 (*a*), 128). This condition is known as 'eversion' and is the very opposite to the inverted or thick-roofed forebrains that are found in the lung-fishes, close to the line of tetrapod descent (p. 278). The whole of this forebrain is reached by olfactory fibres, and there is little evidence that fibres from other receptor centres reach forward to it; it is mostly a smell brain. Extirpation of the telencephalon from various teleosts has not produced changes in locomotion, balance, or vision; there may be slight changes in general activity and social behaviour. No movements have been seen following electrical stimulation of it.

The diencephalon is not large, since most of the optic fibres end not here, but in the midbrain. The roof is everted to form a pineal body, and this and other parts of the diencephalon may contain receptors sensitive to light. The minnow *Phoxinus* has a transparent patch on the head in this region, and it has been found possible to train the fish to give appropriate responses to changes of illumination even after removal of the paired eyes and the pineal body. Evidently there are light-sensitive cells in other parts of the walls of the diencephalon, besides those that become evaginated to form the eyes. Experiments on lampreys also showed the presence of such cells (p. 105). The hypothalamus is well developed and receives large tracts from the forebrain. Below and behind it is a large saccus vasculosus in some forms (p. 169).

The midbrain is often the largest part of the brain. The cells spread out over its roof (tectum opticum) are not all collected round the ventricle but have migrated away to make an elaborately layered system. Into this midbrain cortex there pass not only the great optic tracts but also ascending tracts from the sensory regions of the spinal cord, lateral line system, gustatory systems, and cerebellum. Large motor tracts pass back towards the spinal cord; the details of their endings have not been traced, but they certainly exercise control over motor functions. Electrical stimulation of the optic lobes produces well-coordinated movements of local groups of muscles, for instance those of the eyes or fins. It can hardly be doubted that this well-developed midbrain apparatus thus controls much of the behaviour of the fish and is able to mediate quite elaborate acts of learning and other forms of more complex behaviour. After removal of the tectum of one side a minnow is blind in the opposite eye. Each part of the retina is mapped on to a distinct area of the tectum and if the optic tract is cut and allowed to regenerate this projection is exactly replaced. When a goldfish is trained to respond to some visual stimulus the learning process occurs in the midbrain and continues unaffected after removal of the forebrain. Conversely, olfactory learning takes place in the latter and is undisturbed by injury to the tectum opticum.

The base of the midbrain (tegmentum) contains motor centres. Electrical stimulation here produces abrupt and massive responses of the locomotor apparatus, very different from the sequences of co-ordinated movements that appear after stimulation of the roof of the tectum.

The cerebellum is very large in teleosts, especially in the more active swimmers, and a forwardly directed lobe of it, the valvula cerebelli, extends under the midbrain. Various disorders of movement have been reported after removal of the cerebellum, such as swaying when moving quickly. Presumably it plays an important part, as in

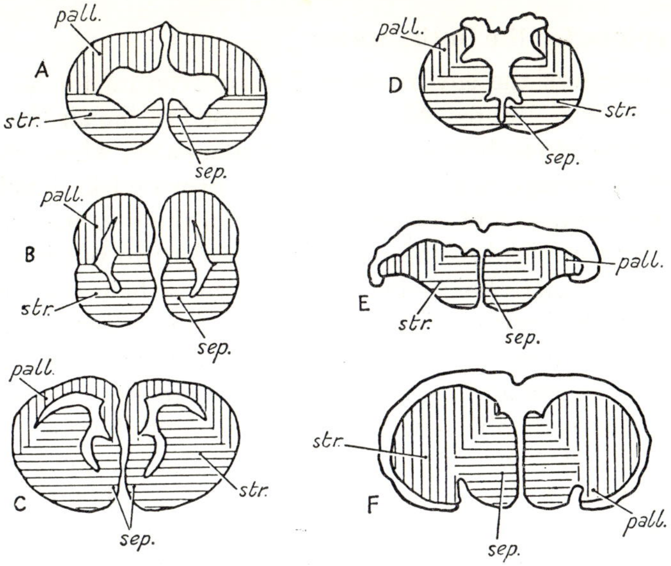

FIG. 128. Transverse sections of forebrain in various vertebrates to show the condition of inversion (thick roof) in A, B, and C and eversion (thin roof) in D, E, and F.
A, lamprey; B, frog; C, chelonian; D, chimaera; E. sturgeon; F, teleostean; *pall.* pallium; *sep.* septum; *str.* striatum. (From Kappers, Huber, and Crosby.)

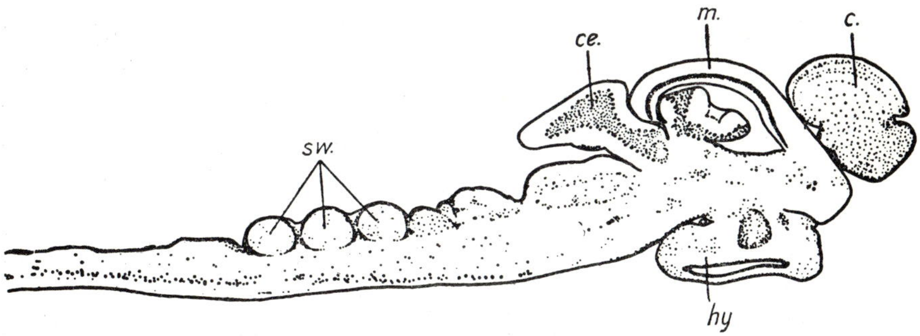

FIG. 129. Sagittal section of brain of the gurnard, showing the swellings in the spinal cord at the point of entry of the nerves from the fin.
c. cerebral hemisphere; *ce.* cerebellum; *hy.* hypothalamus; *m.* midbrain; *sw.* swellings of spinal cord; *v.* valvula. (From Scharrer, *Z. verg. Physiol.* **22.**)

other vertebrates, in producing precise and correctly timed movements. It is enormous in the Mormyridae, where it may assist in direction-finding by electrical pulses (p. 253), perhaps acting as a timing device.

The medulla oblongata is also well developed, having special lobes connected with the entry of the lateral line nerves and gustatory fibres of the cranial nerves. In the gurnard, *Trigla*, there are chemical receptors in the elongated fins. These are innervated from spinal nerves, and there are swellings of the dorsal part of the spinal cord at the points where these nerves enter (Fig. 129).

13. Receptors for life in the water

The features of the environment that are relevant for life are very different in air and water. Man is so well used to the air that it is not easy to appreciate fully the conditions underwater, where changes of illumination, though obviously important, provide less detailed evidence of the sequence of distant events than they do in air. We can say that light carries less information for a fish; to put it in another way, fewer distinct choices between alternative behaviour pathways are made on the basis of visual clues by a fish than by a man.

On the other hand, the water around the fish provides mechanical stimuli both at low and high frequency that are more closely related to distant events than is generally true in air. Both hearing and touch are of great importance in the water and the lateral line system provides a system of 'distant touch' that is perhaps wholly outside our experience. Localization of distant objects by such a sense, perhaps assisted by echo-location by water movements, provides the fish with many relevant clues. It is interesting that these receptors are connected with a very large cerebellar system, perhaps concerned with measuring time differences.

Chemical changes in the water also provide much information and both taste and smell are well developed. That smell is analysed by a distinct system in the forebrain, not directly related to the cerebellar system, is one of the fundamental principles of control of vertebrate behaviour. Distant chemical changes provide the first clue to the presence of food, a mate, or an enemy, whereas the detailed finding of these involves eyes, ears, touch, and an accurate timing system. There thus arises the distinction between the systems for initiation of action in the forebrain ('emotive') and for its fulfilment (executive) by centres farther back.

14. Eyes

An animal provided with suitable receptors can obtain much information about the environment from the changes in illumination. Control of the whole physiology to follow the rhythm of day and night may have been the original reason for the development of photosensitivity in the diencephalon (see p. 105). At the stage of evolution reached by teleosts information is gained from the fact that light varies in frequency (*colour*) and intensity (*brightness*) and that it is reflected from many substances, revealing their *movement* and *shape*. The greatest sensitivity of the fish eye is in the yellow-green, which is the wavelength that penetrates farthest into the water. In order to extract the maximum of information at high as well as low intensities it is necessary to adjust the sensitivity, and hence the signal/noise ratio. For this purpose teleosts have developed retinas with distinct rods, cones, and twin-cones and in some there is a fovea composed of numerous thin cones (e.g. in *Blennius*). The pupil usually varies little in diameter, and adjustment of sensitivity is by migration of pigment

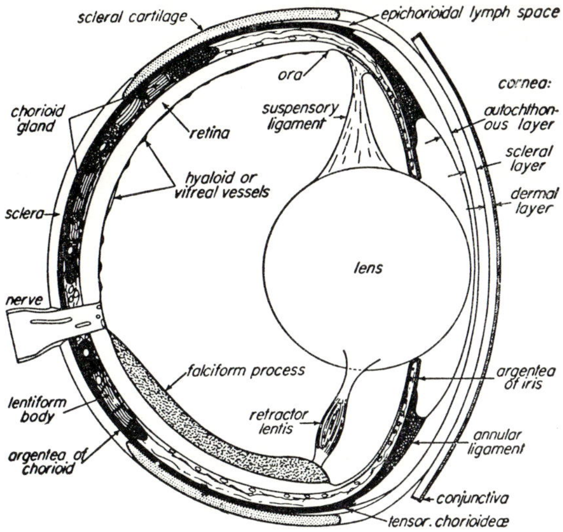

FIG. 130. Diagrammatic vertical section of a typical teleostean eye. Not all the structures here shown are found in all species. (From Walls, *The Vertebrate Eye*.)

between the receptors and contraction of a 'myoid' segment of the latter. In bright light the pigment expands, the cones contract forward, towards the light and the rods contract back, beneath the pigment. These photo-mechanical changes thus serve the same end as changes of pupil diameter in other vertebrates.

The photochemical change in the rods of marine fishes is the same as that of land vertebrates, namely the breakdown of the rose-coloured 'visual purple' (rhodopsin) first to the yellow retinene and then to colourless vitamin A_1. In freshwater fishes there is a different pigment porphyropsin, or visual violet, which breaks down to vitamin A_2. Intermediates between these may be found.

In all fishes there is a very large, dense, spherical lens, to which is attached a retractor muscle (campanula Halleri) inserted on to a falciform ligament, which occupies the persistent choroidal fissure in the retina (Fig. 130). The eye is usually said to be myopic at rest and to be accommodated for distant vision by pulling the lens nearer to the retina. However, this has recently been disputed by Verrier, who denies that the campanula is muscular and believes the eye to be hypermetropic at rest and accommodated, if at all, by fibres of the ciliary body, as in other vertebrates. It may be that the fixed focus is already sufficiently deep and the campanula perhaps serves mainly to steady the lens.

It is unwise, however, to generalize about teleostean eyes, for they are very varied. Whereas the trout, like most, has a round pupil, which varies little if at all in size, other fishes, whose eyes are more exposed to light from above, have a more mobile iris. In flat-fishes and the angler-fishes, such as *Lophius*, and the Mediterranean *Uranoscopus*, the star gazer, the iris has an 'operculum' and is very muscular; its movements are controlled by nerves and not, as in selachians, by the direct effect of light. The sympathetic system sends branches into the head in these animals (Fig. 138) and its fibres cause contraction of the sphincter of the iris, whereas fibres in the oculomotor nerve cause contraction of the dilatator, the opposite arrangement to that in mammals. In the eel the pupil is also capable of wide changes of diameter, but here the control is mainly by the direct response of the circular sphincter iridis muscle to light incident upon it. The pupil of the isolated eye of an eel closes when illuminated and reopens again in darkness (Fig. 131). Presumably because of its lack of nervous control this iris is not affected by many of the usual 'autonomic' drugs. For instance, closure will occur in the presence of atropine and the dark-adapted pupil remains unchanged when placed in a solution as strong as 1 per cent. pilocarpine, but then closes immediately on illumination (Fig. 132). The isolated pupil of *Uranoscopus*, however, closes when pilocarpine is applied (Fig. 133), and in this case the sphincter muscle is innervated by sympathetic nerve-fibres. Adrenaline also causes the sphincter to contract and acetyl choline in moderate concentrations causes dilatation

The eyes may be small or absent in fishes living in caves, muddy waters, or the deep sea. In this last habitat, however, many have exceptionally large eyes, with, apparently, high acuity as well as sensitivity. They may be elongated ('telescopic') and with large binocular fields and a fovea of 'rods'. In *Bathylagus* the rods reach a density of 800,000 mm^2 and are arranged in six superimposed layers, which presumably come into action successively as an object approaches. The cells of the deeper layers are less closely packed.

The tropical fish *Anableps* lives with the head half out of water and the eyes are adapted for use in both media. The upper part of the

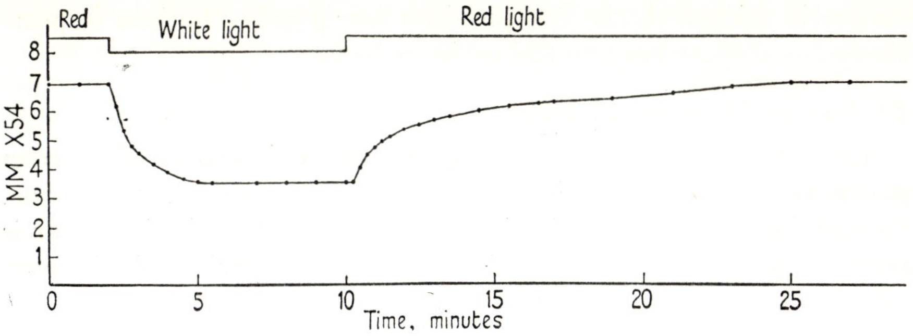

FIG. 131. Closure of the pupil of the isolated eel's eye, followed by plotting the movement of its margin with a camera lucida. The movements are shown magnified 54×. Time in minutes.

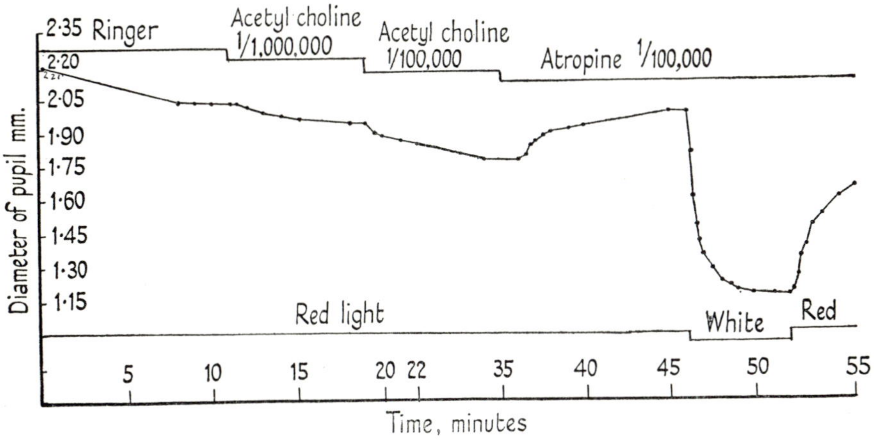

FIG. 132. Changes of diameter of isolated eel's iris in Ringer's solution with the addition of various drugs. Acetyl choline produces some closure, atropine some opening. Light still produces closure after application of atropine.

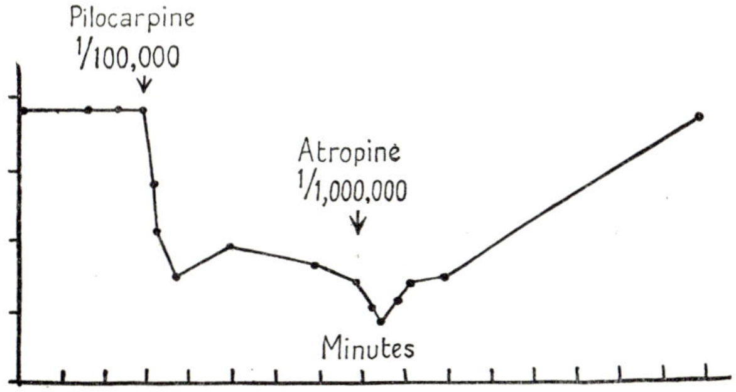

FIG. 133. Movements of margin of pupil in isolated iris of *Uranoscopus* in isotonic solution. Pilocarpine produces closure and atropine opening of this pupil whose sphincter muscle is innervated by *sympathetic* nerve-fibres. (From Young, *Proc. Roy. Soc.* B. **107**.)

cornea is thickened, the iris provides two pupils, the lens is pear-shaped, and there are two retinas in each eye.

15. Ear and hearing of fishes

The ear provides receptors that ensure the maintenance of a correct position of the fish in relation to gravity and to angular accelerations. In addition, in many species it serves for hearing. The inner ear is completely enclosed in the otic bones. There is a perilymphatic space only in those species that hear well.

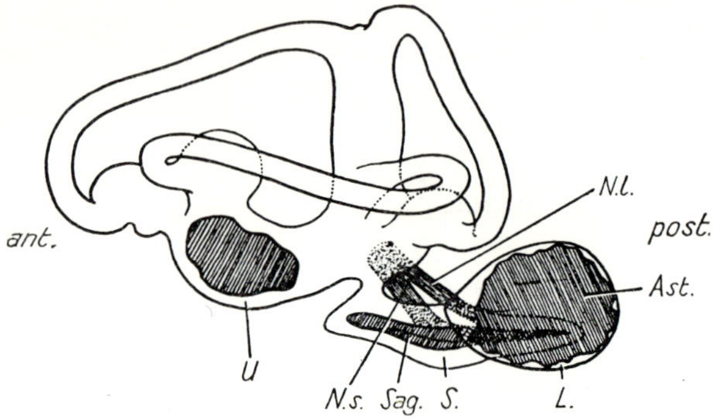

FIG. 134. Diagram of ear of the minnow *Phoxinus*. *Ast.* asteriscus; *L.* lagena; *N.l.*, *N.s.* nerves of lagena and saccule; *S.* saccule; *Sag.* sagitta; *U.* *utricle*. (From V. Frisch, *Z. vergl. Physiol.* **25**.)

Each ear sac is subdivided into three semicircular canals and three other chambers, the utriculus, sacculus, and lagena (Fig. 134). In each chamber is carried an ear stone (otolith) and these are given special names, the lapillus, sagitta, and asteriscus, occupying the above three chambers respectively.

The sensitive macula of the utricle lies horizontally, with the lapillus resting upon it, whereas the maculae of the saccule and lagena are vertical. These receptors with otoliths have double or triple functions. At rest they act as static receptors, signalling the position of the fish in relation to gravity and setting the fins and eyes in appropriate positions. In movement, together with the semicircular canals, they signal angular accelerations, initiating compensatory movements. Thirdly, some of the otolith organs respond to sonic vibrations.

In the fishes that hear well there is a connexion between the air bladder and the ear. This may be either direct, by means of a sac extending forwards (in Clupeidae and others) or indirectly by a chain of modified vertebrae, the Weberian ossicles (Fig. 135). This latter arrangement is found in the freshwater Ostariophysi, which hear particularly well. In these fishes the receptors are in the inferior part of the ear (saccule and lagena, Fig. 134). The sagitta carries a special wing projecting into the cavity and is so suspended as to serve to amplify vibrations. Near to it is a thin portion of the wall of the sac, which would favour the passage of variations of pressure transmitted to the endolymph by the ossicles.

These Ostariophysi respond to sounds between about 60–6,000 vibrations/sec. After removal of the pars inferior responses continue only up to 120/sec. If the air bladder is punctured a minnow can still respond, but only up to 3,000/sec and with a sensitivity diminished by more than fifty times.

Minnows can be trained to discriminate between warbled notes separated by $\frac{1}{4}$ tone. Non-ostariophyse fishes have mostly a much lower upper limit of hearing and lower capacity for discrimination. The Mormyridae (p. 254), however, approach the minnows in this respect and here there is a special isolated portion of the air bladder within the otic bone.

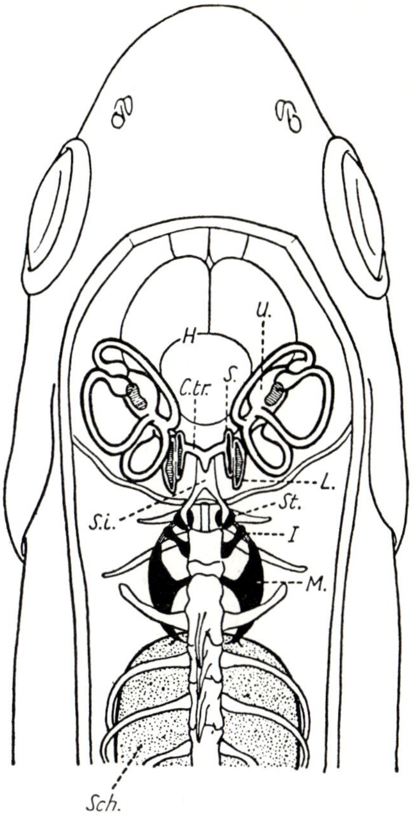

FIG. 135. Position of ear in Ostariophysi and its relation to the Weberian ossicles, which are shown in black.

C.tr. transverse canal between the two sacculi; *H.* brain; *I.* 'incus'; *L.* lagena; *M.* 'malleus'; *S.* sacculus; *Sch.* swim-bladder; *S.i.* sinus impar (perilymphatic space); *St.* 'stapes'; *U.* utriculus. (From V. Frisch.)

In the best cases the sense of hearing of fishes thus approaches that of man, in spite of the absence of a coiled cochlea and basilar membrane with fibres of different lengths. Clearly the discrimination of tones cannot here depend upon differential resonance as the theory of Helmholtz requires. In spite of the considerable powers of pitch discrimination there is little evidence of capacity to localize sounds except when they are loud and near.

16. Sound production in fishes

A surprisingly large number of fishes can produce sounds audible to ourselves, and these noises are used by the fishes either for shoaling, or to bring the sexes together, or to warn or startle enemies. Some fish may use the sounds they produce for echo-location. Among the loudest of the sounds is that produced by the drum-fish (*Pogonias*) of the Eastern Atlantic. The 'whistling' and other noises of the 'maigre' (*Sciaena*) are supposed to be the origin of the song of the Sirens, since they can easily be heard above the water. In both these fishes the sounds are made mostly if not wholly in the breeding season. In others, such as siluroids and *Diodon*, the noise is associated with the presence of spines and may be a warning. In *Congiopodus* the nerves that innervate the muscles of sound production also supply muscles that raise the spines (Packard).

The mechanism for sound production is very varied, involving either stridulation by the vertebrae (some siluroids), operculum (*Cottus*, the bull-head), pectoral girdle (trigger-fishes), teeth (some mackerel and sun-fish), or phonation by the air bladder. The latter may be involved either by its use for 'breathing' sounds in physostomatous forms (p. 261) or as a resonator. Noise production is common in some families (Triglidae, Sciaenidae, Siluridae) but almost absent from others. The advantages to be obtained from sound production underwater have led to parallel evolution of similar mechanisms in several different groups.

17. The lateral line organs of fishes

The lateral line organs occur partly as rows of distinct pits, partly in canals that communicate with the surface through pores in the scales. Besides the main canal running down the body and served by the lateral line branch of the tenth cranial nerve, there are also lines following a definite pattern on the head, namely, supra- and sub-orbital lines, a line on the lower jaw, and a temporal line across the back of the skull. The canals on the head are innervated mainly from the seventh, partly from the ninth cranial nerve. The nerve-fibres enter the very large acoustico-lateral centres of the medulla and valvula cerebelli.

Fishes possess the capacity to react to an object moving some distance away in the water ('distant touch sense') and this is reduced or absent after section of the lateral line nerve. Presumably the moving object sets up currents in the water, which move the fluid (or mucus) in the canals. It has also been suggested that the canals serve to record displacements produced by the swimming movements of the fish itself, but this has not been proved. Fishes deprived of the lateral line show no muscular incoordination, although if blind they collide frequently with solid objects. It has often been suggested that these organs serve for hearing, perhaps at low frequencies, but this is probably not so.

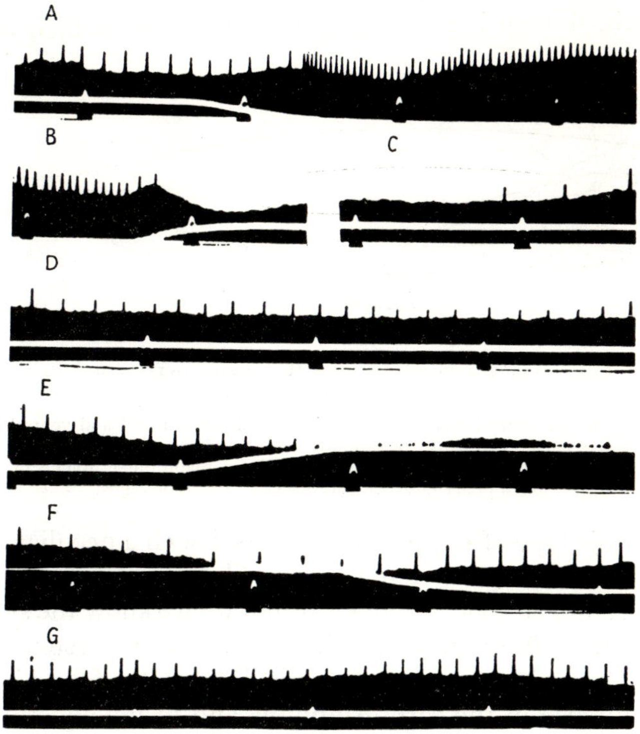

FIG. 136. Responses of a single end organ in a lateral canal of a ray, shown with an oscillograph after amplification. Time signal 1·0 sec. intervals. The movements of the continuous white line show A, the beginning of a headward flow, increasing the frequency of discharge; B, the end of this flow; C, return of spontaneous discharge after an interval of 28 sec.; D, spontaneous discharge 60 sec. later; E, beginning of a tailward perfusion, inhibiting the discharge; F, the end of this perfusion; G, the spontaneous discharge 10 sec. later. (From Sand, *Proc. Roy. Soc.* B. **123**.)

Study of the electrical activities of these organs in rays has shown that many of them discharge impulses all the time, even when not under the influence of any external stimulation (Fig. 136). By passing currents of water along the tubes Sand showed that a tailward flow checks and a headward flow accelerates this 'spontaneous' discharge of impulses. Such changes in the streams of impulses arriving at the brain could, no doubt, form the basis for initiation of movements of the fish. This, however, still leaves open the question of what agency initiates movement of the fluid in the canals during life. It has been shown that when small streams of water are directed against the side of the tail some fish make escaping movements, but that these no longer appear when the lateral line nerve has been cut. The lateral line organs thus provide signals when agitation of the water causes pressure changes. In fact they provide the animal with a kind of water touch, though it is not certain whether this is their only func-

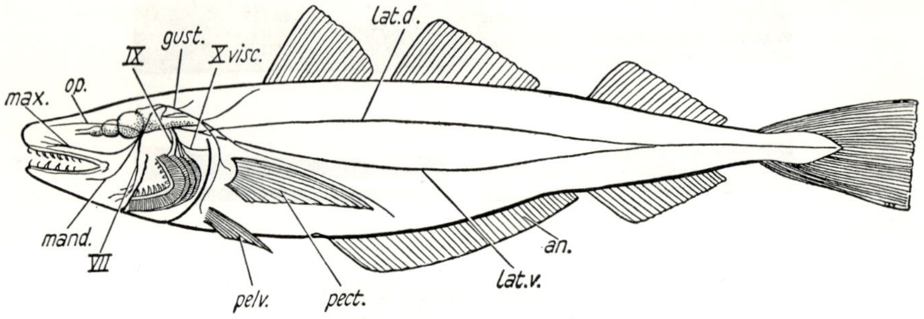

FIG. 137. Dissection of whiting to show the cranial nerves, and especially the nerves for the taste-buds.

an. anal fin; *gust.* gustatory branch of facial; *lat.d.* and *lat.v.* dorsal and ventral lateral line nerves of vagus; *mand.* and *max.* mandibular and maxillary divisions of trigeminal; *op.* ophthalmic; *pect.* pectoral fin; *pelv.* pelvic fin; *VII*, hyomandibular branch of facial; *IX*, glossopharyngeal; *X.visc.* visceral branch of vagus.

tion. Why this type of receptor should need such a peculiar apparatus of canals, rather than a system of nerve-fibres in the skin, innervated by the spinal dorsal roots, is not clear, nor do we know the significance of the pattern of lines on the head. The lateral line system must certainly be of great importance in aquatic life, for it is found in all types of fishes and also in the early Amphibia and in the aquatic larvae of modern members of that group. The distant touch receptors could obviously be used in many ways, not only to locate moving objects and water currents but to serve for echo-location, by computing the time relation of reflected waves set up by the fish itself.

18. Chemoreceptors. Taste and smell

As in all vertebrates, there are two separate chemical senses, taste and smell. The former serves mainly to produce appropriate reactions to food near the body, such as snapping, swallowing, or movements of rejection. Smell, on the other hand, is a 'distance sense', by which the whole animal is steered. The distinction between the two types of receptor is somewhat obscured in bony fishes by the fact that taste-buds are not restricted as they are in mammals to the tongue and pharynx but may occur on the whiskers and all over the body. They are innervated by branches of the seventh, ninth, and tenth cranial nerves, which may reach far backwards (Fig. 137). In some species it has been shown that the fish is able to turn and snap at a piece of food placed near the tail. This power is lost if the branches from the cranial nerves are cut. In mammals taste-buds serve to discriminate only four qualities (salt, sour, bitter, and sweet), most of our so-called 'tasting' being in reality the smelling of the food in the mouth. In fishes, also, the four taste qualities are discriminated by the taste-bud system, and it has been shown that the minnow (*Phoxinus*) continues to make such discriminations after the forebrain has been removed. Other chemical discriminations are made by the nose, however, and can only be performed with an intact forebrain. Thus *Phoxinus* tastes and smells the same classes of substances as man does. The taste-buds are exceedingly sensitive, the threshold for sweet substances being 500 times and for salt 200 times lower than in man. On the other hand, some substances that are very bitter for us produce little reaction in *Phoxinus*.

In many fishes the nose is one of the chief receptors (macrosmatic). There are two nostrils on each side, allowing for the sampling of a stream of water (Fig. 119). The nose does not communicate with the mouth, except in a few fishes that live buried in the sand (*Astroscopus*).

The sense of smell is used to find food and for recognition of the sex of members of the same species. Minnows can be trained to give distinct reactions to extracts made from the skin of other species of fish living in fresh water. In the presence of 'alarm substances' produced by damaged skin of a member of the same species, minnows (and other fishes) show a 'fright reaction', scattering and refusing food.

The state of development of the nose is very varied. It is large in macrosmatic solitary predators such as *Anguilla* and in many schooling species that also have well-developed eyes (*Phoxinus*, *Gobio*). Daylight predators, on the other hand, are microsmatic (*Esox*, *Gasterosteus*). Other evidence shows that fishes can discriminate between the smells of water plants and between the waters of different streams. It is likely that this provides part of the mechanism by which salmon return to the stream in which they were born, having been conditioned as fry to the smell of its water. It has been suggested that they might be decoyed to return to a stream other than that where they were hatched by conditioning them as fry to a substance such as morpholene to which they have a high sensitivity although it is neither an attractant nor repellant.

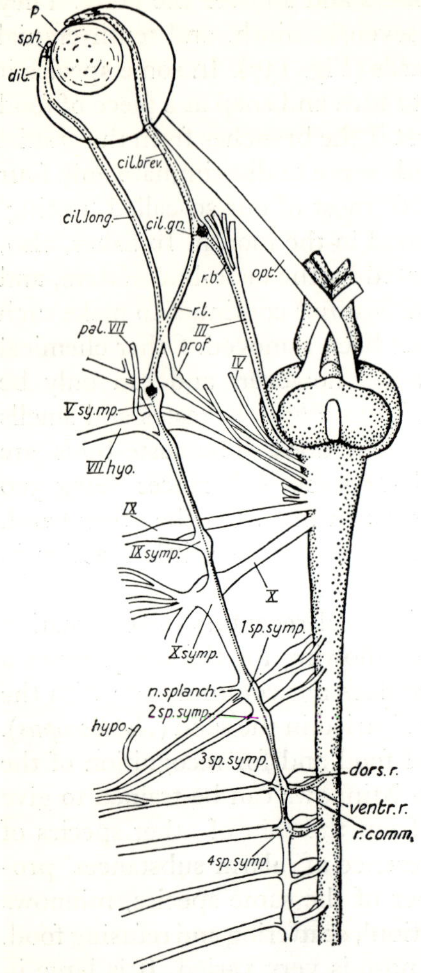

FIG. 138. Diagram of ventral view of the sympathetic system of the front part of the body of *Uranoscopus*, showing the fibres in the sympathetic and oculomotor that are responsible for the light reflex. (From Young.)

cil.brev. cil.long., short and long ciliary nerves; ***cil.gn.***, ciliary ganglion; ***dors.r.*** dorsal root; ***dil.*** dilatator muscle; ***hypo.*** hypoglossal; ***n.splanch.*** splanchnic nerve; ***opt.*** optic nerve; ***p.*** pupil; ***pal.VII***, palatine branch of facial; ***prof.*** profundus; ***r.b.*** short root of ciliary ganglion; ***r.comm.*** ramus communicans; ***r.l.*** long root of ciliary ganglion; ***sph.*** sphincter muscle; ***ventr.r.*** ventral root; *III–X*, cranial nerves with their sympathetic ganglia (*V.symp.* &c.); *VII hyo.* hyomandibular branch of facial; *1–4 sp.symp.* spinal sympathetic ganglia.

19. Touch

Touch is, of course, well developed in fishes, and in many species there are special sensory filaments, which presumably serve this sense. They are usually developed around the mouth, as in the catfish; in other fishes they are modifications of the fins, for instance, the pectoral fins of gurnards, which also contain chemoreceptors.

20. Autonomic nervous system

The autonomic nervous system of bony fishes is organized on a plan rather different from that both of elasmobranchs and of land animals. There is a chain of sympathetic ganglia, extending from the level of the trigeminal nerve backwards, a ganglion being found in connexion with each of the cranial dorsal roots (Figs. 138 and 139). These ganglia do not receive pre-ganglionic fibres from the segments in which they lie, but by fibres that run out in the ventral roots of the trunk region and thence forwards in the sympathetic chain. This emergence of the pre-ganglionic fibres for the head in the trunk region recalls the arrangement in land animals.

Each trunk sympathetic ganglion, besides receiving a white ramus communicans of pre-ganglionic fibres from its spinal nerve, also sends a grey ramus back to that nerve, this ramus carrying post-ganglionic fibres to the skin. Some of these fibres control the melanophores, causing them to contract (p. 259). In elasmobranchs there are no grey rami communicantes and no sympathetic system in the head (p. 173); the differences between the two groups are therefore very striking.

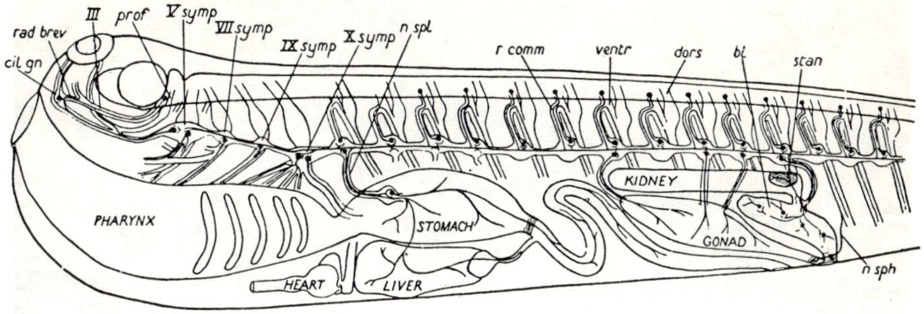

FIG. 139. Diagram of the autonomic nervous system of *Uranoscopus* seen from the side. ***bl.*** mesonephric bladder; ***cil.gn.*** ciliary ganglion; ***dors.*** dorsal root; ***n.sph.*** nerve to anal sphincter; ***n.spl.*** splanchnic nerve; ***prof.*** nervus ophthalmicus profundus; ***rad. brev.*** short root of ciliary ganglion; ***r.comm.*** ramus communicans (including both white and grey fibres); ***stan.*** 'corpuscle of Stannius' (adrenal cortical tissue?); ***ventr.*** ventral root; ***III,*** oculomotor nerve; ***V–X symp.*** sympathetic ganglia associated with the cranial nerves. (From Young, *Quart. J. Micr. Sci.* **75**.)

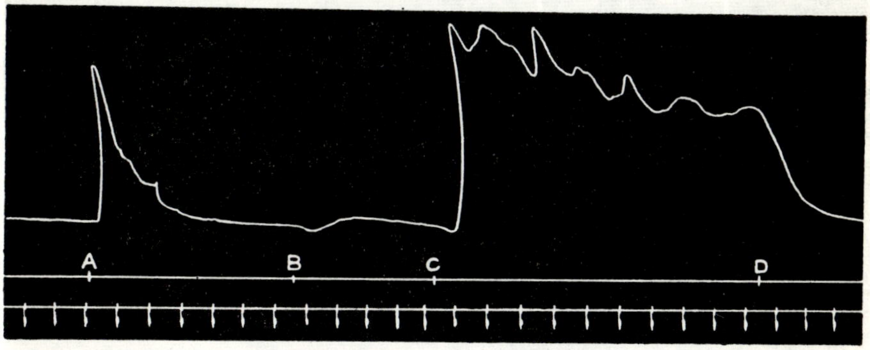

FIG. 140. Tracing of the contractions of a strip of the stomach muscle of the angler-fish, *Lophius*, attached to a lever. Time in minutes. At A, faradic stimulation of the vagus nerve. Drugs then added to the solution to make, at B acetyl choline 1/1,000,000; at C acetyl choline 1/100,000; at D adrenaline 1/100,000. (From Young, *Proc. Roy. Soc.* 120.)

Little is known about the parasympathetic system of bony fishes. The oculomotor nerve carries fibres to the iris, which work in the opposite direction to fibres from the sympathetic (p. 214). There is also a well-developed vagal system, but so far as is known no parasympathetic fibres in other cranial nerves and probably no sacral parasympathetic system. Electrical stimulation of the vagus nerve produces movements of the stomach but not of the intestine; the latter, however, shows movements when the splanchnic nerve is stimulated. In most of the viscera acetyl choline causes initiation of

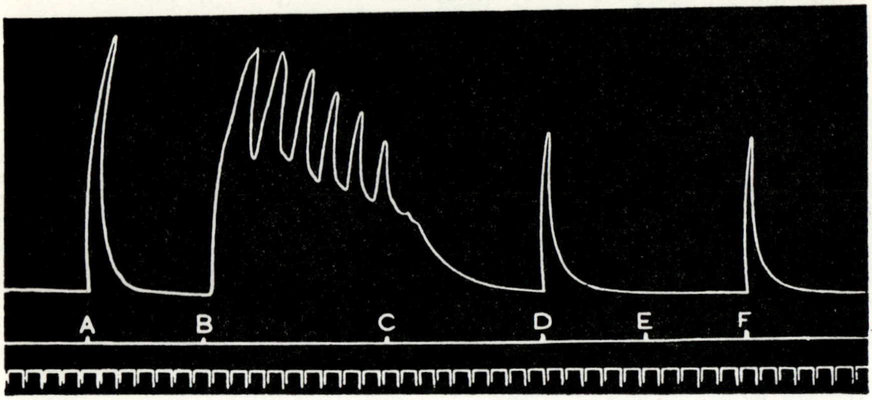

FIG. 141. Tracing of contractions of the muscle of the urinary bladder of *Lophius*, attached to a lever. At A, D, and F faradic stimulation of vesicular nerve. Drugs added to make, at B acetyl choline 1/2,000,000; at C adrenaline 1/500,000; at E ergotoxine 1/50,000. Time, minutes. (From Young.)

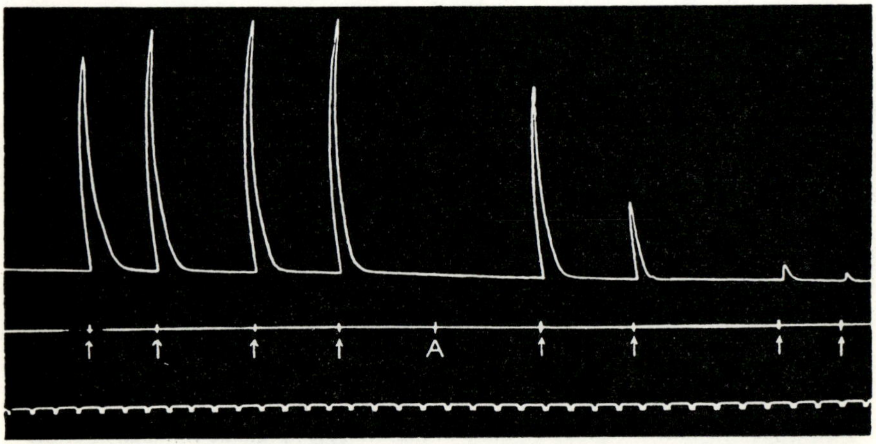

FIG. 142. Tracing to show effect of atropine 1/50,000 added at A, on the contractions of the bladder of *Lophius* produced by faradic stimulation of the vesicular sympathetic nerve. Time, minutes. (From Young.)

rhythmic contractions and these are inhibited by adrenaline (Figs. 140 and 141). In *Lophius* this is true of the stomach, with motor-fibres from the vagus, intestine, with sympathetic motor-fibres, and of the muscles of the bladder, which contract on stimulation of the hinder sympathetic ganglia. However, in the trout adrenaline causes contraction of the stomach (Burnstock, 1958). The effect of the nerves to the bladder is prevented by atropine (Fig. 142) but not by ergotoxine (Fig. 141), though the latter is the drug that in mammals often inhibits sympathetic motor-fibres. In these fishes, therefore, it is not possible to divide up the autonomic nervous system into sympathetic and parasympathetic divisions by either anatomical, physiological, or pharmacological criteria. Presumably the two 'antagonistic' systems found in mammals are a late development, allowing for a delicate balancing of activities for the maintenance of homeostasis.

21. Behaviour patterns of fishes

The well-developed receptors and brain of the teleostean fishes constitute perhaps the most important of all factors in giving them

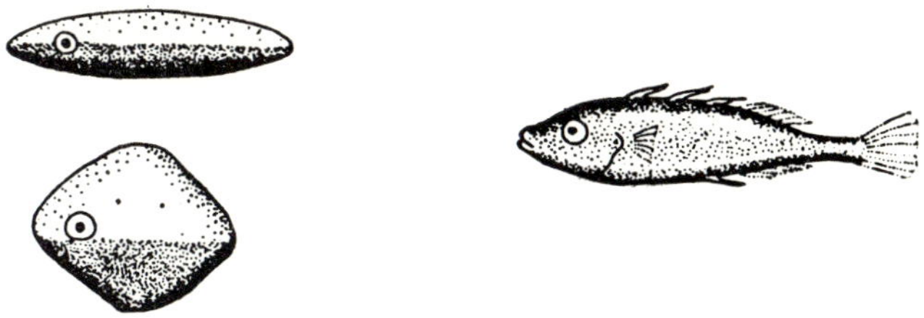

FIG. 143. The red belly of the stickleback releases attacking behaviour in other males and following by females. Of the above models only the two on the left acted as releasers. (From Tinbergen, *Wilson Bulletin*, 1948.)

their great success. Varied habits and quick actions enable the fish to make full use of the possibilities provided by the special features of their structure—the air bladder, mouth, and so on. The receptors and brain make it possible for the fish to learn to react appropriately to many features of its surroundings. Thus the eyes besides orientating the fish to movements in the visual field allow the discrimination of wavelengths and distinct reactions to differing shapes (see Bull, 1957).

The social behaviour of many species includes the development of special 'releasers', shapes, colours, or postures that are displayed by one individual and elicit specific reactions in another (Fig. 143).

There is no doubt that fishes possess great powers of learning. They can form conditioned reflexes involving discrimination of tones, also second-order conditioned reflexes, in which after the animal has learnt to give a certain behaviour in response to a visual stimulus it is then taught to associate the latter with an olfactory stimulus. There are many other examples of such powers, but unfortunately we have as yet little information as to the way in which they are brought about by the brain. Nor have the naturalists provided us with very clear examples of the use of these powers by fishes in nature. There are many tales of carp coming to be fed at the ringing of a bell, and similar powers of association must play a part in the life of fishes in more natural situations. Bull has shown that fishes can be trained to discriminate between very small differences of water flow, temperature, salinity or pH, and no doubt it is by means of such powers that they normally find a suitable habitat.

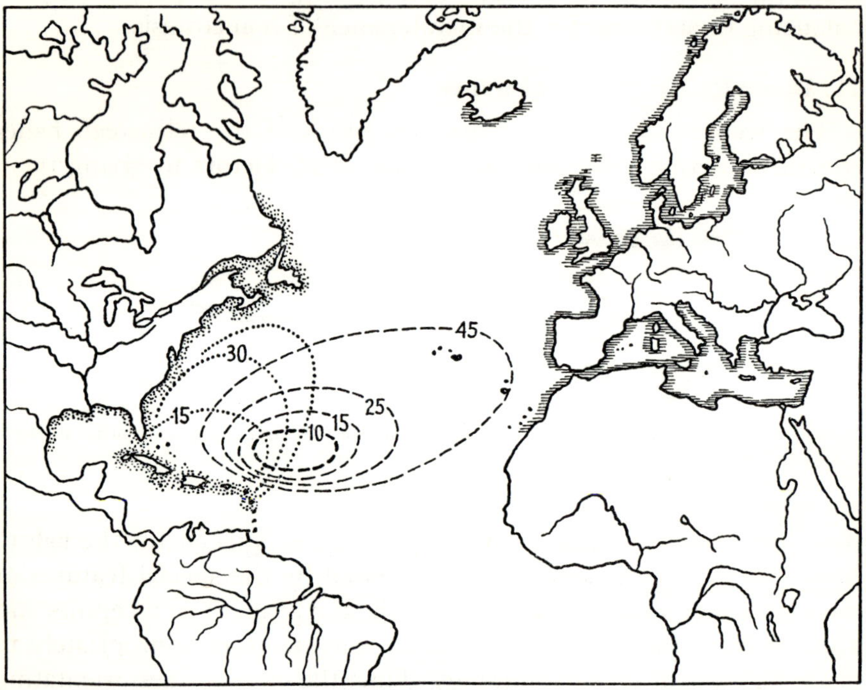

FIG. 144. Migrations of the eels. The European species (*A. anguilla*) occurs along the coasts outlined with lines, the American species (*A. rostrata*) where there are dots. The curved lines show where larvae of the lengths indicated (in millimetres) are taken. (After Norman.)

The migrations of fishes have attracted much attention, but are still imperfectly understood. They vary from the 'catadromous' downward migration of young animals to the sea and the reverse 'anadromous' movement to breed, to the astounding journeys of the eels, 3,000 miles westwards from Europe or eastwards from America to their breeding-place in the Sargasso Sea (off Bermuda) and the return of the elvers to the homes of their parents (Fig. 144). No one has yet discovered the factors that direct these movements, currents may play a part, but can hardly be the only influence. Indeed it has been suggested that European eels never complete the journey but die in their own continental waters. The populations of so-called European eels (*Anguilla anguilla*) would then be maintained by reinforcements of larvae of the American *A. rostrata*, the differences between the two being due to temperature and other factors (Tucker, 1959).

Social behaviour is marked in many species and shoals of some fishes may contain many thousands of individuals. The animals are presumably kept together in most species by visual stimuli, though sounds may play a part. Shoaling gives protection to small fishes, and in some species the animals come together in shoals to breed (herrings). There may also be some advantage for the finding of suitable feeding conditions, but on all these points we can do little more than speculate and hope for further information.

VIII

THE EVOLUTION OF BONY FISHES

1. Classification

Class Actinopterygii
  Superorder 1. Chondrostei
    Order 1. Palaeoniscoidei. Devonian–Recent
      **Cheirolepis; *Palaeoniscus; *Amphicentrum; *Platysomus, *Dorypterus; *Cleithrolepis; *Tarrasius; Polypterus*, bichir
    Order 2. Acipenseroidei. Jurassic–Recent
      **Chondrosteus; Acipenser*, sturgeon; *Polyodon*, paddle-fish
    Order 3. Subholostei. Triassic–Jurassic
      **Ptycholepis*
  Superorder 2. Holostei. Triassic–Recent
      **Acentrophorus; *Lepidotes; *Dapedius; *Microdon; Amia*, bowfin; *Lepisosteus*, gar-pike
  Superorder 3. Teleostei. Jurassic–Recent
    Order 1. Isospondyli
      **Leptolepis; *Portheus; Clupea*, herring; *Salmo*, trout
    Order 2. Ostariophysi
      Cyprinus, carp; *Tinca*, tench; *Silurus*, catfish
    Order 3. Apodes
      Anguilla, eel; *Conger*, conger eel
    Order 4. Mesichthyes
      Esox, pike; *Belone; Exocoetus*, flying fish; *Gasterosteus*, stickleback; *Syngnathus*, pipe-fish; *Hippocampus*, seahorse
    Order 5. Acanthopterygii
      **Hoplopteryx; Zeus*, John Dory; *Perca*, perch; *Labrus*, wrasse; *Uranoscopus*, star gazer; *Blennius*, blenny; *Gadus*, whiting; *Pleuronectes*, plaice; *Solea*, sole; *Lophius*, angler-fish

2. Order 1. Palaeoniscoidei

The actinopterygian stock has been distinct since Devonian times. The early representatives lacked many of the specializations that we find in the successful bony fishes today and showed features of similarity to the Crossopterygii. These Devonian Actinopterygii had not yet acquired the striking signs of full mastery of the waters, which are so characteristic of the group today. They resembled their ancestors the placoderms and their cousins the crossopterygians in being rather clumsy, heavily armoured creatures. From this early type many lines have been derived and can be followed with some completeness to their extinction or modern descendants. Various classifications have been suggested. The one used here is simple but for that very reason obscures the multiplicity of parallel lines. A recent classification recognizes fifty-two orders of Actinopterygii (Grassé).

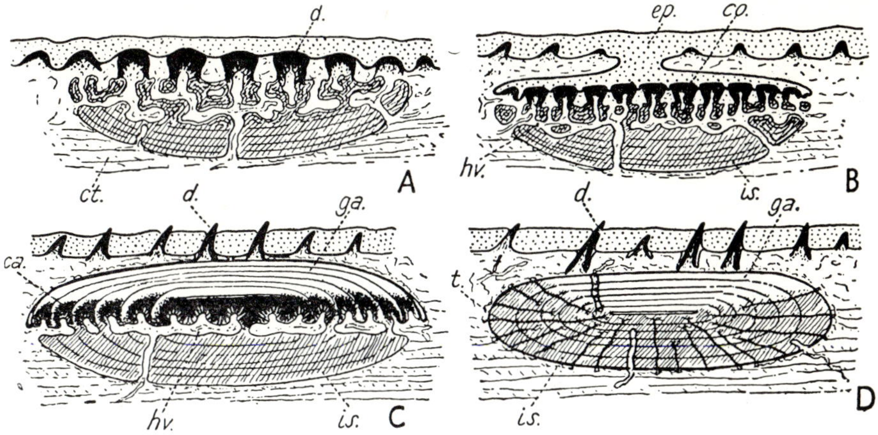

FIG. 145. Scales of some early fishes.

A, hypothetical condition with denticle-like substance (*d.*) attached to a basal bony plate lying in the connective tissue (*ct.*); B, 'cosmoid' scale of early crossopterygians, showing the cosmine layer (*co.*); epidermis (*ep.*); vascular canals (*hv.*) and underlying 'isopedin' (*is.*); C, palaeoniscoid scale with layers of 'ganoin' (*ga.*); D, lepidosteoid scale of the gar-fish with tubules (*t.*). (From Goodrich, *Vertebrata*, A. & C. Black, Ltd.)

The Devonian and Carboniferous forms are grouped together in the order Palaeoniscoidei, and animals of similar type survive today as *Polypterus*, the bichir of African rivers, which though showing some specializations remains in its general organization near the palaeoniscid level.

A typical Palaeozoic palaeoniscid such as **Cheirolepis* was a long-bodied creature (Figs. 146 and 147) with a heterocercal tail, single dorsal fin, and pelvic fins placed far back on the body. The pectoral and pelvic fins had broad bases and the radials fanned out from a small muscular lobe, present in all early actinopterygians but lost in later forms. The body was covered with thick rhomboidal scales very similar to those of acanthodians. They articulated by peg and socket joints and have a structure known as palaeoniscoid (Fig. 145). The scale is deeply embedded and grows by addition both to the bony or isopedin portion and to the shiny surface-layer, the ganoin, which thus becomes very thick. There is a middle layer of 'pulp' corresponding to the cosmine layer of the cosmoid scale of Crossopterygii

(p. 269) and the two types have obvious similarities, though it is not clear how they are related.

The skull was built on a distinctly different plan from that of Crossopterygii, in that there was no joint such as was present in those fishes to allow the front part to flex on the hind. The jaw support was amphistylic, in the sense that the palatoquadrate was attached to the neurocranium by a basal process, but the otic process did not reach the skull and the hind end of the jaw was supported by the hyomandibula. There were even more dermal bones than are found in modern Actinopterygii, arranged so as to form a complete covering for the chondrocranium and jaws. These bones were derived from the original scaly covering of the head and the naming and comparing them with the bones of other forms is a matter of some difficulty. Some of the main bones resemble in appearance and shape those found in tetrapods, but there are others for which no such homologues can be found, and sometimes there is considerable difficulty in recognizing even the main outlines of the pattern. The problem is that we have no rigid criterion by which to set about giving names to the skull bones. No system yet discovered is wholly satisfactory, and we must admit to insufficient knowledge of the factors that determine that bone shall be laid down in certain areas and that sutures shall separate these from each other. However, some of the dermal bones lie in relation to the lateral line canals (or rows of neuromasts), which latter may provide the stimulus to bone formation. The lines are remarkably constant, perhaps because of their function in detecting water movements in relation to swimming, and this is the factor that determines the position of many of the bones. Others fill in the spaces between (anamesic bones). Yet others may be differentiated in relation to the teeth. However, the number of bones along any one line may vary greatly even in one species (e.g. in *Amia*). The whole pattern is more variable in fishes than in higher vertebrates, but it is usual to consider that the bones of early Actinopterygii resemble those of Crossopterygii and of the early amphibians (Fig. 194).

The roof of the skull usually shows a large pair of parietals between the eyes, and post-parietals behind these. Between the parietals and the nostrils there are frontal bones and the front of the head usually also carries a number of rostral bones, not found in higher forms. Behind the post-parietals in the midline is a series of extrascapular bones.

The side of the skull of palaeoniscids is covered by numerous bones, including a series of pre- and post-frontals, post-orbitals, and jugals around the eyes. The outer margin of the upper jaw is covered by premaxillae and maxillae, which are the main tooth-bearing bones. Behind the orbital series of bones the cheek is very variable. Sometimes there is a large bone identifiable as a pre-opercular, with a series of opercular bones behind it. The lower portion of the throat was covered by a series of gular plates. The spiracle in these early forms opened above the opercular bones. The pectoral girdle was attached to the back of the skull by a supracleithrum, below which a cleithrum and clavicle made a series of dermal bones behind the gills, covering the cartilaginous girdle. The roof of the mouth contained a median parasphenoid, with paired prevomers in front of it, and a series of pterygoid bones occupied the space between it and the edge of the jaws, the palatine, ectopterygoid, pterygoid, and sometimes others. Finally the lower jaw, besides the main dentary carrying the teeth, shows many small bones such as the pre-articular and coronoid on the inner surface; splenial, angular, and surangular on the outside.

It will be clear that this skull of **Cheirolepis* may be closely compared with the skull of a crossopterygian or a modern teleostean. The general plan is related to that of the lateral line organs arranged along occipital, supratemporal, and infra-orbital lines. The numerous small bones are evidently similar in the different groups, though it is not easy to assign a suitable name to every one of the more numerous bones of the earlier forms.

These palaeoniscids from the Middle Devonian were rather rare freshwater fishes; they had sharp teeth and probably lived on invertebrates. We have no information about their internal anatomy, but it seems not unlikely that the air-bladder possessed a wide opening to the pharynx (as it still does in *Polypterus*, descended from this stock) and that they breathed air, as did other Devonian fishes. However, they did not have internal nostrils, which are found in the old crossopterygians.

During the Carboniferous and Permian the palaeoniscids were numerous, mostly as small, sharp-toothed fishes. Several distinct lines became laterally flattened and acquired an outwardly symmetrical tail and blunt crushing teeth (Fig. 147). These characteristics probably indicate a habit of feeding in calm waters, perhaps mainly on corals, and they have appeared several times in the actinopterygian stock (p. 241). Palaeoniscids of this type were formerly placed together in a family Platysomidae, but it is now considered probable that the type arose independently several times; thus **Amphicentrum* is found in the Carboniferous, **Platysomus* and **Dorypterus* in the Permian,

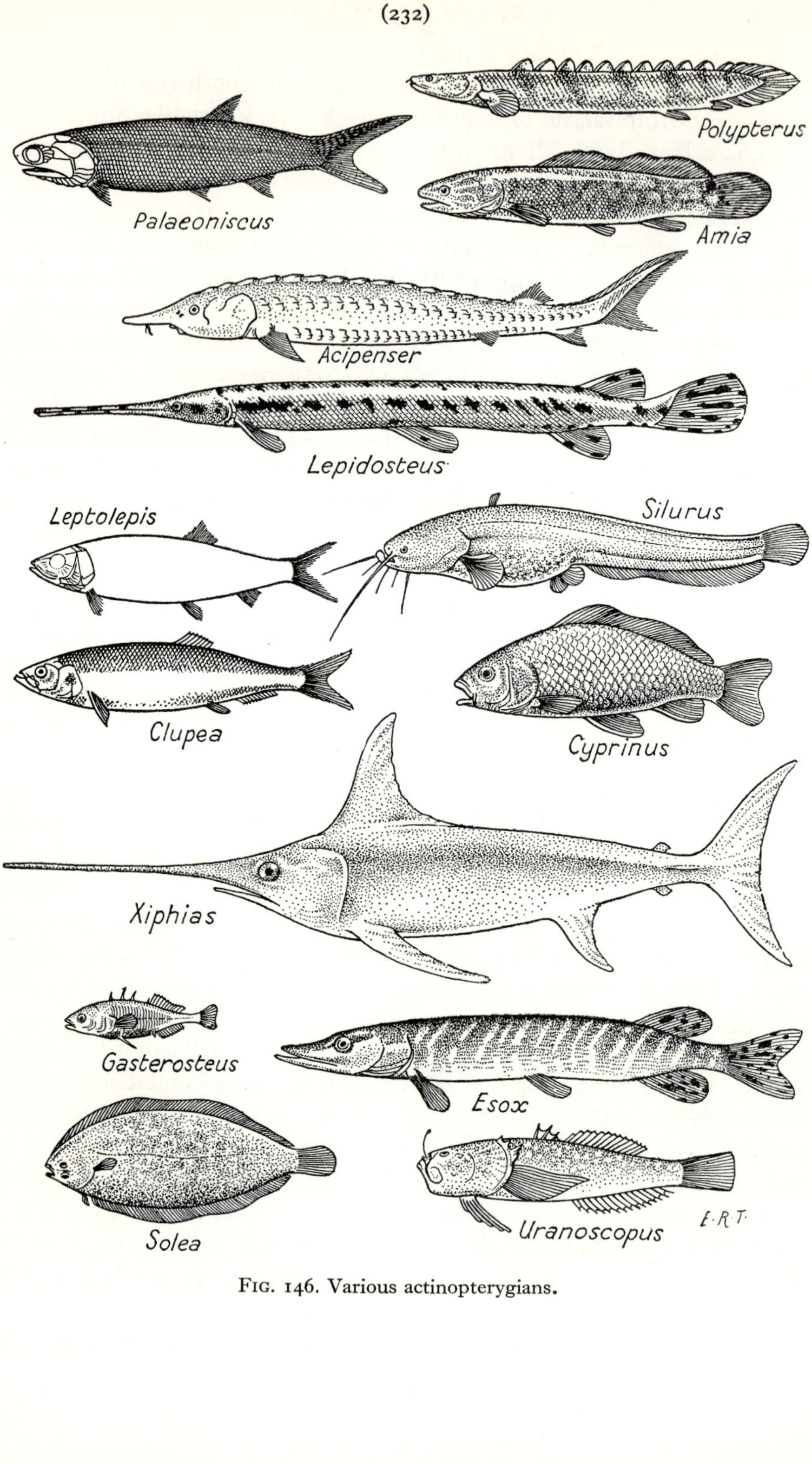

FIG. 146. Various actinopterygians.

**Cleithrolepis* in the Triassic. Similar forms arose again later among the holosteans and teleosteans and we have therefore evidence that this type of animal organization tends to evolve into deep-bodied creatures. **Dorypterus* further resembles modern teleosteans in a great reduction of its scales and in the forward movement of the pelvic fins.

Towards the end of the Triassic animals of typical palaeoniscid type became rare; they were replaced by their more active and speedy, mainly marine descendants, the Holostei (p. 234). Certain of the lines that branched off in the Palaeozoic have, however, survived to the present time, and in spite of subsequent specializations they give us some idea of the characteristics of these early Actinopterygii. Perhaps the most interesting of these survivals are *Polypterus*, the bichir, and the related *Calamoichthys*, both inhabiting rivers in Africa. The air-bladder shows some similarity to a lung. It forms a pair of sacs lying ventrally below the intestine and opening to the pharynx by a median ventral 'glottis' (Fig. 157). This is the arrangement found in lung-fishes (except *Ceratodus*) and in tetrapods, and it seems reasonable to suppose that it has survived in *Polypterus* from Palaeozoic times. However, it is not certain to what extent the air-bladder is still used as a lung, for *Polypterus* cannot survive out of the water.

This fish shows many other ancient characteristics. The covering of thick rhomboidal scales, hardly overlapping, gives the animal an archaic appearance; the structure of the scales is 'palaeoniscoid'. In the skin there is a layer of denticles outside the scales. The presence of a spiracle, the arrangement of the skull bones, and many other features suggest that *Polypterus* is essentially a palaeoniscid surviving to the present day. In the intestine there is a spiral valve, which appears to have been present in the early Crossopterygii and Actinopterygii (as judged from fossilized 'coprolites') and occurs today not only in the Dipnoi but also in sturgeons and, though much reduced, in *Lepisosteus* and *Amia*. There is a single pyloric caecum in *Polypterus* (the caeca are well developed in sturgeons, *Lepisosteus*, and *Amia*). The tail of *Polypterus* is no longer markedly heterocercal, but shows distinct signs of that condition. We can even find a parallel among Carboniferous palaeoniscids for some of the special features of *Polypterus*. The long body and dorsal fin are found in the fossil **Tarrasius*, which may have been close to the ancestry of *Polypterus*, though it lacks the covering of scales. The pectoral fin in **Tarrasius*, as in *Polypterus*, has a peculiar lobed form, which has been compared with the 'archipterygial' pattern (p. 269) and hence held to show that these animals are related to the Crossoptergyii. The resemblance is, however, only superficial and the plan of the fin is essentially actinopterygian. In the brain there is a thin pallium, thick corpus striatum, and a valvula cerebelli. The pituitary is remarkable in that the hypophysial sac remains open to the mouth. In this and other features (persistent pronephros) there are signs of neoteny.

3. Order 2. Acipenseroidei

The sturgeons are a rather isolated line descending from the palaeoniscids and characterized by reduction of bone. This was already apparent in the Jurassic **Chondrosteus*. *Acipenser* and other modern sturgeons live in the sea but migrate up the river to breed. They may reach a very large size (1,000 kg) and since a tenth of this is caviar they are exceedingly valuable. They feed on invertebrates, which they collect from mud stirred up from the bottom by a long snout. This is flattened into a pear-shaped structure in *Polyodon*, the purely freshwater paddle fish of the Mississippi and in *Psephurus* in China. The mouth of all sturgeons is small and the jaws weak and without teeth. In *Polyodon* there is a filtering arrangement of gill-rakers in the pharynx. The jaws of sturgeons hang free from the hyomandibular and symplectic, and can be swung downward and forward during feeding. The skull and skeleton is almost wholly cartilaginous and the dermal skeleton much reduced. The tail is covered with rhomboidal scales, but on the front of the body there are five lines of bony plates bearing spines, with the skin in between carrying structures similar to denticles. There is an open spiracle. The internal anatomy of the sturgeons shows various features that have been held to show affinity with the elasmobranchs; for instance, besides the spiral valve there is a conus arteriosus in the heart and a single pericardio-peritoneal canal. However, there can be no doubt that they are descended from an early offshoot from the actinopterygian line. They retain some features lost by most members of the line, but resemble the Teleostei in other characters, for instance a thin roof to the cerebral hemispheres.

The palaeoniscids and sturgeons may be grouped together in a Superorder Chondrostei and placed with them is a third Order Subholostei, probably a mixed group, including forms that resemble palaeoniscids, but show various trends towards the holostean grade of organization (**Ptycholepis*).

4. Superorder 2. Holostei

During the later Permian period the palaeoniscids gave rise to fishes of a different type, which replaced their ancestors almost completely during the Triassic and flourished greatly in the Jurassic. We may group together the fishes of this type as Holostei but the term is used variously by different authors and includes several lines, whose relationships are not clear. The earliest holostean, **Acentrophorus* from the upper Permian, is much like a paleoniscid but with a small mouth, shorter, deeper body and slightly upturned tail. This 'abbreviated heterocercal' tail was presumably made possible by the changed swimming habits resulting from the use of the air-bladder as a hydrostatic organ. If the fish floats passively there is no need for a heterocercal tail to direct the head downwards (p. 140). Similarly, the head does not need to be flattened to produce an upward lift. The development of the air-bladder has thus made possible the lateral flattening and shortening of the body so characteristic of later Actinopterygii. The body of holosteans was at first covered with thick ganoid scales, but these became thinner in later types. The jaw suspension is characteristic, the maxilla being freed from the pre-opercular. As a result the lower jaw could now be protruded forwards in front of the upper and a 'sucking' action, characteristic of teleosts was evolved, the prey being drawn into the mouth from a distance (Gardiner, 1960). By a change in the insertion of the adductor mandibulae muscle a more powerful jaw action then became possible. Some of the holosteans achieved crushing teeth and replaced the dipnoans in the early Mesozoic. There are various smaller distinctive holostean features, such as the loss of the clavicle.

We do not know whether fishes of this type arose from a single palaeoniscid stock; it is very likely that the change occurred several times, and that throughout the Triassic and Jurassic there were several lines with these holostean characteristics, evolving separately. During the Cretaceous they became fewer, being replaced by their teleostean descendants, but two holosteans survive today, *Lepisosteus* the gar-pike (often written *Lepidosteus*) and *Amia* the bow-fin. These are freshwater fishes, living in the American Great Lakes and other parts of eastern North America, but the group is mainly a marine one, having taken to the sea in the Trias at a time when other groups were doing the same (palaeoniscids, coelacanths, elasmobranchs). The basic cause of this movement is not known, but perhaps there was an increase of planktonic and invertebrate life on which the fish depended.

Lepisosteus shows a rather primitive structure and must have remained at approximately the Triassic stage. With its complete armour of thick scales (Fig. 146) it presents all the appearance of a primitive fish. The air-bladder opens to the pharynx and the gar-pikes come to the surface to gulp air. On the other hand, it has developed certain special features, especially the long jaws, with which it catches other fishes, and the nearly symmetrical tail. Fossils similar to the modern gar-pike are found in the Eocene.

Some of the later holosteans became deep- and short-bodied and developed a small mouth with flat crushing teeth or a beak, for instance, **Lepidotes* (Trias to Cretaceous), and **Dapedius* (Jurassic). They probably browsed on corals, like the modern parrot fishes (Scaridae). **Microdon* and other 'pycnodonts' became laterally flattened, like some palaeoniscids and the modern sea butterflies (Chaetodontidae).

Another line of holostean evolution, developing from the original stock, retained the streamlined body and from these both the modern teleosteans and the amioids were evolved (Fig. 147). **Caturus* (Trias to Cretaceous) was covered with thick scales, but in **Pachycormus* (Cretaceous) they are thinner; these were active pelagic predators. In *Amia* the scales became reduced to single bony cycloid scales, as in Teleostei. Meanwhile other changes took place, the tail fin becoming externally completely symmetrical and the maxilla and other cheek bones reduced. *Amia* has nearly reached the teleostean stage but retains certain primitive features in the skeleton and the small eggs with holoblastic cleavage.

5. Superorder 3. Teleostei

The groups so far considered have been nearly completely replaced by the Teleostei, fishes derived from a holostean stock, which have carried still farther the tendencies to shortening and symmetry of the tail, reduction of the scales, and various changes in the skull, such as reduction of the maxilla. The type apparently arose in the sea in late Triassic times, but remained rare until the Cretaceous, by which time several different lines of evolution had already begun.**Pholidophorus* from the Trias still carried an armour of thick scales but may well have given rise to **Leptolepis* from the Jurassic and Cretaceous, which is generally considered to be close to the ancestry of all Teleostei and may be placed close to the order Isospondyli, many of which are still alive. **Leptolepis* was a long-bodied fish with the pelvic fins placed far back, a skull with a full complement of bones, and a large maxilla. The scales still show traces of the ganoin layer.

From some fish like these leptolepids have been derived the 20,000 or more species of bony fish found today. It is natural that in any group of animals that has evolved relatively recently classification will be difficult, because the separate twigs of the evolutionary bush will show little difference from each other and there may be much parallel evolution. It is only when intermediate forms have become extinct that clear-cut major groups appear. It is therefore not easy to find useful subdivisions of Teleostei; we may divide them among five orders but most classifications require many more.

The first, Isospondyli, fish with soft rays, show primitive features in the large maxilla, which forms the posterior margin of the upper jaw, the persistence of an open duct to the air-bladder, and the posterior position of the pelvic fins. Fishes of essentially similar type are known as far back as the Cretaceous (***Portheus*). Several familiar fishes are of this type, including the salmon and trout (*Salmo*) and herrings (*Clupea*). The order Ostariophysi is a large group of fresh-water fishes, related to the Isospondyli. The anterior vertebrae are modified to form a chain of bones, the Weberian ossicles, joining the swim-bladder to the ear (p. 217). Here belong the carp and gold-fish (*Cyprinus*), roach (*Leuciscus*), and cat-fishes (*Silurus*). The eels (order Apodes) are rather isolated teleosts that diverged early from the main stock and retain many primitive features. The fourth order, the Mesichthyes, includes fishes such as the pike (*Esox*) and stickle-back (*Gasterosteus*) of structure intermediate between that of the more primitive forms and the latest spiny-finned teleosts. The pipe-fishes (*Syngnathus*) and sea-horses (*Hippocampus*) are probably related to the sticklebacks. The flying-fish (*Exocoetus*) also belongs in this group.

The members of the order Acanthopterygii are the most highly developed fishes, characterized by the stiff spines at the front of the dorsal and anal fins. The maxilla is short, the duct of the air-bladder is closed, the body shortened, and the pelvic fins far forward. Fishes of this type already existed in the Cretaceous (***Hoplopteryx*) and the condition may have been evolved along several different lines; the group includes a vast array of modern types. Here belong the perches (*Perca*), mullet (*Mugil*), wrasse (*Labrus*), John Dories (*Zeus*), blennies (*Blennius*) as well as the gadid fishes such as the whiting (*Gadus*). Anglers (*Lophius*), gurnards (*Trigla*), and the flat-fishes, the plaice (*Pleuronectes*) and sole (*Solea*) and others, are further members of this very large order.

6. Analysis of evolution of the Actinopterygii

Our knowledge of the history of the Actinopterygii is sufficiently complete for us to be able to state more definite conclusions about the process of evolution than has been possible from consideration of the more ancient and less perfectly known groups of fishes. First of all we may emphasize the persistence of change. No actinopterygian fish living in the Devonian is to be found today. *Polypterus* may be regarded as a living palaeoniscid, showing features that were common in the Carboniferous, but it has undergone many changes since that time. Similarly the living sturgeons show the stage of organization present in the Jurassic Chondrostei, but they also have changed much. *Lepisosteus* is in general structure similar to a Triassic holostean such as **Lepidotes* and *Amia* to a Jurassic one such as **Caturus*, but both have their own more recent specializations. It is not until we come to the Tertiary history of fishes that types belonging to recognizable modern genera can be found.

The Actinopterygii have therefore been changing slowly, but continuously, throughout the period of their existence. Was this change dictated in some way by a change in their surroundings? Unfortunately we cannot answer this question very clearly. The sea is a relatively constant medium, though not as 'unchanging' as is sometimes supposed. In particular the relative extent of sea and freshwater changes frequently and perhaps because of such a change the early freshwater actinopterygians took to the sea. Probably the life in the sea is not constant over long periods. Almost certainly the available nitrogen, phosphorus, and other essential elements change in amount. We have reason to suspect that there was an increase in the extent and productivity of the sea during the Triassic period. It may be that such gradual changes in the life in the water, depending ultimately on climate or inorganic changes, have been responsible for the continual change of the fish population. However, it cannot be said that we can detect evidence of any such relationship; there is no clear proof that the changes in the fish populations follow changes in the environment. For the present we can only note the fact that change occurs, even in animals living in the relatively constant sea.

Very striking is the fact that as evolution proceeds not merely does each genus change but whole types disappear and are replaced by others. Thus the palaeoniscid type of organization had disappeared almost completely by the Trias and become replaced by the holostean. A few members retained the old organization and still survive today as *Polypterus* and the sturgeons. Similarly the Holostei, with their abbreviated heterocercal tails, hardly survived into the Tertiary, but were replaced by the Teleostei, only *Lepisosteus* and *Amia* remaining to show the earlier organization.

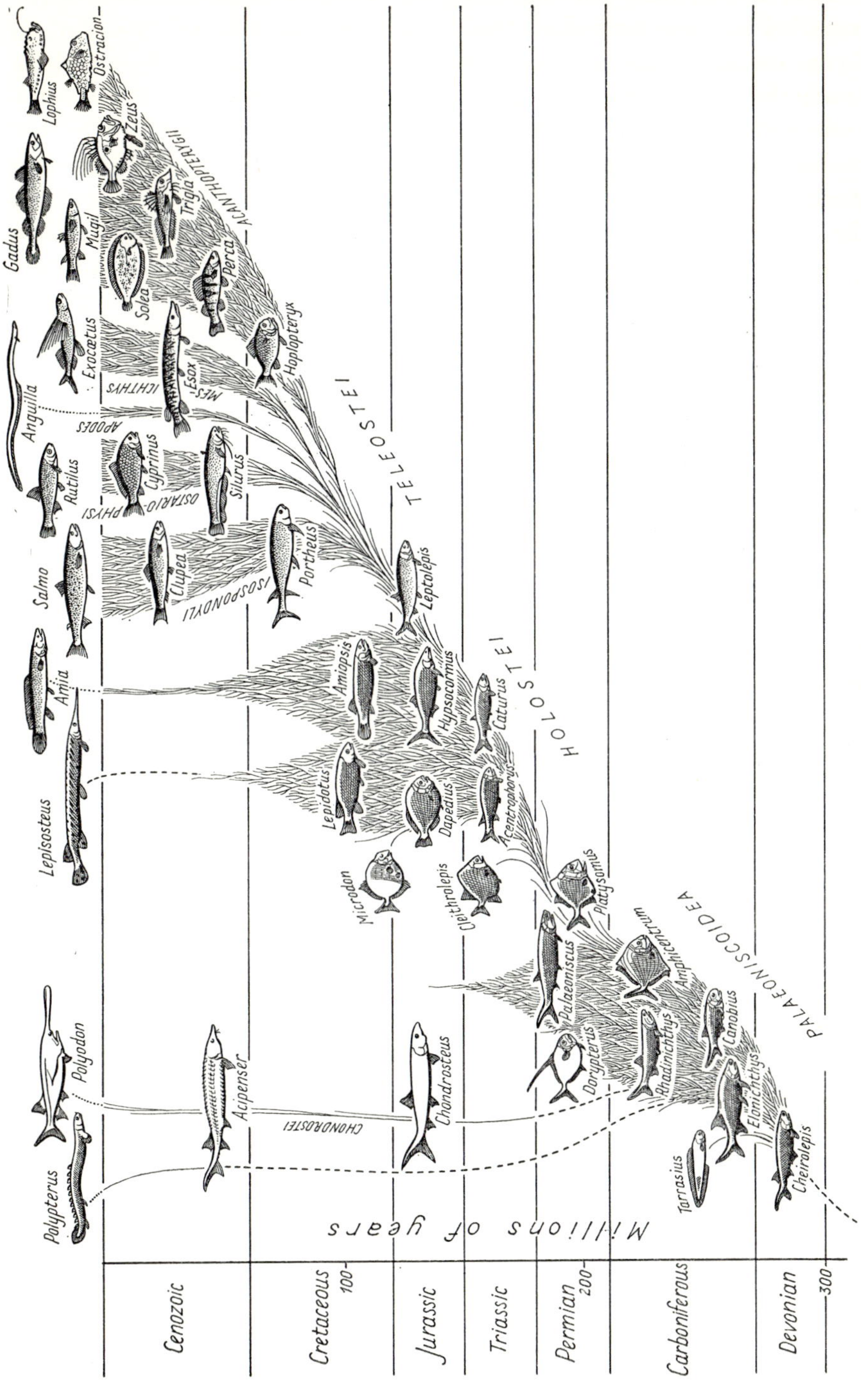

Fig. 147. Evolution of the Actinopterygii.

This replacement of one type by another appears much more remarkable when we reflect that by a 'type', say the palaeoniscid, we do not mean a homogeneous set of similar organisms all interbreeding. Quite the contrary; the palaeoniscids included many separate lines, each with its own peculiarities. When one 'type' therefore is thus replaced by another it must mean either that some one of these many stocks gives rise to a specially successful new population, which ousts the old ones or that *all* the members of the stock are changing their type together. It is not easy with a record such as that of the fish, which is far from continuous, to say which of these is true in any particular case, but we have sufficient evidence to be sure that either of them is possible.

There is no certain example in the Actinopterygii of a single new type replacing all the former ones, but the origin of the Teleostei may perhaps show a case of this sort. It is possible that the Teleostei is a monophyletic group, arising from a single type such as **Leptolepis* and proving so successful that nearly all creatures of previous types soon disappeared.

The Actinopterygii provide also examples of the other and perhaps even more interesting process, the parallel evolution of a number of different lines. There can be no doubt that from Palaeozoic times onwards several independent lines of fishes have shown similar changes. The tail has become shorter and more nearly symmetrical, the body has become flattened and deepened dorso-ventrally, while the pelvic fins have moved forward and the scaly armour has been reduced, all of these being signs of a more effective swimming and steering system (p. 244). As we come closer to modern times and the geological record becomes more complete we obtain more and more critical evidence that such changes have occurred in separate populations. Thus in the descendants of the holosteans we can recognize at least three such lines, that leading to the round-shaped **Microdon*, another leading through **Caturus* to *Amia*, and a third through **Leptolepis* to the Teleostei; probably there were many others. The important point is that although each line possessed peculiar specializations of its own, they all showed some shortening, development of symmetry of the tail, and thinning of the scales.

It is a considerable advance to be able to recognize such tendencies within a group. We begin to see the possibilities of a general statement on the matter. Instead of examining a heterogeneous mass of creatures called Actinopterygii we can recognize an initial palaeoniscid type and state that in subsequent ages this has become changed in certain specified ways and even at a specified rate. Imperfect though our knowledge still is, it enables us to approach towards the aim of our study, to 'have in mind' all the fishes of actinopterygian type.

This is to make the most of our knowledge: there remains a vast ignorance. We cannot certainly correlate this tendency of the fishes to change with any other natural phenomena. Put in another way, we do not know why these changes have occurred. The sea certainly did not stay the same, but it does not seem likely that its changes have been responsible for those in the fishes. It would be very valuable to be able to make a more certain pronouncement on this point, for the case is one of crucial importance. On land the conditions are constantly changing, and therefore we often find reason to suspect that changes in the animals are following environmental changes. But can this be so in the water?

The evolutionary changes in the Actinopterygii certainly involve a definite difference in the whole life. By development of the air-bladder as a hydrostatic organ the animals have become able to remain at rest at any level of the water, and thus, by suitable modification of the shape of the body and fins, to dash about with remarkable agility in pursuit of prey or avoidance of enemies. This has enabled them to dispense with the heavy armour and thus further to increase their mobility. But what made it *necessary* to adopt these changes? Not surely any actual change in the sea itself. We must look then for some factor imposed on the situation by the fishes themselves or the neighbouring animals that constituted their biotic environment. Is it the pressure of competition that has been responsible for the change in fish form? It may well be that the presence of an excess of fishes has led them continually to search for food more and more actively, and in new places, with the result that those types showing the greatest ability have survived. Given the initial genetic make-up of the palaeoniscids, further agility is most easily acquired by those fishes in which competition tended to produce shorter tails, thinner scales, and the other characteristics towards which the animals of this group tend.

The fact that the same set of changes can be produced independently from several different populations of approximately similar type (and presumably genetic composition) is strikingly shown by the specialized creatures evolved for life in coral reefs. Animals with rounded bodies and small mouths, sometimes with grinding teeth, have appeared independently several times; in the Carboniferous, **Amphicentrum*; Permian, **Platysomus* and **Dorypterus*; Triassic

**Cleithrolepis*; Jurassic **Microdon* and **Dapedius*, and in some modern teleosteans such as parrot fishes (Scaridae) and butterfly fishes (Chaetodontidae). This is very valuable evidence of the way in which a common stimulus can work on genetical constitutions that are similar but not identical. In this example the stimulus is a particular set of environmental conditions; in other cases a similar effect may be produced by the stimulus of competition between animals, which was probably the 'cause' of the common changes that affected so many descendants of the palaeoniscids.

The history of these fishes therefore gives plausible ground for the belief that the driving 'forces' that have produced evolutionary change are the tendencies of living things to do three things: (1) to survive and maintain themselves, (2) to grow and reproduce, (3) to vary from their ancestors, all of these operating under the further stress of any slow change in the environment.

Finally we must consider whether this change in the fishes can in any way be considered to be an advance. Several times we have found ourselves implying that this is so, that the later teleosts are 'higher' than their Devonian ancestors. We shall be wise to suspect this judgement as a glorification of the present of which we are part. However, perhaps this danger is less marked when we are dealing with fishes not ancestral to ourselves, whose 'advance' does not therefore bring them nearer to man. The judgement can be put into quite specific terms: the later Actinopterygii are 'higher' than the earlier ones because they are more mobile, quicker, and can live free in the water with lesser expenditure of energy than their ancestors. Unfortunately we have no means of estimating the total amount or biomass of fish matter that is supported by the teleostean organization, but it seems possible that it is absolutely greater than that of any previous type, say the holostean or palaeoniscid. If this is true, the change in plan of structure has perhaps led to an increase not only in fish biomass but in the total biomass of all life in the sea.

The teleostean plan has certainly allowed for the development of a great range of specializations, fitting the animals to all sorts of situations in the sea and fresh water. We must therefore not forget this adaptability in judging the status of the group: it seems likely that modern teleosts are more varied than any of their ancestors. This power to enter a wide range of habitats not previously occupied is perhaps the clearest sign of all that a group has 'advanced', and we have already suggested that it is in this sense of suiting animals to new modes of life that there has been a progress in evolution. It is true that the sea and fresh water have been in existence relatively unchanged throughout the period that we are considering: in a sense the fishes have not found a 'new' environment. But they have found endless new ways of living in the water.

IX

THE ADAPTIVE RADIATION OF BONY FISHES

THE variety of Actinopterygii is so great that it would be impossible to try to give a complete idea of it and the best that we can do is to consider various functions in more detail and specify some of the ways in which the animals have become specially modified.

1. Swimming and locomotion

The teleosts have perfected in various ways the process of swimming by the propagation of waves of contraction along the body. The situation is different from that of elasmobranchs on account of the presence of the air-bladder, serving to maintain the fish steadily at any given level in the water. The stabilization of the animal during locomotion has therefore become a wholly different problem, and the fins are correspondingly changed. In the sharks the pectoral fins serve to correct a continual tendency to forward pitching and by adjustment of their position they are used to steer the animal upwards or downwards in the water.

With an air-bladder the fishes have become freed from the tendency to remain at the bottom, which was prevalent in the more primitive forms and is still so common in sharks that it has several times produced wholly bottom-living ray-like types. A fish with an air-bladder needs only very little fin movement to maintain it at a constant depth or to change its depth. As Harris puts it, 'the elaborate mechanism of pectoral "aerofoils" and a lifting heterocercal tail is no longer needed for the maintenance of a constant horizontal cruising plane. Concomitant with the loss of the heterocercal tail in evolution occurs a rapid and tremendous adaptive radiation of the pectoral fin in form and function.' A stage in this process seems to have been the use of the paired fins to produce oscillating movements during hovering, and this is still found in *Amia* and *Lepisosteus*, fishes that remain relatively slow and clumsy.

Many of the lower teleosteans are relatively poor swimmers and some of them, like so many elasmobranchs, have become bottom-living. Thus in the catfishes there is a large anal fin, acting, like a heterocercal tail, to give lift and negative pitch. The pectoral fins are used to balance this tendency, very much as in sharks.

In the more specialized teleosteans, however, the pectorals are placed high up on the body and are used as brakes (Fig. 148). The plane of the fins' expansion is vertical and they thus produce a large drag force and a small lift force. This lift, of course, tends to make the fish rise in the water when stopping, and there is also a pitching

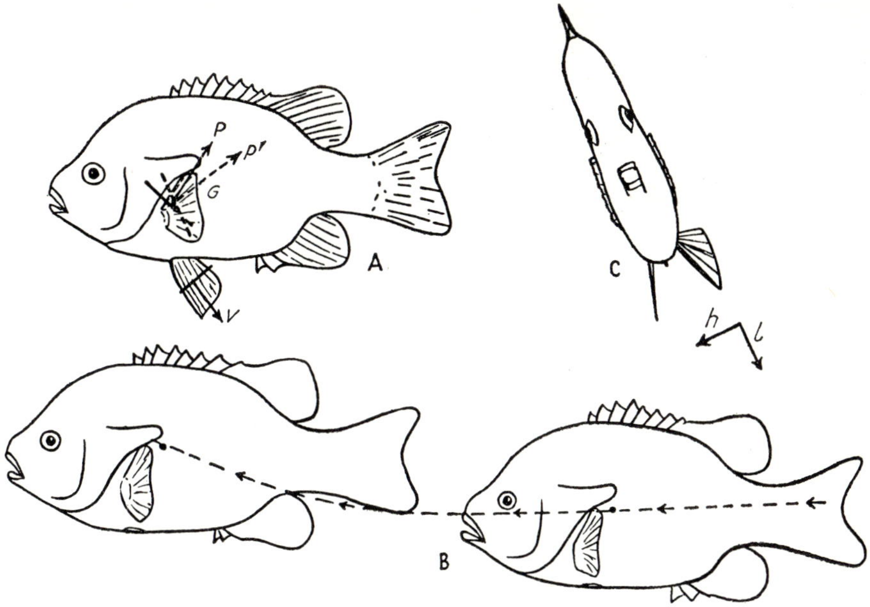

FIG. 148. Use of the paired fins for braking. A. Forces produced by the fins of *Lepomis* during deceleration. The pectoral and pelvic fin planes are represented by the heavy lines. *P* and *V*, the resultant forces on the pectoral and pelvic fin respectively. Dotted line and force *P'*, condition during action of pectoral fins only, pelvic fins being held in 'neutral' position. *G*, position of centre of gravity. B. Sun-fish stopping by extending pectorals. Pelvic fins amputated. Although body remains horizontal, the fish rises during the stop. C. Front view of sun-fish producing a rolling moment by the action of one pelvic fin. *h* and *l*, horizontal and lateral forces. (From Harris, *J. exp. Biol.* **15**.)

moment, depending on the position of the fin in relation to the centre of gravity, usually positive. That the fish does not rise in the water, or pitch, when it stops is apparently due to the anterior position of the pelvic fins, so characteristic of higher Actinopterygii, which has puzzled many morphologists. Experiments on the sun-fish (*Lepomis*) have shown that after amputation of the pelvic fins the fish rises in the water when stopping and raises its head (positive pitch). In fact the pelvic fins are able to produce a downward moment and they tilt the nose downwards. By alterations in their position they can be used to control the rising or diving movements and turning one of them outwards produces rolling (Fig. 148). It has been suggested that the pelvic fins function as keels to prevent rolling, but their amputation in *Lepomis* does not produce excessive rolling. Stability in the transverse plane is presumably assured by the dorsal and anal fins. The

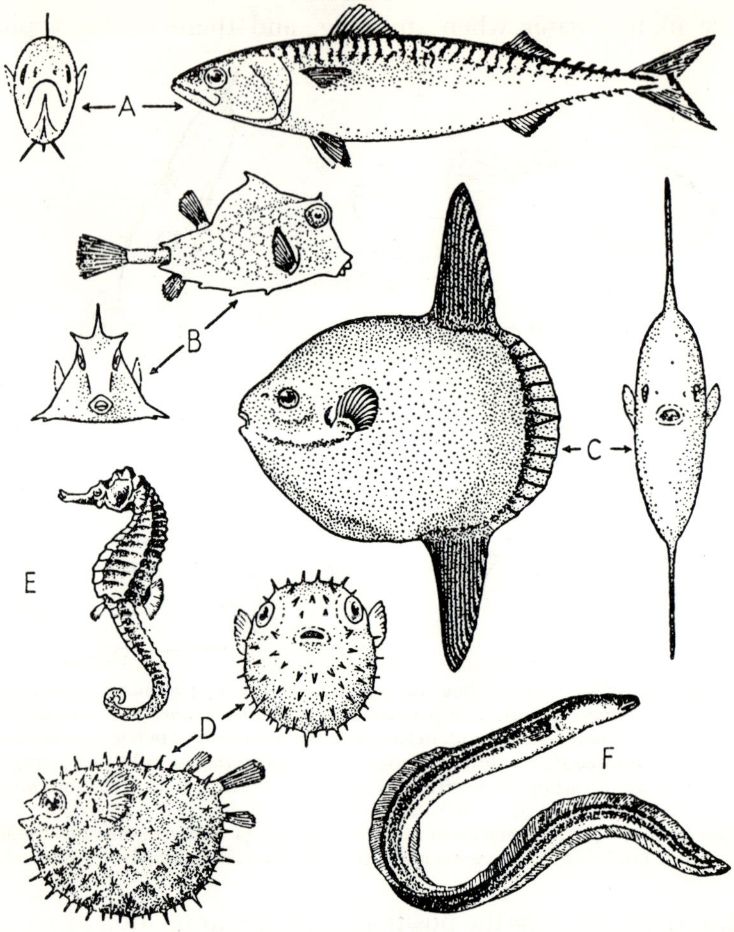

FIG. 149. Differences in form of fishes.

A, mackerel (*Scomber*); B, trunk-fish (*Ostracion*); C, sun-fish (*Mola*); D, globe-fish (*Chilomycterus*); E, sea-horse (*Hippocampus*); F, eel (*Anguilla*). (From Norman.)

use of the fins for stopping was also developed in some Mesozoic fishes, for instance, in the Triassic coelacanth, **Laugia*, which possessed high pectorals, and pelvic fins in the anterior position.

In the fishes with high pectoral fins, therefore, the pelvics are usually found far forward. In the flying-fish (*Exocoetus*), however (Fig. 151), high pectoral fins are found with posterior pelvics. In this position the pelvics would tend to help rather than hinder any tendency by the pectorals to produce a rise; the condition is exactly that which would be expected in a flying-fish.

With these increased opportunities for delicate control of movement without the devices of flattening of the front part of the body and a heterocercal tail, the bony fishes have also been able to make many other improvements in the efficiency of their swimming. The caudal fin has, of course, adopted its symmetrical shape and is used to increase the efficiency in turning. After its amputation a fish is not able to turn in its own length as normally.

With shortening of the body and its lateral flattening all sorts of new factors in streamlining the body are developed, but the details are difficult to understand. The nature of the turbulence produced by the movements of such a complicated structure is far from clear, but it seems probable that the shape of the higher fishes is such as to reduce the total skin-friction and to increase the efficiency of swimming. Other factors such as the flexibility, which has an important influence on the efficiency of the propulsive mechanism, have also been changed, again in ways not fully understood, by the special developments of the vertebral column and ribs. The speed that can be reached increases with the length of the fish. Cruising speeds, which are maintained for hours, are of the order of three to six times the body-length per second, the relationship varying with the species. During sudden bursts the speed may be much greater. Thus Bainbridge found that 10 L/sec could be maintained only for one second, 5 L/sec for 10 sec, and 4 L/sec for 20 sec (in dace, goldfish, and trout).

The locomotion of each type of fish is adapted to its habits. Most freshwater fishes are 'sprinters' but there are varying degrees of staying power. Thus we may distinguish (1) typical sprinters (pike and perch), (2) sneakers (eel) with some staying power, (3) crawlers (rudd, bream), with considerable staying powers for escape, (4) stayers, either for migration (salmon) or for feeding (carp). In the fish with staying powers there is a lateral strip of narrow red muscle fibres in addition to the characteristic broad white fibres of fish muscles.

Bathypelagic fishes, living in deep waters, below the thermocline at about 75 m, encounter special problems. Here currents and turbulence are low, but since the water is cold it is very viscous, making swimming difficult but sinking slow. Many deep-sea fishes have elaborate lures, often phosphorescent. They may be described as 'floating fish traps'. They often have no swim-bladder and achieve an almost neutral buoyancy by great reductions of the skeleton and muscles (e.g. *Ceratias*). The only parts to be well ossified are the jaws. On the other hand, deep-sea fishes that retain the swim-bladder have a well-developed skeleton and powerful muscles (Marshall, 1960).

2. Various body forms and swimming habits in teleosts

Departures from the streamlined body form typical of pelagic fishes have been very numerous; in nearly every case they are associated with a reduction in the efficiency of swimming as such and the development of some compensating protective mechanism (Fig. 149). Lateral flattening, which is already a feature of all teleostean organization, is carried to extremes in many types. Thus the angel-fish,

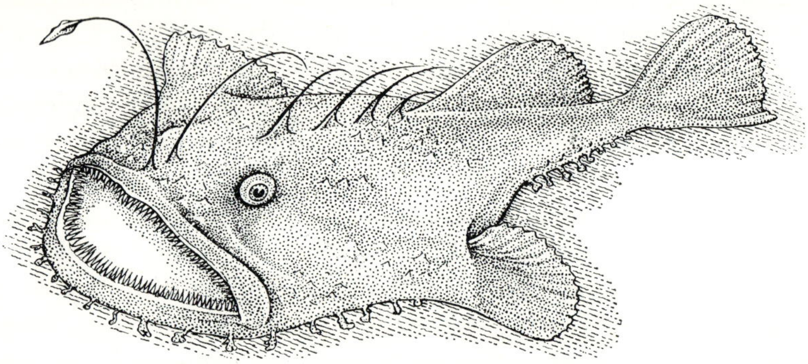

FIG. 150. The angler-fish, *Lophius*.

Pterophyllum, often seen in aquaria, is provided with long filaments and a brilliant coloration, which, in its natural habitat (rivers of South America), give it a protective resemblance to plants, among which it slowly moves. The flat-fishes (plaice, sole, halibut, &c.) have carried this flattening to extreme lengths. They feed on molluscs and other invertebrates on the sea bottom and lie always on one side. The upper side becomes darker and protectively coloured, the lower side white. In order to have the use of both eyes the whole head is twisted during the post-larval period. These forms are mostly poor swimmers, but their coloration gives them a remarkable protective resemblance to the background (p. 257).

The John Dory (*Zeus faber*) has made a different use of lateral flattening. The fish is so thin that its swimming is very slow, but being inconspicuous when seen from in front it can approach close to its prey, which it then catches by shooting out its jaws.

Flattening in the dorso-ventral plane is less common among teleosts than selachians. The flattened forms are mostly angler-fishes, of which there are several different sorts; *Lophius piscatorius* (Fig. 150) is common in British waters. It is much flattened, with a huge head and mouth and short tail. It 'angles' by means of a dorsal fin, modified to form a long filament with a lump at the end, which hangs over the mouth. Swimming, though vigorous, is slow, and protection (both for attack and defence) is obtained by sharp spines, protective coloration, and flaps of skin down the sides of the body, which break up the outline. There is even a special fold of pigmented skin over the lower jaw, serving to cover the white inside of the mouth.

Other anglers are the star-gazers, *Uranoscopus* of the Mediterranean and *Astroscopus* from the Western Atlantic seaboard. Their lure is a red process attached to the floor of the mouth and they lie in wait buried in the sand, with the mouth opening upwards and only the eyes showing. The colour is protective, there are poison spines, and in *Astroscopus* there are electrical organs located near the most vulnerable spot, the eyes, and formed from modified eye-muscles (p. 253).

Other fishes abandon the swift-moving habit for the protection afforded by the development of heavy armour, such as that of the trunk-fish (*Ostracion*) and the globe-fish (*Chilomycterus*) (Fig. 149). Special spinous dorsal rays, such as those of the sword-fish (*Xiphias*) may be developed, without loss of the swift-moving habit; indeed these fish are among the fastest swimmers. There are many groups in which an elongated body form like that of the eel has been developed. In *Anguilla* itself this is associated with the habit of moving over land. The Syngnathidae, sea-horses and pipe-fishes, no longer swim with the typical fish motion but by passing waves along the dorsal fins. The long and often grotesquely cut-up body form gives a strong protective resemblance to the weeds among which they live and on which they feed. The tail of the sea-horses has lost its caudal fin and is used as a prehensile organ, being wrapped around the stems of sea-weeds for attachment.

Evidently the mastery that the Actinopterygii have acquired in the water has depended to a large extent on the freedom given by the use of the air-bladder as a hydrostatic organ. This gives special interest to the question of how this use first began. If we are right in supposing that the bladder was first a respiratory diverticulum of the pharynx, can we suppose that its value as a hydrostatic organ depended on any exertion of effort of the fishes, or was this a case in which those born with the organ better developed found themselves with an advantage? It would seem that the latter orthodox Darwinian interpretation is the more likely; we can hardly imagine the fishes striving to make their bladders bigger. But it must be remembered that only those that were active swimmers, continually venturing into new waters, would be able to make full use of the new organ.

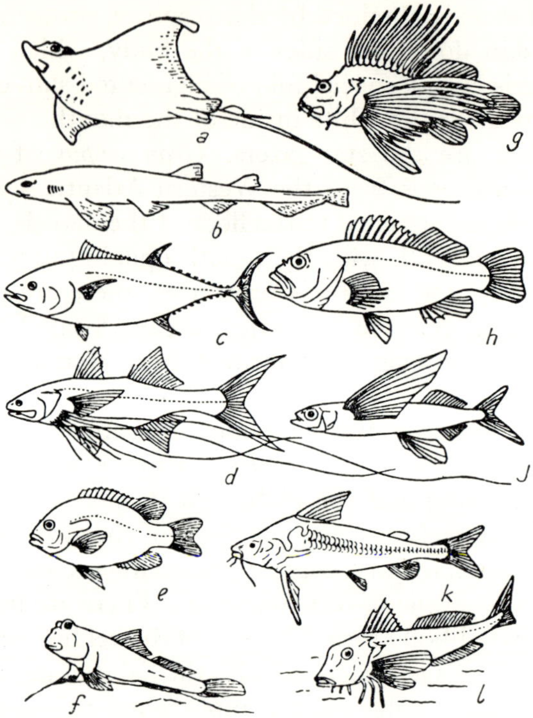

FIG. 151. Various fishes showing special conditions of the pectoral fins.

A, eagle-ray (*Myliobatis*); B, dog-fish (*Scyliorhinus*); C, tunny (*Thynnus*); D, thread fin (*Polynemus*); E, sun-fish (*Lepomis*); F, mud-skipper (*Periophthalmus*); G, scorpion fish (*Pterois*); H, cirrhitid fish (*Paracirrhites*); J, flying-fish (*Exocoetus*); K, catfish (*Doras*); L, gurnard (*Trigla*). (From Norman.)

Many different fishes are able to jump out of the water, presumably to escape enemies. Salmon and tarpon can jump to 8 or 9 ft above the water. The flying-fishes have special structures to assist in such jumps. In *Exocoetus* (Fig. 151) the enlarged pectoral fins serve for gliding for distances up to 400 metres, but in the flying gurnards (*Dactylopterus*) they are actually fluttered up and down, though the flight is feeble.

Several types of fish have the pectoral fin modified to allow 'walking'. The gurnards (*Trigla*) move in this way over the sea bottom (Fig. 151), and the mud-skipper (*Periophthalmus*) chases about catching crustacea and insects on land, using the pectoral fins as levers, provided with special anterior and posterior muscles.

Fishes that live in situations from which they are likely to be carried away develop suckers. Thus in the gobies, found between tide-marks, the pelvic fins form a sucker. The cling-fishes (*Lepadogaster*) are another group with the same habit. The remoras have developed a sucking-plate from the first dorsal fin and by means of this they attach themselves to sharks and other large fish. In order to catch their food they leave the transporting host, though they also feed on its ectoparasites.

3. Structure of mouth and feeding-habits of bony fishes

Although perhaps the majority of fishes are carnivorous, there are species with all sorts of other methods of feeding. The more active predators have strong jaws and sharp teeth, such as those of the pike (*Esox*), cod (*Gadus*), and very many others. The teeth on the edge of the jaw serve to bite and catch the prey, those on the walls of the pharynx to prevent its escape if, as is often the case, it is swallowed whole. The teeth can often be first lowered to allow entrance of the prey and then raised to prevent its exit (e.g. in *Lophius*). In connexion with this habit the walls of the oesophagus and even stomach are often composed of striped muscle, capable of quick and powerful contraction.

Many carnivorous fishes are very fierce. For instance, the blue fish (*Pomatomus*) of the Atlantic move in shoals, cutting up every fish they meet, making a trail of blood in the sea. The barracuda (*Sphyraena*) of tropical waters may attack man. They are said to chase shoals of fish into shallow waters and to keep them there to serve for food as required.

Other fishes feed on invertebrates and are then usually bottom-feeders. Thus the plaice (*Pleuronectes*) has developed chisel-like teeth on the jaws and flattened crushing teeth in the pharynx; it feeds largely on molluscs. The Labridae (wrasses) also have blunt teeth and eat molluscs and crabs. The sole (*Solea*) has a weaker dentition and eats mostly small crustacea and worms. Fish such as the herring (*Clupea*) that live on the minute organisms of the plankton have small teeth and weak mouths, but are provided with a filtering system of branched gill-rakers, making a gauze-like net, comparable with the filtering system found in basking sharks (p. 181), paddle-fish (p. 234), and whale-bone whales (p. 669).

Herbivorous and coral-eating fishes have crushing teeth similar to those of the mollusc-eaters; indeed, many forms with such dentition will take either form of food. The parrot-fishes (Scaridae) have a beak and a grinding mill of flattened plates in the pharynx. With this they break up the corals, rejecting the inorganic part from the anus as a calcareous cloud. The Cyprinidae, including many of our commonest freshwater fishes (goldfish, carp, perch, and minnow), have no teeth on the edge of the jaw, hence the name 'leather-mouths'. There are, however, teeth on the pharyngeal floor, biting against a horny pad on the floor of the skull. These fishes are mainly vegetarians, but many take mouthfuls of mud and extract nourishment from the plants and invertebrates it contains.

4. Protective mechanisms of bony fishes

In general teleosts depend for protection against their enemies on swift swimming, powerful jaws, good receptors, and brain. The majority of them have thus been able to abandon the heavy armour of their Palaeozoic ancestors. In many cases, however, subsidiary protective mechanisms have been developed, and are especially prominent in fishes that have given up the fast-swimming habit and taken either to moving slowly among weeds or to life on the bottom. These developments are a striking example of the way in which, following adoption of a particular mode of life, appropriate subsidiary modifications take place, presumably by selection of those varieties of structure that are suited to the actions of the animal.

These protective devices may be classified as follows:

1. Protective armour of the surface of the body.
2. Sharp spines and poison glands.
3. Electric organs.
4. Luminous organs.
5. Coloration.

5. Scales and other surface armour

The typical cycloid teleostean scales have already been described. They form a covering of thin overlapping bony plates, providing some measure of protection, but not interfering with movement. The hinder edges of the scales are sometimes provided with rows of spines, and are then said to be ctenoid. In many fishes the scales bear upstanding spines and possess a pulp cavity, which recalls that of denticles. In the tropical globe-fishes (or puffers) and porcupine-fishes (*Diodon*) these spines are very long and sharp and the puffers are able to inflate themselves and cause the spines to project outwards, a very effective protective device. In a few fishes the scales have become developed to form a bony armour even more complete than that of the Palaeozoic fishes. Thus in the trunk- or the coffer-fishes (*Ostracion*) the scales are enlarged and thickened into a rigid box, from which only the pectoral fins and tail emerge as movable structures, the former apparently assisting the respiration, the latter the swimming. These fishes live on the bottom of coral pools and have a narrow beak with which they browse on the polyps.

6. Spines and poison glands

Sharp protective spines are often found in teleosts, especially on the operculum and dorsal fins. These may be provided with modified dermal glands that inject poison into the wound. Thus the European weever (*Trachinus*) lives buried in the sand and has poison spines on the operculum and the dorsal fins. It is suggested that the dark colour of the fins serves as a warning. Some catfishes, scorpion fishes, and toad-fishes also have poison spines. Spines may be effective even if not poisonous; the stargazer, *Uranoscopus*, of the Mediterranean and tropical waters has powerful spines on the operculum, which inflict a most unpleasant wound if the animal is disturbed by hand or foot while lying in the sand angling for its prey (p. 249). *Lophius*, the angler, is also armed with dangerous spines. Several species of catfish have large spines, sometimes serrated. In the trigger-fishes (Balistidae and related families) of the tropics one or more of the fins is modified to make a spine that can be raised and locked in that position. These fishes have very brilliant coloration, but since some of them live in the highly coloured surroundings of coral reefs it cannot be considered certain that the colours serve as a warning.

7. Electric organs

The power to produce electric discharges has been developed independently in four distinct families of teleosts, as well as in torpedoes and rays. The electric organs arise bilaterally from modified muscle fibres, the cells of which are plate-like and arranged in rows, the electroplaques. Each plate is innervated on only one surface by motor neurons whose activity is controlled from the forebrain, in some fish there is a controlling nucleus located in the medulla. The physiological properties of transmission at the nerve endings with the electroplaques are similar to those of motor end plates. Unlike other electrogenic tissues such as muscle or nerve, electric organs can develop appreciable voltages in the surrounding fluid, up to 550 volts in *Electrophorus*, the electric eel of the Amazon. These voltages are achieved by series summation of the electromotive forces generated by the individual cells.

The columnar array of several hundreds or thousands of electroplaques in series in the strongly electric fish, *Electrophorus*, *Malapterurus* and *Torpedo* are paralleled so that the electric organs of these fish can generate considerable current at high voltage. A maximum peak power of up to 600 watts has been observed in *T. nobiliana*. The electric organs form the major part of the body of the strongly electric fish. The discharges are used for offence and defence. The weakly electric fish (Gymnotidae, Mormyridae) have only a few columns of series arrays, and relatively few electroplaques in each column. However, many of the species emit pulses of low voltages more or less continuously and regularly (60–400/sec). These pulses probably serve as the power components in an electrical guidance system. All species of the continuously emitting fish are sensitive to changes in the conductance of the water. Presumably the fish sense the altered electric field of their discharges; although the receptors have not yet been identified specialized lateral line organs (mormyromasts) are often present.

8. Luminous organs

Fishes of many different families live at great depths and 95 per cent of individuals caught below 100 fathoms are luminescent. The development of luminescent organs is therefore a further example of parallel evolution. The organs usually show as rows of shining beads of various colours on the sides and ventral surface of the fish.

In many species the light is due to organs containing luminous bacteria, whose appearance may be controlled by the movement of a fold of skin, or of the whole organ, or of chromatophores. Some teleosts, however, have self-luminous photophores and these are also found in *Spinax* and a few other Squalidae. They are formed from modified mucous glands, and may be provided with reflectors and even lenses. They can be flashed on and off, probably by sympathetic stimulation.

The luminous organs probably often serve for recognition of the sexes and often show distinctive patterns. They may serve to startle attackers and in a few cases to illuminate the prey. In the deep-sea anglers (*Ceratias*) the luminous tip of the fin is used as a lure.

9. Colours of fishes

The bony fishes show perhaps the most brilliant and varied coloration of any animals, rivalling even the Lepidoptera and Cephalopoda in this respect. The enormous range of colour and pattern provides an excellent example of the detailed adjustment of the structure and powers of animals to enable them to survive. A great difficulty is introduced into the study of animal coloration by the fact that we are usually ignorant of the capacity for visual discrimination possessed by the animals likely to act as predators. Moreover, it is very difficult for us to obtain this information. When we examine any two objects we are able to say not merely that they are different but that one is red and the other green. A person or animal that is colour-blind may also be able to detect a difference, but yet remain unaware of any distinction of colour; the objects appear to him only as differing in brightness. In order to decide whether animals are able to distinguish between light of two wavelengths we must present them with objects of different colour but the same brightness.

We are therefore faced with the possibility that some of the colours that appear to us so brilliant are to other animals merely differences of tone, and animals to us conspicuous because coloured, when seen in monochrome, may be protected. Some of the colours of fishes may be only a means of producing a pattern of protective greys, as seen through the eyes of an attacker. However, there is no doubt that some fishes are able to discriminate between illuminated bodies which though of different wavelength reflect light of equal brightness. In the subsequent description of fish coloration we shall not be able to consider predators further, but shall describe the colours as they appear to the eye of a normal man.

The colour of fishes is produced by cells in the dermis, (*a*) the chromatophores and (*b*) the reflecting cells or iridocytes (Fig. 152). The chromatophores are branched cells containing pigment, which may be either black (melanin) or red, orange, or yellow (carotenoids or flavines). The iridocytes contain crystals of guanin, making them opaque and able to reflect light so as to produce, where no chromatophores are present, either a white or a silvery appearance. This material is used in the manufacture of artificial pearls, the scales of the cyprinoid *Alburnus lucidus* (the bleak) being used for the purpose. The iridocytes may be either outside the scales, when they produce an iridescent appearance, or inside them, giving a layer, the argenteum, that produces a dead white or silvery colour. By a combination of the chromatophores, and of these with the iridocytes to produce interference effects, a wide range of colour is produced. Thus by mixing yellow and black either brown or green is produced. Blue is usually an interference colour.

The use of the colour by the fish may be classified, according to the scheme introduced by Poulton, as cryptic or concealing, sematic or warning patterns, and epigamic or sex coloration. Cryptic coloration

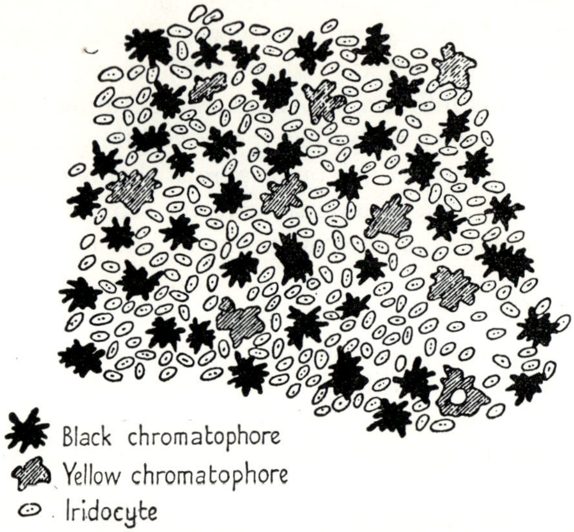

FIG. 152. Coloration elements in the skin of the upper side of a flounder (*Platichthys*). (After Norman.)

may be achieved in various ways and may be subdivided into two main types: (1) assimilation with the background, (2) breaking up the outline of the fish. Assimilation is common, but is often associated with some degree of disruption of outline. The absence of all pigmentation in pelagic fishes, for instance the Leptocephalus larvae of eels, is an example of assimilation. Fishes living among weeds, such as the sea-horses and pipe-fishes, or *Lophius* the angler, often resemble the weeds in colour, and in addition develop 'leaf-like' processes. The colour of many familiar fishes, such as the green of the tench, may be said to resemble that of the surroundings by assimilation. When we consider the much more numerous examples of patterns involving several colours the distinction between assimilation and disruption is more difficult to draw. Many free-swimming pelagic fishes have the upper side dark and striped with green or blue, whereas the under-side is white, the beautiful pattern that is seen in the mackerel (*Scomber*). This gives them protection from above and below, the striping probably making the animal less conspicuous in disturbed water than it would be if of uniform colour. The white under-side also serves to lessen any shadows, an important factor for animals that live in shallow water; similar shading is used by land animals.

FIG. 153. Colour patterns of various tropical fishes.
A, Muraena (*Gymnothorax*); B, bat-fish (*Platax*); C, butterfly fish (*Holacanthus*); D, butterfly fish (*Chaetodon*); E, perch (*Grammistes*). (From Norman.)

Devices of spots and stripes are found on fishes that live against a variegated background (Fig. 153). The beautiful red and brown markings of a trout are a good example. Flat fishes, living on sandy or gravelly bottoms, adopt a spotted pattern, which gives them a high degree of protection, and we shall see later that they are able to change colour to suit the ground on which they rest. The brilliant colours of many tropical fishes probably serve mainly to break up the outline, though no doubt the surroundings in which they live are also brilliant. Great variety of colours may be found on a single fish, especially in the trunk-fishes (*Ostracion*), one species of which is described as having a green body, yellow belly, and orange tail, while across the body are bands of brilliant blue, edged with chocolate-brown. Moreover, the female has another colour scheme and was for long considered as a different species!

Colour differences between the sexes are frequent in fishes, the male being usually the brighter. Thus in the little millions fish, *Lebistes*, there are numerous 'races' of males with distinctive colours, but the females are all of a single drab coloration. The genetic factors that produce the various types of male are carried in the Y chromosome. Presumably the colour of the males acts as an aphrodisiac as a part of the mating display, but the significance of the different races is not known.

Sematic or warning coloration involves the adoption of some striking pattern that does not conceal but *reveals* the animal. This type of colouring is found in animals that have some special defence or unpleasant taste (such as the sting of the wasp), and its use implies that animals likely to attack are able to remember the pattern and the unpleasant effects previously associated with it. It is not easy to be certain when colours are used in this way, but it is possible that the conspicuous spots on the electric *Torpedo ocellata* have this function. Among teleosts there is the black fin of the weevers (*Trachinus*), possibly a warning of their poison spines, and the spiny trigger-fishes and globe-fishes (p. 253) also have conspicuous colours.

10. Colour change in teleosts

In spite of the reputation of the chameleon the teleosts are the vertebrates that change their colour most quickly and completely. The melanophores are provided with nerve-fibres (Fig. 154), and these cause contraction of the pigment and hence a paling of the skin colour. The processes of the cells themselves are not withdrawn, the colour change is produced by a movement of pigment within them.

The nerve-fibres in question are post-ganglionic sympathetic fibres, leaving the ganglia in the grey rami communicantes (Fig. 139) to all the cranial and spinal nerves. The pre-ganglionic fibres that operate them, however, emerge only in a few segments in the middle of the body (Fig. 235), so that severance of a few spinal roots will affect the colour of the whole body. When a nerve to any part of the skin is cut the chromatophores in that region at first expand, making a dark area (Fig. 155). After a few days, however, the skin involved gradually becomes lighter, the process, it is alleged, beginning at the edges and moving inwards. Parker, who has made a careful study of these phenomena, believes that they indicate the presence of melanophore-expanding nerve-fibres (said to be of 'parasympathetic' nature). Following the cut these fibres are supposed to be stimulated to repetitive discharge and hence to make the dark band. Later the band pales

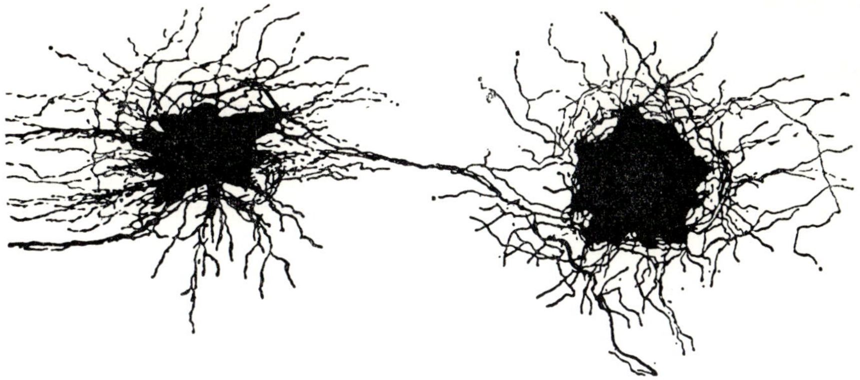

FIG. 154. Nerves of the melanophores of a perch. (From Ballowitz.)

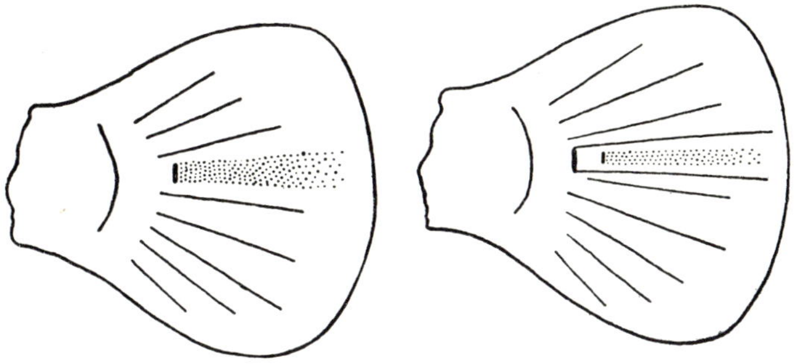

FIG. 155. Diagrams showing, *left,* a cut in the tail of the fish *Fundulus* producing a band of dark melanophores; *right,* when the dark band has faded a second cut makes the melanophores again dark. (After Parker, *Quart. Rev. Biol.* **13**.)

because the stimulating substances ('neurohumors') produced at the nerve-endings of the melanophore-contracting fibres in the neighbouring areas diffuse in gradually.

This hypothesis involves two physiological propositions which are so novel that they would require detailed evidence for acceptance. Firstly the act of cutting is presumed to set up a discharge of impulses lasting for several days, which is unlikely. Moreover, the discharge is presumed to be only in the melanophore-expanding fibres and not in those that produce contraction. Secondly, electrical stimulation of nerves in teleosts always produces *paling* and never darkening (except after the use of ergotamine). We must assume, therefore, that this form of stimulation has the opposite effect to section and only stimulates the melanophore-contracting nerve-fibres. Since neither of these propositions is adequately demonstrated, we must reject the hypothesis and say that there is not sufficient evidence of melanophore-expanding nerve-fibres.

There is, however, another agent that causes expansion of melanophores in a wide variety of vertebrates, namely the posterior lobe of the pituitary(see p. 206), and there is evidence that this works also in teleosts. Hypophysectomized specimens of the Atlantic minnow *Fundulus* are nearly always lighter than normal individuals, especially when on a dark background. Injection of posterior pituitary extracts, or placing of isolated scales in the extract, causes expansion of the chromatophores of any teleost. We may conclude that colour change is produced by the nerve-fibres tending to make the animals pale and secretion of the posterior pituitary to make them dark. Adrenaline induces contraction of chromatophores, and is presumably similar to the sympathetic transmitter.

It is more difficult to decide how external influences are linked with this internal mechanism. Fishes mostly become pale in colour on a light background and vice versa, and the effect is produced predominantly through the eyes. There may also be a slight direct effect of light on the chromatophores. The change in colour begins rapidly, but its completion may take many days. Analysis of the rates of change in normal fishes and in those with anterior and posterior pituitary removal has led to the suggestion that the anterior lobe produces a substance tending to make the fish lighter in colour, at least in the eel. A similar hypothesis has been fully worked out by Hogben and his colleagues in amphibia (p. 300). The colour is also influenced by the pseudobranch, a secretory tissue in the first gill arch. After removal of this a fish becomes dark and the choroid gland of the eye, which receives blood from the pseudobranch, degenerates. It is suggested that the pseudobranch produces a hormone, whose entry into the circulation is controlled by the choroid gland.

The value of the colour change in bringing the animals to the same tint as their surroundings is considerable. Fishes kept on a light background are very conspicuous for the first few minutes when transferred to a dark one. The fisherman acknowledges this by painting the inside of his minnow-can white, to make the bait conspicuous. In the flat fishes, living on the sand, the protection assured by the colour change is of special importance. It has been suggested that it is possible for the fish to assume a pattern similar to that of the ground on which it lies, but it is probable that the degree of expansion of the chromatophores is adjusted to suit the amount of light reflected from the ground; by increasing or decreasing the areas of dark skin, effects approximately appropriate to various backgrounds are produced.

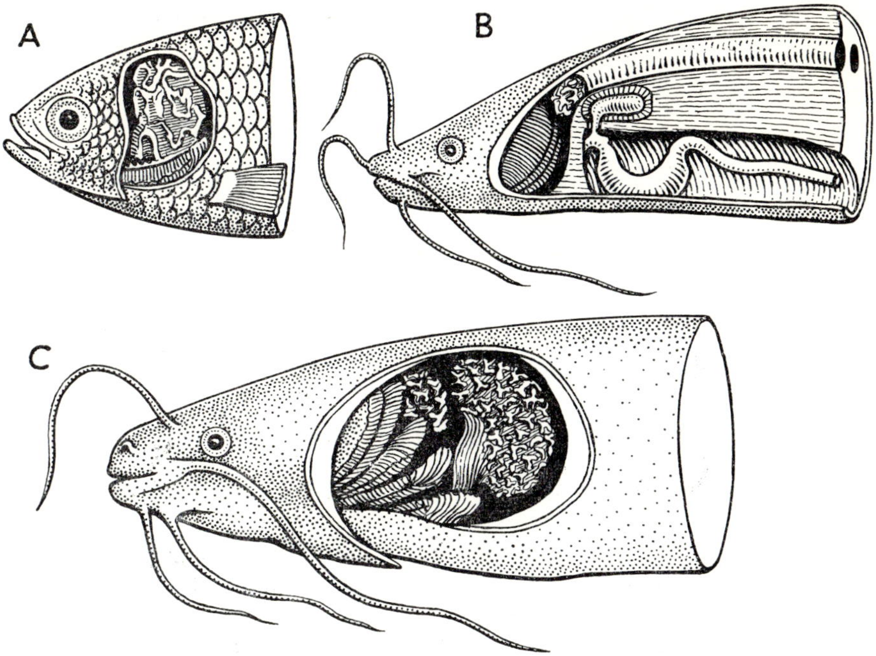

FIG. 156. Special respiratory apparatus, A, in climbing perch (*Anabas*); B, Indian catfish (*Saccobranchus*); C, African catfish (*Clarias*). (From Norman.)

11. Aerial respiration and the air-bladder

Many fishes are able to live outside the water. The excursions on to the land vary from the wriggling of the eel through damp grass to the life of the Indian climbing perch (*Anabas*) spent almost entirely on land. In the eel there is no special apparatus for breathing air (though oxygen may be taken in through the skin). The climbing perch is provided with special air chambers above the gills (Fig. 156) and even when in water it comes to the surface to gulp air and will 'drown' if prevented from doing so, even though it is placed in well-oxygenated water.

Many other fishes gulp air, especially those living in shallow tropical waters, which readily become deoxygenated. There may be other special mechanisms for gaseous interchange. In the Indian catfish

Saccobranchus there are large air sacs growing a long way down the body from the gill chambers (Fig. 156).

The air-bladder, which has contributed so largely to the success of the later teleosts, may have arisen as an accessory respiratory organ,

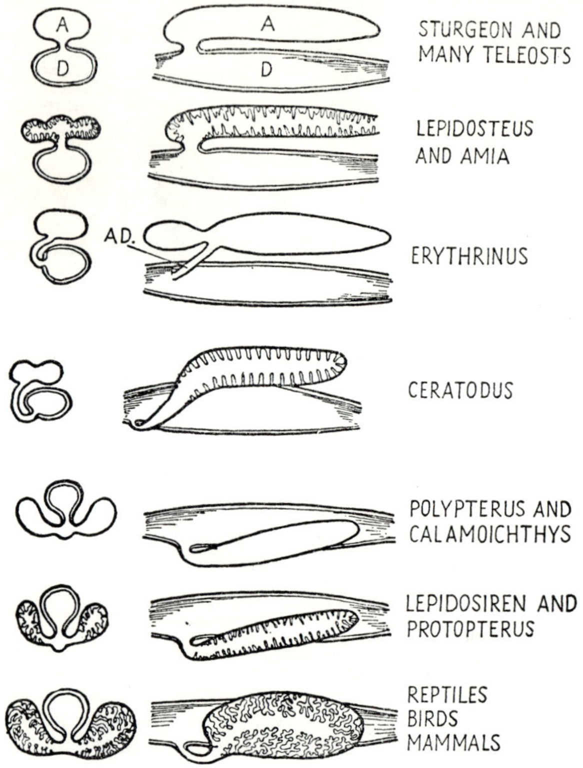

FIG. 157. Air-bladder of various fishes, seen from in front and from the left side. A, air- or swim-bladder; AD, air-duct; D, digestive tract. (From Dean, *Fishes, Living & Fossil,* The Macmillan Co., after Wilder.)

used in the same way as those described above. In all the more primitive teleosts (Isospondyli) the air-bladder preserves in the adult its opening to the pharynx ('physostomatous'), whereas in higher forms it becomes completely separated ('physoclistous'). Survivals of still earlier Actinopterygii have the opening especially well developed, though it varies from group to group (Fig. 157). Thus in the sturgeons there is a wide opening into the dorsal side of the pharynx. In *Amia* and *Lepisosteus* the opening is also dorsal and the walls of the sac are much folded and used for respiration. In *Polypterus* the opening is ventral and the bladder has the form of a pair of lobes below the gut. This arrangement recalls that of the tetrapod lungs and is also found in the modern lung-fishes and presumably in their Devonian ancestors, from which we may suppose that the tetrapods arose (p. 276). This ventral position of the air-bladder was one of the features that for a long time led zoologists to suppose that *Polypterus*

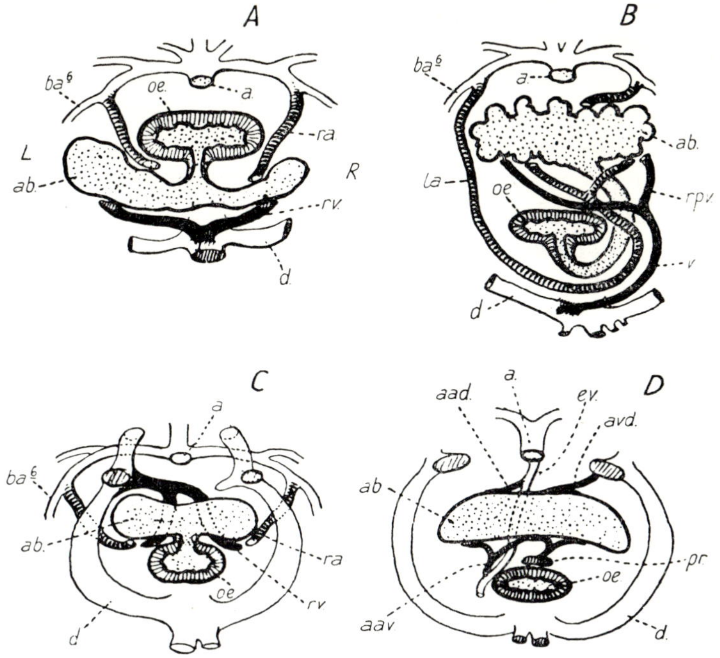

FIG. 158. Diagrams illustrating the blood-supply of the air-bladder in A, *Polypterus*, B, *Ceratodus*, C, *Amia*, and D, a teleost. The blood-vessels are seen from behind, and cut short in transverse section.

a. dorsal aorta; *aad.* anterior dorsal artery from the coeliac; *aav.* ant. ventral artery; *ab.* air-bladder; *avd.* anterior dorsal vein to the cardinal; ba^6, 4th aortic arch (6th of the series); *d.* ductus Cuvieri; *ev.* coeliac artery; *la.* left pulmonary artery; *oe.* oesophagus; *pr.* portal vein receiving posterior vein from air-bladder; *ra.* right 'pulmonary' artery; *rpv.* right (branch of) 'pulmonary' vein; *rv.* right vein from air-bladder; *v.* left 'pulmonary' vein. (From Goodrich, *Vertebrata*, A. & C. Black, Ltd.)

was a member of the crossopterygian line of fishes. It is probable, however, that the affinity is only that which persists between all primitive members of both Actinopterygii and Crossopterygii and is to be taken as an indication that the air-bladder was originally a widely open respiratory sac, or perhaps pair of sacs. Once the power to produce a pharyngeal diverticulum had been developed it is easy to imagine that the actual position of the opening might shift either dorsally, as in the later Actinopterygii, or ventrally, as in the tetrapods.

The blood-supply of the air-bladder should provide some indications both of its origin and function. In *Polypterus* and Dipnoi there are pulmonary arteries springing from the last (sixth) branchial arch and presumably containing venous blood (Fig. 158). Blood returns to the heart by pulmonary veins. Essentially the same arrangement is found in *Amia*, but in all other Actinopterygii oxygenated blood is supplied to the bladder from the dorsal aorta (or sometimes from the coeliac artery). Probably, then, the original function of the air-bladder was respiratory, and this may still be its main function not only in *Amia* and *Lepisosteus* but also in some of the physostomatous Teleostei. However, in the majority of teleosts its dorsal position, closed duct, and arterial blood-supply show that it has some other function and it has long been supposed that this is concerned in some way with flotation. The air-bladder is absent from bottom-living forms, such as flat-fishes, *Lophius* and *Uranoscopus*, though it may be present in their pelagic larvae.

The bladder is provided with special glands by which it is filled and the gas they secrete is mostly oxygen; only in some physostomatous forms is the bladder filled by gulping air. Nitrogen and carbon dioxide are also present and the former even constitutes the main gas in some freshwater fishes at great depths. In the more primitive forms gas is secreted all over the surface of the bladder, but later there develop special anterior oxygen-secreting and posterior oxygen-absorbing regions. The former, known as the red gland, has a special apparatus of blood-vessels, the rete mirabile, and the latter, or 'oval', which may be developed from the closed end of the pneumatic duct, has a special sphincter by means of which it can be closed off.

The pressure of the gases in the swim-bladder is adjusted to make the fish neutrally buoyant, which is achieved when the bladder occupies 7–10 per cent of the total volume in fresh-water and 5 per cent in marine fishes. This may involve partial pressures of oxygen, carbon dioxide, and nitrogen many times greater than those in the blood. The mechanism by which the gases are secreted against a diffusion gradient of several atmospheres has been much discussed. Carbonic anhydrase is present in the gas gland and the oxyhaemoglobin of fish blood is especially sensitive to carbon dioxide, giving up its oxygen even at high oxygen concentration.

If weights or floats are attached to a fish it maintains its position in mid-water by swimming while gas is secreted or absorbed. The receptors concerned are therefore not activated by the tension in the bladder but perhaps by the movements that are necessary when the fish is not in equilibrium. The bladder is innervated by the vagus and sympathetic nerves and after severing the former gas secretion ceases.

The various diverticula connecting the bladder with the ear (and the Weberian ossicles, p. 217) may be associated with pressure receptors that assist in the control of the bladder. Loaches are famous as fish barometers, whose behaviour can be used to predict weather changes.

12. Special reproductive mechanisms in teleosts

The teleosts show great variation in breeding habits, the eggs being sometimes left to develop entirely by themselves, in other cases looked after by one or both parents, while in a few species they develop

FIG. 159. Deep-sea fish *Photocorynus* with parasite male attached. (From Norman.)

viviparously within the mother. Hermaphrodite individuals are not uncommon and in some species of Sparidae and Serranidae are invariably monoecius and self-fertilizing. The method of association of the sexes is correspondingly varied and there are numerous devices for bringing sexes together, such as colour differences, sound production, and the liberation of stimulating substances into the water. In some deep-sea fishes the male is much smaller than the female, to which it remains permanently attached (Fig. 159). Breeding is often preceded by a migration of the fishes to suitable situations and the association into large shoals.

The eggs may be classified as either *pelagic*, if they float, or *demersal*, if they sink to the bottom. In the former case they are sometimes provided with an oil globule and are exceedingly numerous. Thus a single female turbot has been calculated to contain nearly 10 million eggs, a cod 7 million, and a ling 28 million, whereas the herring, whose eggs sink to the bottom, probably does not lay more than

50,000 eggs. The large numbers laid by the pelagic species are presumably an insurance against failure of fertilization and especially against random elimination of the eggs and young. The greater the care devoted to the young by the parents the smaller the number of eggs produced (see, however, p. 283).

Demersal eggs, especially of freshwater animals, are usually laid with some special sticky covering, by means of which they are attached to each other and to the bottom or to stones, weeds, &c. Thus the eggs of many cyprinids (carp, &c.) are attached to weeds. The eggs of salmon and trout, however, though demersal, are not sticky. From depositing eggs on weeds it is only a short step to the building of a nest and guarding of the eggs by one or both parents. Thus the sand goby (*Gobius minutus*) lays its eggs in some protected spot, where they are guarded by the male, who aerates them by his movements. Quite elaborate nests may be built, as by the sticklebacks (*Gasterosteus*), where pairs remain together throughout the breeding-season. A still further development is the retention of the young within the body. In some catfishes they develop within the mouth of either parent. In pipe-fishes and sea-horses the males are provided with special pouches for the young.

Although external fertilization is usual, various teleosts show internal fertilization and the young then develop within the ovary (*Zoarces*, *Gambusia*, *Lebistes*). The mechanisms by which mating and the nutrition of the embryos are assured in these cases show some interesting parallels with the conditions in mammals, including the formation of placentae or nutritive material. In *Lebistes* the female adopts a special position of readiness for copulation, and this has been shown to depend partly on an internal factor in the female and partly on a substance secreted into the water by the male. The embryos are not attached to the wall of the ovary but develop free in the sac, feeding upon an 'embryotrophic' material, apparently produced by the discharged ovarian follicles, which become highly vascular and remain throughout the several months of 'pregnancy'.

Rhodeus amarus, the bitterling (Cyprinidae), shows somewhat similar conditions (Fig. 160). Association of the sexes at mating is here made necessary by the fact that the eggs are laid within the siphon of a swan mussel. For this purpose the cloaca of the female develops into a tubular ovipositor. This development takes place under the influence of a hormone produced by the ovary. Addition of progesterone and related substances to the water containing the fish causes growth of the ovipositor.

The full growth of the ovipositor and preparation of the female for spawning depends on the presence in the water of the male and also of the swan mussel. Water in which males have been kept stimulates growth of the ovipositor. When the female is ready to deposit the eggs she adopts a vertical position in the water and the spawning male, in

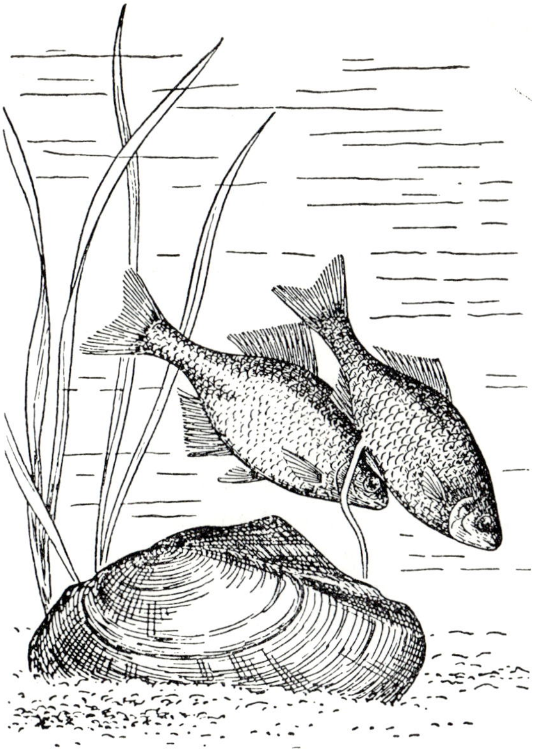

FIG. 160. Male and female bitterling (*Rhodeus*) with swan mussel in which eggs are about to be deposited. (From Norman.)

full nuptial coloration, swims around her. An egg passes into the oviduct and erection of the ovipositor is produced by pressure of the urine, produced by contraction of the walls of the urinary bladder, the exit being blocked by the egg. The extended ovipositor is thus able to place the egg within the siphon of the mussel and the male then immediately thereafter sheds his sperms over the opening and they are presumably carried in by the current. The whole process shows the elaborate interplay of internal devices and external stimuli necessary for the perfection of this remarkable method of caring for the young. Yet the various features are all developments of systems found in other vertebrates.

X

LUNG-FISHES

1. Classification

Class Crossopterygii
Order 1. Rhipidistia. Devonian–Recent.
Suborder 1. *Osteolepidoti. Devonian–Carboniferous
**Osteolepis; *Sauripterus; *Diplopterax; *Eusthenopteron*
Suborder 2. Coelacanthini (= Actinistia). Devonian–Recent
**Coelacanthus; *Undina; Latimeria*
Order 2. Dipnoi. Devonian–Recent.
**Dipterus; *Ceratodus; Neoceratodus; Protopterus; Lepidosiren*

2. Crossopterygians

Although the lung-fishes and their allies are here considered last of all the groups of fishes, because they lead on to the amphibia, it is important to realize that in many features they stand close to the ancestral stock of gnathostomes. It is a mistake to consider them as 'higher' animals than, say, the elasmobranchs or actinopterygians. Only four genera belonging to this group are found at the present time, *Neoceratodus*, *Lepidosiren*, *Protopterus*, the lung-fishes of Australia, South America, and Africa respectively, and *Latimeria*, recently discovered off the east coast of South Africa and the Comoro Islands off Madagascar. These are relics of a group that can be traced back with relatively little change to the Devonian, and there is little doubt that at about that period the first amphibia arose from some similar line. The characters of the modern crossopterygians are therefore of extraordinary interest, because they show an approach to the condition of the ancestors of all tetrapods.

3. Osteolepids

Osteolepis itself (Fig. 161), from the middle Devonian, was the earliest and most primitive member of the group. In appearance it shows an obvious similarity both to palaeoniscids and to early Dipnoi and it was probably close to the line of descent from some placoderm ancestor to both of these groups and to the amphibia. The body was long and the tail heterocercal. A feature distinguishing all early Crossopterygii from early Actinoptergyii was the presence of two dorsal fins in the former. The paired fins have a characteristic scaly lobed form, from which the group derives its name, and the skeleton of the pectoral fin contained a basal element attached to the girdle and a branching arrangement at the tip (Fig. 180). This plan is distinctively different from that of the rayed fin of the Actinopterygii, but could also easily have led to the evolution of a tetrapod limb (p. 307).

The body was covered with thick, pitted rhomboidal scales, with an appearance very similar to that of the palaeoniscid scale. These scales have, however, a characteristic structure known as cosmoid. Each scale (Fig. 145) may be considered as composed of an upper layer of dentine (the cosmine), with a hard covering of shiny vitrodentine, possibly comparable with enamel. Below the cosmine is a 'vascular layer' consisting of pulp cavities in which lay odontoblasts whose processes made the dentine. This layer in turn rests on a bony layer of isopedin. The structure appears to have some relation to that of placoid scales and no doubt the morphogenetic processes that give rise to the isolated pulp cavities of placoid scales are similar to those that produce the cosmoid plates. It is usual to suggest that the latter are formed of 'fused denticles', but this is of course only a manner of speaking. Denticles do not fuse, but morphogenetic processes may occur in such a way as to produce flat plates of dentine. Indeed, it is possible that the evolution occurred in the other direction, that is to say that the placoid scale is a special case of the cosmoid. The condition in which a substance is formed nearly uniformly all over the surface of the body is a more general one than that in which such formation occurs only in isolated areas. Indeed, the discontinuous arrangement is a very remarkable condition, for which we have at present no explanation. The relationship of the cosmoid to the ganoid scale of early Actinopterygii is not quite clear, but the ganoid type seems to show a reduction of the pulp cavities and development of the shiny surface-layer. The early Dipnoi of the Devonian possessed cosmoid scales. In later osteolepids and Dipnoi there has been a thinning of the scales, as among the Actinopterygii, so that the later Dipnoi are covered with thin, overlapping, 'cycloid' scales.

The skull of osteolepids (Fig. 161) was well ossified; there was a series of bony plates arranged according to a pattern with a general similarity to that of palaeoniscids (p. 230) and which might well have been ancestral to that of amphibia. There was, however, a joint across the top of the skull between the parietal and post-parietal bones, and an unossified gap in the base of the skull. A movable joint at this level persists in the living coelacanth (p. 272).

A most important feature, common to all crossopterygians, was that the attachment of the jaws was autostylic, that is to say, similar to the arrangement in amphibia and remotely similar to the earliest gnathostomes (aphetohyoids) but different from that of modern elas-

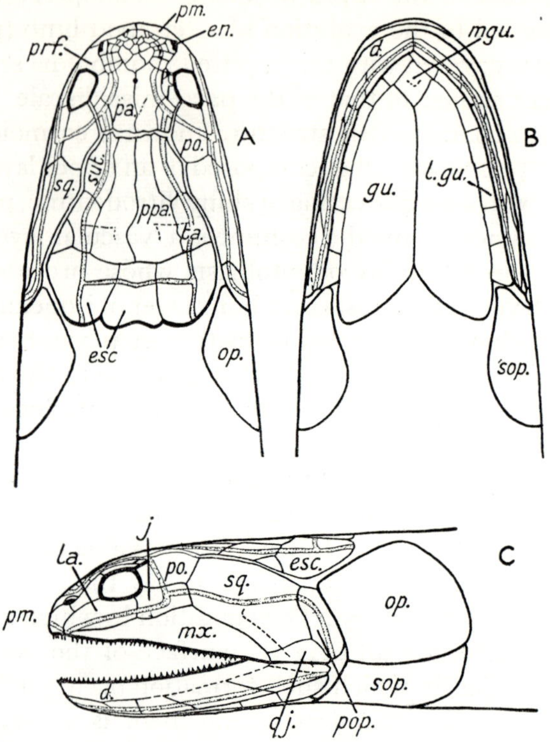

FIG. 161. Skull of *Osteolepis*.

d. dentary; *en.* external nostril; *esc.* extrascapulars; *gu.* gular; *j.* jugal; *la.* lachrymal; *l.gu.* lateral gular; *mgu.* median gular; *mx.* maxilla; *op.* opercular; *pa.* parietal; *pm.* premaxilla; *po.* postorbital; *pop.* preopercular; *ppa.* postparietal; *prf.* prefrontal; *qj.* quadratojugal; *sop.* subopercular; *sq.* squamosal; *sut.* supratemporal; *ta.* tabular. (After Säve-Söderbergh and Westoll, *Biol. Rev.* 1943.)

mobranchs and actinopterygians. The teeth of osteolepids were simple cones, not flattened plates such as are characteristic of Dipnoi, but the teeth on the palate show a somewhat broad folded surface and each tooth is replaced by another growing up near it, both of these being features found in the earliest amphibia. Sections of the teeth show a peculiar infolding of the enamel to make a labyrinthine structure, which is not found in other fishes but is characteristic of the teeth of the first amphibians (p. 327).

There was only one pair of nostrils on the surface of the head and there are gaps, which are considered to be internal nostrils, at the front end of the palate, bordered by the premaxillae, maxillae, palatines, and prevomers. These fishes may have breathed air; they certainly also possessed gills, covered with an operculum.

Animals of this sort seem to have been abundant in Devonian waters and by the end of that period had diverged into several different lines. It is interesting that the tendencies shown by these lines are similar to those that we discovered in the evolution of the Actinopterygii. Some of the later osteolepids became shorter in body, the tails tended to become symmetrical (diphycercal) and the scales to

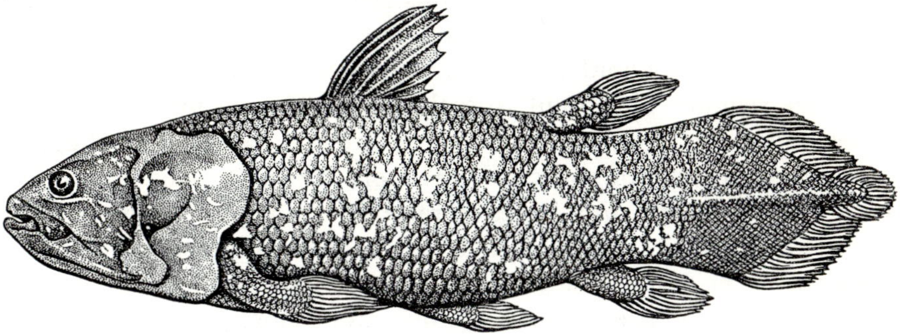

FIG. 162. Coelacanth, *Latimeria chalumnae*, female. Length 142 cm. Caught 1954 near Anjouan. (After Grassé.)

become thinner and overlapping. *Diplopterax* and *Eusthenopteron* represent separate lines from the late Devonian, both showing these characters. Probably the development of these features depends on the use of the air-bladder as a hydrostatic organ and the associated changes in the method of swimming.

4. Coelacanths

The osteolepids became rare in the Carboniferous and disappeared after the early Permian, but a line descended directly from them remained common through the Mesozoic and still survives today. These coelacanths (Fig. 162) show certain very characteristic features, which enabled the strange fish brought to the museum at East London, South Africa, to be recognized immediately as belonging to the group. They are rather deep-bodied animals, with a characteristic three-lobed diphycercal tail. The type first appeared in the late Devonian and was obviously derived from osteolepid ancestry having two dorsal fins, diphycercal tail, lobed fins, and a rhipidistian pattern of skull bones, including in most forms a fronto-parietal joint. There was a calcified air-bladder. *Coelacanthus* and other Carboniferous forms lived in fresh water, but **Undina* and other Jurassic and Cretaceous types lived in the sea.

The first living specimen of the group was fished off the east coast of South Africa and eleven others have since been caught around the Comoro Islands (Madagascar) (Fig. 162). All are referred to the genus *Latimeria*. They have been caught near the bottom at considerable depths (150–400 m). Unlike most of their fossil ancestors they are large fishes, weighing up to 80 kg; they are dull blue in colour. The whole body is covered with heavy cosmoid scales.

The notochord is a massive unconstricted rod. The skull possesses a well-marked joint between a condyle on the hind end of the basisphenoid and a glenoid cavity on the front of the base of the oto-occipital region. This joint, together with fibrous unions between other bones allows of movement of the front part of the head on the hind. A large pair of muscles runs from the parasphenoid up and back to the pro-otic and serves to raise the front part of the head on the hind. Coraco-mandibular muscles attached to the palato-quadrate have the reverse action and the movement is presumably concerned with catching the prey. There are numerous small teeth on the jaws and palate. *Latimeria* lives on other fishes, apparently swallowed whole by the powerful oesophagus. There is a well-developed spiral intestine.

The 'air-bladder' arises by a ventral opening from the oesophagus and proceeds backwards and dorsally for the whole length of the abdominal cavity. The lumen is very small and the organ is 95 per cent fat. It may serve to reduce the specific gravity. Respiration is by the gills. The heart shows a linear 'embryonic' condition, with the sinus venosus and auricle behind the ventricle. There are four rows of valves in the conus. The red cells are large, as in elasmobranchs, Dipnoi, and Amphibia. Nothing is known of the development, except that the eggs are large.

The brain lies far back in the cranium, of which it occupies less than one-hundredth part, the rest being filled with fat. Its structure is somewhat like that of a teleostean, with a thin fore-brain roof, and large striatum, but without eversion. There is no valvula to the cerebellum. The pituitary cleft is large and the gland remains in continuity with the roof of the mouth.

There are anterior and posterior nares but both open on the surface of the head and they have nothing to do with respiration. The rostral organ is a large median sac opening to the surface by three pairs of canals and richly innervated by the superficial ophthalmic nerve.

A similar sac occurs in fossil coelacanths back to the Devonian but its function is quite unknown. The eye, inner-ear, and lateral line system are well developed.

It is hard to see what features have enabled *Latimeria* to survive with little change since the Jurassic or earlier (see p. 771). It clearly cannot be by special development of the brain or receptors. Its habitat is isolated, but not especially protected and its population seems to be small since even by exceptional efforts so few specimens have been found. Perhaps they are more numerous in deeper waters. In some of its features it shows developments parallel to those of the Teleostei rather than to the Dipnoi, whose remote ancestry it shares. Several of its characteristics are paedomorphic. These can hardly be alone responsible for such a long survival, but some of them also appear in the other survivors from the Paleozoic, *Polypterus*, sturgeons, and Dipnoi.

5. Fossil Dipnoi

The Devonian Dipnoi were more like their osteolepid relatives than are the surviving modern forms (Fig. 163). The early members of this group, such as **Dipterus* (Fig. 211), showed the typical elongated body, thick cosmoid scales, heterocercal tail, lobed fins, and well-ossified skull. The pattern of the bones was obscured by a seasonal deposit of cosmine, this being periodically absorbed to allow of growth.

The individual bones have a certain similarity to those of osteolepids, but there are extra bones that are difficult to name. There was no premaxilla or maxilla, nor any teeth along the edge of the jaw; instead broad, ridged tooth-plates were developed on the palate and inside of the lower jaw, presumably as an adaptation for eating molluscs and other invertebrates. These crushing-plates are characteristic of the Dipnoi and preclude even the earliest of them from being the actual ancestors of the amphibia. By the end of the Devonian the Dipnoi were showing changes similar to those of the osteolepids and palaeoniscids. The body became shorter, the first dorsal fin disappeared, the tail became diphycercal, and the scales lost their shiny surface-layer and became thin. The teeth of **Ceratodus* appear in the Triassic and were known to geologists long before the related living animal was discovered. There has been very little change in this animal in more than 150 million years, though the recent members are placed in a distinct genus *Neoceratodus*.

The evolution of Dipnoi is especially interesting because the rate of change has actually been measured (Westoll, 1949). Twenty-six

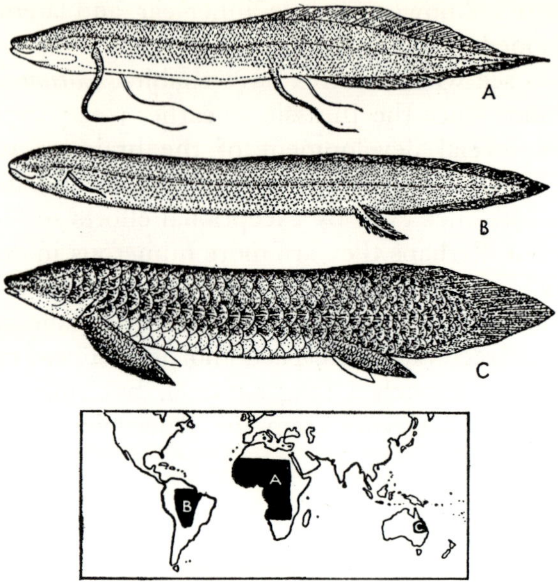

FIG. 163. The three living lung-fishes and their distribution. A, *Protopterus*; B, *Lepidosiren*; C, *Neoceratodus*. (From Norman.)

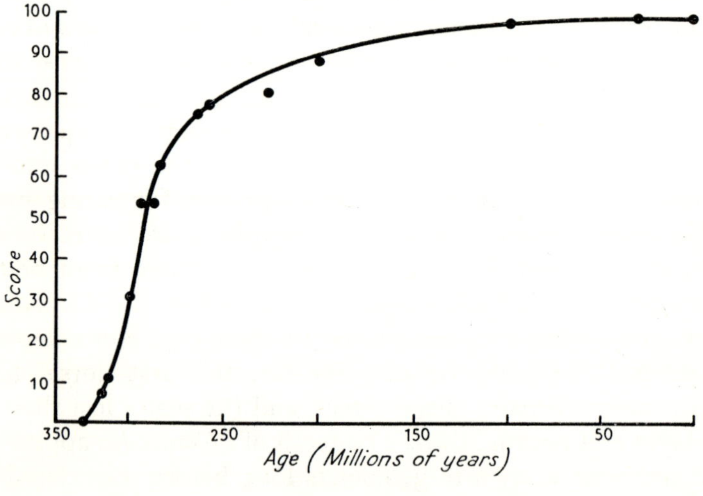

FIG. 164. Rate of evolution in lung-fishes. Each point represents the index of a single genus obtained by taking twenty-six characters and rating them with grades of structure. The lowest value was given for the most primitive condition, the highest for the most modern one. (After Westoll and Simpson.)

different characteristics such as proportion of skull, nature of dermal bones or dentition were divided into 3–8 grades. Each fossil genus was thus given a score, the minimum being for characters appearing in the oldest lung-fish, the hypothetical ancestor being 0. Actual scores ranged from 4 for the earliest to 100 for two of the living genera. Plotting against time we see that there was an early acceleration of evolution followed by very slow or zero change in the last 150 million years (Fig. 164). It has also been shown that a similar method applied to the evolution of coelacanths gives a curve of similar shape (Schaeffer, 1952).

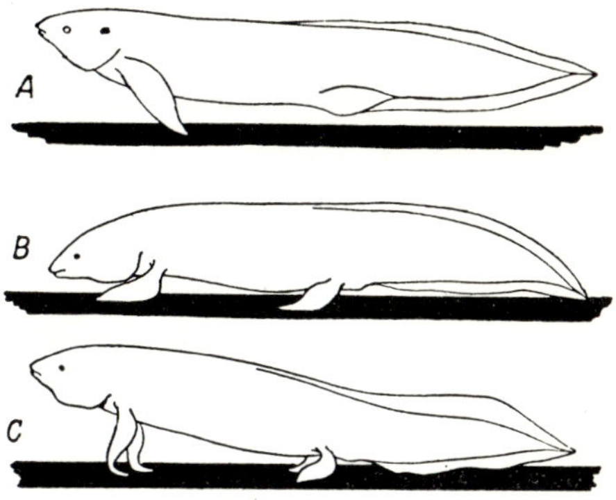

FIG. 165. *Neoceratodus*, showing the method of walking on the bottom. *A*, forwards; *B*, backwards; *C*, resting. (From Ihle, after Dean.)

6. Modern lung-fishes

The three surviving genera of lung-fishes (Fig. 163) are mainly inhabitants of rivers (though *Protopterus* lives in large lakes) and they all breathe air. *Neoceratodus* lives only in the Burnett and Mary rivers in Queensland, the pools of which become very low and stagnant in summer. *Lepidosiren* from the rivers of tropical South America, and *Protopterus* from tropical Africa, can survive when the rivers dry up completely. They dig into the mud, leaving a small opening for breathing, and can remain in this state for at least six months. Remains of cylindrical burrows found associated with dipnoan bones show that this habit of aestivation has been adopted by the group at least since Permian times. The three survivors all show similar deviations from the conditions found in **Dipterus*, but *Neoceratodus* has diverged less than the other two. The tail fin is symmetrical (diphycercal) in all three, with no trace of separate dorsal fins. The paired fins are of 'archipterygial' type in *Neoceratodus*, with an axis and two rows of radials (Fig. 165). The scapula is covered by clavicles, cleithra, and post-temporals, the latter articulating with the skull. The scales are reduced to bony plates.

The vertebrae are cartilaginous arches, the notochord remaining as an unconstricted rod. In the skull there is also a great reduction of ossification, the dorsal bones consisting of a few bony plates, forming a pattern not obviously comparable with that of other forms (Fig.

166). The jaw-suspension is autostylic. The food consists of small invertebrates and decaying vegetable matter, which is eaten in large amounts. In the gut there is no stomach and the intestine is ciliated. There are no hepatic caeca, but a well-developed spiral valve is present.

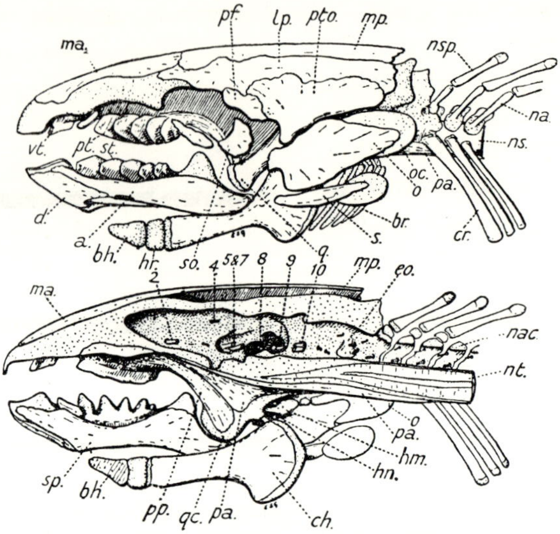

FIG. 166. Skull of *Neoceratodus*. Above, Lateral view; below, view of medial surface of right half.

a. angular; ***bh.*** basihyal; ***br.*** fifth branchial arch; ***ch.*** ceratohyal; ***cr.*** 'cranial' rib; ***d.*** dentary; ***eo.*** 'exoccipital'; ***hm.*** hyomandibula; ***hn.*** hyomandibular nerve; ***hr.*** hypohyal; ***lp.*** lateral plate; ***ma.*** median anterior, and ***mp.*** median posterior plate; ***na.*** neural arch; ***nac.*** cartilage of neural arch; ***ns.*** notochordal sheath; ***nsp.*** neural spine; ***nt.*** notochord; ***o.*** opercular, and ***oc.*** its cartilage; ***pa.*** parasphenoid; ***pf.*** post-frontal; ***pp.*** pterygopalatine; ***pt.*** palatine tooth; ***pto.*** pterotic (?), and ***g.*** its downward process covering the quadrate cartilage, ***qc.***; ***s.*** sub-opercular; ***so.*** suborbital; ***sp.*** splenial; ***st.*** splenial tooth; ***vt.*** vomerine tooth. (From Goodrich.)

The external nostrils lie just at the edge of the mouth and the internal nostrils open into its roof. The air-bladder is developed into a definitely lung-like structure (there is one in *Neoceratodus*, a pair in each of the others), divided into many chambers. *Neoceratodus* has been observed to come to the surface to breathe air and is said to be able to survive in foul water that kills other fishes, but it cannot live out of the water. *Lepidosiren* and *Protopterus* have been shown to obtain 98 per cent of their oxygen from the air. The wall of the air-bladder of all forms contains muscle-fibres and the cavity is subdivided into a number of pouches or alveoli (Fig. 157). In *Protopterus* and *Lepidosiren* the edges of the slit-like glottis are controlled by muscles and there is an epiglottis. The lung is supplied with blood from the last branchial arch in *Neoceratodus* (Fig. 167), but in the other Dipnoi there is a more elaborate arrangement. The second and third gill-arches bear no lamellae and their afferent and efferent branchial vessels are directly continuous, so that blood flows from the ventral to the dorsal aorta and carotids. The pulmonary artery springs from the dorsal aorta. Blood returns in a special pulmonary vein to the partly

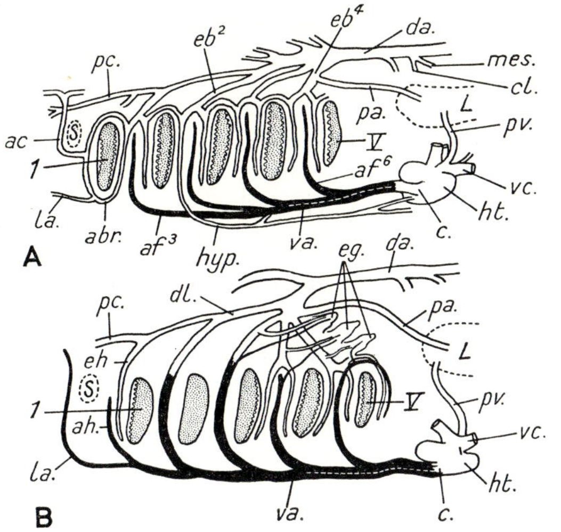

FIG. 167. Branchial circulation of A *Neoceratodus* and B *Protopterus*.

abr. anterior efferent vessel; *ac.* anterior carotid. *af*$^{3-6}$. four afferent vessels (corresponding to the original arches 4–6); *ah.* afferent hyoidean; *c.* conus; *cl.* coeliac artery; *da.* dorsal aorta; *dl.* dorso-lateral aorta; *eb*$^{2-4}$. second and fourth epibranchial arteries; *eg.* external gills; *eh.* efferent hyoidean; *ht.* heart; *hyp.* median hypobranchial; *L.* air-bladder; *la.* lingual artery; *mes.* mesenteric artery; *pa.* pulmonary artery; *pc.* posterior carotid; *pv.* pulmonary vein; *S.* position of closed spiracle; *va.* ventral artery; *vx.* vena cava inferior; *I–V* five branchial slits.

The gills are represented on the hyoid and next four branchial arches. (From Goodrich.)

separated left side of the sinus venosus. The auricle is partly divided into two and the ventricle is almost completely divided by a ridge and a series of muscular trabeculae. The ventral aorta is shortened into a spirally twisted truncus arteriosus, provided with a system of valves such that the blood from the left side of the auricle is directed mostly into the first two branchial arches, that from the right side into the last two. In this way some separation of pulmonary and systemic circulations is achieved. Indeed, although there is every reason to believe that the mechanism has been in existence for nearly 300 million years, it shows us most clearly a possible intermediate stage between aquatic and pulmonary respiration. It seems likely that the earliest amphibia employed a similar system. There is a coronary artery arising from the anterior efferent branchial arches (Fig. 167).

A further amphibian feature of the blood-vascular system is the presence of an inferior vena cava, a vessel collecting blood from the kidneys and reaching to the heart by passing round to the right of the gut in the mesentery. The more dorsal cardinal veins, joining the ductus Cuvieri, remain present, however, and there is a renal portal system.

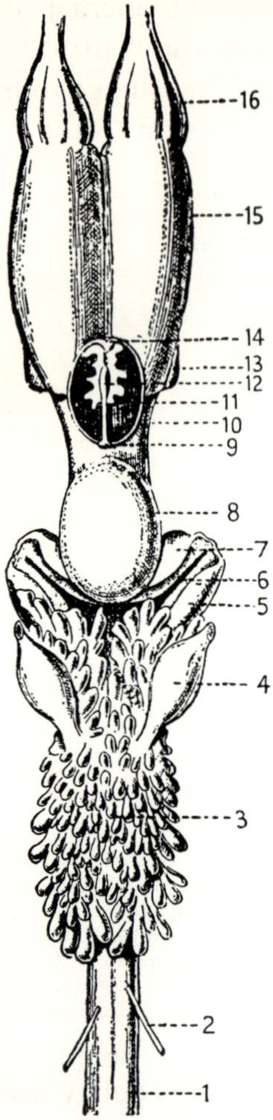

FIG. 168. Dorsal view of the brain of *Protopterus*.

1, spinal cord; 2, dorsal root of first spinal nerve; 3, diverticula of 4, the saccus endolymphaticus; 5, medulla oblongata; 6, fourth ventricle; 7, cerebellum; 8, mesencephalon (fused optic lobes); 9, stalk of pineal body; 10, thalamencephalon; 11, velum transversum; 12, pineal body; 13, lobus hippocampi; 14, choroid plexus; 15, cerebral hemisphere; 16, olfactory lobe. (After Burckhardt, from Goodrich.)

The adrenals of Dipnoi are represented by two separate masses of tissue. The perirenal tissue of *Protopterus* is a considerable mass of material around the kidney, containing lipid, steroid, and round-cell (lymphoid) tissues, as well as endothelial and pigment cells. The steroid tissue shows histochemical properties similar to those of mammalian adrenal cortex and undergoes changes after injection of mammalian ACTH. This tissue thus shows a collection of functions, haematopoietic, phagocytic, storage, endocrine, and pigmentary, which may show the starting point of the evolution of the tetrapod adrenal cortex. Cells that give the chrome reaction, the adrenal medullary tissue, lie in the walls of the intercostal branches of the dorsal aorta. This condition could have given rise to that of amphibia.

The arrangement of the urinogenital system is similar to that of amphibia and is probably closer to that of the ancestral gnathostome than in any living elasmobranch or actinopterygian. In the male there are vasa efferentia by which sperms are passed through the excretory portion of the mesonephros. In the female eggs are shed free into the coelom and carried out by a Müllerian duct, whose opening lies far forward. An interesting feature is that the Müllerian duct is very well developed in the male. This is one of several details (lack of ossification, unconstricted notochord) which raise the suspicion that the living Dipnoi have acquired their special characters by a process of paedomorphosis or partial neoteny, that is to say, becoming sexually mature in an early stage of morphogenesis.

The development, again, shows similarity to that of amphibia and dissimilarity from the other groups of fishes in that cleavage is total and gastrulation takes place to form a yolk plug. There is therefore no blastoderm or extra-embryonic yolk sac. However, the cells of the vegetative pole contribute little to the shape of the embryo and indeed may form a partially separate yolk sac. The larvae show distinct similarity to those of amphibia, especially the larvae of *Lepidosiren* and *Protopterus*, in which there is a sucker and external gills.

The nervous system shows the same affinities as the rest of the organization (Fig. 168). The forebrain is evaginated into a well-marked pair of cerebral hemispheres. The roof of these, though not very thick, is nervous and therefore is definitely of the inverted type, not everted as in Actinopterygii. The optic lobes are little developed, the mesencephalon being hardly wider than the diencephalon. The cerebellum is small. A peculiar feature is the development of the inner ear to form a special lobed saccus endolymphaticus, lying above the medulla oblongata. The significance of this is not known, but it is interesting that similar backward extensions of the ear are found in amphibia.

In many features, then, the Dipnoi differ from modern fishes and resemble the amphibia, which evolved from the same stock. The early crossopterygians probably competed somewhat precariously with other fishes, as do the Dipnoi, and indeed many urodeles, today. It was only after tens of millions of years of evolution during the Carboniferous and Permian times that numerous land animals arose, in the form of the later amphibian and reptilian types (Fig. 211).

XI

FISHES AND MAN

MAN'S influence, which has so altered the face of the land, also reaches out into the water. During the last half-million years or so fishing has grown continually, until at the present time in fresh and shallow sea waters man has become the most important source of mortality of fish. However, he still leaves vast regions of the sea, as of the land, almost untouched. It may be that methods of catching the fish that swim above deep waters will eventually produce changes there too, providing a substantial extension of the food-supplies of man and effecting great alteration in the population of the sea.

Man has caught fish from his earliest times and he uses many parts of their bodies. The skin, especially of elasmobranchs, makes a useful leather and also a polishing material. The scales of the bleak (*Alburnus*) yield a substance that when coated on to the inside of glass beads makes artificial pearls. The lining of the air-bladder, especially of sturgeons, makes isinglass, a shiny powder used for various purposes as an adsorbent (i.e. in wine-making). Fish-glue is obtained from the connective tissue of the skin and other parts. Fish oils are used as food, and are also valuable in the manufacture of soap and other things. Besides their direct use as human food, fish products may be fed to animals, and the liver makes excellent manure.

Fish therefore provide the raw material for many human activities, but it is, of course, principally for food that they are caught. From a fish diet one can obtain not only abundant calorific value, especially if the fish be fat, but also various proteins, often the fat-soluble vitamins A and D, and usually considerable amounts of combined phosphorus and other elements. Herring and mackerel are probably the cheapest source of protein available to many people in Britain. Fish are undoubtedly good food, and what is perhaps even more important they are esteemed as such; most people think that they taste good.

It is not possible to give an accurate estimate of the amount of fish caught every year; 20 million tons, of value £200 million, is certainly not an overestimate for the catch of the whole world in each of the years between the Great Wars. An appreciable fraction of the nutriment of the human race is derived from fish. The annual catch was estimated at 26 million metric tons in 1951. The greatest national catch was by Japan (3·8 million), followed by U.S.S.R. (2·5), U.S.A. (2·3), and Great Britain (1·1).

By far the greater part of all fishing is in the sea. In Great Britain the total catch in fresh water annually is reckoned at only 2,000 tons and the entire stock at 8,000 tons. However, there are important fishing industries in the American Great Lakes, the African lakes, and in many parts of the world carp are raised on fish farms, the water being manured to yield a good crop of the freshwater plants and invertebrates on which the carp feed. The milk fish (*Chanos chanos*), which feeds directly on algae, provides high yields in many tropical and subtropical fish-ponds.

Fishing in the sea is limited mainly by two considerations. First, the labour is only really profitable when the fish population is dense. Secondly, fish do not keep well unless they are treated in some special way and they are therefore best caught near to their market. For these reasons the big commercial fisheries are mostly in the relatively shallow waters of the continental shelf (down to 200 fathoms) and in the northern hemisphere. However, by packing the fish in ice or cooling by other means, the vessels are now able to go longer distances than formerly. Japanese and American vessels now fish for tuna and salmon in the open ocean.

For catching purposes fish may be divided into those that swim away from the bottom, the pelagic forms, such as herring or mackerel, and the bottom-living or demersal fishes, such as the skates and rays, flat-fishes and angler-fishes. For the first the gear employed is usually drift-net. This is of narrow mesh and is shot and left overnight, drifting vertically near the surface, often making a barrier two or three miles in length. The fish, such as herrings, swim into them, and become enmeshed, and are removed as the nets are hauled in.

The success of such fishing with drift-nets depends on laying the nets in the path of large shoals of fish and it is therefore practised especially in regions where fish congregate seasonally. For instance, in the North Sea between East Anglia and Holland the herring congregate in the autumn; at certain times, for unknown reasons, all the fish in the area begin to move, making 'the swim', and they are then caught in great numbers. Herring are usually caught at slack water. At the October full moon, slack water occurs just after dusk and just before dawn and catches are higher then. Other drift-net fisheries depend on intercepting the fish when rising to feed. Evidently these methods demand a close knowledge of the habits and migrations of the fish. The fishermen have learned to know when and where to expect the larger and nearer aggregations, but there is still much to be done in tracing the migration and behaviour of pelagic fishes. It is not impossible that other pelagic fisheries could be developed that would be at least as advantageous as the herring fishery. Besides those eaten fresh, others are preserved as kippers by salting and smoking, or are heavily salted and smoked to make 'red herrings' for which there has been a large export market.

Lining is a method that can be used for either pelagic or demersal fish. For instance, the cod fishery of Newfoundland uses lines with several thousand baited hooks. Inshore fishing may be accomplished by beach *seining*, a process of enclosure of the fish from the shore. There are also various forms of trap into which the fish swim and cannot escape. *Danish seining*, or *purse seining*, completely surrounding the fish and sometimes then also drawing the bottom of the net together, is another effective method.

The greatest quantities of fish are caught, however, by the various forms of *trawling*. These depend essentially on dragging a bag along the bottom of the sea, and the different types adopt various means of keeping the bag open. The earlier way of doing this was by means of a rigid bar and these beam trawls sometimes used poles over 50 ft long. More recently otter-trawling has replaced the beam trawl, the otter boards being flat wooden structures attached at each end, so as to sheer away when they are dragged through the water, thus opening the net. Various devices are used to stir up the fish from the bottom; in particular the otter boards are now usually separated from the net by long wires and a 'tickler' chain is attached in front of the mouth.

With such methods very large amounts of fish can be taken from the sea bottom. In 1860 sailing-ships landed 500 tons of fish at Grimsby; by 1901, 176,000 tons were landed from the steam traw lers 50,000 tons of plaice alone are taken annually from the North Sea. There is evidence that the taking of such amounts of fish from relatively confined waters has a large influence on the population. For instance, it is estimated that 45 per cent of the stock of plaice is caught each year in the North Sea. Under these conditions man provides the main source of mortality for the fishes. It is less clear exactly how such mortality influences the total population of the fish in the area. There is reliable evidence that intensive fishing reduces the productivity of fishing effort in some cases, but there is also evidence that by reducing competition fishing may produce faster growth of the fish. The problem is worth some further discussion since, besides its economic importance, it gives us an insight into the life of the vast populations of the sea and a glimpse of the factors that control the numbers of each animal species.

It used to be held that one of the chief dangers of fishing was that it would destroy the breeding-stock, on the grounds that it is axiomatic that something is wrong if fish are destroyed before they have an opportunity to breed. We may detect something of anthropocentric sentimentality here, for reflection shows at once that it is impossible for every fish that is hatched to survive and, further, that the total annual supply of fish can be provided by relatively few adults. The number of fish in the sea, indeed, bears no obvious relation to the number of eggs. A cod may shed as many as 8 million eggs (usually fewer), which float at the sea surface and hatch into planktonic young. Yet herrings, which are much more numerous as adults, lay fewer eggs, usually less than 30,000 each. Presumably the risks of life are less for the herring, but we cannot see clearly why; the herring's eggs are sticky and become attached to gravel or seaweeds, which may give them some protection; they are known to be eaten by other fishes, perhaps on a smaller scale than planktonic eggs. Both herring and cod fry are planktonic and feed on the nauplius and other larval stages of *Calanus* and other copepods. The cod, like other bottom-fishes, has to undergo the risks of metamorphosis, but this hardly seems sufficient to explain the much greater abundance of herring, especially since many fishes feed on the latter, including the cod themselves! As Graham puts it, 'no one really knows why the herring, which nearly everything eats, should be able to manage with a less rate of reproduction than the cod, which eats nearly everything'. The greater part of the mortality of pelagic fish larvae is due to predation, including that by members of the same species. This probably provides a method of regulation, the number of each species eaten being dependent on their frequency.

This brings us to face the difficult question of the pressure, force, or potential that ensures and controls the number of animals. Reproduction is only a part of the source of this pressure, the other element being the feeding and growth of the animals as a result of the skill and persistence with which they seek and consume their food. It is somewhat easier to study these questions in a marine than in a land community, the whole population being enclosed in one vast bath, the additions and subtractions to which can be known.

We must not forget, however, that conditions are far from constant, even in the 'unchanging sea'. For instance, the extent to which fish-fry hatch and successfully overcome the hazards of the early

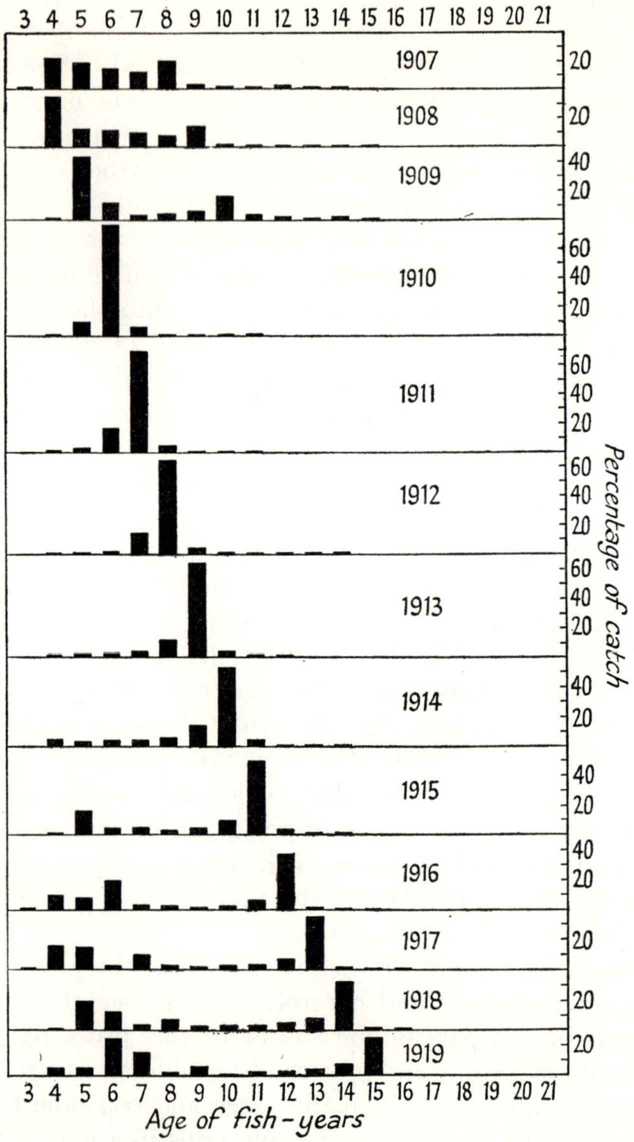

FIG. 169. Age composition of Norwegian herring catches 1907–19, showing the preponderance of particular year-classes. (After Hjort.)

stages of life varies most mysteriously from year to year. By study of the markings on the scales, bones, and otoliths it is possible to show the age composition of a fish population and it is found that the hatches of certain years predominate. Thus the herring hatched in 1904 dominated the population in some parts of the North Sea for years afterwards (Fig. 169). Subsequent good herring years occurred in 1913 and 1918. Similarly there were good cod years in 1904 and 1912. Haddock broods in the North Sea follow a rather regular

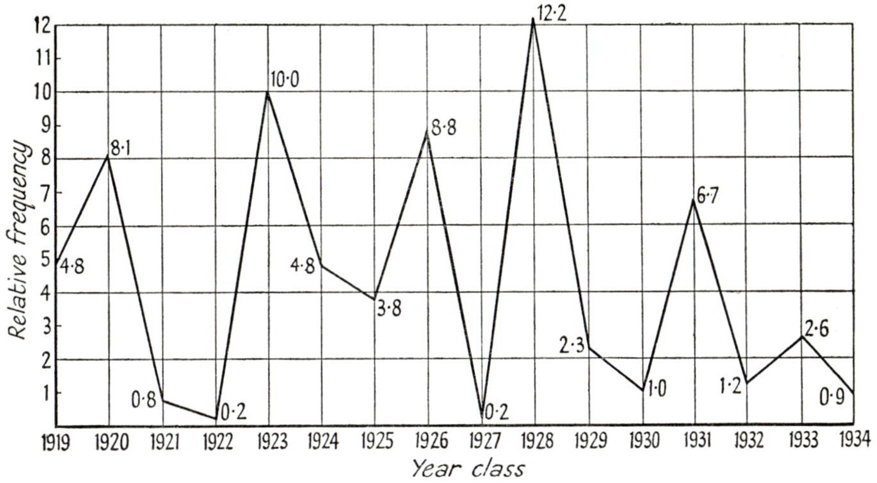

FIG. 170. Relative strengths of haddock broods in the North Sea, showing cyclical fluctuations. (After Raitt, from Russell, *The Overfishing Problem.*)

rhythm with a period of about three years (Fig. 170). These cyclical changes in hatching do not appear to produce fluctuations in the total population of that species, though we have inadequate data on the point. It seems likely that there is a limit set to the total of fish of the species, irrespective of age-grouping.

In addition to such cyclical changes there are also larger ones, such as the increase of cod in the far north around Bear Island in recent years, apparently due to a northerly extension of warm water, the reason for which is not known. Since so many changes are going on in the eternal sea is it to be wondered at that the animals in it are changing? We might be surprised rather at the slowness of their evolution.

All the fish-life in the sea depends ultimately on the diatoms, flagellates, and other green organisms of the plankton, reaching the fish usually after passage through one or more animals, especially copepods. Bottom-living fishes also depend on plankton, in this case after it has died and passed through a bottom-living invertebrate. Thus plaice feed on lamellibranchs and worms, soles on worms alone, haddock on star-fish, molluscs, and worms. The phytoplankton is, of course, built up by photosynthesis from carbon dioxide, water, nitrates, phosphates, and small quantities of other substances. The limiting factors in the plankton growth, granted adequate temperature and sunlight, may be nitrate, ammonia, and phosphate, the latter present only to the extent of less than one part in 20 million. During the spring and early summer nearly all the phosphate in the water may be taken up into the phytoplankton, which increases rapidly. The zooplankton (*Calanus*, &c.) increases somewhat later and by midsummer grazes down the phytoplankton to a considerable extent. It is at this midsummer period that the pelagic fishes feed most easily, and a little later it is the turn of the bottom-living invertebrates and fish, as some of the zooplankton dies and falls to the bottom. At the end of the season there may be a second increase in the phytoplankton, the recovery being due to breakdown of the discontinuity layer (see below), and then with the onset of winter the plants die off and most of the phosphate is returned to the inorganic condition.

Thus to take the case of the cod, whose food chain is one of the longest: the activity of the plants turns carbon dioxide, nitrate, and phosphate into organic matter: the activity of the copepods makes this available to the herrings and the cod eat the herrings. There are, of course, many intermediate effects: *Sagitta* eat some of the copepods, fish-fry and large cod eat younger cod, and so on, but in general we shall not go wrong if we assume that the cycle depends on the activity of diatoms, calanoids, herrings, and codfish—all seeking, eating, growing, reproducing, and dying.

The pressure of these various activities produces a complicated set of interrelations that may without undue extravagance of fancy be called a macro-organism. The activities of all the members contribute to the balance that is set up. Sometimes shortage of materials, such as phosphorus, at key-points may determine the whole cycle.

The phosphate in the sea is increased to a small extent by drainage from the land and, further, in a restricted area such as the North Sea, by the influx of water from more open areas, in that case the Atlantic. Not all the phosphate in a given area of sea is available for organic growth, because there is usually a separation between upper and lower layers owing to the upper layer becoming warmer and less dense. The phosphate in the water below this limiting discontinuity layer, usually lying 20–100 metres down, is thus not available, though it may be made available in certain areas by particular conditions that lead to up-currents and mixing of the layers.

With these facts in mind we can now make further inquiry into the effects on the population of the sea of removal by man of large quantities of organic matter in the shape of fish. There are enough data to give us hints of the changes that follow, and the hope that with further study we may eventually understand the Great Sea Beast sufficiently well to be able, if we can control ourselves, to regulate its growth to our advantage.

In a few cases it has been possible to follow the course of a fishery from its early beginnings, with few vessels and simple apparatus, through stages in which the fishing was progressively more intense. The statistics are seldom adequate to provide us with wholly satisfactory conclusions, but they suggest that (1) as fishing becomes more intense, the yield per unit of fishing power declines; that is to say, the industry becomes relatively less profitable; (2) in spite of this, the total yield may remain constant, or somewhat increase, even though (3) the average size of the fish caught decreases.

For example, fishing for plaice in the Barents Sea, north of Norway, was begun in 1905, and the almost 'virgin' population in 1907 showed great preponderance of large mature fish; whereas in the North Sea, which had already been fished for many years, the average size was much smaller and few of the plaice were mature. The fish taken in the Barents Sea were old and had grown slowly. Transplantation of fish of this stock to the North Sea, however, proved that they were able to grow as rapidly as the local fish. Some measure of the intensity of fishing is given by the number of days' absence of vessels fishing in a given area. No doubt the measure is inexact, particularly if the methods used are changed, but it provides the best estimate available. The landing of plaice taken from the Barents Sea per day's absence was 34·7 cwt in 1906, 50·4 cwt in 1909, and then fell to 46·3, 33·7, and 20·5 cwt in the following years. The total yield showed some decline. The sequence of events is typical of what happens when a stock of mature fish is first exploited.

Graham described a good example of a similar situation in Lake Victoria-Nyanza, where he found two stocks of the carp-like fish *Tilapia*, one fished heavily and the other with primitive devices only. The less heavily fished stock contained five times as many fish as the other and they were much older. Yet by the intensive fishing the poorer stock was being made to yield a ten times greater weight than the richer stock. This desirable state of affairs for man in general is considered unsatisfactory by the fishermen, since the yield *per unit of effort* is much greater when the richer stock is fished.

The yield of haddock in the North Sea per day's absence averaged 5·8 cwt between 1906 and 1913, rose to 15·8 cwt in 1919, averaged 6·5 cwt in 1922–1929, and 3·7 cwt in 1930–7. The fish are divided into large, medium, and small categories, and the percentage of the last category was 50 before the war, 70 in 1922–9, and 85 in 1930–7; evidently the fish were becoming smaller throughout this period, yet the total haddock catch changed only slightly, from 121 million kg (1910–13) to 138 million kg (1922–9) and 94 million kg (1930–6). Fishing seems to have made the population consist of smaller fish, giving a smaller yield per unit of fishing effort. This is a true index of stock and its decline shows the extent of reduction.

The figures available for plaice and cod taken from the North Sea show much the same tendencies. Over a long period fishermen have been afflicted by the consequences of the decline in yield per unit effort indicated by such statistics. Already in the last century those giving evidence before Royal Commissions expressed an uneasy feeling that the profitability of the industry was declining. This situation has been met by continual improvement in methods of fishing and the exploitation of new grounds. The proportion of the total English catch that is obtained from the nearer waters of the North Sea is now much less than formerly, although the catch is no greater. We have already seen that steam trawling and better gear all mean increased labour for the crews, at greater distances from port. With all the improvements in technique and discovery of new grounds we succeed in obtaining from the sea little more fish than at the end of the last century; and there is no greater profit for those engaged in the industry.

The 'Great Law of Fishing', as Graham calls it, 'that fisheries that are unlimited become unprofitable' has been tested on at least three occasions. During the two Great Wars the intensity of fishing in the North Sea was reduced and thereafter both of the expected changes were seen: (*a*) fishing became more profitable, (*b*) there were found to be more old fishes in the stock. Perhaps more encouraging is a case where the intensity of fishing has been regulated not by war but by law. The halibut fishery of the Pacific coast of North America showed the typical history of increasing effort and decreasing yield. Fishermen went farther and farther afield in larger and more expensive boats, setting ever more and more lines, but yet brought home the same or a smaller amount of fish. Thus in 1907 there were 1,800,000 sets of lines used, in 1930, 6,400,000 sets, but instead of the 50 million pounds of fish caught in the first year there were only 23 million pounds in 1930! Then in 1932 a limitation of 23 million pounds was imposed on the total catch allowed. Thereafter the stocks rapidly increased and it was found possible to collect the standard catch in five months instead of nine, with greatly increased profit.

Other attempts to regulate fisheries have been made; in particular by the mesh regulations, designed to avoid taking small fishes. The aim of regulation is nowadays not so much to save the stock from undue reduction or extinction but rather to *crop it in such a way as shall make fishing profitable.* Whatever we do it is unlikely that we shall destroy all the stocks of any species, but there is reason to think that the rate at which the stocks grow varies with the intensity of fishing. If this is so, it may be possible to find an optimum that suits both the fish and the fisherman.

In order to regulate a fishery effectively it is necessary to express its characteristic parameters in a comprehensive equation. This has been done by fisheries research workers who have provided mathematical models such as are so often used in operational research (Beverton and Holt in Graham, 1956). These equations show the effects that are likely to follow variation in such a factor as the size of the mesh of the cod end of the trawl net, which is one of the means used to regulate a fishery. Fishing is a form of hunting, not of agriculture, and if we cannot improve the yield by cultivation we may be able to do so by working out the best way to fish.

Four primary factors are considered in the model, recruitment (R), growth (W), mortality due to natural causes (M), and mortality due to fishing (F). The relation of these factors to the yield by weight of the fishery (Yw) is considered and a theoretical equation for Yw is derived. If the theory is correct the equation should be able to describe the yield of various fisheries in the past. It proves to do this well and forecasts that improvements in yield could be obtained by changing methods of fishing in the future. The data for testing the theory come from statistics of the fishes caught, which give catches per unit effort in each year class and their lengths. These are available for several fisheries and we may consider the plaice in the North Sea for the years 1929–38. Fish are said to be recruited when they first enter the fished area of deeper water at 2–3 years old, being much less than 1 per cent of the original batch of eggs. The value R for a given mesh size is obtained from the number of fishes in the youngest year class in the catch, with corrections to allow for the fact that

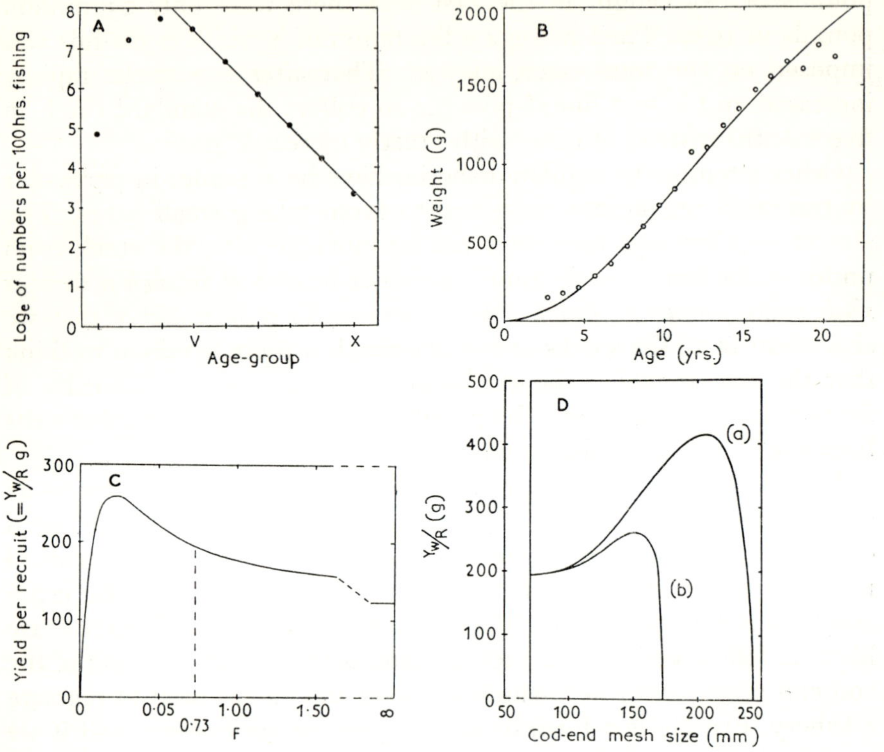

FIG. 171A. Mortality rate of plaice. Natural logarithms of average number caught of each age-group of North Sea plaice per 100 hours' fishing, 1929–38. The slope of the line gives the estimate $(F+M) = 0{\cdot}83$. The first three points are lowered by absence of fish from the exploited area (some individuals still remaining in the nursery areas) and, possibly, by rejection at sea.

FIG. 171B. Growth in weight of plaice. Average weight of fish of each age-group, 1929–38, and the fitted von Bertalanffy equation for growth in weight.

FIG. 171C. Annual steady yield of plaice plotted against fishing intensity. Yield of plaice per recruit (Y_w/R) as a function of fishing mortality coefficient (F) with $t_p' = 3{\cdot}72$ years, corresponding to a 70 mm gauge mesh in trawls, double twine. The vertical line at $F = 0{\cdot}73$ corresponds to the average pre-war fishing intensity.

FIG. 171D. Effect of growth rate changing with density. The yield of plaice per recruit (Y_w/R) is shown as a function of mesh size with fishing intensity constant at $F = 0{\cdot}73$, but in calculating curve (*b*) the growth-rate was assumed to be reduced progressively as the density of stock increases. The differences from the use of constant growth-rate, curve (*a*), are considerable. Not only is the benefit from increase of mesh estimated to be less, but the maximum yield is reached at a considerably smaller mesh size.

recruitment is not simultaneous or sudden. For the plaice in the North Sea, with 70 mm mesh, R has been computed at 280 million fish for the area considered, with mean age of 3·7 years. A check from consideration of numbers and mortality of eggs gave 320 million. Theoretically, we also require to know the length of the life of the fish. Few plaice are caught older than 15 years and almost none over 20 years. Since there are very few in these older age-classes the upper limit is unimportant.

The coefficients of mortality M and F are estimated from the numbers (N_t) of the original recruits R that survive at time t, $N_t = R_e - M\ (t - t_p)$, where e is the natural base of logarithms and t_p the age at entry of the recruits. The total mortality ($F+M$) (due to natural causes (M) and to fishing (F)) is estimated from the catches (Fig. 171A). The part due to fishing alone is found by consideration of how the catch varies when there are known differences in the fishing effort, for example in the number of hours fished. In the extreme case, we have the war-time periods, when there was no fishing. Confirmatory evidence of the likelihood that a fish will be caught can be obtained from marking experiments, giving the time between release and recapture. For the plaice ($F+M$) has been estimated over the years 1929–38 as 0·83. This is the mean for all the age-classes together, though for some purposes it can be calculated for each separately. That the rate is indeed constant is shown by the close fit of Fig. 171A. The first three points do not fit because recruitment of these classes was still incomplete. This method of treatment, in which the contributions of year classes of the same age recruited in different years are considered together, has the effect of averaging out this variation in recruitment, which would be difficult to estimate.

The value of the natural mortality of plaice cannot be obtained for these years from variations in fishing effort, since these were only slight. However, since there was almost no fishing in the southern part of the North Sea from 1940 to 1945, we can compare the age groups V and VI of samples taken before the war with their surviving fellows in the years after 1945. This gives a value for M of 0·1, and subtracting this from 0·83 as the total mortality we obtain 0·73 as the mortality due to fishing. Marking experiments give a value of 0·69, which is a satisfactory agreement considering that the conditions were not entirely similar. Moreover, the conclusions from marking involve other problems, such as damage to the fish and rate of movement away from the point of release. The conclusion that fishing is the most important source of mortality in such areas is not new, but is obviously of very great importance. If man is the chief predator, then change in his activities will greatly influence the populations.

Knowing the number of fishes available and the rates at which they are removed by natural causes and fishing in a given case, we still have the even more difficult task of formulating the way in which the population grows and then deciding whether some other method of fishing would be more profitable. Probably there is a maximum total biomass of fish of any given sort that can be supported in any area, determined ultimately by the supply of inorganic salts. In the absence of fishing this biomass is carried mainly in the form of large fish, whose presence makes the growth of all fish in the population slow. The effect of fishing is to remove mainly these larger animals, with a resultant increased growth from all younger groups. The curve showing the relationship between the yield per recruit and the intensity of fishing effort shows a maximum (Fig. 171C). It should be possible to define a fishing mortality rate at which the decrease in numbers is balanced by the increments in weight of the survivors.

There is every reason to hope that with further study of growth, mortality, reproductive potential, and utilization of food by fish the yield could be increased and the effort of getting it reduced, making fishing more profitable to the fisherman and providing the maximum amount of food. At present the economic conditions and psychology of the fishing populations interact with the factors limiting the stocks and the growth of the fish to produce a complicated system of interrelations that is unstable and unsatisfactory to all parties.

The increase in weight of the population is perhaps the most difficult feature of the pattern to express mathematically. Weight plotted against age shows an asymmetrical sigmoid shape, with an inflexion (Fig. 171B). The curve fitted is arbitrarily chosen, being that deduced from the hypothesis of Bertalanffy, that the weight is subject to opposing forces of anabolism and catabolism, taken as proportional to the absorbing surfaces, that is, to the squares of the linear dimensions. As before, the fit is poor for the lower points, probably because there are incompletely recruited classes. Elsewhere the fit is good but the important point for us is the inflexion, since it indicates that the growth-rate decreases in the later part of life. The older plaice are from this point of view inefficient in providing more biomass. Further, the longer a fish has to be kept alive in the sea before it is eaten, the more of the limited raw materials are devoted to this end and the less to providing human food. With present methods it is not possible to give full weight to all of these factors in deciding what is the best way to fish. However, the theoretical equation derived by Beverton and Holt gives us the yield Y_w in terms of the parameters already discussed with in addition the maximum weight W_∞ and a factor K related to the catabolism

$$\frac{dY_w}{dt} = FN_tW_t. \qquad (1)$$

The number N_t at the time t is given by

$$N_t = R_e^{-(F+M)(t-t_p)}, \qquad (2)$$

and the weight W_t by

$$W_t = W_\infty(1-e^{-K(t-t_0)^3}). \qquad (3)$$

This last is best handled as a cubic of the form

$$W_t = W_\infty \sum_{n=0}^{3} \Omega_n e^{-nK(t-t_0)}.$$

Substituting (2) for N_t in (1), and (3) for W_t we obtain

$$\frac{dY_w}{dt} = FRW_\infty e^{-(F+M)(t-t_p)} \sum_{n=0}^{3} \Omega_n e^{-nK(t-t_0)}.$$

This provides the basic equation with which forecasts of effects of changing the various factors are made.

Empirical values can be obtained for the yield per recruit ($Y_w/_R$) for various values of F, with a mesh of 70 mm. Such a yield/intensity curve of North Sea plaice is shown in Fig. 171C. The graph shows that with infinite effort all fish would be caught at recruitment and would yield their initial weight of 123 g. At the pre-war fishing mortality of 0·73 the yield was 200 g. But the curve has a clear maximum at over 250 g, with a fishing mortality of only 0·22. Therefore if these predictions are correct, a lesser intensity of fishing should provide a greater yield.

One possible way of reducing fishing intensity is to increase the size of the mesh of the cod net. If the age at recruitment were increased to ten years the yield per recruit would be as high as 400 g (Fig. 171D, curve a). Beyond this maximum, the yield falls because of the death of fish by natural causes before entering the exploited phase. However, this curve assumes that the growth-rate is independent of density and it ignores the complex problem of the competition of old and young fishes for a limited supply of raw materials. Curve b of Fig. 171D has assumed a reduced growth-rate with increasing density and it will be seen that the advantage of increasing mesh size is much reduced.

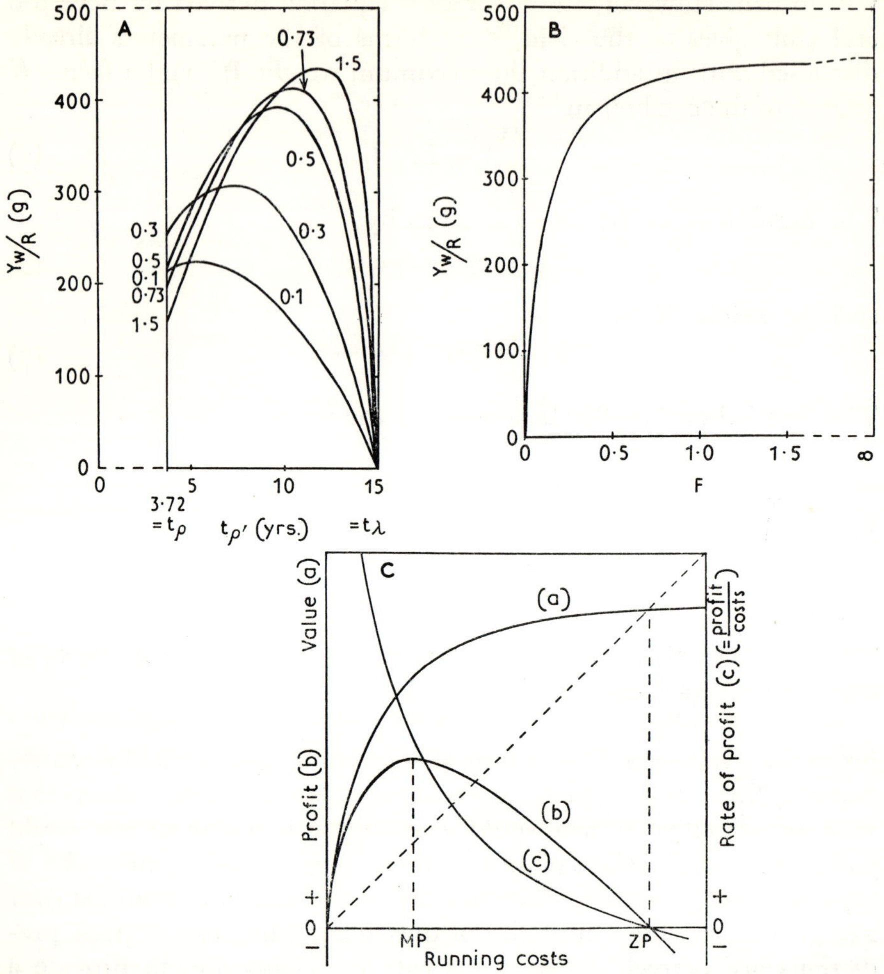

FIG. 172A. Annual steady yield of plaice plotted against mesh at various fishing intensities. The yield per recruit (Y_w/R) as a function of t_p' for various values of F, to show how the height and location of the maximum changes with F.

FIG. 172B. Eumetric Yield curve for plaice. If a change of mesh accompanies a change in fishing intensity in such a way that each value of F is matched by the mesh that would give the maximum steady yield at that value, the resultant curve of steady annual yield per recruit (Y_w/R) for plaice differs widely from that of Fig. 171*c*, particularly in having no maximum.

FIG. 172C. An exercise in bionomics. Curve (*a*) is a eumetric curve in which steady yield instead of being expressed in weight is now in money value, which is plotted against the economic equivalent of intensity of fishing, namely, running costs. From those are derived (*b*) the annual profit and (*c*) the profit expressed as a rate on running costs, which in many situations bear a constant relation to capital outlay.

MP = maximum profit point; ZP = zero profit point.

In practice it is much more economical to reduce the fishing mortality by lowering the number of hours fished than by increasing the mesh size. We require, therefore, to know the optimum mesh size for each level of fishing effort. This we can obtain by plotting, as in Fig. 172A, a series of yield/mesh curves for different sizes of mesh, that is, ages of recruitment t_p. From the maximum of these, plotted against the corresponding fishing mortality, we obtain Fig. 172B, known as the eumetric yield curve. This shows an increasing yield towards an asymptote as fishing increases, but without a maximum.

Evidently the greatest yield is obtained at the highest fishing intensity, but this could be achieved only at a prohibitive cost. In order to obtain the greatest possible yield, we have to consider the cost of unit effort of fishing and the *value* of the yield. This has been done, making certain assumptions, in Fig. 172C. Here the eumetric yield curve is expressed by plotting value of catch against running costs. Another line shows the profit (total value minus total cost) whose maximum might be said in one sense to be the 'best' level of fishing. This curve also crosses the x-axis, where there is no profit in fishing—this being the condition which an uncontrolled fishery tends to approach.

Of course the actual 'best' level of fishing for any given situation may be affected by many social and political factors. However, from studies such as these it begins to be possible to understand the variables that are affecting a fishery and to express them precisely. The value of such work is shown by the fact that international regulation of some fisheries has been agreed. For example, since 1954 the white-fish populations of the North Sea have been regulated by control of the size of mesh used for fishing.

XII

TERRESTRIAL VERTEBRATES: AMPHIBIA

1. Classification

Class Amphibia
*Subclass 1. *Stegocephalia. Devonian–Trias
Order 1. *Labyrinthodontia. Devonian–Trias
Suborder 1. *Ichthyostegalia. Upper Devonian
**Ichthyostega*; **Elpistostege*
Suborder 2. *Embolomeri. Carboniferous
**Eogyrinus*; **Loxomma*
Suborder 3. *Rhachitomi. Carboniferous–Triassic
**Eryops*, Lower Permian; **Cacops*, Lower Permian
Suborder 4. *Stereospondyli. Triassic
**Capitosaurus*, Upper Triassic; **Buettneria*, Upper Triassic
Order 2. *Phyllospondyli. Carboniferous–Lower Permian
**Branchiosaurus*, Lower Permian
Order 3. *Lepospondyli. Carboniferous–Permian
**Dolichosoma*, Carboniferous; **Diplocaulus*, Permian; **Microbrachis*, Permian
Order 4. *Adelospondyli. Carboniferous–Lower Permian
**Lysorophus*, Carboniferous
Subclass 2. Urodela (= Caudata). Jurassic–Recent
Molge; *Salamandra*; *Ambystoma*; *Necturus*
Subclass 3. Anura (= Salientia). Carboniferous–Recent
**Miobatrachus*; **Protobatrachus*;
Rana; *Bufo*; *Hyla*; *Pipa*
Subclass 4. Apoda (= Gymnophiona = Caecilia). Recent
Ichthyophis; *Typhlonectes*

2. Amphibia

During the later part of the Devonian period a population of lungfishes lived in the pools and there is every reason to suppose that some of these animals, first crawling from pool to pool and then spending more time on the land, gave rise to the terrestrial populations that we distinguish as amphibia. No doubt the early efforts at land life were crude. The whole locomotory and skeletal system comes under a completely new set of forces when the support of the water is withdrawn and the effects of gravity become insistent. At the same time the skin must be changed to resist desiccation, the respiratory system adapted to use gaseous oxygen, the receptors to signal the strange new configurations of stimuli. It is not surprising that these new conditions produced greater changes in vertebrate organization than had occurred in tens of millions of years previously. Nevertheless, so slow is the pace of evolution, the only known Devonian amphibia, and many of the Carboniferous ones too, still looked and presumably behaved very like fishes. Animals of this sort (e.g. **Eogyrinus*) floundered about on land for 30 million years or more before producing definitely terrestrial types such as the Permian **Eryops*.

Of all the features that arose at this time in connexion with the new life on land the presence of pentadactyl limbs is perhaps the most conspicuous. It is appropriate that this should be marked in zoological nomenclature: the amphibia are the first of the great group of land vertebrates, the Tetrapoda.

All existing amphibia have been much modified since their Devonian ancestry, yet they retain many features that show how the transition from water to land was produced. These modern forms are by no means a precariously existing remnant but are quite numerous and successful in the ecological niches that they occupy; they form an important element in many food chains. There are some 2,000 species at present recognized, placed in 250 genera. However, contrasting this with the numerous species of teleosts, of birds, and of mammals we shall see that the amphibians, though well adapted for certain situations, do not succeed in maintaining themselves in many different types of habitat. Broadly speaking they are unable to survive for long except in the proximity of water. There are desert toads, such as *Chiroleptes* of Australia, but these survive by burrowing and by special abilities, such as the power to hold large amounts of water, associated with loss of the glomeruli of the kidneys.

Modern amphibia belong to three sharply separated subclasses. Urodela (newts and salamanders) retain the original long-bodied, partly fish-like form. The Anura (frogs and toads) have lost the tail and become specialized as jumpers. The Apoda are limbless, blind, burrowing animals found in the tropics. The urodeles and anurans are found as fossils back to the Cretaceous and Trias respectively, but we have only scanty information about their connexion with the earlier amphibians, which are grouped loosely together as the Stegocephalia. These are found in rocks about 275–160 million years old, that is to say from the late Devonian to the Trias (Fig. 211).

3. The frogs

Perhaps the most successful of all amphibia are those belonging to the genus *Rana*, abundant in every part of the world except in the south of South America, on oceanic islands, and New Zealand. Ranid frogs are typical of the highly specialized subclass Anura, whose members usually inhabit damp places such as marshes or ditches, living for most of their life in the grass or undergrowth and feeding by catching flies and other insects with their tongue. They are preyed upon by birds, fishes, and especially snakes, and escape from these by their hind legs, used either for jumping or swimming. The young develop as tadpoles in the water, where they are omnivorous. The various species differ in size and small points of colour, though there are also some that depart widely from the usual habits, e.g. *R. fossor* which burrows. *R. temporaria* is the species found in Great Britain, *R. esculenta* is a slightly larger form found on the continent of Europe and occasionally in east England, *R. pipiens* is the common small North American frog; *R. catesbiana* the giant bull-frog, whose body is up to 9 in. long, also lives in North America. *R. goliath* of the Cameroons is over a foot long, but is mainly aquatic.

4. Skin of Amphibia

The earliest amphibia possessed the scales of their fish ancestors, but these were soon lost in most lines, though retained in some Apoda; perhaps they were too heavy to be worth while for creatures contending for the first time with gravity, unaided by water. Some frogs carry dermal plates on the back, however, fused to the neural spines (*Brachycephalus* of Brazil). Amphibia differ from reptiles in that the skin is moist and used for respiration; on the other hand, the skin also shows a character typical of land animals in having heavily cornified outer layers. The epidermis therefore consists of several layers in the adult frog and is renewed at intervals by a process of moulting. The moult is under the control of the pituitary and thyroid glands and does not occur if either of these be removed, the keratinized cells merely accumulating in those circumstances as a thick skin. Local thickenings of the epidermis often occur in amphibia, for instance to form the horny teeth by which the larva feeds. Such thickenings are also a conspicuous feature of the warty skin of the toads (*Bufo*), which mostly have a drier skin and are more fully terrestrial than are the frogs. The fact that the epidermis of amphibia can produce local thickenings is of interest in considering the origin of feathers and hairs. In larval amphibians the skin is ciliated.

The glands of the skin are more highly developed than in fishes, and are of two types, mucous and poison glands. Both of these consist of little sacs of gland-cells, derived from the epidermis. The mucus serves to keep the skin moist, this being essential if the skin is to respire; the secretion may perhaps also serve for temperature regulation. The problem of regulation of temperature is important for all terrestial animals, since air conducts heat much less well than water and therefore violent changes of temperature are met. Evaporation produces large influences on temperature and no doubt it was the adjustment of these effects that led to the development of temperature-regulating mechanisms in birds and mammals. Frogs in dry air are always found to be colder than their environment, the difference being sometimes as much as 5° C. It is probable that in some circumstances use is made of this cooling, since tree-frogs (*Hyla*) may be found fully exposed to tropical sunlight, which would be expected to raise their temperatures to a lethal level. On the other hand, the loss of water involved by evaporation in this way would presumably soon become serious.

The poison glands or granular glands are less developed in *Rana* than in *Bufo* ('the envenom'd toad') where they are collected into masses, the parotoid glands. The effect of the poison on man is to produce an irritation of the eyes and nose; only rarely does it affect the skin of the hands. When swallowed it produces nausea and has a digitalis-like action on the heart. The poison of *Dendrobates* of Colombia is used on arrows; it acts on the nervous system.

Some amphibia have characteristic smells, produced by secretions, and these are probably used to attract the sexes to each other. In some male newts (Plethodontidae) there are special collections of these gland-cells below the chin.

Another use of glandular secretions is to keep the eyes and nostrils free from obstruction. The demands of terrestrial life require the production of numerous such special devices and lead to the complexity that we recognize as an attribute of these 'higher' animals.

5. Colours of Amphibia

The use of colour is also highly developed in amphibia. The animals are often greenish and the colour is produced by three layers of pigment cells, melanophores lying deepest, guanophores, full of granules, which by diffraction produce a blue-green colour, and yellow lipophores, overlying these and filtering out the blue. Change of colour is produced by expansion of the pigment in the melanophores under the action of the secretion of the pituitary gland (Fig. 173). Movements in the other chromatophores can also affect the colour, yellow being produced by disarrangement of the guanophores and so on. Other colours may contribute to the patterns, blue (though rarely) by the absence of the lipophores, red by pigment in the lipophores.

Changes in the melanophores may be of two sorts, primary or direct and secondary or visual. The primary response depends on the

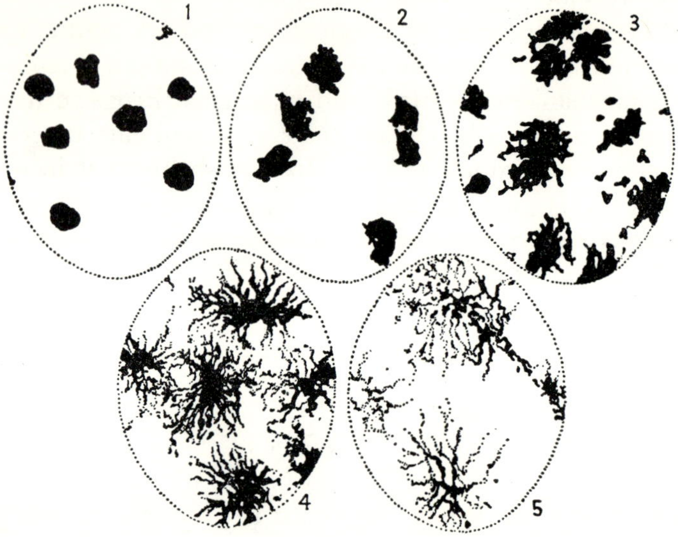

FIG. 173. Stages of dispersal of pigment in the melanophores in the web of the frog *Xenopus*, as used by Hogben to assess the melanophore index. (After Hogben and Slome, *Proc. Roy. Soc.* B. **108**.)

direct effect of light on the skin, causing expansion. The secondary effect consists in contraction of the pigment if the animal is illuminated on a light-scattering surface (light background) but expansion (and hence darkening of the animal) when it is illuminated from above on a light-absorbing (dark) background. There are, therefore, distinct responses from different parts of the retina. Illumination of the dorsal part produces contraction, of the ventral part expansion of the melanophores.

Hogben and his co-workers have shown that the control of the colour change of amphibians is mediated by variation in the secretion of the pituitary gland; there is no direct nervous control of the melanophores. There is still some doubt whether the pituitary produces its effects by means of one hormone or two. The most fully known influence is that of the posterior lobe, producing a B substance, also known as intermedin, which makes the melanophores expand. Extracts of the pituitary of mammals (or other vertebrates) produce this effect when injected into frogs, and after removal of the pituitary a frog becomes pale in colour. There is also some evidence for secretion by the anterior lobe of a W substance that causes paling. After removal of the whole pituitary the melanophores are found to be not in the wholly contracted stage 1 of Hogben's melanophore index (Fig. 173) but in a state (stage 2 or 3) intermediate between this and full expansion (stage 5). If, however, the posterior lobe alone is removed the animal becomes completely pale (stage 1). This certainly suggests

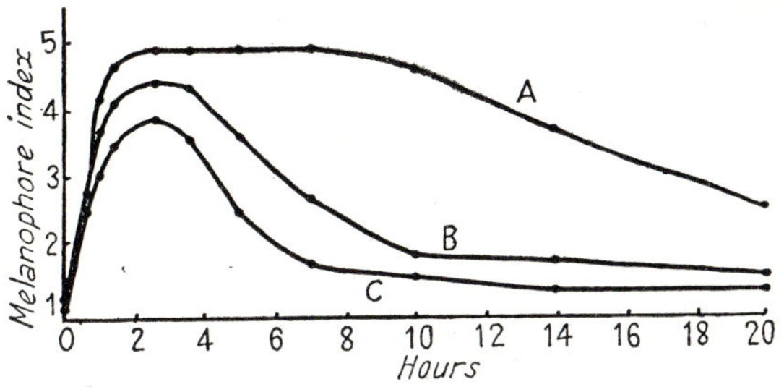

FIG. 174. Effect of an extract containing the B (melanophore-expanding) substance on three groups of *Xenopus*. All were of the same weight and received the same dose. A, whole pituitary removed; B, intact animals; C, posterior lobe only removed. (From Waring, after Hogben and Slome.)

the secretion of a W substance by the pars anterior. There are also differences in the response to injection of B-containing extracts after total and partial removal of the pituitary (Fig. 174). The position is complicated by the fact that extracts of mammalian pineal or adrenal medulla will cause contraction of amphibian melanophores, though it is uncertain whether these effects have any physiological significance.

In amphibia there is no direct control of the pigment cells by nerve-fibres such as are present in bony fishes (p. 259). The colour change is therefore rather slow. After removal of the pituitary the melanophores still show slight changes correlated with change of incident illumination, indicating a small degree of direct response as independent effectors. Temperature and humidity also influence the colour in many amphibia. In frogs contact with water accentuates the black-background response and in darkness produces expansion. On the other hand, drying induces contraction of the melanophores, even upon a black background.

The colour patterns adopted are usually cryptic or concealing in their effect, but the colour also has an important influence on the temperature and varies with it and with the humidity, as well as with the incident illumination. The uniform brilliant green of tree-frogs makes them very difficult to see among the leaves; on the other hand, *R. temporaria* and other species living among grass show a pattern of dark marks, which breaks up their outline. In other amphibians, however, the colour makes the animal conspicuous, for instance the black and yellow markings of *Salamandra maculosa*. Conspicuous colour is often associated with great development of the poisonous parotoid glands and is therefore presumably sematic or warning coloration, allowing recognition by possible attackers. This correlation is not always found, however; the toad *Ceratophrys americana* is dull coloured but poisonous, whereas *C. dorsata* has a bright pattern but is harmless.

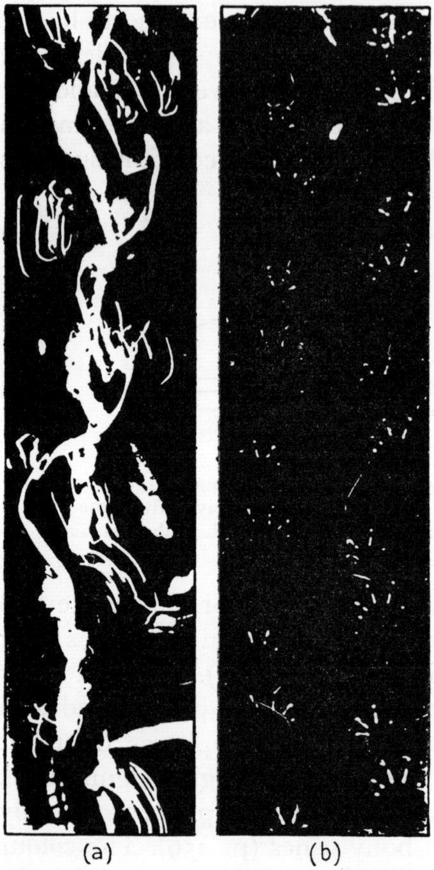

FIG. 175. Record of the movements of *Ambystoma* walking on a smoked drum, A, in rapid locomotion (with the body on the ground); B, in slow locomotion (raised up on the legs). (From Evans, *Anat. Rec.* 95.)

Many frogs make a sudden exposure of brightly coloured patches on the thighs when they jump. This presumably serves to startle the attacker and such colours may be called dymantic or startling. A similar use of colour is made by the cuttlefish (*Sepia*), which may suddenly produce two black spots when alarmed, and also by some Lepidoptera. It is interesting that the colour used in this way so often takes the form of black spots ('eye-spots'), which have an especially striking quality. In some anurans these colours are irregular dark marks, but in *Mantipus ocellatus* they take the form of definite 'eye-spots'.

It must not be forgotten that the presence of pigment serves to protect the organs from the effects of light, which may cause contraction when it falls directly upon muscles. Dark colour may also assist in the absorption of heat, both in the adults and in the eggs.

6. Vertebral column of Amphibia

The general build of the body is essentially fish-like in stegocephalian and urodele amphibians. Such forms have two means of locomotion. When they are frightened and move fast they wriggle along with the belly on the ground, the effective agent being serial contraction of the segmentally arranged myotomal musculature, by means of which the animal as it were 'swims on land', with the legs hardly touching the ground (Fig. 175). When moving deliberately, on the other hand,

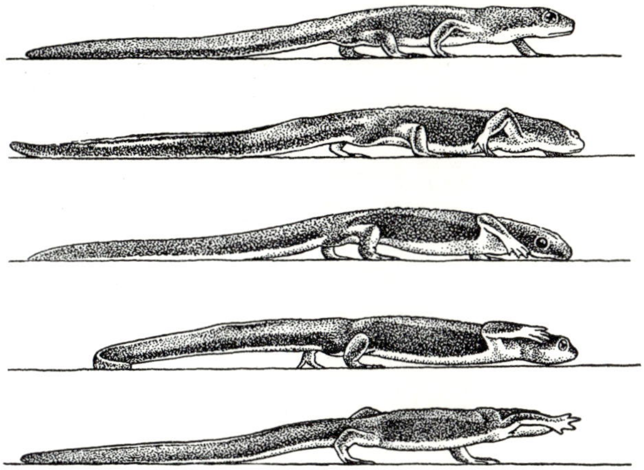

FIG. 176. Drawings made from photographs of a newt (*Triturus*) in slow locomotion. (After Evans, *Anat. Rec.* **95**.)

a newt raises up its body on the legs, which then propel it along as movable levers, the main part of the action being produced by drawing back the humerus or femur, the more distal muscles of the limbs serving to maintain the digits pressed against the ground (Fig. 176).

The carrying of the weight on four legs places an entirely new set of stresses on the vertebral column. Instead of being mainly a compression member as it is in fishes it comes to act as a girder, carrying the weight of the body and transmitting it to the legs. This new function produces a column whose parts are largely bony and articulated together, flexibility becoming less important than strength. The new types of strain involve new muscle attachments and the development of special processes and parts of the vertebrae (p. 307). These changes, however, have not proceeded very far in the amphibians; many urodeles spend much time in the water and their vertebrae often show a lack of ossification, parts of the notochord persist and provide the main compression member required for swimming.

In the anurans the entire skeletal and muscular system has become specialized for the peculiar swimming and jumping methods of locomotion, by means of extensor thrusts of both hind limbs, acting

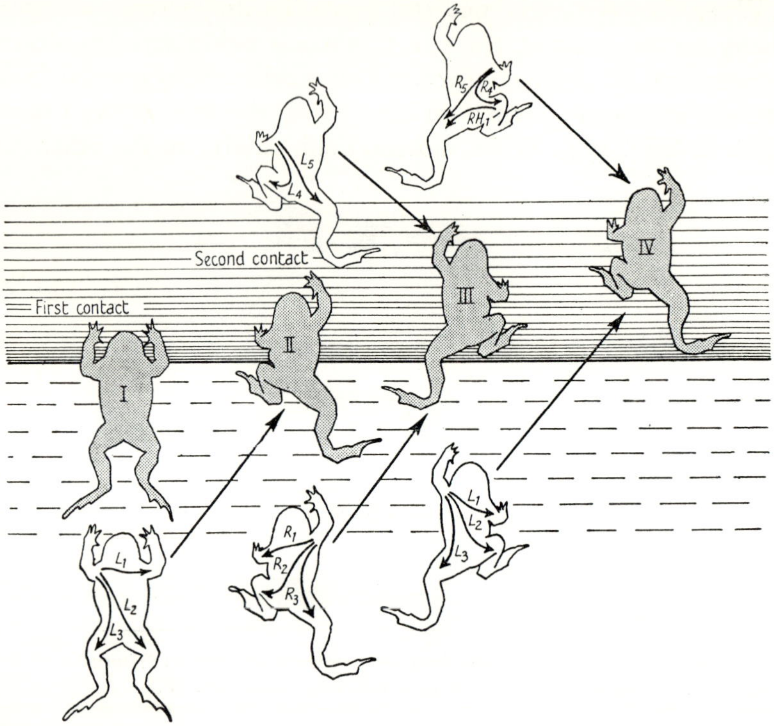

FIG. 177. Reflexes associated with the transition from swimming to walking in toads. The shaded outlines show successive positions as the animal emerges on to solid ground. The first effective contact is by the left fore-limb whose retraction and extension elicits a crossed protraction reflex in the right fore-limb (L_1), a diagonal extensor response in the right hind-limb (L_2), and a placing response in the left hind-limb (L_3). The right fore-limb then touches the ground and produces corresponding responses R_{1-3}. The left fore-limb in response to stretch of its protractor muscles swings forward and this produces retraction of the left hind-limb (L_4) and protraction of the right hind-limb (L_5). Fixation of the right hind foot then produces a crossed flexor response (RH_1). (From Gray, *J. exp. Biol.* **23**.)

together. Frogs, and especially toads, also walk on land, bringing into play a set of myotactic (proprioceptor) reflexes that depend on the contraction of the muscles against an external resistance (Fig. 177).

The actions of jumping and walking are possible because of profound changes in the arrangement of the skeleton and muscles. The myotomal muscles no longer perform their primitive function of producing metachronal waves of contraction, and accordingly the vertebral column (Fig. 178) has lost its original flexibility. Instead, it is

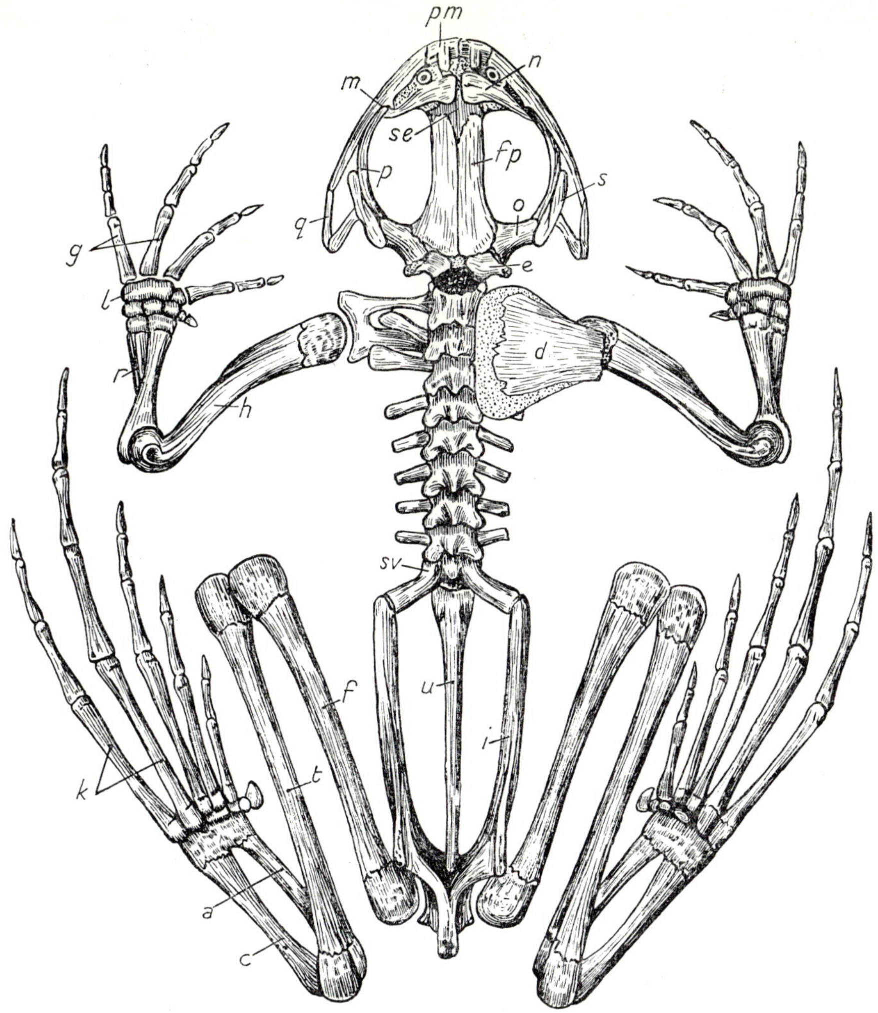

FIG. 178. The skeleton of the frog, seen from the dorsal surface; the left suprascapular and scapular have been removed.

a. astragalus; *c.* calcaneum; *d.* suprascapular; *e.* exoccipital; *f.* femur; *fp.* frontoparietal; *g.* metacarpals; *h.* humerus; *i.* ilium; *k.* metatarsals; *l.* carpus; *m.* maxilla; *n.* nasal; *o.* prootic; *p.* pterygoid; *pm.* premaxilla; *q.* 'quadratojugal'; *r.* radio-ulna; *s.* squamosal; *se.* sphenethmoid; *s.v.* sacral vertebra; *t.* tibio-fibula; *u.* urostyle.
(After Marshall, *The Frog*, Macmillan.)

attached to the pelvic girdle and acts as a support by which the movement of the hind limbs is transmitted to the rest of the body. There is no longer any sinuous motion and the number of vertebrae is very low (nine in the adult *Rana*), and behind them is an unsegmented rod of 'hypochordal' bone, the urostyle. Shortening of the body is a characteristic feature of the change from aquatic to terrestrial

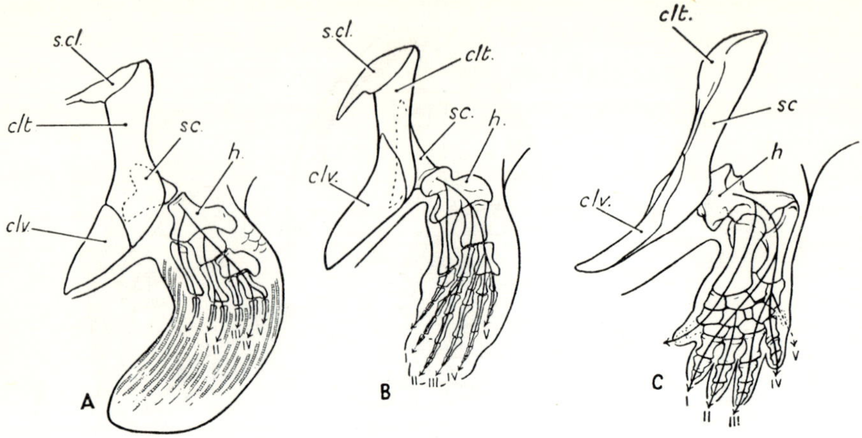

FIG. 179. Transformation of crossopterygian pectoral girdle and fin into pentadactyle limb. Oblique front view of left side.

A, **Eusthenopteron*; B, **Eogyrinus*; C, **Eryops*; *clt.* cleithrum; *clv.* clavicle; *h.* humerus; *p.* pubis; *sc.* scapula; *s.cl.* supracleithrum. (Modified from Gregory and Raven, *Ann. N.Y. Acad. Sci.* **42**.)

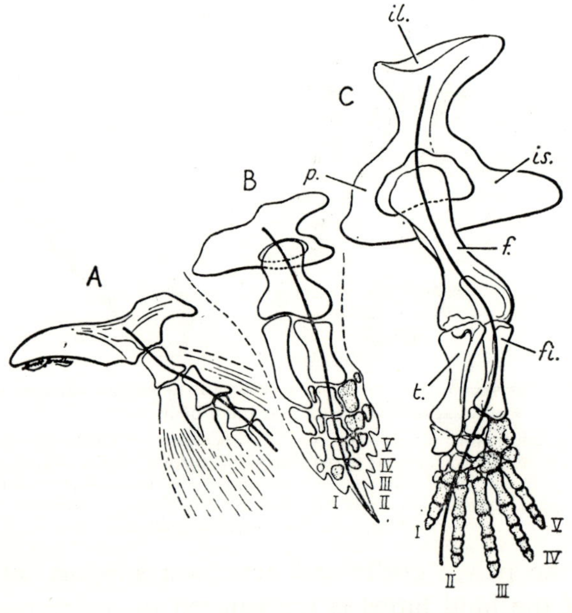

FIG. 180. Transformation of crossopterygian pelvic fin into tetrapod limb.

A, **Eusthenopteron*; B, hypothetical; C, **Trematops*. *f.* femur; *fi.* fibula; *il.* ilium; *is.* ischium; *t.* tibia. (Modified from Gregory and Raven.)

life, and is seen in many lines of amphibian and reptilian evolution. It has proceeded farther in the frogs than in any other tetrapods.

The second to eighth vertebrae of *Rana* are concave in front, convex behind (procoelous), and have large transverse processes. In other amphibians they may be amphicoelous or opisthocoelous. They fit together by complex zygapophyses. The first vertebra has two concave facets for articulation with the two condyles of the skull; its centrum and transverse processes are much reduced. The ninth

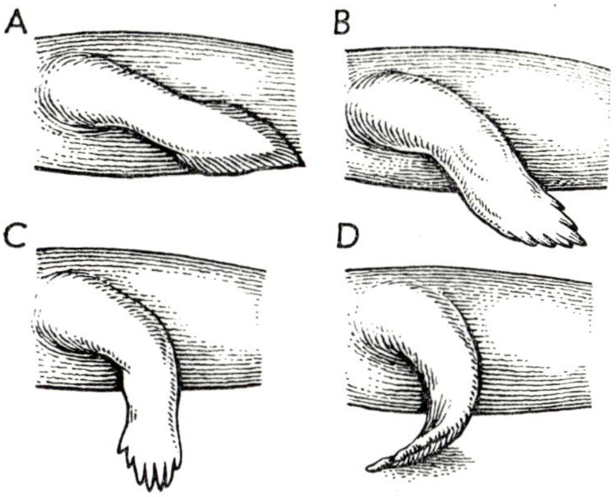

FIG. 181. Diagrams illustrating the probable changes in position during the evolution of a pelvic fin into a tetrapod limb. A as in *Ceratodus*; B, double flexure to give knee and ankle joints, leaving foot direct backwards; C and D, rotation of tarsus and digits turning foot forward. (From Gregory and Raven, after Romer and Byrne.)

(sacral) vertebra has large transverse processes, which articulate with the ilia of the pelvic girdle. There are free ribs in the primitive frogs *Ascaphus* and *Leiopelma*.

7. Evolution and plan of the limbs of Amphibia

The girdles of the paired limbs have become much changed from their fish-like condition (Figs. 179 and 180). Their basic pattern is similar in the two limbs and has been retained throughout the whole tetrapod series. Whereas in fishes the girdles are rather small cartilages and bones, the pelvic girdle being restricted to the ventral region of the body, in amphibians they become enlarged in connexion with the weight-bearing function of the limbs.

The details of the sequence of stages by which a tetrapod limb arose from a fish fin are still somewhat disputed. It is probable that the ancestral crossopterygian possessed a lobed fin, rather like that seen in **Eusthenopteron* (Fig. 179). As the fishes came on land the fin would be used as a lever, giving greater effect to the wave-like motions by which the creature 'swam on land'. The muscles of the limb, contracting in a serial manner, would tend to move it backwards and forwards relative to the body, thus assisting in locomotion. At first the limb perhaps carried only little weight, but as tetrapod evolution proceeded the limbs became elongated and turned under the body, raising it off the ground. To work effectively in this way the limbs came to be held bent down at elbow and knee (Fig. 181) and a firm application to the ground was produced by bending outwards at wrist and ankle. Finally the limbs were brought in to the side of the

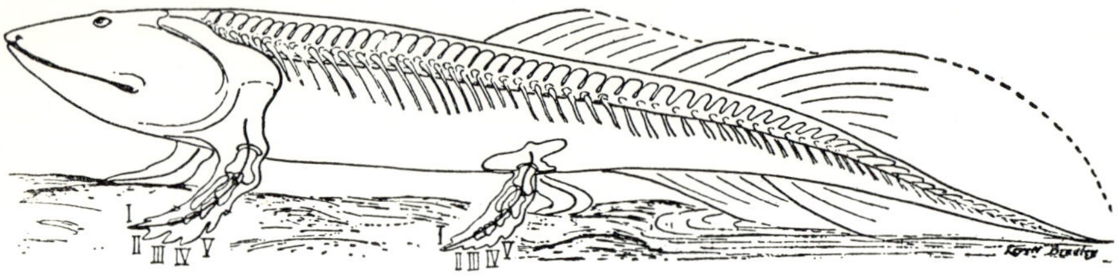

FIG. 182. Suggested protetrapod stage, between crossopterygian and labyrinthodont. (From Gregory and Raven.)

body by rotation, such that the elbow pointed backward and the knee forward.

These are the changes that must have occurred at some time to produce the full tetrapod condition, but we cannot follow exactly the order in which they took place. Their effect is to convert a paddle-like fin, whose main movements were up and down, and were used for stabilization in the horizontal swimming plane, into an elongated jointed strut, on which the animal can balance, and which can be moved as a lever to produce locomotion.

The limbs and girdles and their muscles show a remarkable constancy of pattern throughout the tetrapods. The muscles of the fins of fishes are concerned mainly with lowering and raising (Fig. 192), and they run from a girdle in the body wall to the basal radials in the fin, and between the radials. After the animals came on land the muscles served not only to raise and lower the limbs but also to draw them forwards and backwards; indeed, many fishes already make such movements, including *Protopterus*. The muscles therefore become arranged around the shoulder and hip joints into groups serving as adjustable braces, by which the body is balanced on its legs and by whose contraction the latter are moved. Those muscles that draw the limb towards and away from the mid-ventral line can be called medial and lateral braces (adductors and abductors) and the muscles drawing the leg backwards and forwards are posterior and anterior (retractor and protractor) braces. For the attachment of these muscles proximally the pectoral and pelvic girdles, small in fishes, become expanded into plates (Figs. 180 to 182), and these are divided into a number of characteristic pieces, though the mechanical reason for the division is not clear.

8. Shoulder girdle of Amphibia

The earliest labyrinthodonts (e.g. **Eogyrinus*) inherited a shoulder girdle almost exactly like that of their osteolepid ancestors except that a new dermal element, the interclavicle, was added to the ventral surface. Although the presence of a sternum has never actually been recorded, it is generally assumed that a cartilaginous structure of this type was present between the hindermost margins of the epicoracoid cartilages. As in gnathostomes generally (except elasmobranchs) the shoulder girdle was a dual structure consisting of (*a*) a primary or endochondral component evolved from the basal fin elements of the ancestral fish form and serving to provide an articulatory surface for the limb as well as points of attachment for the limb musculature, and (*b*) a dermal ring of bony elements (skin scales) which had sunk inwards and applied themselves to the ventro-anterior surfaces of the endochondral girdle which, consequently, they braced and supported.

The endochondral girdle consisted of two half rings, which overlapped in the ventral midline. Each half was a single unit but, by topographical comparison with girdles of later tetrapods, it is often arbitrarily divided into two regions, a dorsal scapula and a ventral coracoid. Between these two regions a screw-shaped glenoid received the humerus. The one endochondral ossification is usually homologized with the scapula of amniotes (Watson, 1917). Later forms (e.g. **Seymouria*, **Diadectes*) possessed a second bony element which is generally interpreted as a precoracoid. The endochondral girdle was small in the earliest amphibia (e.g. **Eogyrinus*). In later genera its size progressively increased, presumably to withstand the greater thrust transmitted by the larger limbs of these forms and to provide attachment for the increased mass of brachial musculature.

The dermal girdle consisted, typically, of paired cleithra, clavicles, and interclavicle. The latter, a new element, lay between and often beneath the clavicles and, together with the sternum, probably formed a locking mechanism preventing the complete separation of the epicoracoid cartilages. In the earliest rhachitomes (e.g. **Eogyrinus*) the dermal girdle was attached to the post-temporal region of the skull, as in bony fish. This connexion was soon lost in later forms, presumably to permit greater mobility of the head. This foreshadowed the reduction and loss that was the subsequent fate of the dermal, shoulder girdle elements in tetrapod evolution.

Of the modern amphibia, the Salientia most nearly approach the condition of the fossil forms and they, alone, of recent tetrapods, have retained a cleithrum. Each half of the endochondral girdle consists of a dorsal, bony scapula with a cartilaginous suprascapula, and a ventral coracoid bone connected to an anterior precoracoid cartilage by a mesial epicoracoid cartilage. The precoracoids are invested by the clavicles and, as in all modern amphibians, the interclavicle is absent.

Anuran shoulder girdles may be divided into two broad categories according to whether the two epicoracoid cartilages are fused mesially (*a*) along their entire lengths (firmisternal condition) or (*b*) along their anterior edges only (arciferal girdles). The latter occurs typically in 'walking', toad-like Anura (e.g. Bufonidae, Pelobatidae) and in the aquatic xenopids. The clavicles are the main struts for keeping the glenoids apart and, consequently, they are well developed and never lost. The coracoids, on the other hand, may only be moderately well developed. Immediately behind their point of fusion the epicoracoid cartilages diverge and overlap and their posterior margins are continued as epicoracoid horns, which run in lateral grooves on each side of the sternum. The posterior tip of each horn has a muscle attachment connecting with abdominal recti. This type of sternum/epicoracoid system permits a certain degree of independent movement of the girdle halves whilst, at the same time, preventing the epicoracoid cartilages from being forced too far apart. The mechanism clearly facilitates the independent arm movements characteristic of locomotion in the arciferal frogs.

The firmisternal girdle is a rigid structure allowing no independent movement of the two halves (Fig. 183). It occurs typically in frogs with a jumping habit (e.g. Ranidae, Microhylidae) and provides an excellent landing mechanism. The glenoids are braced apart by the large coracoids. The clavicles and precoracoids are thus deprived of their strutting function and frequently become reduced or even completely lost. No epicoracoid horns are present and the sternum, no longer involved in locking the girdle halves, serves principally for the attachment of pectoral muscles. This function is also performed, in some frogs, by a prezonal (omosternal) element, which is really an

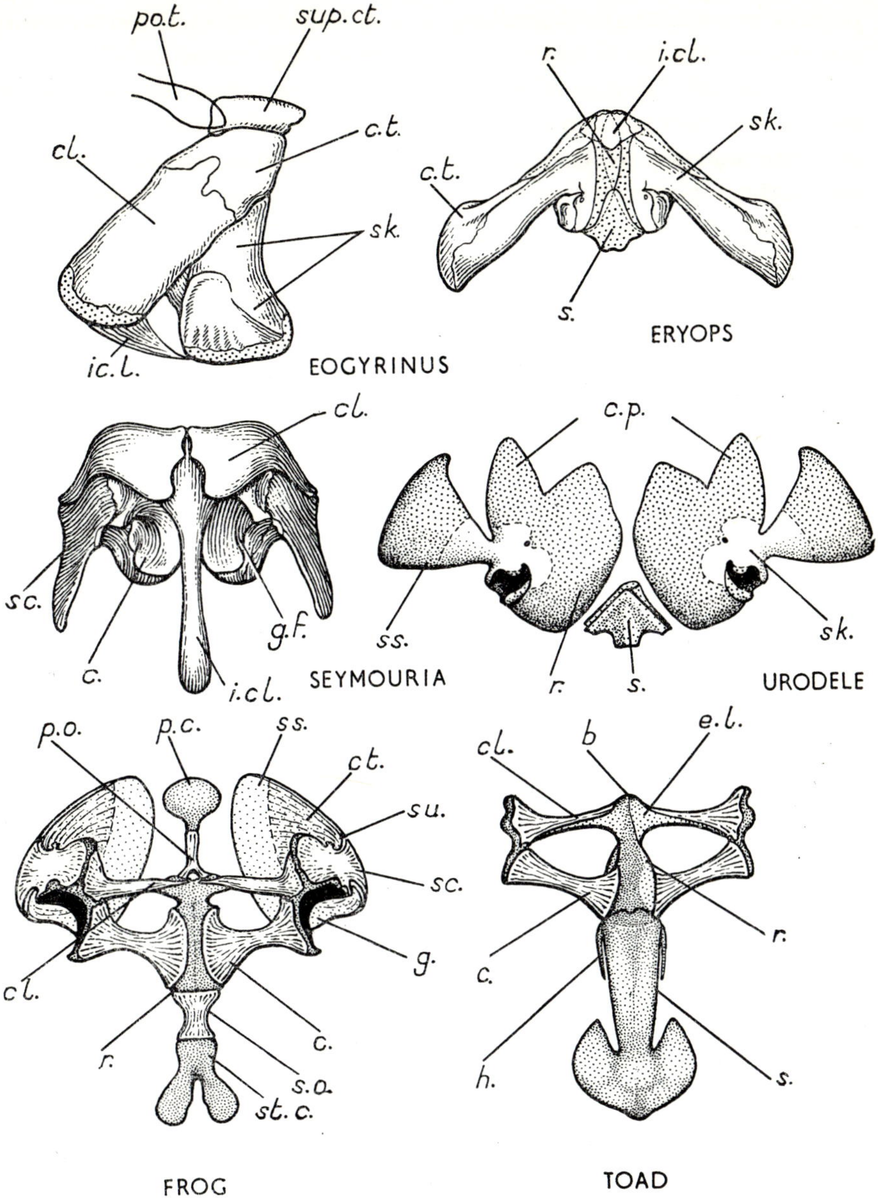

FIG. 183. Amphibian pectoral girdles.

b. precoracoid bridge; *c.* coracoid; *cl.* clavicle; *c.p.* coracoid process; *c.t.* cleithrum; *d.b.* dorsal blade; *e.l.* precoracoid cartilage; *e.m.* epicoracoid muscle; *g.* glenoid; *g.f.* glenoid fossa; *h.* epicoracoid horn; *i.cl.* interclavicle; *l.* left epicoracoid cartilage; *pc.* prezonal cartilage; *p.o.* prezonal bone; *po.t.* posterior temporal; *r.* epicoracoid cartilage; *s.* sternum; *sc.* scapula; *sk.* scapulo-coracoid; *s.o.* sternal bone; *s.r.* ventral blade; *ss.* suprascapular cartilage; *st.c.* sternal cartilage; *su.* coraco-cleithral suture; *sup.ct.* supra-cleithrum.

extension of the precoracoid cartilages. This structure, although present in some arciferal girdles (e.g. Leptodactylidae), is more usually associated with the firmisternal pattern and a jumping habit.

The shoulder girdles of modern urodeles are greatly simplified, the only ossification being a scapulo-coracoid encircling the glenoid. The two epicoracoids overlap broadly, and anteriorly are quite free of each other; posteriorly they are usually rather weakly locked by a

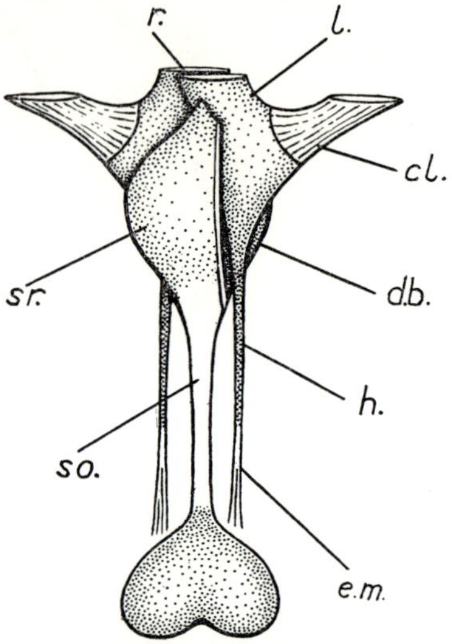

FIG. 183(*a*). *Leptodactylus pragnathus*. Ventral view showing sternal articulation with girdle. The left half of the ventral sternal blade has been removed. Labelling as in Fig. 183.

cartilaginous sternum. The Apoda, of course, retain no vestiges of either limbs or limb girdles.

9. Pelvic girdle of Amphibia

The pelvic girdle is much larger in land animals than the small ventral cartilages found in fishes. It is formed of three main cartilage bones in all tetrapods (Fig. 184), but it is not clear how these originated, nor whether the division has mechanical significance. The dorsal ilium becomes attached to specially modified transverse processes of one or more sacral vertebrae. This ilium can be regarded mechanically as the ossification along a line of compression stress due to the weight-bearing.

The ventral portion of the girdle consists of an anterior pubis and posterior ischium, the three bones meeting at the acetabulum, where the femur articulates. The girdle thus provides a plate to which the muscles that brace the limb can be attached in such a way as to balance the body on the leg.

In urodeles the pelvic, like the pectoral, girdle becomes reduced and mainly cartilaginous. The pelvic girdle of anurans is highly specialized and unlike that of any other vertebrate. The ilia are very long and directed forward to articulate with the transverse processes of the single pair of sacral vertebrae. The base of the ilium is expanded to make the

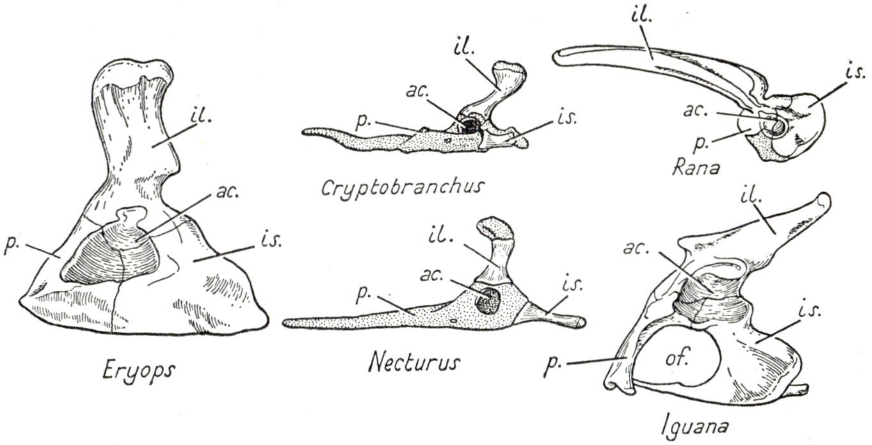

FIG. 184. Pelvic girdles of lower tetrapods. Regions mainly cartilaginous are stippled. *ac.* acetabulum; *il.* ilium; *is.* ischium; *of.* obturator foramen; *p.* pubis. (After Evans.)

dorsal portion of a disk, of which the pubis is the anterior, the ischium the posterior part, with the acetabulum at the centre. The girdle is thus developed into a long lever for transferring force from the limb to the vertebral column during jumping.

Considerable movement is possible at the ilio-sacral joints, at least in Salientia (Whiting, 1961). In *Rana* the ilia may rotate through an angle of over 90° on the sacral ribs in the vertical plane. This movement is used during a strong leap. In *Discoglossus* the sacrum can be turned laterally on the pelvis through 20°. The movement is used both in turning to take food and in locomotion. In *Xenopus* the sacrum can slide backwards and forwards on the pelvis, producing a considerable shortening and lengthening of the whole animal. This movement is probably used in driving into the mud.

10. The limbs of Amphibia

The pattern of bones and muscles in fore and hind limbs of tetrapods is surprisingly constant in spite of the various uses to which the limbs are put. Evidently similar morphogenetic processes are at work in both limbs. There are nearly always three main joints in each limb, at shoulder (hip), elbow (knee), and wrist (ankle). The hand and foot provide basically similar five-rayed levers, with several joints in the digits (Figs. 185 and 186).

The bones of the limbs can be plausibly derived from those of a crossopterygian fin, and indeed the condition in **Eusthenopteron* already distinctly suggests that of the limb of an early amphibian (Figs. 179 and 180). We know less about the origin of the hind than of the front leg, but the two are so similar that they may be treated together for elementary analysis. There is a basal humerus (femur), articulating distally with two bones in each case, a more anterior (pre-axial) radius (tibia) and a posterior (post-axial) ulna (fibula). These bones articulate at the wrist or ankle with a carpus or tarsus, consisting, in the fully developed condition, of three rows of little bones, namely 3 in the proximal row, about 3 centrals, and 5 distals. Each of the latter carries a digit, composed of numerous jointed phalanges. In naming these bones of the carpus and tarsus it is convenient to call the proximal carpals by their position radiale, intermedium, and ulnare and the tarsals tibiale, intermedium, and fibulare. The centrals and distal carpals may then be numbered beginning with 1 at the pre-axial border in each case. Unfortunately other less explicit systems of naming are in use, as shown in the following table.

Plan of the Tetrapod Carpus and Tarsus. (*The names used for the bones in man are shown in brackets.*)

		CARPUS		
	Pre-axial		*Post-axial*	
Proximal	radiale (scaphoid)	intermedium (lunate)	ulnare (triquetral)	
Central		centrale (tubercle of scaphoid)		
Distal	carpal 1 (trapezium)	2 (trapezoid)	3 (capitate)	4 and 5 (hamate)

		TARSUS		
	Pre-axial		*Post-axial*	
Proximal	tibiale (talus or astragalus)	intermedium (os trigonum)	fibulare (calcaneum)	
Central		centrale (navicular)		
Distal	tarsal 1 (medial cuneiform)	2 (intermediate cuneiform)	3 (lateral cuneiform)	4 and 5 (cuboid)

The plan of the carpals and tarsals can well be imagined to have been derived from that of a fin such as is seen in the fish **Eusthenopteron* (Fig. 179), which might be said to have humerus, radius, and ulna,

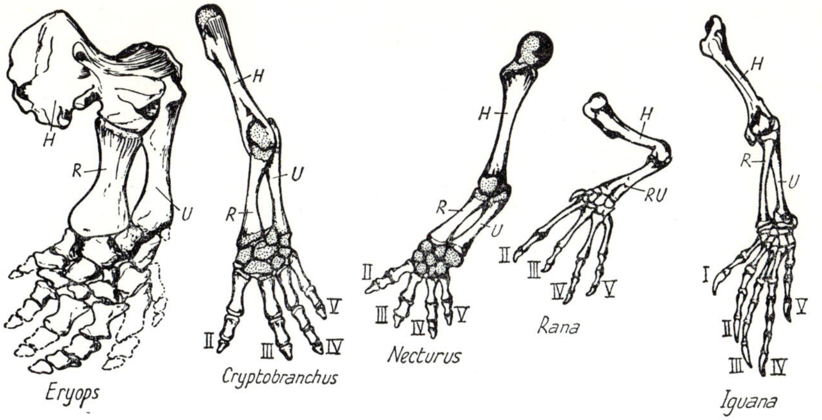

FIG. 185. Front legs of various lower tetrapods.

H. humerus; R. radius; U. ulna. (Modified from Evans, *J. Morph.* **74**.)

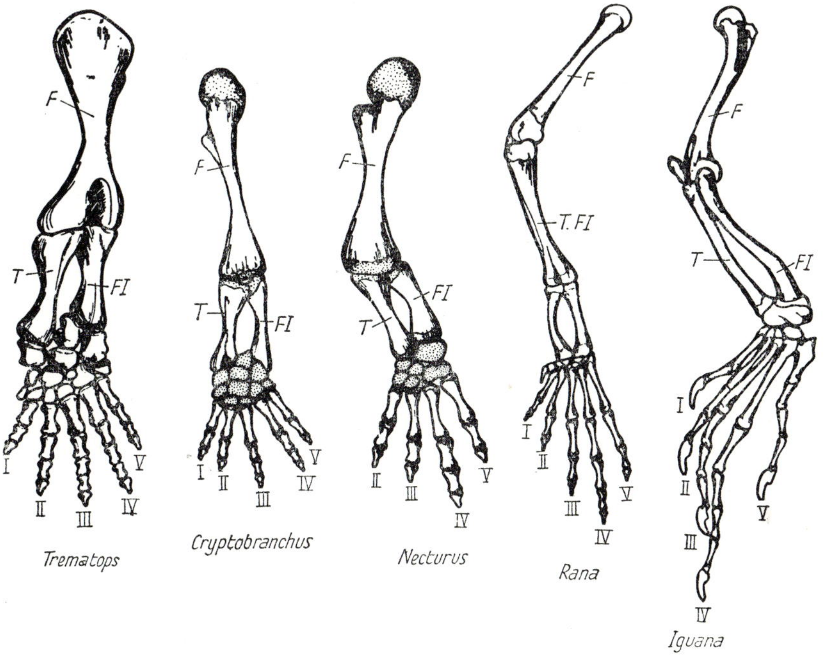

FIG. 186. Hind legs of various tetrapods.

F. femur; FI. fibula; T. tibia. (Modified from Evans.)

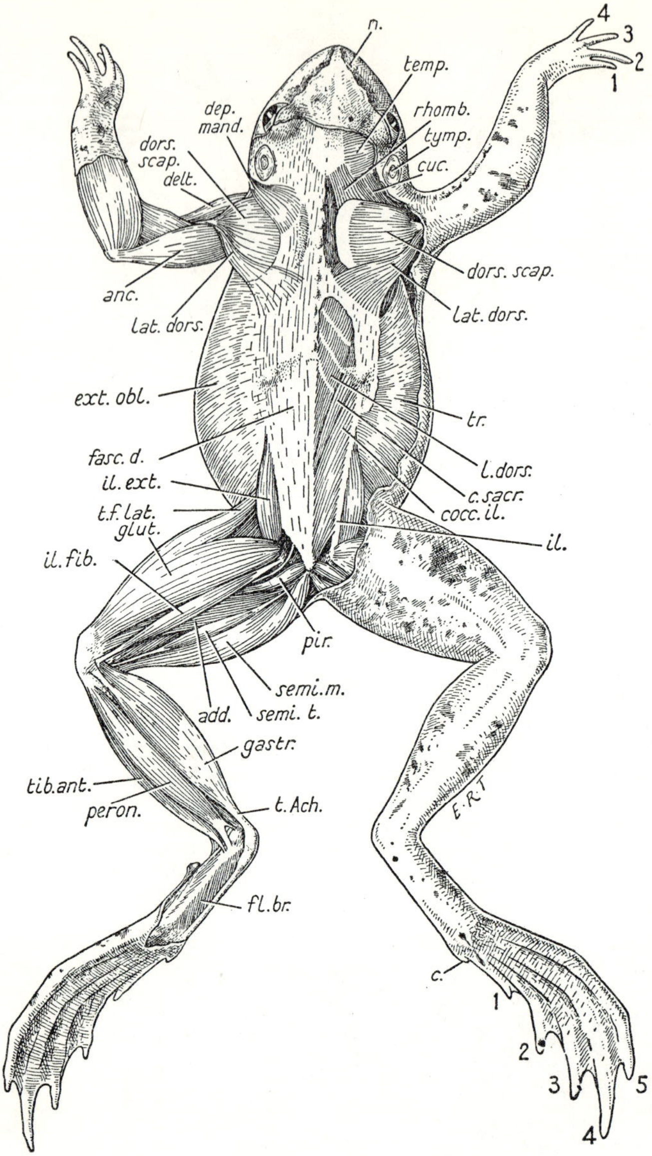

FIG. 187. *R. temporaria* dissected from the back.

add. adductor; *anc.* anconeus ('triceps'); *c.* calcar; *c.sacr.* coccygeo-sacralis; *cocc.il.* coccygeo-iliacus; *cuc.* cucullaris; *delt.* deltoid; *dep. mand.* depressor mandibulae; *dors.sc.* dorsalis scapulae; *ext.obl.* obliquus externus abdominis; *fasc. dors.* dorsal fascia; *fl.br.* flexor brevis; *gastr.* gastrocnemius; *glut.* gluteus; *il.* ilium; *il.ext.* iliacus externus; *il.fib.* ilio-fibularis; *lat.dors.* latissimus dorsi; *l.dors.* longissimus dorsi; *n.* nostril; *peron.* peroneus; *pir.* piriformis; *rhomb.* rhomboideus; *semi.m.* semimembranosus; *semi.t.* semitendinosus; *t.Ach.* tendo Achillis; *t.f.lat.* tensor fasciae latae; *tib.ant.* tibialis anterior; *tr.* transversus abdominis; *tymp.* tympanum. (Partly after Gaupp.)

carpals, and 7 or 8 digits. In the amphibian **Eryops* most of the digits radiate from the radius, in later forms mostly from the ulna. Moreover, in the hand of **Eryops* there seem to have been six digits and it is usually stated that the first of these is a pre-pollex 'not comparable with the pollex of higher forms'.

The effect of this system is to provide a lever that can be held firmly against the ground while it is moved by the muscles running from the girdles to the humerus or femur. In addition the lever is itself extensible by means of its own muscles. Whatever may have been their origin in fishes these muscles in tetrapods work in such a way as to bend each segment up and down. The shoulder and thigh joints usually allow movement in several planes, both towards and away from the midline (adduction and abduction), and forwards and backwards (protraction and retraction). As we have seen, the animal balances at these joints by muscles arranged round them. Movements of rotation are also possible at these, and sometimes at other joints, the distal bone turning about its own axis on the proximal one. Such movements may be very important for the proper placing of the limbs in walking. Pronation is the rotation of the radius about the ulnar bone, so that the manus is directed caudally, supination being the opposite movement. The terms flexion and extension are convenient at certain joints (e.g. the elbow), but have no consistent meaning with reference to the main axes of the body.

The limbs of the earlier amphibians were ponderous affairs, with large bones and widely expanded hands and feet (Figs. 181 and 182). It is not certain exactly how they were used; probably they were held out sideways, giving a wide base on which the somewhat precarious balance was maintained, the body being often slumped on to the ground. In modern urodeles the limbs retain the full pattern of parts, but with imperfect ossification, as would be expected since they carry little weight.

In frogs, specialized for jumping, the radius and ulna are united and the carpals are reduced in number. There are only four true digits, the first digit (thumb or pollex) being reduced. There is, however, a small extra ossification, the pre-pollex, which becomes well developed as a copulatory organ in the male and may be compared with a similar digit found in some stegocephalians. It is to be expected that in a system of repeated parts, such as a tetrapod limb, multiplications and reductions will be common. It can be imagined that they can be produced by changes in the rhythm of morphogenetic processes, and it is surprising that there is such constancy in number of digits.

The hind legs of frogs are long, giving a good leverage in jumping. The tibia and fibula are united and the proximal row of tarsals is reduced to two, greatly elongated and known as the tibiale (astragalus or talus) and fibulare (calcaneum). The distal tarsals are reduced to a total of three, bearing five 'true' digits and an extra one, the calcar or prehallux.

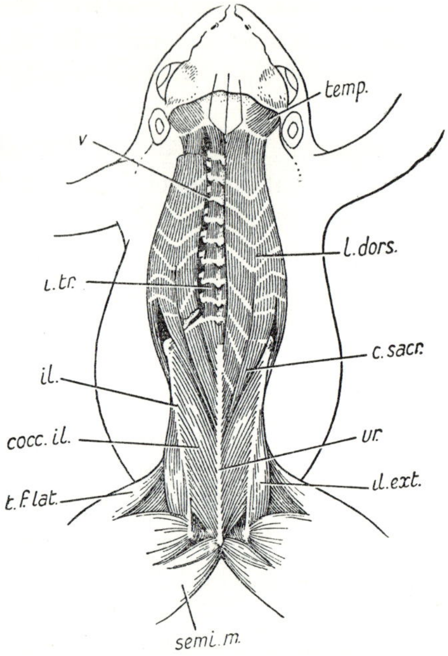

FIG. 188. Deeper dissection of muscles of back of frog.

i.tr. inter-transversarii; *ur.* urostyle; *v.* vertebra; other letters as Fig. 187. (Partly after Gaupp.)

11. The back and belly muscles of Amphibia

With the change in the method of locomotion the muscular system has become greatly modified from that found in fishes. In urodeles, which still use the old method and hence may be said to swim on land, the dorsal musculature is well developed (Fig. 189), but in anurans the dorsal portions of the myotomes, the epaxial musculature, no longer have to produce the locomotory effect by lateral flexion. They remain in frogs only as muscles that bend the body dorsally, serving to brace the vertebral column on the sacrum (Figs. 187 and 188). Short muscles run between the vertebrae, and dorsal to these is a continuous sheet of longitudinally arranged fibres, the longissimus dorsi muscle, running from head to sacral vertebra and urostyle. This muscle, though forming a continuous band, is crossed by a tendinous intersection, showing its segmental origin. At the hind end the coccygeo-sacralis and coccygeo-iliacus muscles brace the urostyle on the pelvic girdle.

The pectoral girdle is attached to the axial skeleton by a series of muscles. Rhomboid and levator scapulae muscles run from the suprascapula to the vertebrae and skull. The cucullaris muscle corresponds

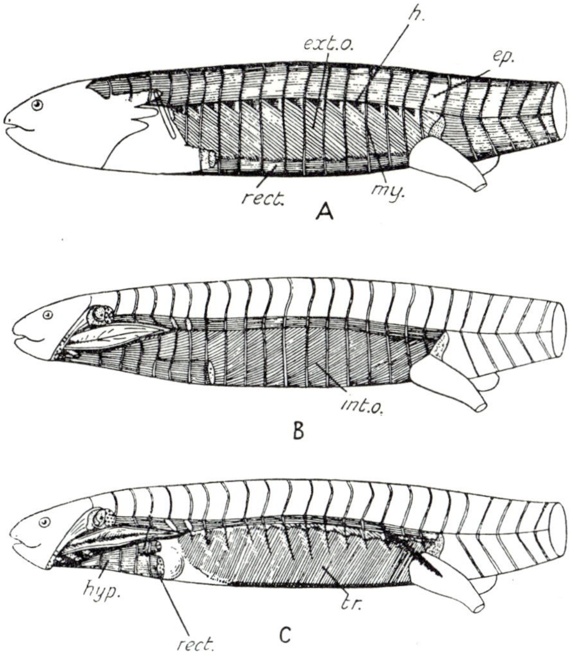

FIG. 189. Muscles of larval *Ambystoma*. A, B, and C show successive layers.
ep. epaxial muscles; *ext.o.* external oblique; *h.* horizontal septum; *hyp.* hypobranchial muscles; *int.o.* internal oblique; *my.* myocomma; *rect.* rectus abdominis; *tr.* transversus abdominis. (From Ihle, after Maurer.)

to the mammalian sternomastoid, running from the skull to the suprascapula; it is derived from lateral plate musculature and innervated by the vagus. The naming of these muscles of the scapula, and indeed all amphibian muscles, meets the difficulty that many of the bundles of fibres are similar in their general course to muscles found in mammals and yet differ sufficiently to raise serious doubts about the wisdom of using the mammalian names. The similarity of arrangement of the limb muscles is so striking throughout the tetrapods that there is probably no harm in keeping to the well-established system of names, but we know so little of the hereditary or mechanical

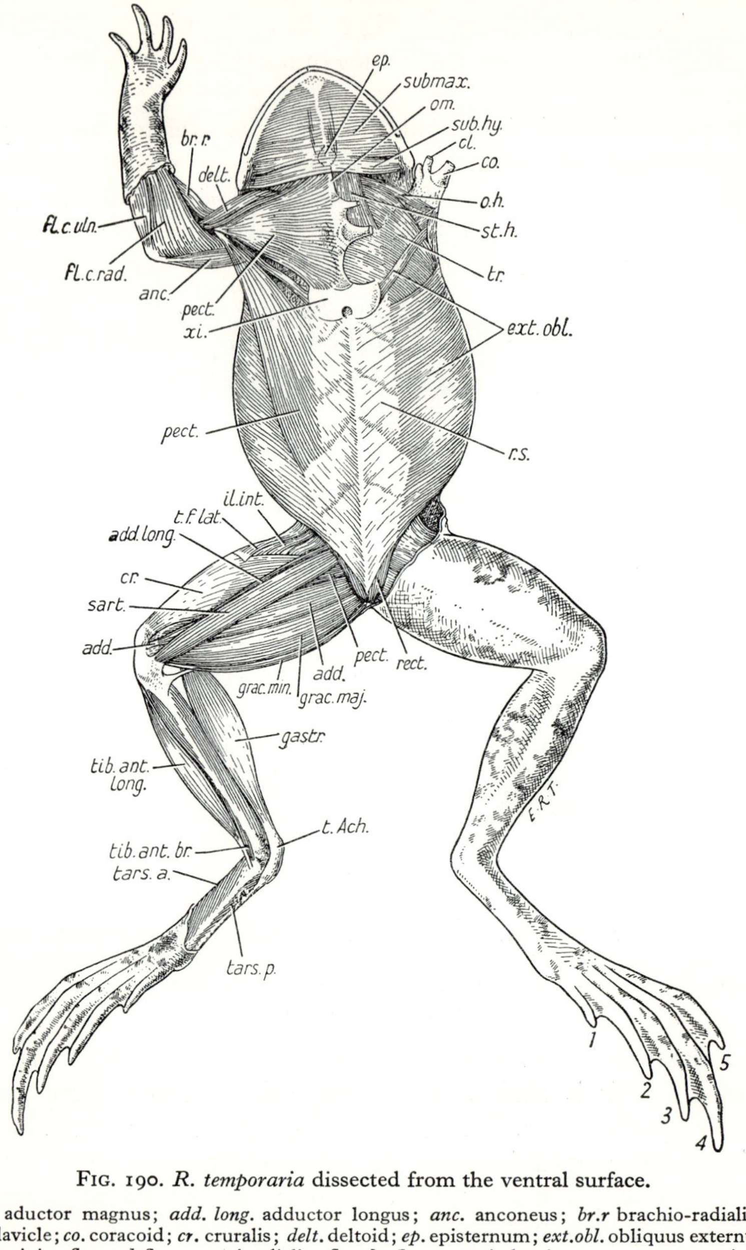

FIG. 190. *R. temporaria* dissected from the ventral surface.

add. aductor magnus; ***add. long.*** adductor longus; ***anc.*** anconeus; ***br.r*** brachio-radialis; ***cl.*** clavicle; ***co.*** coracoid; ***cr.*** cruralis; ***delt.*** deltoid; ***ep.*** episternum; ***ext.obl.*** obliquus externus abdominis; ***fl.c. rad.*** flexor carpi radialis; ***fl.c.uln.*** flexor carpi ulnaris; ***gastr.*** gastrocnemius; ***grac. maj.*** and ***min.*** gracilis major and minor; ***il.int.*** iliacus internus; ***o.h.*** omohyoid; ***om.*** omosternum; ***pect.*** pectoralis; ***r.s.*** rectus sheath; ***rect.*** rectus abdominis; ***sart.*** sartorius; ***st.h.*** sterno-hyoid; ***submax.*** submaxillary; ***sub.hy.*** subhyoid; ***t.Ach.*** tendo Achillis; ***t.f.lat.*** tensor fasciae latae; ***tars.a.*** and ***tars.p.*** tarsalis anterior and posterior; ***tib.ant.br. and long.*** tibialis anterior brevis and longus; ***tr.*** transversus abdominis; ***xi.*** xiphisternum. (Partly after Gaupp.)

factors that control the arrangement of muscle-fibres into 'muscles' that discussion of homologies is difficult.

The hypaxial musculature, formed from the more ventral portions of the myotomes, is more developed than in fishes and differentiated into several parts, for the purpose of slinging the viscera, which of course need support in air in a way that is unnecessary in water.

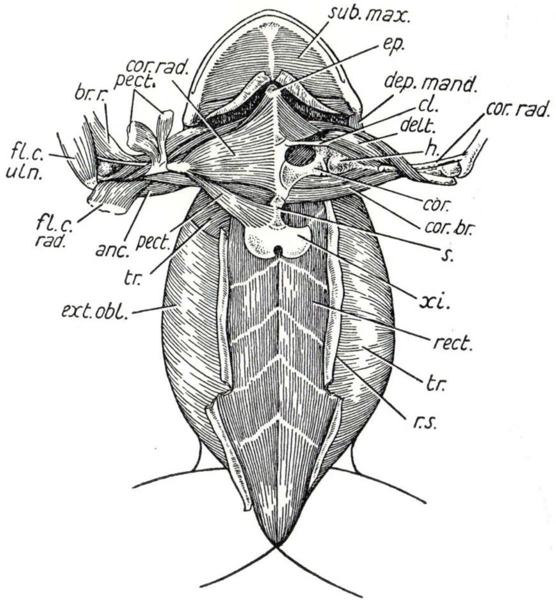

FIG. 191. Dissection of muscles of frog from ventral surface.
cor.br. coraco-brachialis; *cor.rad.* coraco-radialis; *dep.mand.* depressor mandibulae; *h.* head of humerus; *s.* sternum; other letters as Fig. 190. (Partly after Gaupp.)

These muscles are differentiated into layers whose fibres run in different directions. The plan found, with modifications, in all tetrapods is seen in amphibian larvae and includes four sets of fibres. The external obliques run caudally and ventrally; inside this layer is the internal oblique, running in the opposite direction, and within this again the transversus abdominis running approximately dorso-ventrally (Fig. 189). The rectus abdominis consists of fibres in the midline running antero-posteriorly.

In the adult frog three of these sets of fibres can be recognized. In the mid-ventral region (Fig. 191) are the longitudinally arranged fibres of the rectus abdominis, making a sling between the sternum and the pubis. These fibres are interrupted at intervals by transverse fibrous tendinous inscriptions, giving an appearance of segmentation. In the mid-ventral line is the tendinous linea alba. The sling formed by the rectus abdominis is supported laterally by thin sheets of muscle-fibres running up to the vertebral column, the obliquus externus and transversus abdominis (Fig. 191).

In the anterior region the hypaxial muscles have become restricted to the throat, where they form the hyoid musculature, which by raising

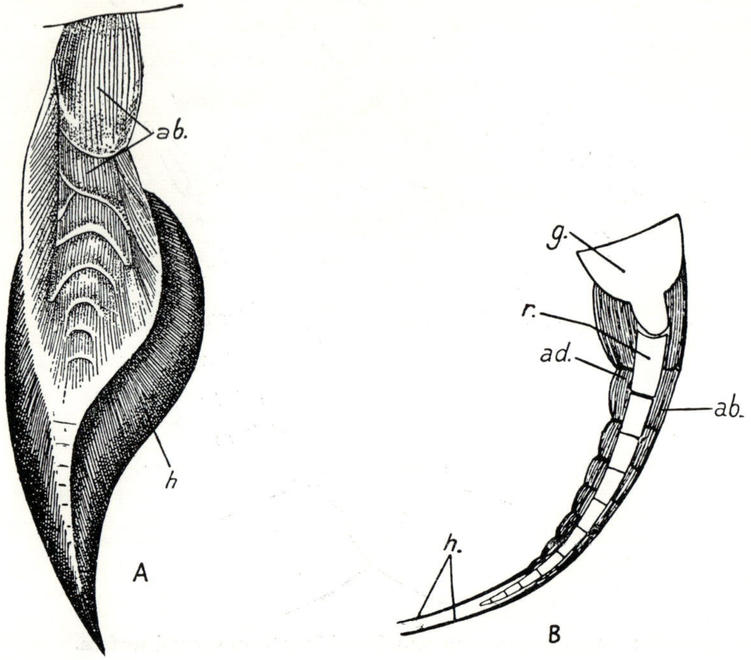

FIG. 192. A. Lateral surface of the fin of *Neoceratodus*, showing the abductor muscle-bundles. B. Section through the fin in the transverse plane, showing the arrangement of the muscle-bundles as abductors and adductors. *ab.* abductor muscles; *ad.* adductor muscles; *g.* pectoral girdle; *h.* horny fin rays; *r.* radials. (From Ihle, after Braus.)

and lowering the floor of the mouth is the main agent of breathing. The submaxillary muscle runs transversely between the rami of the jaw. Deep to this lie other muscles, including the sternohyoid, close to the midline, which is a forward continuation of the rectus abdominis.

12. The limb muscles of Amphibia

The muscles of the limbs were presumably derived from the radial muscles that moved the fins of fishes. These are formed from the myotomes and they are mainly arranged so as to raise and lower the fin (Fig. 192). In modern amphibia the limb musculature is still partly formed from myotomes (Griffiths, 1959). The segmental origin of the limbs is also shown by the fact that they are innervated by branches of the spinal nerves of several segments (2 for the fore-limb, 4 for the hind-limb in the frog). Presumably the original arrangement was such as to move the limbs in association with the waves of contraction passing down the body. In modern urodeles the limb is brought forward and its joints flexed as the epaxial muscles at the level of its front end contract, and then passes back and extends as the wave of contraction moves past. This may have been the primitive movement, making the limb more useful as a lever during the early attempts to 'swim on land' (Fig. 176).

The muscles of the limbs of tetrapods are presumably derived from those that raise and lower the fins of fishes, modified, as we have seen, to brace the limbs and move them, allowing standing and walking. The muscles that run from the girdles to the humerus and femur are therefore able to draw the leg forward and backward, as well as to raise and lower it in the transverse plane. The actions of the various bundles are of course not confined to a single plane: all the muscles running from the back to the humerus can raise (abduct) the upper limb, but the more anterior members also protract, the more posterior retract it. Similarly there is a ventral series whose anterior members work with the anterior dorsal muscles as protractors, although they antagonize the action of raising the whole limb. Moreover, many of the muscles have a rotating action on the humerus and femur. It is, however, possible to consider the muscles of the arm and leg in two great groups; first a more anterior and ventral ('ventro-lateral') set serving to draw the limb mainly forward and towards the midline (protraction and adduction) and to flex its more distal joints, second a more posterior, dorsal ('dorso-medial') mass serving mainly to draw the limb backwards and away from the body (retraction and abduction) and to extend its joints.

In the fore-limb the proximal members of the ventral group make a sheet of fibres running transversely to the main body axis and attached to the sternum and hypaxial muscles at one end and to the humerus at the other (Fig. 190). Within this sheet can be recognized the deltoideus, pectoralis, coraco-radialis, and coraco-brachialis muscles. In the limb itself this group is continued, there being, roughly speaking, a set of muscles in each segment that serves to flex it on the next. Thus the brachio-radialis flexes the elbow joint and in the forearm the flexor carpi radialis and flexor carpi ulnaris flex the wrist. The flexor digitorum longus muscle arises from the medial epicondyle of the humerus and is inserted by tendons to the carpus and terminal phalanges. Flexor digitorum brevis muscles arise from this tendon for insertion on the digits.

Of the dorsal muscle mass (Fig. 187) the latissimus dorsi and dorsalis scapulae are the most proximal, running from the middle of the back to the humerus and serving to abduct and draw back the whole limb. The triceps (anconeus) serves to extend the elbow; in the forearm are extensor carpi ulnaris and radialis and extensors for the fingers.

According to this plan protractor (flexor) muscles lie mainly anterior to retractors (extensors), corresponding to the ancient movement by which the limb was drawn first forward then back as a swimming wave passed down the body. In all tetrapods flexor muscles are in general innervated by spinal roots anterior to those for the extensors. The locomotory movements of the limbs therefore still show the passage of an excitation wave backwards along the spinal cord, a relic of the swimming rhythm of fishes. However, the changes that have taken place in the relative positions of the parts of the limbs make it difficult to follow out this simple pattern in detail. It must also continually be remembered that many muscles produce rotation as well as movement in the main planes of the body.

In the hind limb, muscles of the same two general types can be recognized, namely anterior muscles, which draw the limb forward and flex and adduct its joints, and posterior ones, which draw it back and extend and abduct. The specialization of the main muscle masses has gone much farther, however, so that more individual muscles are found, especially round the hip joint, each serving to move the limb in a special way.

In the thigh (Figs. 187 and 190) the muscles of the anterior group, lying on the ventral surface, are the pectineus and the adductors, running from the pelvic girdle to the femur and thus serving to move the whole limb inwards (adduction). The sartorius, biceps, semimembranosus, and semitendinosus are two-joint muscles mainly producing flexion at the knee as well as at the hip.

The more posterior and dorsal group of muscles includes the gluteus and tensor fascia lata from girdle to femur, extending the thigh joint, and the very large cruralis (including the rectus femoris and triceps femoris) running from girdle and femur to tibia. This is the main extensor of the knee, being helped by gracilis and semimembranosus. This extension is obviously an important part of the jumping movement of the frog.

In the shank the arrangement of the flexors and extensors into the anterior and posterior groups is much modified. The more conspicuous muscles are the tibialis anterior and peroneus running from the femur to the tarsus so as to flex the ankle joint. Long and short flexors move the toes, as in the fore-limb. At the back of the tibiofibula the gastrocnemius (plantaris longus) runs from the femur to be attached by the tendo Achillis to the tarsus. Its main action is to extend the ankle in the movements of jumping and swimming. Tibialis posterior runs from the tibia to the tarsus. Within the foot there is an elaborate system of small muscles for bending and stretching the toes and abducting them away from each other, so as to expand the web for swimming.

The whole system is designed to produce the characteristic sudden simultaneous extension movement of all the joints of both hind limbs, by which the frog moves both in water and on land. The hind limbs can also be used for alternate walking movements, especially in toads (Fig. 177).

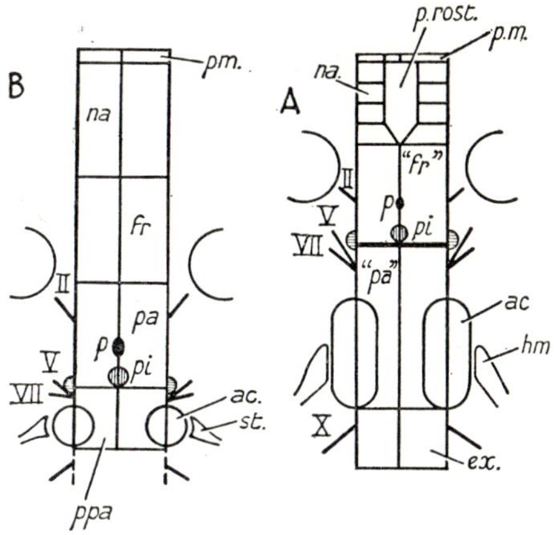

FIG. 193. Diagram of skull bones and other structures. A, an osteolepid; B, a stegocephalian.

Letters for this and Fig. 194: *ac.* auditory capsule; *ex.* extrascapular; *fr.* frontal; *hm.* hyomandibula; *it.* intertemporal; *j.* jugal; *l.* lachrimal; *na.* nasal; *mx.* maxilla; *p.* pineal; *pa.* parietal; *pi.* pituitary; *pm.* pre-maxilla; *po.* post-orbital; *pof.* post-frontal; *ppa.* post-parietal; *prf.* pre-frontal; *p.rost.* post-rostrals; *qj.* quadratojugal; *sq.* squamosal; *st.* stapes; *sut.* supratemporal; *t.* tabular.
(After Westoll.)

13. The skull of Stegocephalia

The skull of the Devonian and Carboniferous amphibia was essentially like that of the osteolepid fishes in the arrangement of the bones, but the proportions had been altered so that the pre-optic region was relatively large and the more posterior 'table' of the skull short (Fig. 193).

The nasals and frontals, which were small in crossopterygians, were quite long in stegocephalians, whereas the parietals were shorter and the post-parietals absent altogether in the later forms. The difference is so marked that for a long time people were deceived in identification of the bones and it was said that the pineal opening lay between the frontal bones in fishes but between the parietal bones in tetrapods. The bones identified as 'frontal' in the fish types were, of course, parietals, whereas the 'parietals' were the post parietals, which have gone completely from most amphibians, though still present in the earliest Devonian forms (Fig. 194). This is an excellent example of how study of changes of proportion can clear up morphological difficulties.

The opercular apparatus covering the gills was lost early in amphibian evolution; perhaps the reduction of the whole posterior part of the head was effected by a single morphogenetic change. In

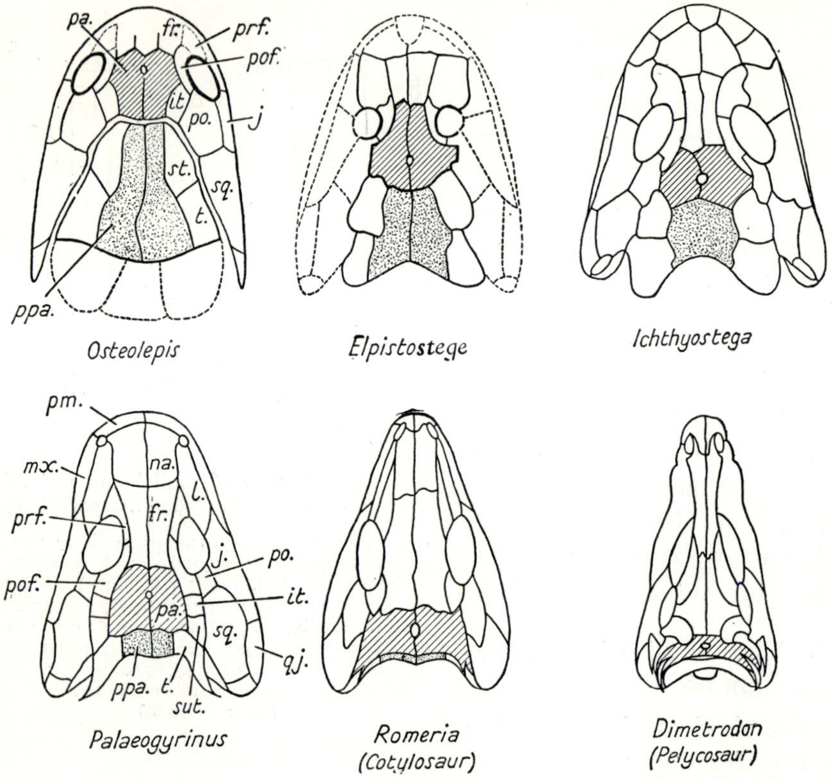

FIG. 194. Skulls of a crossopterygian and various early tetrapods to show the shortening of the posterior region. Lettering as Fig. 193. (After Westoll.)

modern amphibia the skull is much flattened and its ossification reduced, so that large spaces are left; in the earlier forms, however, the skull was of the more usual domed shape and the roof and jaws were covered by a complete set of dermal bones. Presumably the loss of bone was another development producing a reduction of weight advantageous to a terrestrial animal.

Lateral line organs are present in aquatic amphibians and their position is marked on the bones of the fossil skulls by rows of pits. By using these lines as reference marks it is possible to compare the pattern of the bones on osteolepid and early amphibian skulls and to confirm the remarkable similarity. The main new development found in the skull of early amphibians was correlated with the modification of the Eustachian tube in connexion with the sense of hearing, and the need for a sensitive resonator to pick up the air vibrations. Already in the earliest amphibians the opercular coverings of the gills were lost (there was a small pre-opercular bone in **Ichthyostega*) and the spiracular opening thus uncovered acquired a tympanic membrane. The hyomandibular cartilage, no longer concerned (if it ever had been) with supporting the jaw, was modified to form the columella auris, serving to carry vibrations across to the inner ear. At first, however, there was no trace of the fenestra ovalis, the hole in the auditory capsule into which the columella fits in the frog. In modern urodeles the whole ear apparatus is much modified, there being no tympanum. Instead the columella is fused to the squamosal and the ear thus receives its vibrations from the ground.

Other small changes in the skull in passing from the fish to the amphibian stage include the increase in size of the lachrymal bone, which also came to have a hole to carry the tear duct, draining the orbit. A series of small bones surrounds the orbit in early amphibians, as in fishes; large squamosals and quadratojugals support the quadrate. At the back of the skull these stegocephalians possessed various of the small bones that are found in fishes but not in modern amphibians, the supratemporal and intertemporal, post-parietal (much smaller than in crossopterygians) and tabulars. In fact there are numerous small bones, arranged in a pattern clearly recalling that of the fish ancestor, but showing some reductions and less variation than in those very variable fish skulls. This simplification (which was later carried farther), together with some changes in the shape, are the chief transformations that have converted the fish skull into the amphibian skull.

The palate of the early amphibians also resembled that of crossopterygians, showing a complete plate made of vomer, palatines, pterygoids, and ecto-pterygoids. These bones, as well as the premaxillae and maxillae, often carried folded teeth (hence 'labyrinthodonts'), with a pit for a replacing tooth beside each one, an arrangement similar to that of their fish ancestors (p. 270). The internal nostril opened far forward, through the palate. The lower jaw was covered by a number of dermal bones (Fig. 208), but the actual jaw articulation was made between cartilage bones, the upper quadrate, and the lower articular.

14. The skull of modern Amphibia

Modern amphibia share several cranial features that distinguish them from typical labyrinthodonts. The number, extent, and thickness of the dermal elements are greatly reduced so that the otic

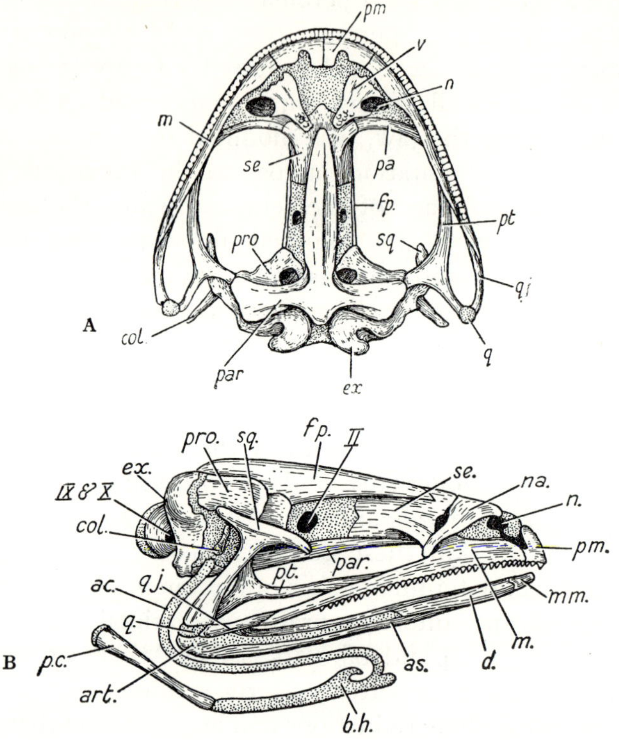

FIG. 195. Skull of the frog. A. Ventral view. B. Side view.

ac. anterior cornu of hyoid; *art.* articular; *as.* angulosplenial; *col.* columella auris; *d.* dentary; *ex.* exoccipital; *fp.* fronto-parietal; *m.* maxilla; *m.m.* mento-Meckelian; *n.* nostril; *na.* nasal; *pa.* palatine; *par.* parasphenoid; *pc.* posterior cornu of hyoid; *pm.* premaxilla; *pro.* pro-otic; *pt.* pterygoid; *q.* quadrate; *qj.* quadrato-jugal; *se.* sphenethmoid; *sq.* squamosal; *v.* vomer; *II*, *IX*, and *X*, nerve foramina. (After Marshall, *The Frog*, Macmillan.)

capsules are generally exposed. The orbits and interpterygoid vacuities are large, the mandibular ramus is short and the skull as a whole much flattened. The occiput is shortened so that the hypoglossal nerve emerges behind the skull and (with the few exceptions noted below) the parietal foramen has been lost.

The skull of the frog (Fig. 195) shows great reduction and specialization from the early amphibian type. It may be considered as consisting of a series of cartilaginous boxes or capsules, in whose walls some ossifications occur, partly covered by dermal bones. The cartilaginous

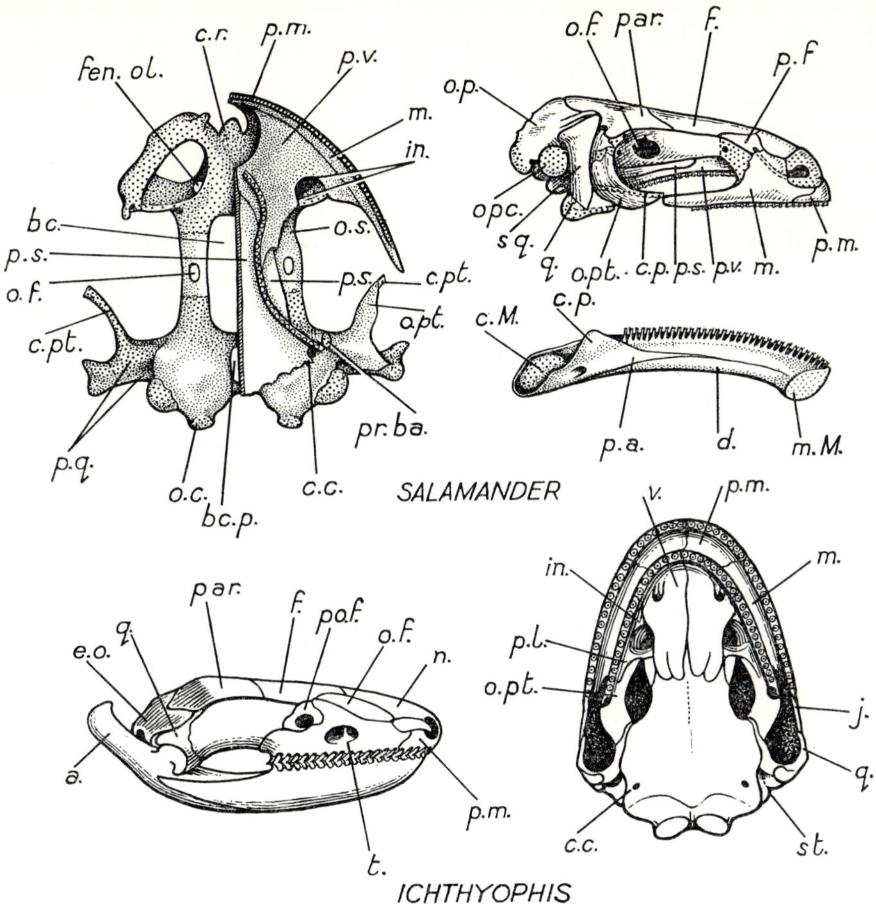

FIG. 196. Skulls of amphibians.

a., articular; *bc.*, basicranial fenestra; *bc.p.*, posterior basicranial fenestra; *c.c.*, carotid canal; *c.M.*, Meckel's cartilage; *c.p.* coronoid process; *c.pt.* pterygoid cartilage; *c.r.* process of internasal plate; *d.* dentary; *e.o.* exoccipital; *f.* frontal; *fen.ol.* olfactory fenestra; *in.* internal naris; *j.* jugal; *m.* maxillary; *m.M.* mento-Meckelian; *n.* nasal; *o.c.* occipital condyle; *o.f.* optic foramen; *o.p.* occipito-petrosal; *opc.* operculum; *o.pt.* pterygoid bone; *o.s.* orbito-sphenoid; *p.a.* pre-articular; *par.* parietal; *p.f.* prefrontal; *p.l.* palatine; *p.m.* premaxilla; *po.f.* postfrontal (enclosing orbit); *p.q.* palato-quadrate; *pr.ba.* basal process; *p.s.* para-sphenoid; *p.v.* prevoma; *q.* quadrate bone; *sq.* squamosal; *st.* stapes; *t.* tentacular groove; *v.* vomer.

boxes, well seen in a tadpole's skull, are the central neurocranium around the brain, and the olfactory and auditory capsules. Ossifications occur especially at the points of compression stress, namely, around the foramen magnum (the exoccipitals), where the auditory capsule joins the cranium (the pro-otic), and at the base of the nasal capsules (the mesethmoid). The paired occipital condyles are found only in modern amphibia, and are formed by the failure of the basioccipital to become ossified. Paired occipital condyles have also arisen, independently, in the mammal-like reptiles.

The dermal bones covering the roof of the skull are the nasals and frontoparietals, while on the floor is the large dagger bone, the parasphenoid, and a small tooth-bearing vomer. The remains of the cartilaginous palato-pterygo-quadrate bar can be recognized as a rod, covered in front by premaxillae and maxillae, and dividing behind into an otic process fixing it to the skull (autostylic) and a cartilaginous quadrate region articulating with the lower jaw. This region is covered by the pterygoid ventrally, the quadrato-jugal laterally, and the squamosal dorsally. The palatines are membrane bones forming the anterior wall of the orbit. The upper jaw is thus supported by struts formed from the nasals and palatines in front and the squamosal and pterygoid behind, an arrangement that gives a large mouth for respiration and eating insects, combined with the advantages of strength, mobility of the lower jaw, and lightness in weight.

The lower jaw consists of Meckel's cartilage, covered on its outer surface by a dentary and on its inner by an angulo-splenial bone. The anterior tips of the cartilages ossify as the mento-Meckelian bones.

The visceral arches are well formed in the tadpole but are much modified in the adult frog. In the tadpole the skeleton of the hyoid arch consists of a large pair of ceratohyals attached to a basal hypohyal. As a result of subsequent metamorphosis the ceratohyals later form the long anterior cornu of the hyoid, attached to the pro-otic bone. The body of the hyoid is a plate lying in the floor of the mouth and formed from the hypohyal and from the hypobranchial plate at the base of the remaining arches. The posterior cornua support the floor of the mouth and the whole apparatus assists in respiration. The sixth and seventh of the series of branchial arches give rise respectively to the arytenoid and cricoid cartilages of the larynx.

The lateral plate muscles of the branchial arches are well developed only as the muscles of the jaws. Certain muscles of the scapula (the cucullaris (p. 319) and interscapularis) are innervated from the vagus and recall the sternomastoid and other muscles innervated by the spinal accessory nerve in mammals.

The muscles of the hyoid arch, innervated by the facial nerves, remain mainly as the depressor mandibulae (Fig. 197) running from the back to the angle of the jaw and serving to lower the floor of the mouth. The jaw-closing muscles, m.m. adductor mandibulae, belong to the mandibular segment and are innervated by the trigeminus.

They run from the hind end of the jaw to the surface of the skull and squamosal.

The skull and jaws of the frog thus constitute a protection for the brain and special sense-organs, a feeding apparatus, and a means of respiration. The heavy protection afforded by the dermal bones of fishes and early amphibians has been largely dispensed with, probably for lightness. The front part of the skull, concerned with the nose, eyes, and brain, has become increased in size and the hind part, originally concerned with the gills and pharynx, greatly reduced.

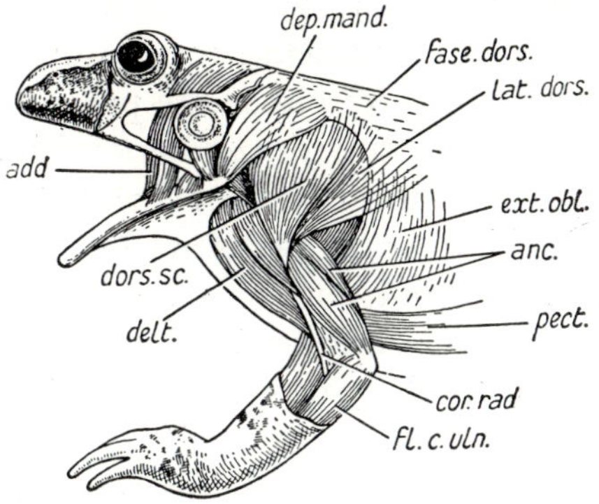

FIG. 197. Muscles of head and neck of frog dissected from the side. *add.* adductor mandibulae. Other letters as in Figs. 187 and 191. (Partly after Gaupp.)

These changes, carried to extremes in frogs, have been in progress throughout the evolution of amphibia. It is not difficult to imagine that they have been the result of rather simple genetic changes, affecting the relative growth of various parts of the skull. We are still far from the knowledge necessary to say exactly what developmental changes have occurred, but we know enough to imagine how selection through millions of years has changed the quantities of certain substances so as to produce gradually less bony and shorter heads, such as enabled their possessors to maintain sufficient mobility to hold a place in a world peopled by the reptiles and other still more active descendants of the early amphibians.

The preceding account, particularly with regard to the osteology, should not be regarded as diagnostic of all anurans. Bufonid skulls are completely devoid of teeth but they possess a supratemporal bone, which fuses with the squamosal and roofs the otic capsule. Hylids frequently develop secondary dermal ossifications to form expanded helmets; this trend also occurs in leptodactylids (e.g. *Calyptocephalus*), where the skull may be so completely roofed and sculptured as to simulate the condition of the extinct branchiosaurs. Pseudoteeth (serrations of the jaw elements) frequently occur on the dentary and pre-articular (e.g. *Amphodus*) but the only modern form to possess true teeth on the lower jaw is *Amphignathodon*. No recent frog retains the large parietal foramen so typical of the fossil amphibia but some leptodactylids and the aquatic xenopids have a small canal perforating the fronto-parietal, through which runs a fibro-nervous tract from the pineal organ to the habenular ganglion (Griffiths, 1954). The anuran skull is always easily distinguished from those of all other Amphibia by the fact that the frontals are fused with the parietals.

Urodele skulls are, in some respects, less specialized than those of Anura. The frontals and the parietals remain discrete and in certain species both lacrimals and prefrontals are present. In other respects they are clearly more degenerate (or paedomorphic?). No urodele has either a jugal or quadrato-jugal (except *Tylotriton*) and in perennibranchs even the maxillaries and nasals are lost. Urodeles are further distinguished from frogs (but not from caecilians) by the great size of the prevomers (each consisting really of a prevomer+palatine) and by the possession of a tooth-bearing coronoid, as well as a dentary and a prearticular.

The apodan skull is a much more rigid structure than that of either of the above subclasses and, at first sight, approaches more closely to the ancestral pattern. The number of bones present, however, is no greater than in any of the other modern groups. The overall compactness is effected particularly by the expansion of the nasals and of the marginal elements of the upper jaw and is probably correlated with the burrowing habits of the group. Lower as well as upper jaws carry teeth and a toothed coronoid is present in the mandible.

15. Respiration in Amphibia

The new problems presented by life on land have led to the production of very varied means of respiration among amphibia. In a terrestrial habitat oxygen is available in plenty; the difficulty is evidently to arrange for a regular interchange of air in contact with adequately moistened surfaces. The interchange is provided for in most cases by modifications of the apparatus used in fishes, but pumping air presents new problems and it seems that these are not easily solved, since in many amphibians the skin is used as an accessory respiratory mechanism. The retention of moisture becomes more difficult as the ventilation becomes efficient; probably for this reason air is often only transferred to the lungs after it has remained for some time in the mouth. We see again that the new way of life, in a medium remote from water, makes it necessary to possess more complicated methods of self-maintenance.

16. Respiration in the frog

The lungs of the frog are paired sacs, opening to a short laryngeal chamber, which communicates with the pharynx by a median aperture, the glottis. The glottis and laryngeal chamber are supported by the arytenoid and cricoid cartilages. The arytenoids guard the opening of the glottis and are moved by special muscles. During breathing the mouth is kept tightly closed, the lips being so arranged as to make an air-tight junction. Air is sucked in through the nostrils by lowering the floor of the mouth by means of the hypoglossal musculature, and can then either be breathed out again or forced into the lungs by raising the floor. The external nares are closed by a special pad on the anterior angle of the lower jaw, supported by the mento-Meckelian bones. This pad is thrust upwards and pushes the premaxillaries apart, so altering the position of the nasal cartilage that the nostrils are closed. This is a special mechanism, found only among anurans. In urodeles the nostrils are closed by valves provided with smooth muscles. Such valves are present in the frog but are said to be functionless.

The movements of the floor of the pharynx are not continuously of the same amplitude. After a period of relatively slight movements the nostrils are kept closed while the throat is lowered. Air is thus drawn from the lungs and then again returned to them once or twice before the nostrils are reopened. The whole procedure presumably ensures the maximum gaseous interchange for the minimum water-loss.

This method of taking in air is clearly derived from the movements of the floor of the mouth of fishes, by which water is passed over the gills. In amphibian larvae water is pumped in this way and there is direct continuity between the mechanism of larva and adult. The basic rhythmic mechanism, centred on the nerve-cells of the medulla oblongata, is no doubt the same throughout, but the anurans have improved upon it by the addition of special features, requiring intricate coordination of the muscles of the larynx and the apparatus for closing the nostrils.

The skin is very vascular, and especially so in the buccal cavity. It plays a large part in respiration, actually serving to remove more carbon dioxide than do the lungs. There is, however, little power to vary the amount of exchange through the skin, which is therefore constant throughout the year. There is considerable regulation of the exchange in the lungs. The rate of breathing depends, as in mammals, on the effect of the carbon dioxide tension of the blood on a respiratory centre in the medulla. There is also a vasomotor control of the blood-supply to the lungs and, through the vagus nerve, of the state of contraction of the latter. By such means the rate of respiratory exchange is greatly increased during the breeding season, and made to vary with the activity of the animal.

17. Respiratory adaptations in various amphibians

The skin and the lungs show many variations according to the habitat of the species, special devices being adopted to enable the animals to live in particular environments. The lungs vary from the well vascularized sacs with a highly folded surface found in the frogs, and especially in the drier-skinned toads, to small simple sacs in some stream-living amphibia. The lung will serve to lift the animal in the water; for this reason it is reduced in the frog *Ascaphus*, which lives in mountain streams in the eastern U.S.A. In newts this hydrostatic function of the lungs is predominant and the inner surface is often quite simple. The lung is entirely lost in stream-living salamanders, such as the European alpine *S. atra*. The coldness of the water reduces activity and lowers the need for respiratory exchange to a level at which it can be fully met by the skin. The skin shows increased vascularity in these forms with reduced lungs, capillaries reaching nearly to the outermost layers of the epidermis. In the African frog *Astylosternus*, in which the lungs are vestigial, the male develops vascular papillae on the waist and thighs during the breeding season.

Gills are present in amphibian larvae, and also in certain adult urodeles that may be considered as larvae that have failed to undergo metamorphosis (p. 364). The gills are extensions of the branchial arches, and carry branched villi, richly supplied with blood. Where the main trunk is long the gill projects and is 'external', whereas in other cases, as in the later frog tadpole, the filaments are directly attached to the arches and are called 'internal'. There are no profound differences between the two types.

18. Vocal apparatus

Sound is produced as a protective (fear) response and by the male frog as a call to attract the female. Both sexes have vocal organs, but those of the female are much the smaller. The noise is produced by the vibration of the elastic edges of a pair of folds of epithelium of the laryngeal chamber, the vocal cords. Air is passed backwards and

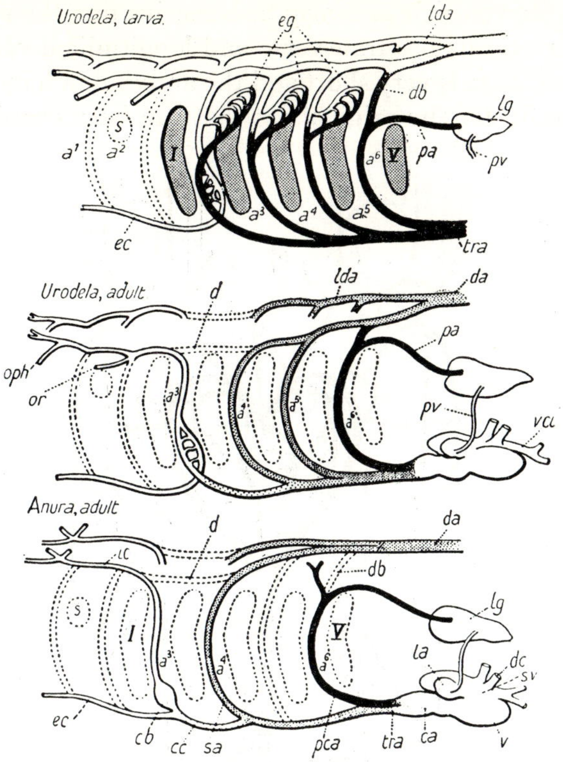

FIG. 198. Diagrams illustrating development and fate of aortic arches in Amphibia, left-side view completed. Vessels carrying most arterial blood white, most venous blood black, and mixed blood stippled.

a^{1-6}, Primary arterial arches; *ca.* conus arteriosus; *cb.* carotid gland; *cc.* common carotid; *da.* median dorsal aorta; *db.* ductus Botalli; *dc.* left ductus Cuvieri; *ec.* external carotid; *eg.* blood-supply to external gill; *ic.* internal carotid; *la.* left auricle; *lda.* lateral dorsal aorta (*d.* obliterated part, ductus caroticus); *lg.* lung; *oph.* ophthalmic; *or.* orbital; *pa.* pulmonary artery; *pca.* pulmo-cutaneous arch; *pv.* pulmonary vein; *s.* closed spiracular slit; *sa.* systemic arch; *sv.* sinus venosus; *tra.* truncus arteriosus (ventral aorta); *v.* ventricle; *vci.* vena cava inferior. (From Goodrich.)

forwards between the lungs and a large pair of sacs (or a single median sac), the vocal pouches, formed below the mouth. These also serve as resonators, and are developed only in the male.

19. Circulatory system of Amphibia

The venous and arterial systems are less fully separated in Amphibia than in lung-fishes. The auricles are completely divided by an inter-auricular septum, venous blood returning to the right, arterial to the left auricle. There is only a single ventricle, but this is provided with spongy projections of its wall, which may prevent the mixing of the blood. The ventral aorta (conus arteriosus), springs from the right side of the ventricle and may thus receive first the venous blood. The conus arteriosus has transverse and longitudinal valves.

The ventral aorta is very short and the arches much modified in the adult (Fig. 198). Of the original six that can be recognized the first

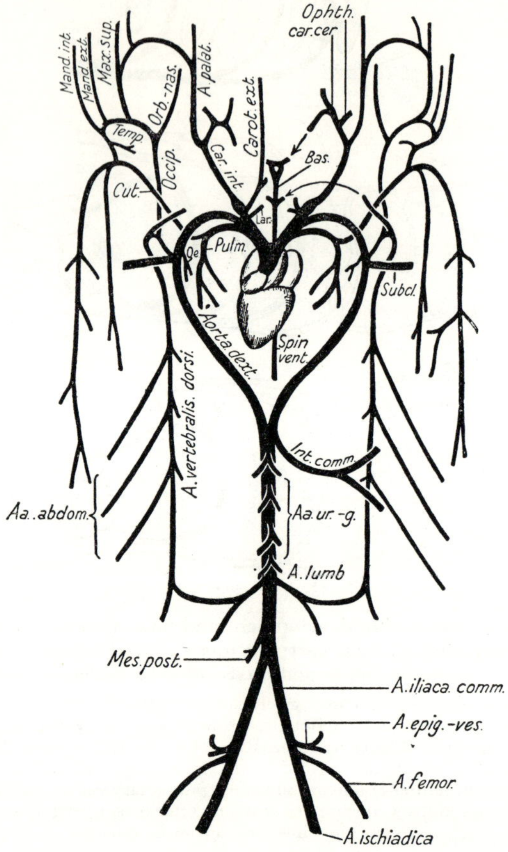

FIG. 199. Diagram to show the chief arteries and their anastomoses in the frog. (After Gaupp.)

two disappear, the third on each side gives rise to the carotid artery, the fourth remains complete and forms the systemic arch. The fifth remains also in some urodeles, but disappears in anurans. The sixth arch becomes the pulmonary artery and loses its connexion with the dorsal aorta: special 'cutaneous arteries' carry de-oxygenated blood from this arch to the skin (Fig. 199). These pulmonary arches probably offer a lesser resistance than do the systemic and carotid ones; the pressure in the latter is said to be increased by a special network, the 'carotid gland', though it may well be that this organ is a receptor, connected with regulation of the blood-pressure. However, it is claimed that the first blood leaving as the ventricle contracts flows to the lungs. In anurans this separation may be further assisted by an arrangement such that the pulmonary arteries join at their base and

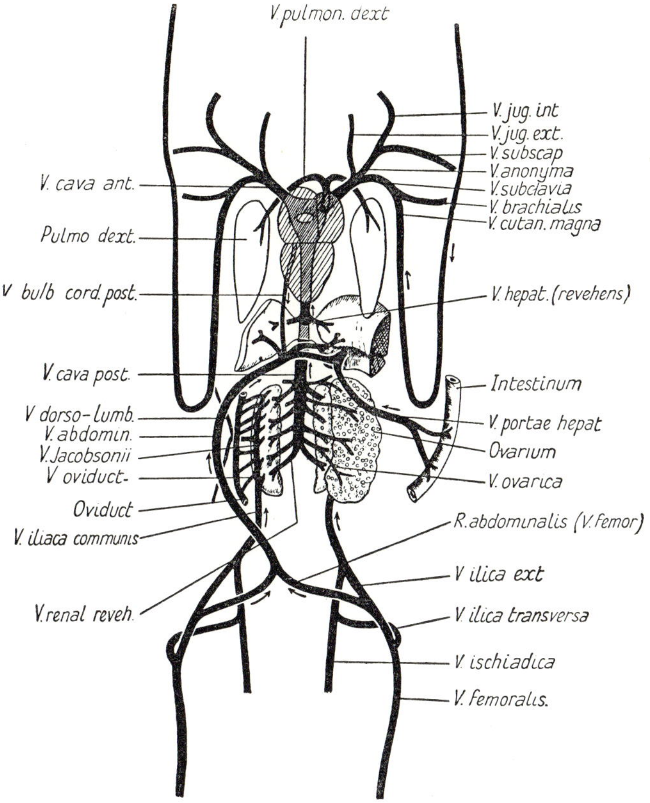

FIG. 200. Diagram to show the chief veins of the frog. (After Gaupp.) The r. abdominalis is often called the pelvic and the v. iliaca communis the renal portal vein.

open to the dorsal part of the truncus arteriosus (cavum pulmocutaneum), which is partly separated from the more ventral cavum aorticum, leading to the carotid and aortic arches. The classical view of this system is that as the pressure rises the truncus contracts and the spiral valve moves in such a way as to force all the blood that leaves the ventricle during the later part of its contraction into the ventral portion and hence to the systemic and carotid arches. In this way a separation of blood from the right and left auricles would be achieved. The view that the heart allows such a separation has, however, been challenged on the basis of experiments made by injection of X-ray opaque material, allowing the course of the circulation to be watched. It is stated that by this method it can be shown that blood returning to either auricle reaches all parts of the ventricle and that no separation occurs. Since the blood from the skin returns to the right auricle (Fig. 200) it is not clear that a separation of the streams would be advantageous. It may be that the undivided condition of the ventricle in amphibians is a secondary development, perhaps not present in the earlier forms such as **Eogyrinus*, which reached a larger size (Foxon, 1955). The spongy walls of the ventricle may allow metabolic exchange since the heart is provided only with very small coronary arteries.

The venous system (Fig. 200) is based on the same plan as that of Dipnoi. The posterior cardinal veins are replaced early in life by a vena cava inferior. Most of the blood from the hind limbs passes through the renal portal system, but there is an alternative path through pelvic veins and a median anterior abdominal vein, which breaks into capillaries in the liver.

The blood-pressure is regulated by the extrinsic nerves of the heart, fibres from the vagus tending to slow and from the sympathetic nervus accelerans tending to speed the beat. The latter nerve is a new development, there being no sympathetic innervation of the heart in fishes (the condition in Dipnoi is unknown). The diameter of the arteries throughout the body is also under control from sympathetic vasoconstrictor and perhaps also vasodilatator nerves. The arterioles in the web of the foot can be seen to constrict when the medulla oblongata is stimulated. Substances extracted from the posterior lobe of the pituitary and from the adrenal medulla also serve to cause constriction of the arteries and perhaps also of the capillaries.

There is therefore a complex mechanism for ensuring that the pressure of the blood is maintained and the flow directed into the part of the body that requires it for the time being.

20. Lymphatic system of Amphibia

The transfer of substances between the cells and the blood-stream is effected in any vertebrate by a transudation through the walls of the capillaries into the tissue fluids. Under the pressure of the heart-beat water and solutes leave the capillaries, passing through their walls, while proteins remain behind. The blood passing into the venous ends of the capillaries therefore has a high colloid osmotic pressure and this serves to suck back fluid from the tissues. In this way a circulation from the capillaries into the spaces around the cells is produced. Clearly, however, it is essential for this mechanism that the pressure of the ventricular heart-beat shall exceed the colloid osmotic pressure of the blood. This it does by about three times in the frog.

The lymphatic system consists of a set of spaces which, in the frog at least, communicate with the tissue spaces around the capillaries. Injection of gum into the lymphatic system, by increasing the colloid osmotic pressure in the tissue spaces, prevents the back suck of fluid into the venules and hence leads to swelling of the part injected. The lymph spaces in the tissues join to form larger channels and great sinuses, such as that below the loose skin of the back of the frog. The lymph is kept circulating by the action of lymph hearts. In the frog there are anterior and posterior pairs of these, opening into veins. The more posterior pair lies on either side of the coccyx and can be seen if the skin is removed. The lymphatic vessels also assist in the process of repair. If, after injury, red cells come to lie in the tissues the lymphatics send out sprouts for as far as $\frac{1}{10}$ mm to pick them up and return them to the blood-stream.

21. The blood of Amphibia

The red corpuscles of amphibia are much larger than those of mammals, reaching in the urodele *Amphiuma* the immense size of 70 μ; they nearly always exceed 20 μ. The red cells are formed mainly in the kidney, and are destroyed, after a life of about 100 days, by the spleen and liver. The bone-marrow is a source of red cell formation in *Rana temporaria* but not, except during the breeding-season, in *R. pipiens*. A process of breaking up of the red cells occurs after they have entered the blood-stream, giving a number of enucleated fragments, and this, when the part remaining with the nucleus is small, produces a result like the extrusion of the nucleus during the development of the red cell of mammals. In *Rana* only small portions of the cytoplasm are broken off in this way, but in *Batrachoseps* a large proportion of enucleated corpuscles is produced.

The haemoglobin of the frog has a lower affinity for oxygen than that of mammals, even when both are considered at the same temperature, and in this respect is notably less efficient. Also, although the power of the blood to combine with carbon dioxide is great, there is a less delicate regulation of the reaction of the blood than in mammals.

The white cells of amphibia are of three types, lymphocytes, with a large nucleus and small cytoplasm, monocytes, which are larger phagocytic macrophages, and polymorphonuclear granulocytes.

These last may be neutro-, eosino-, or basiphil and are migratory and phagocytic. Thus the white cell picture with which we are familiar in mammals was evidently established a very long time ago.

There is a globular spleen near the tail of the pancreas.

The blood of frogs also contains numerous small platelets (thrombocytes), which probably break down when in contact with foreign surfaces to produce the thrombin that combines with the fibrinogen of the blood-plasma to produce clotting.

22. Urinogenital system of Amphibia

The excretory organs of adult amphibia are always the tubules of the mesonephros. In *Rana*, where there is a general shortening of the body, these extend over only a small number of segments and the kidneys are compact. In urodeles and the primitive frog *Ascaphus* the kidneys are elongated and show some evidence of their segmental nature. The mesonephros consists essentially of a series of tubules leading from the nephric funnels to the Wolffian duct. In the frog the funnels do not open into the tubules, however, but into the veins; moreover, they form independently of the rest of the tubule. In the adult there are some 2,000 glomeruli, from each of which a short ciliated tube leads to the proximal convoluted tubule. There follows a second short ciliated region, corresponding in position to the Henle's loop of mammals, and leading to a distal convoluted tubule, which joins the Wolffian duct.

The blood-supply of the kidney differs from that of mammals in that blood arrives from two distinct sources; the branches of the renal artery run mainly to the glomeruli, those of the renal portal vein to the tubules. This corresponds to the functions now well established for those two parts, namely that the glomerulus filters off water and crystalloids, some of which are then reabsorbed by the tubule. Many details of this process are not clear, however, for instance how the urea concentration in the urine is raised many times above that of the blood.

The frog, having a moist skin, is presumably in constant danger of osmotic flooding with water when it is submerged, and of desiccation when on land. The flooding is prevented by the efficient functioning of the glomeruli; they allow the frog to excrete as much as one-third of its weight of water per day (man 1/50th). The mechanisms for resistance to desiccation are less perfect. There is no long water reabsorbing segment, the part of the tubule corresponding to Henle's loop being short. There is, however, a large cloacal (allantoic) bladder (to be distinguished from the mesodermal bladder of fishes) from which water can be reabsorbed. Certain desert amphibia (*Chiroleptes*) conserve water by losing the glomeruli altogether. *Rana cancrivora* is euryhaline and may have 2·9% of urea in the blood (Gordon 1961).

The Müllerian duct, by which eggs are carried to the exterior, develops separately from the Wolffian system in the frog, but arises from the latter during development in urodeles. In this, as in many other features, the frog shows a greater degree of specialization of its developmental processes. The ovaries are mere folds of the peritoneum, having no solid stroma such as is found in mammals. There are, however, follicle cells around each egg; these presumably produce the ovarian hormones. Sections of an ovary show eggs in various stages of development, but not all those that begin complete their maturation; many degenerating, atretic eggs are found. Ripening of the eggs proceeds under the influence of a hormone produced by the anterior lobe of the pituitary. This in turn is controlled by external environmental factors to ensure breeding in the spring. Suitable injections of mammalian anterior pituitary extracts will ensure ripening of the ovaries and ovulation at any time of year. The 'prolans' excreted in the urine of pregnant women have a similar effect, and the production of ovulation in *Xenopus* is used as a test for the diagnosis of human pregnancy.

Having left the ovary the eggs find their way to the mouths of the oviducts mainly by ciliary action of the latter. The walls of the oviduct are glandular and secrete the albumen; they are dilated at the lower end to form uterine sacs, in which the eggs are stored until laid.

The testes discharge directly through the mesonephros by special ducts, the vasa efferentia, formed by outgrowths from the mesonephros into the gonad. This is presumably a secondary development from the original vertebrate condition in which the sperms were carried away by the nephrostomes. The fact that the sperms pass through the kidney emphasizes that the amphibia have diverged at a very early stage of the evolution of the vertebrate stock, and remain still in many respects at a lower level of evolution than the modern fishes, all of which have acquired separate urinary and genital ducts. In *Alytes*, in many ways primitive, the sperms do not, however, pass through the kidney!

In some frogs (*R. temporaria*) there is a special diverticulum, the vesicula seminalis, leading by several small channels to the lower end of the Wolffian duct. It contains spermatozoa during the breeding-season and its appearance suggests a secretory activity.

Most of the amphibia have failed to effect the complete transfer to land life: they return each year to the water to breed. Special modifications of the reproductive system for land life are therefore not found. Secondary sexual differences are marked in many species. In frogs the males precede the females to the water and then attract the latter by their vocal apparatus. The male clings to the back of the female by means of a 'nuptial pad', developed as an extra digit, prepollex, on the hand (p. 317). Injection of male hormones or implantation of testis will cause this organ to develop in young female frogs.

In newts fertilization is ensured by an elaborate courtship. Sperms are made into spermatophores by special pelvic and cloacal glands and there are also abdominal glands, which produce a secretion attractive to the female. After a courtship ceremony the spermatophores are picked up by the cloaca of the female and stored in a spermathecal chamber.

23. Digestive system of Amphibia

Nearly all adult amphibia feed on invertebrates, mainly insects, partly also worms, slugs and snails, spiders and millipedes. The larval stages are usually omnivorous, but they may be cannibalistic, feeding on the tadpoles of the same or other species—an interesting form of provision for the next generation by excess productivity of the mother. There are only minor modifications of particular species in relation to their diet; as regards their food amphibia occupy a generalized or 'easy' habitat. The fact that they are not particular in choice of diet has no doubt been part of the secret of their success.

The tongue is the characteristic organ for catching the food and is one of the special features required for terrestrial life, being reduced in aquatic amphibia. In *Rana* it is attached to the floor of the mouth anteriorly and flicked outwards by its muscles. To keep it moist and sticky a special inter-maxillary gland is found. From the shape of the premaxillae it can be deduced that this gland was present in labyrinthodonts. The saliva contains a weak amylase and some protease. It is suggested that these serve to release sufficient substances for tasting. Special tracts of cilia carry the secretion from the intermaxillary glands to the vomero-nasal organ and palatal taste-buds (Francis, 1961).

Another feature made necessary by terrestrial life is the presence of cilia to keep the fluids moving over the oral surfaces. These cilia are absent in aquatic amphibia.

The teeth on the premaxillae, maxillae, and vomers are used only to prevent the escape of the prey; few amphibia bite. Biting teeth are present, however, in the adult *Ceratophrys ornata*, whose larvae also have powerful jaws and are cannibalistic. The South American tree-frog *Amphignathodon* has teeth in the lower as well as the upper jaw and presumably has redeveloped them, a remarkable case of the reversal of evolution. Teeth are also present on the lower jaw of most urodeles.

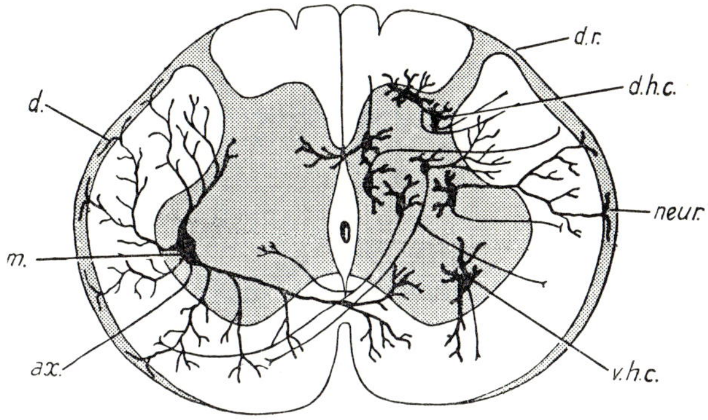

FIG. 201. Transverse section of the spinal cord of a frog, showing cells in the grey matter with their axons and dendrites spreading into the 'white' matter.

ax. axon of ventral horn cell, leaving cord in ventral root; *d.* dendrite of ventral horn cell; *dhc.* small cells of dorsal horn; *dr.* dorsal root entry; *m.* cell body of ventral horn cell; *neur.* neuropil at periphery of spinal cord; *vhc.* small cell of ventral horn. (After Gaupp.)

The oesophagus is not sharply marked off from either mouth or stomach and the latter is a simple tube. Its lining epithelium of mucus-secreting cells is folded and simple tubular glands open at the base of the folds. These glands, unlike those of mammals, are composed of only a single type of cell, which secretes both the acid and the pepsin found in the stomach.

The intestine is marked off from the stomach by a pyloric sphincter. It is relatively short and dilates into a large intestine, there being a valve interposed in the frogs, though not in all amphibia. The liver and pancreas have the structure common to all vertebrates and produce juices of the usual type. The intestine of the omnivorous tadpole is more coiled than that of the adult frog. The type of food taken depends on what is available, most species of amphibia are not particular feeders. However, they can learn with only one or two trials to avoid distasteful insects. Frogs and toads devour large numbers of insects. If the common insects available are pests the amphibian's part in controlling their number works to the advantage of man.

24. Nervous system of Amphibia

The organization of the nervous system of amphibia might be said to be essentially similar to that of fishes. In both groups there are highly developed special centres in the brain, each centre related to a

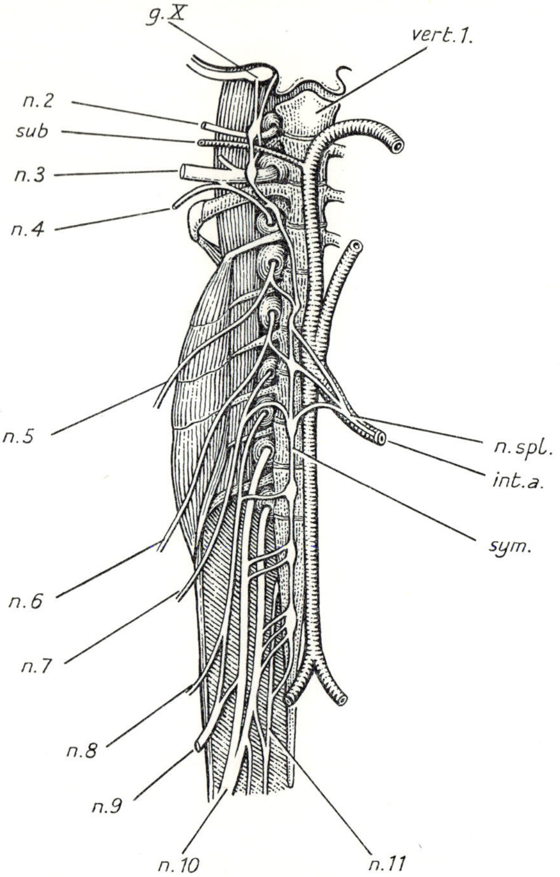

FIG. 202. Ventral branches of the spinal nerves (2–11) of the frog. The sympathetic chain is also shown.

g.X, vagus ganglion; *int.a.* intestinal artery; *n.spl.* splanchnic nerve; *n.* (2–11), spinal nerves; *sub.* subclavier artery. *sym.* sympathetic chain; *vert. 1*, 1st vertebra. (After Gaupp.)

special receptor system. In neither group is there a dominant part, integrating the activity of the whole, as does the cerebral cortex in mammals.

The plan of the spinal cord is like that of fishes, but well-marked dorsal and ventral horns are present. The large motor-cells of the cord have dendrites that spread widely in the white matter, where their synaptic connexions are made in a complicated 'neuropil' (Fig. 201). This is a simpler arrangement than is found within the grey matter of the mammalian cord.

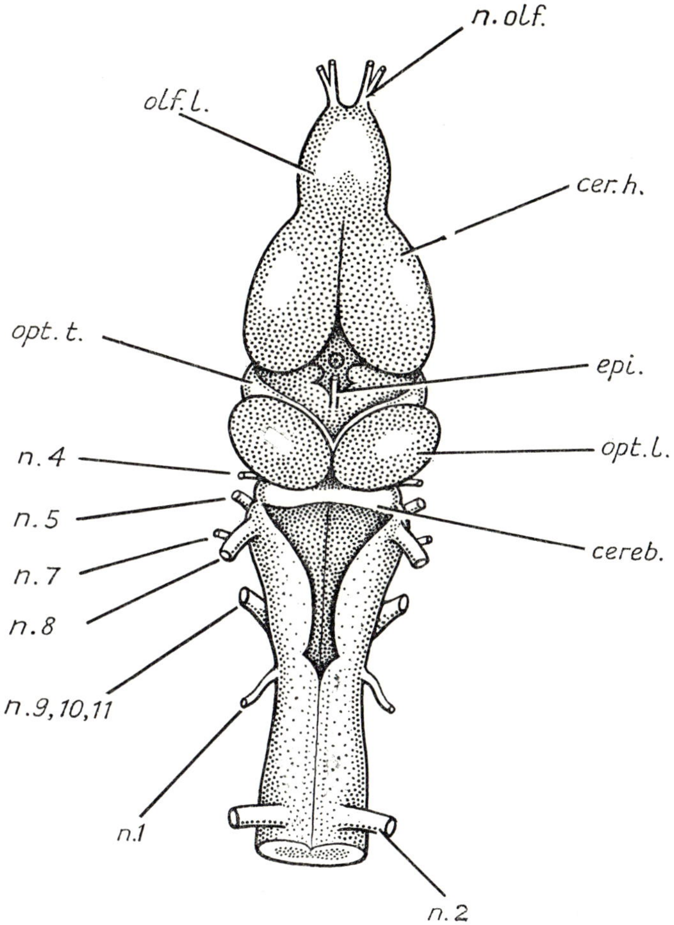

FIG. 203. The brain of *Rana*, dorsal view.

cer.h. cerebral hemisphere; *cereb.* cerebellum; *epi.* epiphysis; *hth.* hypothalamus; *hyp.* pituitary gland; *lam.t.* lamina terminalis; *olf.l.* olfactory lobe; *n.olf.* olfactory nerve; *n. 3–11* cranial nerves; *n.* (1 and 2), spinal nerves; *opt. c.* optic chiasma; *opt.l.* optic lobe; *opt.t.* optic tract. (Modified from Gaupp.)

The arrangement of the spinal nerves is much modified by the development of the limbs. Ten spinal nerves are found but since an embryonic first one is missing they are sometimes numbered 2–11 (Fig. 202). Two spinal segments contribute to the brachial and four to the sciatic plexuses in the frog. From these plexuses fibres are distributed to the muscles and skin of the limbs (Fig. 202).

The brain (Figs. 203 and 204) resembles that of Dipnoi very strikingly. The prosencephalon is based on an inverted plan (p. 211); the large evaginated cerebral hemispheres therefore have a thick nervous roof as well as floor. In the frog there is only a short unpaired region of the forebrain (diencephalon) but this is longer in urodeles. The walls of each hemisphere may be divided into a dorsal pallium, medial ventral septum, and latero-ventral striatum (Fig. 205). The cell bodies lie around the ventricle in all parts of the hemisphere and there are several layers of them. The cells are pyramidal in shape and the connexions are made in the outer 'white' matter.

Nearly all parts of the hemisphere are reached by olfactory tract fibres, the axons of the mitral cells of the olfactory bulb (Fig. 205). In the frog there are regions at the hind end of the hemispheres that receive forwardly directed fibres, some probably connected with tactile and others certainly with optic impulses. There is therefore some opportunity for the hemispheres to act as correlating centres, but we have little information as to the functions performed in them. Their backward projections are made by means of two large tracts, the lateral and medial forebrain bundles, but these reach only to the thalamus, hypothalamus, and midbrain, not back to the cord. Electrical stimulation of the forebrain does not produce movements of the animal; presumably such a crude method, though it may excite a few neurons, cannot imitate the more subtle patterns in which they are normally active.

Removal of the cerebral hemispheres is said to have little influence on the normal feeding and other reactions of the frog. After this operation the animals are said to be more sluggish, to show less 'spontaneity', and to learn less well. If the latter is true it shows a considerable advance in the functioning of the hemispheres over the stage reached in fishes, whose learning can certainly take place in other parts of the brain, and is apparently little affected by removal of the forebrain (p. 210).

Some indication of the function of the cerebral hemispheres is given by the fact that by placing electrodes connected with a suitable amplifier upon them, rhythmical changes of potential can be recorded (Fig. 206). These are most marked in the olfactory bulb and probably propagate backwards along the hemisphere. The rhythms continue even in a brain that has been removed from the head. They are therefore a sign of some intrinsic activity of the brain, rather than of response to peripheral stimulation.

The diencephalon is interesting chiefly for the considerable number

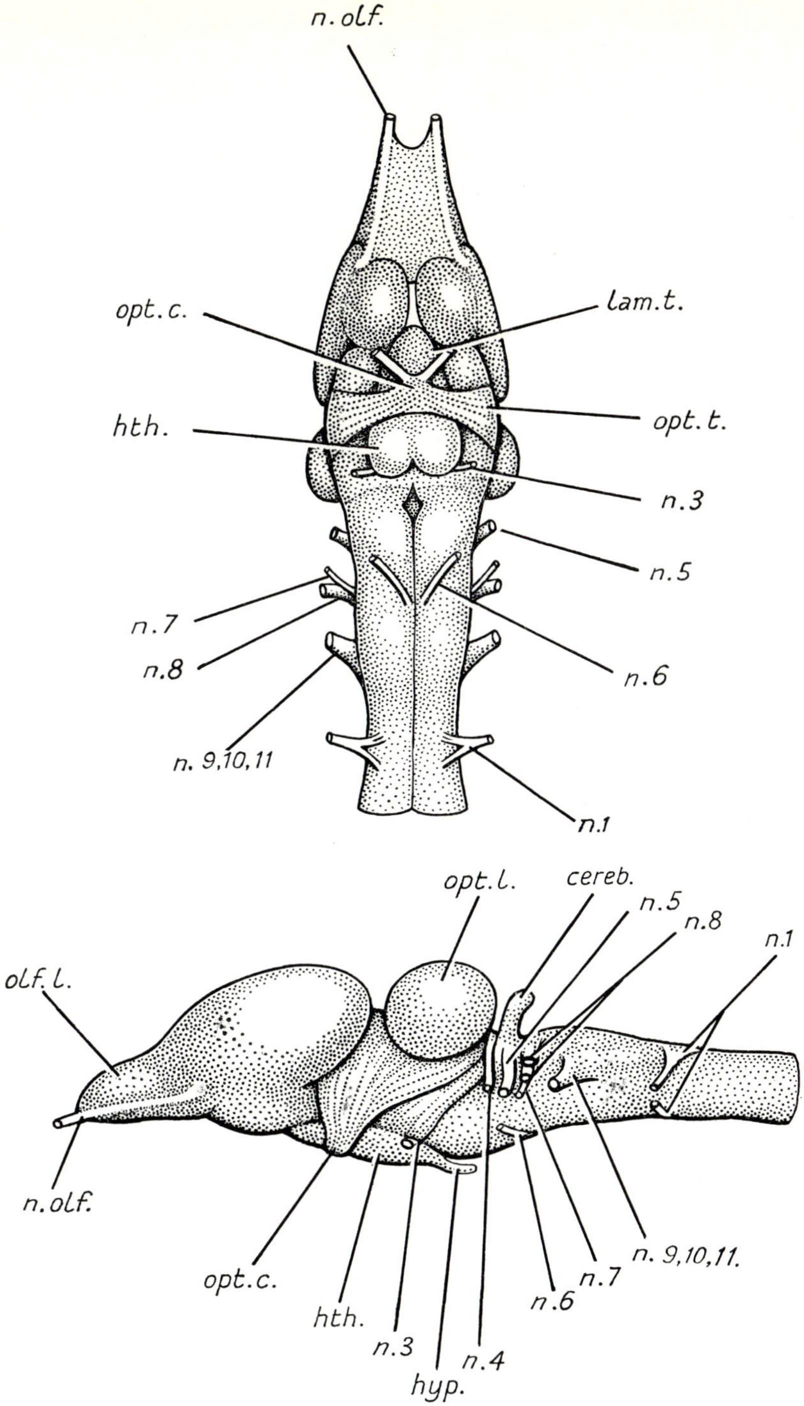

FIG. 204. Two further views of the brain of *Rana*.
Above: ventral view. Below: lateral view. (Modified after Gaupp.)
Lettering as Fig. 203.

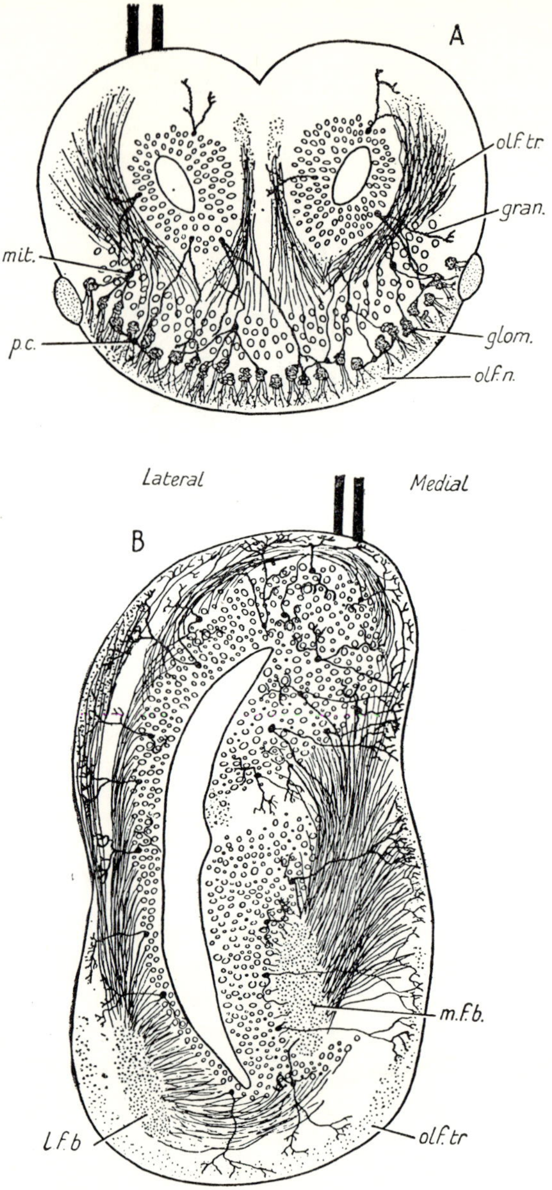

FIG. 205. Diagrams of the structure and probable cell connexions in, A, the olfactory bulb, and B, the cerebral hemisphere of the frog.

glom. glomerulus in which fibres of olfactory nerve make contact with dendrites of mitral cells; *gran.* granule cell (cell without any axon); *l.f.b.* lateral forebrain bundle; *m.f.b.* medial forebrain bundle; *mit.* mitral cell; *olf.n.* fibres of olfactory nerve; *olf.tr.* olfactory tracts; *p.c.* periglomerular cell. The electrodes are shown as they would be placed for recording the potentials shown in Fig. 206. (From Gerard and Young.)

of optic tract fibres that end in its walls; other sensory projections also reach here. In anurans, but not in urodeles, there is a partial division into separate thalamic sensory nuclei, such as are found in mammals, for touch, sight, and other receptor modalities.

The pineal organ shows evidence of a retina in a few amphibia; in most it is a simple sac.

The pituitary body is well developed and the usual partes, anterior, intermedia, nervosa, and tuberalis can be recognized, though they are not in the same relative position as in mammals.

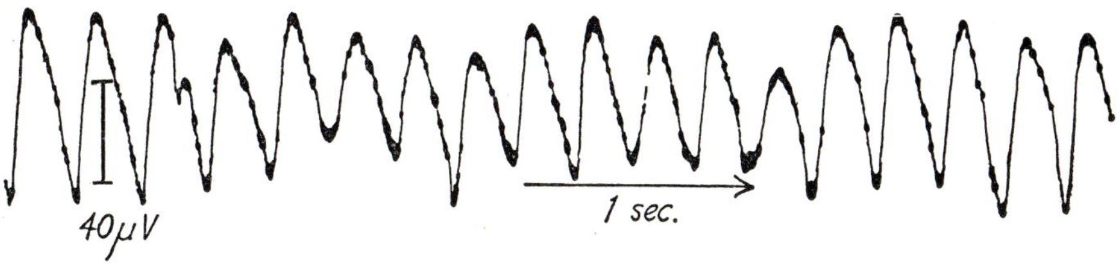

FIG. 206. Rhythmical changes of potential between electrodes placed on the surface of the olfactory bulb of the frog as in Fig. 205. (After Gerard and Young.)

The midbrain is very well developed and shows many similarities to that of fishes. The cells it contains do not all lie round the ventricle, many have moved out to make an elaborate system of cortical layers. Electrical stimulation of various parts of the optic tectum produces movements of the limb and other muscles; there can be no doubt that this region plays a dominant part in behaviour. Most of the fibres of the optic tract end here, and there are also other pathways from the olfactory, auditory, medullary (gustatory?), and spinal regions. Efferent fibres leaving the tectum pass to the midbrain base, medulla, and perhaps back into the cord. This region therefore has wider connexions than any other part of the nervous system and thus nearly reaches the status of a dominant integrating organ.

The cerebellum of amphibia, on the other hand, is very small; perhaps because these are mostly animals that do not have to adjust themselves freely in space during locomotion, they move mainly in a single plane. There is little need for control of speed or distance of movement, except of the head and tongue, which are controlled by the tectum.

25. Skin receptors

Lateral line organs are present in the skin of all aquatic amphibian larvae and in some aquatic adults, such as those of the anuran family Pipidae. They are of simple form, consisting of groups of cells in an open pit. In newts they are present in the larvae, which are aquatic, but are covered by epidermal layers during the first post-larval stage during which the newt lives on land. In the final aquatic adult stage the organs reappear.

The skin, of course, also contains tactile organs, and in addition is often sensitive to chemical stimuli. This chemical sense is mediated by fibres running in the spinal nerves, not by special elements such as the taste-buds found spread out over the body in fishes. The skin is also sensitive to heat and cold, and there is some evidence that these senses are served by fibres different from those that mediate touch, pain, or the chemical senses. Histologically, however, there is little sign of the development of the special sensory corpuscles that are so conspicuous in the skin of birds and mammals. All the nerve-endings are of the type known as 'free nerve-endings', except for a few touch corpuscles on special regions such as the feet. In this the amphibia again resemble the fishes and show less differentiation than do the higher animals.

The taste-buds on the tongue and palate are probably able to respond to the presence of only two of the four types of substance that are discriminated by mammals. Applications to the tongue of the frog and recordings of nerve-impulses in its nerves show that there are chemoreceptors present able to respond to salt and sour substances, but that no reaction is given to substances that in mammals are classed as sweet or bitter.

The olfactory organ functions both on land and in the water, special mucous glands being present to keep it moist when in air. A continual circulation of water or air is maintained over the olfactory epithelium by cilia or the movements of respiration. The internal nostril may have originally developed from the double nostril of fishes, in order to make a circulation around the olfactory receptors possible. Jacobson's organ is a special diverticulum of the olfactory chamber, serving to test the 'smell' of food in the mouth.

The Apoda, being blind, have a great development of the sense of smell, including a hollow tentacle or olfactory tube.

26. The eyes of Amphibia

Provided that certain requirements are met the air gives more scope for the use of photoreceptors than does the water. Light is transported with less disturbance through the air and image formation is facilitated by the refraction of the air-corneal surface. The amphibia have exploited these advantages and sight has become the dominant sense of most of the forms. For clear vision it is essential that the surface of the eye be protected, kept moist and free of particles, and for these purposes the eyelids and lachrymal glands are present. The upper lid is fixed, but the lower is very mobile and folded to make a transparent structure, the nicitating membrane, able to move rapidly across the surface of the eye.

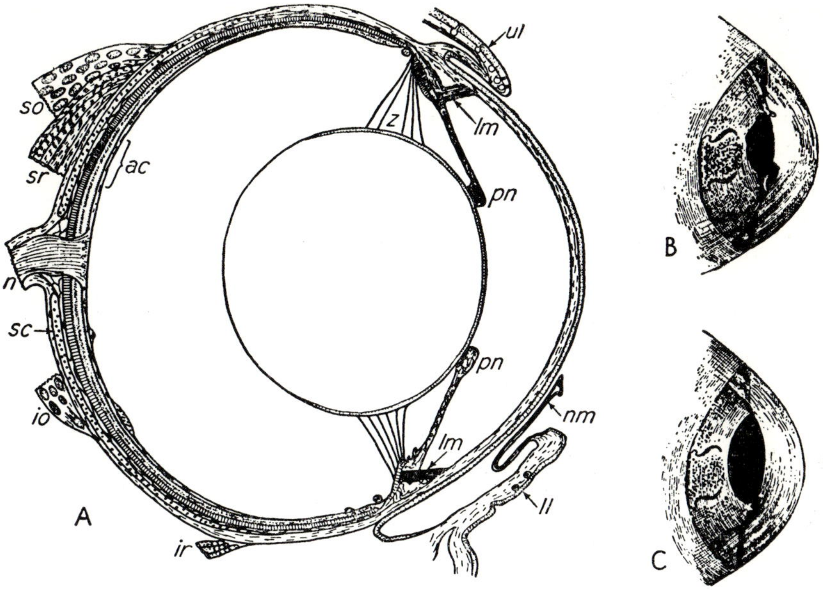

FIG. 207. The amphibian eye and its accommodation.

A, anuran eye in vertical section. *ac.* area centralis; *io.* inferior oblique; *ir.* inferior rectus; *ll.* lower lid; *lm.* lens muscles (protractors); *n.* optic nerve; *nm.* nictitating membrane; *pn.* pupillary nodules; *sc.* scleral cartilage; *so.* superior oblique; *sr.* superior rectus; *ul.* upper lid; *z.* zonula Zinnii. B, anterior segment of *Bufo* in relaxation. C, in accommodation; note forward movement of lens. (From Walls, after Franz and Beer.)

The eyeball is almost spherical, with a rounded cornea. The lens is farther from the cornea than in fishes and is flattened, more so in anurans than in urodeles. These modifications allow focusing of a more distant image. There is an iris, with a rapidly moving aperture, operated by powerful circular (sphincter) and radial (dilatator) muscles. Although these muscles are partly actuated by a nervous mechanism they are also directly sensitive to light, and the pupil of the isolated eye of the frog shows wide excursions with change of illumination.

Accommodation is effected by protractor lentis muscles, attached to the fibres by which the lens is supported (Fig. 207). These muscles move the lens forward, whereas the muscles of the lens of teleostean fishes move it backwards. Other fibres, the musculus tensor chorioideae, run radially and around the lens. They may help the protractors, and are probably the ancestors of the ciliary muscles of higher forms.

In amphibia living in the water the eye is based much more on the fish plan and the lens is rounder. There are then no lids or lachrymal glands and the eye is enabled to make an image, in spite of the absence of the air-corneal interface, by a thickening of the inside of the cornea.

Rods and cones are present in the retina, the former containing visual purple, which may be red or greenish. The two sets of receptor are apparently found throughout the retina in urodeles, but in *Rana* there is a macular region in which the cones are in excess and this is still further developed in *Bufo*. Study of the impulses in the optic nerves of *Rana* shows that six types of detector operate upon the information provided by the rods and cones. (1) Contrast detectors give a sustained response when a sharp edge moves into the visual field; (2) convexity detectors respond to objects that are curved, the discharge being greater the more curved (smaller) they are. These two types together may be called 'on' fibres; (3) moving-edge detectors ('on/off' fibres) respond with a frequency proportional to the velocity of movement; (4) dimming detectors respond on reduction of illumination ('off' fibres); (5) darkness detectors fire with frequency inversely proportional to illumination. These types of fibre project to different depths in the tectum as sheets of endings, and the arrangement of the retina is accurately reproduced there although the fibres are interwoven in the nerve (perhaps to prevent 'cross-talk'). Moreover, if the nerve is severed the fibres regenerate in such a way as to reconstitute the map. The sixth type of fibre is sensitive to blue light and is connected with the thalamus.

These operations serve to provide reports of the types of change relevant to the animal. Thus the second type might be called 'insect detectors', responding when a small dark object enters the field and moves about intermittently. More complex visual discriminations are also possible, for example toads can distinguish between shapes.

The skin is probably sensitive to light in all amphibians: frogs react to light even after removal of the eyes and cerebral hemispheres. This skin sense is especially developed in certain cave-living urodeles, *Proteus*, in which the eyes are not functional. A similar degeneration also occurs in Apoda.

27. The ear of Amphibia

The inner ear is divided into a utricle, from which the semicircular canals arise, and a saccule, from which there is a diverticulum, the lagena, part of whose receptor surface is covered with a tectorial membrane somewhat similar to that of mammals. There is, however, no coiled cochlea. The middle ear of the frog consists of a funnel-shaped tympanic cavity communicating with the pharynx and closed externally by a tympanum supported by a tympanic ring. Sound waves are transmitted across the cavity by a rod, the columella, fitting by an expanded foot, the otostapes, into the fenestra ovalis, a hole in the wall of the auditory capsule. This hole is also partly occupied by a second plate, the operculum, which is joined to the scapula by a special opercular muscle. The operculum and otostapes develop within the wall of the auditory capsule and the middle part of the columella (mediostapes) forms as an outgrowth from the otostapes. The outer part of the columella (extra-columella) and the tympanic ring develop close to the quadrate and probably from its cartilage.

The columella, therefore, shows no developmental relationships to the hyoid arch. The tympanic cavity is developed from the spiracular cleft, after a strange series of changes. The original cleft degenerates six days after hatching but about six of its lining cells persist and at the end of the tadpole stage form a tympanic vesicle, which becomes connected with the pharynx by a rod of cells. This rod then degenerates again and an open air passage to the vesicle of the drum is not established until some thirty days after emergence from the water, when a pouch from the pharynx joins the tympanic cavity. These events show the complexities that may result from the modification of developmental processes, and they emphasize the difficulty in assigning 'homologies'. It is still debated to what extent the middle ear of the frog can be compared with that of amniotes. The hyomandibular nerve, which divides above the middle ear of amniotes (and above the spiracle of the dogfish) lies behind the tympanic cavity of the frog and branches below it.

The arrangement for conveying vibrations to the ear varies considerably among amphibians. In urodeles there is no tympanum. In some of them the columella is attached to the squamosal, perhaps in connexion with a semi-aquatic or burrowing habit. A similar arrangement may have been present in the earliest amphibians, which have a columella but no oval window. In other urodeles (Plethodontidae) the columella is attached to the quadrate and there may be a second ossicle, the operculum, working in parallel, with its inner end in the oval window caudal to the columella and its outer end attached by a muscle to the scapula. In terrestrial forms the columella becomes fused with the window at metamorphosis and its function is taken over by the operculum, probably receiving vibrations from the fore-legs. The more aquatic forms (*Cryptobranchus*) retain the larval condition and never develop an operculum. The tympanum and columella are also reduced in some terrestrial anurans (*Bombinator*) but in the aquatic *Xenopus* and *Pipa* the operculum and its muscle are lost, perhaps a paedomorphic feature.

The sense of hearing is certainly well developed, especially in Anura, which respond to vibrations from 50 to 10,000 a second. The hearing is used especially in the breeding-season, when the croaking serves to attract both sexes to the water. The prey may also be located by sound. Urodeles have been shown to give no response to the ringing of a bell suspended from the ceiling, which, however, produces reactions in *Rana* and *Bufo*.

A peculiar feature of many Anura is an immense backward development of the perilymphatic space of the inner ear, forming a sac extending above the brain and on either side of the spinal cord as far back as the sacrum. Portions of this sac emerge between the vertebrae, showing as whitish masses on account of the granules of chalk they contain. The calcium salts in these sacs diminish greatly during metamorphosis and they then refill. The system may serve as a calcium reserve also for the adult.

28. Behaviour of Amphibia

The habits of amphibia, like their special structures, enable them to deal with the various emergencies that threaten the continuation of life on land. Frogs and toads have a strong sense of place and they show distinct 'homing' reactions. They are able to learn to find their way out of mazes and to remember the way for periods of at least thirty days.

Complex migrations are made by many species; nearly all migrate to the water in spring. In this migration the males usually precede the females, then attract the latter by their calling. The receptors for the orientation towards the water are known in the osmoreceptors in the mouth of the frog. This orientation is particularly clear in urodeles, in which sound plays no part in the migration. The power to find water is obviously of first importance for any animal living on land, and further study of the receptors involved would be interesting.

The search for food and the avoidance of enemies are not in principle more difficult on land than in the water, but they probably demand new mechanisms. For example, the greater range of visibility can be a disadvantage, especially when it is exploited by one's successful and predatory descendants. A hawk or owl makes fuller use of its opportunities in this respect than does the frog, who can only remain safe from them by behaviour that keeps it concealed. Similarly there are dangers in certain situations, for instance of desiccation, which are additional to those that are met by an animal in the water.

In the emergence of the first land vertebrates we thus see a conspicuous example of the invasion by living things of a medium far different from themselves. This produces a situation that calls forth all the powers of the race to produce new types of individual, and necessitates that the individuals make full use of their capacities. New patterns of structure and behaviour are developed as the various possible situations emerge. The types of organization that at first manage to survive gradually give place to others, still more complex or 'higher'. Some traces of the organization of the early venturers can still be seen in the amphibia, which today exploit the damper situations on the earth.

XIII

EVOLUTION AND ADAPTIVE RADIATION OF AMPHIBIA

1. The earliest Amphibia

THERE are such close resemblances between the skulls of the earliest amphibians and those of the Devonian crossopterygian fishes that there can be no doubt of the relationship (Fig. 194). At present there is, however, no detailed fossil evidence of the stages of transition from the one type to the other. The fossils that appear to be closest to the possible tetrapod ancestor are the osteolepids of the Lower and Middle Devonian periods, about 375 million years ago. These were definitely fishes, though they may have breathed air. **Elpistostege* is a single Upper Devonian skull intermediate between such fishes and the earliest undoubted tetrapods, **Ichthyostega* and similar forms, found recently in freshwater beds of Greenland. These are dated as very late Devonian or early Carboniferous, that is to say about 350 million years ago. They are the oldest members yet found of the great group of Stegocephalia, which, throughout the succeeding 100 million years of the Carboniferous and Permian periods, flourished and developed many different lines, one giving rise to the reptiles and others to the modern amphibia. The term Stegocephalia is convenient to cover the whole group of palaeozoic amphibia, all probably of common descent. At least seven types can be recognized (p. 296), but attempts to group these have not been altogether successful; the nomenclature remains confused. The Labyrinthodontia were the central stock and were in the main terrestrial forms, giving off at intervals lines that returned to the water. A characteristic labyrinthodont feature is a folded pattern of the teeth, similar to that of their crossopterygian ancestors.

The earliest Stegocephalia were definitely tetrapods and already showed sharp changes from the fish type. **Ichthyostega* is known chiefly from the skull (Fig. 208), which shows all the characteristic amphibian features, but retained traces of fish ancestry in its shape, with a short, wide snout and long posterior region (table), and presence of a preopercular bone. The nostril lies on the very edge of the upper lip, apparently partly divided by a flange of the maxilla into internal and external openings.

2. Terrestrial Palaeozoic Amphibia. Embolomeri and Rhachitomi

We possess more complete information about the slightly later forms, the Embolomeri, such as **Eogyrinus*, from the Lower Carboniferous (Fig. 211). These were long-bodied animals, rather newt-like, and their small limbs cannot have made very effective progress on land. They probably lived mostly in the water, eating fish. The

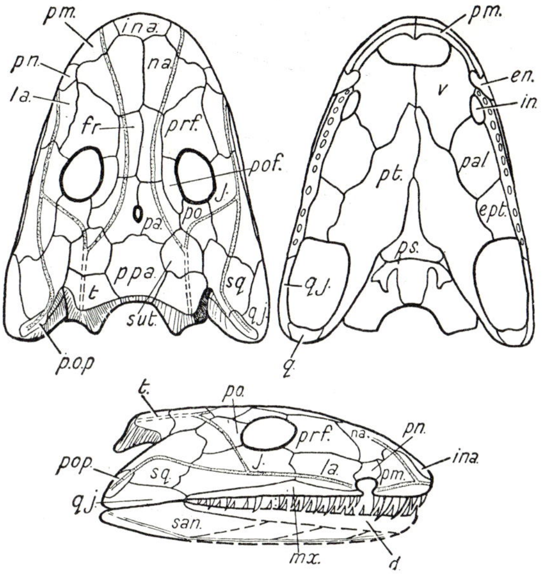

FIG. 208. Skull of *Ichthyostega*. (From Westoll, after Säve-Söderbergh.)

d. dentary; *en.* external nostril; *ept.* ectopterygoid; *fr.* frontal; *in.* internal nostril; *ina.* internasal; *j.* jugal; *la.* lachrymal; *mx.* maxilla; *na.* nasal; *pa.* parietal; *pal.* palatine; *pm.* premaxilla; *pn.* postnarial; *po.* postorbital; *pof.* postfrontal; *pop.* preopercular; *ppa.* postparietal; *prf.* prefrontal; *ps.* parasphenoid; *pt.* pterygoid; *q.* quadrate; *qj.* quadratojugal; *san.* sur-angular; *sq.* squamosal; *sut.* supratemporal; *t.* tabular; *v.* vomer.

pectoral girdle was still joined to the skull by a process of the tabular bone, as in fishes. The pelvic girdle did not form a full articulation with the sacral vertebrae but was apparently attached only by ligaments.

The structure of the vertebrae has given rise to much controversy. In the earliest amphibia we find three elements, a more dorsal neuropophysis and a centrum composed of two parts, pleurocentrum and hypocentrum, the latter associated with a ventral arch and rib. These elements can be identified in the vertebrae of crossopterygians but it is still not clear what relationship, if any, they have to the two pairs of vertebral 'arches' alleged to be present in elasmobranchs (see Williams, 1959).

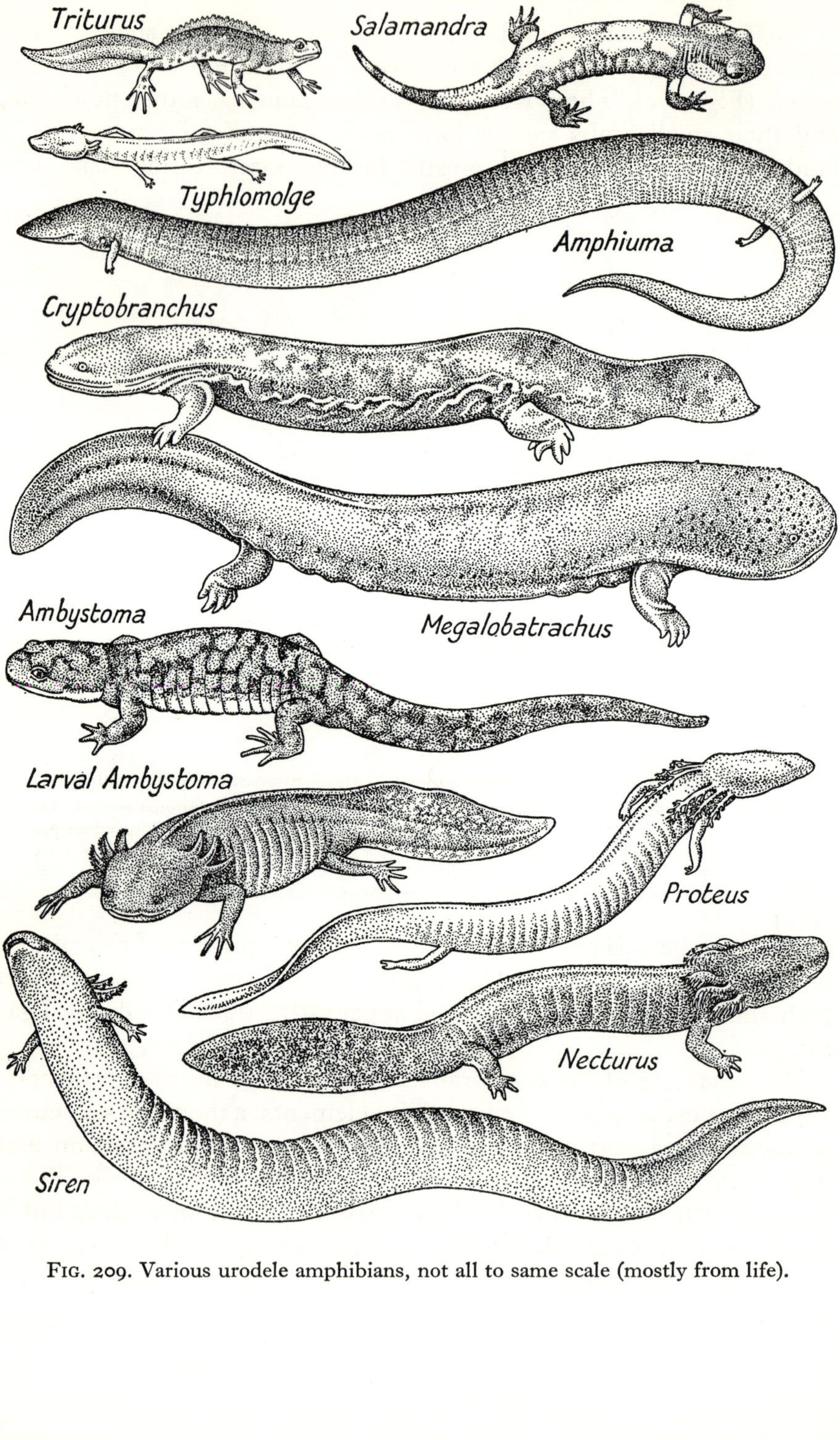

FIG. 209. Various urodele amphibians, not all to same scale (mostly from life).

The skull shows the full series of bones that we have already discussed; there is therefore no reason to suppose that these animals represent a secondarily 'degenerate' branch, which had returned to the water. They were probably very close to the ancestors of all tetrapods. They were numerous in the Carboniferous swamps, but disappeared early in the Permian. So close were these Embolomeri to the ancestry of the reptiles that many workers classify them near the Permian **Seymouria*, which we shall consider as a cotylosaurian reptile (p. 386).

Throughout the Carboniferous, Permian, and Triassic there were abundant amphibia of partly terrestrial habit, the Rhachitomi, in which both vertebral centra were present, the pleurocentrum being the larger. **Eryops* (Fig. 211) was a typical form living in the Permian, about 250 million years ago. The animals were 5 ft or more in length, rather like crocodiles, relatively shorter in body and tail than **Eogyrinus* and with stronger limbs. Nevertheless they probably lived partly in the water and may have been fish-eaters. The skull was long and narrow in the front and short in the 'table' behind the eyes, continuing the previous tendency. A characteristic feature of later labyrinthodonts now began to appear, namely a dorso-ventral flattening of the skull. The pectoral girdle was no longer attached to the skull, but there was a joint between the pelvic girdle and the sacrum. Some Permian Rhachitomi became still more completely terrestrial than **Eryops*, for instance **Cacops* had very large limbs and protective plates along its back.

3. Aquatic Amphibia of the later Palaeozoic

Other lines of amphibia, however, show an accentuation of the tendency to return to the water. In the vertebrae the anterior hypocentrum became large, while the pleurocentrum disappeared. At the same time the skull became very flattened and the limbs weak. Several stages are known leading from the Rhachitomi to these fully aquatic forms of the Trias, which are placed in the suborder Stereospondyli. **Capitosaurus* and **Buettneria* are typical of the group, which remained numerous until near the end of the Triassic period, about 150 million years ago. Probably this change from rhachitomous to stereospondylous condition occurred on several independent lines of descent. Thus amphibia, after becoming semi-terrestrial in the Carboniferous and then probably giving rise to the early reptiles, later returned to the water.

We also have record of various other secondarily aquatic amphi-

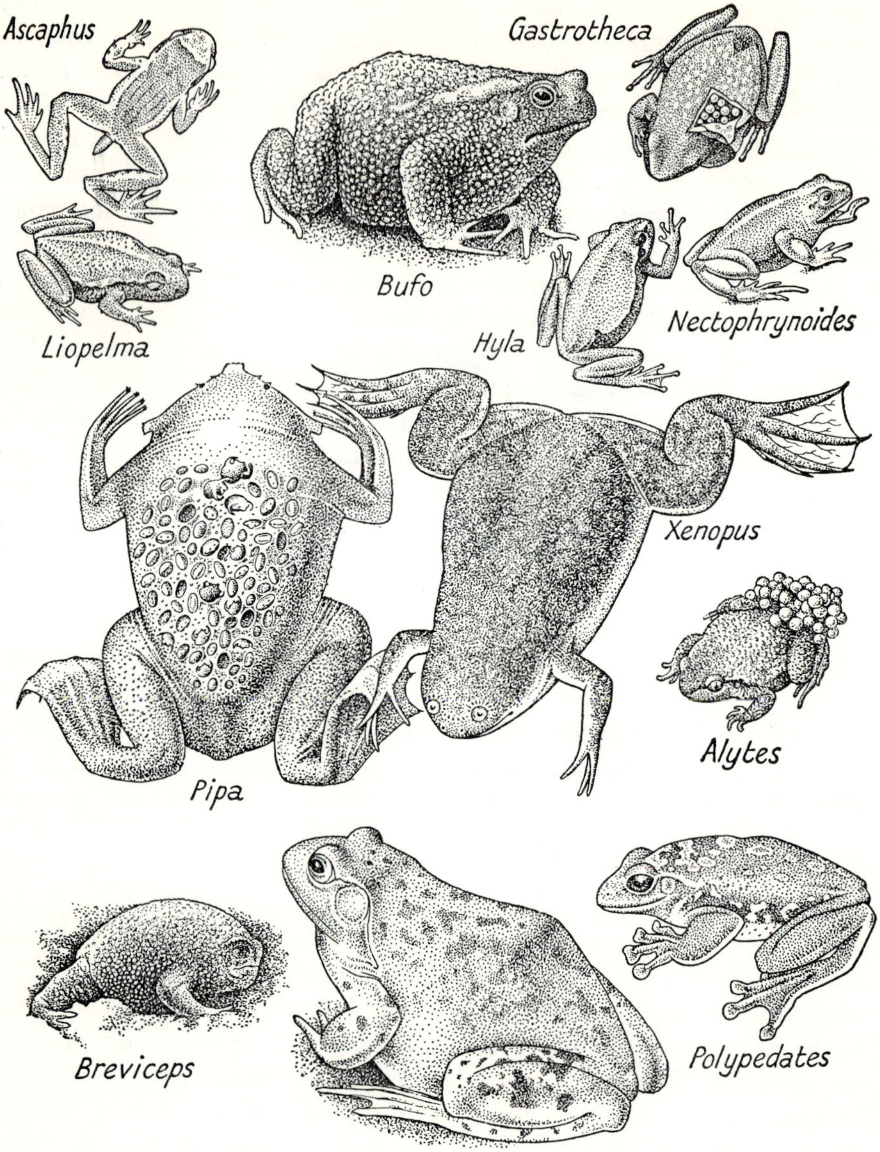

FIG. 210. Various anuran amphibians, not all to one scale.
(*Ascaphus*, *Nectophrynoides*, and *Polypedates* after Noble, *Breviceps* after Thompson, *Gastrotheca* after *B.M. Guide*, others from life.)

bians whose affinities are less certain. They have in common a reduction of ossification in general and in particular in the centra, which seem not to be formed from separate cartilaginous elements as in labyrinthodonts but as thin continuous sheets of bone. Some of these animals classed as Phyllospondyli or 'Branchiosaurs' were almost certainly larval Rhachitomi; the external gills can be recognized and stages found connecting them with known adults of that group. It is necessary, however, to retain the order for the present.

Other early aquatic amphibia are less easy to classify and are grouped for convenience as an order Lepospondyli, all having vertebrae composed of a single piece, and a continuous notochord. They show, however, at least three distinct lines, probably separate offshoots from the main labyrinthodont stock. **Dolichosoma* and other forms from the Carboniferous were like snakes and had lost the limbs. **Diplocaulus* from the Permian possessed remarkable horned skulls. These creatures with broad flat heads and upward-looking eyes and small limbs were presumably bottom-dwellers and the development parallels that of the Stereospondyli. A third aberrant group placed here, the Microsaurs, such as **Microbrachis*, were animals with long bodies and small limbs, presumably aquatic, but showing many similarities to the reptiles in the skull.

The order Adelospondyli has been created for a further collection of presumably secondarily aquatic amphibia such as **Lysorophus*, in which the neural arch and centrum are not fused but articulate by means of a jagged suture. The skull shows reduction and variation of the bones, and for this and other reasons it has been suggested that the urodeles may have arisen from an adelospondylous line.

The relationship of the modern amphibia to these palaeozoic stegocephalians remains uncertain. The earliest anuran is **Protobatrachus* of the Triassic, possessing ribs and a tail, but with elongated ilia and an anuran type of skull. It is probably a larva in metamorphosis. **Miobatrachus* of the Carboniferous is a form in which the posterior portion of the skull is shortened and the temporal bones lost. In other respects the skull is like that of a rhachitome and suggests that the frogs diverged from the labyrinthodonts at this very early stage.

The urodeles can be traced back only to the Jurassic. It is often suggested that both they and the Apoda have arisen from aquatic Lepospondyli, such as the microsaurs, but there is no real evidence of this.

4. Tendencies in the evolution of fossil Amphibia

The changes in the form of amphibia can be followed from the beginning of the Carboniferous to the end of the Triassic period, and, indeed, in the form of their reptilian descendants far beyond. There are signs, however, of very many distinct lines (as we should expect), and it is not possible to trace details of the history or fate of particular populations. Two distinct tendencies appear over this period: (1) to become fully terrestrial, (2) to return to the water. The terrestrial forms became very gradually shorter in body and stronger in leg. The skull remained fairly high and domed and the otic notch became deeper, as a more effective tympanum developed. In the vertebrae the hypocentrum became reduced as muscles developed attached to the pleurocentrum.

Return to the water led to animals of two distinct types, (*a*) snake-like or (*b*) flattened, but in both there was a reduction of limbs and a secondary lengthening of the body, with return to the sinuous movements of fish-like locomotion. In the bottom-living forms, such as some Stereospondyli and **Diplocaulus* among Lepospondyli, the skull became flattened, with the eyes looking upwards, the otic notch being shallow. The snake-like **Dolichosoma* retained the more normal skull shape but became immensely elongated and lost the limbs altogether.

These observed tendencies can be understood to result from the situation that developed as the vertebrates first colonized the land. The earlier amphibia, such as **Eogyrinus*, were partly aquatic, by force, one might say, of inexperience. Throughout the Carboniferous various lines of them became more fully equipped for terrestrial life, moving faster, seeing and hearing better, and so on. The competition and predatory attacks of these more successful lines then drove others back into the water and so the process continued, until later the earlier reptilian lines, themselves driven back to the water by their own descendants, removed most of the amphibians from the waters as well as from the land, leaving only some few remaining populations, from which the modern orders have evolved.

It certainly does not seem necessary to postulate any special directive force to explain all this. We could wish, of course, for much more information, but it seems reasonable to imagine that these changes were produced by the action of the animals with each other and with the environment, supposing that the animals continually strive to feed, grow, and reproduce others rather, but not quite, like themselves.

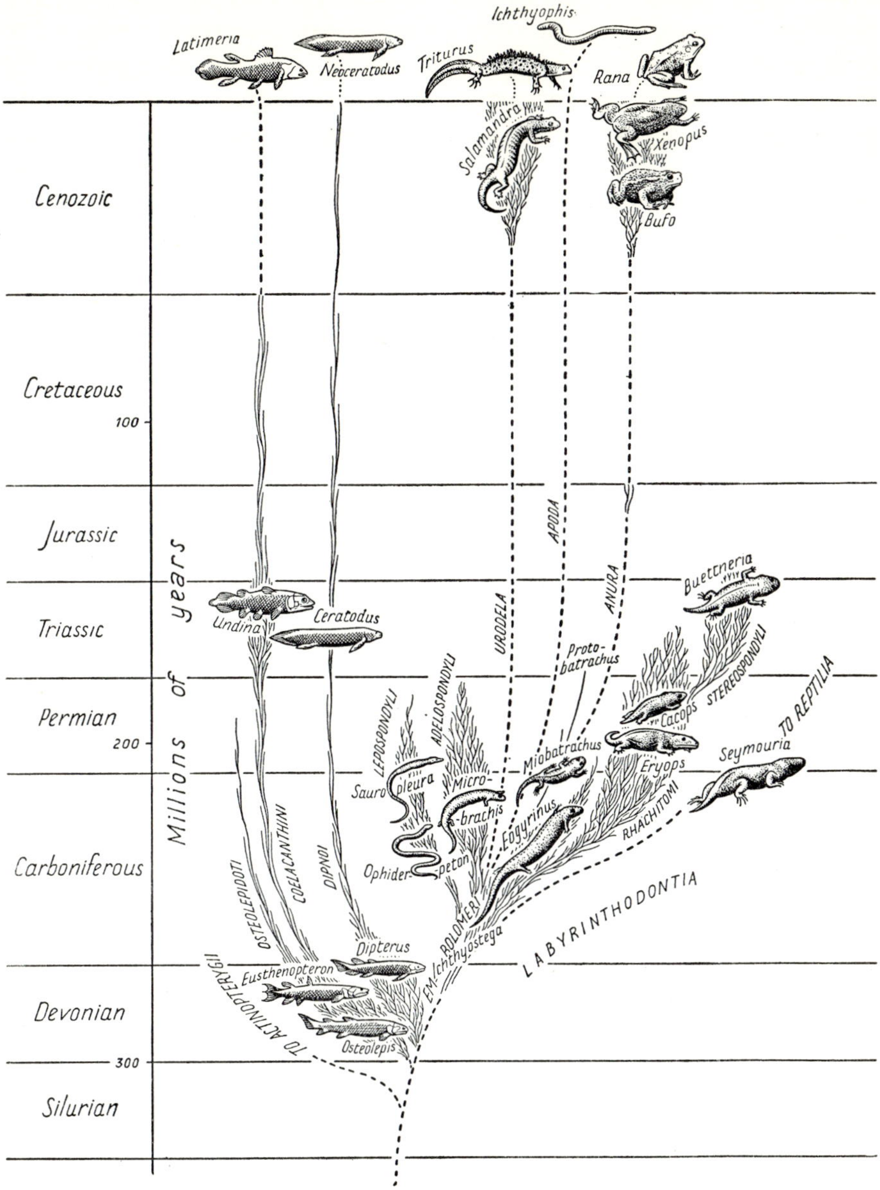

FIG. 211. Chart of evolution of amphibia.

It must be stated, however, that Watson, who has contributed more than anyone to knowledge of amphibian evolution, believes that it is possible to recognize a number of non-adaptive trends, which are independent of environmental influences. Changes suggested by Watson and others as non-adaptive include flattening of the skull, doubling of the occipital condyles, reduction of the number of roofing bones and loss of ossification in the neurocranium, the changes in the vertebrae already mentioned, and many other features. It is not entirely clear how the 'non-adaptive' nature of these features is established. Severtzoff has suggested that the flattening of the head is connected with the development of a large mouth for buccal respiration. Palaeontology necessarily deals with small points of structure, whose significance for the animal may be difficult to determine, but it does not follow because we are not able to discern the significance of a part that it therefore has none. It is not at all easy in biology to hold the balance between credulous acceptance of a function for every character and a sceptical attitude that insists on regarding the organic world as a jumble of unrelated substances. The only safe rule is to search continually for signs of regular recurrence of similarities of structure and action, and then to make hypotheses about function, which can be tested by experiment.

5. Newts and Salamanders. Subclass Urodela

The urodeles, also called Caudata or tailed amphibia, show less deviation from the general form and habitats of the amphibia as a whole than do the specialized anurans (Fig. 209). The adult and larval urodeles differ little from each other, and characters suitable for aquatic life are frequently found in the adult. Indeed, all stages of suitability for land occur, from the terrestrial salamanders, such as *Salamandra maculosa*, the European salamander, which is viviparous, to the fully aquatic forms, for instance *Necturus*, the mud-puppy of North America. In many of the aquatic animals there is a tendency to retain in the adult characters usually found in larvae. This process of paedomorphosis has developed to various extents, and independently in several groups. Thus the giant salamander *Megalobatrachus*, 5½ ft in length, in China and Japan, has no eyelids, but loses its gills in the adult. In *Cryptobranchus*, the hell bender of the United States, the spiracle remains open and is used for the outlet of water during respiration.

Amphiuma, also from the southern U.S.A., is a very elongate form, with absurdly small legs, no eyelids, and four branchial arches. In the still more modified forms, such as *Necturus*, external gills are present and the lung is so reduced that the animals can live walking along the bottom. *Proteus* from European caves is a blind urodele with external gills and no pigment. *Siren* shows almost entirely larval characteristics and has no hind limbs.

The more typical terrestrial newts are of several sorts. In North America the common genus *Ambystoma* (usually written *Amblystoma*) has eleven species, many adapted to special habits, including *A. mexicanum*, in which some races become mature without metamorphosis, because of lack of iodine in the water, whereas others, the axolotls of Mexico, are genetically neotenous.

The common British newt *Triturus vulgaris* is a typical example of the more definitely terrestrial urodeles, though it is not able to live in very dry situations. However, the limbs support much of the weight of the body, and their soles are applied to the ground and turned forwards. The tail shows various degrees of reduction to a circular organ, but in the breeding-season, when both sexes return to the water, it develops a large fin, especially in the male. The common newts of America form a distinct family, including *Plethodon* and many specialized forms, such as the blind *Typhlomolge*, inhabiting the waters of caves.

6. Frogs and Toads. Subclass Anura

Among the frogs and toads are very many suited for special modes of life, and it must again be emphasized that this is far from being a static and precariously surviving group. We have already mentioned the frog *Ascaphus*, which lives in mountain streams in the north-west of the United States and has reduced lungs, showing a combination of specialized and primitive features. Internal fertilization is assured by a penis-like extension of the cloaca. In this genus and the New Zealand *Leiopelma* (Fig. 210) there are several primitive features, including tail muscles (absent in all other anurans), amphicoelous vertebrae, free ribs, abdominal ribs, and persistent posterior cardinal veins.

In *Alytes*, the midwife toad of Europe, the male carries the eggs wrapped round the legs. *Pipa* is a related and still more specialized aquatic frog from South America; it has no tongue, and, curiously enough, has developed an elaborate arrangement by which the young are carried in pits on the back. *Xenopus* of Africa is related to *Pipa*, but without the habit of carrying its young (Fig. 210).

The bufonid toads are among the most successful of all amphibian groups and are more fully adapted than most for a terrestrial life, but return nearly always to the water to breed. *Bufo* itself is found in almost all possible parts except in Australia and Madagascar; related genera, many of them with special features, are found all over the world. Curiously enough only one genus, *Nectophrynoides* from East Africa, is viviparous, the young being in that case provided with a long vascular tail, by means of which they maintain contact with the wall of the 'uterus', even though embedded in a mass of embryos.

Hyla and other tree-frogs, very widely distributed, are similar to the bufonids but have pads on the toes by which they climb, and many other adaptations to arboreal life. *Gastrotheca* (= *Nototrema*), the marsupial frog, is a genus in which the young develop in a sac on the back of the female, this sac being in one species protected by special calcareous plates. *Rana* and its allies, the true frogs, are also cosmopolitan. A number of frogs related to *Rana* have taken to a tree-living habit, developing pads on the toes. *Polypedates* is a widespread genus and there are several others, each independently derived from ranids. This is therefore a striking illustration of parallel evolution—the hylid tree-frogs having arisen from bufonids and probably several sorts of polypedatids from ranids.

Burrowing with the legs has also been evolved several times by anurans. In *Breviceps* (Fig. 210), which digs for ants, there is a large snout, as in other anteaters.

7. Subclass Apoda (= Gymnophiona = Caecilia)

These (such as *Ichthyophis*) are burrowing, limbless creatures living like earthworms, in the tropics. They show several interesting primitive features, including the retention of small scales in the skin. They are specialized, however, in having a very short tail and some features suited to their terrestrial life, such as copulatory organs. The animals are blind, the place of the eyes being taken by special sensory tentacles. The eggs are large and yolky and cleavage is meroblastic; they are laid on land and the embryos develop around the yolk sac, but often have long, plumed gills. Vivipary is common, including in the aquatic form *Typhlonectes*.

8. Adaptive radiation and parallel evolution in modern Amphibia

Even this superficial study of the 250 genera and about 2,000 species of modern amphibians shows that the features we have already recognized in fish evolution are found also in evolution on land. It is difficult in a short time to gain an impression of the very great variety that is characteristic of any group of animals when closely studied. Besides the main types that can be distinguished, countless lesser variations will be found, and one realizes that the characteristics of the populations are still today in process of continual and perhaps rapid change. Anyone trying to discover the relationships of the various derivatives of ranids or bufonids must be impressed by the presence of series of parallel lines of development, so that it is impossible to disentangle the relationships. Evolution viewed at close quarters by the student of abundant modern animals looks very different from the simple picture seen by the lucky collector of a few rare fossils, who can arrange his types in genealogical trees and is apt to forget that they represent only an infinitesimally small sample of abundant and varied populations.

We can perhaps find certain tendencies in the modern amphibian populations that are similar to the tendencies of the fossil series. Many return to the water, especially among the urodeles. Others become more fully terrestrial, either by climbing trees or by burrowing into the earth. Both these habits have been independently adopted many times by recently evolved lines and, no doubt, still more often in the past by creatures that have died out, leaving no trace.

9. Can Amphibia be said to be higher animals than fishes?

It is not easy to decide whether there is a clear sense in which amphibians can be said to have advanced over their fish ancestors. They have moved from the water into environments that are in a sense less suitable for life. In order to maintain a watery system, such as a frog or toad, outside the water, various special structures and methods of behaviour have been evolved. The presence of such additional systems can be said to add complexity to the organization. It is difficult to make a count of the number of 'parts' involved in the organization of any animal. Amphibia possess many special devices, for instance, for respiration without loss of moisture, for control of water intake and water loss, for return to water to breed, and so on. Even without making a proper quantitative computation it seems reasonable to say that these add up to make an organization more complicated than that of a fish. The integration of the action of so many parts requires an elaborate nervous system, and there is evidently some connexion between the increased size and importance of the nervous system and the development of this more complicated organization that enables life to continue in a different environment.

Considering the matter in this way it is hardly sensible to ask the question 'Are the amphibians more efficient than the fishes?', the work that they do in maintaining life is so different that a comparison of 'efficiency' is fallacious. One method of assessing living efficiency might be to judge each animal organization by the extent to which it maintains its constancy—by its power of homeostasis. Data about the fluctuations of the internal environment are so scanty among lower vertebrates that we cannot proceed very far on these lines. It is probable that the blood of fishes shows greater fluctuations, for instance in osmotic pressure or lactic acid content, than does that of amphibians, such fluctuations being perhaps even an advantage in allowing life in waters of differing salinity. In fact, to say that the whole mechanism of homeostasis becomes more complicated in land animals is only to say over again that they are 'higher' because they have more special work to do to maintain themselves in a difficult environment. Almost every part of the body shows signs of this greater complexity; the central nervous system becomes larger, the autonomic nervous system develops more elaborate control of the viscera. The endocrine glands become more numerous and differentiated, the muscular system shows more distinct parts, enabling the animal to act in new ways.

However difficult such comparisons may be it is hardly possible to deny them some validity. Amphibian organization differs from that of fishes and may be said to be 'higher' in the sense that it is more elaborate and allows life in conditions that the fish organization cannot tolerate.

XIV

LIFE ON LAND: THE REPTILES

1. Classification

Class. Reptilia

Subclass 1. Anapsida

Order 1. *Cotylosauria. Carboniferous–Trias

**Seymouria; *Captorhinus; *Diadectes*

Order 2. Chelonia. Permian–Recent

**Eunotosaurus; *Triassochelys; Chelys; Emys; Chelone; Testudo*

Subclass 2. *Synaptosauria

Order 1. *Protorosauria. Permian–Trias

**Araeoscelis; *Tanystropheus*

Order 2. *Sauropterygia. Trias–Cretaceous

**Lariosaurus; *Pliosaurus; *Plesiosaurus; *Placodus*

Subclass 3. *Ichthyopterygia

Order 1. *Ichthyosauria. Trias–Cretaceous

**Mixosaurus; *Ichthyosaurus*

Subclass 4. Lepidosauria

Order 1. *Eosuchia. Permian–Eocene

**Youngina; *Prolacerta*

Order 2. Rhynchocephalia. Trias–Recent

**Homoesaurus; *Rhynchosaurus; Sphenodon* (= *Hatteria*)

Order 3. Squamata. Trias–Recent

Suborder 1. Lacertilia (= Sauria). Trias–Recent

Infraorder 1. Gekkota. Mainly Recent

Gecko

Infraorder 2. Iguania. Cretaceous–Recent

Iguana; Anolis; Phrynosoma; Draco; Lyriocephalus; Agama; Chamaeleo

Infraorder 3. Scincomorpha. Eocene–Recent

Lacerta; Scincus; Amphisbaena

Infraorder 4. Anguimorpha. Cretaceous–Recent

**Dolichosaurus; *Aigialosaurus; *Tylosaurus; Varanus; Lanthanotus; Anguis*

1. Classification (*cont.*)

Suborder 2. Ophidia (= Serpentes). Cretaceous–Recent
**Palaeophis; Python; Natrix; Naja; Vipera*

Subclass 5. Archosauria

Order 1. *Pseudosuchia (= *Thecodontia). Trias
**Euparkeria; *Saltoposuchus*

Order 2. *Phytosauria. Trias
**Phytosaurus; *Mystriosuchus*

Order 3. Crocodilia. Trias–Recent
**Protosuchus; Crocodilus; Alligator; Caiman; Gavialis*

Order 4. *Saurischia. Trias–Cretaceous

Suborder 1. *Theropoda
**Compsognathus; *Ornitholestes; *Allosaurus; *Tyrannosaurus; *Struthiomimus*

Suborder 2. *Sauropoda
**Apatosaurus* (= **Brontosaurus*); **Diplodocus; *Yaleosaurus; *Plateosaurus; Brachiosaurus*

Order 5. *Ornithischia. Trias–Cretaceous

Suborder 1. *Ornithopoda
**Camptosaurus; *Iguanodon; *Hadrosaurus*

Suborder 2. *Stegosauria
**Stegosaurus*

Suborder 3. *Ankylosauria
**Ankylosaurus; *Nodosaurus*

Suborder 4. *Ceratopsia
**Triceratops*

Order 6. *Pterosauria. Jurassic–Cretaceous
**Rhamphorhynchus; *Pteranodon*

Subclass 6. *Synapsida. Carboniferous–Permian

Order 1. *Pelycosauria (= *Theromorpha)
**Varanosaurus; *Edaphosaurus; *Dimetrodon*

Order 2. *Therapsida. Permian–Jurassic
**Scymnognathus; *Cynognathus; *Bauria; *Dromatherium; *Dicynodon*

Order 3. *Mesosauria (= Proganosauria). Permian
**Mesosaurus*

2. Reptilia

Towards the end of the Devonian period, say 350 million years ago, the vertebrate organization produced a population of amphibian creatures and from this has been derived not only various modern groups classed as amphibia but also the more fully terrestrial populations that do not need to breed in water—the Amniota. Since that time many divergent lines have evolved from this stock, including the birds and the mammals, and it is evident, therefore, that it is likely to be difficult to specify what is meant by a reptile, as distinct from an amphibian or a bird or mammal. The term does not define a single vertical line of development or branch of an evolutionary tree, but is rather a horizontal division, marking a band on the evolutionary bush, specifying a level of organization beyond that of an amphibian but before that of either bird or mammal. Attempts have been made to divide the reptiles vertically into sauropsidan (bird-like) and theropsidan (mammal-like) lines, but such a division, although it has some foundation, obscures the fact that their bush-like evolutionary radiation has produced not two but many types.

The existing reptiles belong to four out of the dozen or more main lines that have existed. The most successful modern forms are placed in the order Squamata, the lizards and snakes, the latter being of relatively recent appearance in their present state. Secondly, the tuatara, *Sphenodon*, of New Zealand is a relic surviving with little change from the Triassic beginnings of this group. Thirdly, the crocodiles are an older offshoot from the stock from which the modern birds were derived. Finally the tortoises and turtles (Chelonia) have retained in some respects the organization of still earlier times, perhaps through the special protection of their shells. Though they are much modified in some ways, they still show us several characteristics of the earliest Permian reptiles.

These four modern types are all that remain of the reptiles that flourished throughout the Mesozoic, culminating in the giant dinosaurs of the Jurassic and Cretaceous. Evidently a profound change affected the population of the world, including the sea, between the end of the Cretaceous and the Eocene. This change will be discussed further in Chapter XXI, but we must briefly discuss here the possible relation of the decline of the reptile populations to the rise of their descendants, the birds and mammals. It can hardly have been only the more efficient organization due to the warm blood that gave these their opportunity, for there were forms in the Trias so similar to mammals in their skeletons that we may reasonably (though not certainly) suppose them to have been warm-blooded. There were birds with feathers in the Jurassic, and it is probable that they also already had warm blood. However, as a working hypothesis, we may suppose that the climate, which had been suitable for reptiles in the Mesozoic, became less so in the early Tertiary, and the most obvious suggestion is that colder conditions developed all over the earth's surface. The modern reptiles for the most part live in the temperate and tropical zones, indeed they flourish only in the latter. However, it must be remembered that climate fluctuates continually (p. 13); it is dangerous to make generalizations about conditions over such long periods as the Cretaceous.

3. The organization of reptiles

The organization we call reptilian is, generally speaking, suitable for life in warm countries, though two species, the common lizard (*Lacerta vivipara*) and the adder (*Vipera berus*), are found as far north as the Arctic Circle. No doubt the distribution of reptiles is limited largely by the fact that they cannot maintain a temperature above that of the surroundings by production of heat from within. The widespread idea that reptiles have no means of regulating their body temperature, however, has been overemphasized. Bogert and his collaborators in the U.S.A. have shown that in the wild (though not as a rule under laboratory conditions) reptiles are often able by suitable behaviour to maintain their body temperatures at a remarkably high and constant level throughout much of the day, by varying their exposure to the available sources of heat. When they get cold they bask in the sun or rest on warm rocks; when they get too hot they shelter under vegetation or in holes. In some species, too, colour change plays a part in temperature control, the animals becoming darker or lighter in colour, according to whether heat absorption or reflection is the appropriate response.

It has also been shown that each species of reptile has an optimum range of temperature, below which the animals become inactive and above which they quickly die. In some desert lizards the upper limit is above 40° C. The range tends, as one would expect, to be higher in diurnal than nocturnal forms, and is in general higher in lizards than it is in snakes or alligators.

The reptilian method of temperature control differs essentially from that of mammals in that it depends on the availability of external sources of heat such as the sun, rather than on the ability to conserve or lose heat generated within the body. For this reason reptiles are sometimes termed 'ectothermic' and mammals 'endothermic'. These terms are perhaps preferable to 'poikilothermic' (having a variable temperature) and 'homoiothermic' (having a constant temperature) which are in more general use.

The ectothermic method of temperature control presupposes some sensitive mechanism for registering slight changes in the temperature of the surroundings. There is evidence that the pineal complex is the receptor and that the hypothalamus may be involved in thermal homeostasis.

It remains true to say, however, that no reptile can retain an independent body temperature for a long period. For this reason, reptiles living in temperate climes must hibernate during the winter, while in warm countries some, conversely, aestivate during the hottest months.

4. Skin of reptiles

The skin is characteristically dry; unlike the skin of amphibians and mammals it contains few or no glands. The Malphigian layer of the epidermis produces the horny scales, which are periodically shed in flakes, or, as in snakes, cast as a single slough. Beneath the horny scales many reptiles (some lizards, crocodiles, some dinosaurs) develop bony plates in the dermis (called osteoderms). These may be restricted to the head, where they lie superficial to the skull bones, or may cover most of the body. The tortoise's shell contains both horny (epidermal) and bony (dermal) components (p. 394). The horny scales are often modified to form crests, spines and other appendages.

Many reptiles, particularly lizards and snakes, have bold and elaborate colour patterns. These may play a part in concealment (though they often seem conspicuous in captive specimens away from their normal terrain). In some forms, especially lizards, there are marked colour differences between the sexes (see p. 407). The well-known phenomenon of colour change, which is much more marked in certain lizards than in any other known reptiles, is discussed on p. 410.

5. Posture, locomotion, and skeleton

The elongated body and small laterally projecting legs of many reptiles recall those of a urodele, and the method of locomotion is in general similar in the two groups. Many retain the primitive five digits in both hand and foot. With the similarity of movement goes a general similarity in plan of the skeleton: there are, however, certain most significant features, characteristic of the reptiles. The head is carried off the ground, on a well-developed neck. The two first cervical vertebrae are modified to form the atlas and axis. The atlas is a ring of bone without centrum, but with a facet in front for the occipital condyle and one behind for the odontoid process, a peg attached to the front of the axis but derived in development from the centrum of the atlas segment.

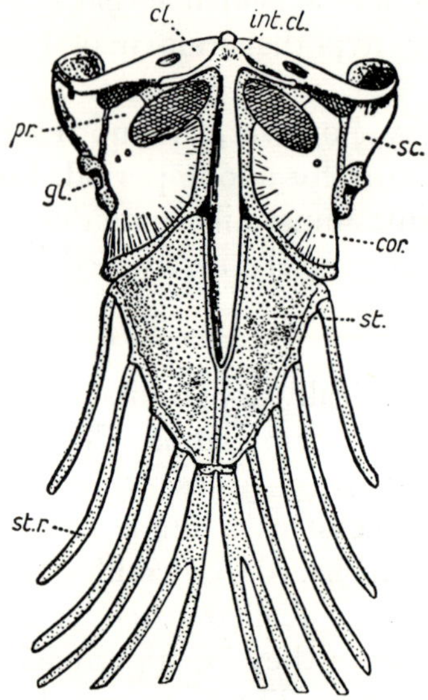

FIG. 212. Shoulder girdle and sternum of a lizard (*Iguana*).
cl. clavicle; *cor.* coracoid; *gl.* glenoid; *int.cl.* interclavicle; *pr.* procoracoid process; *sc.* scapula; *st.r.* sternal rib; *st.* sternum. (From Reynolds, after Parker.)

The vertebrae articulate with each other by a system of interlocking processes much more elaborate than that found in fish-like vertebrates and presumably serving to allow the column to carry weight. As a rule, each centrum is concave in front, covering the convex hind end of the vertebra next to it, a condition known as procoelous. In aquatic vertebrates the centra articulate by flat surfaces and this condition was retained in the amphicoelous vertebrae of many primitive reptiles. Besides this articulation of the centra the vertebrae are also united by the zygapophyses, facets on the neural arches, so arranged that the upwardly facing surfaces of the anterior zygapophyses slide over the down-facing surfaces of the posterior zygapophyses, an arrangement that is found throughout the amniotes.

Ribs are well developed in the middle or trunk region; each articulates with the body of the vertebra by a single capitular facet.

The vertebrae can often be divided into four sets. The cervical vertebrae, which are very variable in number, have short ribs, not reaching the sternum. Sternal ribs occur in the thoraco-lumbar segments (Fig. 212). The ribs of the two sacral vertebrae are short and broad and articulate with the ilia. The numerous caudal vertebrae show reduction of all parts, especially towards the tip of the tail. The chevron bones are ossicles attached to the caudal centra and representing the reduced intercentra.

The girdles and limbs (Figs. 183–6 and 212) show the same general structural and functional features as those of amphibia. The limbs form the main locomotor system, the metachronal contraction of the myotomes playing a lesser part than in urodele amphibians. The humerus and femur are normally held in such a position that their outer ends lie higher than the inner, that is to say, in a position of abduction. The radius and ulna and tibia and fibula proceed downwards towards the ground (at right angles to the proximal bones) and the hand and foot are turned outwards at right angles, to rest on the ground. The main muscles thus draw the humerus and femur backwards and forwards as well as downwards, and the ventral regions of the girdles are large and flattened to receive these muscles; in mammals, with a different system of progression, the more dorsal parts of the girdles have become developed.

The pectoral girdle (Figs. 183, 212) consists of a dorsal scapula and a large ventral coracoid, which may be fenestrated. Distinct pro- and post-coracoid elements are probably only found in the extinct mammal-like reptiles. The dermal components are represented by the paired clavicles and median interclavicle. A cleithrum is found in a few very primitive forms.

In the pelvic girdle (Fig. 184) the usual dorsal ilium, anterior pubis, and posterior ischium are found, the last two meeting their fellows in midline symphysis.

The characteristic modifications of the reptilian skull are discussed on p. 391. The general plan is similar to that of primitive amphibians, but in all except the most primitive reptiles there is a development of holes (fossae) in the temporal region to provide space for the bulging temporal muscles. The skull roof is in some respects more primitive than that in modern amphibians (Figs. 213 and 214). It is made up of a large series of dermal bones, including the nasals, prefrontals, frontals, supra-orbitals, and parietals. The side of the skull is usually less complete, composed of the tooth-bearing premaxilla and maxilla, lacrymal, jugal, post-orbital, squamosal, supratemporal and quadrate. The naming of some of the smaller bones round the orbit and above the quadrate is a matter of controversy.

The margins of the palate are formed by flanges of the premaxillae

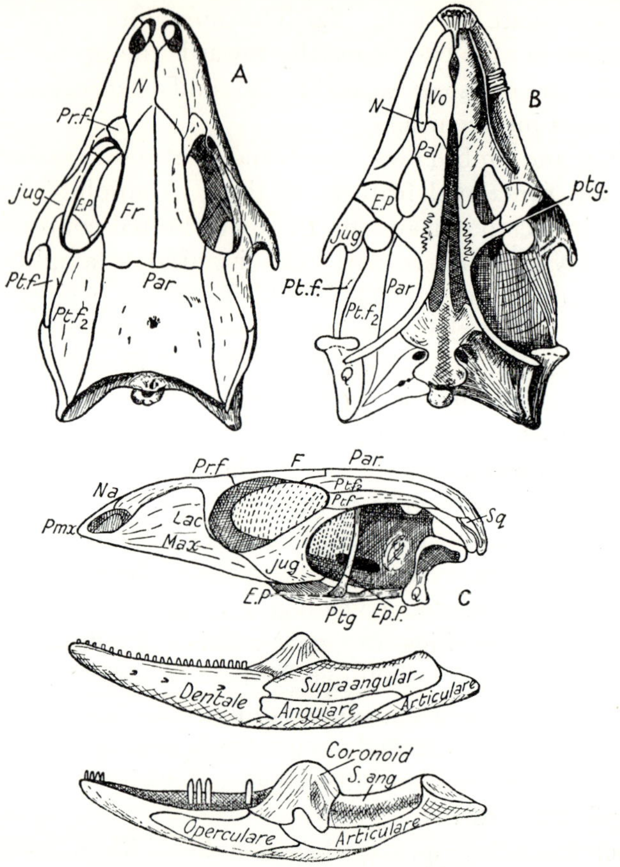

FIG. 213. Skull and lower jaw of *Lacerta*.

A, dorsal view; B, ventral view; C, from left side; D, right half of lower jaw from inner side, showing the pleurodont arrangement of the teeth. *E.P.* ectopterygoid; *Ep.P.* epipterygoid; *F, Fr.* frontal; *jug.* jugal; *Lac.* lachrymal; *Max.* maxillary; *N, Na.* nasal; *N,* in B, inner narial opening; *Pal.* palatine; *Par.* parietal; *Pmx.* premaxillary; *Pr.f.* prefrontal; *Pt.f.* post-orbital; *Pt.f*$_2$. post-frontal; *Ptg.* pterygoid; *Q.* quadrate; *S.ang.* supra-angular; *Sq.* squamosal; *Vo.* vomer. The regions of persistent cartilage are not shown in detail. (After Gadow.)

and maxillae and the small ectopterygoids. The internal nostrils usually lie forwards between the maxillae, vomers, and palatines. More posteriorly the floor of the skull is made up mainly by pterygoid bones and the parasphenoid, which is partly fused with the lower surface of the basisphenoid. Occipital bones surround the foramen magnum and make up the single occipital condyle, which in some forms is indented to form three partly distinct lobes. In many reptiles there is an epipterygoid bone on either side of the brain-case behind the orbits; this is regarded as an ossification in the ascending process of the palato-quadrate. The lower jaw usually consists of six bones, the articular forming the joint with the quadrate, and the dentary carrying teeth.

The anterior part of the chondrocranium, surrounding the front of the brain, and the nasal capsule, remain more or less unossified, and

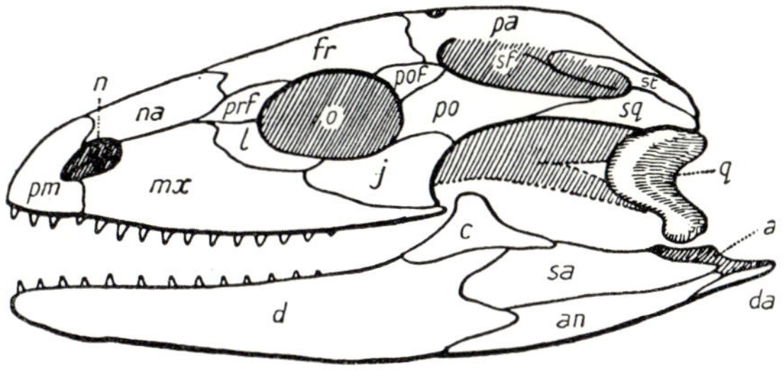

FIG. 214. Diagram of the skull of lizard to show temporal fossa.

a. articular; *an.* angular; *bo.* basioccipital; *bs.* basisphenoid; *c.* coronoid; *d.* dentary; *da.* dermal articular; *do.* dermal supraoccipital; *ept.* ectopterygoid; *eo.* exoccipital; *fr.* frontal; *j.* jugal; *l.* lachrymal; *mx.* maxilla; *n.* nostril; *na.* nasal; *o.* orbit; *op.* opisthotic; *pa.* parietal; *pal.* palatine; *pv.* prevomer; *pm.* premaxilla; *po.* post-orbital; *pof.* post-frontal; *pr.* pro-otic *pra.* prearticular; *prf.* prefrontal; *ps.* presphenoid; *pt.* pterygoid; *q.* quadrate; *qj.* quadrato-jugal; *sa.* surangular; *sf.* upper temporal fossa (this is shown diagrammatically, as it occurs in many lizards; in *Lacerta* it is largely covered by an extension of the post-frontal—see Fig. 213); *so.* supraoccipital; *sp.* splenial; *sq.* squamosal; *st.* supratemporal; *v.* vomer. (From Goodrich.)

in places may be membranous. There may, however, be small ossified orbitosphenoids and farther back pleuro- or laterosphenoids, which develop in the pila pro-otica uniting the orbital cartilage with the otic capsules. Between the eyes there is in most reptiles a thin sheet of cartilage known as the interorbital septum, which may be partly ossified by small presphenoid elements. The posterior part of the chondrocranium ossifies to form the following bones; occipital complex, basisphenoid, and the ossifications in the otic capsule (pro-otic, opisthotic, &c.).

In many reptiles the upper jaw and front part of the skull can move to some extent in relation to the occipital region and cranial base, such movement being termed kinesis (p. 405). This is often associated with mobility of the quadrate, as in lizards, snakes, and certain dinosaurs. Kinesis helps to widen the gape and may provide a shock-absorbing effect when the jaws are snapped together.

The postmandibular visceral arches play no part in jaw support but are incorporated into the ear and hyoid apparatus. There is a rod-like columella auris with a small cartilaginous element (extra-columella) at its outer end. The columellar system usually conducts vibrations from the tympanum, lying behind the quadrate, to the fenestra ovalis and inner ear. In some forms, e.g. snakes, however, the tympanum is absent and the outer end of the columella is applied to the quadrate. These animals may be deaf to air-borne sounds but sensitive to ground vibrations, transmitted through the bones of the jaw.

The hyoid apparatus consists of a basal plate, which projects into the tongue, and three pairs of ascending horns. These represent the remains of the hyoid and branchial arches.

6. Feeding and digestion

Food is seized either by the teeth or, in some specialized lizards such as the chameleon, with the elongated tongue. The teeth are situated along the edges of the jaws and often also on some of the bones of the palate. Typically, they are all of the same conical shape, but may be slightly serrated, or modified to form crushing plates, poison fangs, and other devices. As a rule, tooth succession is continuous throughout life, though exceptions to this are found among the lizards. Salivary glands are well developed in some forms; in snakes and one genus of lizards (*Heloderma*) some of them are modified to form poison-glands. The tongue is very variable, being hardly movable in some reptiles (e.g. crocodiles) but long, forked, and highly mobile in others (e.g. snakes).

Digestion proper begins in the stomach. The alimentary canal is built on the typical vertebrate plan, with a tubular stomach, rather short small intestine, and wider large intestine, leading to a short caecum.

There is a well-marked cloacal chamber in all reptiles, subdivided into a coprodaeum for the faeces, and a urodaeum for the products of the kidneys and genital organs. These two chambers open into a final common proctodaeum, closed by a cloacal sphincter. This division of the cloaca is associated with the necessity for the retention of water, the cloacal chambers serving for water reabsorption from both the faeces and urinary excreta (p. 380).

7. Respiration, circulation, and excretion

The typical method of respiration is a backward movement of the ribs, produced by the muscles attached to them. There is no complete separation of the thorax from the abdomen, but a partial diaphragm may be present. The glottis is a slit at the back of the mouth and leads into a larynx with supporting cricoid and arytenoid cartilages. Many reptiles are able to produce small sounds, but the voice-box is less developed than in either amphibia or birds.

The lungs are sacs whose walls are folded into ridges, separating a number of chambers or bronchioles. The hinder part of the lung is

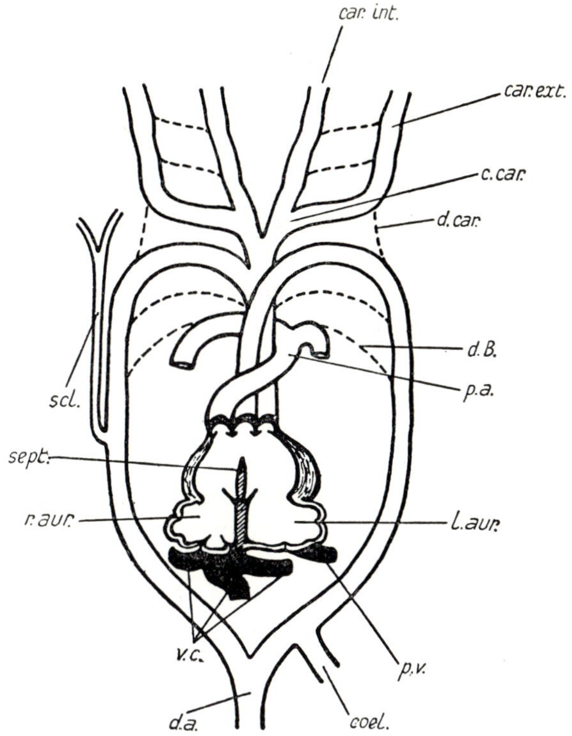

FIG. 215. Diagram of heart and arteries of *Lacerta*.

car.ext. external carotid; *car.int.* internal carotid; *c.car.* common carotid; *coel.* coeliac artery; *d.a.* dorsal aorta; *d.B.* ductus Botalli (arteriosus); *d.car.* ductus caroticus; *l.aur.* left auricle; *p.a.* pulmonary artery; *p.v.* pulmonary vein; *r.aur.* right auricle; *scl.* subclavian artery; *sept.* interventricular septum; *v.c.* caval veins. (From Ihle, after Goodrich.)

nearly smooth and in some lizards, as in birds, it becomes developed into characteristic air sacs. The tendency is for the more anterior portion of the lung to become the effective vascular and respiratory region, allowing the air-stream to be drawn across it at both inspiration and expiration.

The circulation (Fig. 215) of lizards shows a partial separation of venous and arterial blood; there are two auricles, but only one ventricle, this being partly divided by a septum into right and left sides. Three arterial trunks arise directly from the ventricle, these being the right and left aortae, and the pulmonary trunk. The opening of the latter lies opposite the right side of the ventricle and receives predominantly venous blood. The left systemic arch opens opposite to the incomplete ventricular septum and receives mixed blood, whereas the right systemic arch opens from the left side of the ventricle and carries almost pure arterial blood. The carotid arteries of both sides arise from the right systemic arch. This 'classical' view of the circulation of the blood in the reptilian heart has recently been confirmed by the radiographical studies of Foxon and his colleagues on the green lizard. Variations may be found, however, in different species.

The venous system is based on the same plan as that of the frog, with pelvic veins receiving blood from the tail and hind legs and returning it to the heart through either an anterior abdominal vein or renal portals, and the inferior vena cava.

In the urinogenital system is seen another feature characteristic of amniotes, the development of a posterior region, the metanephros, concerned solely with excretion, leaving the mesonephric (Wolffian) duct to function as the vas deferens in the male. There is sometimes an endodermal (allantoic) bladder.

The waste nitrogen is largely excreted as uric acid and this allows the reabsorption of much of the water in the urodaeum, with precipitation of the organic matter as a chalky white mass of urates. The advantage of this method of excretion is that it allows for a greater economy of water than would be possible if the end product was the more soluble urea. There is, however, some variation in the mode of excretion among certain members of the group and this may depend on the manner of life of the species and the necessity for water conservation. Thus among Chelonia the more aquatic forms (*Emys*) produce considerable amounts of ammonia and urea, but relatively little uric acid, whereas the last is the main excretory product of the fully terrestrial types, such as the Grecian tortoise (*Testudo graeca*), which can live under almost desert conditions.

8. Reproduction of reptiles

Fertilization has become internal, and in all modern reptiles except *Sphenodon* special organs of copulation derived from the cloacal wall are developed in the male. In crocodiles and tortoises there is a single median penis, but in lizards and snakes there are a pair of these structures, though only one is inserted at a time. The mechanism of erection involves both muscular action and vascular engorgement. The sperms pass from the vasa deferentia into the urodaeum, and after traversing this region they are carried into a groove along each penis. In snakes the sperms may survive within the female for long periods, and instances are known of isolated individuals laying fertile eggs after months, sometimes even years, in captivity.

Some of the most serious difficulties in the colonization of the land are concerned with reproduction, and these problems have been largely solved in the reptiles, allowing the animals to reproduce without returning, as many amphibia must do, to the water.

The eggs of oviparous reptiles are always laid on land. They therefore require a firm physical support and protection against desiccation, as well as an adequate supply of food and special means of gaseous exchange and storage of waste products. These requirements are met by the development of a shell, secreted by the walls of the oviduct and often hardened by lime impregnation, by the formation of special embryonic membranes, the amnion and allantois, and by the provision of a large quantity of yolk enclosed in a bag, the yolk-sac. The method of embryonic cleavage is affected by the great amount of yolk, and as in birds is only partial. An albumen or egg-white layer is present in the eggs of crocodiles and tortoises and presumably serves as a reservoir of water; in the eggs of lizards and snakes, however, the albuminous layer is poorly developed or absent.

The formation of the amnion and allantois is one of the most remarkable features of the development of reptiles; it is characteristic of all higher vertebrates, distinguishing them sharply from the lower types. The amnion is developed from folds, which cover the embryo and enclose a sac filled with fluid, where development can proceed in the absence of the pond that was necessary for the earlier vertebrates. The allantois began as an enlarged bladder, serving for the reception of the waste products during the life within the shell. Coming close to the surface and fusing with the chorion, it then becomes the vehicle for the transport of oxygen to the embryo.

The evolution of eggs and embryonic membranes of the kind described must have been an event of critical importance in tetrapod history. Romer has suggested that this advance took place under climatic conditions of alternate drought and flooding, so that eggs laid above the high-water mark had the best chance of survival. Since many of the early reptiles are thought to have spent much of their time in the water, it is possible that the egg preceded the adult in the process of adaptation to terrestrial life.

Most reptiles lay their eggs, but in many lizards and snakes these are retained within the oviduct until the young are ready or nearly ready to hatch (e.g. *Lacerta vivipara*, *Anguis fragilis*, *Vipera berus*). This method of reproduction is termed ovoviviparous; in forms that practise it the eggshell is reduced to a thin membrane or is lost altogether. In some species (e.g. certain skinks and other lizards, sea-snakes) a placenta is developed from the chorio-allantois or the yolk-sac or both. The placenta may, as in *Lacerta vivipara*, serve only for the transfer of water and gases, but in the more advanced forms it probably provides a means of transport for food (supplementing the yolk) and excretory products.

Young born alive are perhaps less susceptible to the hazards of weather than those left to hatch in the sun or among rotting vegetation, and it is interesting that most, if not all, of the few reptiles that live in places where the climate is really severe are ovoviviparous.

Young reptiles have special devices to assist their escape from the egg. In *Sphenodon*, Chelonia, and Crocodilia, as in birds, there is a horny epidermal egg-breaker on top of the snout tip, called the egg-caruncle. In the Squamata, a true egg-tooth, projecting from the front of the upper jaw, has the same function. The egg-tooth is present, though sometimes rudimentary, in ovoviviparous forms.

Some reptiles make a simple nest but the group is not noted for maternal care, usually abandoning their new-laid eggs or newborn young. There are, however, some exceptions to this; female pythons and certain other snakes and lizards brood their eggs, and female alligators are said to guard their nests.

Many reptiles exhibit well-marked courtship and display phenomena during the breeding-season, the males fighting and displaying, either to intimidate each other or to evoke a suitable response from the female. This is particularly striking in certain lizards, notably those of the iguanid and agamid groups, where the males are often brightly coloured and may be adorned with crests and distensible fans under the throat. In these lizards bobbing movements of the head and front part of the body, often accompanied by colour change, form an important part of the display. As in birds, courtship may be associated with territory, a male holding an area of ground on which females, but not rival males, are tolerated. Breeding behaviour and sexual coloration are, of course, under the control of the endocrine system, especially the anterior pituitary and the gonads, and may be modified by castration. The onset of the breeding season is also influenced by climatic conditions; most reptiles breed only once or twice a year, but a few species living in warm stable climates may breed at intervals nearly all the year round.

9. Nervous system and receptors of reptiles

All the modifications of structure that fit the reptiles for life on land would be useless without the development of appropriate behaviour. This in turn depends on suitable structure and function of

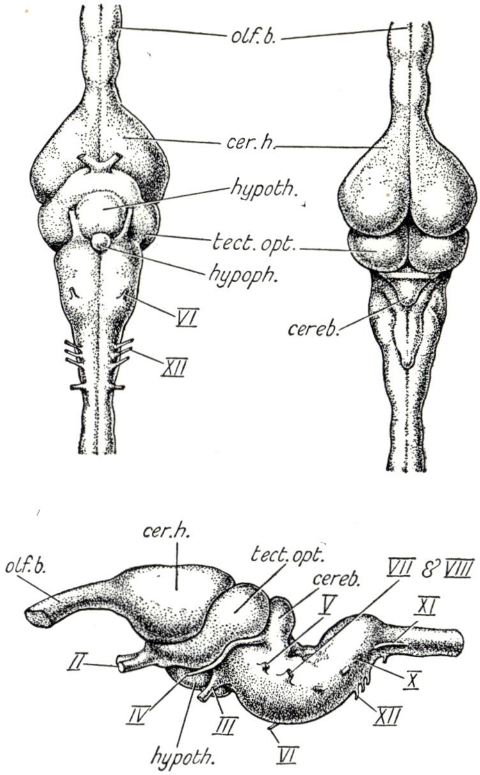

FIG. 216. Three views of the brain of a lizard.
cereb. cerebellum; *cer.h.* cerebral hemispheres; *hypoph.* hypophysis; *hypoth.* hypothalamus; *olf.b.* olfactory bulb; *tect.opt.* tectum opticum; *II–XII,* cranial nerves. (After Frederikse.)

the nervous system. The brain accordingly shows some interesting developments. The cerebral hemispheres are relatively larger in reptiles than in amphibians (Fig. 216). The increased bulk lies mainly in the basal parts of the hemisphere (the corpus striatum), as in birds (Fig. 217). The roof (pallium) is little developed and lacks the elaborate cortical differentiation found in mammals. The thalamus is well developed and receives connexions from the optic tracts, which no longer run mainly to the midbrain. There are also many fibres from the thalamus to the cerebral hemispheres. This, together with the other features mentioned, may be evidence for the transfer of many nervous functions from lower levels of the nervous system to the cerebral hemispheres, a process that has been carried much farther in mammals. In most respects, however, the brains of modern reptiles are more like those of birds than of mammals.

The behaviour of reptiles, despite the elaborate courtship of some forms, remains of a relatively stereotyped character. The eyes are usually the main exteroceptive sense-organs and are usually provided

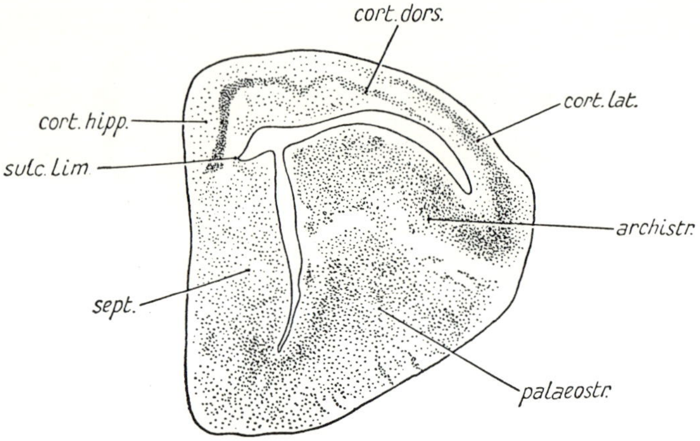

Fig. 217. Transverse section through forebrain of *Lacerta*.

archistr. archistriatum (upper part of corpus striatum); *cort.dors.* dorsal cortex; *cort.hipp.* medial (hippocampal) cortex; *cort.lat.* lateral (pyriform) cortex; *palaeostr.* palaeostriatum (basal part of corpus striatum); *sept.* septum; *sulc.lim.* sulcus limitans. (After de Lange and Kappers.)

with movable eyelids, including a third eyelid or nictitating membrane. The lacrymal and Harderian glands provide secretions that keep the surface of the cornea moist. The eye is supported in most reptiles by a scleral cartilage and a ring of bony scleral plates. Accommodation is produced by the striated ciliary muscles, so arranged that they cause the ciliary process to squeeze the lens, making its anterior surface more rounded (Fig. 218). In many reptiles the retina possesses both rods and cones, the latter predominating in diurnal types.

The pineal complex is often well developed, and in *Sphenodon* and many lizards a 'pineal' or parietal eye is present with lens-like and retina-like components. In such forms there is a pineal foramen in the parietal bone near the fronto-parietal suture. Similar foramina are found in many fossil reptiles, especially the more primitive types. The function of the reptilian pineal is still rather obscure, but there is evidence that in lizards it registers solar radiations, and, perhaps by the secretion of hormones, influences the animal's thermoregulatory behaviour in exposing itself to sunlight. It is also possible that the pineal complex plays some part in the control of reproduction.

In the majority of reptiles the olfactory region of the nose is quite well developed, but, except in crocodiles, there is only a single nasal concha. The organ of Jacobson (vomero-nasal organ), a specialized and sometimes separate region of the nose, innervated by a separate

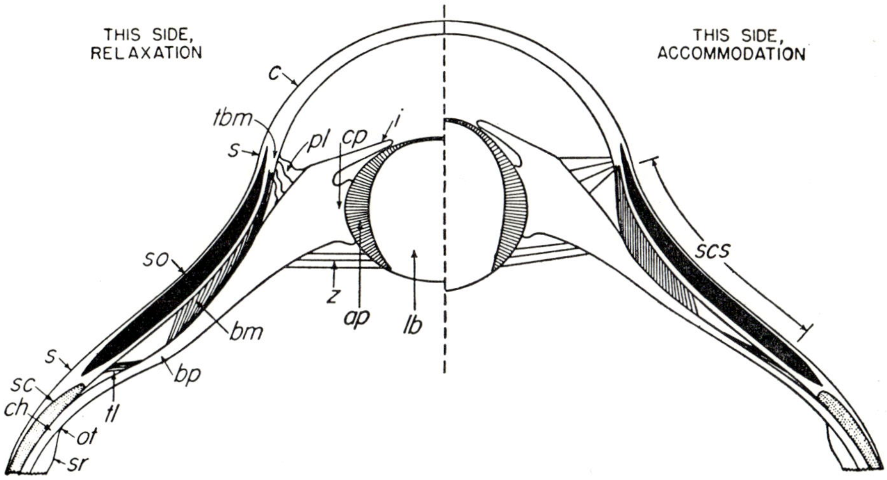

FIG. 218. Diagram to show the mechanism of accommodation in the eye of reptiles.

ap. annular pad of lens; *bm.* Brucke's muscle; *bp.* base plate of ciliary body; *c.* cornea; *ch.* chorioid; *cp.* ciliary process; *i.* iris; *lb.* lens body; *ot.* ora terminalis; *pl.* pectinate ligament; *s.s.* sclera; *sc.* scleral cartilage; *scs.* sclerocorneal sulcus; *so.* scleral ossicle; *sr.* sensory retina; *tbm.* tendon of Brucke's muscle (continuous with inner layers of corneal substantia propria); *tl.* tenacular ligament; *z.* zonule. (From Walls, *The Vertebrate Eye.*)

branch of the olfactory nerve, is present in turtles, *Sphenodon*, and Squamata. In the latter it is usually very highly developed (see p. 405).

The tympanum when present lies at the back of the jaws, sunk a little below the surface. The range of response to sound waves is not known in lizards, but the ears of certain tortoises are very sensitive to sound over a narrow range of about 110 cycles per second; apparently there is some resonating mechanism, perhaps the columella auris, which vibrates at this frequency. Generally speaking, the sense of hearing is best developed among reptiles in the Crocodilia and certain lizards.

XV

EVOLUTION OF THE REPTILES

1. The earliest reptile populations, Anapsida

THE organization of a reptile is well suited to maintain life on land. Many features show a considerable advance in this respect over the amphibia, for example, the dryness of the skin, the method of reproduction, and the devices for economizing in the use of water. The immense radiation of the reptiles into every sort of land habitat during the Mesozoic period shows the efficiency of these mechanisms, which were probably present, at least in imperfect form, in the earliest Carboniferous and Permian offshoots from the ancestral Stegocephali (p. 356).

We have sufficient knowledge to be able to trace the early stages of reptilian evolution with considerable certainty. An animal known as **Seymouria*, found in the lower Permian of Texas (perhaps 250 million years old), is of critical importance in our understanding of reptile origins (Figs. 219, 220). It was a lizard-like creature, about 2 ft long, probably living on insects and perhaps some larger animals. Its characteristics are so exactly intermediate between those of amphibians and reptiles that it is not possible to place it definitely with either group; many zoologists class it with the Amphibia. This intermediacy is shown in almost every structure of the body and is often a subtle matter of the shape or size of the parts. Although a list of anatomical features is apt to give an unreal picture of any living organization it is the only method available to us in the absence of any more ingenious calculus, and we may therefore give first some of the characteristically reptilian features of **Seymouria*. (1) The neural arches of the vertebrae were convex dorsally so that they have a 'swollen' appearance which is also seen in early reptiles; (2) a canal for the lachrymal duct was present; (3) the occipital condyle was single; (4) the pectoral girdle possessed a long interclavicle; (5) the pelvic girdle was attached to the vertebral column by two sacral vertebrae; (6) the blade of the ilium was expanded for the attachment of the large muscles used in walking; (7) there were five digits in the hand instead of four as in many labyrinthodonts and in living amphibia; (8) the phalangeal formula was 2:3:4:5:3 or 4, approximating to the reptilian, rather than the usual labyrinthodont condition, in which the phalanges are less numerous.

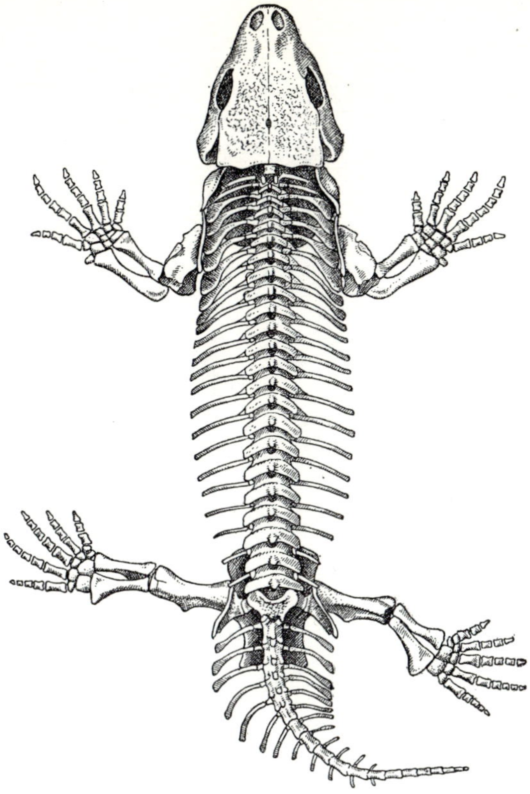

FIG. 219. Skeleton of *Seymouria*. Actual length 20 in.
(From Williston, *Osteology of the Reptiles*, Harvard University Press.)

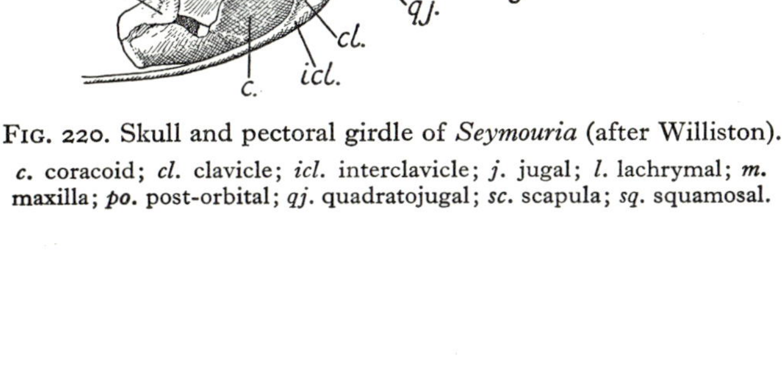

FIG. 220. Skull and pectoral girdle of *Seymouria* (after Williston).

c. coracoid; *cl.* clavicle; *icl.* interclavicle; *j.* jugal; *l.* lachrymal; *m.* maxilla; *po.* post-orbital; *qj.* quadratojugal; *sc.* scapula; *sq.* squamosal.

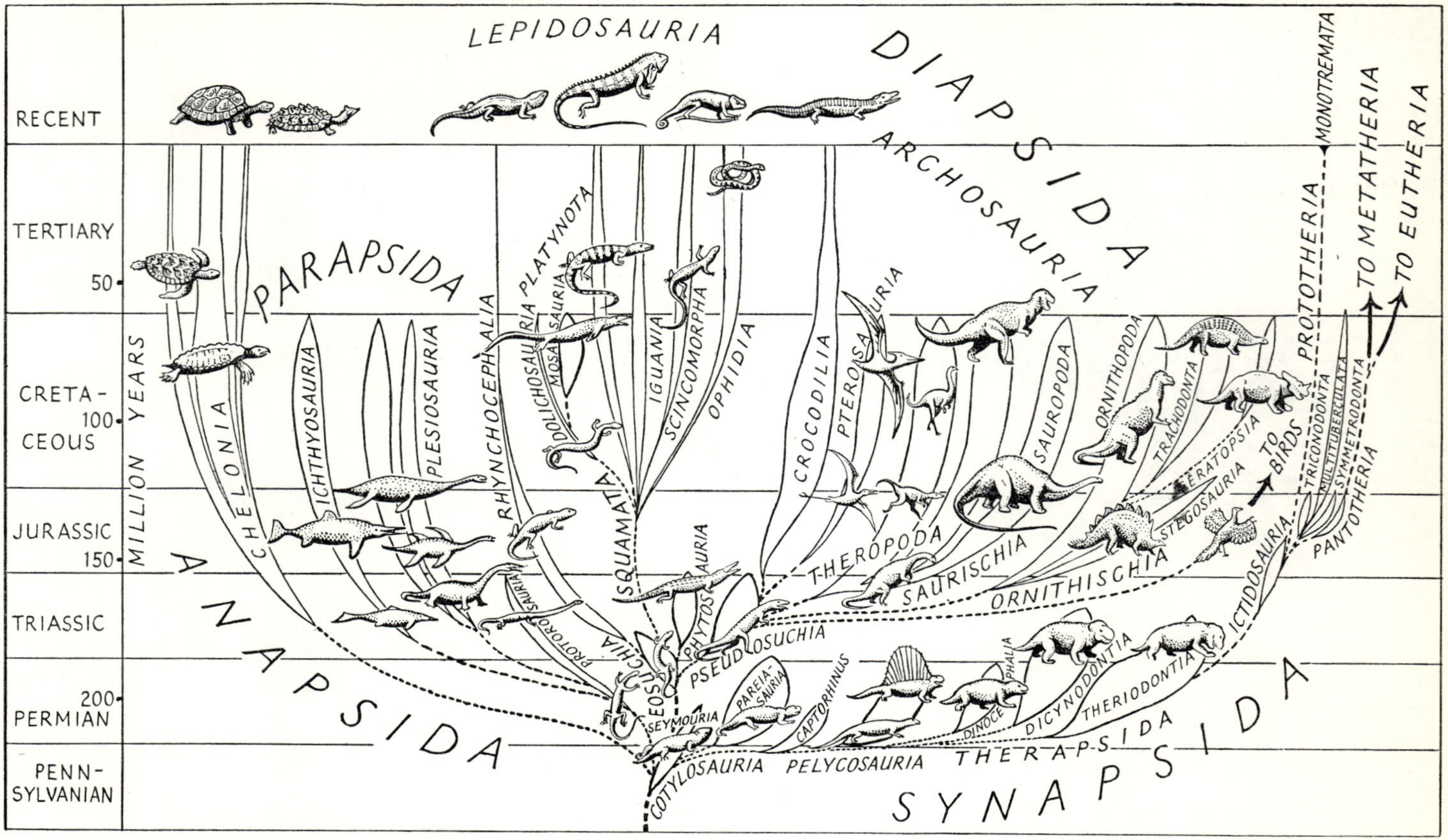

FIG. 221. Chart to show the course of evolution of the reptiles.

On the other hand, there are many considerations that would lead one to classify this fossil as an amphibian. (1) The skull bones were conspicuously pitted or 'sculptured'; (2) the pattern of the bones was very like that of early amphibians; for instance an intertemporal bone was present; (3) the teeth still showed a labyrinthine structure and the palatal teeth were distributed in pairs as in many labyrinthodonts; (4) the structure of the otic notch, across which the tympanic membrane was stretched, and certain other features of the auditory apparatus suggest amphibian rather than reptilian affinities; (5) other fossils are known which, though clearly related to **Seymouria*, show trends such as flattening and reduced ossification of the skull that are characteristic of late labyrinthodonts rather than early reptiles; (6) perhaps the most significant point of all is that some adult specimens show signs of the presence of lateral line canals. Since the structure of the skeleton of **Seymouria* suggests terrestrial habits, the presence of these canals suggests that the animal may have passed through an aquatic larval stage in which they were functional. Hence the case for classifying **Seymouria* as an amphibian becomes very strong.

It must be added that **Seymouria* may show other features characteristic of both early amphibians and primitive reptiles. The neck, for example, was short, with the pectoral girdle lying close behind the skull. There was little differentiation between the vertebrae, all those in the cervical region bore ribs. The ribs were double-headed, a feature that has been retained by some more advanced reptiles.

**Seymouria* itself existed too late to have been a direct ancestor of the more advanced groups of reptiles, since some of the latter had already appeared by the early Permian. Whether it should actually be classified as an amphibian or a reptile is uncertain; in Romer's recent *Osteology of the Reptiles* it and its allies are placed in the latter group. It may be regarded as a most interesting link between the labyrinthodonts and the primitive stem-reptiles (Cotylosauria), from which soon arose a great variety of descendants, which came to dominate not only the land but also the sea and air throughout the subsequent Mesozoic period.

The cotylosaurs must have existed throughout the later part of the Carboniferous, though they did not become prominent before the beginning of the Permian. From the Red Beds of Texas we have, besides **Seymouria*, forms such as **Limnoscelis*, **Captorhinus*, and **Labidosaurus*, all with the rather high narrow skulls and pointed nose characteristic of reptiles rather than amphibians, but differing from the latter in the absence of the otic notch. This and other features suggest that these animals were related to the early mammal-like reptiles (p. 539). *Limnoscelis*, however, was a particularly primitive form, and may have been partly aquatic in habits.

Other rather later cotylosaurs were larger forms, such as **Diadectes* (Fig. 222) and **Bradysaurus* and other 'pareiasaurs' from the Permian and Triassic of Europe, Africa, and America. These were up to 10 ft

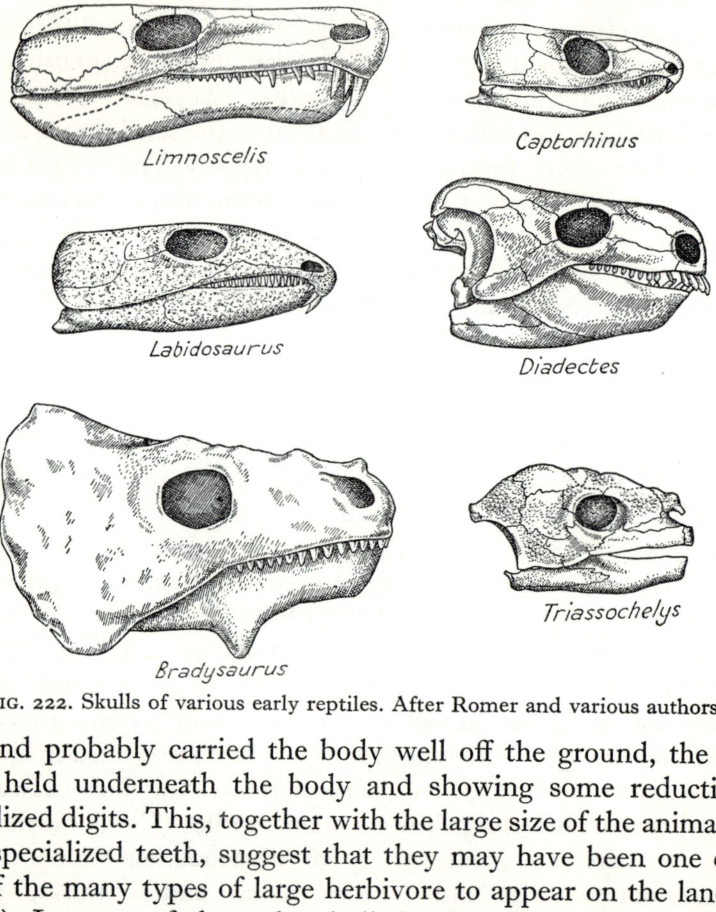

FIG. 222. Skulls of various early reptiles. After Romer and various authors.

long and probably carried the body well off the ground, the limbs being held underneath the body and showing some reduction of specialized digits. This, together with the large size of the animals and their specialized teeth, suggest that they may have been one of the first of the many types of large herbivore to appear on the land (see p. 429). In some of them the skull developed grotesque protective protuberances, a feature recalling similar later developments in reptiles (Ceratopsia, p. 426) and mammals (amblypods, p. 717). A characteristic structural feature was the presence of an otic notch low down on the side of the skull. This distinguishes the pareiasaurs from the more mammal-like forms and suggests affinity with some of the other reptilian descendants of the early cotylosaurs. These cotylosaurs multiplied and became very diversified throughout the 45 million years of the Permian period, by which time the main reptilian types had appeared. The individual reptilian orders nearly all became established during the subsequent 45 million years of the Trias and most of them reached their maximum development in the Jurassic and Cretaceous.

2. Classification of reptiles

Since our knowledge of reptiles depends mainly on fossil remains it is convenient to classify them by means of the skull into four great

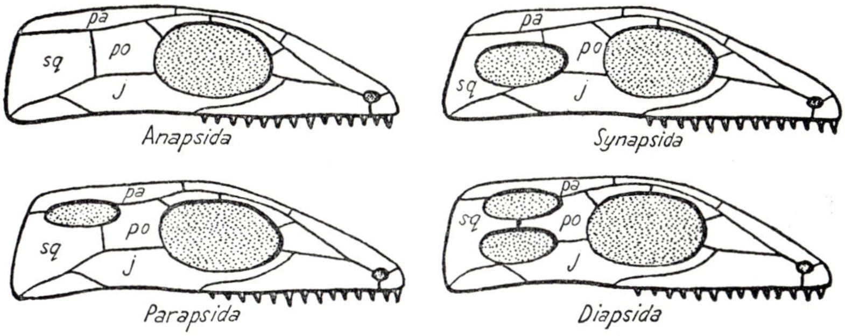

FIG. 223. Diagrams of reptilian skulls to show arrangement of the temporal openings. Anapsida, no opening. Synapsida, a lower opening, with post-orbital and squamosal meeting above it. Parapsida, an upper opening with post-orbital and squamosal meeting below it. Diapsida, two openings, separated by a bar.

.jugal; *pa.* parietal; *po.* post-orbital; *sq.* squamosal. (From Romer, *Vertebrate Paleontology*, Chicago University Press.)

groups (Fig. 223). Such a classification is in some ways artificial, but it serves to indicate in a broad way the main lines of evolution within the class.

In the cotylosaurs the dermal bones of the temporal region of the skull presented an unbroken surface and there were no temporal fossae. There were therefore no arches or 'apses' of bone in the temporal region. Such forms are placed in the subclass Anapsida. The jaw muscles took origin from the deep surface of the temporal side wall, between it and the brain-case, and they passed down through holes in the palate to be inserted on the lower jaw. This represents the most primitive condition found in reptiles, and resembles that in the early amphibians. It is still seen today, though often in a modified form, in the Chelonia, which are hence placed in the anapsid subclass.

In more advanced groups of reptiles fossae bounded by bony arches appear in the temple region, enabling the jaw muscles to extend through them on to the outer surface of the skull, an arrangement that increases their mechanical advantage.

In many reptiles, two such fossae appeared, the condition being termed diapsid. This is seen in the subclasses Lepidosauria and Archosauria, perhaps the most successful groups of reptiles. In lepidosaurs of the order Squamata, however, the lower temporal arch is always incomplete, having no quadrato-jugal bone and the jugal separated from the squamosal. In some lizards and in snakes the upper arch is also lost.

In other groups only a single fossa and arch is present. When this is situated high on the skull the condition is known as parapsid. Parapsid skulls are seen in the subclasses Ichthyopterygia (icthyosaurs) and Synaptosauria (plesiosaurs, &c.). Formerly, these two subclasses were placed together in a group known as the Parapsida, as is shown in Fig. 221, but this classification is now regarded as artificial, since the ichthyosaurs and sauropterygians are not closely related; in fact a careful analysis shows that the bony relationships of their single temporal fossae were rather different.

In the remaining subclass, the Synapsida, there is also a single fossa, but in the earlier forms at least it is placed low down, and is bounded below by the jugal and squamosal. The term synapsid, meaning 'fused arch', is actually a misnomer, due to the fact that early workers believed, wrongly, that the single arch was formed from the fusion of the two seen in diapsids.

The synapsids comprise the mammal-like reptiles, but in the later members of the group, such as **Cynognathus*, and in their descendants the mammals, the temporal fossa has greatly enlarged, and has lost its primitive relationships.

3. Order 1. Chelonia

Shut away in their boxes the tortoises and turtles have retained some of the features of the earliest anapsid reptiles. Even today they are a not unsuccessful and quite varied and widespread group, with more than 200 species. These include terrestrial animals, such as *Testudo graeca*, the Grecian tortoise of south Europe, which is herbivorous; the freshwater tortoises, such as *Chrysemys* and other American terrapins, and *Emys* the European water-tortoise, all of which are carnivorous. The marine Chelonia, usually known as turtles, are often very large. *Dermochelys*, the leathery turtle, which has no horny shell, is over 6 ft long and weighs half a ton. *Chelone mydas*, the green or edible turtle, is over 3 ft long.

The characteristic of chelonian organization is the shortening and broadening of the body, together with the development of bony plates, forming a box into which the head and limbs can be withdrawn. The total number of segments is only about 8 in the neck, 10 in the trunk, and a series of reduced caudals; the body is therefore

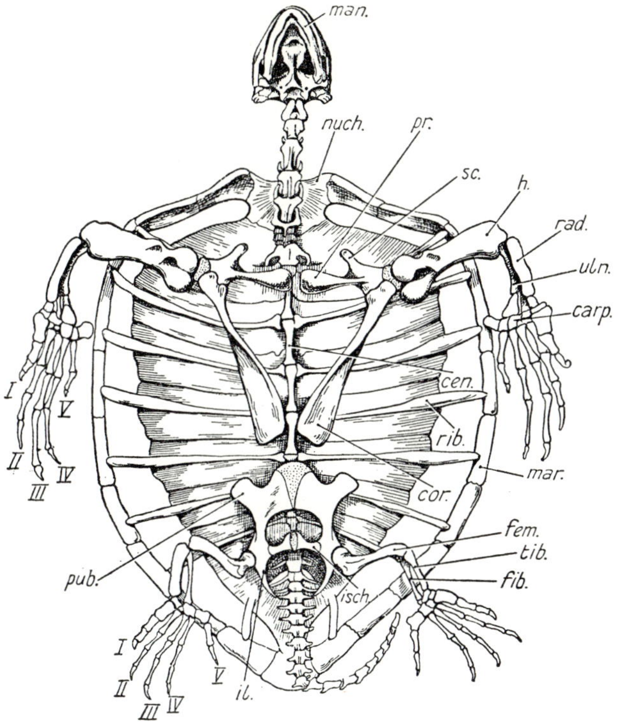

FIG. 224. Skeleton of turtle (*Chelone*).

carp. carpus (note hook-shaped 5th metacarpal); *cen.* centrum of vertebra; *cor.* coracoid; *fem.* femur; *fib.* fibula; *h.* humerus; *il.* ilium; *isch.* ischium; *mar.* marginal plate; *nuch.* nucal plate; *pr.* 'proscapular' process or acromion; *pub.* pubis; *rad.* radius; *rib,* rib, partly fused with costal shell-plate; *sc.* scapula (foreshortened); *tib.* tibia; *uln.* ulna. (After Shipley and McBride and Reynolds.)

morphologically shorter than in any other vertebrate except the frog. Probably this shortening and broadening is the result of some quite simple change in morphogenesis.

The shell is usually considered to include a dorsal carapace and ventral plastron. Each of these is made up of inner plates of bone, covered by separate outer plates of horny material, comparable to the scales of other reptiles. The carapace includes five rows of bony plates, namely, median neurals, and paired costals and marginals (Fig. 225). These plates are ossifications in the dermis, attached to the vertebrae and ribs, but not actually formed from the latter. The plastron is developed from the expanded dermal bones of the pectoral girdle, together with dermal ossifications comparable to the abdominal ribs found in crocodiles and other reptiles. The whole is covered in most chelonians by rows of special smooth epidermal plates forming the 'tortoise shell' (Fig. 225). A new and larger layer is added to each of these plates each year, the old one remaining above it, thus making a number of 'growth rings' from which some indication of the age of the

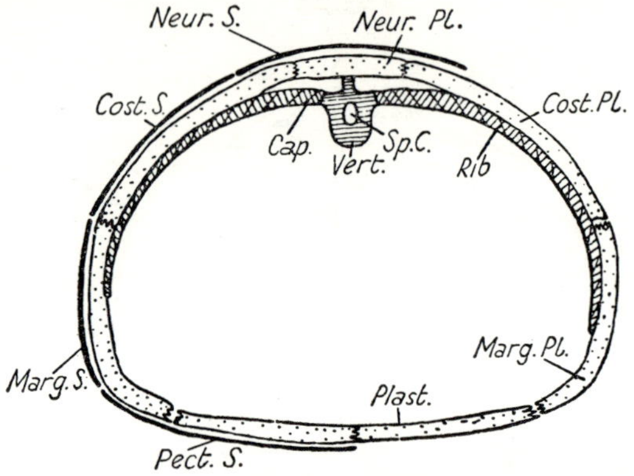

FIG. 225. Diagram of the arrangement of the shell of the tortoise (*Testudo*). The horny shields are shown only on the left.

cap. capitulum of rib; *cost.pl.* costal plate; *cost.s.* costal shield; *marg. pl.* marginal plate; *marg.s.* marginal shield; *neur.pl.* neural plate; *neur.s.* neural shield; *pect.s.* pectoral shield; *plast.* plastron; *sp.c.* spinal cord. (After Gadow).

tortoise can be calculated, though the outer members often become rubbed off.

In order to support this box the limb girdles have become much modified and lie inside the encircling ribs. The pectoral girdle has three prongs, a scapula that meets the carapace dorsally and carries a long 'acromial process' and a backwardly directed coracoid, the two last being attached by ligaments to the plastron. The ilia are attached to two sacral vertebrae and the ischia and pubis are broad. The limbs are stout, but otherwise typically reptilian, with five digits in each. In the marine turtles they are transformed into paddles.

The interpretation of the skull is still somewhat doubtful, but it seems not unlikely that the turtle, *Chelone*, shows the simplest case, namely, the original anapsid condition (Fig. 226). Here the roofing is complete, the dermal bones being widely separated from the brain-case, forming a tunnel for the jaw muscles and those producing retraction of the neck. The tympanum is stretched across a sort of otic notch, bounded by the squamosal, quadrato-jugal and quadrate, the columella auris articulating with the latter. In other groups of Chelonia, however, the dermal roofing has been reduced or 'emarginated', presumably to give still better attachment for the jaw and neck muscles

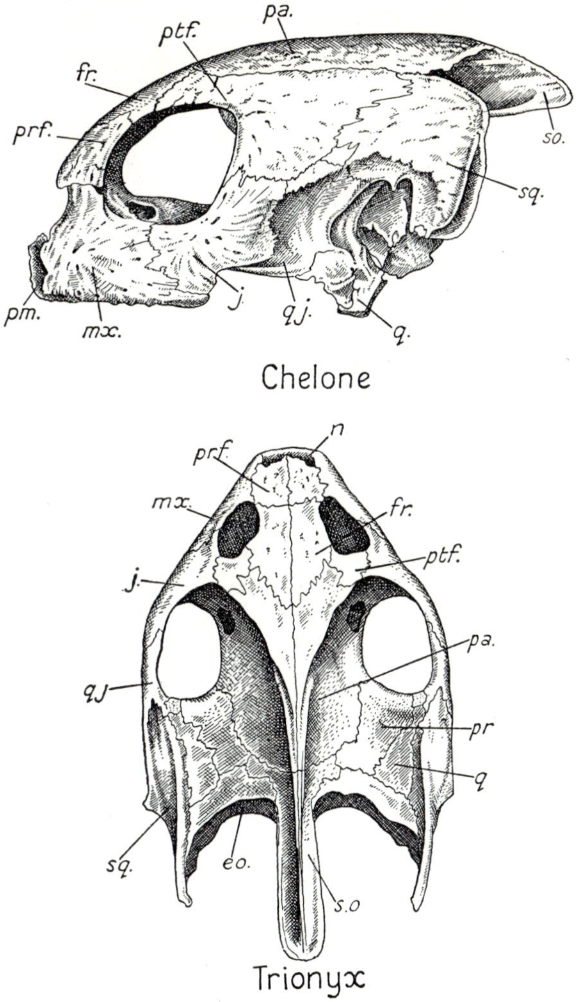

FIG. 226. Skulls of *Chelone* and *Trionyx*. Lettering as Fig. 214, p. 377. (After Goodrich from Parker and Haswell, and Zittel.)

(Fig. 226). It has been argued that the condition in *Chelone* is secondary, but there is no evidence of true temporal fossae in any chelonian, and since the early form **Triassochelys* also had a fully roofed skull there seems no reason for denying that we have here essentially an anapsid condition.

A peculiarity of recent Chelonia is the entire absence of teeth, alike

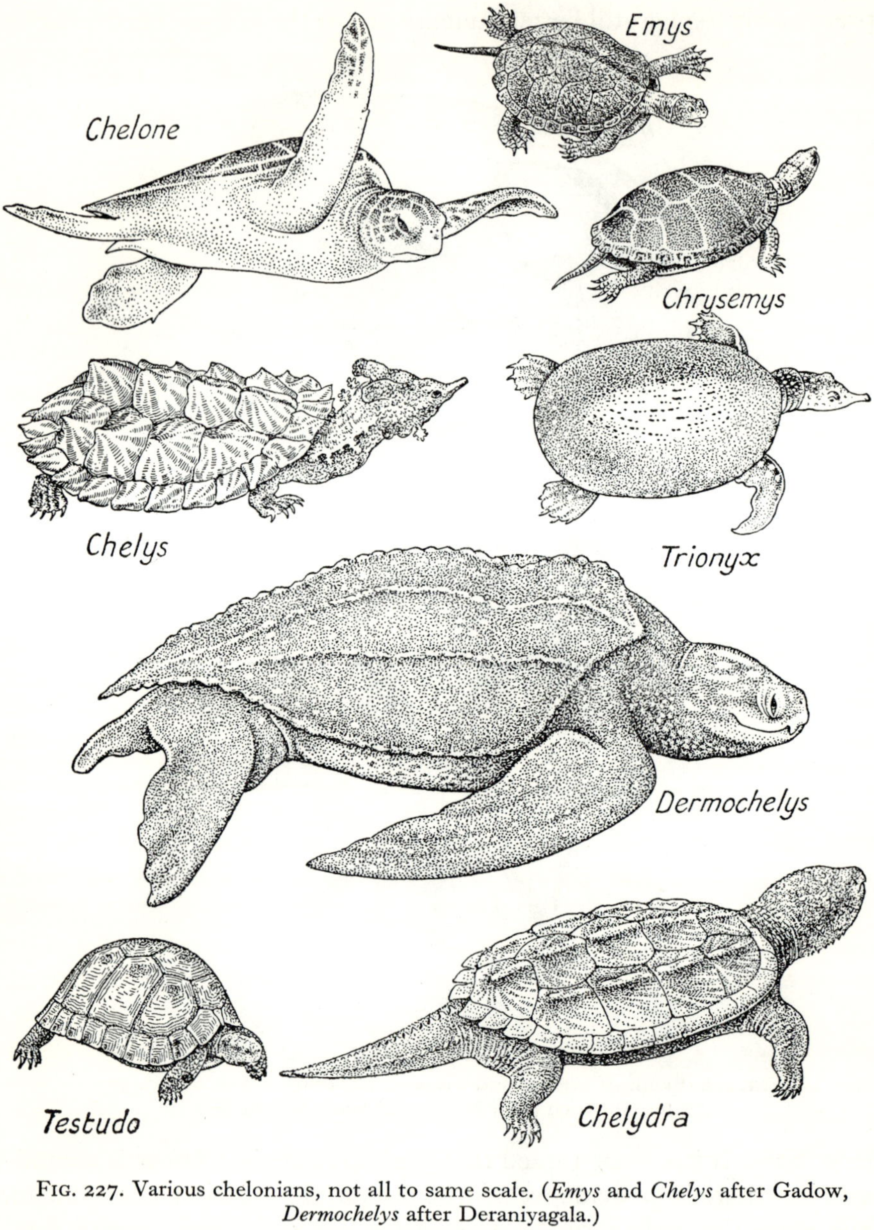

FIG. 227. Various chelonians, not all to same scale. (*Emys* and *Chelys* after Gadow, *Dermochelys* after Deraniyagala.)

in the herbivorous and carnivorous forms. The edges of the jaws form sharp ridges, covered with a formidable horny beak.

The similarity of the soft parts of Chelonia to those of other surviving reptiles (which are all diapsids) suggests that the general organization of the group has changed little since the Permian. The heart possesses a partly divided ventricle and there are two equal aortic arches (Fig. 244). Respiration is modified by the rigidity of the body wall; the lungs are spongy structures attached to the dorsal surface of the shell, sometimes enclosed in a separate pleural cavity (*Testudo*). Breathing is mainly brought about by the contraction of the modified abdominal muscles, which function in a manner comparable with that of the mammalian diaphragm, and by means of pumping movements of the pharynx. Some aquatic forms (*Emys*) also respire by taking water into special vascular sacs, diverticuli of the urodaeum. The metabolism of tortoises is slow and they can remain for long periods without breathing. In temperate climates all species hibernate regularly.

The kidney is metanephric and the nitrogenous excreta are mainly uric acid, there being a typical subdivision of the cloaca and reabsorption of water to form a solid whitish excretory product (see p. 380). There is a single copulatory organ and the eggs are whitish, with hard or soft shells. Like other aquatic reptiles that lay eggs, turtles all come ashore to breed. Thus the marine *Chelone* breeds in the West Indies, in the Straits of Malacca, and on the coast of West Africa; they are caught as they come ashore and made into turtle soup. On the Amazon chelonian eggs are (or at least were) so plentiful that large numbers were eaten. The eggs are usually carefully placed in holes made by boring with the tail and scooping with the feet (*Emys*). The traces of the nest are then covered, often with considerable success. Nevertheless Bates reckoned that at least 48 million eggs were taken annually on the upper Amazon. The chief enemies of the young are vultures and alligators, and these were presumably the ultimate losers when collection by man began, though the numbers of turtles have also decreased as a result of the human depredations.

The proverbial slowness of the tortoise is a necessary corollary of its heavy armour, but the nervous organization and behaviour is more complex than is sometimes supposed. The brain shows well-developed cerebral hemispheres, with not only the basal regions but also the pallium quite large. This was therefore probably true also of the earliest reptiles, as it is of amphibians. In the mammals, also derived from cotylosaur ancestors, there has been still further development of the dorsal regions of the hemispheres to form the cerebral cortex, whereas in the remaining reptile groups, and in the birds, the ventral portion has become large, the dorsal thin. The eyes are probably the chief receptors of chelonians, but the nose is also well developed and the animals are very sensitive to vibration. The tympanum is often covered with ordinary skin and hearing is probably not acute (p. 385). The voice is also small.

The various species show many special habits, some of them complicated and ingenious, especially among the aquatic forms (Fig. 227). For instance, the snapping turtles (*Chelydra*) and alligator turtles (*Macroclemys*) of North America and *Emys* in Europe show considerable care and skill in stalking and capturing not only fish but also young ducks and other birds. Similarly the smaller turtles, such as *Chrysemys picta*, the painted terrapin, with bright yellow, black, and red colours, feed not only on insect larvae, but also on flies, which they catch near the water surface. Many observers have shown that the common Grecian tortoise has a marked sense of locality, returning to a favourite spot even after hibernation.

Our knowledge of the geological history of the Chelonia extends back to the Trias. **Triassochelys* (Fig. 222) was an early turtle, with a shell like that of modern forms, but still possessing teeth on the palate. The skull was anapsid and the pectoral girdle contained interclavicles, clavicles, and perhaps cleithra; these dermal bones were already somewhat enlarged and incorporated in the plastron. The head, tail, and limbs could not be withdrawn into the shell and were protected by spines. In the later evolution of the Chelonia retraction of the head became possible by one of two methods. In the suborder Pleurodira, 'side-neck turtles', the neck is folded sideways. The group is known from the Cretaceous and survives today in tropical Africa (*Chelys*), South America, and Australia. The more successful group is the suborder Cryptodira, in which the neck is curved in a dorsoventral plane. This type is also known from the Cretaceous and includes most of the modern types. The aquatic chelonians show various modifications and it is probable that several lines have independently returned to the water. As a result of this habit the bony shell is often reduced, presumably in the interests of lightness and because of absence of enemies. This had occurred already in **Archelon* of the Cretaceous, which is very similar to the modern *Chelone*. *Dermochelys*, the leathery turtle, has a curious 'carapace' consisting only of a mosaic of small bony plates beneath its leathery skin, and *Trionyx* is a freshwater turtle with a soft shell and no horny plates. All of these forms are best considered as aberrant cryptodirans.

We can trace the history of the Chelonia back rather satisfactorily to the Triassic, but unfortunately there is little to show how they evolved to that stage from some Carboniferous cotylosaurian ancestor. **Eunotosaurus* from the Permian of South Africa had a small number of vertebrae, with very broad, expanded ribs. This perhaps suggests some affinity with Chelonia, though in the latter the ribs themselves are not expanded. There is therefore little to tell us how, when, or why one of the early reptilian populations shortened its bodies and covered them with armour for protection against the hazards of the land they had recently invaded.

4. Subclass *Synaptosauria

*Order *Protorosauria*

All the Synaptosauria characteristically possessed a single temporal fossa in the upper or parapsid position. The earliest forms were small terrestrial lizard-like creatures such as **Araeoscelis* from the lower Permian (Fig. 228). A few more specialized forms, including the remarkable Triassic **Tanystropheus* with a long neck and short body, are known, but the protorosaurs seem never to have been an important element of the early reptilian fauna. Their relationships are not well understood but it is possible that they gave rise to the sauropterygians. The theory that the Squamata were derived from protorosaurs by the emargination of the lower temporal region is now held to be unlikely.

*Order *Sauropterygia*

This was a very successful line of marine reptiles, extending from the Trias to the end of the Cretaceous. The earlier nothosaurs, such as **Lariosaurus* (Fig. 228) from marine Triassic deposits, were small (3 ft long) and had a long neck, and limbs partly converted into paddles. The upper temporal fossa was enlarged and the nostrils lay rather far back, as in many water reptiles.

All of these features were further developed in the plesiosaurs of the Jurassic and Cretaceous, such as **Muraenosaurus* (Fig. 228). In some the neck became very long, presumably for catching fish; 76 cervical vertebrae have been recorded. In others the neck was shorter and the skull longer. The limbs were developed into huge paddles, the ventral portions of the girdles being large for the attachment of muscle masses inserted on the flattened humerus and femur. The dorsal portion of the girdles, so well developed in terrestrial reptiles, was here small: the ilium hardly articulates with the sacral vertebrae. The hands and feet were enlarged by increase in number of joints (hyperphalangy), but there was no increase of digits (hyper-

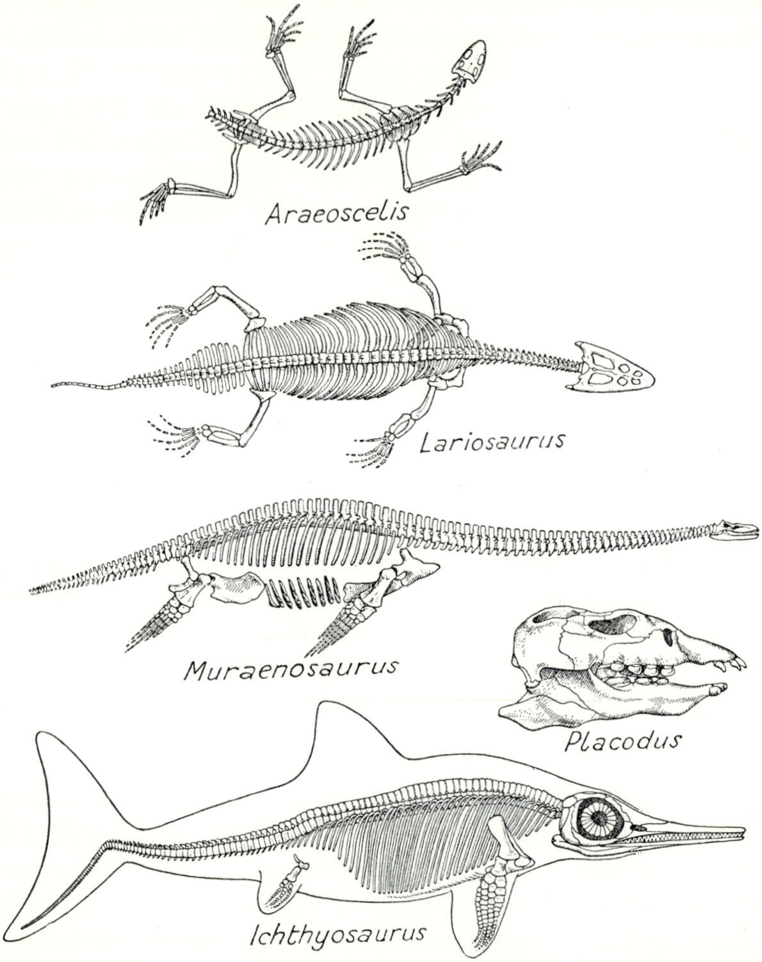

FIG. 228. Icthyosaurs, plesiosaurs, and their allies. (Partly after Romer.)

dactyly) such as is seen in ichthyosaurs. The skull was as in nothosaurs, but with the nostril still further displaced on to the upper surface.

These animals were numerous in Jurassic and Cretaceous seas and some of them reached 50 ft in length. They were obviously fish-eaters, but little is known of their habits. It is not known whether they were viviparous or came ashore to lay eggs. *Placodus and related

Triassic forms were related to the plesiosaurs and were specialized for mollusc-eating by the development of large grinding teeth on the jaws and palate.

5. Order *Ichthyopterygia

These animals, found mainly in the Triassic and Jurassic seas, were even more modified for aquatic life than the plesiosaurs (Fig. 228). They occupied a position comparable to that of the dolphins and whales during the Tertiary period. The body possessed a streamlined fish shape and swimming was by lateral undulatory movements. The vertebrae were amphicoelous disks and there were large dorsal and caudal fins, with the vertebral column apparently continued into the lower lobe of the latter. The paired fins were small and presumably used as stabilizing and steering agents. The pelvic girdle did not articulate with the backbone. In the limbs the number of digits was often greater or less than the usual five, and there was often hyperphalangy. Evidently this type of skeleton gives better support for a fish-like paddle than does the pentadactyl tetrapod type and it is interesting to find it evolved again in vertebrate stocks that returned to the water. This seems in a sense to be a case of reversal of evolution.

The head was much modified for aquatic life, with a very long snout armed with sharp teeth, and nostrils set far back. The eyes were large and surrounded by a ring of sclerotic bony plates. The temporal fossa, though in the parapsid position, had boundaries different from that in the synaptosaurians.

The Triassic ichthyosaurs were already greatly modified and we have no trace of the origin of the group. Romer has suggested a possible derivation from cotylosaurs related to the ancestors of mammal-like reptiles. The ichthyosaurs were more highly adapted to aquatic life than any other reptiles known. They seem to have been viviparous, since the skeletons of small specimens have been found within larger ones. Like the plesiosaurs they developed a special mollusc-eating type, **Omphalosaurus*, in the Triassic.

6. Subclass Lepidosauria

Most of the animals popularly considered as characteristic of the period of reptilian dominance have a two-arched or diapsid skull. This condition, or some modification of it, is found in all the surviving reptiles except the Chelonia, and in the birds. Formerly all the two-arched reptiles were placed in a single subclass, the Diapsida, but it is now customary to divide them into two subclasses, the Lepidosauria, which includes *Sphenodon*, the lizards and snakes, and the Archosauria, including the crocodiles, dinosaurs, pterosaurs, and the ancestors of birds. It is not known if the archosaurs were derived from primitive lepidosaurs such as **Youngina*, or whether the two groups arose independently from cotylosaurian ancestors.

*Order *Eosuchia*

The earliest lepidosaurs belong to the order Eosuchia. The best known of these, **Youngina* (Figs. 229 and 230), was a lizard-like creature, found in the Upper Permian of South Africa, and retaining many cotylosaurian features, for instance, teeth on the palate and no opening between the bones of the snout (antorbital vacuity). The two fossae at the back of the skull immediately show the affinity with other diapsids. Little is known of the post-cranial skeleton. The fifth metatarsal does not show the hooked shape that is found in other diapsids and also in Chelonia. It is difficult, however, to ascribe very great weight to this single point, as against the general features of the skull, which indicate that **Youngina* could have given rise to the later two-arched reptiles and, by loss of the lower margin of the lower temporal fossa, also to the lizards and snakes. **Prolacerta*, from the Lower Trias, shows how this may have come about; it is so like **Youngina* that it is classed as an eosuchian, but there is a gap in the lower temporal arch, suggesting that the animal may have been near the ancestry of lizards.

7. Order Rhynchocephalia

Sphenodon (= *Hatteria*), the tuatara of New Zealand, is the oldest surviving lepidosaurian reptile; it still remains in essentially the eosuchian condition. Very similar Mesozoic fossils (e.g. **Homoeosaurus* from the Jurassic) show the continuity of the type (Fig. 221). Among the many primitive features that this race has preserved unchanged for 200 million years are the two complete temporal fossae (Fig. 231), the well-developed pineal eye (the pineal foramen is marked in the early diapsid fossil skulls), and the amphicoelous vertebrae with intercentra. *Sphenodon*, alone of surviving reptiles, has no copulatory organ. The large wedge-shaped front teeth are among the few specialized characters.

The tuatara was once widespread throughout New Zealand but became much reduced and in danger of extinction. Recently rigid conservation measures seem to have allowed it to recover in numbers on some small northern islands. The animals are up to 2 feet long and

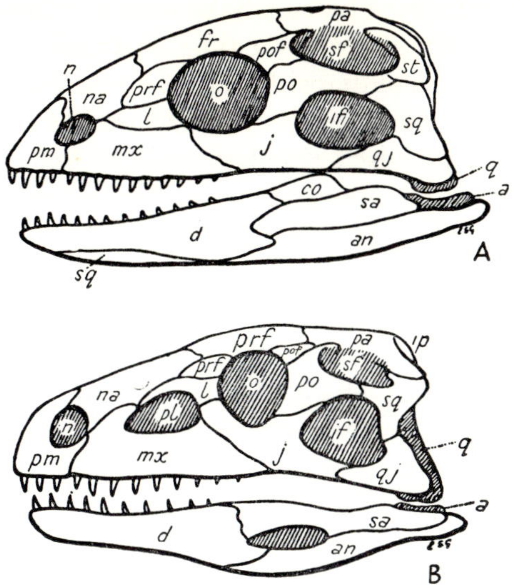

FIG. 229. Diagrams of skulls of A, eosuchian, B, pseudosuchian, showing the plan of the diapsid skull.

Lettering as Fig. 214. *lf.* lower temporal fossa; *pl.* preorbital fossa; *sf.* upper temporal fossa. (From Goodrich.)

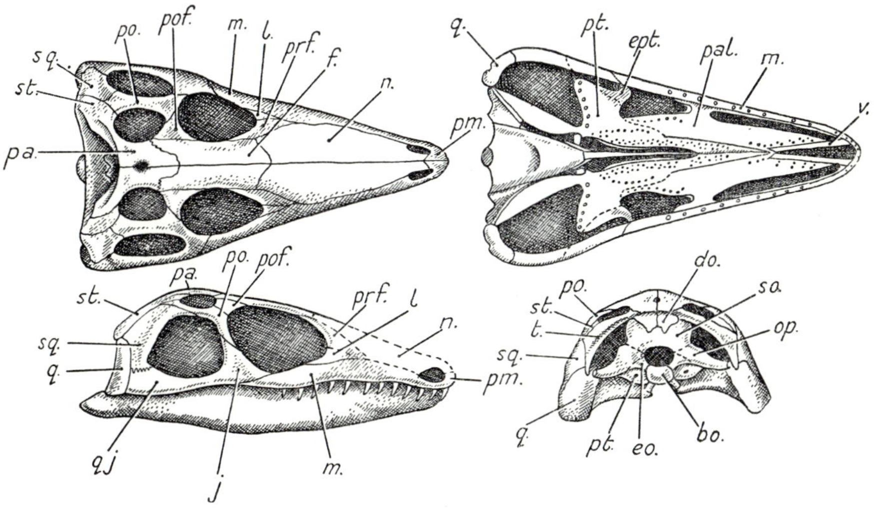

FIG. 230. Skull of *Youngina*. Lettering as Fig. 214. (After Romer, *Vertebrate Paleontology*, Chicago University Press.)

are insectivorous and carnivorous. They live in burrows, often in association with petrels. The eggs take over a year to hatch.

Sphenodon evidently shows us a type that has departed relatively little from the condition of diapsids in the late Permian. Yet its appearance, habits, and soft parts are very like those of lizards, and provide us with evidence that these animals remain in essentials rather close to the original amniote populations. A few other extinct rhyncho-

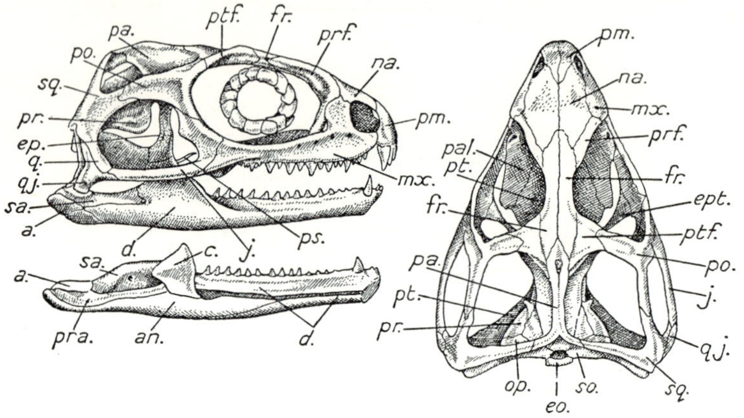

FIG. 231. Skull of *Sphenodon*. Lettering as Fig. 214. (After Romer, *Vertebrate Paleontology*, Chicago University Press.)

cephalians are known; these include the rhynchosaurs, in which the tip of the upper jaw had a hooked beak-like appearance.

8. Order Squamata

The lizards and snakes are the most successful of modern reptiles, numbering between them over 5,000 species. Probably the groups arose from eosuchians related to **Prolacerta*. Such forms would also have been close to the Rhynchocephalia, differing from them, however, in a tendency to lose the lower temporal arch and to develop a movable quadrate.

It has now been shown that the lizards are a more ancient group than was formerly supposed, and had appeared by the end of the Triassic. Furthermore, some of the early forms were already considerably specialized. Our knowledge of the early lacertilian radiation is still incomplete, however, and none of the existing lizard families are known much before the Cretaceous. The earliest undoubted snake occurs in the upper Cretaceous and the group does not seem to have become abundant until the Oligocene.

Although typical lizards preserve a number of primitive reptilian features, the Squamata as a group show several interesting specializations that are absent in *Sphenodon*, and to which their success may be partly attributed. In the majority of forms, especially in the snakes, the skull is highly kinetic, having a freely movable quadrate, which imparts its motion to the bones of the upper jaw. Paired copulatory organs of a unique type are present in the male. There is a widespread tendency towards limb reduction, which has apparently occurred

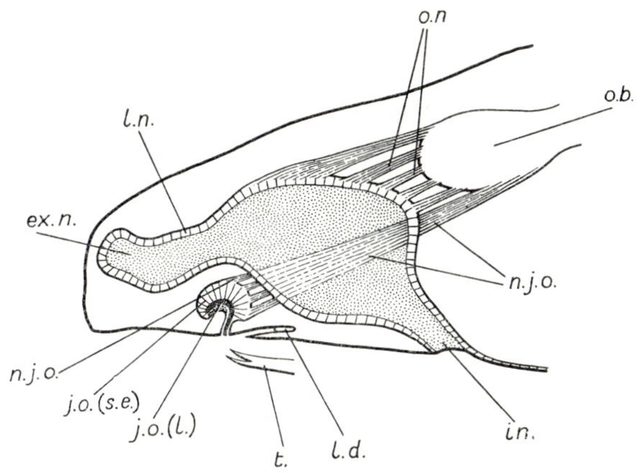

FIG. 232. Jacobson's organ. Diagram of reconstructed L.S. of snout of lizard showing nose and organ of Jacobson, both seen in section from the lateral side.

ex.n. region of external nostril; *in.* internal nostril; *j.o.* (*l.*) Jacobson's organ lumen; *j.o.* (*s.e.*) Jacobson's organ sensory epithelium; *l.d.* lachrymal duct, front end, cut; *l.n.* lining of nose; *n.j.o.* nerves to Jacobson's organ; *o.b.* olfactory bulb; *o.n.* olfactory nerves; *t.* tongue.

independently in members of about half the existing families of lizards, and in snakes.

The paired organs of Jacobson are highly elaborated and of great functional importance. In snakes and in most lizards these organs (Fig. 232) are hollow domed structures above the front of the palate, each opening into the mouth by means of a slender duct. The lachrymal duct opens into or near the duct of Jacobson's organ, instead of into the nose, suggesting that the secretions of the eye glands may have some special function related to that of the organ. Odorous particles are carried to the ducts of Jacobson's organs, or to the immediate neighbourhood of them, by the tongue tip, which is forked in snakes and many lizards. The lumen of the organ is partly lined by sensory epithelium, supplied by a separate branch of the olfactory nerve. Experiments by Noble and others have shown that the organs assist in such functions as sex recognition and following trails left by prey.

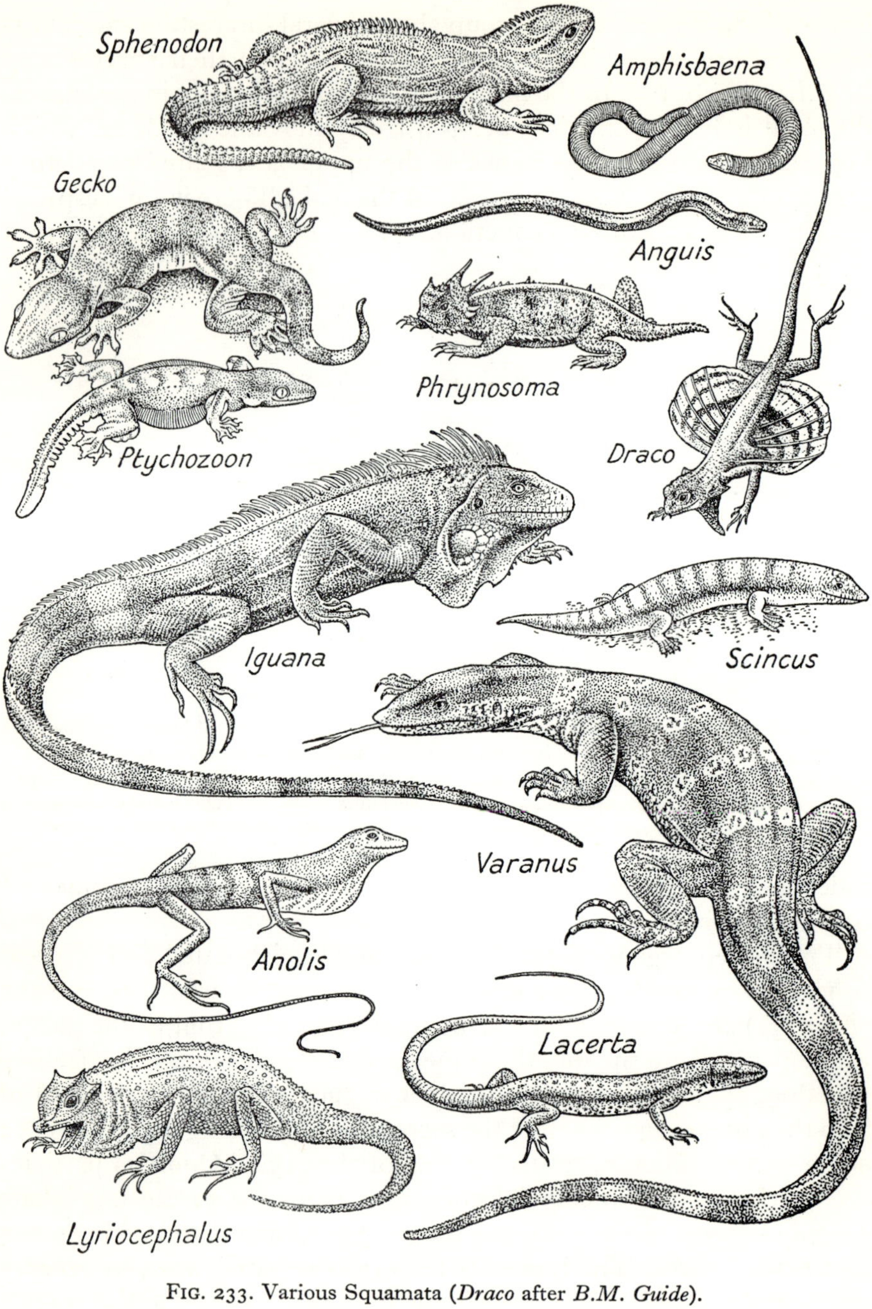

FIG. 233. Various Squamata (*Draco* after *B.M. Guide*).

In some lizards and all snakes, the eyelids are modified to form an immovable transparent spectacle covering the cornea, with the loss of the nictitating membrane. The adaptive significance of this, which is foreshadowed in some lizards by the development of a window in the lower eyelid, is not always clear, since although it is found in many that burrow or live in sand, it also occurs in arboreal forms.

9. Suborder Lacertilia

The modern lizards show extensive adaptive radiation (Fig. 233) and include terrestrial, arboreal, burrowing, and aquatic forms. The majority are carnivorous but there are some herbivores. It is difficult to say which of the twenty or so living families is the most primitive, and the grouping of these into infraorders is a matter of some difficulty.

The Gekkota contains the geckos and a small group of Australasian limbless forms, the pygopodids. Geckos are mainly small nocturnal and arboreal insectivorous lizards of warm climates, with ridged pads on the toes and sharp claws that enable them to climb an almost smooth surface. Some species have taken to living in houses. The tree-gecko *Ptychozoon* has webs of skin on the limbs and along the sides of the body, which perhaps act as a parachute to break its fall. Many geckos live in colonies and unlike most lizards are extremely vocal, making clicking and cheeping sounds. Their hearing is probably acute. The endolymphatic ducts of the inner ear are greatly expanded to form sacs in the neck containing calcareous deposits, but the functional significance of this is obscure. Their eyes are, as a rule, covered by a spectacle. Geckos are the only Squamata that lay hard-shelled eggs, those of other forms being leathery in texture.

The Iguania is a large group comprising the agamids, iguanids, and chameleons. The first two include terrestrial, arboreal, and amphibious types, sometimes of large size, and often furnished with crests, dewlaps, expansible throat-fans and other appendages that play a part in rivalry and courtship. The males are often brightly coloured and the whole group is characterized by a visually dominant behaviour pattern. In some arboreal forms, as in the chameleons, the sensory parts of the nose and the organ of Jacobson are reduced.

The agamids are found in the Old World and Australasia and include such well-known types as the oriental 'blood-sucker' (*Calotes*), so-called because of the red colour of the throat, the spiny lizards (*Uromastix*) of the north African and Indian deserts, and the Australian frilled lizard (*Chlamydosaurus*) (Fig. 240). The Indo–Malayan genus *Draco* has a large lateral web, supported by the ribs, which can be spread out and used for gliding, though it is not moved in true flight. *Lyriocephalus* from Ceylon is an agamid with a remarkable convergent similarity to the chameleons. Some agamids (and other

FIG. 234. Chameleon catching a fly, showing its changes in colour.

A, cream with yellow patches, the usual night colour. B, grey-green with darker patches. C, dark brown patches and yellow spots. D, reaction produced by pinching tail, inflation and darkening of all spots. (After Gadow.)

lizards) can run on their hind legs when they are in a hurry (Fig. 240). In agamids the teeth are set squarely on the summit of the jaw, as in *Sphenodon*; this condition is termed acrodont. In most other lizards the teeth are attached obliquely to the inner side of the jaw (pleurodont).

The iguanids are found mainly in the New World and parallel the agamids in many ways. *Anolis* is a small, common North American form. *Iguana* from south and central America reaches 6 ft in length.

Amblyrhynchus, found only in the Galapagos Islands, is remarkable as the only existing marine lizard, though it spends much of its time on shore, basking and feeding on the sea-weed left at high tide. *Phrynosoma*, the horned toad, with spikes on the head and back, is found in the deserts of North America and burrows in the sand. This is one of the few ovoviviparous iguanids.

The chameleons are highly modified arboreal lizards from Africa, Madagascar, and India. Some species have casques on the head, or one or more horns on the snout. The tail is prehensile and the digits are arranged in groups of two and three so as to be opposable and allow the grasping of branches. Chameleons live on insects, caught by means of the very long tongue (Fig. 234), which has an adhesive clubbed tip and is projected by a remarkable muscular mechanism. Their movements are slow and deliberate, but they show considerable care in stalking their prey; as they approach it their eyes, which normally move independently, converge so as to bring the prey into binocular vision, presumably serving to judge its distance. Their powers of colour change are described on p. 410.

The Scincomorpha are another large assemblage, including *Lacerta*, common in Europe and North Africa, and the skinks, many of which are modified for burrowing, sometimes in the sand. Many skinks have well-developed limbs, but others show all degrees of limb reduction, either the fore- or hind-limbs, or both, being lost. The most highly fossorial of all lizards, the worm-lizards or Amphisbaenidae, may also belong to the Scincomorpha, though they show many remarkable specializations. Their eyes are rudimentary, their tails are blunt and resemble the head, and they are able to move freely in either backward or forward direction. External limbs are usually absent.

The Anguimorpha contains the anguids, of which the European slow-worm *Anguis* is a familiar limbless example, and the monitor lizards (*Varanus*) and their allies, which are placed in the superfamily Platynota. The monitors of the Old World and Australia include the largest of existing lizards, one species, the Komodo Dragon, growing to at least 10 ft long. They are carnivorous, killing vertebrates as well as insects, and are often semi-aquatic. Three related groups, now extinct, occurred in the Cretaceous. The aigialosaurs and dolichosaurs were amphibious lizards of moderate size, but the later mosasaurs, such as **Tylosaurus*, were huge creatures, sometimes 30 ft long, and were highly adapted for marine life, with long jaws and paddle-like limbs showing some hyperphalangy. The strikingly coloured helodermsfrom North and Central America are also playtnotids and are of special interest as the only known poisonous lizards. Another allied form is the rare earless monitor, *Lanthanotus*, from Borneo, which seems to be a survivor of the primitive platynotid stock.

Many lizards are able to change colour, the chameleons, *Anolis* (often called 'American chameleons'), and certain agamids being the most notable for this. The colour may change with the environment, serving the obvious purpose of concealment. Special colour patterns

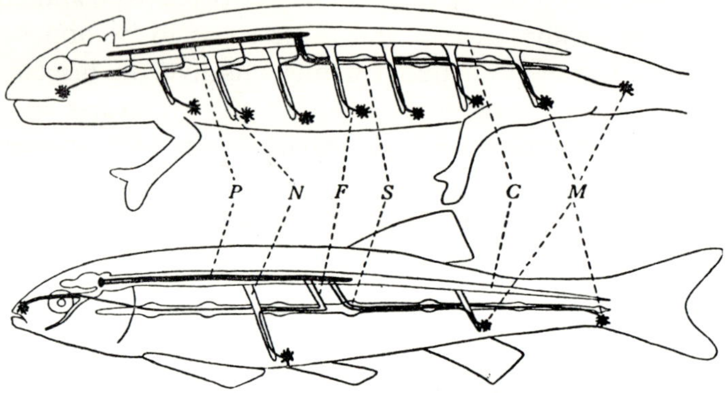

FIG. 235. Diagram of the nervous control of the melanophores in the chameleon (above) and minnow (below).

C, spinal cord; F, pathway of pigmentomotor fibres (the synapse in the sympathetic ganglia is omitted); M, melanophore; N, spinal nerve; P, pigmentomotor centre; S, sympathetic. (From Sand, minnow after v. Frisch.)

are displayed in courtship or threat, and colour change may also occur in response to temperature and other environmental changes. The physiological mechanism of colour control varies in different reptiles. In *Anolis* there are probably no nerves to the melanophores, which are controlled by hormones produced by the posterior pituitary and possibly other glands. In chameleons, however, the melanophores are controlled partly or entirely by the autonomic nervous system (Fig. 235).

Many lizards are able to break off their tails when threatened or seized by a predator; this ability, known as autotomy, is due to the presence of special planes of weakness through the bodies of the caudal vertebrae. Such fracture planes are also found in *Sphenodon*, but not in snakes. After autotomy the tail regenerates, but the new member is not a replica of the normal. The vertebrae, for instance, are not regenerated, their place being taken by an unsegmented tube of cartilage. It has been shown experimentally that in *Anolis* regeneration will not occur if the motor-nerves are prevented from growing back from the stump. The new sensory innervation of the skin is derived entirely from the surviving dorsal roots, whose cells become greatly enlarged.

10. Suborder Ophidia

The snakes are obviously descended from lizards of some kind, but their precise mode of origin is obscure. Some workers believe that their nearest living relatives are the platynotid lizards (monitors, &c.). There is evidence that the snakes passed through a burrowing stage in their early history, although no known platynotids show marked fossorial adaptation. A burrowing ancestry is particularly suggested by the structure of the eye, which, as Walls has pointed out, differs widely from that in typical lizards (Fig. 238). Thus there are no scleral ossicles or cartilages in snakes, and accommodation is brought about in a manner unusual for reptiles, involving displacement of the lens. The visual cells include cones of a peculiar type, which have apparently been derived from rods. The yellow retinal droplets that serve to protect lacertilian retinae from excessive light are absent, and instead some diurnal snakes protect their retinas by a yellow-tinted lens. These features can all be interpreted on the supposition that the ophidian eye was once drastically reduced, but has subsequently been refurbished in response to the needs of life above ground. Other characters that seem to point in the same direction include the structure of the ears which, as in many burrowing lizards, have apparently degenerated. The ear drums, tympanic cavities, and Eustachian tubes are absent, and the columella auris articulates with the quadrate. It seems unlikely that snakes can hear airborne sounds at all well, though doubtless they are sensitive to ground vibrations transmitted through the bones of the jaw.

The snakes show many other interesting peculiarities, the most obvious being the complete absence of limbs. Only in a few of the more primitive forms such as the boas and pythons can rudiments of the hind limbs and their girdles be found; in these snakes claws may be present externally on either side of the cloaca and are said to play a part in coitus.

Locomotion is produced by the lateral undulation of the body, which exerts pressure on surrounding objects and pushes the snake forwards; the enlarged transverse ventral scales of most species help to prevent slipping. A few snakes (e.g. some boas and vipers) can also progress by muscular movements of the ventral scales, with their bodies stretched out almost in a straight line. The spine is strengthened by additional intervertebral articulations known as the zygantra and zygosphenes.

The skull is highly modified, permitting, in all except a few burrowing forms, an enormous gape and the swallowing whole of large prey. The premaxilla is small and usually toothless, and the bones of the upper jaw are loosely attached to the rest of the skull. The two halves of the lower jaw are united only by ligaments. The sharp recurved teeth are carried on the palate bones as well as on the maxilla

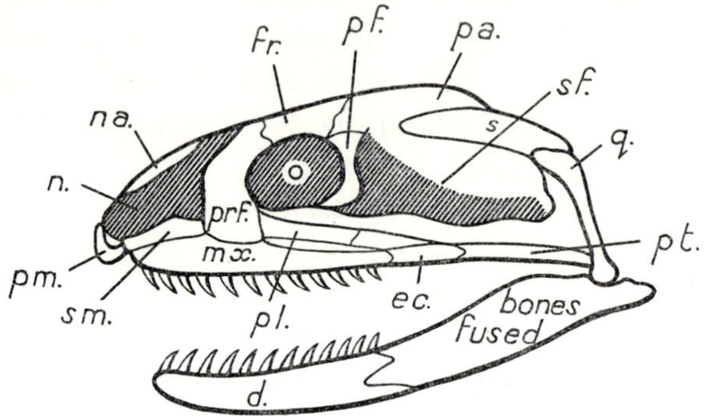

FIG. 236. Diagram of skull in ophidians.
d. dentary; *ec.* ectopterygoid; *fr.* frontal; *mx.* maxilla; *n.* nostril; *na.* nasal; *o.* orbit; *pa.* parietal; *pf.* postfrontal; *pl.* palatine; *pm.* premaxilla; *prf.* prefrontal; *pt.* pterygoid; *q.* quadrate; *s.* squamosal or supratemporal; *sf.* upper temporal fossa (this is shown diagrammatically, as it occurs in many lizards; in *Lacerta* it is largely covered by an extension of the postfrontal—see Fig. 213); *sm.* septomaxilla. (Modified from Goodrich.)

and dentary. The brain-case is strong and compact, the brain being protected from mechanical injury during swallowing by the massive parasphenoid and by flanges of the frontals and parietals, which lie between the orbits, so that there is no interorbital septum.

In the normal ophidian kinetic mechanism the upper jaw as a whole is raised as the result of forward rotation of the lower end of the freely mobile quadrate, which is attached to the back of the pterygoid. The well-developed protractor muscles of the pterygoid and quadrate play an important part in the process. In the viperid snakes a further elaboration of this mechanism is seen, the maxilla being very short and able to rotate on the prefrontal so that the fangs can be erected (Fig. 237). A slip from one of the muscles is attached to the poison gland and helps to expel the venom as the snake bites.

The respiratory system and viscera of snakes are also much modified. The glottis can be protruded so as to keep the airway clear while prey is being swallowed, and in some forms a part of the trachea is specialized for respiration as a tracheal lung. The left of the two paired lungs is usually reduced, often to a rudiment, as in some limbless lizards, and the other paired viscera tend to lie at different levels on the two sides. The heart usually lies a quarter to a third of the way down the body, and the carotid arches are asymmetrical, the right common carotid artery tending to be suppressed.

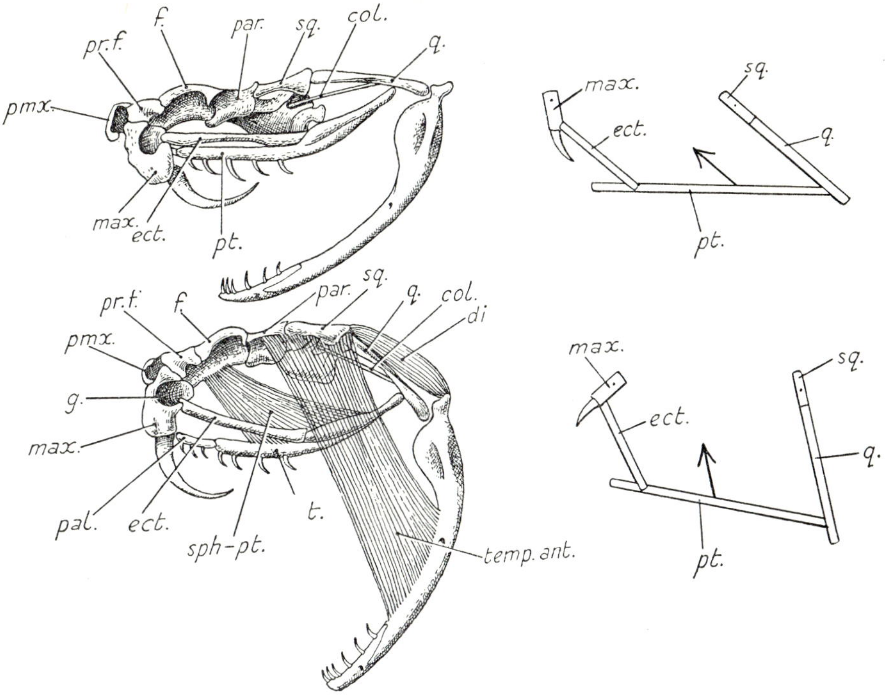

FIG. 237. Skull of rattle-snake (*Crotalus*) with jaws partly and fully opened.

Lettering as Fig. 214; *sph-pt.* the protractor-pterygoid muscle, which pulls the pterygoid forward, causing it to push the ectopterygoid, which rotates the maxilla and erects the fang; *di.* the digastric muscle that assists in opening the jaw; *temp.ant.* the anterior temporal which shuts the mouth. The diagrams at the right show the actions of the levers that erect the fang; *g* is the groove characteristic of crotaline snakes. (Modified after Gadow.)

The snakes show nearly as much adaptive radiation as the lizards, though there is less structural variation among them. The more primitive forms, with pelvic rudiments, include a number of small burrowers such as *Typhlops*, as well as the large boas and pythons of the family Boidae, which tend to be arboreal and amphibious in habits and kill their prey by constriction. In general, the pythons lay eggs, whereas the boas are ovoviviparous.

The majority of living snakes belong to the family Colubridae, which contains many medium-sized harmless snakes such as the grass-snake (*Natrix*) and some moderately poisonous ones with grooved fangs at

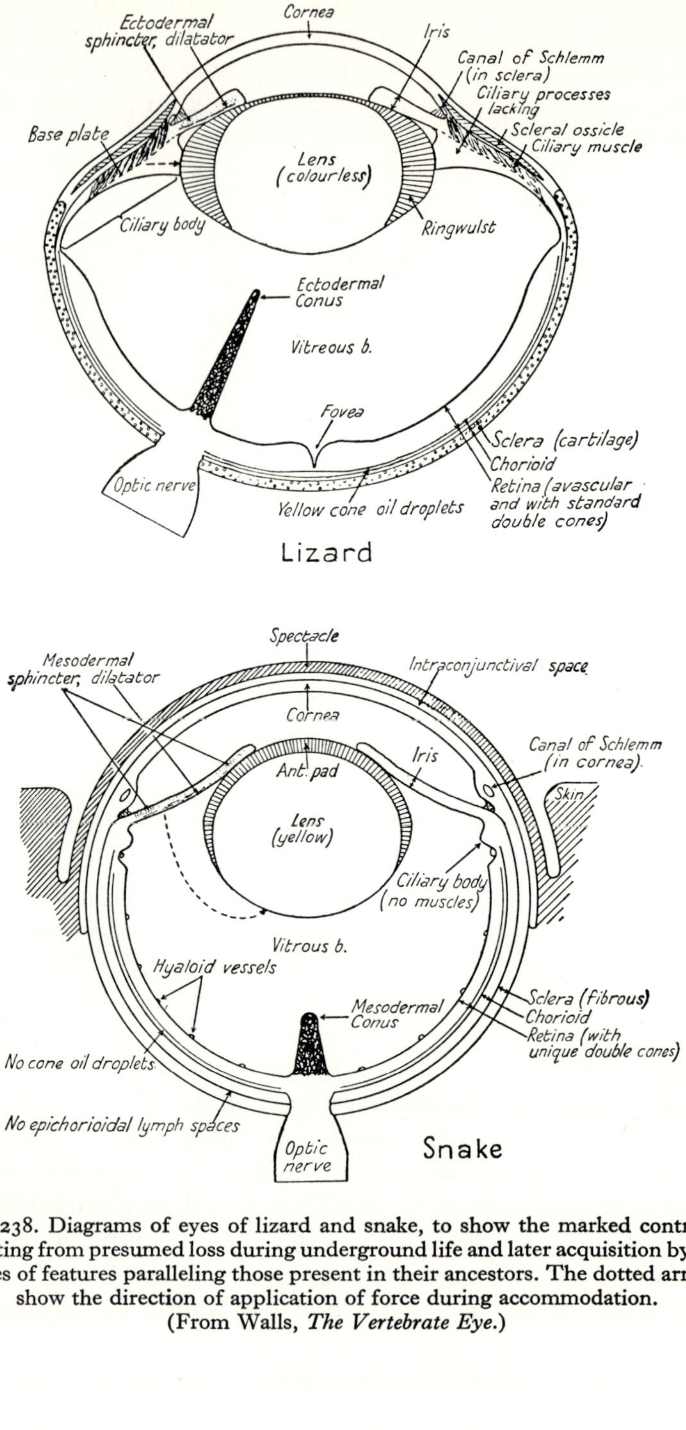

FIG. 238. Diagrams of eyes of lizard and snake, to show the marked contrasts resulting from presumed loss during underground life and later acquisition by the snakes of features paralleling those present in their ancestors. The dotted arrows show the direction of application of force during accommodation. (From Walls, *The Vertebrate Eye.*)

the back of the maxillae; these are known as back-fang snakes or opisthoglyphs. The South African boomslang (*Dispholidus*) is one of the few whose bite may be lethal in man. *Dasypeltis*, the egg-eating snake, is also a member of this group; it swallows the eggs whole and crushes them with special tooth-like processes of the neck vertebrae.

The family Elapidae contains the cobras, kraits, and coral snakes, all highly poisonous with quite small and relatively non-movable fangs at the front of the maxilla, and a venom predominantly neuro-toxic in

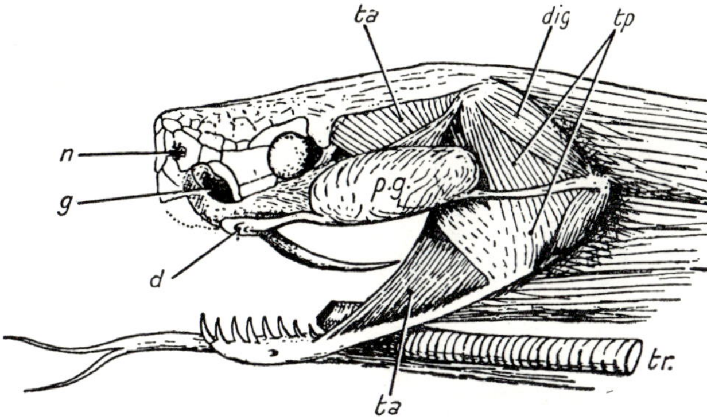

FIG. 239. Head of crotaline snake (*Lachesis*) after removal of skin.
d. duct of poison gland, bending at base of fang; *dig.* digastric muscle; *g.* sensory pit or groove; *n.* nostril; *p.g.* poison gland; *ta.* and *tp.* anterior and posterior portions of temporalis muscle; *tr.* trachea. (From Gadow.)

action. All the poisonous Australian snakes belong to this group. The hood of the cobra is expanded by the long cervical ribs and probably has a warning (sematic) function. The king cobra (*Hamadryas*) is the largest poisonous snake and reaches 18 feet in length.

The very poisonous sea-snakes (Hydrophiidae) are related to the elapids. Their tails are vertically compressed for swimming; some species can hardly move on land. Like many freshwater snakes they are (with a few exceptions) ovoviviparous.

The family Viperidae consists of the vipers of the Old World and the rattle-snakes and pit-vipers, mainly from the New World. The two latter groups, placed in the subfamily Crotalinae, are distinguished by the presence of a remarkable sensory pit on each side of the head between eye and nostril (Fig. 239). This is highly sensitive to temperature changes and helps the snake to detect warm-blooded prey. The rattle-snakes are, of course, also noted for their caudal appendages, which are composed of articulated rings and modified skin. One ring is formed at each moult, though the older and most posterior ones break off periodically. The rattle is vibrated voluntarily as a warning and perhaps prevents the snake from being trodden on by large mammals.

Most of the Viperidae are highly poisonous, though the bite of the European adder (*Vipera berus*) is seldom fatal to man. The venom is predominantly haemolytic in action. The fangs are canalized, the canal having apparently being evolved by the progressive deepening of a groove until its margins have come into apposition. The fact that the fangs are erected when the snake strikes and can be folded back along the roof of the mouth when not in use, makes it possible

FIG. 240. Drawings of three frilled lizards (*Chlamydosaurus*) and a *Grammatophora* (at right) to show the bipedal habit. (Drawings made by Heilmann from photographs of the lizards running at full speed, taken by Saville Kent.)

for these structures to be very long, about 1 inch in the case of a large puff adder.

Some of the American pit-vipers are very large, the dreaded bush-master (*Lachesis*) reaching about 10 ft. The majority of the Viperidae bear their young alive and the finding of late embryos within the bodies of female adders and rattle-snakes may have given rise to the tale and that these reptiles temporarily hide their young by swallowing them in the face of danger.

11. Superorder Archosauria

We have seen that about 130 million years ago the diapsid stock produced the most successful modern reptile group, the Squamata (Fig. 221). Much earlier an even more successful type had developed from the Eosuchia, having as its outstanding feature the habit of walking on the hind legs. Creatures of this type were the dominant land animals of the later Mesozoic, and they include the dinosaurs and pterosaurs. Crocodiles are the only living descendants of the group that have remained at the reptilian level. They have, of course, abandoned the bipedal habit and survive as a specialized amphibious remnant. The birds, which are also undoubtedly descendants of this archosaurian group, give us in some ways a better idea of the characteristic structure than do the crocodiles.

All the lines of archosaurs are characterized by certain common tendencies, mostly associated with bipedalism, which is possible also in some lizards (Fig. 240); features barely indicated in the earlier forms become developed in the later. In all archosaurs the hind legs were much longer than the front and the acetabulum formed a cup, open below, so that the legs were held vertically below the body. At the same time the ischium and pubis became elongated, presumably to allow for the attachment of muscles producing a fore-and-aft movement (see p. 375). In later forms the ilium became fused with several sacral vertebrae. The femur has a lateral head and the tibia becomes long and strong and sometimes fused with the proximal tarsals; the distal tarsals may fuse with the metatarsals as in birds, and the digits are reduced, usually to three long ones turned forward while the first is turned back. The skull is typically diapsid, but tends to have certain modifications, such as the development of antorbital vacuities behind the nostrils and other spaces in the palate, presumably serving to give lightness without loss of strength.

12. Order *Pseudosuchia

The earliest archosaurs were the Triassic pseudosuchians, creatures evidently not far removed from the Permian eosuchians. These animals (**Saltoposuchus*) can be visualized as lizards that ran on their hind legs (Fig. 241). They were small and carnivorous, having sharp teeth set in sockets along the edges of the jaws (hence 'thecodont'). The skeleton showed all the archosaur characters in a most interesting incipient form. Thus the bones of the pelvis were still plate-like, but arranged in the characteristic triradiate manner. The front legs were already much shorter than the hind. Antorbital vacuities were present and there was no pineal foramen.

13. Order *Phytosauria

Even in the Triassic at least one line, the phytosaurs, abandoned the bipedal habit, becoming amphibious. These creatures were not actually ancestral to the crocodiles, but show remarkable parallelism to them in the elongated jaws and general build (Fig. 241). However the nostrils were set far back. There can be no doubt that the phytosaurs were derived from pseudosuchians and the two groups are often placed together in an order Thecodontia.

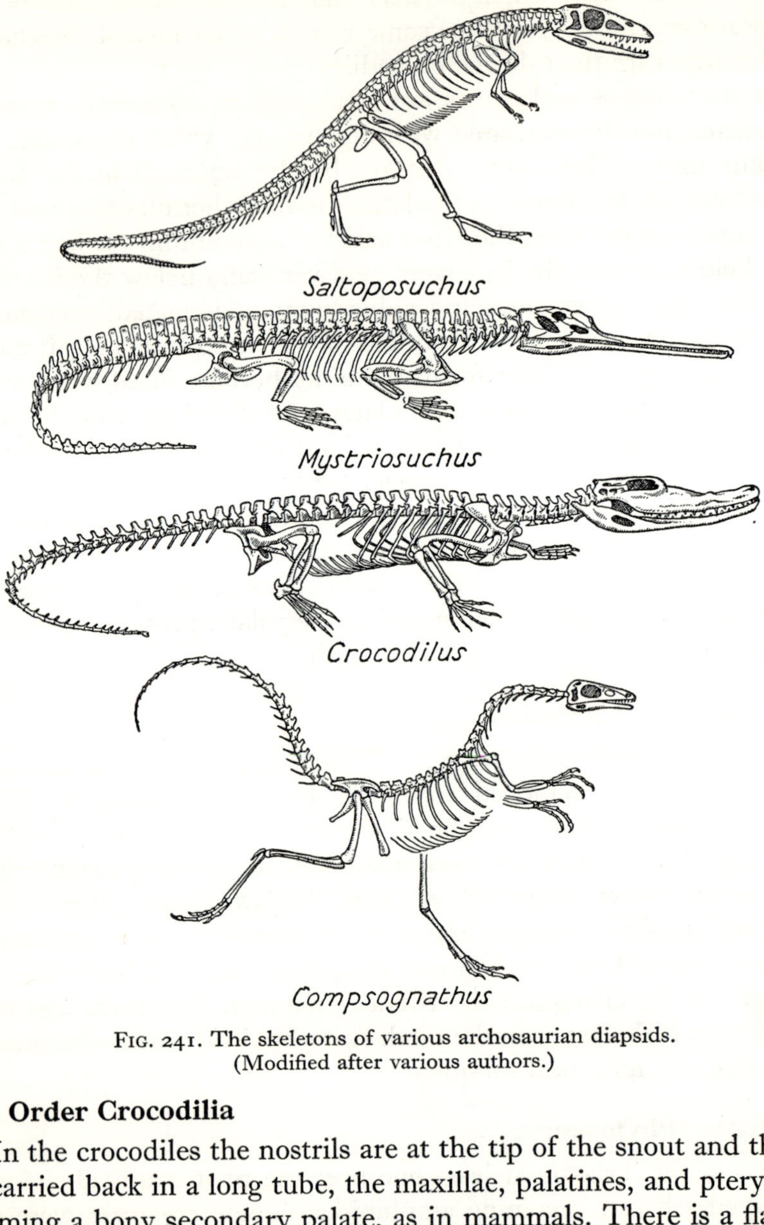

FIG. 241. The skeletons of various archosaurian diapsids. (Modified after various authors.)

14. Order Crocodilia

In the crocodiles the nostrils are at the tip of the snout and the air is carried back in a long tube, the maxillae, palatines, and pterygoids forming a bony secondary palate, as in mammals. There is a flap on the hind end of the tongue, which, with a fold of the palate, enables the mouth to be closed off from the respiratory passage and hence kept open under water. The nostrils can also be closed by a special set of muscles, and the ear-drums are protected by scaly movable flaps. The Eustachian tubes are very complicated, and parts of the skull are pneumatized by extensions from the middle ear cavity, as in birds.

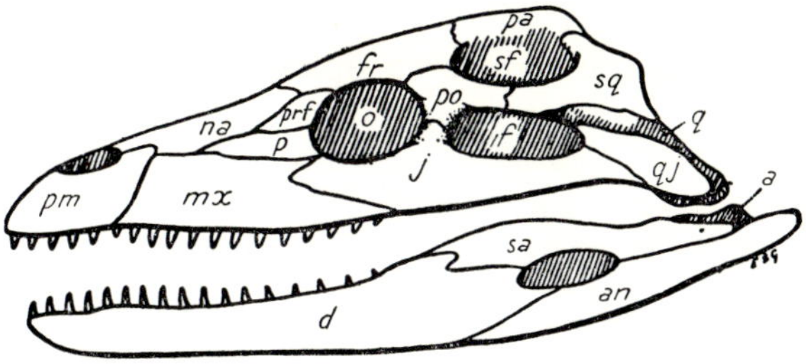

FIG. 242. Diagram of the skull of Crocodilia. Lettering as Fig. 214, p. 377. (From Goodrich.)

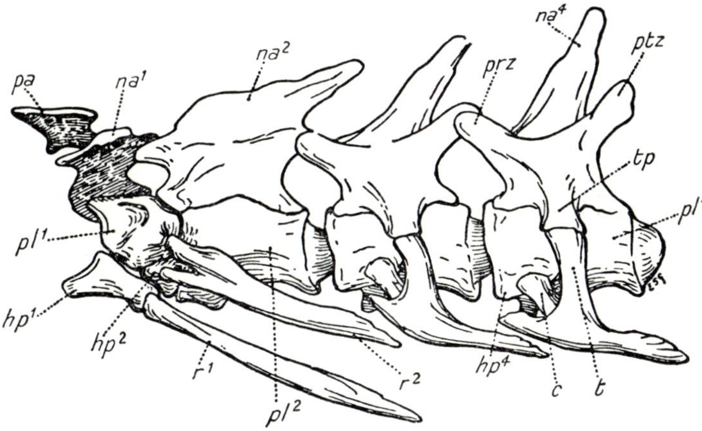

FIG. 243. Anterior cervical vertebrae of *Crocodilus*.

c. capitulum; *hp.* hypocentrum; na^{1-4}, neural arches; *pa.* pro-atlas; *pl.* 1–4, pleurocentra; *prz.* prezygapophysis; *ptz.* postzygapophysis; *r.* rib; *t.* tuberculum; *tp.* transverse process. The first neural arch and the pro-atlas of the left side have been removed to show the first pleuro-centrum (pl^1) which is the odontoid process. (From Goodrich.)

The crocodiles use all four limbs in walking, but the front are shorter than the hind, indicating bipedal ancestry. The pelvis of the crocodiles shows signs of the typical triradiate structure, but there are only two sacral vertebrae. Rapid swimming is produced by lateral movements of the tail, but when moving slowly the partly webbed feet are used to push the animal along. The ribs (Fig. 243) are two-headed and there is a proatlas element between the skull and atlas. The scutes of the back and, in some forms, of the belly, are reinforced by osteoderms, and there are well-developed abdominal ribs.

The soft parts of the crocodiles are of special interest because crocodiles are, except the birds, the only living creatures closely related to the great group of dinosaurs. The heart (Fig. 245) shows a complete division of the ventricle, but there are still two aortic arches. The truncus arteriosus is divided in a spiral manner to its base, so that the aortae cross and the right arch opens from the thick-walled left

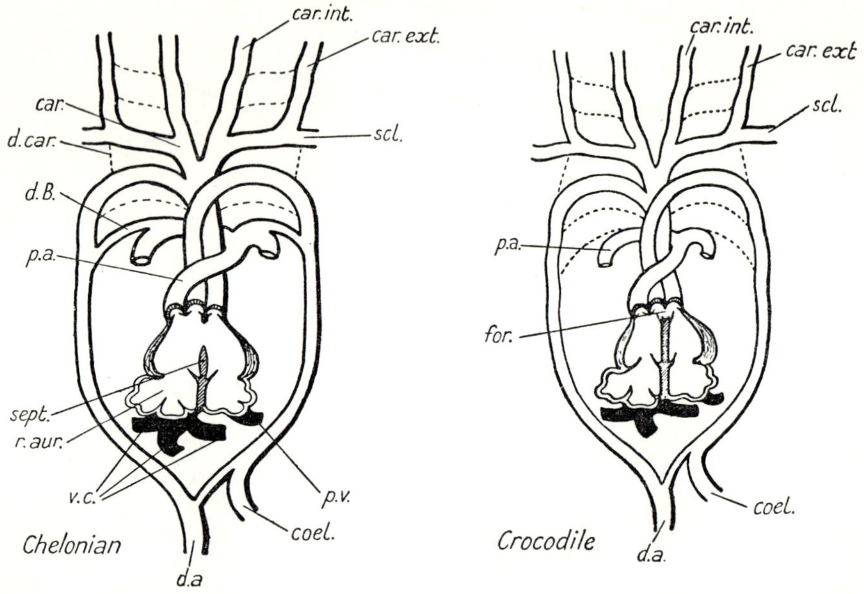

FIG. 244. FIG. 245.

Diagram of heart and arterial arches of a chelonian and of a crocodile, seen from below. Lettering as Fig. 215, p. 379. (From Ihle, after Goodrich.)

ventricle, while the left opens with the pulmonary arteries from the weaker right ventricle. The left arch would therefore contain venous blood, but an aperture, the foramen of Panizza, connects the two arches near the base and presumably the higher pressure in the left ventricle ensures that the left arch receives at least some oxygenated blood. Possibly, however, the pressure in the right ventricle is increased when the crocodile dives and the blood flows through the foramen from right to left.

The lungs are well developed, having a system of tubes ending in sacs. A transverse partition separates off a thoracic from the main abdominal cavity. This 'diaphragm' is not itself muscular, but is continued into a diaphragmatic muscle attached to the abdominal sternal plates. This muscle, innervated by abdominal spinal nerves, presumably assists in respiration. It is a development for this purpose quite distinct from the mammalian diaphragm. It is not impossible that the dinosaurs possessed further developments of this arrangement of the heart and lungs, and that they owed some of their success to this mechanism.

The modern crocodiles represent only the survivors of a once much more abundant group. *Crocodilus* is the most widespread genus, occurring in Central America, Africa and Asia, Malay and East Indies, and North Australia. *Alligator*, with each fourth lower tooth penetrating into a hole in the maxilla, is found in North America and in China. *Caiman* of Central and South America is related to *Alligator*. The length of the snout varies considerably in different species, and is extremely long and slender in the fish-eating *Gavialis*, Indian gharial, and *Tomistoma* of the East Indies. Crocodiles lay hard-shelled eggs in large clutches, depositing them in the sand or in nests composed of vegetation. The crocodiles seem to have changed little since they first appeared in the late Triassic, perhaps 190 million years ago. **Protosuchus* of that time had a pelvis like that of crocodiles but was otherwise very like a pseudosuchian. There were numerous types of crocodile in the Jurassic and Cretaceous, living both in fresh water and in the sea. In these forms the palate was closed only as far back as the palatine bones; the addition of flanges of the pterygoids took place only in the Eocene crocodiles, which were numerous in many parts of the world, including northern continental regions that today are too cold for such animals. In spite of their specializations for aquatic life, the crocodiles show us many features that were present in the earliest archosaurs and they therefore give some idea of the characteristics of the ancestors of the pterodactyls, dinosaurs, and birds.

15. The 'Terrible Lizards', Dinosaurs

In the 10 million or so years at the end of the Triassic some of the descendants of the pseudosuchians became very successful and numerous and many of them were very large. The large size was not a characteristic only of one line but of two quite distinct ones, each with several sub-divisions. The term dinosaur is applied to all of them, but the two main lines have little in common beyond the characters common to all archosaurs. The desire to explain this extraordinary exuberance of reptiles has attracted much attention to these giants.

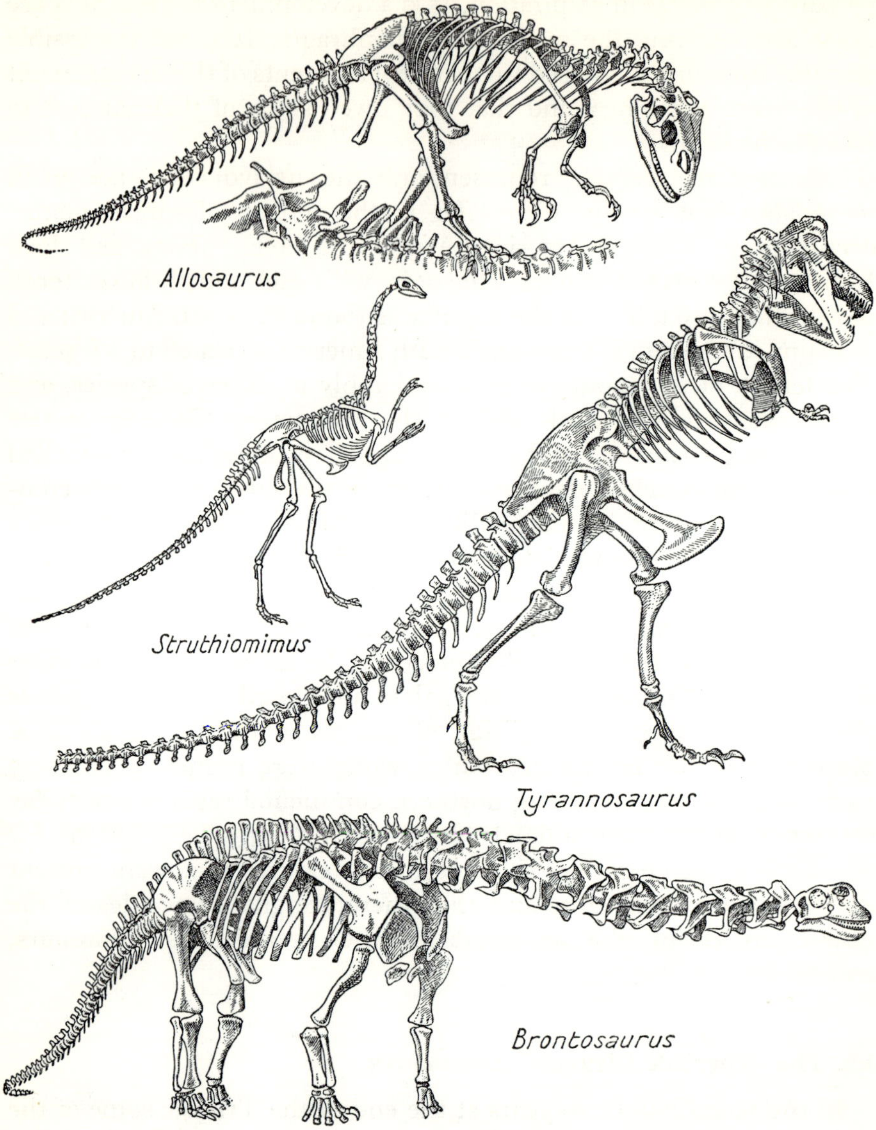

FIG. 246. The skeletons of various saurischian dinosaurs. (Modified after various authors.)

16. Order *Saurischia

These include forms with a triradiate pelvis, very like that of the pseudosuchians. The earlier types, like their ancestors, were bipedal carnivores of no great size, such as **Compsognathus* from the Jurassic of Europe and **Ornitholestes* from that of North America (Fig. 241). The front legs were short, with 4 or 3 digits, provided with claws; the pectoral girdle was reduced to scapula and small coracoid, with no trace of clavicles. Some members of this line, the theropods, soon developed into large carnivores, such as **Allosaurus* (Fig. 246), over 30 ft long (Jurassic, North America). These animals apparently swallowed their food whole and to help with this the quadrate was movable and there was a joint between the frontals and parietals, as in many lizards. In other respects the skull was very similar to that of the pseudosuchians.

At the end of the Cretaceous this theropod line produced the largest carnivores that have appeared on the earth, such as **Tyrannosaurus rex*, nearly 50 ft long and 20 ft high, from North America. All the previously mentioned tendencies were here accentuated, producing creatures with bipedal habit, very powerful head and jaws, and much-reduced fore-limbs. They presumably preyed upon the large herbivorous dinosaurs of the Cretaceous and became extinct with their prey, either from a common inability to meet the rigours of the climate or in competition with the mammals and birds. Throughout most of the Jurassic and Cretaceous the theropods were the dominant carnivores of the world, taking the place occupied earlier by the synapsid reptiles (p. 540) and later again by descendants of the synapsids, the carnivorous mammals.

In the Cretaceous the organization of this saurischian line also produced some exceedingly bird-like forms, **Struthiomimus* and **Ornithomimus*, walking on three toes and having three also in the hand, one opposable and used for grasping. The skull became very lightly built and the teeth disappeared, possibly in connexion with an egg-eating habit (Fig. 246).

All these carnivorous, bipedal saurichians may be grouped into a suborder Theropoda. Another line of organization, starting from bipedal, carnivorous Triassic theropods, adopted a herbivorous diet and reverted to the quadrupedal habit. These animals, the suborder Sauropoda, culminated in the immense Jurassic forms, **Apatosaurus* (= **Brontosaurus*) and **Diplodocus*, the largest of all terrestrial vertebrates. Several stages of the transition from bipedal to quadrupedal habit can be traced. **Yaleosaurus* from the Trias was a bipedal creature 6 ft long but with rather long front and short hind legs. **Plateosaurus*, also of the Trias, was 20 ft long, but still bipedal. Soon the front limbs became larger and more used for walking, though the disparity always remained. The neck was immensely elongated and the head very small, with a lightly built skull. The nostrils lay on the top of the head and in **Diplodocus* formed a single opening. This seems to indicate that the animals were aquatic or amphibious, as would in any case be suspected from the very large size, making it unlikely that the legs could bear the full weight. **Diplodocus* and **Brachiosaurus* were over 80 ft long and the weight of the latter must have been nearly 50 tons. However, the structure of the vertebral column shows that much weight was carried on the legs, for the vertebrae are strong, though hollowed in places. Footprints of the animals have been found. One or more of the digits bore claws. The skull became relatively short and broad, and among the many puzzling features of these giant animals is the weakness of the jaws and small size of the teeth, mostly crowded towards the front of the mouth. These teeth would have served well enough for cropping, but there are no teeth on the hind part of the jaws and no provision for grinding the food. Animals of large size can only have been supported by this feeble apparatus if some very nutritious food was readily available. This perhaps agrees with the small size of the brain, which was several times smaller than the lumbar enlargement of the cord.

17. Order *Ornithischia

The second main group of dinosaurs appeared later than the sauropods and possessed a 4-radiate pelvis, with the pubis directed backwards and an extra pre-pubic bone pointing forwards. The teeth were restricted to the hind part of the jaws, the front bearing a beak. At the front end of the lower jaw there was an extra bone (predentary). These were herbivorous forms and they appeared in the Jurassic and achieved their maximum in the Cretaceous, by which time the sauropods had become less common. The earliest of the ornithischians were bipedal animals, included in a suborder Ornithopoda, from the Jurassic and Cretaceous. These animals, such as **Iguanodon*, were built on the same general lines as the pseudosuchians, from which they were presumably derived. The skull was heavily built and adapted for a herbivorous diet, with powerful muscles attached to a coronoid process of the lower jaw. The bipedalism was less marked than in saurischians and the fore-limbs less reduced. Several separate lines then reverted to a quadrupedal habit. The trachodonts (**Hadrosaurus*) were a very successful group of amphibious forms in the Cretaceous, with webbed feet. The teeth were suited for grinding, parallel rows being present, making as many as 2,000 teeth in one animal. In several types of hadrosaur the top of the head was pro-

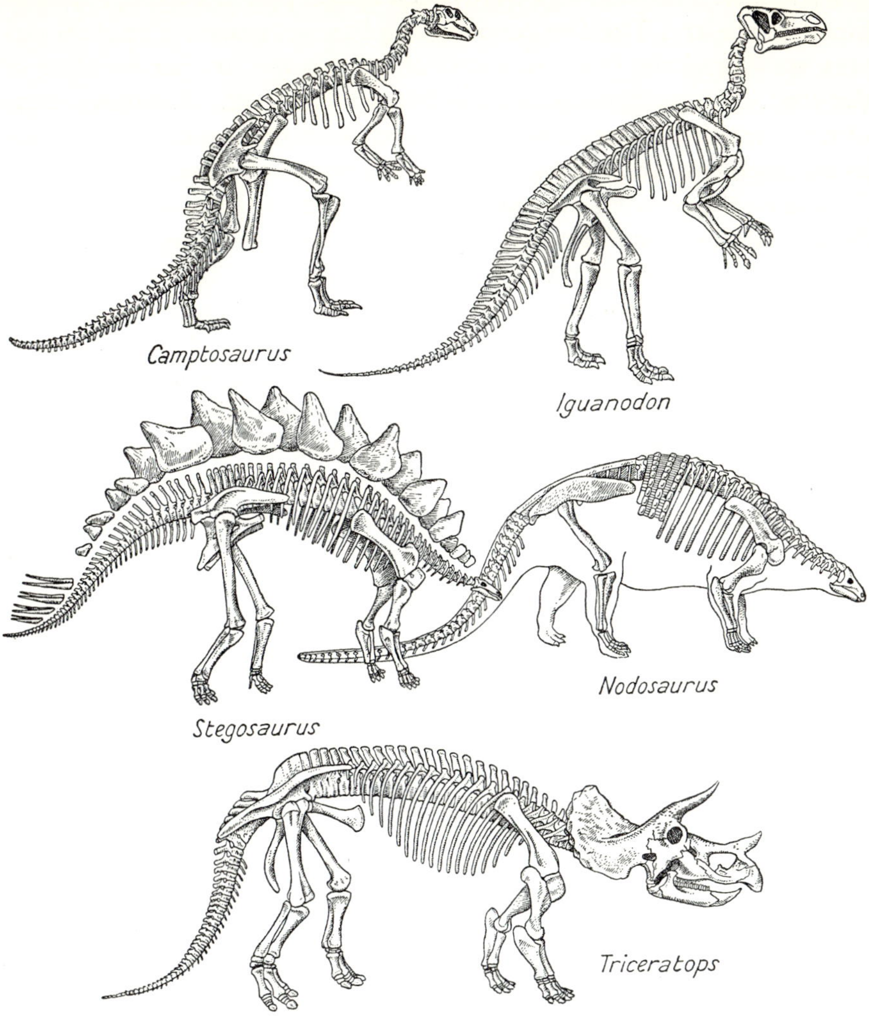

FIG. 247. The skeletons of various ornithischian dinosaurs. (Modified after various authors.)

longed in various ways, giving a structure that perhaps allowed the nostrils to remain above water while the animal was feeding below. These animals reached 30 ft in length and may have supplanted the sauropods as marsh-living forms, possibly when the soft foods gave place to harder plants.

Other lines of ornithischians became more fully terrestrial and quadrupedal and were mostly heavily armoured. Thus the stegosaurs of the Jurassic carried immense spines on the back and the tail bore sharp spikes. The hind legs were much longer than the front, a relic of bipedal ancestry. The feet carried hoof-like structures. The skull was very small and the brain much smaller than the lumbar swelling of the cord. The teeth were in a single row and small. The ankylosaurs of the Cretaceous were covered all over with bony plates, somewhat in the manner of the mammalian glyptodonts (**Nodosaurus*, Fig. 247).

Finally, the ceratopsians, such as **Triceratops* of the late Cretaceous, developed enormous heads, with huge horns and a large bony frill, formed by extension of the parietals and squamosals to cover the neck. These later Cretaceous animals appear to have lived on dry land and to have walked on all fours, although the bipedal ancestry is shown in the shortness of the front legs. There are several indications that the climate at the close of the Cretaceous was becoming drier and the organization of the giant reptiles became modified accordingly. They survived successfully for a while, but were ultimately replaced by the mammals, perhaps as a result of still further change in the climate (see p. 538).

18. Order *Pterosauria

The Triassic archosaurian reptiles gave rise to two independent stocks that took to the air, the pterodactyls and the birds. Both of these appear first in the Jurassic as animals already well equipped for flight, although obviously basically of archosaurian structure. We cannot therefore say anything about the steps by which their flight was evolved and can only speculate about the influences that drove them to take to the air. The early archosaurs were bipedal animals, and the fore-limbs were therefore free and available for use as wings. There has been much speculation about the intermediate stages by which flight was produced. Other reptiles, such as *Draco*, the flying lizard (p. 407), develop a membrane between the limbs and the body to assist them in making soaring jumps. The flight of pterodactyls and birds may have originated thus or, as suggested by Nopcsa, by the flapping of the fore-limbs during rapid running on the ground, the animals then becoming airborne for longer and longer periods.

The stages of the evolution of flight may have been different in the two cases, for whereas the birds are obviously bipedal animals and the similarity to such reptiles as **Struthiomimus* and **Ornithomimus* is obvious, the pterodactyls probably could not walk on their hind legs and may have used the wing more for soaring than for flapping flight. In spite of great differences there are interesting parallelisms in the structure of the fully evolved fliers of the two groups, for instance the limb bones became light, the skull bones fused, and the jaws toothless and beaked. This parallelism in lines known to be distinct, although of remote common origin, is similar to that which we have noticed before in aquatic animals, and it can be interpreted as showing that

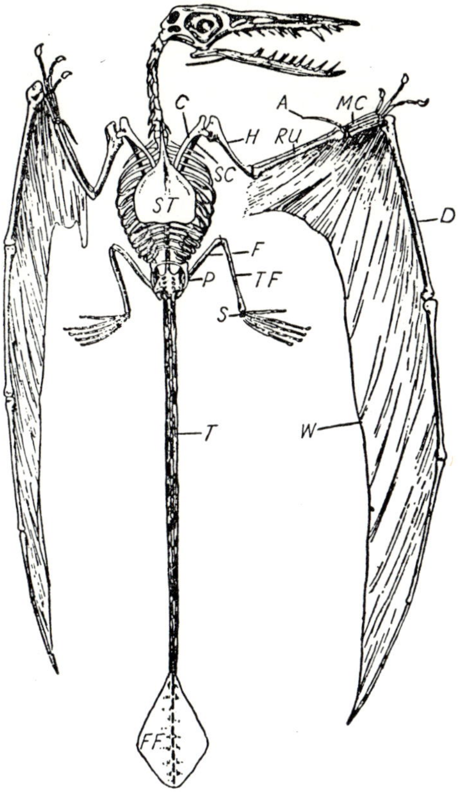

FIG. 248. The skeleton of a pterodactyl.

A, extra wrist bone; *C*, coracoid; *D*, elongated digit; *F*, femur; *FF*, fin; *H*, humerus; *MC*, metacarpal; *P*, pelvis; *RU*, radio-ulna; *SC*, scapula; *ST*, sternum; *T*, tail; *TF*, tibio-fibula; *W*, wing. (From Thompson, *The Biology of Birds*, Sidgwick & Jackson, Ltd.)

populations with similar genotypes will respond to similar environmental stimuli in the same way.

The pterodactyls are most commonly found in the Jurassic strata, less often in the Cretaceous. Many specimens have been found in marine deposits and seem to have been fish-eaters. The characteristic features that have produced the pterodactyl structure from a thecodont ancestry may be described as a lengthening of the head and neck, shortening of the body and ultimately of the tail, lengthening of the arms and especially of the fourth digit, shortening of the legs, and development of the ventral parts of the limb girdles. These are the changes that can be recognized in the bony parts available for study; no doubt there were many others in the soft parts also paralleling the evolution of birds, for instance the animals may have been warm-blooded. However, there is no evidence that they possessed feathers; the wing was a membrane (patagium).

**Rhamphorhynchus* of the Jurassic is still recognizably of archosaurian structure, especially in the skull, which has two fossae and large forward-sloping teeth (Fig. 248). The fore-limb was elongated, but the carpus still short, with an extra 'pteroid' bone in front, presumably to support the wing. The first three digits were short and hooked, the fourth long, supporting the wing, and the fifth absent. The hind-limb was slender, with five hooked digits. There was a long tail, ending in an expanded 'fin'. Both girdles had well-developed ventral regions and there was a large 'sternum', keeled in front. The scapula articulated directly with the vertebral column.

**Pteranodon*, of the Cretaceous, showed further modifications. The trunk became shortened to ten or fewer segments and the fore-limb further lengthened, the carpus being long and the fourth digit much longer than the other three. The hind-limb remained small and the tail became very short. The very large and elongated head gradually lost its teeth, presumably acquiring a horny beak. In the latest forms the skull was drawn out backwards into an extraordinary process. Some earlier related forms were only a few inches long, but **Pteranodon* itself, of the late Cretaceous, had a wing-span of 25 ft.

Zoologists have not yet succeeded in reconstructing the life of these animals, and it is hard to see how they could have walked on land. The membrane, which stretched between both legs and the body, and perhaps also included the head, must have been easily torn. The feathers of birds can be ruffled without breaking and the loss of a few does no great harm: the bat's wing can be torn, but at least it is supported by many digits, whereas that of the pterodactyl was a huge continuous membrane supported by a single finger. Again it is difficult to see how the animals can have perched; if they hung, as the claws suggest, was it with the front or with the hind legs? And how can they have staged a take-off, which in birds is greatly helped by the jump of the hind legs? It is possible that they always came to rest hanging from cliffs, which they could leave by soaring. Even the flight itself presents many difficulties. Although there is a sternum and a strong humerus, neither suggests the presence of muscles sufficiently strong to carry a creature as large as **Pteranodon*. We cannot solve any of these mysteries, but one clue is that the biggest pterodactyls were mostly, if not all, marine. The largest flying birds alive today are the albatrosses, which use their great weight to gain height with the increasing velocity of the wind a few feet above the sea (p. 460). It is possible that the pterodactyls used a similar method of soaring. They were presumably unable to compete with the birds, however, and died out at the end of the Cretaceous, along with so many other reptiles.

19. Conclusions from study of evolution of the reptiles

Many of the conclusions that have been drawn from study of vertebrate evolution in the water also apply to the forms that have come on land. The fossil record leaves no doubt that almost all the populations have changed very markedly. Few forms of reptile alive today are closely similar to any found in the Permian or Triassic periods. *Sphenodon* has shown relatively less change than most others; it may be significant that it is found in an isolated island region (but see p. 772).

The data are not sufficient to show the rate of evolutionary change. We cannot be sure whether it has been constant or even continuous, but particular types are found only from a limited range of strata and there is little evidence that any terrestrial form remains unchanged for more than a few million years, at most. Each type is successful for a while and then the niche that it fills becomes occupied by another type, either descended from the first or, more usually, from some related stock. Thus the earliest large land herbivores were probably the pareiasaurs; these were replaced by other reptilian types such as the herbivorous mammal-like reptiles, and later the sauropods (in so far as these were terrestrial) and various types of ornithischians; then perhaps by the hadrosaurs in the more watery habitats and the stegosaurs, ankylosaurs, and ceratopsia on drier ground. Finally, all these gave place to the earliest mammalian herbivores, which were in turn replaced by others (p. 776).

Throughout early tetrapod evolution there is a tendency to return to the water, perhaps under some pressure of competition from descendants on land. This is marked among reptiles, where besides the chelonians and ichthyosaurs and plesiosaurs there are the phytosaurs and crocodiles, and among Squamata the mosasaurs and tylosaurs, not to mention the sea-snakes.

The large size of many reptiles has been one of their most striking features, but it is, of course, not true to say that there is a strong tendency for size to increase in all reptile groups. While many have become enormous, others, such as the lizards, have produced probably as great a biomass spread over a large number of small individuals. Large size in a reptile may help to conserve heat (p. 372), but could also endanger the animal from overheating, since the ratio of surface area to volume decreases as the absolute size increases, and heat cannot be lost so readily through the skin. Up to a point size may be a protection, but it involves the dangers of those who place all the eggs in one basket; incidentally, the actual eggs of these large animals must have provided formidable physical problems for their support.

Parallel evolution of several lines descended from a single stock is as common among reptiles as among other groups of vertebrates. Thus the bipedal habit, with hind legs longer than the front, has been adopted independently by a number of diapsids; again, elongated jaws are found among fish-eaters, whether ichthyosaurs, plesiosaurs, phytosaurs, crocodiles, or mosasaurs.

Although it is difficult to see in all this any persistent tendency except to change, yet the very fact that each type is so rapidly replaced suggests that descendants in some way more efficient are continually appearing. In the case of the reptiles the more interesting of these are the birds and mammals, and we shall therefore leave the problem of serial replacement among amniotes for later discussion. Meanwhile we may note once again that the reptiles surviving today, although not of larger size nor obviously better suited for life than their mesozoic ancestors, yet exist in considerable numbers alongside and even in competition with the birds and the mammals.

XVI

LIFE IN THE AIR: THE BIRDS

1. Features of bird life

THE quality we define as 'life' is perhaps more fully represented in birds than in other vertebrates, or indeed in any animals whatsoever. It is difficult to find units by which accurate comparisons can be made of such matters, but there is a meaning in the statement that the life of a bird is more intense than that of, say, a reptile or a fish. Following out our definition of the life of a species as the total of the activities by which that particular type of organization is preserved, we shall find that the birds have many and very varied activities, by means of which a great deal of matter is collected into the bird type of organization. Moreover, this is achieved under conditions remote from those in which life first arose; the birds get a living by moving in the air, the most difficult medium of all.

Flight is of course the characteristic that gives us most fully the feeling that the birds are active animals; it impresses us as a technical marvel and as a means by which the animals obtain a most enviable and valuable freedom, enabling them to avoid their enemies and to seek new habitats. Almost equally important items in the active life of birds are the high and constant temperature and large brain. These features have been acquired independently by birds and mammals, and have led to profound changes in behaviour. A homoiothermic animal does not need to change its activity with the changes in environmental temperature; it can be continuously active, and, perhaps even more important, its steady continuity of life makes possible the accurate recording of past experience in the memory. Probably only with a high and constant temperature can full use be made of the possibilities of delicate balance of activities within large masses of nervous tissue. In homoiothermic birds and mammals we find larger brains and more elaborate social and family habits than in any other animals.

2. Bird numbers and variety

Flight necessitates a large surface-weight ratio, therefore birds do not become so large as some mammals; nevertheless, an immense biomass is produced by their very large numbers. Any attempt to enumerate the bird population is largely guess-work, but the density of breeding birds in different habitats in Britain has been estimated, and varies from 200 per 10 acres in woodland to 20 on agricultural land and 10 or less on moorland. Calculating from such figures Fisher estimates that there may be 100 million land birds in Great Britain and 100,000 million birds in the world altogether, including sea birds. This is perhaps a low estimate; it would represent a total biomass of the same order as that of 3,000 million human beings. With all their activity, therefore, the birds organize less matter into themselves than do the mammals.

One of the most striking features of bird life is that although the basic organization remains fairly constant differing types show a great variety of special features, fitting them for numerous habitats. Besides differences in behaviour, in body form, and in powers of flight there are found others in the shape of the bill, and hence of food habits, and in the details of many other parts, such as the feet, that make fascinating studies in adaptation to environment.

3. The skin and feathers

The skin of birds differs from that of mammals in being thin, loose, and dry; there are no sweat glands, indeed the only cutaneous gland present is the uropygial gland or preen gland at the base of the tail. The bird cleans its feathers with its beak, obtaining oil from this gland, which is especially well developed in aquatic birds.

The keratin-producing powers of the skin are of course mostly devoted to producing feathers, but scales like those of reptiles are present on the legs and feet and sometimes elsewhere. The bill (p. 466) and claws are also specialized scale-like structures and are sometimes moulted.

Nerve-endings are present throughout the skin, and the cere at the base of the bill is perhaps an organ of touch. The bill may itself have special endings, such as the corpuscles of Grandry found in the ducks.

The feathers of modern birds provide a covering whose uses vary from heat insulation and flight to protective coloration and sexual display. It is likely that in evolutionary history the function of heat regulation came first. The two main functions of heat conservation and flight are indeed today performed by feathers of different types (Fig. 249). The down-feathers or plumules, which form the covering of the nestling and may be present also in the adult, are simpler than the contour feathers or pennae, and the elaborate flattened flight feathers. Filoplumes are a third type, being very fine, hair-like feathers. Usually several generations of feathers are produced; first the nestling feathers (neoptiles), then one or more generations of juvenile feathers, which

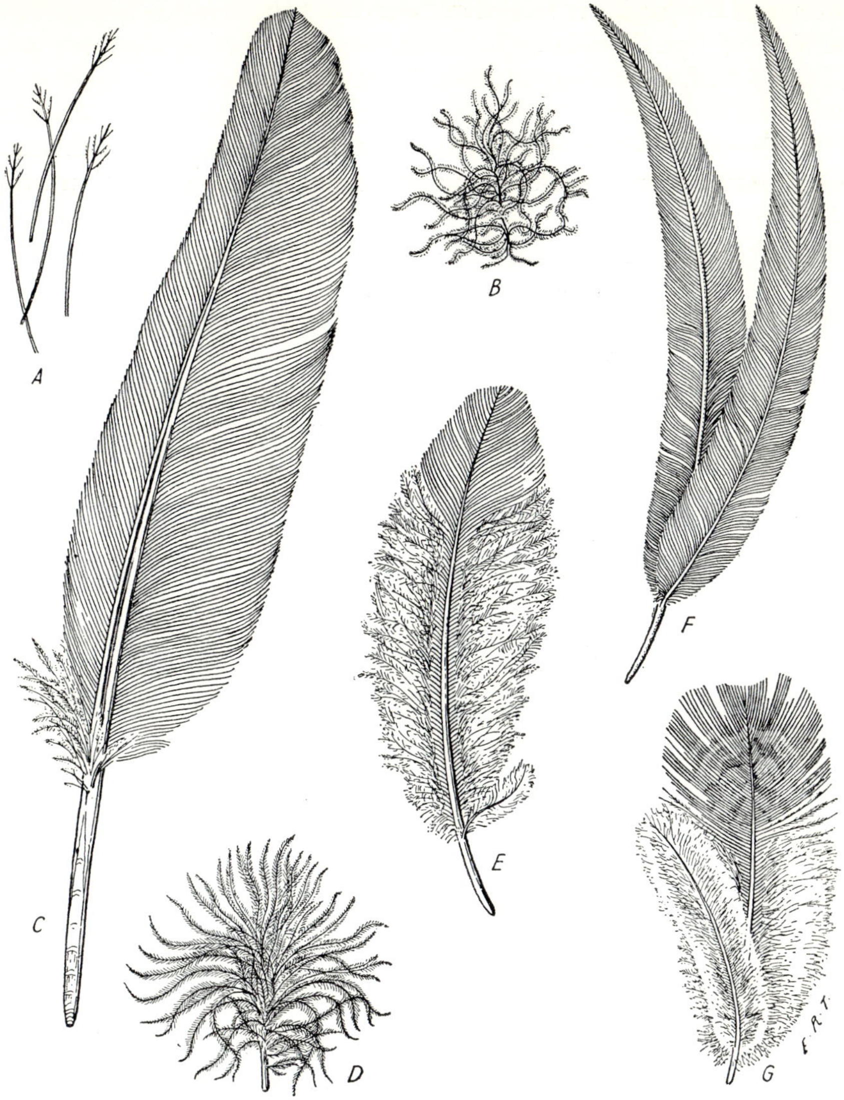

FIG. 249. Various types of feather.

A, filoplumes; B, nestling down-feather; C, primary wing feather of pigeon; D, permanent down feather; E, feather with free barbs; F, emu's feather with long aftershaft; G, contour feather of pheasant with aftershaft. (Partly after Thompson, *The Biology of Birds*, Sidgwick & Jackson, Ltd.)

may be of various types, prefiloplumes, preplumules, and prepennae; finally, the adult feathers (teleoptiles).

Each feather, of whatever type, is formed from a dermal papilla or follicle, over whose surface keratin is produced. In down-feathers the surface of the papilla is ridged all round and the result is to produce a number of fine threads or barbs of keratin, covering the body with a coat of fluff, which acts as a heat insulator by preventing air circulation.

Feathers, like other epidermal structures, are moulted, either at a certain stage in the life-cycle or seasonally, a new generation being

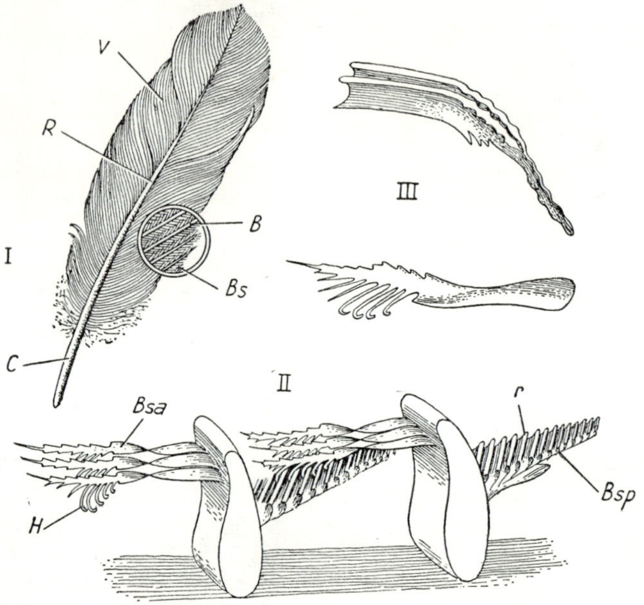

FIG. 250. The structure of a feather.

I. Whole feather showing calamus (quill), C; rhachis (shaft), *R*; and vane, *V*. On the right side a small area is shown as it appears under a lens; *B*, barb; *Bs*, barbule. II. A section cut at right angles to two barbs in the plane of the barbules of the anterior series (*Bsa*). Note how the hamuli, *H*, of the anterior barbs interlock with ridges (*r*) on the posterior barbules (*Bsp*). III. Shows one anterior and two posterior barbules isolated. (After Pycraft, *A History of Birds*, Methuen & Co., Ltd.)

produced from the old papillae. Most birds moult after the breeding-season, some a second time during the year. The down-feathers of the nestling are partly replaced by contour feathers; the follicle, instead of producing equal barbs, now forms two large ones at one side, which together become the central axis (rhachis), carrying a series of further barbs that spread at right angles to it to form the vane (vexillum). Each feather (Fig. 250) thus consists of a central rhachis, forming the hollow calamus or quill below and carrying the barbs, which make the vane. The calamus opens at the base by the inferior umbilicus, the entrance of the mesodermal papilla, and at the beginning of the vane there is a second hole, the superior umbilicus. At this point there is often a loose tuft of barbs or an extra shaft, the aftershaft, perhaps in some way representing the down-feather.

The barbs or rami that make up the vane are held together by rows of barbules (radii) running nearly at right angles to the barbs and carrying hooks (hamuli) by which the barbules of one radius become fixed to grooves in those of the next (Fig. 250). Anyone who has played with a feather knows that these connexions can be broken down so that the barbs become separate, but can be joined again by 'preening' the whole feather.

The feathers are provided with muscles at the base and the control of their position is important for the regulation of heat loss, for flight, and in many other activities, for instance sexual display. Like the hairs of mammals the feathers are also used as organs for the sensation of touch, nerve-fibres being wound round the base of the papilla. In owls and other night-birds special vibrissae, analogous to those of mammals, are present. Various specialized feathers are used for eyelashes, ornament, and other purposes, and in some birds patches of special feathers without rhachis break up to make a greasy 'powder down'.

The feathers are not spread uniformly over the body but are localized to certain tracts, the pterylae, separated by bare areas, apteria. Among the contour feathers it is usual to recognize the remiges of the wing and rectrices of the 'tail'. The former are divided into primaries on the hand and secondaries on the forearm (Fig. 256). Each large feather, whether in wing or tail, is usually covered above and below by several rows of upper and under coverts. In many birds there is a peculiar gap in the secondary feathers of the wing, the fifth remex feather being absent (diastataxis); the condition in which this feather is present is called eutaxis. The feathers have a remarkably flexible structure, so that they adopt different shapes with different positions of the wing. The shape of the quill and barbs varies between feathers and parts of a feather, for instance the barbs at the tip of the primary feathers provide a stream-lined cross-section, like that of certain aeroplane propellers (p. 453). The small covert feathers at the front of the wing stand up vertically, but have a right-angle bend, thus providing the wing camber.

The rectrices vary greatly, being almost absent in birds that live near to the ground, such as the wrens, but very large in fast-moving birds that change direction quickly (swallows). In these latter the outer rectrices are enlarged for steering purposes. The rectrices may be put to special uses, as in the woodpeckers, where they make a rigid brace, or in the peacocks, whose display feathers are the tail coverts.

4. Colours of birds

Birds possess colour patterns more vivid than those of any other vertebrates, using them not only for concealment but also as the chief means of recognition and sexual stimulation and hence as the basis of their social life. Like other animals that live far from the ground and move fast (primates) the birds have a poor sense of smell, often none at all, but they have very good vision, and in many species the turning of discriminating eyes by one sex upon the other has led to the development of a very gorgeous covering. The feathers alter the appearance of the bird so completely that it is not fantastic to compare their effect with that of clothing in man.

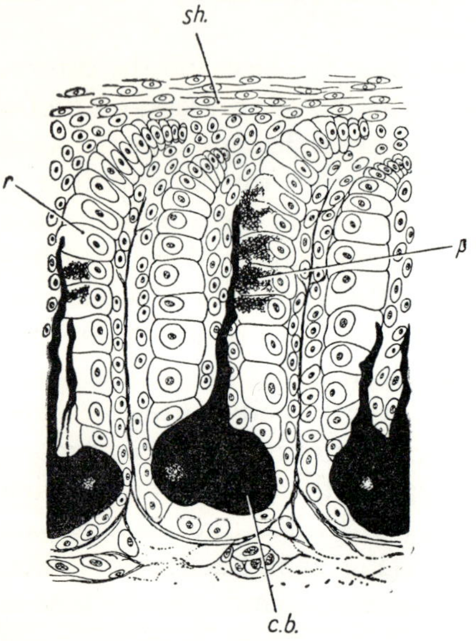

FIG. 251. Deposition of pigment in feather germ. Transverse section through a developing arm feather.

c.b. cell body of pigment cell; *p.* process of pigment cell; *r.* cells forming a radius; *sh.* sheath of feather germ. (After Strong, from Streseman.)

As in other animal groups the colours are produced partly by pigments and partly by reflection and diffraction effects (structural coloration). The most common pigments are melanins, ranging from black through brown to yellow, and laid down in the feathers by special cells in the papilla. The processes of these amoeboid chromatophores convey pigment to the epidermal cells (Fig. 251). Carotenoid pigments (soluble in organic solvents) are also found, such as the yellow xanthophyll of the duck's bill and feet and the red astaxanthin of pheasant wattles. White is usually given by reflection. In blue colours incident light is reflected from a turbid porous layer overlying a deposit of melanin pigment. In iridescent feathers interference of light in thin surface films gives colours like those of soap bubbles. The more specialized iridescent feathers produce Newton's rings, with colours of the second and even third orders. The turacos or plantain-eaters of Africa contain two very peculiar pigments, a copper-containing red porphyrin turacin, which is soluble in weakly alkaline water and dissolves out in the rain, and the green, iron-containing turacoverdin.

The actual colour patterns vary with the habits of the bird. Concealing (cryptic) coloration is very common; even the brighter colours may serve this purpose, by breaking up the outline of the bird when at rest or in motion. Most birds are dark above and white below. The feathers often show mottled or speckled patterns rather than a homogeneous colour. Finches and other birds living in the sunlit upper branches show bright yellow, yellow-green, and blue colours, either singly or combined. Birds living in thickets, such as the thrush and blackbird, are usually duller brown or black. An example of disruptive coloration that is easy to observe is the white patch on the throat of a thrush. If the nest is approached while the bird is sitting the head is held rigidly still with the beak upwards; the white mark on the neck breaks the outline and instead of an obvious bird's head there appear only the meaningless shapes of the sides of the jaws. In most species coloration is a compromise between concealment and conspicuousness. Sometimes selection has acted so that the female is cryptic, the male conspicuous (e.g. ducks). In hole-nesting shelducks both sexes are conspicuous. In other birds bright colours are concealed most of the time (e.g. the robin's red breast is underneath, many waders have conspicuous colours under their wings).

Some colour patterns seem to make the bird conspicuous and may be a warning of a distasteful quality. The black and white pattern shown by the magpie may be an example of such sematic coloration; certainly this bird is seldom preyed upon, no doubt partly because of its large size. The conspicuous black of rooks and starlings may be connected with their social life, making it desirable that the birds should easily follow each other, the group being protected by the combined receptors of its many members and the quick response of all to escape movements by any one.

The protective functions of the colour often give place in one or both sexes to garments used for communication between individuals, for such purposes as pair formation, aggression between males, nest site selection, or rearing the young.

5. The skeleton of the bird. Sacral and sternal girders

The arrangement of the whole locomotor apparatus is based on the plan of the bipedal archosaurian reptiles, modified and simplified for the purposes of flight and balancing and walking on two legs. The bones are very light and often of tubular form, but sometimes with internal strutting well suited to the stresses they must bear (Fig. 252). Many of the bones contain extensions of the air-sacs; even the wing and leg bones are pneumatized in this way in very good fliers, such as some birds of prey and the albatross. Fusion of bones has proceeded so far that the skeleton consists of a few hollow girders and large plates of special shape (p. 441). This result is achieved by limiting the joints at which movement occurs and simplifying the muscular system. The long bones ossify from a single diaphysis, there are no epiphyses at the ends.

The skeleton of the backbone and limb girdles is so modified as to allow the weight of the body to be carried in two quite distinct ways,

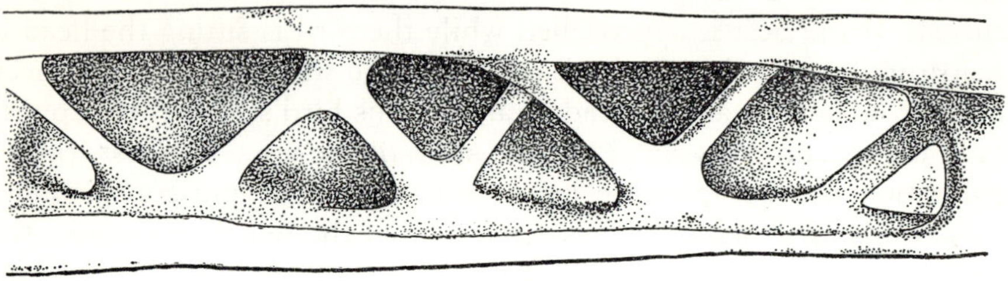

FIG. 252. Metacarpal bone from the wing of a vulture, sectioned to show the arrangement of the struts similar to that known to the engineer as a Warren's truss, such as is often used in aeroplane wings. (After Prochnow and D'Arcy Thompson.)

on the wings or on the legs. For this purpose there are two plate-like girders, the sternum and the synsacrum, curved in opposite directions. The muscles around the shoulder and hip joints balance the weight on these girders and produce propulsion. The main thrusts come from the pectoralis major in flying and from the leg retractors in walking. Perhaps no other animals are suited so perfectly for locomotion by two distinct means, and of course many birds can swim as well as fly and walk.

The whole axis of a bird is morphologically shorter than that of any other vertebrate except a frog or a tortoise (Fig. 253): only the neck remains a long and mobile structure. The number of cervical vertebrae varies and is greater in the birds with longer necks; there are fourteen in the pigeon, if we include two that bear ribs not articulating with the sternum. The cervical centra have saddle-shaped surfaces, the concavity running from side to side on the front and up and down behind, allowing great mobility in all directions.

There are four or five thoracic vertebrae, all except the last united into a single mass. The ribs are large, double-headed, and jointed to the vertebrae. They bear uncinate processes on their vertebral portions, hook-like projections overlapping the rib next behind and thus strengthening the whole thoracic cage. There is a well-marked joint between the vertebral and sternal portions of the ribs. The latter are bony, not cartilaginous as in mammals, and are jointed to the sternum, which is a very large keeled structure in all flying birds, serving to

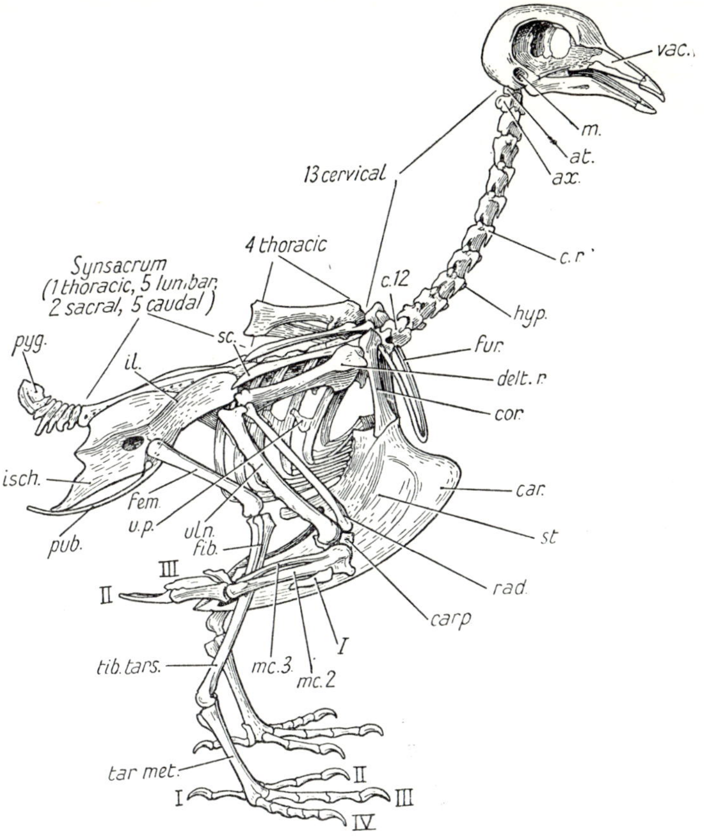

FIG. 253. Skeleton of the pigeon (*Columba*).

at. atlas; *ax.* axis; *C12*, 12th cervical vertebra; *car.* keel of sternum; *carp.* carpus; *cor.* coracoid; *c.r.* cervical rib; *delt.r.* deltoid ridge of humerus; *fem.* femur; *fib.* fibula; *fur.* furcula; *hyp.* hypapophysis; *il.* ilium; *isch.* ischium; *m.* auditory meatus; *mc. 2* and *3*, metacarpals; *pub.* pubis; *pyg.* pygostyle; *rad.* radius; *sc.* scapula; *st.* body of sternum; *tar.met.* tarsometatarsus; *tib.tar.* tibio-tarsus; *uln.* ulna; *up.* uncinate process; *vac.* vacuity in side of skull.

carry the weight of the body to the wings by the attachment of the main wing muscles (Fig. 254). The pectoralis major, which depresses the wing in flight, is attached to the edge of the sternum and the great depth of the keel serves to increase the length and mechanical advantage of the fibres of the muscle and also, by its shape, to strengthen the sternum. When the bird is in the air the sternum is carrying a large part of the weight. By this arrangement the centre of gravity is kept well below the centre of pressure, giving great stability.

The last thoracic (rib-bearing) vertebra is united with about five that can be regarded as lumbars, two sacrals and five caudals to make a synsacrum, which is also fused with the ilium. This produces a very thin plate-like structure, whose ridged shape gives it sufficient strength to carry the bird's weight. Finally, there is a short bony tail of about six free caudal vertebrae, carrying four that are fused together to form the upturned pygostyle, supporting the tail feathers.

The joints of the vertebral column are therefore reduced so as to allow movement only in the cervical region, between the thorax and

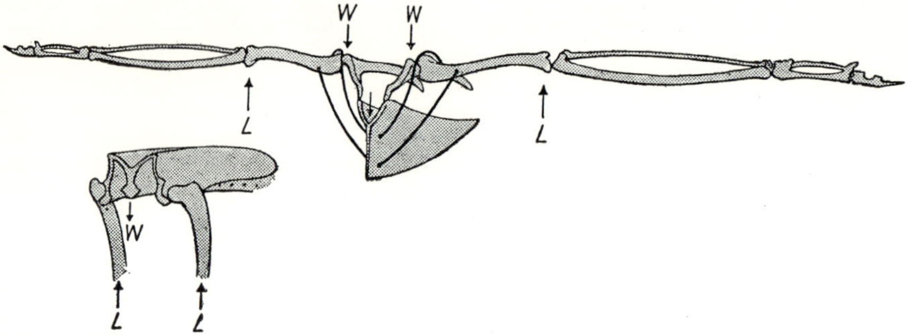

FIG. 254. Diagrams of the pectoral and pelvic girdles of an eagle, to show the methods of support in flying and walking. In each case the weight is carried on an arch, the strength of which is obtained by the peculiar kinked shape of the thin sheets of bone.

synsacrum and in the tail. The axial muscles have been correspondingly reduced. Those of the neck are large and the hinder cervical and the thoracic vertebrae have special ventral hypapophyses for attachment of the flexor muscles of the neck. The other back muscles, except those of the tail, are reduced and the whole back forms a single rigid strut, carrying the weight of the breast and viscera through the ribs and the abdominal muscles either to the pelvic girdle or to the sternum. In flying this weight is suspended on the wings and there is therefore a compression stress throughout the ribs, and this no doubt accounts for the ossification of their ventral parts. The weight of the bird when resting on its wings (Fig. 254) is thus carried by the pectoralis major as a tension member, through the plate-like sternum; the ribs, and especially the coracoid, act as compression members. The last-named bone lies nearly in the plane of the pectoralis major and is very strongly built.

6. The sacral girder and legs

In standing, perching, and walking the weight is balanced on two legs. To achieve this posture the type of girder found in the vertebral column of other terrestrial vertebrates has been abandoned, and with it the system of braces (back muscles) holding up the weight of the forepart of the body. Instead the whole axis is so shortened that the centre of gravity lies far back, low, and over the feet. This is not

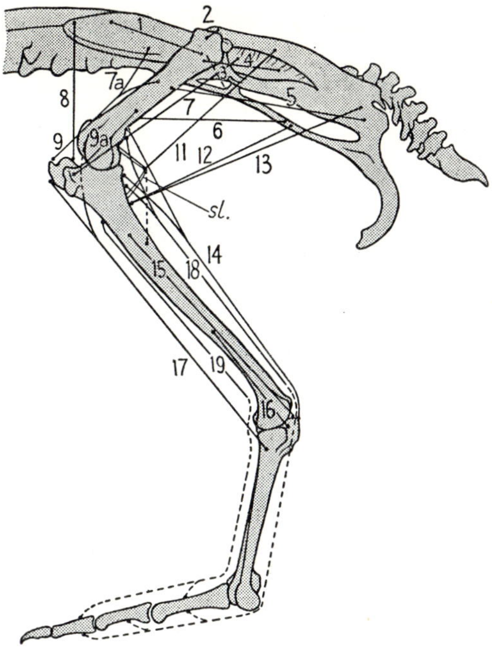

FIG. 255. Diagram of muscles of the hind leg of a bird. Tendons shown dotted.

1, Mm. ilio-trochanterici; 2, M. ilio-femoralis; 3, M. obturator; 4, M. ischio-femoralis; 5, M. caud-ilio-femoralis; 6, M. pub-ischio-femoralis; 7, M. ilio-tibialis posterior; *7a*, M. ilio-tibialis anterior; 8, M. sartorius (ilio-tibialis internus); 9, M. femoro-tibialis medius; 9*a*, M. femoro-tibialis externus; 11, M. ilio-fibularis; 12, M. ischio-flexorius; 13, M. caud-ilio-flexorius; 14, M. gastrocnemius; 15, M. peroneus superficialis; 16, M. peroneus profundus; 17, M. tibialis anterior; 18, Mm. flexores digitorum; 19, Mm. extensores digitorum. *sl.* sling for M. ilio-fibularis. (After Stolpe.)

apparent from Fig. 255, which is not in a normal perched position. Birds whose feet are placed far back for swimming must hold the body nearly upright to achieve a stable position with the centre of gravity over the feet (auks, penguins). The ribs and abdominal muscles transfer the weight to the greatly elongated ilia, which are fused to the vertebrae, making a long girder of approximately parabolic form. Though this is composed of bone of almost paper thinness, it is strengthened by longitudinal ridges (Fig. 254). Its strength, like that of the sternum, lies not in its arched shape in the transverse plane, but in the distribution of weight that is achieved by its longitudinal curve and peculiar kinked shape. The whole pelvic girdle is modified to allow

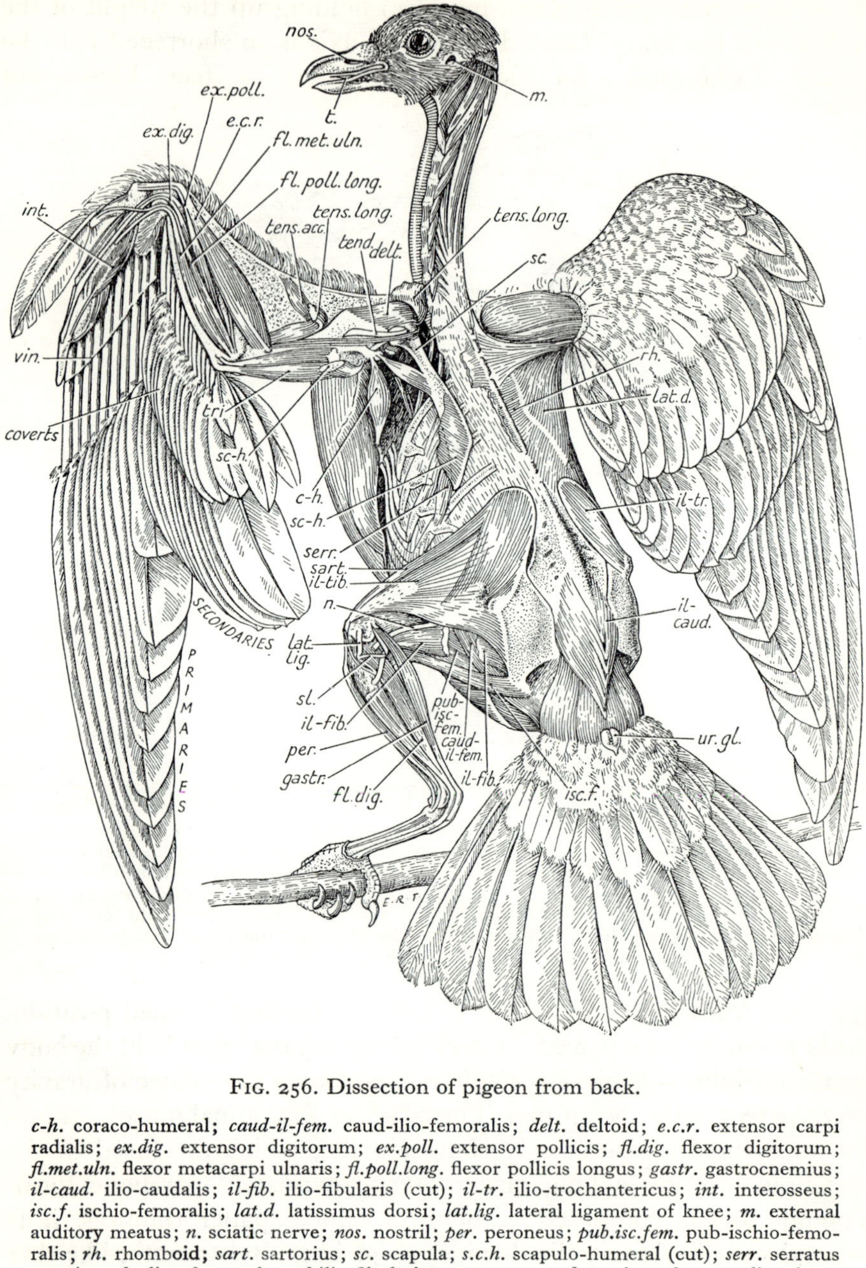

FIG. 256. Dissection of pigeon from back.

c-h. coraco-humeral; ***caud-il-fem.*** caud-ilio-femoralis; ***delt.*** deltoid; ***e.c.r.*** extensor carpi radialis; ***ex.dig.*** extensor digitorum; ***ex.poll.*** extensor pollicis; ***fl.dig.*** flexor digitorum; ***fl.met.uln.*** flexor metacarpi ulnaris; ***fl.poll.long.*** flexor pollicis longus; ***gastr.*** gastrocnemius; ***il-caud.*** ilio-caudalis; ***il-fib.*** ilio-fibularis (cut); ***il-tr.*** ilio-trochantericus; ***int.*** interosseus; ***isc.f.*** ischio-femoralis; ***lat.d.*** latissimus dorsi; ***lat.lig.*** lateral ligament of knee; ***m.*** external auditory meatus; ***n.*** sciatic nerve; ***nos.*** nostril; ***per.*** peroneus; ***pub.isc.fem.*** pub-ischio-femoralis; ***rh.*** rhomboid; ***sart.*** sartorius; ***sc.*** scapula; ***s.c.h.*** scapulo-humeral (cut); ***serr.*** serratus anterior; ***sl.*** sling for tendon of ilio-fibularis; ***t.*** tongue; ***tend.*** tendon of pectoralis minor; ***tens.acc.*** tensor accessorius; ***tens.long.*** tensor longus; ***tri.*** triceps; ***ur.gl.*** uropygial gland; ***vin.*** vinculum elasticum.

this arrangement. The ischium and pubis are directed backwards and do not meet in a symphysis, which would prevent the underslinging of the viscera.

The legs are used for balance and walking or hopping in ways that show interesting similarities and differences from those of man. The femur is turned under the body and articulates with the acetabulum in such a way that movement is almost restricted to the antero-posterior direction. The bird balances on its hips only in the sagittal plane; there are no movements of abduction and adduction such as are found in man. Abduction of the leg, or the falling medially of the bird's body when standing on one leg, is prevented by the fact that besides the ball and socket articulation of the femoral head there is also a second joint surface between the trochanter and an anti-trochanter of the ilium. The ligaments across the top of this joint are very strong and they limit abduction movements, while movements of adduction are restricted by a strong ligamentum teres attached to the femoral head.

In life the femur is held nearly horizontal, bringing the legs well forward. The bird replaces the movements of abduction and adduction, which we make at the hip during walking in order to prevent falling over while only one leg is on the ground, by movements of rotation at the knee. The muscles around the hip joint form a system of braces allowing balancing and locomotion much as in man, but they are well developed only anteriorly and posteriorly; the lateral and medial (abductor and adductor) elements are weak (Figs. 255–8). The anterior group (protractors) includes a sartorius (ilio-tibialis internus) running from the ilium to the tibia, an ilio-femoral, and a large anterior ilio-tibial inserted through a patella to a ridge on the front of the tibia. Associated with this muscle, which crosses both hip and knee joints, there are also, as in man, femoro-tibial muscles, making up with the longer muscles, the extensor system of the knee.

The lateral side of the hip joint is supported by rather small abductor braces, the ilio-trochanteric muscles, corresponding to our glutei, and acting mainly as medial rotators, opposed by obturator and ischio-femoral muscles, which work as lateral rotators. The main locomotor muscles are the posterior braces or retractors, lying behind the hip joint and including muscles known as the posterior ilio-tibial, ilio-fibular, caud-ilio-flexorius, pub-ischio-femoral, ischio-femoral, and caud-ischio-femoral. Some of these also act with the obturator muscle as lateral rotators, and those placed more medially function as adductors or medial braces, so far as such are required.

The femur articulates with both tibia and fibula at the knee. The fibula is distinct at its upper end, fused with the tibia below. The joints of the foot are greatly simplified by the union of the proximal

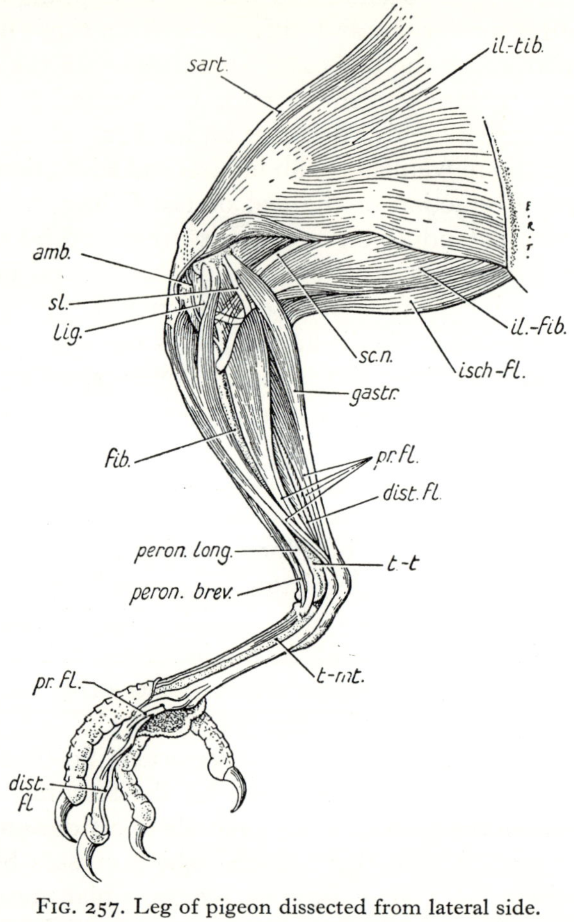

FIG. 257. Leg of pigeon dissected from lateral side.

amb. ambiens tendon; ***dist.fl.*** flexors of distal phalanges; ***fib.*** fibula; ***gastr.*** gastrocnemius; ***il-fib.*** ilio-fibularis; ***il-tib.*** ilio-tibialis; ***isch-fl.*** ischio-flexorius; ***lig.*** lateral ligament of knee; ***peron.brev.*** peroneus brevis; ***peron.long.*** peroneus longus; ***pr.fl.*** flexors of proximal phalanges; ***sart.*** sartorius; ***sc.n.*** sciatic nerve; ***sl.*** sling for tendon of ilio-fibularis; ***t.t.*** tibio-tarsus; ***t.mt.*** tarso-metatarsus.

tarsals with the tibia to make a tibio-tarsus, articulating at an intertarsal joint with the remaining three tarsal and metatarsal bones, fused to make a single tarso-metatarsus. There are usually four digits articulating with the tarso-metatarsus; three directed forwards and one backwards. In standing, the weight is usually balanced in tripod fashion on three of the four points provided by the front and back portions of the feet.

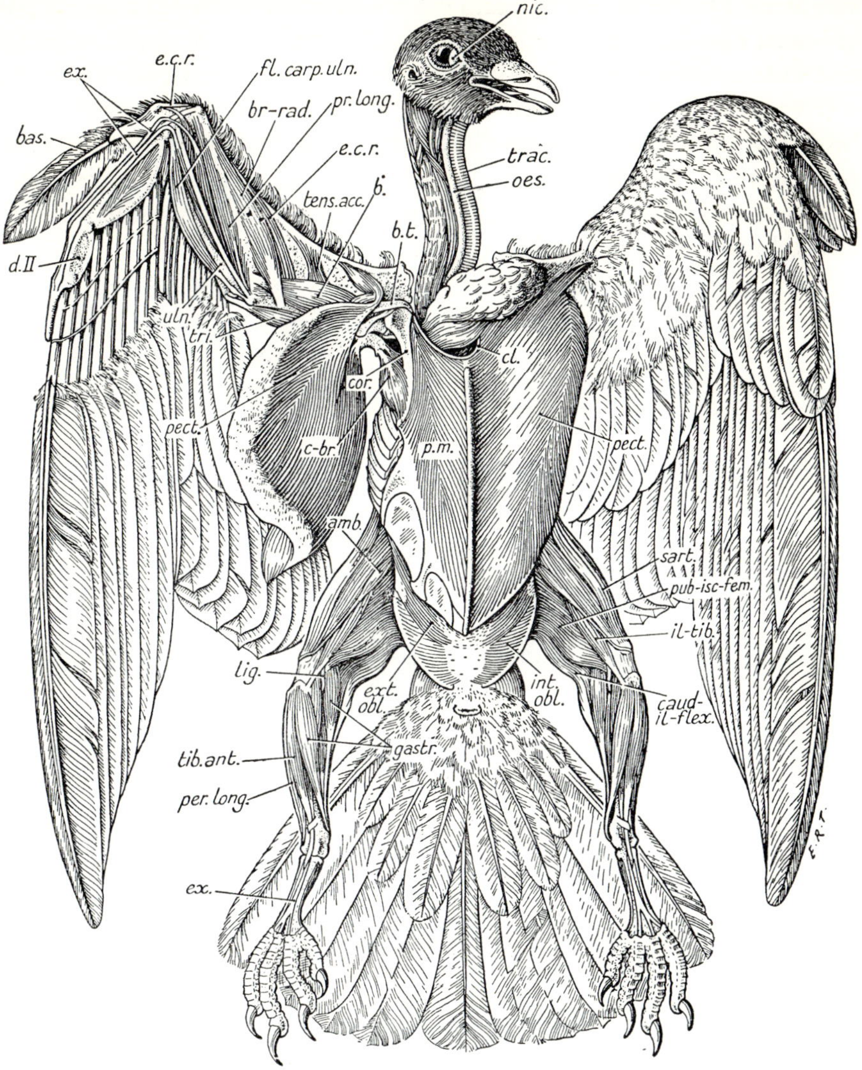

FIG. 258. Pigeon dissected from ventral surface.

amb. ambiens; ***bas.*** bastard wing; ***br-rad.*** brachio-radialis; ***b.*** biceps; ***b.t.*** biceps tendon; ***caud-il-flex.*** caud-ilio-flexorius; ***cl.*** clavicle; ***c.br.*** coraco-brachialis; ***cor.*** coracoid; ***d.II.*** 2nd digit; ***e.c.r.*** extensor carpi radialis; ***ex.*** extensor; ***ext.obl.*** external oblique; ***fl.carp.uln.*** flexor carpi ulnaris; ***gastr.*** gastrocnemius; ***oes.*** oesophagus; ***il-tib.*** ilio-tibialis; ***int.obl.*** internal oblique; ***lig.*** lateral ligament of knee; ***nic.*** nictitating membrane; ***pect.*** pectoralis major; ***p.m.*** pectoralis minor (supracoracoideus); ***per.long.*** peroneus longus; ***pr.long.*** pronator longus; ***pub-isc-fem.*** pub-ischio-femoralis; ***sart.*** sartorius; ***tens.acc.*** tensor accessorius; ***tib.ant.*** tibialis anterior; ***trac.*** trachea; ***tri.*** triceps; ***uln.*** ulnar.

The knee joint has some remarkable similarities to that of man. It is stabilized by lateral, medial, and cruciate ligaments and contains a pair of lunate cartilages or 'menisci'. The joint allows movements of flexion and extension and the femur as it extends on the tibia in walking rotates laterally because of the arrangement of the joint surfaces. The bird thus balances in the medio-lateral plane by rotation at the knee, somewhat as we do by abduction–adduction at the hip (Fig. 259). When it makes a step forward the weight is brought by this rotation at the knee over the leg that remains on the ground.

FIG. 259. Drawings from photographs of a goose, A, standing; B, stepping. The centre of gravity *S* is brought over the foot on the ground by lateral rotation of the femur on the tibia. Note the position of the tail in B. (After Heinroth, from Stolpe.)

The intertarsal joint allows mainly movements of flexion and extension. It is largely supported by ligaments and has a very strong capsule and lateral and cruciate ligaments rather like those of the knee; there is even a meniscus on the lateral side. The back of the tibia is occupied by the gastrocnemius and the flexor muscles of the toes and at the front there is a tibialis anterior acting across the inter-tarsal joint, and also extensors of the toes. The calf muscles are mainly concerned with producing flexion of the toes in the act of perching and they form an elaborate system of tendons attached to the phalanges. These tendons often act as a single unit, and there is an arrangement by which the flexion is passively maintained by the weight of the body, even during sleep. Many of the muscles are specially arranged to allow support of the joint whether in the flexed or extended position. The iliofibular muscle passes through a conspicuous sling for this purpose (Figs. 255 and 257). The flexor muscles of the toes are inserted largely above the knee and thus tend to tighten as the bird sinks. In this they may be assisted by the ambiens, a muscle found in reptiles and some birds, which takes origin from the ilium. The muscle belly lies on the medial side of the thigh, and its tendon runs beneath the patella on to the lateral surface of the lower leg, where it is attached to the upper end of the muscles that flex the toes. This arrangement provides a single string crossing hip, knee, and ankle and allows the weight of the body to flex the toes as the joints bend.

The second mechanism for maintaining the bird on its perch is a locking device that holds the toes flexed. The under-surface of the flexor tendon is ridged at the metatarso-phalangeal joint, where the weight of the body presses it against a branch. The upper side of the tendon sheath is also ribbed and as the bird settles on its perch the two sets of ridges interlock.

The feet show a wide variety of adaptations for special habitats (Fig. 260). In the cursorial and walking birds there are often long digits in front and behind to give a long base for balance, but the number may be reduced—to two in running birds, such as the ostrich. Hopping is used by small birds on the ground and in the trees and produces quick movement. It is expensive because of the large displacements of the centre of gravity, and for long distances or large animals walking is more efficient. Many different groups of birds have acquired webbed feet for swimming. In birds exposed to cold the digits may be enclosed in a coat of feathers. Birds of prey develop long raptorial talons. Throughout the great group of perching birds one digit is directed backwards, allowing firm grasp of a branch. In climbing birds the fourth digit is often directed backwards as well as the first, so that the foot forms a sort of pincer, with long curved claws.

7. Skeleton of the wings

The wing is designed to have a minimum moment of inertia about an axis parallel to the sagittal plane and passing through the shoulder joint. Movements are produced by muscles lying either outside the arm or in its proximal part, with long tendons. The wing feathers are carried along the post-axial border of the humerus, ulna, and hand, and the shape of the wing depends on the position in which the feathers are held by their muscles, as well as on membranes, the pre- and post-patagia, developed where the limb joins the body. The active movements of flight are produced mainly by the pectoral muscles; the joints and muscles of the wing itself serve to spread the wing and to adjust its shape during each beat. The humerus is short and broad with a large head and an expanded surface for attachment of the pectoral muscles. Radius and ulna are both large, especially the latter. There are only two free proximal carpals and the remainder of the wrist is formed of three metacarpals, one short and two long and fused. Only one digit, probably representing the second, is well developed, having two broad phalanges; the third and the first digits consist of single rods, the

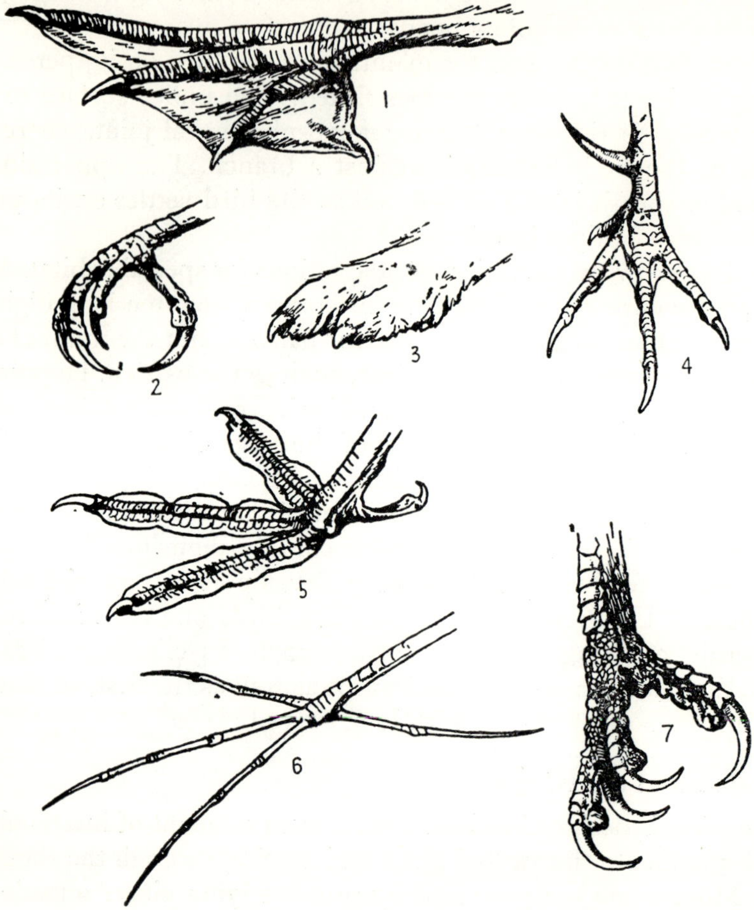

FIG. 260. Various types of feet in birds.

1, shag (swimming); 2, crow (perching, lifting); 3, ptarmigan (stockinged by feathers); 4, jungle fowl (walking, scraping); 5, coot (lobate, swimming); 6, jacana (suited for walking on floating plants); 7, sea-eagle (raptorial). (From Thompson, *Biology of Birds*, Sidgwick & Jackson, Ltd.)

latter, standing somewhat apart at the front of the base of the hand, is capable of independent movement; it carries the bastard wing (alula or ala spuria).

The glenoid cavity is formed at the union of a blade-like scapula and a stout coracoid. The former lies horizontally and is attached by muscles to the vertebral column and ribs. The coracoid holds the wing away from the sternum, with which it makes a joint. The furcula, probably consisting of the combined clavicles and interclavicles, is loosely attached to the sternum and carries the origin of muscles that rotate the humerus about its long axis.

8. Wing muscles

Depression of the wing is produced mainly by a single mass of muscle, the huge pectoralis major, making up as much as one-fifth of the whole weight of the body. It runs from the sternum and furcula to the under side of the humerus, to which it is attached, at some distance from the joint, by a complicated tendon of insertion. The fibres of this muscle are very red in strongly flying birds and often contain numerous lipoid inclusions. In the fowl the fibres are white and contain glycogen, but little lipoid. Elevation of the wing is produced by a muscle also attached to the sternum, lying deep to the pectoralis major and often called the pectoralis minor, but more properly supracoracoideus. Its tendon passes through the foramen triosseum, between the furcula, scapula, and coracoid, to be inserted on the upper side of the humerus. It is assisted by latissimus dorsi and deltoid muscles.

The chief muscles of the shoulder are thus a massive set serving to raise and lower the wing. There is little development of the other muscles such as are present in other vertebrates for the purpose of balance and drawing the limb backwards and forwards for standing and locomotion. Such a system of braces all round the joint is unnecessary; the bird balances on its wings mainly by the action of the pectoralis major as the chief brace, between the sternum and the humerus, with the coracoid as a compression member between. Stresses must of course arise in other directions besides those tending to produce a vertical fall and these are met by the muscles that produce rotation of the humerus and various other movements of the wing, especially a pronation, depressing the leading edge. The muscles used in other tetrapods to sling the weight of the body to the pectoral girdle and fore-limb are little developed. The scapula is held to the vertebral column by small rhomboid muscles and there is a short series of slips attached to the ribs, the serratus anterior.

Other muscles running from the body to the humerus produce rotation of the humerus at the glenoid and adjustments of the patagia, movements that are very important in flight. From the outer surface of the scapula arises a scapulo-humeral muscle, inserted in such a way as to produce adduction and lateral rotation of the humerus, raising the hinder edge of the wing. The coraco-humeral muscle is a compact bundle attached near to the last and producing the opposite effect of abduction and medial rotation, lowering the hinder aspect of the wing.

The deltoid muscle is divided into several parts and besides its main abductor action on the wing also has slips inserted into the skin of the anterior patagium, muscles known as the long and short tensors of that membrane. There is also a tensor accessorius, running from the surface of the biceps to the skin of the leading edge of the wing.

The muscles within the arm itself serve to extend or fold the whole wing and to alter the positions of the parts, especially by pronation and supination during flight. Large triceps and smaller biceps muscles act at the elbow. In the forearm there is a large extensor carpi radialis and an extensor carpi ulnaris, serving to keep the wing extended at the wrist. Flexor carpi ulnaris folds the wing. There are also two large pronators, brevis and longus (brachio-cradialis), rotating the radius medially and lifting the back of the wing. A system of digital flexors and extensors, inserted into the distal phalanx of the main digit, keeps the wing tip spread out or folds it. The position of individual feathers is controlled by an elaborate system of tendons and muscles along the back of the hand. The first digit is moved independently by abductor and adductor pollicis muscles, controlling the position of the bastard wing, which increases the angle of stall and thus allows slow flying speeds in take-off and landing.

9. Principles of bird flight

A plane surface moved through the air in a direction inclined at an angle to this plane is known as an aerofoil. The forces generated can be resolved into a lift force acting upwards and a drag force tending to stop the motion. On this fact depends the power of supporting weight in the air that is possessed by birds and human heavier-than-air machines. Both lift and drag forces are proportional to the square of the speed, and the requirement for sustained flight in still air is that the object shall have sufficient speed to generate a lift force equal to its weight.

The flow of air over the upper surface of the wing reduces the pressure there and provides the main portion of the lift (Figs. 261–2). By tilting the wing (increasing the 'angle of attack') the pressure on the underside can be increased, but the air flow now tends not to follow the upper surface but to become turbulent, especially at the hind edge, destroying the lift (Figs. 261–2). When an aerofoil falls below this critical speed it stalls; that is to say, drops suddenly, being no longer supported. The smooth flow of air over the wing tends to be especially

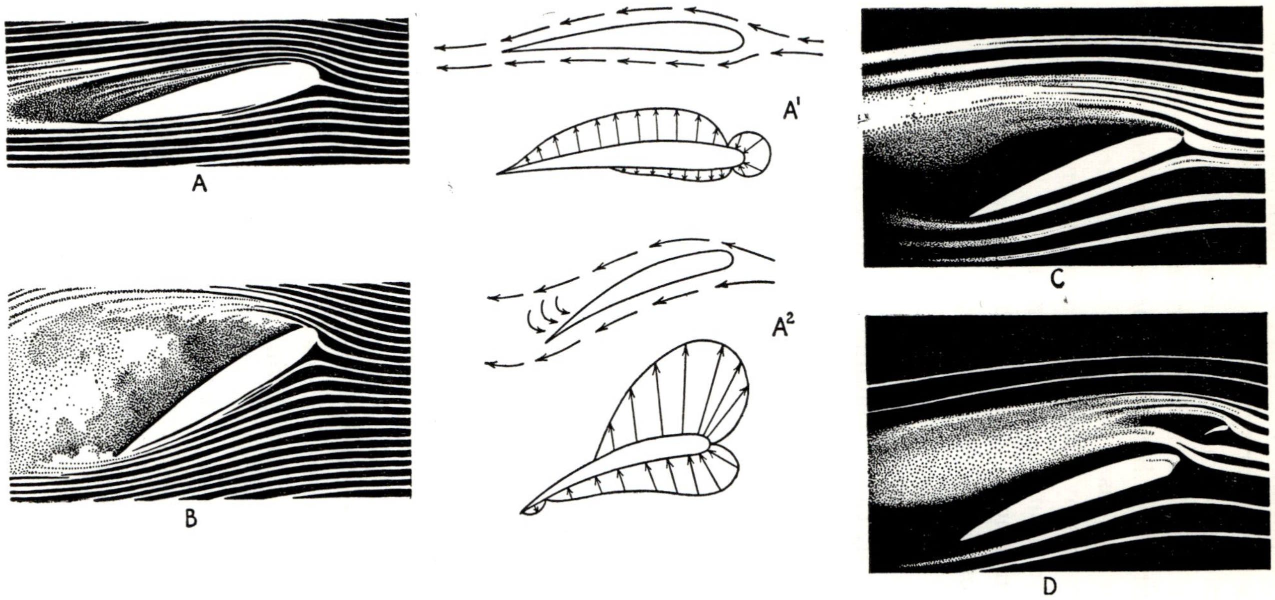

FIGS. 261–2. Effect of an air stream flowing past a wing under various conditions in a wind tunnel. The white lines are jets of smoke. In A they are close together below the wing and farther apart above it. These increased and reduced pressures produce the lift and the arrows in A^1 and A^2 show the actual pressures registered by gauges, with different angles of attack. In B and C the angle of attack is so large that the stream does not follow the upper surface, turbulence develops, the stream produces drag instead of lift, and the wing stalls. D shows how this is prevented by slotting. (After Storer from official photographs of the National Advisory Committee for Aeronautics and from Jones.)

disturbed at its hinder ('trailing') edge and by eddies round the end ('tip vortex'). The proportion of length to breadth (aspect ratio) in the wing suitable for a particular type of flight depends on the need to provide a sufficiently large undisturbed area.

The shape of the aerofoil is of critical importance in determining its aerodynamic capacities. For birds, as for aeroplanes, there are differing shapes, suitable for various types of flight. To understand them we must classify the means by which birds attain the necessary forward velocity. First and most obvious is flapping flight. Though the details of this are varied and not fully understood, it can be regarded as a screw-like motion of the wings, providing forward and upward components (p. 455).

In still air the only alternative to flapping flight is to glide downwards, which obviously cannot continue indefinitely. Yet some birds, such as the gulls, and especially the albatrosses and the buzzards, condors, and other birds of prey, can be seen to soar for many minutes, gaining height without flapping the wings. Lord Rayleigh showed how they can do this by making use of the fact that the air is seldom still. Theoretically the bird can use three types of air movement: (1) ascending currents, usually thermal; (2) variations in the wind velocity at any one level (gusts); (3) differences in wind velocity at different levels. The first method is that used by human sail-planes and is certainly adopted by many soaring land birds. The gustiness of the wind is probably turned to advantage by gulls, rooks, and many other birds, and the decrease in wind velocity near the sea surface is used by marine soaring birds, notably the albatross.

10. Wing shape

A wing of the shape that allows an albatross or swift to make its superb manœuvres would stall immediately at the speed of flight adopted by a crow. In discussing wing shape the chief factors to be considered are (1) the wing area, (2) the aspect ratio (wing length/breadth), (3) the wing outline and taper, (4) the presence of holes or slots, (5) the camber or curvature of the wing.

11. Wing area and loading

A small wing area is necessary for fast flight, since the drag $\propto$ area $\times$ speed^2, at least for high speeds. For this reason fast aeroplanes and birds have small wings, but in the bird the fact that the wing provides the forward momentum as well as the lift greatly complicates matters. For flapping flight the wings must be moved relatively fast and for this small size is an advantage. On the other hand, a large wing area allows slow flight (lift $\propto$ area$\times$speed^2) and is found in hawks, vultures, storks, and other birds that fly slowly to hunt, especially if they soar on thermal currents, for which a large lift is necessary (p. 458).

The loading of the wing varies considerably. Since the weight increases with the cube but the wing area only with the square of the linear dimensions, it follows that large birds must have relatively larger wings than small. However, the larger birds usually have a heavier loading of the wing, for instance, 10 kg./m.^{2} in the duck (*Anas*), 20 in the swan (*Cygnus*), 1 in the goldcrest (*Regulus*), 3 in the crow (*Corvus*). A considerable 'safety margin' remains in most birds; for instance, pigeons were found to be able to fly until as much as 45 per cent of the wing surface was removed; hawks and owls have an especially high safety margin and they can carry prey almost as heavy as themselves.

12. Aspect ratio

Although a small wing area reduces drag, many fast-flying birds have a large wing-span. The aerodynamic advantages of this allow a low rate of descent when gliding, reducing the expenditure of energy necessary to sustain flight. High aspect ratio is therefore found in birds that fly fast by flapping flight (swifts and swallows) and especially in those, such as the albatross, that glide fast in order to obtain sufficient kinetic energy to convert into altitude. However, these wings with very high aspect ratio stall at relatively high speeds and the birds that soar slowly on thermal up-currents over the land mostly have a low aspect ratio. Some figures for aspect ratios are:

Albatross (*Diomedea*)	25	Vulture (*Neophron*)	6
Gull (*Larus*)	11	Rook (*Corvus*)	6
Swift (*Apus*)	11	Sparrow (*Passer*)	5
Shearwater (*Puffinus*)	10		

13. Wing tips, slots, and camber

A pointed wing tends to stall first at its tip and is therefore only suitable for fast fliers. Such birds show great development of the hand feathers, producing a long narrow wing, whereas birds built for slower flight and manœuvre have a shorter broader wing with long arm feathers (Fig. 263).

The condition of the air around the wing is of first importance for the maintenance of lift; if there is not a smooth stream over the upper and under surfaces the air becomes turbulent, and the aerofoil stalls (Figs. 261–2). This tends to happen either if the speed falls too low or if the angle of the wing relative to the line of motion increases above about

20°. Turbulence is mitigated, however, by the provision of openings, known as slots, which let through part of the air and provide the necessary smooth stream. The spaces that occur between the feathers, especially towards the wing tip, almost certainly function as slots. Probably the arrangement provides a series of such apertures, giving a very efficient high-lift device. Such slots are conspicuous in slow

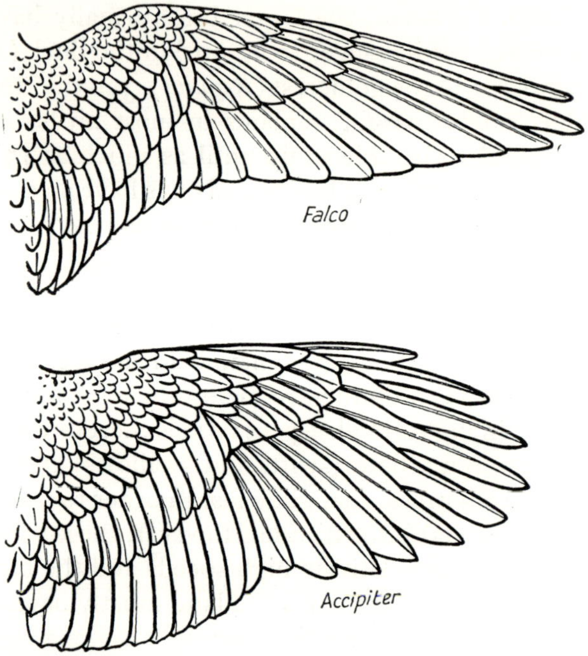

FIG. 263. Wings built for speed (falcon, *Falco*) and for manœuvring (hawk, *Accipiter*). The former is long and narrow, with relatively large hand feathers. The latter is short and broad, the arm feathers being long and the primaries arranged to make slots. (After Fuertes.)

fliers (rooks) and especially in those that soar on thermal up-currents (vultures). The feathers of such birds are often individually tapered (Fig. 263). Slots are also found in the wings of large birds that are fast fliers (pheasants), the wing being liable to stall in certain phases of the down strokes. It is possible that the bastard wing acts as a slotting device; indeed, consideration of it played a part in development of the theory of turbulence and slotting.

The shape of the wing has a very important influence on the air stream. In most birds there is a stiff leading edge and a thinner trailing edge. Nearly all wings are cambered, that is to say, they taper from the leading to the trailing edge, especially in the region of the forearm

(Figs. 261–2). This arrangement directs the air stream over the upper surface of the wing in such a way as to provide an extra lift by creating a 'suction zone' of reduced pressure. However, high camber, like low aspect ratio, reduces the speed of the bird.

14. Flapping flight

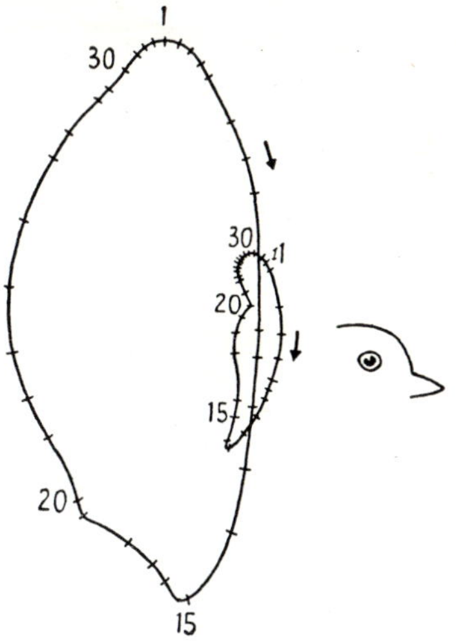

FIG. 264. Pathway of the wing tip and wrist joint relative to the body during free flapping flight of a gull. Equal time-intervals are shown. Note the great speed of the upward beat and that the forearm is raised before the wing tip. (After Demoll.)

Flapping flight involves a complex, screw-like motion of the wing, downwards and forwards then upwards and backwards, more rapid upwards than downwards (Fig. 264). The action of the wings differs during take-off or landing from that in sustained flight (Brown, 1953). In the former conditions, when the speed is slow, forward velocity is provided by backward movement of the wings. During each stroke, beginning with the wings raised, they are first moved downward and then forward, providing lift (Figs. 264, 265). This movement is produced mainly by the pectoralis major. During the upstroke the wing is first adducted, folded and flexed, and supinated at the wrist, by the actions of pectoralis minor and other muscles. A very rapid backward flick then follows, produced by upward and forward rotation of the humerus, extension of the wing, and pronation of the manus. The effect of these movements, produced largely by the triceps and other extensors, is to provide a forward component.

This form of flight involves mainly the primary feathers. It is evidently very tiring and can only be continued for a few seconds. In sustained flight the downstroke is as in slow flight but the upstroke is much simpler, with only a slight backward flick of the primary feathers. The action is such that the inner part of the wing provides lift, the tip propulsion. The upstroke in fast flight is thus mainly passive, produced by the pressure of air against the under surface. The major part of the effort needed to provide lift and forward propulsion is thus provided by the pectoralis major. This muscle weighs as much as one-fifth of the body weight in flappers, such as the lapwing, as little as

FIG. 265. Pigeon (*Columba*) with wings at bottom of stroke during rising. From a photograph taken at very short exposure. Note forward position of wings but upward and backward curled primary feathers. (After Aymar.)

FIG. 266. Drawings from four photographs from film of a take-off by an eagle (*Aquila*). A. The legs are jumping and the upper arms nearly vertical before the wrist joint is extended. B. Wing fully extended with bastard wing spread out, some pronation. C. Marked pronation as the upper arm reaches its lowest point. D. The upper arm is proceeding upwards, although the hand has not yet reached its lowest point. (Drawn from photographs by Knight, from Streseman.)

one-ninth in soarers, such as the gull. It does nearly all the work, the other muscles serving to give extra lift when needed.

The feathers are held by an elaborate system of tendons and in some birds they are allowed to twist only when the wing is being raised and the barbs of the feathers themselves are so arranged that they open like the vanes of a blind when under pressure from above, but close when the pressure is from below (Fig. 268). In other birds, especially those that fly fast with a slow wing beat, such as gulls and swans, the wing is probably rigid on the up as well as on the down strokes, and is twisted so as to produce forward and upward components on the up stroke.

FIG. 267. Bill-fisher leaving its hole in a bank, showing wings half-way through the upstroke. The upper arm has reached its highest point and the forearm is just starting upwards; its primary feathers have opened on the right wing, reducing resistance. (Drawn from photograph by Aymar, *Bird Flight*, published by John Lane, The Bodley Head, Ltd.)

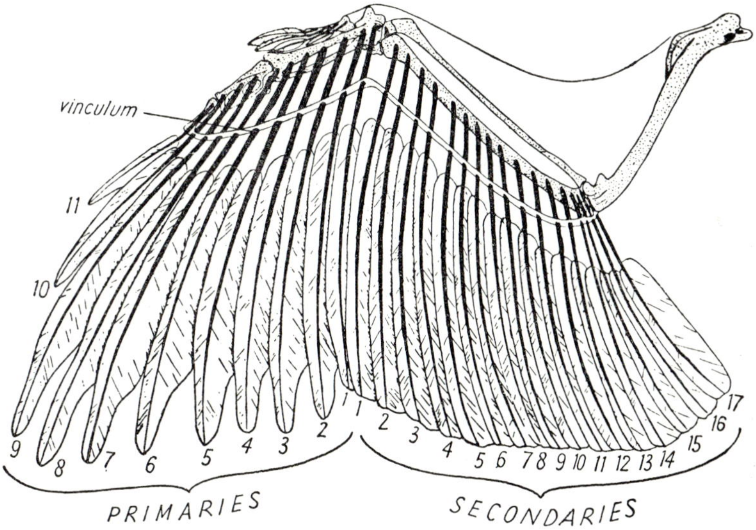

FIG. 268. Diagram of wing to show arrangement of the flight feathers. (From Pycraft, *A History of Birds*, Methuen & Co., Ltd.)

The whole upward movement is usually faster than the downward one. Before the wing tip has reached its highest point the upper arm is already beginning to descend and in this way the line of flight is maintained almost straight and does not follow a wavy path as it would do if the parts of the wing vibrated together. In small birds the wing works more nearly as a whole and the flight differs in several respects from that of larger birds. In general the wing is a very labile system and regulates itself automatically with changes in the aerodynamical forces. This regulation is produced partly by feather plasticity and joint mobility, with participation of reflex muscular adjustments that are little understood.

The whirring flight of some small birds, especially of humming birds, enables them to remain almost in one place in the air, or even to move backwards. The wings beat backwards and forwards (Fig. 269), often as fast as 200 times a second, and the 'pectoralis minor' is almost as large as the major.

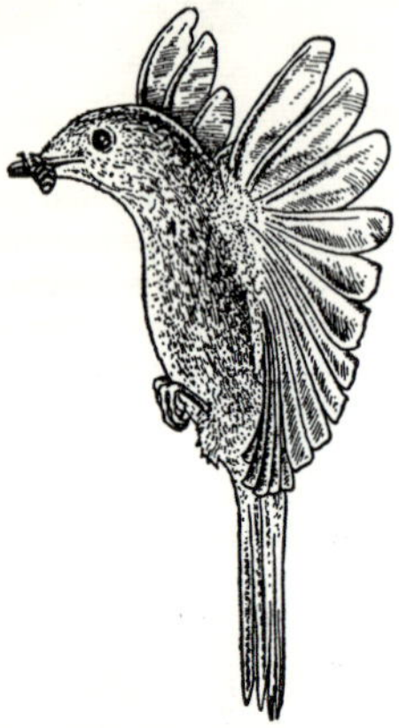

FIG. 269. Spotted fly-catcher hovering. The wings are passing backwards and there are spaces between the feathers. (Drawn from a photograph by E. Hosking, with permission.)

15. Soaring flight

Many birds economize the energy needed for flapping flight by making skilful use of the possibilities presented by movement of the air. All birds glide for short distances, some small birds with wings folded, others with wings outstretched. Sustained gliding and soaring upwards without flapping the wings is found only in large birds, probably because considerable weight is necessary to provide kinetic energy sufficient to ensure continuous flight and efficient use of wind variations. As has been suggested, there are two distinct types of soaring birds: (1) land birds using thermal up-currents, (2) marine birds using variations in wind above sea level.

16. Soaring on up-currents

Up-currents of air arise in the neighbourhood of large objects on the ground (cliffs or even a ship) and particularly from variations in the rate of heating of the earth's surface in the sun, over rocks, vegetation, mountain shadows, &c. Birds using such currents usually proceed upwards in a series of small circles, a behaviour seen in buzzards and other hawks and especially characteristic of vultures, which may ascend in this way above 1,000 ft (Figs. 270–2). The characteristic features of such thermal soarers are large wing area,

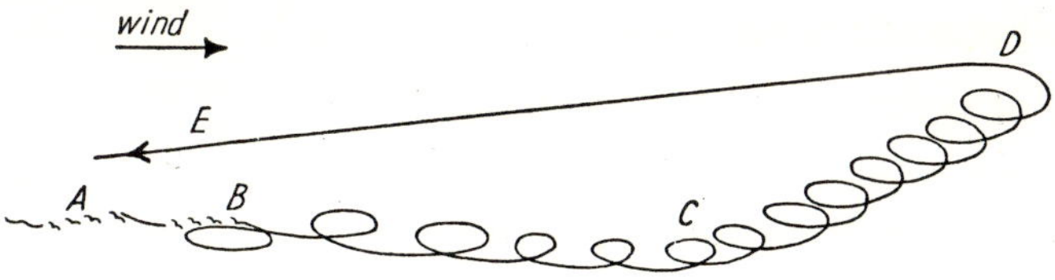

FIG. 270. Flight of an eagle (*Aquila*).

A–B, flapping flight; B–C, soaring at constant height; C–D, soaring in ascending spirals; D–E, gliding. (After Ahlborn.)

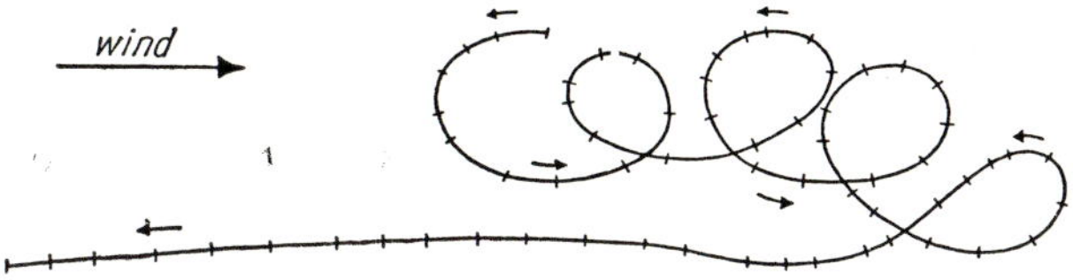

FIG. 271. Soaring flight of kite (*Milvus*).

Losing height downwind and gaining it upwind. Time-marks one second. (After Hankin.)

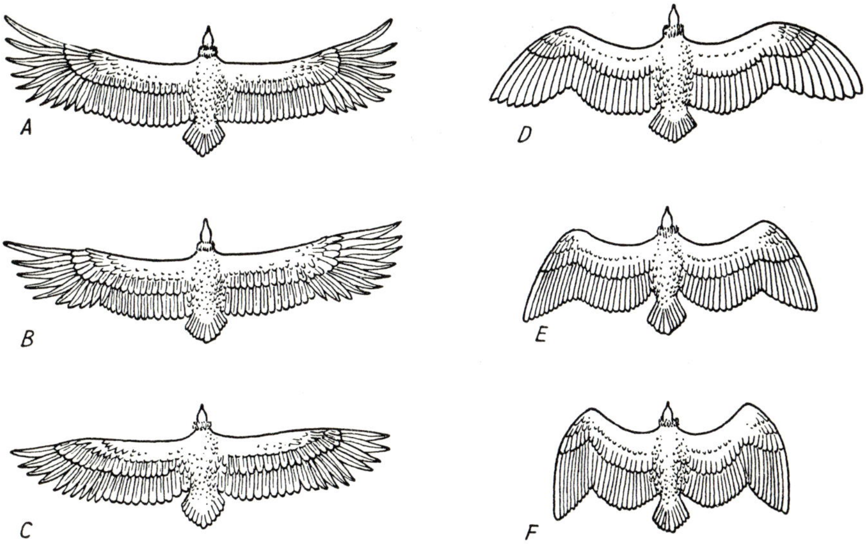

FIG. 272. Wings of vulture (*Gyps*).

A, soaring upwind, gaining height; B, upwind on level; C, downwind, losing height; D–F, gliding flight. (After Ahlborn.)

low aspect ratio, and wings broad at the tip and usually provided with well-marked slots.

17. Use of vertical wind variations

The decreasing effect of surface friction causes the wind to blow faster at greater heights and this phenomenon is used by some birds at sea, where conditions are presumably more uniform than over the

FIG. 273. Wings of the albatross (*Diomedea*), used for soaring flight. Wing very narrow, with long upper arm region. (Drawn from photograph by Aymar.)

land. The albatross (*Diomedea*) (Fig. 273) is the classic example of this type of bird, proceeding in a regular series of movements, without flapping the wings, downwind losing height and gaining speed and then upwind gaining height and passing into a faster-moving layer. Each downwind tack is longer than the upwind one. During the upwind tack the wings are spread forwards, downwind backwards. The albatross remains all the time within 50 ft of the sea surface, because the variations in wind velocity are marked only at low levels. The albatross shows the characteristics suitable for this type of flight, namely, large size (it is the largest of all flying birds), great wing span (11 ft), high aspect ratio (25), and pointed wing tips, without slots.

Other sea birds, such as the gulls, though not so highly specialized, can take advantage of variations in horizontal wind velocity, including gusts at any one level. The bird moves upwards as it meets an accelerating gust and turns when the wind decelerates. Gulls also use up-currents at cliff faces and, no doubt, air movements of all sorts are

FIG. 274. Heron (*Ardea*) leaving its perch. The legs have been used to make a jump and the wings are fully spread. (Drawn from photograph by Aymar.)

widely used, especially by large birds. However, it is evident that the wing equipment only allows the bird a limited range of choice and probably even the slightly different wing shapes of related species depend on the differing conditions they are called upon to meet. A vulture could no more zoom backwards and forwards over the waves than an albatross could circle slowly on a gentle thermal up-current. A pigeon cannot equal a gull at steady gliding and soaring, but the pigeon can rise more steeply or descend more rapidly without stalling.

18. Speed of flight

Estimation of the speed of flight involves distinguishing between air and ground speed. The speed relative to the ground may be very high; there are records of birds covering more than 100 land miles in an hour, with the help of the wind. Racing pigeons can average 40 miles an hour or more for considerable periods. Air speeds of 30–50 m.p.h. can certainly be reached by many birds: swifts are said to reach 100 m.p.h. in still air.

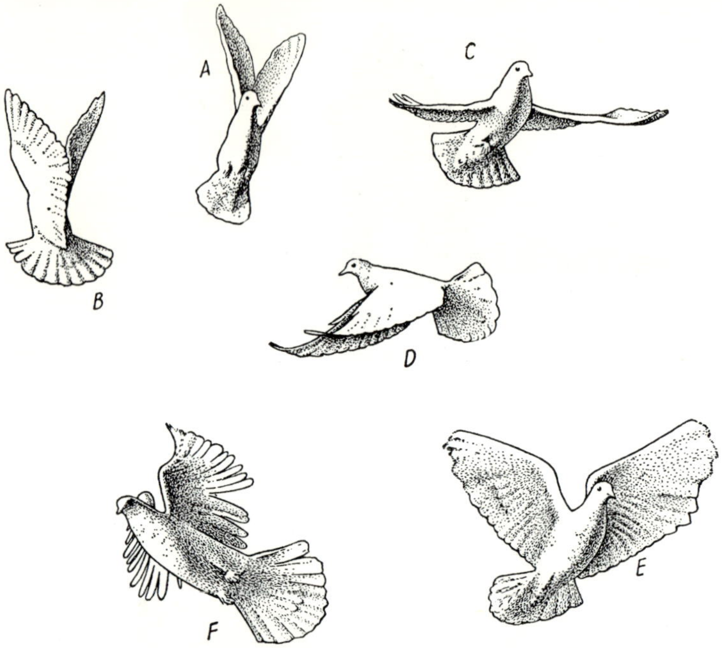

FIG. 275. Pigeons (*Columba*), photographed during take-off, with exposures of 1/825 second. A, front, and B, rear view, with wings together. C, nearly, and D, quite at bottom of downstroke; note pronation and forward movement of wing. E and F, wings during the upstroke; in F the primary feathers have opened; note that the wing moves backwards and that the motion is faster than on the downstroke. (From photographs by Aymar.)

19. Take-off and landing

At the take-off the bird has to acquire sufficient forward momentum to provide lift, and yet must leave the air sufficiently undisturbed for subsequent beats to be effective. In many birds, especially the smaller, the jump provided by the legs is adequate for the take-off (Fig. 274). Large birds must run or swim rapidly to obtain sufficient speed. Eagles are said to be unable to rise without a long run, and many large birds nest on a cliff or tree, which gives them an up-current for the take-off. Swifts usually come to rest high up and can only rise off the ground with difficulty. The albatross is unable to take off from the sea surface in a dead calm.

The first beats are usually very large, beginning with the wings above the back and held at such an angle as to produce a large forward component. The wings may be heard actually clapping together in pigeons (Fig. 275). During rapid ascent, as in the larks, the body is nearly vertical and the wing changes its angle at the shoulder very

FIG. 276. Jackdaw about to land. The wings are fully extended on the downbeat, and the tail is fanned out and bent downwards. (From photograph by Aymar.)

FIG. 277. Hawk striking at a dummy owl. Note long legs and the method of braking. The wings are broad and rounded, giving a large safety factor. (From photograph by Aymar.)

sharply between the downward and recovery strokes. The bastard wing is held in such a position that the beat provides extra forward momentum.

Landing is also a delicate operation, especially since it often involves coming to rest suddenly on a branch (Fig. 276). This is achieved by lowering and fanning out the tail, which thus acts as a flap, providing both lift and braking. The legs are then lowered; often one further wing stroke is given to bring the bird forward to drop onto the perch. The adjustment of braking in such a way as to prevent stalling involves a very special system of coordination (Fig. 277). Other methods of landing are possible, for instance, rooks may make a roll and sideslip to the ground.

20. The skull in birds

The arrangement of the parts of the bird's skull is similar to that of archosaurian reptiles (Fig. 242). Individual bones can be recognized in the young, but they mostly become united in the adult to form a continuous thin-walled structure that encloses the brain and sense-organs and supports the beak (Fig. 278). Most birds are microsmatic; the nasal passages are simple and the turbinals reduced. There is seldom a complete bony secondary palate, such as there is in mammals, instead the internal nostril opens into the mouth relatively far forward. The large size of the brain and reduction of its olfactory portions are responsible for the rounded form of the top of the head, and there are very large orbits at the sides, separated by an ossified septum. The base of the skull is formed by a basioccipital behind, carrying a single occipital condyle. There is a large basisphenoid, covered ventrally by a pair of basitemporals, probably representing the parasphenoid, the front part of which makes a 'basisphenoid rostrum', as in archosaurs.

The jaws are characteristically slender and elongated; in the more advanced birds they have a very special form of support. The upper part of the front of the skull is composed of the enlarged premaxillae, the nostrils lying very far back and the nasal bones being small. The palatines are long and fused far forward with the maxilla, while they articulate movably behind with the pterygoids and base of the skull. The pterygoid is a slender rod, itself movably articulated with the skull and with the quadrate, which is a triangular bone with clearly separate otic and basal articular processes. The upper jaw is thus a long thin bar composed of maxillae, quadrato-jugal, and jugal, and as in many reptiles it is capable of considerable movement ('kinesis'). It is raised when the lower end of the quadrate moves forwards. This mechanism is particularly well developed in parrots, where the beak is freely hinged on the skull. This type of palatal arrangement is known as neognathous. In some birds, such as the flightless ratites, the palatines are shorter, the vomer larger, and the pterygoids less movable, a condition called palaeognathous (p. 514). The lower jaw, also elongated, consists of the articular bone and four membrane bones.

21. The jaws, beak, and feeding mechanisms

There is a complete lower temporal bar, composed of jugal and quadrato-jugal bones. The temporal region is hard to interpret, but

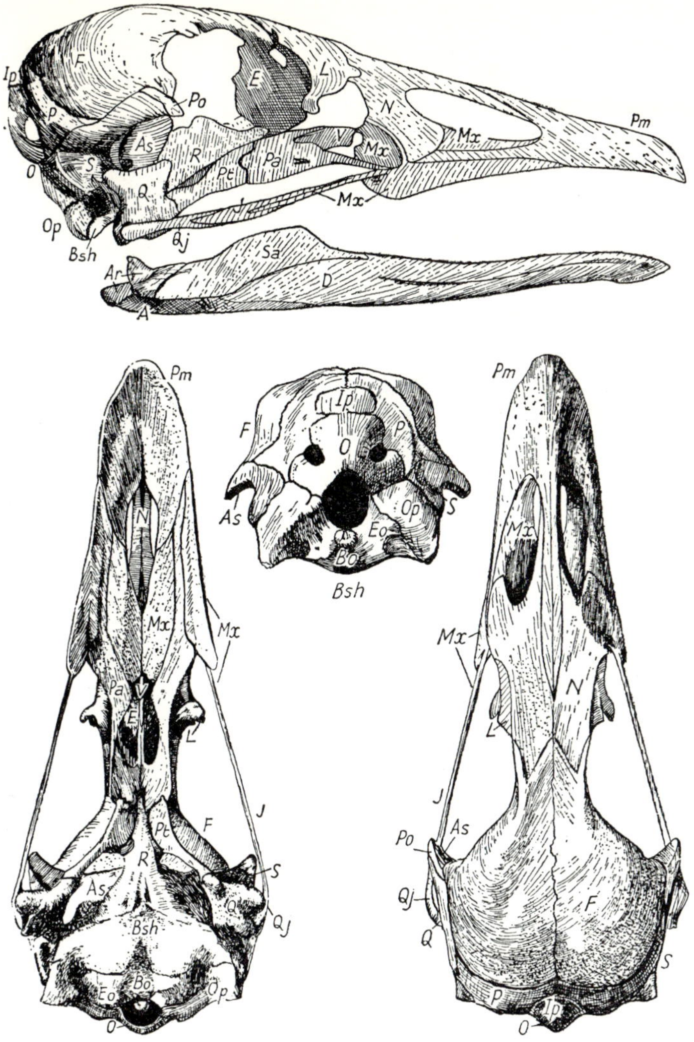

FIG. 278. Skull of young gosling (*Anser*).

A. angular; *Ar.* articular; *As.* alisphenoid; *Bo.* basioccipital; *Bsh.* basisphenoid; *D.* dentary; *E.* ethmoid; *Eo.* exoccipital; *F.* frontal; *Ip.* interparietal; *J.* jugal; *L.* lachrymal; Mx. maxillary; *N.* nasal; *O.* supra-occipital; *Op.* opisthotic; *P.* parietal; *Pa.* palatine; *Pm.* premaxillary; *Po.* postorbital; *Pt.* pterygoid; *Q.* quadrate; *Qj.* quadrato-jugal; *R.* rostrum of basisphenoid; *S.* squamosal; *Sa.* sur-angular; *V.* vomer. (From Heilmann, *The Origin of Birds*, H. F. & C. Witherby, Ltd.)

has presumably been derived from the diapsid archosaurian condition. Typically, there is a single large fossa, communicating with the orbit, but this is often partly subdivided by bony processes; occasionally, there is a complete post-orbital bar (parrots). There are moderately large temporal and pterygoid muscles, but the jaws are not usually

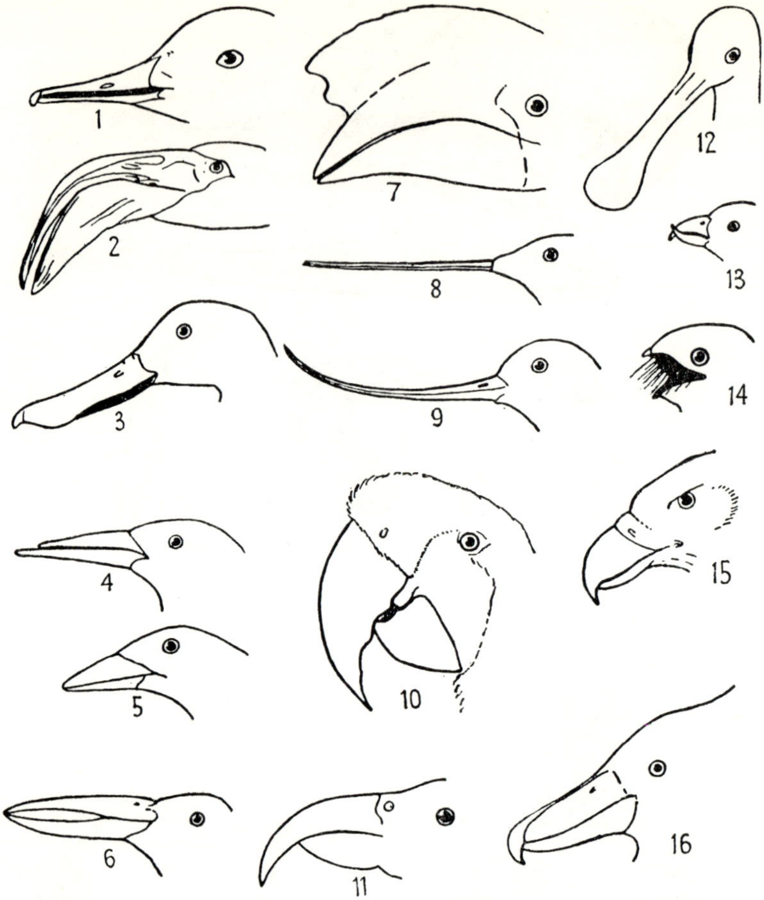

FIG. 279. Various bird beaks.

1, Merganser; 2, Flamingo; 3, Shoveller; 4, Scissor-bill (adult); 5, Scissor-bill (young); 6, *Anastomus*; 7, Hornbill; 8, Humming bird; 9, Avocet; 10, Parrot; 11, Parrot; 12, Spoonbill; 13, Crossbill; 14, Nightjar; 15, Eagle; 16, *Balaeniceps*. (From Pycraft, *A History of Birds*, Methuen & Co., Ltd.)

very powerful, though, of course, formidable in carnivores. Having completely lost the teeth, the birds must rely largely on internal processes to break up the food. The beak is, however, characteristically modified according to the food habits (Fig. 279). There is very great variety in the feeding, as in so much of the life of birds, and though many species keep strictly to one diet others are able to adapt themselves to the food available. The ingenuity and persistence with which birds seek and collect food must be a main factor in their success.

Many birds with a moderately long bill, such as the song-thrush (*Turdus*), can eat either flesh (snails, earthworms, or caterpillars) or fruit. Incidentally we may notice the ingenious behaviour by which the snail's shell is cracked open to obtain the food, by beating it against a stone. Birds that mainly eat seeds, such as the finches, usually have short, thick, strong bills. Large strong bills are present in the hornbills and toucans; they push through dense foliage to obtain the fruit, which may have a hard case. In parrots the beak is moved on the skull, pushed up by the upper jaw when the latter is pulled forward by the digastric muscle.

FIG. 280. The Galapagos woodpecker finch (*Camarhynchus pallidus*) using its stick. (From Lack, drawn by R. Green from photograph by R. Leacock.)

The carnivorous birds, such as most eagles and owls, have short and sharp beaks, whereas fish-eating, as in other vertebrates, results in long jaws. Another widely found arrangement is the flattened bill of some ducks, similar to those of some sturgeons and of the platypus, which also sift out food from water or mud. The long, thin beak of the curlew selects food from mud in a different way, mostly worms and other soft-bodied invertebrates. Lesser flamingos feed on blue-green algae and microscopic phytoplankton, collected by a filter system on the jaws, using a current of water produced by the sucking mouth and piston-like tongue. Some insectivorous birds have long beaks for finding their prey under bark. The woodpeckers have a strong beak like a pick-axe for excavating in wood, and most elaborate special modifications for the purpose of licking up insects; there is an enormously long protusible tongue and special hyoid. The woodpecker finch (*Camarhynchus pallidus*) on the Galapagos Islands probes insects from the bark by means of a cactus spine, a remarkable case of the use of a tool by a bird (Fig. 280). Among the most specialized feeders are the humming-birds, eating nectar, the beak being long or short according to the type of flower visited, and the tongue provided with a special tubular tip.

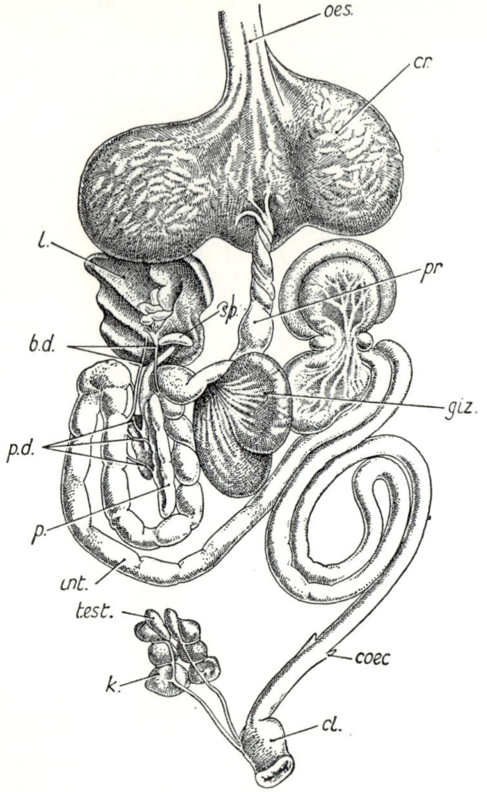

FIG. 281. Dissection of pigeon.
b.d. bile-ducts; *cl.* cloaca; *coec.* coeca; *cr.* crop; *giz.* gizzard; *int.* intestine; *k.* kidney; *l.* liver; *oes.* oesophagus; *p.* pancreas; *p.d.* pancreatic ducts; *pr.* proventriculus; *sp.* spleen; *test.* testis. (After Schimkewitsch and Streseman.)

22. Digestive system of birds

Once the food is in the mouth it is manipulated by the long, thin tongue, moistened with saliva, which usually consists of mucus but is said to contain diastatic enzyme in some seed-eating finches. Food swallowed down the oesophagus may be stored in a large receptacle, the crop, found especially in grain-eating birds; its lining is of oesophageal structure (Fig. 281). The true stomach is divided into two parts, a glandular proventriculus and a muscular gizzard. The structure of the anterior chambers of the gut varies greatly with the diet. In grain-eating birds, such as the pigeon, the crop is large and the seeds are first macerated by storage there. They are then mixed with peptic enzymes in the proventriculus and ground up in the muscular gizzard, which in pigeons has a horny lining and also contains numerous small stones. In insectivores and carnivores the crop is usually smaller or absent, but is very large in some fish-eating birds. In carnivores the gizzard has the character of a more normal stomach. It was stated by John Hunter that herring gulls, normally living on fish, readily take to eating grain, and that after a year or so of this diet the gizzard becomes muscular and has horny walls.

The peptic juice has powerful digestive powers and many carnivorous and fish-eating birds dissolve even the bones of their prey, though in owls these are regurgitated with fur or feathers, making characteristic pellets. The crop of pigeons is also remarkable for the milk it produces to nourish the young. There are special glands for this purpose and they become active in the breeding-season under the influence of a pituitary hormone, prolactin, which has been

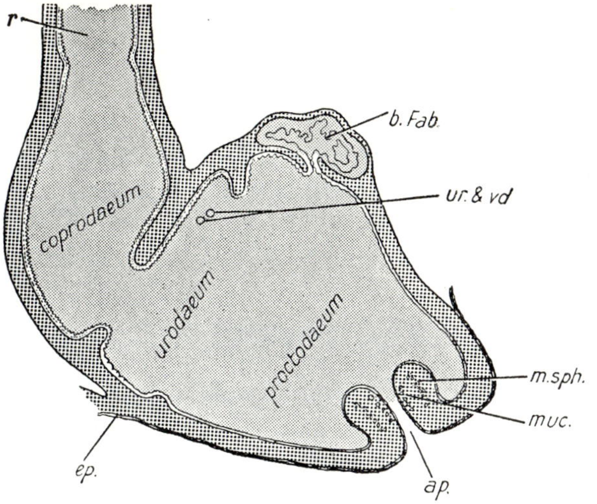

FIG. 282. Diagrammatic section through cloaca of pigeon. *ap.* external aperture; *b.Fab.* bursa Fabricii; *ep.* epidermis; *m.sph.* sphincter muscle; *muc.* mucous glands; *r.* rectum; *ur.* & *vd.* papillae for ureter and vas deferens (or oviduct). (After Clara, from Streseman.)

crystallized and is probably protein in nature. Prolactin causes regression of testes and ovaries and involution of secondary sexual characters, but induces brooding behaviour in the female. Its action is comparable with that of the galactogenic hormone of the mammalian pituitary.

The duodenum and coiled intestine are of characteristic vertebrate type, relatively rather short, though somewhat longer in grain-eating birds. The bile and pancreatic ducts usually open into the distal limb of the duodenum; in pigeons the left bile-duct enters close to the pylorus (Fig. 281). There is a peculiar pair of coeca at the junction of rectum and intestine. The food enters these coeca, but it is not clear what function they perform, possibly it is related to the absorption of water. The arrangements of the cloaca are certainly concerned with this end (Fig. 282). The rectum opens into a coprodaeum and this in turn receives a urodaeum, which is the terminal portion of the urinary and genital ducts. A final chamber, the proctodaeum, opens at the anus. The urinary products are made solid by subtraction of water in the urodaeum and the walls of the other chambers serve a similar purpose. The bursa Fabricii is a blind sac with much lymphoid tissue, opening into the proctodaeum; its function is probably to protect locally against infection and to produce lymphocytes for the bloodstream, hence it has been called a 'cloacal thymus'. Like the thymus, it is prominent in young animals and usually much reduced in the adult.

The large surface area, high temperature, and great activity of birds necessitate a high food intake, especially in the smaller types. This is made possible by rapid passage of food through the gut. Thus a shrike (*Lanius*) is said to digest a mouse in 3 hours, and hens take only 12–24 hours over the most resistant grain. The amount of food taken per day may reach nearly 30 per cent of the body weight (6 g) in the very small goldcrest (*Regulus*) but is about 12 per cent in a starling (*Sturnus*) weighing 75 g.

23. Circulatory system

Many of the features characteristic of birds depend on an efficient circulation, allowing of a high rate of metabolism, and hence a high and constant temperature. It is significant that the birds and mammals are the only vertebrates that have achieved complete separation of the respiratory and systemic circulations, making possible a high arteriolar pressure, which allows materials to reach the tissues rapidly.

The heart shows its sauropsidan characteristics clearly in that the ventral aorta is split to its base into aortic and pulmonary trunks. The former arising from the left ventricle curls round the pulmonary trunk to form a single right aortic arch. The heart has lost the sinus venosus; as in mammals no such extra chamber is necessary to step up the venous return pressure. The ventricles are large, especially the left. The right auricle and ventricle are separated by a flap-like valve, the left side having valves with chordae tendinae, somewhat as in mammals. There are enormous innominate arteries to supply the pectoral muscles. In the venous system there are renal portal veins.

The size of the heart and rate of heart-beat vary with the size and activity of the bird, larger birds having in general relatively smaller and less rapid hearts. In a turkey the rate of beat may be less than 100 per minute, in a hen about 300, and in a sparrow nearly 500.

The red corpuscles of birds differ from those of mammals in being oval and nucleated. They carry a large amount of a haemoglobin that gives up its oxygen suddenly at a relatively high oxygen tension. The red corpuscles are smaller in actively flying birds than in the larger flightless ratites. Haemopoetic tissue is widespread in the young, restricted mainly to the marrow in the adult, although it may also be found in the liver and spleen. The white corpuscles are more numerous than in mammals. They include neutrophils laden with crystals, and thrombocytes, as well as the mammalian types. Lymphatic tissue is dispersed rather than aggregated into nodes. There is a pair of lymph hearts in the sacral region of the embryo and these may persist in the adult. There is a high basal metabolic rate and a temperature considerably higher than that of mammals, usually about 42° C, reaching nearly 45° C in some cases. The means by which this is kept constant in the absence of sweat glands are not known certainly. Heat loss is minimized by the absence of vascularized extremities, the feet being little more than keratin and collagen. The formation of the wing from large avascular surfaces has no doubt been a large part of the secret of the success of birds.

The air-sacs may serve to conserve heat by providing an air cushion for the viscera, with perhaps the alternative possibility of losing heat in this way, by ventilation, when necessary. There is a system of direct arterio-venous connexions in the feet, and elsewhere. The anastomotic regions have powerful muscles, whose contraction closes them and forces the blood through the capillary system. There must be a whole system of nervous pathways for the control of upward and downward temperature regulation, evolved independently of that found in mammals. At least one species (the nightjar) is known to hibernate, and certain humming birds, whose small size render heat loss a serious problem, become temporarily poikilothermic at night.

24. Respiration

Special arrangements are present to provide the large supply of oxygen necessary for the active metabolism and these are based on the plan found in some reptiles. Beyond the respiratory portion of the lung, which is relatively small, there are membranous air-sacs, which are filled at inspiration and then sweep the used air out of the lungs at expiration, thus avoiding the 'dead space' of unrespired air, which is considerable in mammals. When the bird is at rest the air-sacs contain air with a high content of CO_2, but during periods of activity the abdominal air-sacs fill with fresh air containing little CO_2; they then serve not only as a means of ventilating the lungs but also for regulation of the body temperature. The exact direction of the air currents passing through the lungs and the different air-sacs is not fully understood.

The larynx of birds is a small structure guarding the entrance to the trachea. The latter is often long and coiled, perhaps to warm the air. The tracheal rings are bony and complete. The voice is produced in the syrinx, a slight enlargement at the *lower* end of the trachea, containing a pair of semilunar membranes with muscles that alter the pitch of the sound. The apparatus is simple in many birds, but the muscles are very complicated in the singing birds and are especially

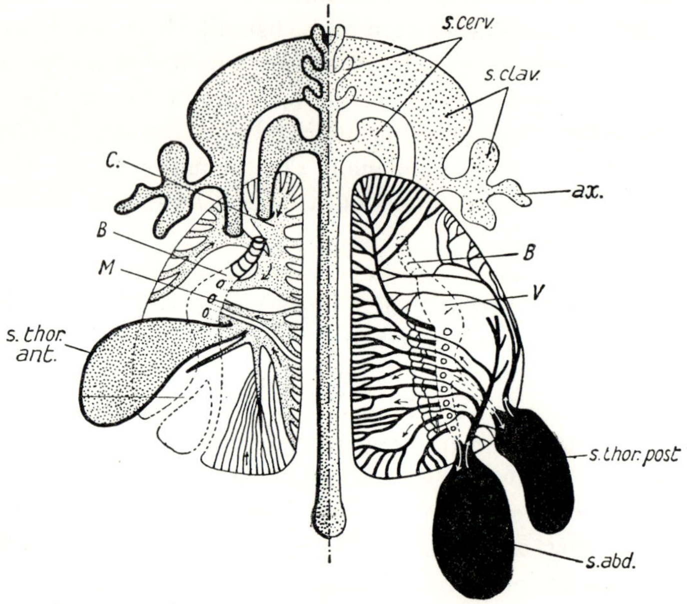

FIG. 283. Diagram of lungs and air-sacs of pigeon, seen from ventral side on left, dorsal on right. On the left side only the ventral surface of the lungs and the expiratory bronchi and air-sacs are shown (dotted). On the right are the inspiratory bronchi and air-sacs (in black).

B. main bronchus; *C.* cervical ventrobronchus; *M.* mesobronchus; *V.* vestibule; *s.abd.* abdominal air-sac; *s.cerv.* cervical air-sac; *s.clav.* clavicular air-sac with diverticulum (*ax.*) in axilla; *s.thor.ant.* and *post.* thoracic air-sacs. (After Brandes and Ihle.)

large in the males. Many varieties of sound are produced, from simple cries appropriate to each sex to elaborate songs. In many species the song is given in its full complexity by individuals that have had no opportunity of hearing others sing, but in some the song is largely learnt by the young and may show considerable local variation. The voice is used for communication in various ways, including, in social birds such as rooks, the giving of warning and the frightening away of intruders. The language may include as many as fifteen sounds used under different circumstances (chaffinch). The more elaborate song of male birds is used in courtship both as a sexual stimulant and as a threat to other birds invading the chosen territory (p. 503).

The lungs are rather small spongy organs, with little elasticity. The air passes backwards in a large main bronchus running through the lungs and giving off branches to the lung substance, but continuing beyond to the inspiratory air-sacs (Figs. 283 and 284). These are thin-walled chambers, divided into two sets, the posterior inspiratory and anterior expiratory. The posterior, inspiratory, air-sacs are the abdominal and posterior thoracic and they are filled by the air rushing into them through the main bronchus. The anterior or expiratory air-sacs include an anterior thoracic, median interclavicular, and cervical, these often communicating with spaces in the bones. At expiration the air passes from the more posterior sacs through the lungs by special recurrent bronchi into the anterior sacs. From these the air may be expelled to the exterior, return to the lungs being prevented by closure of sphincters. In some conditions, however, especially in diving birds, the air may be passed backwards and forwards through the lungs several times, until all its oxygen has been used. The branches of the bronchi in the lungs do not end blindly in alveoli, but make an elaborate system of lung capillaries. Air sweeps through the larger channels at inspiration and expiration, but probably reaches the finer capillaries by diffusion.

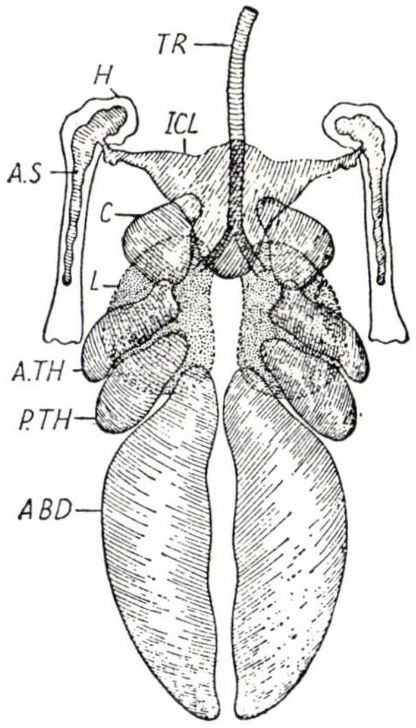

FIG. 284. The air-sacs of a bird. L. right lung; C. cervical air-sac; ICL. interclavicular; A.S. outgrowth into humerus (H.); A.TH. anterior thoracic air-sac; P.TH. posterior thoracic; ABD. abdominal air-sac; TR. trachea. (From Thompson, *Biology of Birds*, Sidgwick & Jackson, Ltd.)

The mechanism by which the ventilation is produced is complicated and depends largely on the movements produced during locomotion. The upper surface of the lung adheres to the ribs, its lower surface is covered by a special membrane derived from the peritoneum and known as the pulmonary aponeurosis (Fig. 285). This is connected with the ribs by costopulmonary muscles. The floor of the thoracic air-sacs, which lie below the lungs, is also covered by a fibrous membrane, the oblique septum, but the walls of the remaining air-sacs are very thin. Quiet respiratory movements are produced by the intercostal (inspiratory) and abdominal (expiratory) muscles, acting upon the thoracic and abdominal cavities so as to enlarge and contract the thorax, drawing air in and out of the air-sacs, through the lungs. During flight the movements of the pectoral muscles provide the ventilation, the sternum moving towards and away from the vertebral column.

25. Excretory system

The kidneys are, of course, metanephric and are relatively large, elongated, and lobulated. They are provided with venous blood by

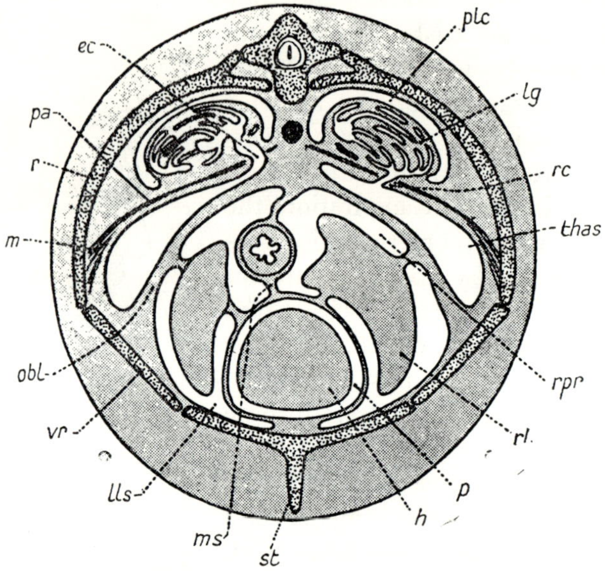

FIG. 285. Diagram of transverse section through the thorax of a bird.

ec. Excurrent passage from lung to air-sac through pulmonary aponeurosis; *h.* heart; *lls.* left liver-sac; *lg.* lung; *m.* muscle; *ms.* mesentery below oesophagus; *obl.* oblique septum; *p.* pericardial coelom; *pa.* pulmonary aponeurosis; *plc.* reduced pleural coelom; *r.* dorsal rib; *rc.* recurrent bronchus from sac to lung; *rl.* right lobe of liver; *rpr.* right pulmonary recess; *st.* sternum; *thas.* posterior thoracic air-sac; *vr.* sternal rib. (From Goodrich.)

the renal portal veins and arterial blood from the renal arteries. The arrangement is essentially as in amphibia and reptiles, with the renal arteries supplying the glomeruli and the portal veins, which break up into inter-lobular branches, sending blood to the renal tubules, whence it is collected into a central intra-lobular vein. It is not certain, however, exactly how the system operates, and it is possible that much of the blood-flow is directly from the renal portal to the renal veins, making little contact with the tubule walls.

The excretory system is highly specialized for water-saving. For this purpose the end product of nitrogenous metabolism is the relatively insoluble uric acid, synthesized in the liver, probably from ammonium lactate. After excretion by the kidney the urine is concentrated in the cloacal chambers and the uric acid precipitates as whitish granules. There is no urinary bladder in the adult bird. More soluble excretory end substances, such as urea, would reach toxic concentrations. The glomeruli are much more numerous and smaller than those of mammals. The urinary tubules effect a considerable concentration of the urine by means of long loops of Henle. The viscous fluid that enters the urodaeum then passes up into the coprodaeum, where further water is abstracted, and the mixed faeces and urinary products are then excreted as the characteristic semi-solid white guano. The water-conservation system is certainly very effective, and some desert-living birds are said to be able to survive for many weeks without water. In this respect the birds have freed themselves from the original aquatic environment to a remarkable degree.

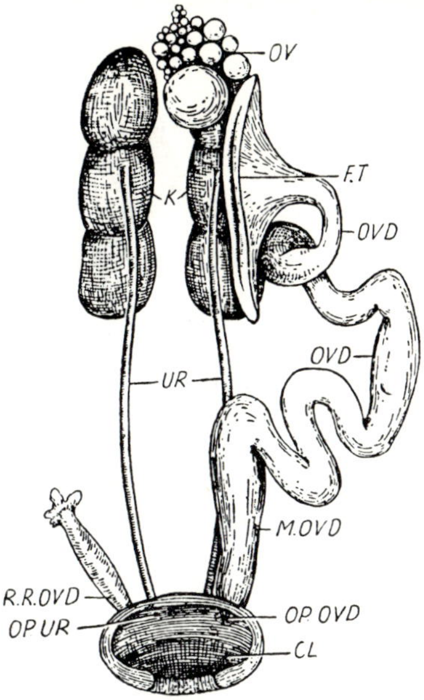

FIG. 286. Female reproductive organs of a hen.

OV. ovary; K. kidneys; F.T. funnel; OVD. oviduct; M.OVD. muscular part of oviduct; OP.OVD. opening of oviduct; UR. ureters; OP.UR. opening of right ureter; R.R.OVD. rudimentary right oviduct; CL. cloaca. (From Thompson, *Biology of Birds*, Sidgwick & Jackson, Ltd.)

26. Reproductive system

The testis consists of coiled tubules of the usual type, joining to form a long epididymis and vas deferens, opening into the urodaeum by an erectile papilla that is the only copulatory organ of most birds. During copulation the proctodaea of male and female are everted and pressed together, so that the sperm is ejaculated direct into the female urodaeum and finds its way up the oviduct. A definite penis (and also clitoris) is found in ratites, anseriformes, and a few other birds. The condition of the testis and its ducts varies greatly with the time of year, the weight of the gland being as much as 1,000 times greater in the breeding-season than it is in the non-breeding, when it contains only spermatogonia.

The provision of material sufficient for the development of a warm-blooded creature is, of course, made possible in birds by the extremely yolky eggs, so large that they allow room for development of only one ovary, nearly always the left (Fig. 286). The right ovary remains present as a rudiment and if the left is destroyed by operation or disease the right is able to differentiate, but then forms not an ovary but a testis. Complete sex reversal can thus occur, at least in some races of domestic fowl, and the transformed bird may acquire cock plumage and tread and fertilize hens (Fig. 287). Sex reversal rarely, if ever, takes place in the opposite direction. We must suppose that there is some switch over in the balance of male and female determining

FIG. 287. Secondary sexual characters of the fowl (*Gallus*). *Left* cocks, *right* hens.
A, normal; B, castrated; C, cock with implanted ovary and hen with implanted testis. (After Zawadowsky.)

processes, taking place relatively early in the case of normal definitive males but later on in life also in 'females', so that all birds become potentially 'male' at the end of their life.

Of the large number of oocytes only few ripen to make the enormous follicles. After each follicle has burst it quickly regresses; there is no 'corpus luteum'.

The egg is taken up by the ciliated and muscular funnel of the left oviduct, and passes down a tube with circular and longitudinal muscles and a glandular, ciliated mucosa. The albumen of the egg is produced by long tubular glands, opening to the lumen. The oviduct has various parts, the upper secreting mainly albumen, the lower producing the shell, and the lowest mucus, to assist the act of laying. The blue background colour of the egg (oocyanin) is produced during shell-formation in the upper part of the tube; spots of red-brown ooporphyrin are added lower down. The pigments are derived from the bile, ultimately from haemoglobin.

As much as a third of the weight of calcium in the whole skeleton is needed for the shells of the two eggs laid by a pigeon. A reserve is collected as the ovarian follicles mature. The oestrogen they produce increases the uptake of calcium from the food and stimulates its deposition in the bones. After ovulation the oestrogen level falls, the calcium is mobilized from the bones, and its concentration in the blood becomes very high, until used by the eggs.

27. The brain of birds

The brain is larger relative to the body in birds than in any other vertebrates except mammals (Fig. 288), and there is no doubt that one result of the high temperature has been to allow opportunity for an elaborate nervous organization and complicated behaviour. Unfortunately we have little information about the way in which the large masses of tissue of the brain function; they are certainly different from anything found in mammals. There are considerable differences in the development of the parts in various birds, for instance, the forebrain is especially large in the rooks and crows (*Corvus*) and in the parrots, the behaviour of which also shows signs of outstanding 'intelligence'.

In the spinal cord the most characteristic feature is the relatively small size of the dorsal funiculi, and their nuclei in the medulla are also small. Evidently the sense of touch is less well developed over the body than it is in mammals, perhaps less than in reptiles. No doubt movement of the feathers provides impulses leading to reflex actions, but it is not surprising that the loose covering does not allow elaborate organization of the sense of touch. The finer senses of birds are restricted to the eyes, ears, and bill. On the other hand, there are large spino-cerebellar tracts, presumably proprioceptive and concerned with the delicate adjustments necessary for flight. The spinal cord is controlled by large efferent tracts from the brain, including cerebello-spinal, vestibulo-spinal, and tecto-spinal pathways. There is no direct tract from the forebrain to the spinal cord, but the influence of the large corpora striata is probably exercised through fibres running to the red nucleus and tegmentum of the midbrain, from which others pass to the cord.

The cerebellum is also large (Fig. 288), a state of affairs perhaps connected with the precise timing and control of movement in all planes of space during flight. Besides large spino-cerebellar and vestibulo-cerebellar pathways there are also tecto-cerebellar and strio-cerebellar tracts, the latter perhaps conducting in both directions. The effect of the cerebellum on other parts of the brain is exercised through cerebellar nuclei, the cells of which give origin to the cerebello-spinal tract.

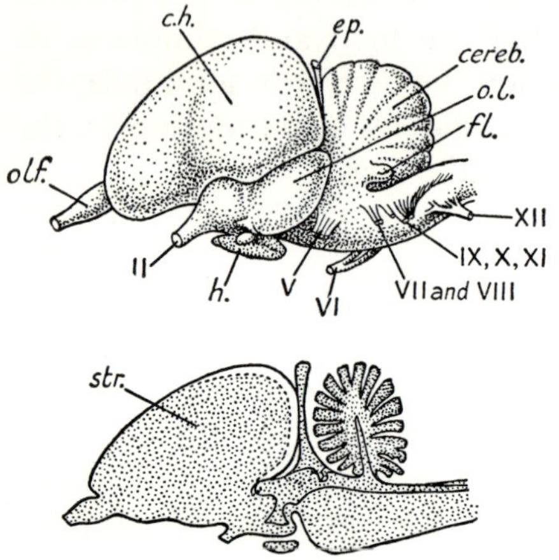

FIG. 288. Brain of a duck (*Anser*)

c.h. cerebral hemisphere; *cereb.* cerebellum; *ep.* epiphysis; *fl.* flocculus; *h.* hypophysis; *o.l.* optic lobe; *olf.* olfactory lobe; *str.* striatum; II–XII, cranial nerves. (After Butschli and Ihle.)

The optic tracts are completely crossed and end mainly in the midbrain, as in lower vertebrates. However, a considerable portion of the optic tracts passes to the thalamus, and the midbrain and thalamus are both highly developed and have intimate and reciprocal connexions with the striata of the cerebral hemispheres. The optic lobes also receive ascending fibres from the trigeminal nuclei and from the spinal cord. Their efferent pathways run to the oculomotor nuclei, to the underlying tegmentum, and to the medulla and spinal cord. Evidently they play a large part in correlating visual with other afferent impulses. The thalamus is large and its dorsal part well differentiated into nuclei. It receives, besides optic fibres, also projections from tactile, pain, temperature, and perhaps auditory sources. There are large thalamo-striatal tracts, probably conducting in both directions. The ventral thalamus receives impulses from the striatum and sends them to the tegmentum, this being the main efferent pathway of the forebrain. The hypothalamus is rather small, probably because of the reduction in the olfactory system.

The cerebral hemispheres are much larger than any other part of the brain and show an exaggeration of the condition found in the lizards (Fig. 289). The ventro-lateral portions are enormously developed, whereas the medial ventral walls are thin and the pallium is quite small, thin, and not folded. The olfactory regions of the brain are small, including the hippocampus.

The corpus striatum is a huge solid mass of tissue, receiving projections forward from the thalamus and sending them back through the latter, to the midbrain roof and floor, to the cerebellum, and thence to the medulla and spinal cord. This very characteristic striatum can be divided into various regions. The part representing the 'original' or lower striatum is called the 'palaeostriatum'; other parts, lying above this, are known as the mesostriatum and hyperstriatum.

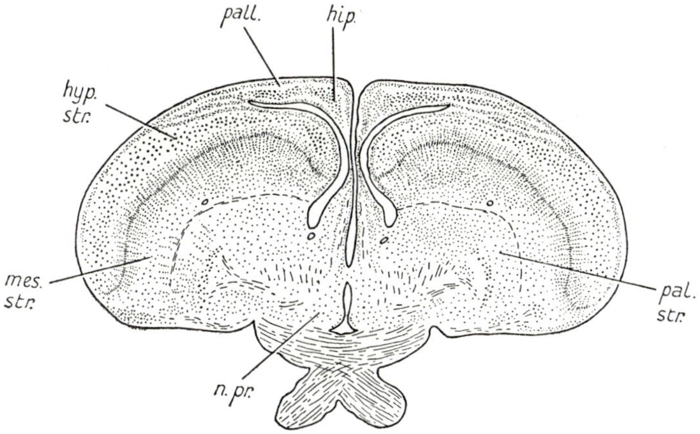

FIG. 289. Transverse section through forebrain of sparrow.
hip. hippocampus; *hyp.str.* hyperstriatum; *mes.str.* mesostriatum; *n.pr.* preoptic nucleus; *pal.str.* palaeostriatum; *pall.* pallium. (Partly after Kappers, Huber, and Crosby.)

28. Functioning of the brain in birds

Loss of one complete hemisphere by a pigeon is not followed by any gross motor defect or asymmetry of movement. This suggests that the corpora striata do not control individual muscle movements, which agrees with the fact that there is no direct pathway from the forebrain to the spinal cord, corresponding to the pyramidal tract of the mammals. Electrical stimulation does not produce movements; the striata are 'silent areas' to stimulation.

Complete removal of both hemispheres does not reduce a pigeon to a helpless state. The animal can still maintain its temperature and its balance and can feed itself if the food is placed near to it. However, a bird so treated is far from normal. It may show a lack of activity, remaining inert for long periods, and then become aimlessly restless for a while. Evidently the normal balance of excitation and inhibition has been upset. Deficiencies in vision can be detected in birds with various portions of the cerebral hemispheres removed, and the mating and nesting behaviour are also affected. Even small removals of the cortex and top of the striatum are said to prevent incubation (though not copulation) and with deeper lesions the whole process of rearing the young becomes impossible.

These observations on the functions of the brain of pigeons may not be applicable to birds in general. The forebrain is larger in many other birds than in the pigeon, and there is some evidence that in the parrots movements and even 'phonation' can be elicited by electrical stimulation of the corpora striata, which are especially large. Removal of one particular area is said to lead to disturbances of 'speech'.

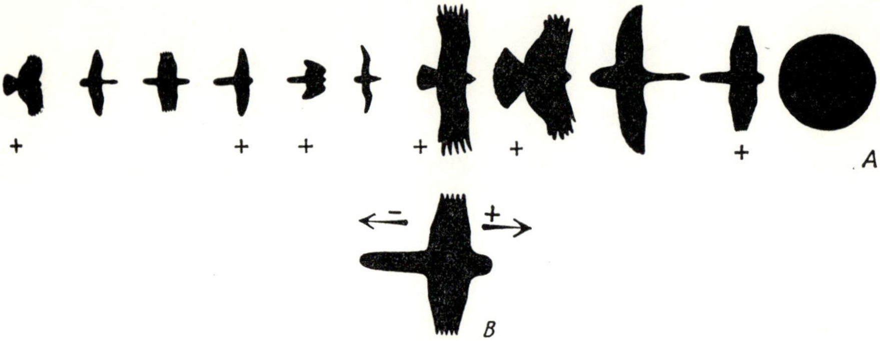

FIG. 290. A. Models used by Lorenz and Tinbergen. Small birds reacted with escape movements to the models marked +. B. The model induced escape reactions when towed to the right ('hawk') but not when towed to left ('goose'). (From Tinbergen.)

It seems, therefore, that the large masses of nerve-cells in the striata are concerned in some way with the elaboration of the more complex acts of behaviour. This is a very vague statement, but is the best that we can give at present. It may be that further investigation of the reciprocal actions of striatum and thalamus will show whether the essentials of their action consist in some reverberating or scanning systems and whether these actions are at all similar to those in the forebrain of mammals. The fact that the striatum consists of solid masses of tissue suggests that the arrangement does not depend, as does the mammalian cerebral cortex, on the projection of patterns of excitation onto an extended surface.

Birds are usually said to show more stereotyped patterns of instinctive behaviour than primitive mammals. Once they have embarked on a line of action, even a complex one like nest-building, they are supposed to pursue it in a given manner, without ability to adapt themselves to unusual happenings. Watching the exploratory behaviour of a robin or a tit it is difficult to feel that the existence of this difference is adequately proved, but whatever distinction exists is presumably a reflection of the difference between the functions of large masses and spread-out sheets of nervous tissue. Certainly birds can count as well as rodents, and their performance in mazes and puzzle boxes is at least comparable to that of most mammals.

There are many signs that a suitable initial stimulus sets off in the bird a whole train of behaviour, organized from within. On the other hand, inappropriate stimuli may sometimes set off a reaction, as if the 'keys' to these cerebral 'locks' were not very elaborate or specific. A

FIG. 291. Pintail ducks (*Anas acuta*). Males displaying dark brown feathers of neck with white bands on either side. (From Tinbergen, after Lorenz.)

robin held in the hand may burst into song, cock ostriches frightened by an aeroplane fall to the ground in their characteristic sexual display. Birds frightened or disturbed may proceed to the actions of bathing, preening, feeding, or drinking, performed in a ritual and cursory manner for a long time. Such displacement activities show that the organization of the bird's nervous system, like that of a mammal, provides for some strange deviations, whose study may reveal much about the method of working of the brain.

Many complex forms of behaviour are responses to only limited parts of the natural stimulus situation. Thus when the models shown in Fig. 290 were towed above certain young birds, only the models marked with a cross induced escape reactions: apparently the configuration of the short neck is the essential feature. Much of the elaborate social life of birds depends on such sign stimuli displayed by one bird (the 'actor') and serving as releasers setting off particular actions or trains of action in another bird (the 'reactor'). Many of the elaborate forms of display evolved by birds (p. 497) are releasers of this sort (Fig. 291), and structures and actions on the part of the young release the appropriate behaviour of the parent. The red breast of the robin (*Erithacus*) is the agent that releases attacks by other birds (Fig. 292). Similar phenomena are known in fishes and other vertebrates (p. 225), and it remains to be shown whether they can be attributed to any single or particular neural basis.

29. The eyes of birds

Birds depend more on their eyes than on the other senses; they are perhaps more fully visual than are any other animals. The eyes are extremely large: those of hawks and owls, for instance, may be absolutely larger than in man. The shape is not spherical, the lens and

FIG. 292. The red tuft of feathers is attacked by male robins holding territory, but the complete juvenile bird (without red) is left alone. (After Lack, from Tinbergen.)

cornea bulge forwards in front of the posterior chamber, this form being maintained by a ring of bony sclerotic plates (Fig. 293). In most birds the whole eye is thus broader than it is deep, but in those with very acute sight it is longer, and in some eagles and crows becomes almost tubular. The great distance between lens and retina allows broadening of the image, thus improving the fine two-point discrimination that is needed by these diurnal birds. The shape of the back of the eye is such that 'the retina lies almost wholly in the image plane, so that all distant objects within the visual angle are sharply focused on the photosensitive cells, whereas in the human eye this is only true of objects lying close to the optic axis' (Pumphrey; Fig. 294). The lens is usually soft and accommodation is effected by changing its shape, and especially the curvature of its anterior surface, by the pressure upon it of the ciliary muscles behind. These, like the iris muscles, are striated, presumably allowing for the quick accommodation necessary in a rapidly moving bird, though it must not be forgotten that these muscles are also striated in lizards. The ciliary muscle is characteristically divided into 'anterior' and 'posterior' portions, the muscles of Crampton and Brücke (Fig. 293). The latter draws the lens forward into the anterior chamber so that since the shape of the eye is fixed by the sclerotic plates, the lens becomes more curved and hence accommodated for near vision; contraction of the iris sphincter assists in the process. Crampton's muscle is so arranged as to pull on the cornea, shortening its radius and further assisting in

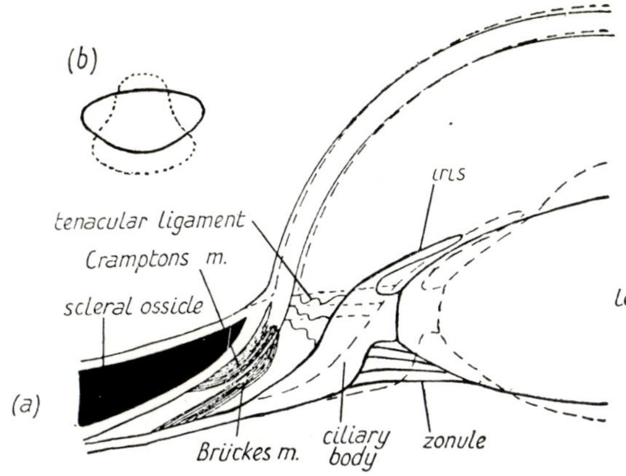

FIG. 293. The mechanism of accommodation in a bird's eye; the positions during near vision are shown dotted. *b*. The lens of the cormorant's eye at rest (full line) and fully accommodated (dotted line). (From Pumphrey, after Franz and Hess.)

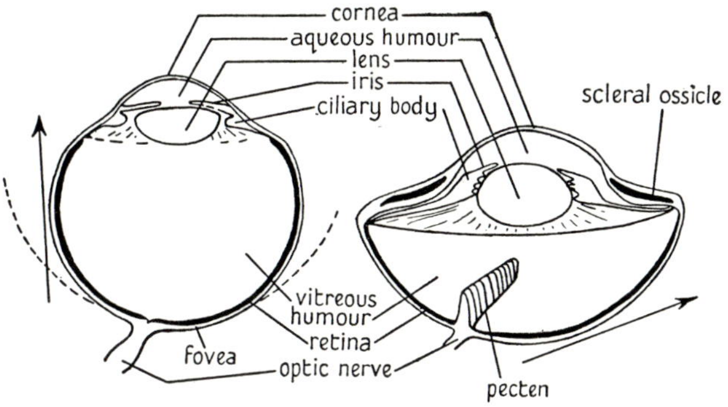

FIG. 294. Diagrams of right eye of man and left eye of swan to show the difference in shape. The position of the image plane in man is shown dotted; it lies behind the retina except near the centre. The arrows point forward. (From Pumphrey.)

accommodation. This double method of active accommodation for near vision is most fully developed in diurnal predators, such as the hawks, less so in night-birds. In aquatic birds Crampton's muscle is reduced, and the cornea is of little importance in image-formation. Special arrangements are found in diving birds, for instance in the cormorants Brücke's muscle is large and there is a very powerful iris muscle, which assists the ciliary muscles to give the great change in shape of the soft lens, allowing accommodation of 40–50 diopters (about

10 in man). The kingfishers are said to possess an amazing arrangement of double foveas, placed at different distances from the lens, so that as the bird dives under water the image is transferred from one fovea to the other without any change in the dioptric apparatus. These details

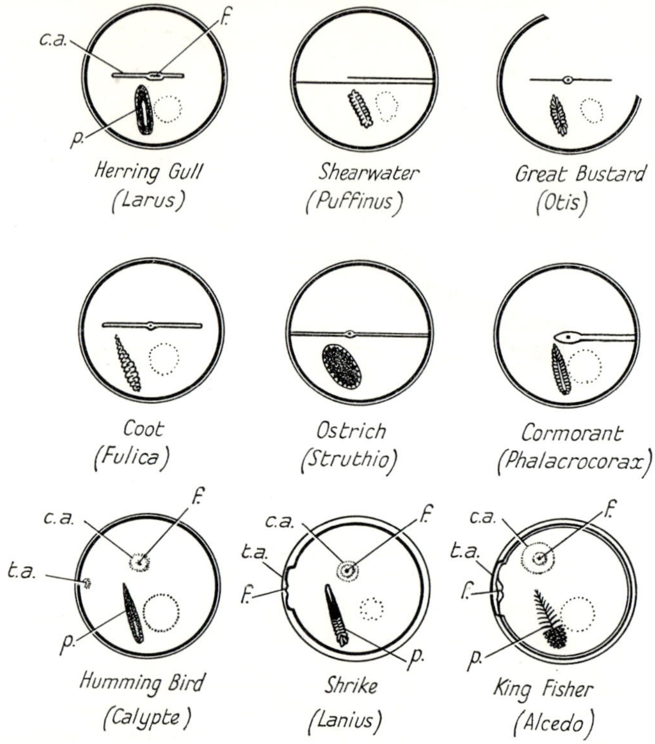

FIG. 295. The appearance of the retina of various birds as seen with an ophthalmoscope through the pupil.
c.a. central area; *f.* fovea; *p.* pecten; *t.a.* temporal area. (After Wood, from Pumphrey.)

of the visual system show, like so many other features of bird anatomy, how readily the structure conforms to special habits of life.

The retina of day-birds consists largely but not wholly of cones; these animals are more fully diurnal than is man. The high resolving power and hence high powers of discrimination and of movement-detection depend on the great density of the cones, as many as 1 million to each square millimetre in the fovea of a hawk, three times denser than in man. Nocturnal birds, on the other hand, have retinas composed mainly or completely of rods, and the differences between the behaviour of these two types of eye, found in birds as in mammals, have been a powerful support for the duplicity theory of vision. There are usually one or more areae, regions of the retina consisting of tightly packed receptors. In birds that live on the sea, in the desert, or other open spaces the area often has the form of an elongated horizontal band (Fig. 295), whereas in tree-living birds it is circular. Some birds have two areae, a central one in the optic axis and a second placed on the temporal surface of the eye, so that the image of objects in front of the head falls on the temporal areae of both eyes. This arrangement is common in birds that follow moving prey (shrike, *Lanius*) or for

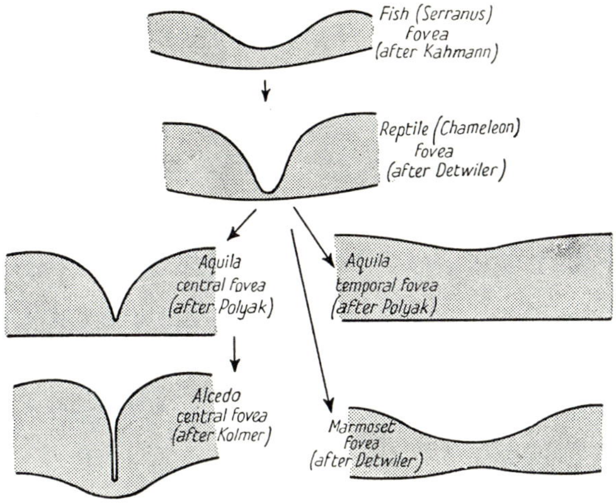

FIG. 296. Forms of the fovea in various vertebrates, showing the development from moderately to sharply convexiclivate types in foveas adapted for detection of movement, but flattening of the fovea where there is binocular vision. (From Pumphrey.)

some other reason require accurate perception of distance (swallows, humming-birds; the latter feed their young on insects caught on the wing). The density of the cones is so high in diurnal birds, even outside the areae, that they probably obtain a good detailed picture in all directions. They do not, therefore, scan the world with the central area of the retina as we do; indeed, the eyes move relatively little. Instead the bird is able to detect very small movements anywhere in its surroundings. The bird's-eye view usually lacks stereoscopic solidity and it is possible that in compensation for this the animals appreciate distance by movements of the intrinsic eye-muscles. The familiar cocking of the head of a bird before pecking may be its means of judging distance.

As in man there is often within the central area of the eye a fovea or pit, and in many birds the sides of this pit are steeply curved (Fig. 296). Walls has suggested that since the vitreous humour and retina differ in refractive index this curvature serves to magnify the image and increase acuity. Pumphrey points out that any such advantage would be counteracted by the aberration introduced and he makes the suggestion that this disturbance of the picture by the 'convexiclivate'

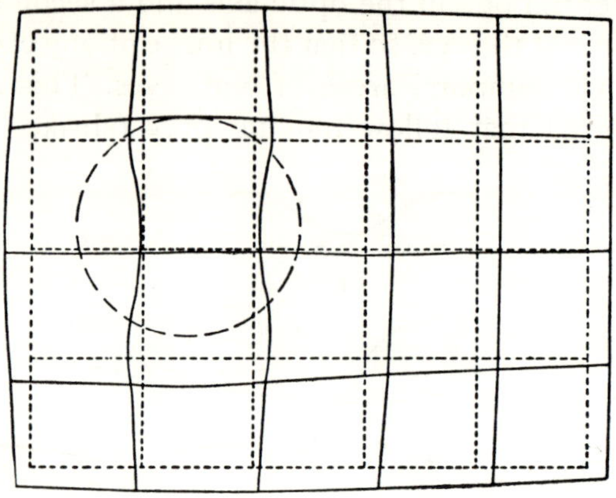

FIG. 297. Effect of refraction produced by the curvature of the fovea of the golden eagle. An image of the form shown with dotted lines is distorted by the fovea to the form shown in solid lines. The circle represents a radius of 10μ at the centre of the fovea. (From Pumphrey.)

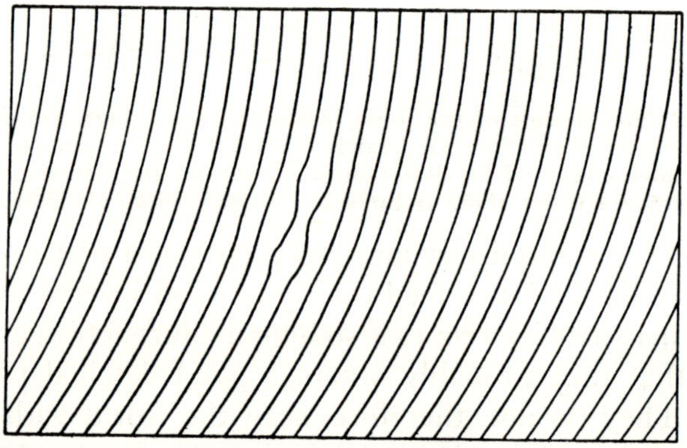

FIG. 298. Distortion by the fovea. The lines represent the successive images at equal time-intervals of the boundary of a regular object when the object moves steadily across the visual field. If this picture is viewed at 7 m the area of irregularity subtends an angle about equal to the angle subtended by the central part of the hawk fovea. It will be found that the irregularity is very evident to the human eye at this distance though the lines are resolvable with difficulty. (From Pumphrey.)

fovea is itself an advantage, improving the power of fixation and sensitivity to movement, at the sacrifice of acuity. Such an arrangement would serve to emphasize angular displacements, transforming a radially symmetrical image into an asymmetrical one, except when there is coincidence between the axes of symmetry of the fovea and of the object (Figs. 297 and 298). Foveas with steep sides are found in birds of prey, kingfishers, and others that have very high powers of detecting movement; a similar but less pronounced arrangement is found in fishes and reptiles. It is probable that the primitive functions of the eyes were fixation and detection of movement, rather than resolution of detail and recognition of patterns. Some birds have one convexiclivate and one flatter fovea (Figs. 295 and 296), the latter being on the temporal surface of the retina and used in binocular vision. The fovea is also flat in the retinae of primates with binocular vision; evidently the optical errors of a curved fovea cannot be tolerated where there is fusion of the two retinal images.

Birds undoubtedly discriminate colours, apparently on a trichromatic basis similar to that of mammals. No other animals, except perhaps primates, show such responsiveness to colour in their surroundings, including the food and other members of the species. In animals that move so freely recognition and attraction of the sexes is more efficiently performed in this way than by touch or odour. The cones of birds often contain red and yellow droplets, which may heighten visual acuity by reducing the effects of chromatic aberration. The droplets in the central area are always yellow. The presence of droplets of various colours in adjacent cones may also increase powers of discrimination. Sometimes the droplets are so arranged as to allow accentuation of different contrasts in the parts of the visual field. The lower part of the pigeon's retina contains red, the upper yellow filters, increasing the contrast of blues and greens respectively, as required for vision against the sky in the one case and the ground in the other. There have been many investigations of the distribution of sensitivity in the retina; probably many birds are rather insensitive to the blue end of the spectrum. There is no truth in the suggestion that the eyes are sensitive to infra-red radiation.

Although the eyes of some birds are directed forwards, so that their fields overlap, they are said not to have binocular vision and decussation of the optic tracts is complete. Perception of distance, a very important function for the bird, must be performed in some other manner. In many birds the eyes are directed sideways, and the fields of view may even overlap behind the head, for example in waders. This may serve to give warning of predators. Many functions have been suggested for the most enigmatic organ of the bird's eye, the pecten, a pleated highly vascular fold, projecting from the retina into the vitreous. It is possible that the irregular shadow cast by this organ provides, as it were, numerous small blind spots and hence by a stroboscopic action increases the number of on-and-off effects produced by a small object in the visual field, increasing contrast and allowing detection of its movement. This, however, is only one of the numerous suggestions about the function of the pecten, and the only real support for the idea is that the body is large and much pleated in predatory birds, which detect minute movements at great distances,

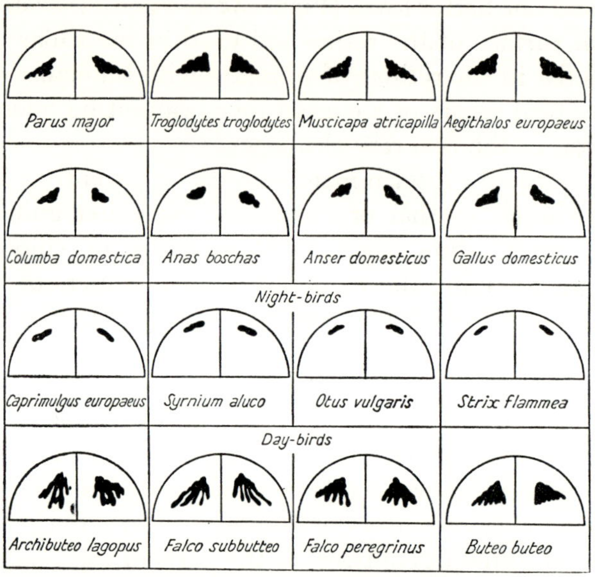

FIG. 299. Tracing of the shadow of the pecten on the retina in various birds. (After Pumphrey and Menner.)

and is small and smooth in nocturnal birds (Fig. 299). However, it is almost certain that the original function of the pecten was to bring nourishment to the vitreous and retina. It has often been suggested that the pecten is in some way connected with accommodation; it is not likely that it actually assists in focusing, for instance, by pressing forward the lens, and no changes have been seen in it during accommodation. However, it might possibly assist by adjusting the intra-ocular pressure, which must be increased by the extensive changes in the lens during accommodation.

30. The ear of birds

Both vestibular and auditory parts of the ear are well developed in birds. The former are not known to possess special peculiarities, but the large connexions with the cerebellum suggest great importance in the operations of flight, presumably especially by the semicircular canals. There is a distinct cochlea, slightly curved and especially well developed in owls and in parrots (Fig. 300). In this there is a basilar membrane, with fibres increasing in length towards the tip and carrying an organ of Corti with hair-cells in contact with a tectorial membrane (Fig. 301), as in mammals. At the tip of the cochlea is a special sensory region, the lagena, similar to that of lower vertebrates and perhaps responsible for reaction to lower notes, the basilar membrane

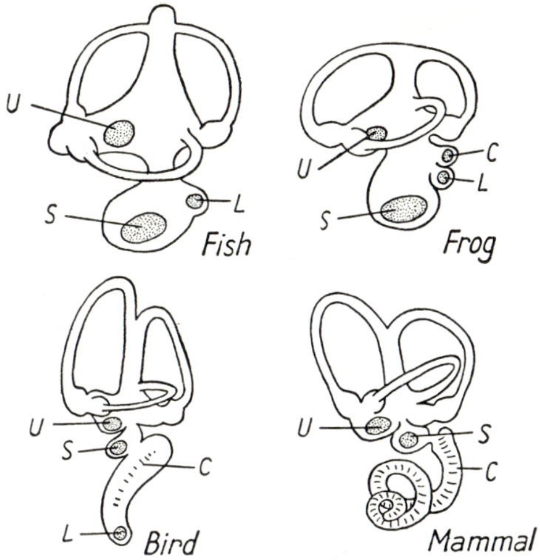

FIG. 300. Labyrinths of various vertebrates, to show varying development of cochlea (C) and lagena (L). S, saccule; U. utricle. (From v. Frisch.)

responding to the higher frequencies. Birds are known to be more sensitive to distant gunfire and other low-frequency vibrations than is man. Transmission of vibration from the tympanum to the inner ear is effected by the columella auris, derived from the cartilages of the hyoid arch. The inner portion of the columella is rod-like (stapes), but the outer end makes contact with the tympanum by means of three processes, of somewhat irregular shape.

Hearing is, of course, acute and the song-birds must be able to discriminate between simple tunes; some of them are surprisingly good mimics. Ability to localize sound is high and owls and other night-birds probably find their prey largely by ear. For the purpose of direction-finding they have developed an asymmetrical arrangement of the ear cavities (*Strix*) or asymmetrical external ears (*Asio*). A few birds that live in caves have the power of avoiding obstacles by echo-location (*Steatornis*, the oil bird, *Collocalia*, swiftlet). They emit up to 5 to 6 clicks a second at 4 to 5 Kc. The rate varies inversely with the amount of light and increases when obstacles are met.

31. Other receptors

The corpuscles of Grandry in the bill of ducks and other birds are probably touch receptors, comparable to Meissner's corpuscles in mammals. The corpuscles of Herbst, found in the dermis elsewhere in

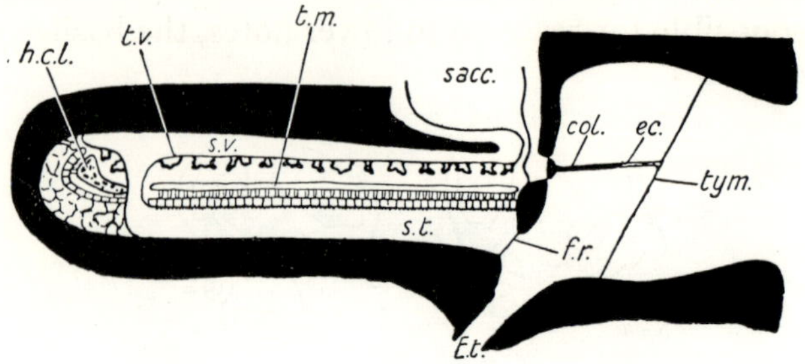

(A)

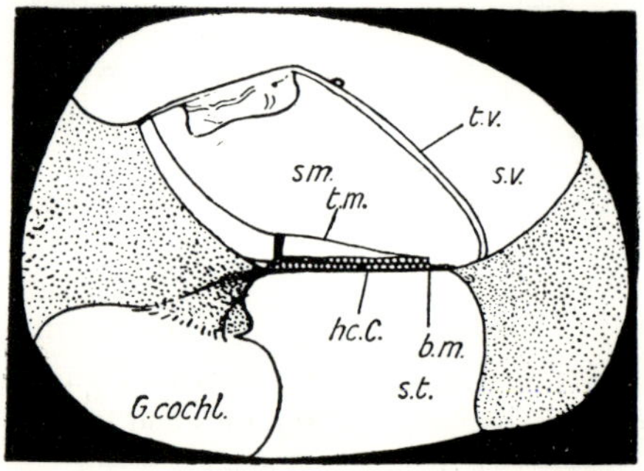

(B)

FIG. 301. A. Diagram of section of bird's ear.

b.m. basilar membrane; *col.* columella; *e.c.* extra-columella; *E.t.* Eustachian tube; *f.r.* round window; *h.c.C.* hair-cells of Corti's organ; *h.c.l.* hair-cells of lagena; *sacc.* sacculus; *s.t.* scala tympani; *s.v.* scala vestibuli; *t.m.* tectorial membrane; *t.v.* tegmentum vasculosum; *tym.* tympanic membrane.

B. Diagram of section across cochlea of a bird.

G.cochl. cochlear ganglion; *s.m.* scala media; other lettering as in A. (From Pumphrey, after Satoh.)

the body, resemble Pacinian corpuscles. They may be receptors for vibration and are numerous in certain situations, for example in the feather follicles, the beak, between the tibia and fibula, and in the tip of the tongue of a woodpecker.

Chemoreceptors for taste and smell are little developed. There are few taste-buds on the tongue. The nasal cavity is large but the olfactory epithelium restricted. It is doubtful whether most birds use the nose as a distance receptor; they may use it to test air coming from the internal nostril. In kiwis, however, which are nocturnal and terrestrial, the olfactory sense is well developed.

XVII

BIRD BEHAVIOUR

1. Habitat selection

THE success of bird life has been largely due to the great variety and ingenuity displayed in finding situations suitable for providing food and allowing reproduction. Their powers of habitat selection allow diversification of species by reducing interspecific competition. Species living in the same area rarely eat the same food, especially if it is scarce. It is often said that birds are creatures of stereotyped habits, yet they have certainly exploited their mobility to the full; they obtain the means of life in most various ways. This mobility makes it difficult to specify the 'environment' of a bird. For instance, a swallow may pass part of its life in the tropics, part near the Arctic Circle. A gull may nest on a rock, eat grain in a field, and then fish in the sea, all within a few hours. Observation of the familiar birds of town and country soon shows that they are at home in a much greater variety of situations than could be supported by most animals, and that within limits they can adapt their behaviour to each situation. Birds show, therefore, in a marked degree, two of the features most commonly used as criteria for the recognition of a higher animal, namely, freedom to move to different conditions and ability to obtain a living in unpromising circumstances.

Nevertheless each species nests in a limited variety of habitats and feeds in a limited variety of habitats. There is evidence that the appropriate habitat is recognized by a relatively small number of conspicuous features, independently of learning.

2. Food selection

Where the food is very specific substitutes will only be accepted under unusual conditions of starvation, but many birds are more catholic in tastes and some of these are among the most common, for instance rooks, starlings, thrushes, blackbirds, and gulls. Both field observation and experiment suggest that birds quickly learn from experience where and how food may be obtained. They will remember to visit an abundant source of supply and there are numerous stories of the ingenuity of such birds as the jackdaws in obtaining it. This short-term memory is probably of great importance in allowing the bird to establish an effective routine for each part of its life, especially as it is combined with the power to explore elsewhere when conditions change.

Food selection thus depends on species-characteristic motor patterns and structures, whose use varies to suit the circumstances. There is an initial responsiveness to a wide range of stimuli, later modified by learning. In young birds these actions are not necessarily related to appropriate objects or situations. Thus young chaffinches or tits peck at spots of many sizes, but only when they are *not* hungry. When they are, they beg from the parents. Young kestrels 'play' at hunting pine cones, even after obtaining food by real hunting (see Hinde, 1959).

This range of response is then narrowed by learning. Objects that provide food are pecked again, those that do not or are distasteful are avoided. The range of learning that is possible must, however, be influenced by the hereditary equipment. Thus chaffinches never use the foot to hold objects but goldfinches and tits do so and can thus learn to pull over a grass stem and peck off insects otherwise out of reach, or indeed to open milk bottles!

3. Recognition and social behaviour

Great mobility has made it necessary for birds to develop specific means for recognizing their fellows, their enemies, and their competitors; from this power an elaborate social life has developed in many species. In spite of their freedom many birds are not individualists, they live much of their lives together in flocks; the unit of life is larger than the 'individual' body. Species feeding on the ground, such as rooks, starlings, and partridges, commonly move about in groups during the winter and obtain the advantage that the alertness of each single bird serves to warn for many. The lack of procryptic coloration in some of these social birds is a measure of the effectiveness of the protection afforded by the society; indeed, it may be advantageous that the birds should be conspicuous to their fellows. Starlings carry the communal life farther by collecting together in large numbers each night to roost. As many as 100,000 may be found in one roost, the birds flying home from their feeding-grounds every night for distances of many miles. Possibly in this way the disadvantage of the conspicuous outline is minimized while roosting. Rooks show somewhat similar behaviour, but it is not found in the protectively coloured partridge.

Many different means are adopted by birds for recognition of other members of the species and of the same and opposite sex; bird-life contains elaborate social and sexual rituals for many occasions. For example, relief of the one bird by the other at the nest is accompanied by a peculiar wing-flapping ceremony in herons and other birds. Greeting ceremonies are common in many species, and there is a host of sexual recognition and courtship rituals, to be considered later, serving the immediate function of regulating the aggressiveness that otherwise arises when two individuals approach each other closely. Such ceremonies prevent attempted copulation with the same sex and ensure it with a member of the opposite sex, and of the right species.

Apart from sexual behaviour birds show many complex mutual and social reactions. Thus in communities of hens or pigeons there is quickly established a rank of 'pecking order', such that each bird is submissive to the one above it, the order changing, however, when age, moulting, or experiment (e.g. sex-hormone injection) alters the state of the birds. More pleasing communal habits are the dances and corporate flights, which are well known in cranes and other species.

4. Bird migration and homing

Among the remarkable devices of birds is their habit of seasonal movements to obtain the advantage of the favourable conditions offered in more northerly regions only during the summer. The most familiar migrations are those north and south (sometimes north-east and south-west) over the land masses of the northern hemisphere, but there are similar movements also in the southern hemisphere, though they are of lesser extent. Some tropical birds migrate to breed in the rainy season in the outer tropics, removing to the central tropics in the dry season. Marine birds also may make extensive migrations. Thus the great shearwater (*Puffinus*) breeds on Tristan da Cunha, but comes as far north as Greenland or Iceland in May, returning again after months of wandering at sea, apparently without making a landfall. The Arctic tern (*Sterna*) breeds in the north temperate zone, and migrates to the Antarctic along both sides of the Atlantic. Penguins make migrations by swimming. Distances up to 6,000 miles from Northern Europe to South Africa have been recorded for swallows and storks, and even farther for the Arctic tern. The factors determining the direction and course of migration are beginning to be known. It has been shown that the power to follow a given course depends partly on the ability to navigate by observation of the position of the sun or stars (Matthews, 1955; Sauer, 1957). Birds certainly do not learn the routes from their elders, indeed the young often leave first. However, juveniles return only approximately to their birthplace, whereas older individuals return to their old nesting sites. Individual memory therefore plays a part. The birds often migrate singly and

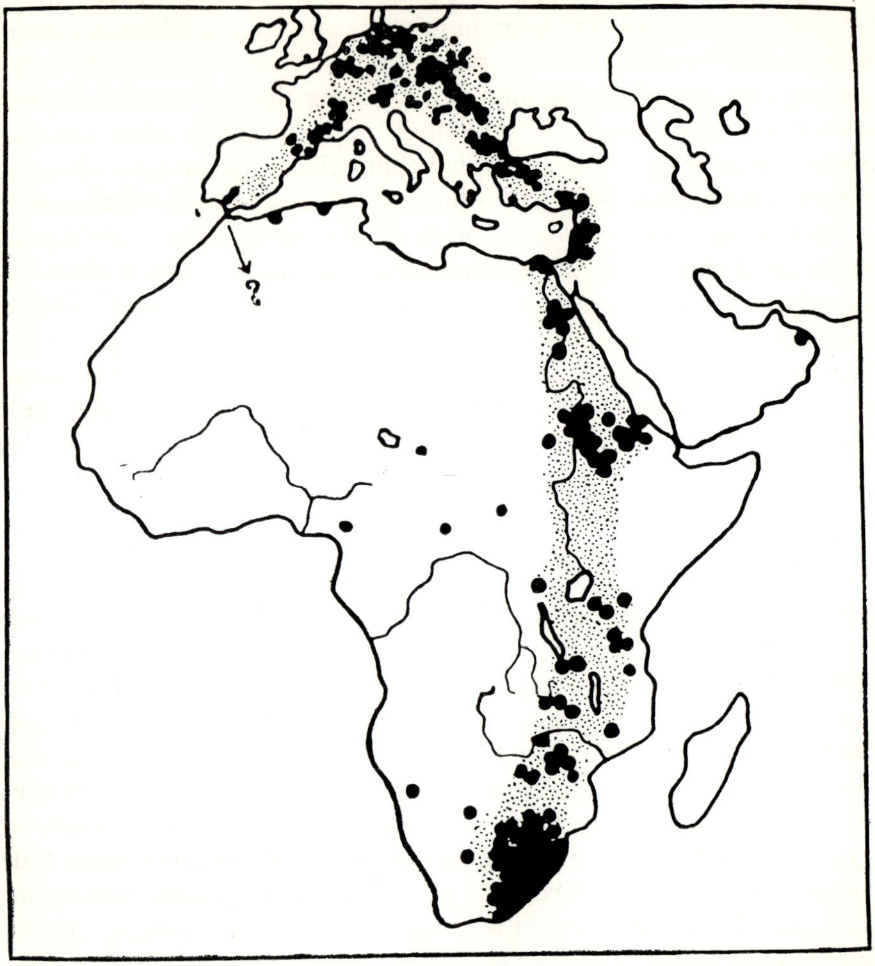

FIG. 302. Direction of migration of the stork (*Ciconia*). The black spots represent points of recovery of birds ringed in Germany. The dotted areas show the pathways produced by ecological and geographical factors. (After Schülz and Stresemann.)

experiments in a planetarium have shown that warblers turn accurately in the appropriate direction as given by the stars.

The available evidence therefore suggests (*a*) that for most species at least there is a preferred direction for migration, which is independent of experience. Hand-reared warblers will orient correctly by the stars without previous experience; (*b*) that they may be deviated by features of the environment such as coastlines or hills; (*c*) that after experience they can do more than just steer a course—they can navigate, that is fix position, calculate the course to steer, and follow it.

Powers of 'homing' are remarkable in birds, altogether apart from migration. It is probable that these feats are performed by the use of visual clues, combined with a tendency to follow coastal outlines and other conspicuous geographical features. Pigeons are trained to 'home' by release at progressively increasing distances and can acquire the ability to return from more than 500 miles.

It has now been satisfactorily proved that pigeons return home after release from a distant point even if they have never been there before, nor have had any previous training in returning from situations out of sight from the loft. Matthews (1955) and others have shown that upon release the birds fly off towards home provided that they can see the sun. This capacity to navigate by the sun must depend upon determination of position on two coordinates by observations of the sun's altitude, azimuth, and/or movement. This implies the use of a very accurate chronometer as a means of determining the difference between home and local time. It is hard to believe that all this could be achieved in the few seconds following release, but the facts demand some such hypothesis. Moreover, experiments designed to upset the 'chronometer' by altering the period of daylight have been claimed to be effective in altering the direction of flight of birds upon release.

Other birds are also able to return home from spectacular distances, an instance being the Manx shearwater removed from its nest (burrow) on Skokholm Island off the Welsh coast and sent to Boston by air: it returned in 12 days, the distance being 3,067 miles across the Atlantic. In another experiment three out of ten untrained terns returned to their nests from a distance of 855 miles, in about 6 days. There are many peculiar and unexplained features about such long journeys.

5. The stimulus to migration

It is probable that the north-to-south migrations of birds in the northern hemisphere take place under some stimulus provided by the internal condition of the gonads, these being themselves affected by the seasonal change. Rowan has made extensive experiments with juncos, birds that are summer visitors to Alberta, Canada. If the birds are caged in the autumn and illuminated to compensate for the shortening day the gonads do not regress as they normally do at that season, so that full breeding song continues on into the middle of winter. Rowan had the interesting idea of releasing these birds in the winter and found that they immediately moved away, perhaps northwards, conditions being then appropriate if anything for that direction. On the other hand, control birds, which had been kept in Alberta but without extra light, showed no tendency to move away if released in mid-winter. Since the gonads of these controls had, of course, by then undergone reduction, the stimulus to migrate south was not felt. This would agree with the fact that only some individuals of species such as the thrush and blackbird migrate. These experiments suggest that seasonal changes in illumination may be an important factor determining migration. However, they leave many points unsettled. For instance, is there a stimulus that starts the bird migrating north again if it has 'wintered' near the equator, where there is little or no seasonal change? It may be that once alteration of the gonads has been started, say, by reduction in light, a cycle will be set up, the gonads developing again with the longer days in the south and thus driving the bird north again, and so on.

6. The breeding-habits of birds

The complexity and variety of bird behaviour show especially in their breeding; perhaps in no other creatures except men is such elaborate behaviour involved in bringing the birds together and caring for the young. In order to ensure adequate provision for the development of a warm-blooded animal and its nourishment until it can fend for itself it is necessary either to keep it within the mother or to provide a means of incubating the eggs. In either case a long period of care is necessary after birth; the young animals, having a large surface area, require a great amount of food to keep warm. Thus young starlings and crows may eat as much as their own weight of food each day.

Mammals and birds set about providing for this warmth and food in different ways. Since a female mammal can move about and get food while pregnant the father can desert her altogether, though frequently he does not do so. In birds the eggs and young cannot be left cold for long and it is therefore especially desirable that the father should help. In birds, therefore, perhaps even more than in mammals, the breeding-habits involve the development of elaborate systems of mutual relations, serving not only to bring the parents together but also to keep them together throughout the period of incubation and while feeding the young. The actual building of the nest may be an intricate business in which both birds collaborate and a further factor is that the pair occupies a territory around the nest, which they defend against others of the same species.

The type of association of the sexes varies greatly. In the ruff and certain game birds (blackcock) there is no pair formation: display and copulation occur at communal display grounds. The wren (*Troglodytes*) and a few other birds are polygamous, each male forming continuous association with several females. The great majority of birds form pairs throughout a single season, occasionally they change mates for the second brood. The same pair may mate in successive years (crows, swifts) and a few birds stay together through the year (ducks).

7. Courtship and display

The breeding of birds is nearly always seasonal, even in the tropics where conditions are apparently almost uniform throughout the year. In temperate latitudes breeding begins in spring as the gonads develop, probably under the influence of increasing illumination. The changes in behaviour with ripening of the gonads vary, of course, greatly with the species. Birds that have been social through the winter, for instance buntings, begin to leave their flocks, and the voice of the male changes from the simple winter notes to the more complex breeding song. The production of the elaborate secondary sexual characters of the plumage and other features used in display is controlled partly by direct genetic effects on the tissues, partly through hormones. Injections of male or female sex hormones or anterior pituitary extracts influence the production of some characters but not others, according to the species. In the majority of birds there is a breeding-season, initiated, at least in many, by the effect of increasing length of day in the spring, acting through the pituitary on the gonads. Other factors such as degree of activity and food taken play their parts, especially near the equator, where there is little seasonal variation.

The song is one feature of the elaborate business of display and courtship. It has somewhat different functions from species to species. In the simplest case the display serves to bring the sexes together, to enable recognition, and at a later stage as a stimulus to copulation. Moreover, in some birds (doves, budgerigars, and canaries), the display serves as part of the stimulus to ovulation. Involved in the display actions, however, is often a threat to other males. The aggressive displays are usually different from the courtship displays, though in many species the male's first response to a potential mate is an aggressive one. When the female does not flee or fight back, as a male would do, he gradually changes over to courtship display. Finally, some forms of courtship, especially those that are mutual, seem to serve to keep the partners together for the period of incubation and feeding.

We may recognize in courtship, therefore, three elements, first sexual stimulation, secondly threat to other males, and thirdly mutual stimulation while rearing a family, but the various types of display are combined in so many different ways that any classification or analysis is bound to be arbitrary and only a few examples that have been thoroughly studied can be given. There are, of course, also begging displays, given by young to parents, various displays given to potential predators, and others.

Song and displays that bring the sexes together are responsible in part at least for many of the pronounced secondary sexual characters in which the male and female differ. The beautiful plumage of the cock pheasant or peacock and the more bizarre combs and wattles of turkeys are displayed before the female in a manner that is clearly an excitant to copulation. The secondary sexual characters by which the sexes are differentiated may affect features as different as the colour, length, and structure of the feathers, the colour of the iris and size of the pupil, the shape and size of the body, the voice, the ornamentation of the head, and the spurs on the feet. The importance of species recognition in leading to differentiation of male plumage is shown by Darwin's finches (p. 524) which, in the isolation of the Galapagos Islands, where there are few other passerine birds, have abandoned the highly coloured male plumage found in other finches. The recognition of individuals of the same species in this case is based on the characteristics of the beak; a male will begin to attack an intruder only when the face is seen.

Display takes place either by one bird to the other or mutually, and has the effect of bringing the animals together and keeping them together for periods varying from a few minutes to several years. Long unions are common in the large birds of prey and are often a result as much of mutual association with the nest as with display by the other bird. Some birds pair in the winter long before the gonads are ripe (ducks), but more usually after two birds pair off the display produces a gradual heightening of tension, leading to nest-building, copulation, and ovulation within a few days. Probably the process of bringing the birds together and ensuring coition requires even more elaborate stimulation in birds than other animals because of their great mobility and the fact that the male cannot grasp the female. Before he can tread her in such a way as to ensure coition she must be brought into a suitably receptive state. The function of the display is certainly largely to induce this state in the female, though much more is involved in addition. The processes of sexual stimulation, nest-building, incubation, and caring for the young necessitate a particular state of excitability that must last a long time, and the display serves to provide this condition in both birds. The reproductive process may be interrupted at any time if the stimuli are inadequate. Thus ovula-

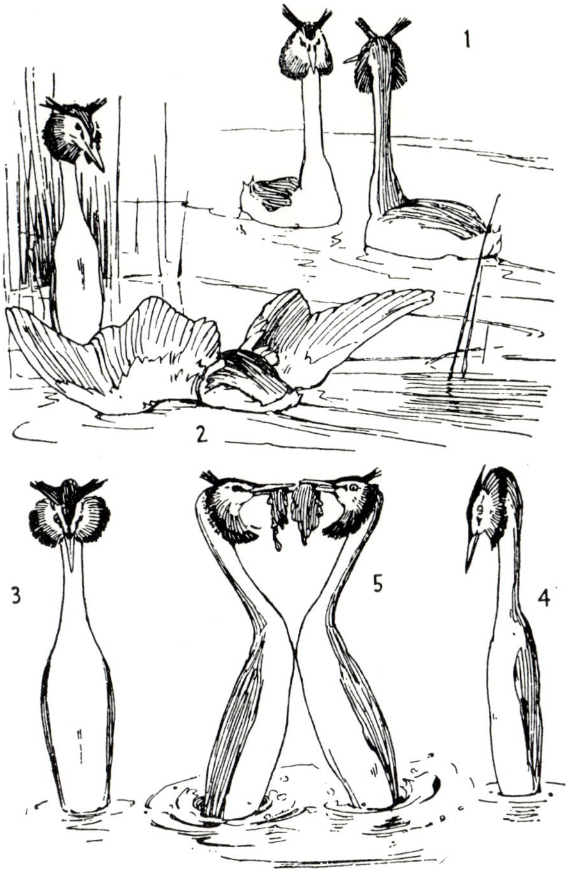

FIG. 303. Incidents in the courtship of the great crested grebe. 1. Mutual head shaking. 2. The female is displaying before the male who has dived and shoots out of the water in front of her. 3 and 4. Further views of the male rising from the water. 5. Both birds have dived and brought up weeds. (From Huxley, *Proc. Zool. Soc.*, 1914, by permission of the Zoological Society.)

tion may depend upon courtship and copulation (pigeons) and eggs are often deserted if the birds are interfered with in any way, or if they fail to stimulate each other.

The actual procedure of display is as varied as any other feature of bird life. When the sexes are alike in colour and shape the performance is usually mutual, as in the great crested grebes (*Podiceps*), which approach each other over the water and go through various actions such as the head-shaking ceremony (Fig. 303). Where the male is more strongly coloured, provided with a special comb, &c., he often displays before the female, who adopts a more passive role. A very common element in the procedure of courtship is the sudden revelation of some

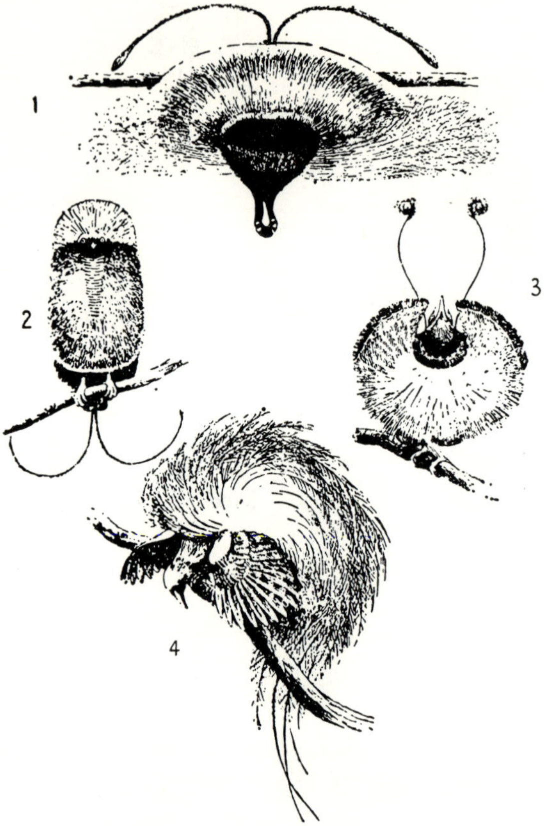

FIG. 304. Display of various birds of paradise.

1 and 4, *Paradisea*; 2, *Diphyllodes*; 3, *Cicinnurus*. (From Streseman, after Seth Smith.)

feature or pattern, serving as it were to arrest and awaken the female and at the same time almost to reduce her to passiveness. The actual movements involved are very various. An excellent example is the peacock, who approaches showing his dull-coloured back and then suddenly turns on his hen, revealing the pattern of spots (which we have noted elsewhere to be an arresting shape), shaking his 'tail' with a rustling noise and himself emitting a scream, in a way that can easily be believed both arresting and fascinating. It is a common characteristic of all bird displays that they involve sights and actions that have a peculiar and exaggerated quality for us and probably also for the partner (Figs. 304, 305). Watching the effects of such appearances on the female one has the impression that this vision acts as it were as a key or 'releaser', opening in her the appropriate course of behaviour. Under its influence she acts as if 'mechanically', moving towards the male and adopting the receptive position, while his display activity passes over into that of mounting and coition. This sort of behaviour we have seen elsewhere in bird life; there must be in the nervous

FIG. 305. Display of the pheasant, *Centrocercus*. 1. Beginning of inflation. 2. Complete display with cervical air-sacs inflated. (From Stresemann, after Horsfall.)

organization a receptive matrix ('the lock') ready to be actuated by the appropriate 'key', which may be the display of structures as bizarre as the wattles and tippets of a turkey.

Many displays are modified versions of everyday actions of the birds and some at least of these seem to be symbolic. Thus, in many birds courtship and coition include ritual feeding of the female by the male. Involved in this is a reversion by the female to an infantile condition. She may beg for food, often with actions similar to those she used as a nestling. The prime significance of this feeding is not in the nourishment provided but, in many cases at least, in the fact that it is the male who provides it. This is proved by the fact that the female (robin, for instance) will not feed herself when food is all around, but will beg the male to give it to her. Female herring gulls bringing back fish may even beg to bc fed by the male, who has not left the rock! There are, of course, many cases of practical feeding of the sitting female by the male, but it is possible that these have been derived from the ritual feeding, rather than vice versa.

The various forms of billing and gaping ceremony probably represent a further degree of abstraction from ritual feeding. A variety of birds touch bills during courtship, for instance, gulls, ravens, great crested grebes, and some finches. The inside of the mouth is sometimes brilliantly coloured in adults, as it is in so many nestlings; it is green in some birds of paradise, yellow in many birds. During display it may be opened suddenly, producing an obvious effect on the mate, who approaches fascinated into a state of passive acceptance by this surprising revelation, perhaps recalling a possibility of satisfaction remembered from childhood.

Other aspects of the courtship may show this reversion to infantile behaviour. Thus, in female sparrows and many other birds one or both wings are held drooping and fluttering during display, as they are by the chick craving food or by the frightened adult; a behaviour known as injury feigning or distraction display. The female hedge-sparrow may quiver with one wing and open her bill to the male at the same time. This injury-feigning also has survival value in distracting the attention of a predator from the nest, as can be well seen in plovers and other ground-nesting birds. Some plovers have different displays for use against different types of nest enemy.

The effects of bird display are by no means restricted to ensuring mutual recognition of males and females and stimulating them to coition, especially important though such functions must be in 'flighty' creatures. The element of threat and even fighting with other males is very common. It is seen in its purest form in such birds as the blackcock and ruff, which are promiscuous. The male ruffs congregate on a chosen 'courting ground' and go through an elaborate series of ritual fights. The females ('reeves') do not take part in this procedure, but at intervals one of them will 'select' a male by fondling him with her bill and then adopt the receptive attitude for copulation. Selous, who observed this display, noted that males with large ruffs were chosen especially often. Such selection of males by females was the basis of Darwin's theory of sexual selection, the supposition being that the males chosen would be the most gorgeous, victorious, and hence most vigorous and effective breeders. Selous recorded that in one area, during a period of 3½ hours, there were 12 copulations, 10 of them with a single male. It must be very difficult when observing birds to establish exactly the actions and relationships of males and females and hence to distinguish between the 'stimulation' of female by male and 'selection' of male by female. Perhaps there is only a verbal difference between the two.

8. Bird territory

The element of threat in singing and display has a further importance in connexion with the territories that many birds establish around their nests. Eliot Howard especially has developed the concept of bird territory as a result of observation mainly of warblers and buntings. For instance, in the warblers (Sylviidae) the males, returning from migration some days earlier than the females, establish themselves on a certain area, singing often from a tall tree or other headquarters near its centre. As other males arrive boundaries develop, so that the region becomes divided up into a number of areas, at first each of about 2 acres, later reducing to 1. When the females arrive they pair off with the males and throughout the whole season the two birds occupy a single territory, driving off other birds that encroach and in this way establishing quite definite boundaries to their area.

Howard supposed that this arrangement was widespread in birds and that it has four desirable effects for the species.

1. Uniform distribution over the habitable area is ensured.

2. Females are assisted to find unmated males.

3. The two birds are kept together and are not distracted by wanderings far from home.

4. It is possible to find adequate food without travelling far from the nest, this being especially important during the period of incubation and rearing of the young. There is no doubt that many birds do remain mostly in the area around their nest and that they may resist invasion. It is probable, however, that the territory is usually less rigid than Howard implied, and there is certainly much variety between different species. According to Lack the territory is often mainly associated with the sexual display of the male; it is his area, part, as it were, of the method he adopts to stimulate the female and to keep the pair together. By establishing a territory he ensures the opportunity to display and copulate without disturbance, a very necessary precaution since he is vulnerable at these times and other individuals may attack a copulating male, trying to displace him. The complicated song, characteristic of so many male birds at the breeding-season, is, on this view, partly an attraction to the female, but largely also a threat to warn off other males. We cannot exclude that it serves as a stimulus to the male himself. The impression is strong when one hears a thrush 'singing for joy' at his headquarters or a lark soaring above his patch of ground.

Territory is therefore, according to Lack, put to various uses in different species. It may be either (1) a mating arena only, as in the ruff and blackcock mentioned above, or (2) it may be a mating station and nest as in the plover and swallow, which birds will not allow others near the nest, although all mix freely for feeding. (3) In sparrows and herring gulls the nest is likewise defended, but is not a mating station. (4) In the warblers investigated by Howard the territory, besides being a mating station, is also a feeding-ground. The significance of this in spacing out the birds remains, however, doubtful. It may limit the effect of predators, by ensuring dispersal. (5) In still other birds, such as the robin, it is a feeding-ground mainly, and therefore it is kept throughout the winter. Examination of the territory concept thus shows that birds have a strong sense of place and that they associate this in various ways with their life, especially during the breeding-season. It is certain that the occupation of territory helps in the initiation and maintenance of the pair, but not yet proved that it serves to limit the breeding density and ensure a food-supply for the young.

9. Mutual courtship

In many birds courtship displays do not necessarily end in coition and may continue long after the eggs have been laid. A classic example of this sort is the great crested grebe, a water-bird watched by J. S. Huxley (Fig. 303). The male and female birds do not differ greatly and the ceremonies are mutual. One bird may dive and come up close to the other and they then approach with necks stretched out on the water, giving a curious ripple pattern that Huxley called the plesiosaur appearance. When they meet the birds come together neck to neck and a period of swaying ensues and may sometimes end by the mounting of one bird by the other, not necessarily the female by the male. At other times the diving bird comes up with pieces of nest-building material and elaborately presents them to its mate. This is apparently a symbolic act; the material is not actually used to make the nest and it is not far-fetched to suppose that such behaviour is an expression of the mutual activity in which the birds are engaged. In so far as it has a biological function it serves, like the rest of the ritual, including post-ovulatory copulation, to keep the two individuals together while rearing the young.

As already mentioned, courtship may include ritual feeding of the other sex during display and often before coition, for instance, in pigeons and gulls, and this again may have a symbolic function. It is perhaps not fantastic to find analogies between behaviour of this sort and the elaborate and prolonged courtship and frequent copulation of man, continuing without cyclical breeding-seasons for as long as the pair remain together and rear their young. In birds, as in man, the 'procreation of children' is not fully accomplished by a single act of fertilization.

10. Nest-building

As in every other aspect of their life we find the nest of birds varied in many ways to suit different manners of life. It is suggested that the habit of making a nest may have arisen from the 'sex-fidgeting' that is commonly seen before, during, and after copulation. This fidgeting may take various forms, including making a 'scrape' in the soil or picking up pieces of grass, &c., after copulation. There is certainly, in many birds, a close connexion between nest-building and copulation. Ritual offering of nest-material is an important element in many courtship displays and in some species (e.g. magpie) male birds may build extra nests.

The nest is therefore often at first a sex site and its position may be chosen by the male. The actual building of the nest is done very variously. Sometimes the male brings the material and the female uses it. She may do the fetching as well, perhaps accompanied by the lazy male; or he may have nothing to do with the whole business. The nest is built by means of a limited number of stereotyped movements, which are characteristic of the species. The integration of these movements into a functional sequence of behaviour depends, however, on experience. Nest building and copulation occur at about the same stage of the reproductive cycle and both can be induced by oestrogen (in canaries).

The complicated forms of nest are found only in passerine birds; in others it is usually simply a hollow in the ground or a heap of sticks. The more elaborate nests show many protective devices; in temperate regions, where predators come largely from below, the nests are often open, but where there are many snakes they are mostly domed or hung from branches or provided with a long tubular entrance (weavers).

In building the nest the bird follows a set pattern, laid down in some way by the method of working of its brain and showing no sign of foreknowledge of the result. Young birds, however, build rougher nests than mature ones. Weaver birds reared by hand for four generations made perfect nests of a type which, of course, they had never seen. On the other hand, it has been claimed that canaries deprived of nest-building materials for some generations then build clumsily at first, though they quickly improve.

The methods used, of course, vary with the materials. Many of them involve most elaborate tying of grasses to the branches, which is done with the beak or the feet or both (Fig. 306). The shape and construction of the nest varies with the habits of the bird. Many sea-birds, nesting safely on the cliffs, make only very simple nests or none at all.

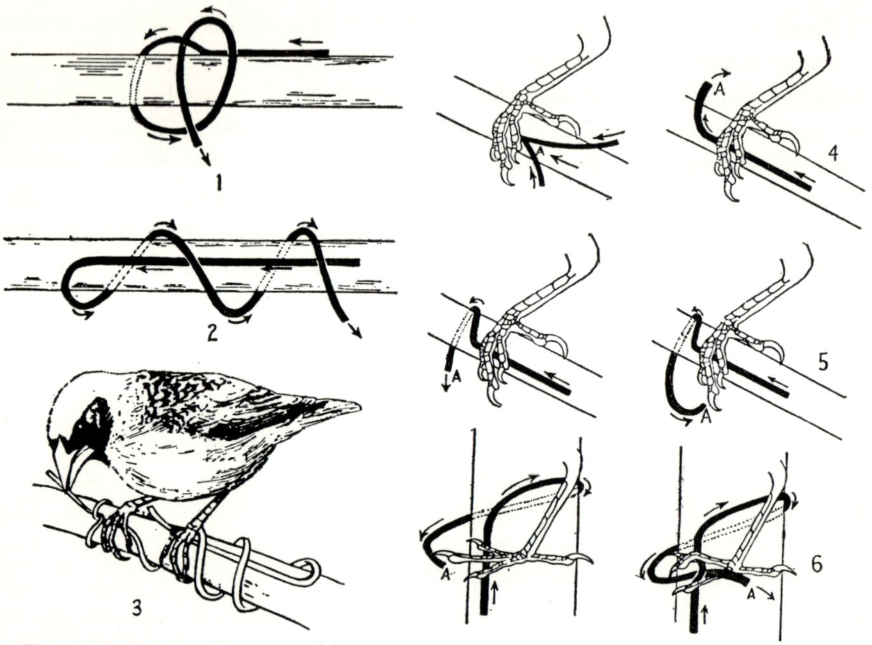

FIG. 306. Process of nest-building by a weaver bird (*Quelea*). The arrows show directions in which the piece is pulled. A, the points of holding by the beak; 4, 5, and 6 show successive stages of co-operative weaving by the foot and beak. (From Stresemann, after Friedmann.)

In such birds as the plovers or larks the nest is a cup in the earth, the eggs being procryptically coloured with a blotched green and brown pattern, so that it is very difficult to see them on the spring ploughland or partly green earth. Nests constructed in trees vary from the simple sticks of rooks or pigeons to the elaborate domed and lined nests of many passerines. The nest is woven from materials brought in with the beak and often lined with moss or in ducks with feathers from the breast of the bird. The thrush lines its nest with mud moistened with saliva and the swift makes nearly the whole nest of saliva. Many birds build roughly and often use the old nests of others, for instance, kestrels are often found using the nests of crows. Others build in burrows, either accidental or, in kingfishers, made by the bird itself. Similarly, holes already in trees may be used (tit-mice), but the woodpeckers drill their own holes. The female hornbill walls herself into a hole in a tree with the help of mud brought by the male and for the three months of incubation she is fed by the male through a small aperture. In the brush turkeys (Megapodes) the eggs are not incubated by the bird but are buried in a mound of decaying material, which provides the necessary heat; the young bird is independent from the moment it hatches.

The bower-birds of Australia and New Guinea are passerines in which the male builds an elaborate structure of plants, ornamented with bright objects. Here he displays to a female and sings. A bower that is left may be destroyed by another male and its decorations stolen. The male constantly refurbishes his bower and gyrates round it, tossing the decorations violently. This seems to be a form of displacement activity. When the female ultimately becomes receptive copulation occurs, the bower often being demolished in the act. The eggs are laid in a nest nearby.

11. Shape and colour of the eggs

Eggs are as varied as the rest of bird structures. The shape is determined by the pressure of the oviducal wall and the blunt end always emerges first. With lesser pressure from behind the egg approaches a spherical shape. The pointed end may serve to prevent the eggs rolling away (guillemot) or to help the eggs to fit together (plover). The pigment is laid down at the end of the travel along the oviduct and is derived from the bile pigments. The coloration is usually procryptic in eggs laid on the ground, whereas those laid in holes are white and birds like the pigeons that do not attempt to protect themselves or their nests by concealment also have light-coloured eggs. The significance of the varied colours of eggs that are not procryptic is obscure. It is likely that they serve as a stimulus to the brooding bird, who will sometimes leave a nest when a wrong-coloured egg is inserted. A further sign of this is that there are various races of cuckoo, each laying eggs appropriate to the nest it parasitizes; but the genetics and behaviour of cuckoos are still very obscure subjects.

12. Brooding and care of the young

Usually it is the hen who broods the eggs, but the cock may assist and in a very few species he does all the brooding. Brooding is not a mere sitting on the eggs, but depends on the development of a vascular response in a part of the skin, the brood spots, produced by a local moult of the feathers. At hatching the parents may assist the action of the caruncle of the young. The care of the nestlings is a very elaborate business in most birds, involving many separate actions. The young are warmed, fed, and occasionally watered, and in many species the nest is kept clean by careful removal of the faeces, which may be produced as pellets enclosed in a skin of mucus; these, being shining white, are eaten by the parents. In warm climates the parents may shield the nestlings from the sun during the heat of the day.

The work of caring for the young birds is often performed by both parents and there are various adaptations to ensure this. The young react strongly to the return of the parents to the nest, usually by opening the beak and displaying the coloured inside of the mouth, an action that strongly stimulates the parent, releasing the feeding behaviour, which varies with the species. The young pigeon thrusts its bill into the throat of the adult to collect the milk secreted by the crop. It is probable that this careful attention by the parents is ensured by a series of somewhat simple stimulus reactions. For example, Eliot Howard showed that a female linnet responds to its own nest rather than to its own young. When its young were put into an abandoned nest and the latter placed near to its own the hen usually returned to its nest and neglected the young, though the male gave them some food. Cuckoos similarly make use of this undiscriminating 'instinctive' behaviour; birds will feed any young that provide the appropriate stimulus and if no young appear they will make little effort to find them.

The rate of development after hatching varies greatly. In the gallinaceous birds, in many ways a primitive group, the young are well-developed at hatching and soon fend for themselves (nidifugal). In nidicolous species, on the other hand, the young is naked and helpless, it is a growing machine, with a large liver and digestive system but little developed nervous system. Yet the birds of these species ultimately have much larger brains and are more 'intelligent' than those that leave the nest soon after hatching.

XVIII

THE ORIGIN AND EVOLUTION OF BIRDS

1. Classification

Class Aves.

*Subclass 1. Archaeornithes. Jurassic

**Archaeopteryx*

Subclass 2. Neornithes

*Superorder 1. Odontognathae. Cretaceous

**Hesperornis; *Ichthyornis*

Superorder 2. Palaeognathae. Ratites. Cretaceous–Recent

*Struthio; Rhea; Dromiceius; Casuarius; *Dinornis; *Aepyornis; Apteryx; Tinamus*

Superorder 3. Impennae. Penguins. Eocene–Recent

Spheniscus; Aptenodytes

Superorder 4. Neognathae. Cretaceous–Recent

Order 1. Gaviiformes. Loons

Gavia, loon

Order 2. Colymbiformes. Grebes

Colymbus (= *Podiceps*), grebe

Order 3. Procellariiformes. Petrels

Fulmarus, petrel; *Puffinus*, shearwater; *Diomedea*, albatross

Order 4. Pelecaniformes. Cormorants, Pelicans, and Gannets

Phalacrocorax, cormorant; *Pelecanus*, pelican; *Sula*, gannet

Order 5. Ciconiiformes. Storks and Herons

Ciconia, stork; *Ardea*, heron; *Phoenicopterus*, flamingo

Order 6. Anseriformes. Ducks

Anas, duck; *Cygnus*, swan

Order 7. Falconiformes. Hawks

Falco, kestrel; *Aquila*, eagle; *Buteo*, buzzard; *Neophron*, vulture; *Milvus*, kite

Order 8. Galliformes. Game birds

Gallus, fowl; *Phasianus*, pheasant; *Perdix*, partridge; *Lagopus*, grouse; *Meleagris*, turkey; *Numida*, guinea fowl; *Pavo*, peacock; *Opisthocomus*, hoatzin

Order 9. Gruiformes. Rails

Fulica, coot; *Gallinula*, moorhen; *Crex*, corn-crake; *Grus*, crane; **Phororhacos; *Diatryma*

1. Classification (*cont.*)

Order 10. Charadriiformes. Waders and Gulls
Numenius, curlew; *Capella*, snipe; *Calidris*, sandpiper; *Vanellus*, lapwing; *Scolopax*, woodcock; *Larus*, gull; *Uria*, guillemot; *Plautus*, little auk

Order 11. Columbiformes. Pigeons
Columba, pigeon; **Raphus*, dodo

Order 12. Cuculiformes. Cuckoos
Cuculus, cuckoo

Order 13. Psittaciformes. Parrots

Order 14. Strigiformes. Owls
Athene, little owl; *Tyto*, farm owl; *Strix*, tawny owl

Order 15. Caprimulgiformes. Nightjars
Caprimulgus, nightjar

Order 16. Micropodiformes. Swifts and humming-birds
Apus, swift, *Trochilus*, humming-bird

Order 17. Coraciiformes. Bee-eaters and kingfishers
Merops, bee-eater; *Alcedo*, kingfisher

Order 18. Piciformes. Woodpeckers
Picus, woodpecker

Order 19. Passeriformes. Perching birds
Corvus, rook; *Sturnus*, starling; *Fringilla*, finch; *Passer*, house-sparrow; *Alauda*, lark; *Anthus*, pipit; *Motacilla*, wagtail; *Certhia*, tree-creeper; *Parus*, tit; *Lanius*, shrike; *Sylvia*, warbler; *Turdus*, thrush; *Erithacus*, British robin; *Luscinia*, nightingale; *Prunella*, hedge-sparrow; *Troglodytes*, wren; *Hirundo*, swallow

2. Origin of the birds

Many characteristics of birds show close resemblance to those of reptiles and in particular to the archosaurian diapsids. Already in the early Triassic period the small pseudosuchians such as **Euparkeria* (p. 417) showed the essential characteristics of the bird group, especially those associated with a bipedal habit. From some such form the birds have almost certainly been derived, by a series of changes parallel in many cases to those found in other descendants of the pseudosuchians, such as the crocodiles, dinosaurs, and pterosaurs.

3. Jurassic birds and the origin of flight

We have no detailed evidence of the stages by which cold-blooded terrestrial reptiles were transformed into warm-blooded flying birds,

FIG. 307. Restored skeleton of **Archaeopteryx*, compared with the skeleton of a pigeon drawn on a more reduced scale.

c. carpal; *cl.* clavicle; *co.* coracoid; *d.* digits; *f.* femur; *fi.* fibula; *h.* humerus; *i.* ilium; *is.* ischium; *mc.* metacarpals; *mt.* metatarsals; *p.* pubis; *py.* pygostyle; *r.* radius; *s.* scapula; *st.* sternum; *tm.* tarso-metatarsus; *tt.* tibiotarsus; *u.* ulna; *v.* ventral ribs; *I–IV*, toes. (From Heilmann, *The Origin of Birds*, H. F. & C. Witherby, Ltd.)

but two fossil specimens from the upper Jurassic rocks of Bavaria show us one intermediate stage on the way (Fig. 307). These **Archaeopteryx* certainly had achieved some powers of flight or gliding, but they were less specialized for the purpose than are modern birds. The whole body axis was still elongated and lizard-like. The vertebrae articulated by simple concave facets as in reptiles, without the saddle-shaped articular facets of the centrum seen in birds. The dorsal vertebrae were not fixed and only about five went to make up the sacrum. There was a long tail, with feathers arranged in parallel rows

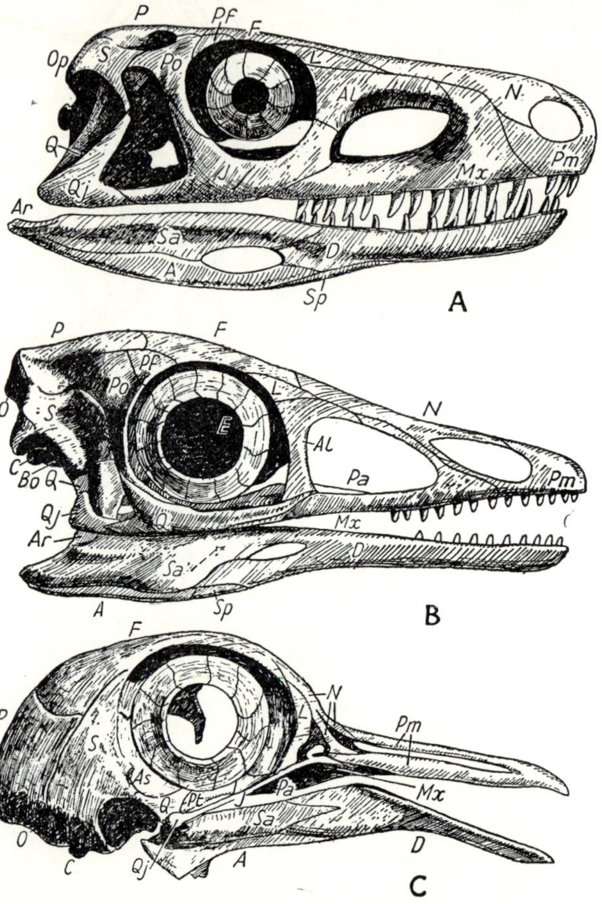

FIG. 308. Skulls of A, **Euparkeria*; B, **Archaeopteryx*; C, *Columba*.

A. angular; *Al.* adlachrymal; *Ar.* articular; *Bo.* basi-occipital; *C.* condyle; *D.* dentary; *E.* eye with sclerotic ring; *F.* frontal; *J.* jugal; *L.* lachrymal; *Mx.* maxilla; *N.* nasal; *O.* occipital; *Op.* opisthotic; *P.* parietal; *Pa.* palatine; *Pf.* post-frontal; *Pm.* premaxilla; *Po.* postorbital; *Pt.* pterygoid; *Q.* quadrate; *Qj.* quadratojugal; *S.* squamosal; *Sa.* sur-angular; *Sp.* splenial. (From Heilmann, *The Origin of Birds*, H. F. & C. Witherby, Ltd.)

along its sides, probably an important organ, as in other animals that live in trees and jump and glide. The fore-limb ended in three clawed digits, with separate metacarpals and phalanges, the hallux being opposable. The limb was used as a wing, for the fossils show feathers on the back of the ulna and hand, but the wing area was small and the shape rounded, like that of the wing of birds that fly for short distances only. There was a furculum and a small sternum. The ribs were slender and had no uncinate processes. The pelvic girdle and hind limb resembled those of archosaurs, with elongated ilium and backwardly directed pubis. Only six vertebrae were fused to form the sacrum (at least eleven in birds). The fibula was complete and the proximal tarsals were free, but the distal ones were united with the metatarsals.

In the skull of **Archaeopteryx* (Fig. 308) there were teeth in both jaws. The shape was more reptilian than bird-like, with rather small eyes and brain, and premaxillae and frontals much smaller than in modern birds. There was a large vacuity in front of the eye and probably there were post-frontal and post-orbital bones. The condition of the temporal region is unfortunately not clear on account of the crushing of the material. The brain-case was large and many of the bones were united, as in modern birds. The bones were not pneumatized. The cerebral hemispheres were elongated as in reptiles and the cerebellum was small.

These very interesting fossils suggest that the birds arose from a race of bipedal arboreal reptiles, living in forests and accustomed to running, jumping, and gliding among the branches (Fig. 309). There has been much controversy about the origin of flight, some maintaining that the earliest birds were terrestrial and used the wings to assist in running, leading eventually to a take-off, perhaps at first for short distances. The claws and long tail of **Archaeopteryx* speak definitely against this view and in favour of a gliding origin for flight.

4. Cretaceous birds. Superorder Odontognathae

These Jurassic fossils are so distinct from other birds that they are placed in a distinct subclass *Archaeornithes. All other known living and extinct birds have a short tail, reduced hand, a sternum, and other characteristics of the subclass Neornithes. A few fossils are known from the upper Cretaceous in which certain reptilian characteristics are still preserved. **Hesperornis* probably possessed teeth. Another Cretaceous bird skull (**Ichthyornis*) was found associated with a toothed jaw, but the latter is now believed to have belonged to a mosasaur. However, the two birds are placed in a superorder *Odontognathae. They were aquatic birds, and the former was a diver that had lost the power of flight.

Already in the Cretaceous there were some birds that had lost the teeth and can be referred to orders found alive today. Birds are not commonly found as fossils, however, and it is not possible to give a detailed history of the evolution of the various orders that are recognized. We do not even know whether bird life first became abundant after the Cretaceous, at the same time as the mammals began to be numerous.

5. Flightless birds. Superorder Palaeognathae

The flightless birds or 'ratites', such as the ostrich, cassowary, and kiwi, with reduced wings and no sternal keel, long legs and curly feathers, have in the past been placed in a distinct group and regarded as primitive. Indeed, it has even been suggested that they diverged so early from the ancestral avian stock that they never passed through a flying stage. Recently, however, de Beer and others have pointed out that certain of their allegedly primitive characters, such as the arrangement of the palate bones, may be regarded as manifestations of neoteny and do not indicate a truly primitive condition. Some neognathous birds pass through a palaeognathous stage during development. The evidence strongly suggests that the 'ratites' have been descended from flying birds and are not a natural group, but represent several different evolutionary lines. The detailed relationship of these is still obscure and for convenience they are retained in a superorder Palaeognathae.

The various ratite birds have been placed in as many as eight distinct orders, but the orders of the ornithologist generally represent lesser degrees of difference than are usual elsewhere in the animal kingdom. The ostriches (*Struthio*) are the largest living birds, now limited to Mesopotamia. The rhea (*Rhea*) occupies the same ecological position in South America and the emu (*Dromiceius*) and cassowary (*Casuarius*) in Australasia. The moas (**Dinornis*) were another type; several species lived in New Zealand until recent times. The elephant-birds (**Aepyornis*) were similar, with several species in Madagascar in the Pleistocene. Some were larger than ostriches, with eggs estimated to weigh more than 10 kg, presumably the largest single cells that have existed!

The kiwis (*Apteryx*) of New Zealand are smaller, terrestrial birds whose relationship to the ratites is doubtful. They are nocturnal and insectivorous or worm-eating, with a long beak and small eyes. The sense of smell and the parts of the brain related to it are better developed than in other birds; it is not clear whether this is the retention of a primitive feature. The palate shows large basipterygoid processes. There is a penis, as in other ratites.

Still more doubtful is the position of the tinamus (*Tinamus*), terrestrial birds rather like hens, of which about fifty species are found

FIG. 309. Restoration of hypothetical proavian. (From Heilmann, *The Origin of Birds*, H. F. & C. Witherby, Ltd.)

throughout South America. They show similarities to the ratites in the palate and other features and may be placed among the Palaeognathae. Probably, like the more typical ratites, they are an early offshoot that has long developed independently.

6. Penguins. Superorder Impennae

The penguins (*Spheniscus*) are birds that early lost the power of flight and became specialized for aquatic life. They may have a common ancestry with the petrels. Unlike most other water-birds they swim chiefly by means of the fore-limbs, modified into flippers; the feet are webbed. The penguins are mainly confined to the southern hemisphere. They come ashore to breed; many make no nests, but sometimes carry the one or two eggs on the feet throughout the incubation period. The emperor penguin breeds in winter on the Antarctic ice and is the only bird that never comes on land. The egg is supported on the feet.

7. Modern birds. Superorder Neognathae

All the remaining birds have the characteristic palate, sternum, and other features already described and are placed in a single group as the superorder Neognathae. Birds of this type probably existed in the Cretaceous and many of the orders are known from Eocene times, but the fossil evidence is not adequate for us to be able to say when they became numerous and differentiated as they are now. The existing birds, as has already been suggested, show very great variety of details of structure and habits, superimposed on a common basic plan. Classification of the vast number of genera involves recognition of over forty distinct orders and even then one of the orders, the Passeriformes, contains about half of all the species. Unfortunately, little can be done in a short space towards describing the great variety of bird life. We can only list the important orders, mentioning a few of the characteristics of some of the more interesting types. Birds are so conspicuous that their species have been very fully described, there are about 25,000 well-defined species and subspecies. The arrangement of the orders adopted for the survey, *Birds of the World*, by J. L. Peters has been used here.

Order 1. Gaviiformes. Loons

The divers are aquatic birds retaining some primitive characteristics. They are birds of open waters, feeding mainly on fishes. Various species of *Gavia* live mostly on the sea, but breed by lakes throughout the holarctic region.

Order 2. Colymbiformes. Grebes

The grebes (*Colymbus* = *Podiceps*) (Fig. 303) are also aquatic birds, almost unable to walk on land. They resemble the divers in some ways, but are perhaps not closely related to them. They nest on lakes, laying a small number of white eggs in a floating nest.

Order 3. Procellariiformes. Petrels

The petrels (*Fulmarus*), shearwaters (*Puffinus*), and albatrosses (*Diomedea*) are birds highly modified for oceanic pelagic life, some of them very large. They lay one white egg, often in burrows. Their long narrow wings are specialized for soaring flight (Fig. 273).

Order 4. Pelecaniformes. Cormorants, Pelicans, and Gannets

This is another order of aquatic birds, much modified for diving and fishing and including the cormorants (*Phalacrocorax*), pelicans (*Pelecanus*), and gannets (*Sula*). They nest in colonies on rocks cr trees; the eggs are usually unspotted and covered with a rough chalky substance. These birds make spectacular dives when fishing; gannets may plunge from more than 50 feet.

Order 5. Ciconiiformes. Storks and Herons

The storks (*Ciconia*), herons (*Ardea*) (Fig. 274), and flamingoes (*Phoenicopterus*) are large, long-legged birds, living mostly in marshes and feeding mainly on fish. They are strong flyers and some of them perform extensive migrations. Nests are usually in colonies and may be used year after year; there are elaborate display ceremonies. Eggs are few and unspotted.

Order 6. Anseriformes. Ducks

The ducks (*Anas*) and swans (*Cygnus*) represent yet another group of birds specialized for aquatic life. The characteristic flattened bill is used to feed on various diets. Some are vegetarians, a few filter-feeders; some eat molluscs, others fish. The numerous eggs are usually white or pale and the nest is built on the ground.

Order 7. Falconiformes. Hawks

This order includes the birds of prey that hunt by day, having sharp, strong, curved bills and powerful feet and claws. The retina contains mainly cones. Many different types are found throughout the world. Most feed on birds or mammals, some on carrion, and a few on fish or reptiles. Typical examples are the kestrel (*Falco*), eagle (*Aquila*), buzzard (*Buteo*), and vulture (*Neophron*). The eggs, few in number, are usually spotted and the nests are generally made on cliffs, tree-tops, or other inaccessible places; some, however, are on the ground.

Order 8. *Galliformes. Game birds*

These are mainly terrestrial, grain-eating birds, capable only of short, rapid flights; some of their structural characters and habits are certainly primitive. The palate differs from both that of ratites and of most modern birds, suggesting an early divergence. There is often a marked difference in plumage, and sometimes in size, between the sexes. The nest, usually made on the ground, is simple and the eggs numerous, white, or spotted. The young develop very quickly after birth. The order contains many successful types and is of world-

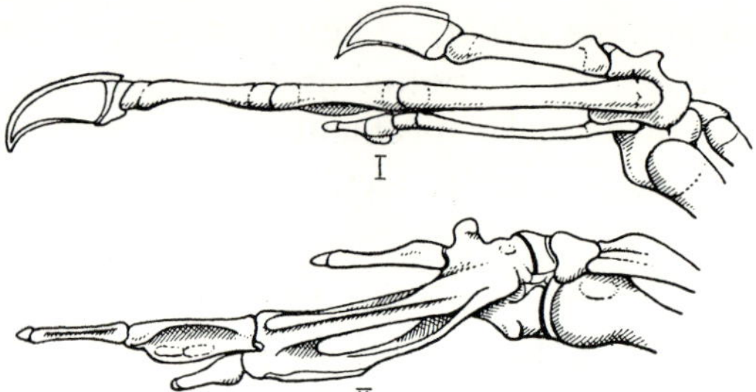

FIG. 310. Claws on hand of the hoatzin (*Opisthocomus*), I, in nestling; II, adult. (After Parker and Heilmann.)

wide distribution. It includes *Gallus*, the jungle-fowl of India, and all its domesticated descendants, also *Phasianus* and other pheasants, *Perdix* (partridge), *Lagopus* (grouse), *Meleagris* (turkey), *Numida* (guinea-fowl) and *Pavo* (peacock). The Megapodes or mound-builders of the Australasian and east Indian regions lay their eggs in mounds of decaying leaves and earth. In *Opisthocomus*, the hoatzins of tropical South America, one of the few tree forms, the young possess well-marked claws on the digits of the wing (Fig. 310), which they use for climbing. These claws are usually considered to be a secondary development; their resemblance to the claws of **Archaeopteryx* is remarkable.

Order 9. *Gruiformes. Rails and Cranes*

The rails are mostly secretive, terrestrial birds, compressed laterally and often living in marshy country and having an omnivorous diet; common British members are the coots (*Fulica*) and moorhens (*Gallinula*). They run, swim, and dive easily, but are poor flyers; they build rather simple nests and lay numerous, often dark-spotted eggs. *Crex* (the corncrake) and other landrails are of more terrestrial habit. The cranes (*Grus*) are long-legged birds found in swamps and probably allied to the rails rather than to the waders as is still often supposed.

Possibly related to the rails are the cariamas of South America, carnivorous birds with very long legs, hardly able to fly, living largely on reptiles. The Miocene **Phororhacos* was a similar bird, reaching 6 feet high; evidently the group was successful in a region free of mammalian carnivores. **Diatryma* was an even larger flightless carnivorous bird, found in the Eocene of Europe and North America and perhaps also related to the early ancestors of the Gruiformes, though usually classified in a separate order, or near the herons.

Order 10. *Charadriiformes. Waders and Gulls*

This is a large order including the wading birds and the gulls, terns, and auks, which have evolved from them. The typical waders are birds that live mainly on the ground, often inhabiting open watery places or marshes. They are usually gregarious out of the breeding-season and are often very numerous on the sea-shore. They often have long legs and long bills and feed chiefly on small invertebrates. The curlews (*Numenius*), snipe (*Capella*), and sandpipers (*Calidris*) are well-known examples. The lapwings (*Vanellus*) and related plovers are birds found on drier land than is usual among other waders; the woodcocks (*Scolopax*) inhabit swampy woods.

The gulls (e.g. *Larus*) are a very important group of birds derived from the waders and adapted to life by and on the sea. Usually they have a grey or white colour, often with black head and wing-tips. The young are usually darker than the adults and mottled with brown. The guillemots (*Uria*) and little auks (*Plautus*) are more fully marine animals, breeding in very large colonies on the cliffs.

Order 11. *Columbiformes. Pigeons*

The pigeons are tree-living, grain- or fruit-eating birds, mostly good flyers but retaining some primitive features. They are of world-wide distribution. There is little sexual dimorphism; the nest is usually simple and the eggs normally one or two and white. The young are born very little developed and are nourished by the 'milk' secreted by the crop (p. 508). The dodo (**Raphus* = **Didus*) was a pigeon that adopted a terrestrial habit in the island of Mauritius and grew to a large size, but was exterminated by man in the seventeenth century.

Order 12. *Cuculiformes. Cuckoos*

The cuckoos include some species that build nests but many lay their eggs in those of other birds. In the common cuckoo (*Cuculus*), any one individual female lays mostly in the nests of a single foster species, in England often the meadow-pipit or hedge-sparrow. She watches the building of the nest and lays her egg on the same day as the foster parent, removing one of the clutch before she does so. Often about twelve eggs are laid in this way, each in a different nest; even more have been recorded. The eggs are usually strongly mimetic with those of the host, variable in colour, more so when varied host nests are available. The young hatch before the host eggs, which are then ejected from the nest by the young cuckoo.

Order 13. *Psittaciformes. Parrots*

The parrots are birds found mainly in warm climates, living among the trees and having many special characteristics. With the crows, they are usually reckoned to be the most 'intelligent' birds and certainly have considerable powers of memory. They are predominantly vegetarian and some, though by no means all, make use of the beak for breaking open hard shells. The eggs are usually laid in holes and are white and round. The period of parental care after hatching is unusually long (2–3 months).

Order 14. *Strigiformes. Owls*

The owls, specialized for hunting at night, resemble the hawks, by convergence, in their beaks, claws, and in other ways. The food is swallowed whole. They probably detect their prey mainly by sound, and show various specializations in the ears. The eyes contain mostly rods and are directed forwards; they are very large and they cannot be moved in the orbits, the movements of the neck compensating for this restriction. The feathers are so arranged as to make very little noise in flight. The eggs are white and laid in holes or in the old nests of other birds, some on the ground. Many genera are recognized from all parts of the world, examples being the barn owls (*Tyto*) and the eared owls (*Asio*).

Order 15. *Caprimulgiformes. Nightjars*

The nightjars (*Caprimulgus*) are a rather isolated group of crepuscular birds, feeding on insects taken on the wing. Two mottled eggs are laid on the bare ground.

Order 16. *Micropodiformes. Swifts and Humming-birds*

The swifts (*Apus*) and humming-birds (*Trochilus*) are perhaps more fully adapted to the air than are any other birds. The wings are very long, composed of a short humerus and long distal segments. The swifts are insectivorous and have very large mouths, adapted for feeding on the wing. The nests are made in holes, the eggs are white, and the young helpless at birth.

Order 17. *Coraciiformes. Bee-eaters and Kingfishers*

This is a large group of birds, including the bee-eaters (*Merops*), mainly tropical and often brightly coloured. The three anterior toes are united (syndactyly). The nests are usually made in holes and the eggs are white. The kingfishers (*Alcedo*) are modified for diving into the water to catch fish.

Order 18. *Piciformes. Woodpeckers*

The woodpeckers (*Picus*) are highly specialized climbing, insectivorous, and wood-boring birds. The bill is very hard and powerful and the tongue long and protrusible and used for removing insects from beneath bark. The tail feathers are used to support the bird as it climbs the tree-trunk. The nest is made in a hole in a tree and the eggs are white.

Order 19. *Passeriformes. Perching birds*

The great order of perching birds contains about half of all the known species. They are birds mostly living close to the ground, rather small, and of very varied habits. There are always four toes arranged to allow the gripping of the perch. The display and nesting behaviour is usually complicated, with a well-developed song in the male. Many species build very complicated nests and the eggs are often brightly coloured and elaborately marked. The young are helpless at birth. Only a few of the many and varied types can be mentioned here.

The rooks and jackdaws (*Corvus*) are the largest passerines and perhaps 'highest' of all birds; they are mostly colonial. The starlings (*Sturnus*) are also partly colonial and nest in holes. The finches (*Fringilla*, &c.) are seed-eating birds with a short, stout, conical bill. The house-sparrows (*Passer*) are closely related to the finches and have become commensals of man all over the world. The larks (*Alauda*) make their nests on the ground. The pipits (*Anthus*) and wagtails (*Motacilla*) are somewhat like the larks, largely terrestrial birds with slender bills. The tree-creepers (*Certhia*) are tree-living, insectivorous birds with long bills, showing some convergent resemblance to woodpeckers. The tits (*Parus*, &c.) are a large group of woodland birds; they chiefly eat insects, also buds and fruits. The shrikes (*Lanius*) are peculiar among passerines in being mainly carnivorous, using their strong bills to eat other birds, amphibia, reptiles, and large insects.

The warblers (*Sylvia*, &c.) make a very large group of woodland birds, living in trees or scrub. Related to them are the thrushes and blackbirds (*Turdus*), the British robins (*Erithacus*), and nightingales (*Luscinia*), mainly eating small invertebrates, also fruits. They have a very wide distribution and are among the most recently evolved and successful members of the whole class. The hedge-sparrows (*Prunella*) are small omnivorous passerines, possibly related to the thrush group. The wrens (*Troglodytes*) are small and mainly insectivorous. The swallows (*Hirundo*) form a very distinct family of passerines, suited for powerful flight and feeding on insects caught in the air. The very long pointed wings and 'forked' tail allow rapid manœuvring in the air and the insects are taken in a wide mouth. These features, together with the elaborate migrations, mark the swallows as among the most specialized of all birds.

8. Tendencies in the evolution of birds

The bird plan of structure, originating in the Jurassic period, perhaps 150 million years ago, has become modified to produce the great variety of modern birds. In trying to discover the factors that have influenced this modification we are handicapped by the poverty of fossil remains; it is not possible to trace out individual lines as it is in other vertebrate groups. It is clear that the process of change has been radical, the later types often completely replacing the earlier ones: no long-tailed or toothed birds remain today.

Our knowledge of direction of the change is largely dependent on study of the variety of birds existing today, which is perhaps more thoroughly known than in any other group of animals. In the reports of those who have studied this variation there are two distinct, indeed opposite, tendencies. Many have observed that adaptive radiation has occurred; birds are found occupying a wide variety of habitats, with modifications appropriate to each way of life. Other workers, recording minor differences between races occurring in different areas, have found difficulty in believing that these have adaptive significance. The existence of such 'subspecies' with a geographical limitation is a striking characteristic, especially conspicuous in widely distributed species such as the chaffinch (*Fringilla*). On continental areas such subspecies usually grade into each other (making 'clines') and are interfertile at the areas where they meet. On the other hand, where a group of individuals becomes isolated on an island, or by some other geographical barrier, it may become infertile with the 'parent' species. If the two groups again come to occupy a common area, then either one eliminates the other or slight modifications of habits enable the two to survive side by side as two distinct 'species'.

Definite cases of formation of new species in this way have been recorded in birds, which are especially suitable for such study. Thus in the Canary Islands, besides a local form of the European chaffinch (*Fringilla coelebs*) there is the blue chaffinch (*F. teydea*), which was probably originally an offshoot from the European form. On the mainland *F. coelebs* inhabits both broad-leaved and coniferous forests, but in Grand Canary *F. teydea* occupies the pine woods, *F. coelebs* the chestnut and other woods. On the island of Palma, however, the blue chaffinch is absent and the European form occupies both habitats. It is presumed that the blue form is more suited to the coniferous woods, but this could not be deduced from its specific characters as recorded by a systematist. The adaptive significance of the differences between groups of animals may not be easy to discover, but it is most unwise to assume that it does not exist until a very thorough study has been made. Detailed observation generally shows small differences in habits and behaviour between animals occupying what seems at first to be a single 'habitat'. As Lack, who has studied this question in detail, puts it, 'A quick walk through the English countryside might suggest that there is a wide ecological overlap between the various song-birds. In fact, close analysis shows that there are extremely few cases in which two species with similar feeding habits are found in the same habitat.' The differences may be in the breeding habitat, as in the case of the meadow, tree, and rock pipits (*Anthus pratensis*, *A. trivialis*, and *A. spinoletta*). The spotted and pied flycatchers (*Muscicapa striata* and *M. hypoleuca*) catch their food in slightly different ways and the chiff-chaff (*Phylloscopus collybita*) feeds higher in the trees than the willow warbler (*P. trochilus*).

A complicating factor is that birds that occupy similar habitats for one part of the year may migrate to different regions, for instance, the tree and meadow pipits and the chiff-chaff and willow warbler.

Birds are able to get their food and to breed in many different ways, and when a race finds a situation occupied it perhaps often survives by a slight change of habits, creating a new 'habitat' not previously occupied. This presumably results from the action of certain individuals whose constitution differs from the mean of the race, enabling them to pioneer. It is not certain to what extent individual birds are able to 'adapt themselves' in this way to new habitats. Probably the majority of the members of any population are limited by their structure and behaviour pattern to a rather narrow habitat range.

9. Darwin's finches

A remarkable example of evolution and adaptive radiation is provided by the birds of the Galapagos Islands. It was these birds and

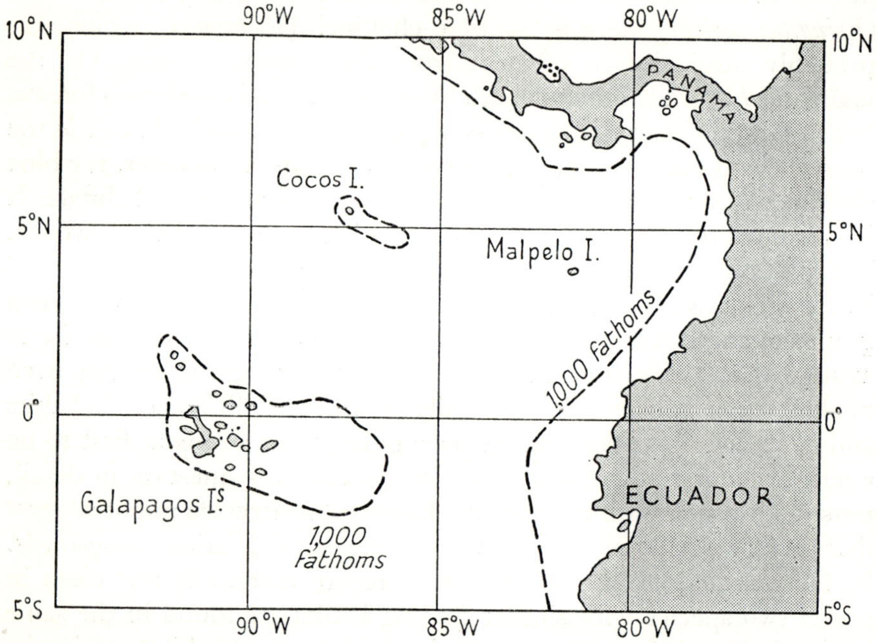

FIG. 311. Position of the Galapagos Islands. (From Lack, *Darwin's Finches*, Cambridge University Press.)

the giant tortoises (after which the islands are named) that started Darwin on his study of evolution. 'In July opened first note-book on "Transmutation of Species". Had been greatly struck from about month of previous March on character of S. American fossils—and species on Galapagos Archipelago—These facts origin (especially latter) of all my views' (Darwin's Diary 1837).

These islands are volcanic and probably of Tertiary (? Miocene) date. They lie on the equator, 600 miles west of Ecuador, the only other nearby land being the island of Cocos, 600 miles north-west (Fig. 311). Though it has been suggested that the islands were connected by a land bridge with South America, it is much more probable that their limited stock of plants and animals has arrived across the sea. Of hundreds of species of land birds on the mainland, descendants of only seven species are found in the Galapagos. The only land mammals are a rat and a bat. The land reptiles include giant tortoises, iguanas, a snake, one lizard, and one gecko. There are no amphibians and only a limited number of land insects and molluscs. There are large gaps in the flora; for instance, no conifers, palms, aroids, or Liliaceae. This fragmentary flora and fauna strongly suggest that the islands have been colonized by chance transportation across the sea, and that, once arrived, the animals and plants have proceeded to settle not only in the habitats they occupied on the mainland but also in others, not filled, as in their homeland, by rivals. Thus the tortoises and iguanas, arriving presumably by chance, have grown to large size, to occupy the ecological position usually taken in other faunas by mammalian herbivores. The composite plant *Scalesia* and the prickly pear, *Opuntia*, have become tall trees in the Galapagos. We have here, therefore, an example of the results of evolution over a relatively limited period of time (perhaps less than 20 million years) from a limited number of initial creatures, and this provides an excellent opportunity for trying to discern the forces that have been at work.

There are thirteen larger islands in the Galapagos group, the largest 80 miles long. They are separated by distances of up to 100 miles and several of the peculiar Galapagos animals have formed island races. The land birds present perhaps the most interesting features of the whole strange fauna. Besides two species of owls and a hawk they consist of five passerine types and a cuckoo, all very close to others found on the South American mainland, and a group of fourteen species of finches, placed in a distinct subfamily, Geospizinae. These finches are related to the family Fringillidae, which is represented in South America, but they cannot be derived from any single species now existing there. Like the other animals in the islands the birds tend to become differentiated into distinct island races, but this process has gone to varying extents. The cuckoo, warbler, martin, and tyrant flycatcher are similar in all the islands and it is significant that they are all very close to species occurring in South America. Presumably they are recent arrivals. The vermilion flycatcher, also a South American species, has three island races. The mocking-bird, *Nesomimus*, is placed in a genus distinct from that on the mainland and has different races on each island, some of them being reckoned as separate species. The extreme of island differentiation is shown by the finches, now reckoned to belong to fourteen species, classed in four genera (Fig. 312), one of these being found on the distant Cocos Island.

Evidently these finches have been in the archipelago for a considerable time and the specially interesting feature is that not only have they formed races recognizably distinct, but they have radiated to form a series of birds that have quite varied habits, many of them very un-finchlike. The main differences are in the form of the beak, which varies greatly with the food habits (Fig. 313). The central

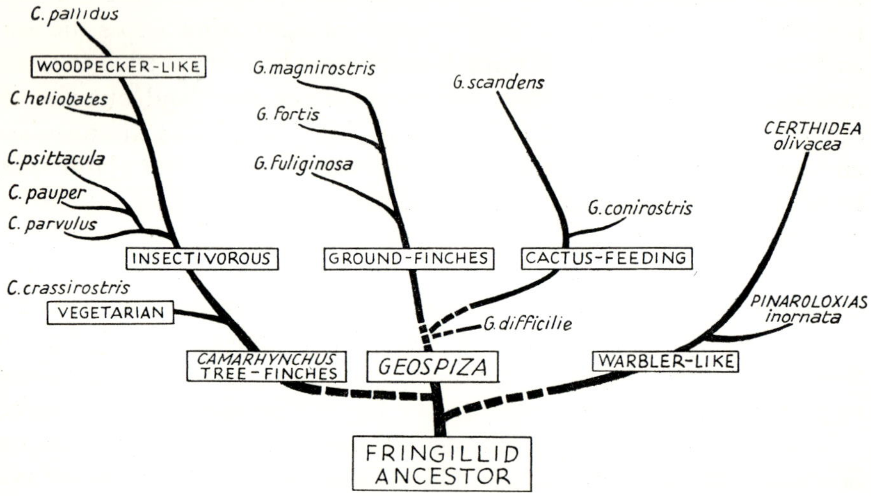

FIG. 312. Suggested evolutionary tree of Darwin's finches. (From Lack.)

species, which are also the nearest to the presumed Fringillid ancestor, are the ground-finches, *Geospiza*, of which there are five species, feeding mainly on seeds. Two further species of *Geospiza* have left the ground and taken to eating cactus plants. The tree-finches, placed in a distinct genus *Camarhynchus*, include one vegetarian and five mainly insectivorous species, one of the latter, *C. pallidus*, having acquired the habit of climbing up the trees like a woodpecker and excavating insects with a stick (Fig. 280). A third group of this remarkable subfamily has acquired a convergent likeness to warblers: *Certhidea* has a long slender beak and eats small soft insects; by many it has been regarded as distinct from the other Galapagos finches, but it has now been shown to resemble them, not only in structure but also in breeding-habits. Nevertheless it probably diverged some time ago and is found on all the islands. Presumably its success is due to the absence of other warbler-like birds, since the true Galapagos warbler is a recent arrival. The fourth genus of the Geospizinae is

Pinaroloxias, the Cocos-finch, found on Cocos Island 600 miles away, and also having warbler-like characteristics.

These birds provide a remarkable example of adaptive radiation 'with seed-eaters, fruit-eaters, cactus-feeders, wood-borers and eaters of small insects. Some feed on the ground, others in the trees—

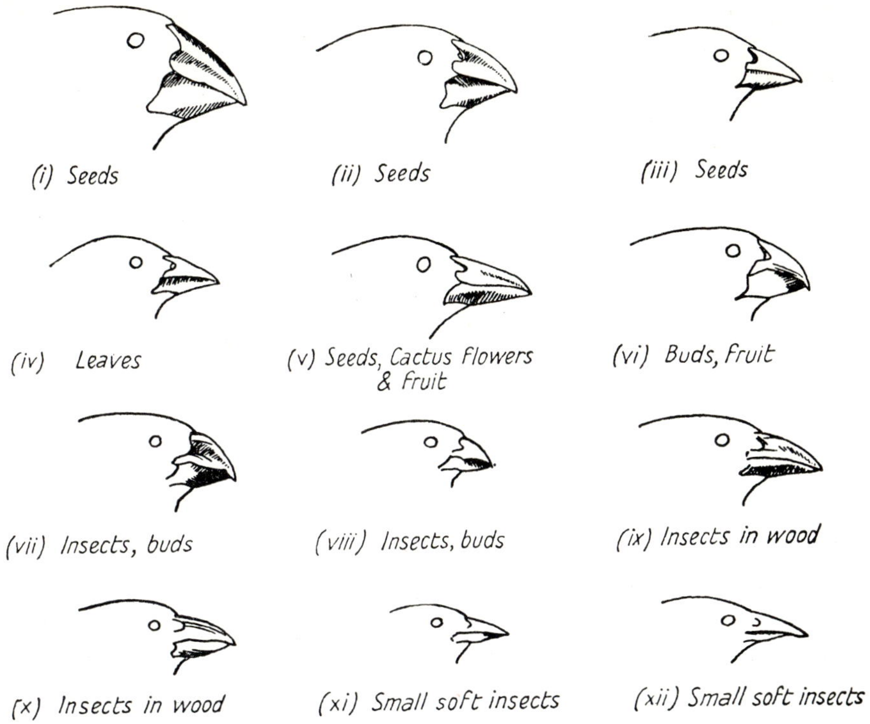

FIG. 313. Beaks of Darwin's finches.

(i) *Geospiza magnirostris*; (ii) *Geospiza fortis*; (iii) *Geospiza fuliginosa*; (iv) *Geospiza difficilis debilirostris*; (v) *Geospiza scandens*; (vi) *Camarhynchus crassirostris*; (vii) *Camarhynchus psittacula*; (viii) *Camarhynchus parvulus*; (ix) *Camarhynchus pallidus*; (x) *Camarhynchus heliobates*; (xi) *Certhidea olivacea*; (xii) *Pinaroloxias inornata*. (After Swarth, from Lack.)

originally finch-like, they have become like tits, like woodpeckers and like warblers' (Lack). This is interesting enough, but more can be learned from this extraordinary natural experiment than from other examples of adaptive radiation. Another most striking feature is that the birds are very variable, and are by no means uniformly distributed over the islands. The outlying islands lack certain species and their place is then taken by variants of others, with corresponding modification of the beak. In some such cases the local subspecies can be clearly seen to have an adaptive significance, but this is by no means always so. Some races found on one or more islands differ in minor features of colour or size. It remains to be shown whether these are themselves significant or connected with other significant factors.

The striking feature is that so many distinct races should appear, even in birds, which could easily cross the distances, mostly less than

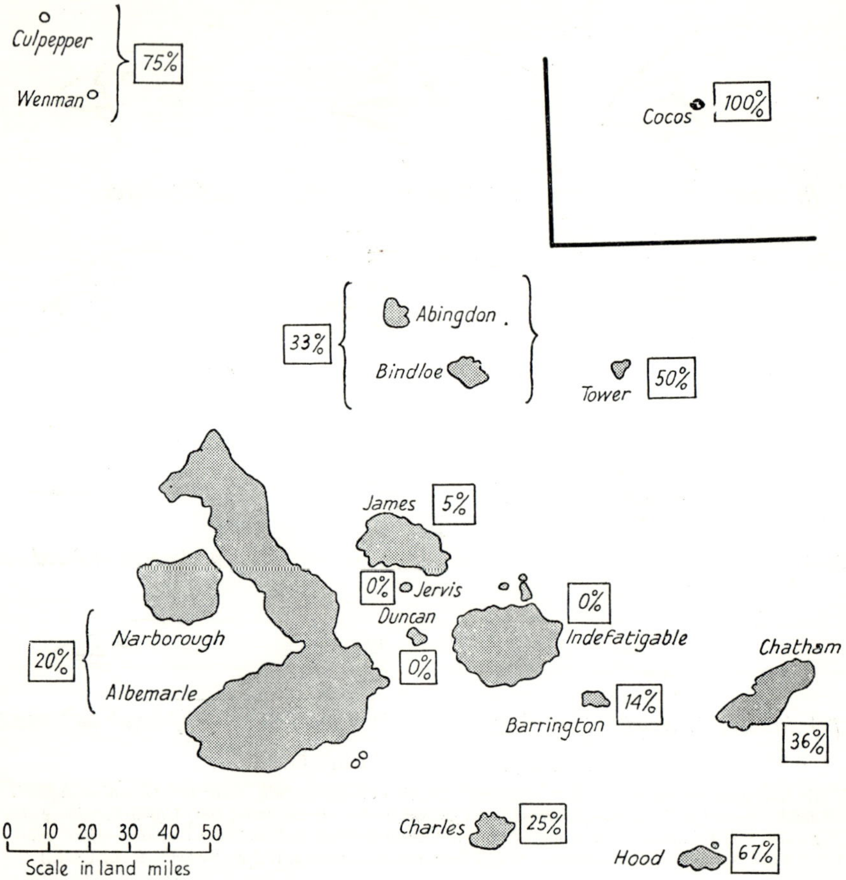

FIG. 314. Percentages of endemic forms of Darwin's finches on each island, showing the effects of isolation. (From Lack.)

50 miles, between the islands. Evidently the birds tend to remain at home, and there is no doubt that degree of geographical separation is a main factor in producing the races. Thus the central islands of the group, lying close together, have no endemic subspecies, whereas the forms in the outlying islands are mostly distinct (Fig. 314). It is evident that isolation is tending to break up the population into a number of distinct units and it is remarkable that, in spite of the large amount of variation, intermediates are rare between species and even

between subspecies. The species, of course, do not cross in nature, and such as have been tested in captivity mate only with their own species.

It is still not possible to give an entirely clear idea of the factors

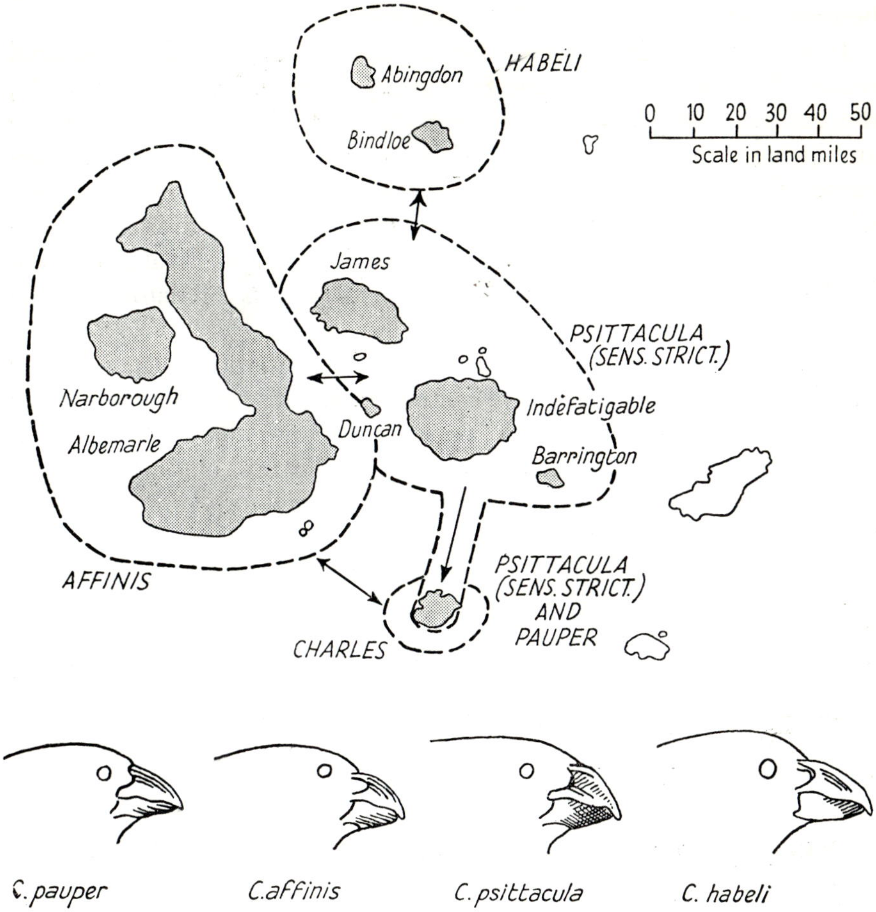

FIG. 315. Development of distinct forms from a single one in the Galapagos finch *Camarhynchus*. Charles Island has two forms derived independently from the central form C. *affinis*. (From Lack, partly after Swarth.)

responsible for the development of varied animals, even in this much simplified case. There is no clear line between subspecific differences, say of the beak that are not obviously adaptive and those with clearly adaptive character. Since favourable mutations will spread through a population the easiest assumption is that all these differences are in fact adaptive, or linked with unseen adaptive characters. The easiest assumption is not necessarily correct, but the demonstrably adaptive character of many of these differences certainly constitutes a case for considering that the others are also of this nature. In very small populations (< 1,000) unfavourable characteristics may become established by chance ('genetic drift', Wright, 1940), and this factor may be responsible for some of the island races.

It is especially interesting that in the Galapagos animals special reasons have made it possible for variety to arise. If, as seems likely, the earliest *Geospizas* were among the first birds to arrive in the archipelago, they found there neither competitors nor enemies. Because of the distance from the mainland this condition has remained with little change ever since. Even today the six passerines not belonging to the Geospizinae do not provide serious rivals, and the local owls and hawks are apparently not serious predators on the finches.

The other great factor that has led to differentiation is the splitting up of the area into a number of isolated units. This has allowed slightly different races to emerge, and we may suppose that if these again came into contact with each other they would find slightly different optimal conditions, and therefore, with partial or complete intersterility, would continue as distinct species. This has almost certainly happened with the Canary chaffinch (p. 523), and in Galapagos there is a similar case in that two species of the large insectivorous tree-finch *Camarhynchus* occur together on Charles Island. Both are derived from a single population, offshoots from which have independently colonized the island (Fig. 315).

10. Birds on other oceanic islands

While the case of Darwin's finches is very striking it is important to recognize that similar radiation from a few species is not found on all oceanic islands. For instance, there are numerous birds on the Azores and they differ little from those of Europe. Whereas migrants are frequent in the Azores they are rare in the Galapagos; evidently there are special factors producing the isolation of the latter.

A radiation similar to that of the Galapagos has, however, occurred in Hawaii, where there are only five passerine forms and one of these, a finch-like bird, the sicklebill, has produced an even greater variety than Darwin's finches (Fig. 316). Some of these sicklebills feed on insects, others on nectar, fruit, or seeds, and the beaks have developed accordingly. One species climbs like a woodpecker, digs for beetles with a short lower mandible, and probes them out with a long curved upper one.

Fig. 316. Adaptive radiation of sicklebills in Hawaii. (From Lack, after Keulemanns.)

11. The development of variety of bird life

This and similar evidence from the study of the relatively recent evolution of birds might be summarized by saying that the production of variety of animal types is largely due to the interplay of biotic factors. The particular characteristics that each population acquires depend partly on physical and geographical conditions, but the stimulus to change, or the check on change, comes from the interaction of the animals and plants with each other.

It has already been suggested that the tendencies to increase and to vary are important among these factors influencing animal evolution and there is evidence that both have been at work in the development of the population of Galapagos finches and other animals. It is safe to say that there are more and varied finches, tortoises, iguanas, and mocking-birds than there were when each arrived. The other factors that we are now able to isolate, by their absence in this case, are the competitors and the predators. In more fully developed continental populations these probably tend to limit the development of variety, allowing only those individuals of a population showing the mean or 'normal' structure and behaviour to survive. Extreme variants that venture to brave the enemies and seek new habitats are eliminated. Variation will arise when these checks are weak. In a crowded habitat this may occur as a result of some peculiar swing of the elaborately balanced interacting system of biotic factors, or by some external physical change. We still do not know which of these is the more important in producing the changes of these complicated mainland populations, but the evolutionary laboratories provided by the volcanoes of the Galapagos and other islands suggest that evolutionary change does not follow only on climatic or other physical changes. A single population will become divided into several distinct ones by its own tendencies to growth and variation, given absence of competitors and predators and some means of isolating the animals in different parts of the range.

XIX

THE ORIGIN OF MAMMALS

1. Classification

Class Reptilia
Subclass *Synapsida
Order 1. *Pelycosauria (= Theromorpha). Carboniferous–Permian.
**Varanosaurus; *Edaphosaurus; *Dimetrodon*
Order 2. *Therapsida. Permian–Jurassic
Suborder 1. *Dicynodontia
**Galepus; *Moschops; *Dicynodon; *Kannemeyeria*
Suborder 2. *Theriodontia
**Cynognathus; *Scymnognathus; *Bauria; *Dromatherium; *Tritylodon*; **Oligokyphus*

Class Mammalia
Subclass 1. Eotheria
*Order Docodonta. Trias–Jurassic
**Morganucodon*, Trias, Europe; **Docodon*, Jurassic, N. America and Europe
Order *inc. sed.* **Diarthrognathus*

Subclass 2. Prototheria
Order Monotremata. Pleistocene–Recent. Australasia
Tachyglossus (= *Echidna*), spiny anteater, Australia; *Zaglossus* (= *Proechidna*), New Guinea; *Ornithorhynchus*, platypus, Australia

*Subclass 3. Allotheria
*Order Multituberculata. Jurassic–Eocene. Europe and N. America
**Plagiaulax*, Jurassic, Europe; **Ptilodus*, Palaeocene, N. America

Subclass 4. Theria
*Infraclass 1. Pantotheria
*Order 1. Eupantotheria. Jurassic. Europe and N. America
**Amphitherium*
*Order 2. Symmetrodonta. Jurassic. Europe and N. America
**Spalacotherium*

1. Classification (*cont.*)

Infraclass 2. Metatheria

Order Marsupialia

Infraclass 3. Eutheria (= Placentalia)

*Order *inc. sed.* Triconodonta. Jurassic. Europe and N. America

**Amphilestes*; **Triconodon*

2. The characteristics of mammals

The idea that the mammals include the highest of animals can easily be ridiculed as a product of human vanity; we shall find, however, that in many aspects of their structure and activities they do indeed stand apart as animals that are 'higher' than others, in the sense in which we have used the word throughout this study. The mammals and the birds are the vertebrates that have become most fully suited for life on land; among them are many species in which the processes of life are carried on under conditions far remote from those in which life first arose. The mammalian organization includes a great number of special features that together enable life to be supported under conditions that seem to be extravagantly improbable or 'difficult'. For example, the surface of the body is waterproofed, and elaborate devices for obtaining water are developed. A camel and the man he is carrying through the desert may perhaps contain more water than is to be found in the air and sandy wastes for miles around. This is only an extreme example of the 'improbability' of mammalian life, which is one of its most characteristic features.

The faculty of maintaining a high and constant temperature has opened to the birds and mammals many habitats that were closed to the reptiles. Besides making life possible under extremely cold conditions, such as those of the polar regions, the warm blood vastly extends the opportunities for life in more temperate climates. Mammalian races, which can feed all through the winter, can of course expand more rapidly than their reptilian cousins, which must hibernate for much of the year, during which period they consume rather than produce living matter.

The success of the mammals in maintaining life in strange environments is largely due to the remarkable powers they possess of keeping their own composition constant. All living things tend to do this, but it seems probable that the mammals maintain a greater constancy than any other animals, except perhaps the birds. Claude Bernard's famous dictum 'La fixité du milieu intérieur c'est la condition de la vie libre' may be doubtful of vertebrates in general, but it can certainly be applied to mammals. They have a life that is more free than that of other groups, in the sense that they can exist and grow in circumstances that other forms of life would not tolerate, and they can do this because of the elaborate mechanism by which their composition is kept constant. Besides the regulation of temperature there is also a regulation of nearly all the components of the blood, which are kept constant within narrow limits. Barcroft has pointed out that the achievement of this constancy has enabled the mammals to develop some parts of their organization in ways not possible in lower forms. For instance, an elaborate pattern of cerebral activities requires that there shall be no disturbances by sudden fluctuations in the blood.

We shall expect to find in the mammals, therefore, even more devices for correcting the possible effects of external change than are found in other groups. Besides means for regulating such features as those mentioned above we shall find that the receptors are especially sensitive and the motor mechanisms able to produce remarkable adjustments of the environment to suit the organism, culminating in man with his astonishing perception of the 'World' around him and his powers of altering the whole fabric of the surface of large parts of the earth to suit his needs.

Such devices for maintaining stability often take peculiar and specialized forms in particular cases. The activity and 'enterprise' of mammals has led many of them to make use of particular structures and tendencies in order to develop very odd specializations, which enable them to occupy peculiar niches. What could be more bizarre than the development of the muscles of the nose until a huge mobile trunk appears, so that the heaviest of four-footed beasts, while using its legs for support, can also handle objects more delicately than almost any other animal?

The mammals have developed along many special lines and many of these have already become extinct; others, especially among rodents and primates, remain among the dominant land-animals today. Warmth, enterprise, ingenuity, and care of the young have been the basis of mammalian success throughout their history. The most characteristic features of the modern mammals are thus seen to be largely in their behaviour and soft structures. Mammalian life is above all else active and exploratory. Mammals might well be defined as highly percipient and mobile animals, with large brains, warm blood, and a waterproofed, usually hairy skin, whose young are born alive. Since it is difficult to recognize such characters as these in fossils we cannot say exactly when they arose, and our technical definition of a mammal must be made on the basis of hard parts.

The whole series of mammal-like forms from Carboniferous anapsids onwards forms a natural unit, and it is only by an arbitrary convention that we separate the reptilian subclass Synapsida from the class Mammalia. The present-day mammals form a distinct group of animals, which we identify superficially by their possession of hair. For instance, the duck-billed platypus is immediately referred because of its hair to the mammalia, and we class it apart from reptiles or birds, even though its internal organization and the fact that it lays eggs show it to have many similarities with reptiles, and its bill is like that of a duck. The technical characteristic of the class Mammalia is conventionally given by the presence of a single dentary bone in the lower jaw, the articular and other bones forming, with the quadrate, part of the mechanism of the middle ear. However, fossils showing intermediate conditions are now known from the upper Trias, say 180 million years ago, and all stages can thus be traced in the jaws. The full mammal-like condition was established by the middle Jurassic period, about 150 million years ago. The reduction of the jaw bones may perhaps have been associated with the habit of chewing the food, in order to obtain the large amounts necessary to maintain a high temperature.

It would not be very rash to suggest that by Jurassic times the synapsids had developed the other mammalian characters, such as active habits, large brain, warm blood, hair, and perhaps also a diaphragm, four-chambered heart, and single left aortic arch. They may even have been viviparous, for the surviving monotremes, which lay eggs, probably diverged at a still earlier period, perhaps in the Trias (p. 556).

3. Mammals of the Mesozoic

In spite of all the uncertainties of the fossil record it is now possible to follow the history of the Mammalia back to their origin from cotylosaurian reptiles of the Permian, more than 225 million years ago. Sufficient information is available for us to be quite sure that some population of early anapsid reptiles, such as **Seymouria* of the Carboniferous and Permian times, besides giving rise to all the modern reptiles, to the dinosaurs, and to the birds, also produced the mammals. The evidence for this connexion rests on a most interesting series of fossils, together with some 'living fossils' such as the monotremes of Australia. The fossil history is not at all times equally clear. The mammalian stock first became distinct in late Carboniferous times, as a special type of cotylosaurian reptile, with a tympanum placed behind the jaw and later a lateral temporal aperture. This stock quickly became very abundant and successful as the theromorphs and therapsids of Permian and Triassic times, but then nearly died out in the Jurassic, during which period we know of the mammals only from fragmentary remains of a few small animals. Then, in the Cretaceous, some of these small forms became more numerous and from them arose, before the Eocene, a variety of different types of mammal, from which the histories of the modern orders can be followed in some detail.

We have, therefore, abundant evidence of the earliest stages of mammalian evolution, say, 20 million years, from fossils in the Permian Red Beds of Texas. For the next following 40 million years or so we have also a rich material, from the Upper Permian and Triassic Karroo beds of South Africa. In these early times the mammal line was a flourishing one, more so indeed than the diapsids or any other of the descendants of the cotylosaurs. The animals were mostly carnivorous, though there were also herbivorous types. Many of these early mammal-like forms became quite large and numerous, in fact this stock dominated the land scene in the Permian. In the later Trias, however, if our fossil evidence is a safe guide, their numbers became reduced, perhaps the carnivorous dinosaurs took their place. However, already at this time many of the essential features of mammalian organization had been developed, so far as these can be judged from the bones. Possibly the soft parts of these Permian forms were also mammal-like, the animals may have been active and 'intelligent', have possessed hair and warm blood. However, their brains were small and reptilian in structure.

Throughout the succeeding 100 million years of Jurassic and Cretaceous times our knowledge of the mammalian stock is dim. Mammals of a somewhat rodent-like type, the multituberculates, became quite numerous and produced some large forms, but then became extinct in the Eocene. Unless we have been singularly unlucky in the preservation of mammalian remains, no other mammal-like animals larger than a polecat existed throughout this long period. Since we possess detailed information, based on numerous fossils, about scores of large and small diapsid reptilian types, it can hardly be only an accident that mammal-like fossils are so rare.

We must conclude that the mammalian organization, after an initial success in the Permian and Triassic, was almost supplanted in the Jurassic and Cretaceous by the various diapsid creatures. However, a few very rare fossils, mostly lower jaw bones, give us some idea of the nature of the animals that carried our type of organization through the Jurassic period. Then from the top of the Cretaceous, about 70 million years ago, come the first fossil remains of mammals similar to those alive today; rare, insectivorous creatures, not widely different from modern hedgehogs. This was the time of the beginning of an astonishing revolution, which completely altered the life on the face of the earth. The descendants of these shrew-like animals multiplied exceedingly in the new conditions, and by the earliest Palaeocene times, 70 million years ago, they had produced recognizable ancestors of most of the modern orders of mammals.

Our knowledge about this revolution is still very dim. Some people claim that it accompanied one of the major geological crises, a period in which the surface of much of the earth became unstable, there were great volcanic outflows and the building of vast mountain chains. It is possible that such upheavals led to the disappearance of the swampy conditions, which had been so suitable for large reptiles, and to the appearance of dry uplands on which the mammals and birds flourished, especially where the climate was cold. We must be careful here, however, not to argue in a circle. Our evidence about the climate is derived from the changes in the populations, as revealed by the fossils. Study of the plant remains, however, confirms the supposition that conditions became drier during this time.

Our knowledge of the origins of mammals derived from fossils is supplemented by certain surviving mammals, whose structure shows them to have diverged rather early from the main stock, especially the monotremes (duck-billed platypus and *Echidna*) and the marsupials (kangaroos, opossums, &c.). Unfortunately it is still not clear exactly how these survivors are related to the main stocks as revealed by the fossils. The monotremes are unknown except as Pleistocene fossils; tantalizingly enough we know only little about the affinities of this ancient group. The characteristics of their bony skeleton show that they must have diverged from other mammals well back in Mesozoic times. For instance, the pelvic and pectoral girdles are very 'reptilian' in structure. It is therefore not wild speculation to use the characters of the soft parts of monotremes to deduce those of the mammalian stock in late Triassic or early Jurassic times, say, 180 million years ago. The marsupials appear as fossils in the late Cretaceous. Some of the earliest forms are very like modern opossums, and we may conclude that the condition of modern marsupials throws some light on the probable condition of the soft parts of mammals 70 or 80 million years ago. However, it has been realized in recent years that many features in marsupial organization are special developments, not 'primitive' characteristics carried over from the ancestral condition.

The piecing together of all this evidence to give a reliable picture of the history of mammals is a valuable exercise in zoological and geological method, as well as a means of becoming familiar with a fascinating story. We may now return to the Carboniferous times to examine the evidence more in detail (see Fig. 221).

4. Mammal-like reptiles, Synapsida

All our evidence about the origin of mammals must of course be based upon the study of hard parts, which can become fossilized, and in particular on the skull. The characteristic feature of the skull in the populations that led to the mammals is the development of a hole in the roof, low down on the side of the temporal region, the lower temporal fossa (Fig. 223). This hole was at first bounded above by the post-orbital and squamosal bones, below by the jugal and squamosal. It was at one time supposed that this single fossa was formed by union of the two present in diapsid reptiles, and thus the whole group acquired the inappropriate name Synapsida. Animals having this characteristic appear in the rocks at the same time as the cotylosaurs, with no hole in the skull roof (Anapsida), from which they were presumably derived. We have only fragmentary knowledge of anapsids from the Carboniferous; most of our information comes from lower Permian forms, such as **Seymouria*; yet there are quite well-preserved synapsids of Upper Carboniferous date. Therefore we have no complete series of successive types to show how the earliest mammal-like types arose; nevertheless we can see something of the probable stages of this progress within the Permian Anapsida. One of the characteristics of the skull of **Seymouria* was the presence of an 'otic notch', in which lay the tympanic membrane, at the back of the skull (Fig. 220). In **Captorhinus* and similar anapsid fossils from the Lower Permian this notch has disappeared (Fig. 222). The tympanum apparently lay behind the quadrate, leaving the whole side of the skull as a rigid support for the jaw. A long series of subsequent evolutionary changes led to one of the most characteristic features of the skeleton of mammals, namely, the conversion of the hinder jaw bones (quadrate and articular) into ossicles for the transmission of vibrations to the ear (p. 550).

**Captorhinus* and its allies were two or more feet in length and showed some development of the limbs towards the mammalian condition, in that the bones were moderately elongated and slender and the limbs perhaps tended to be held vertically under the body. However, there was no great development of a backwardly directed elbow and forward knee. The teeth were considerably specialized, possibly for eating molluscs, and there were several rows of crushing teeth on the edge of the jaw, and long overhanging ones in the premaxillae.

5. Order *Pelycosauria (= Theromorpha)

The synapsid line began when an early offshoot from some such anapsid as **Captorhinus* developed a temporal fossa. This must have

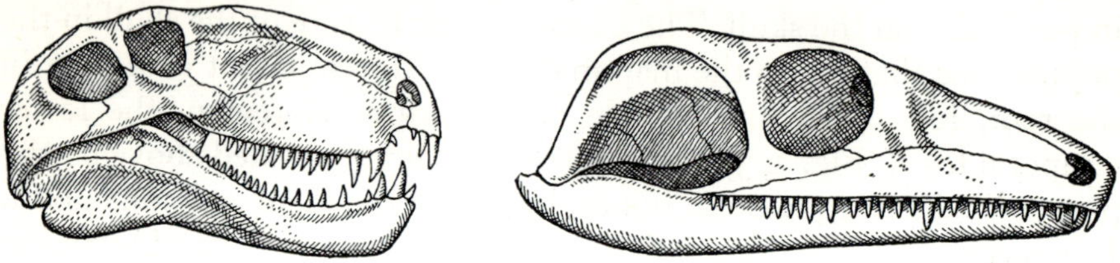

FIG. 317. Skulls of two early synapsids. (After Romer, *Vertebrate Paleontology*, University of Chicago Press, and Abel.)

occurred in the middle of the Carboniferous, nearly 300 million years ago, producing in the later part of that period, and in various Pennsylvanian and Lower Permian strata, especially in North America, these earliest synapsid populations, classified together as pelycosaurs or Theromorpha. **Varanosaurus* (Fig. 317), from the Texas Red Beds, is a typical example, a carnivore, about 3 ft long, showing a lizard-like appearance little different from that of anapsids or primitive diapsids. Intercentra were found all along the vertebral column, as in anapsids, and there were abdominal ribs. The teeth showed the beginnings of the mammalian differentiation in that one or more near the front of the series were elongated as 'canines'. The skull of pelycosaurs also showed tendencies in the mammalian direction in having a long anterior region and a relatively high posterior part, giving a large brain-case and deep jaw.

Other pelycosaurs developed long neural spines, in more than one line, **Edaphosaurus* and its allies being herbivores, some of them as much as 12 ft long, whereas **Dimetrodon* (Fig. 317) was a carnivore. The spines presumably supported a web, perhaps used in temperature regulation. These animals were important in the late Carboniferous and early Permian fauna, preying on the other early reptiles and on rhachitomous amphibia. They did not survive long, however, being apparently replaced by their descendants. From some form similar to **Dimetrodon* arose a whole series of lines classed together as Therapsida and leading on to the mammals.

6. Order *Therapsida

In these animals there was never any trace of the bipedalism that developed in the Archosauria; the fore-limbs were always at least as long as the hind, and tended to be turned under the body and to carry it off the ground. The skull became deeper and its brain-case enlarged; a bony secondary palate developed from flanges of the premaxillae, maxillae, palatines, and pterygoids. The teeth became differentiated for various functions and the bones of the lower jaw, except the dentary, were gradually reduced.

These features seem to have developed in several different lines descended from pelycosaur ancestors; the sorting out of the various genealogies is not yet complete. It is therefore still difficult to decide for certain the interesting question whether parallel evolution occurred, and especially whether similar mammalian features appeared independently in animals of different habits. The fossils are nearly all found in South Africa and have been studied in great detail by R. Broom.

Much of the surface of the southern part of Africa is covered by rocks known as the Karroo system. These consist partly of shales and mud-stones formed by the matter brought down by a large Mesozoic river, and the remainder are sandstones composed of blown sand. Both sorts of rock were particularly favourable for the preservation of the remains of terrestrial animals. Unfortunately the absence of marine fossils makes it difficult to give dates for these rocks. Altogether the strata present a thickness of some 15,000 ft, laid down over a period corresponding probably to that from the Middle Permian to the Upper Trias in Europe, that is to say, from 250 to 180 million years ago.

The therapsids fall into two groups, the mainly herbivorous dicynodonts and the carnivorous theriodonts, probably preying upon the former. **Galepus* (Fig. 318) perhaps shows a stage of evolution of dicynodonts from pelycosaurs. The temporal opening was small, the teeth all alike and the bones at the hind end of the jaw large. The early dicynodonts include the herbivorous 'dinocephalia', such as **Moschops*, which retained many primitive features, but the legs were turned under the body and the phalanges became reduced to the mammalian formula of 2.3.3.3.3. The roof of the head was expanded into a large dome, giving the name to the group. The later dicynodonts were a still more specialized and successful offshoot, the first of the many great tetrapod herbivores. They became the commonest of all reptiles in the later Permian. They were large creatures (**Kannemeyeria* and **Dicynodon*, Fig. 318), some probably living in marshes. The margins of the jaws were covered with a horny beak and the teeth reduced to a single pair of upper 'canine' tusks.

The most interesting of the therapsid reptiles are, however, those placed in the suborder *Theriodontia, which, by the early Trias, had produced a very mammal-like type of organization, apparently in several independent lines. The temporal opening became progressively wider, presumably for the accommodation of larger jaw muscles, so that the parietal bone entered into the margin of the fossa and post-orbital and squamosal bones no longer met above it. Eventually the post-orbital bar itself became incomplete, leading to the typical mammalian condition, with the orbit and temporal fossa confluent. There was a large columella, articulating broadly with the quadrate and perhaps serving to brace the latter as well as to transmit vibrations to the inner ear. A bone of this size could act as an efficient vibrator (Hotton, 1959). The brain-case was high and large and the cerebellum better developed than in modern reptiles. The cerebral hemispheres, however, though large, probably remained mainly olfactory structures, as indeed they were in many Eocene mammals and are in some that survive today.

The ribs were well developed and seem to have formed a cage, which may have been used, with a diaphragm, for respiration. The limbs came to support the body off the ground, and the dorsal parts of the girdles were developed accordingly, the scapula becoming large and bearing an acromial spine for muscle attachments, the coracoids being reduced. The anterior portion of the ilium became large and a hole appeared between the pubis and ischium. The head of the femur lay at the side, and a special knob, the great trochanter, appeared on it for the attachment of muscles running from the front part of the ilium. In the hands and feet the toes show a reduction of the formula to 2.3.3.3.3. The teeth became differentiated into incisors, canines, and cheek teeth, the latter having several cusps in place of the cone of a typical reptile tooth. In most forms the tooth replacement was as in reptiles but in some it was more limited.

These features were little developed in the earliest theriodonts, such as **Scymnognathus* and other Permian forms (Fig. 318). The

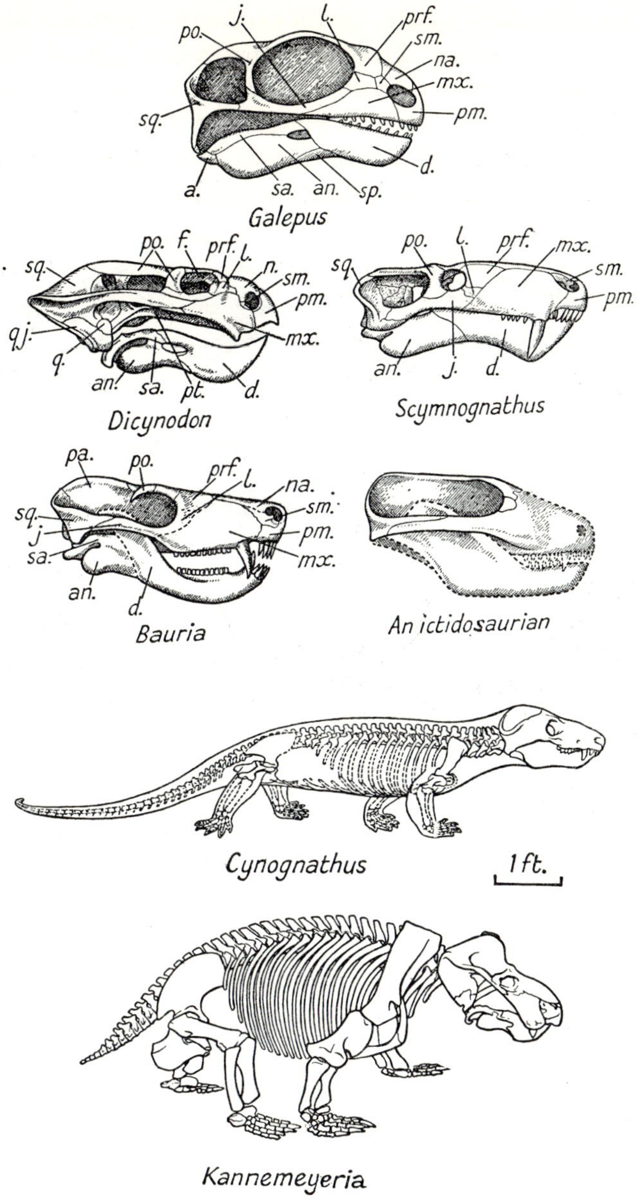

FIG. 318. Skull and skeleton of various synapsid reptiles, abbreviations as Fig. 214, p. 377. (After Romer, *Vertebrate Paleontology*, University of Chicago Press, from Broom, Watson, Gregory, & Pearson.)

temporal fossa still resembled that of pelycosaurs, there was a single condyle, no secondary palate, and a phalangeal formula of 2.3.4.5.3. **Cynognathus* and other Triassic forms were typical theriodonts, showing the above 'mammalian' features. They were carnivores, of distinctly dog-like appearance, although they still retained many signs of a heavy reptilian build. **Bauria* was of another type, still more advanced than **Cynognathus* in that the orbit and temporal fossa were

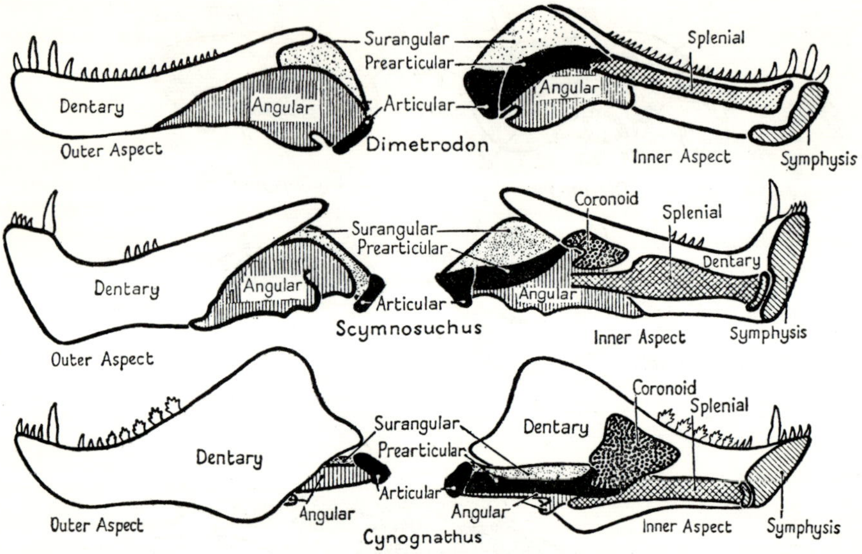

FIG. 319. Reduction of the articular and other bones and increase in the dentary in therapsid reptiles. (From Neal and Rand, *Comparative Anatomy*, The Blakiston Company, after Watson.)

confluent (Fig. 318). The foramina of the maxilla of some of these animals suggest the passage of nerves and blood-vessels for a facial musculature, which is characteristic of mammals but absent in other vertebrates. Other types of theriodont were more specialized, with rodent-like features (**Tritylodon*, **Oligokyphus*).

These *Theriodontia were the dominant carnivores of the early Trias, but by the end of that period they had almost disappeared. The latest synapsids include in some classifications the *Ictidosauria, such as **Diarthrognathus* of the Upper Trias. They are rare fossils, showing almost completely mammal-like structure. There was no post-orbital bar and no prefrontal, post-frontal, or post-orbital bones. A well-developed secondary palate was present. The bones at the hind end of the lower jaw had become very small (Fig. 319). In at least one form the dentary articulates with the squamosal, though there was also a quadrate-articular joint. This fossil has been appropriately named **Diarthrognathus*. It shows that it is not possible to find an absolutely rigid definition of a 'mammal', even by means of the condition of the jaw. We shall classify this fossil as a mammal.

Study of the synapsids, mainly from the Karroo system, shows us, therefore, a series of types, of which the earliest were very like the first reptiles and the latest very like true mammals. There can be no doubt of the general tendency, but the series is not complete enough to enable us to follow the details of the evolution of the populations. These fossils are revealed by denudation and can be given only approximate dates. It is certain that some of the mammal-like features appeared independently in lines whose evolution proceeded separately from a common ancestor. Thus the dicynodonts and later theriodonts all had the mammalian phalangeal numbers, but each of these lines has certainly evolved independently from pelycosaur-like ancestors having a greater number of phalanges.

The influences that produced the evolution of these populations must have been quite complex, since they did not affect all parts of the body at once. For instance, some early therapsids, in spite of their mammalian phalangeal formula, still showed pelycosaur features in the absence of a secondary palate, and presence of a single occipital condyle and small dentary. If the presence of a squamo-dentary articulation is taken as the criterion of 'a mammal' this condition was almost certainly reached independently by several different lines (see Simpson, 1959). We still know too little to be able to specify clearly the conditions controlling such evolutionary changes, but it seems possible that the gradual appearance of terrestrial life and of large herbivores led various animals of a suitable structure and disposition to a carnivorous life. For this purpose certain changes of the ancestral structure would be suitable, leading to parallel evolution in related stocks. However, at present we can hardly do more than pose questions about such matters and resolve to be rigorous in interpretation of the available evidence.

7. Mammals from the Trias to the Cretaceous

The types classified as synapsid reptiles, which we have been considering, are not found later than the early Jurassic, 170 million years ago; mammals of approximately the modern type appear in the late Cretaceous. For the enormous time of more than 90 million years between these dates the mammalian organization maintained itself in the form mostly of small insectivorous animals, perhaps arboreal and nocturnal; few remains of these are found as fossils, but discoveries now made have helped to bridge the gap (Kermack and Mussett, 1958). No doubt many types existed of which we have no fossil remains, but the Mesozoic mammals that we know can be divided into five orders, *Docodonta, *Multituberculata, *Triconodonta, *Symmetrodonta, and *Eupantotheria (= *Trituberculata). The last of these may perhaps be directly related to the animals that gave rise to the modern mammals, the other four lines are specialized offshoots. All of them were true mammals in that the articulation of the lower jaw was between dentary and squamosal, though there is evidence that the articular and other bones remained relatively large in the docodonts. The brain-case (where known) was high and the temporal fossa joined with the orbit. The post-cranial skeleton is little known and indeed many forms are known only from lower jaws. However, the pectoral girdle of docodonts includes the same bones as that of monotremes in an even more reptilian form. This and other features suggest that the monotremes have evolved from animals like the *Docodonta and that they have preserved to the present day many features that characterized the mammalian stock during the Mesozoic period. Some of the 'Mesozoic mammals', however, had advanced beyond the stage of the monotremes, for instance the limb girdles of multituberculates resemble those of modern mammals rather than reptiles.

For purposes of classification we can most conveniently divide the class Mammalia into four subclasses, the Eotheria for the docodonts, Prototheria for the monotremes, Allotheria for the multituberculates, and the Theria for the modern mammalian line. This latter subclass can then be subdivided into three infraclasses, Pantotheria for those 'Mesozoic mammals' that probably led to the rest, Metatheria for the marsupials, and Eutheria for the placentals. The triconodonts must be left as of uncertain affinities. This classification is based upon that provided by G. G. Simpson of the American Museum of Natural History, who has not only greatly increased our knowledge of many orders of mammals but also provided a complete systematic review of the group.

The *Docodonta are known from isolated teeth and jaws in North America and more abundant remains from the latest Triassic of South Wales, which include parts of the skull and shoulder girdle. The molar teeth carry three cusps in a row, a condition known as 'triconodont'. There is a well-marked condyle on the dentary and on the lower margin of the jaw a process (the 'angle'). Above this is a conspicuous trough, quite unlike anything found on the dentary of therian mammals. This is presumed to have contained the articular and perhaps other bones, fragments of which can be seen (Fig. 320). A similar groove appears on the jaws of later theriodonts. Jaws carrying triconodont teeth but

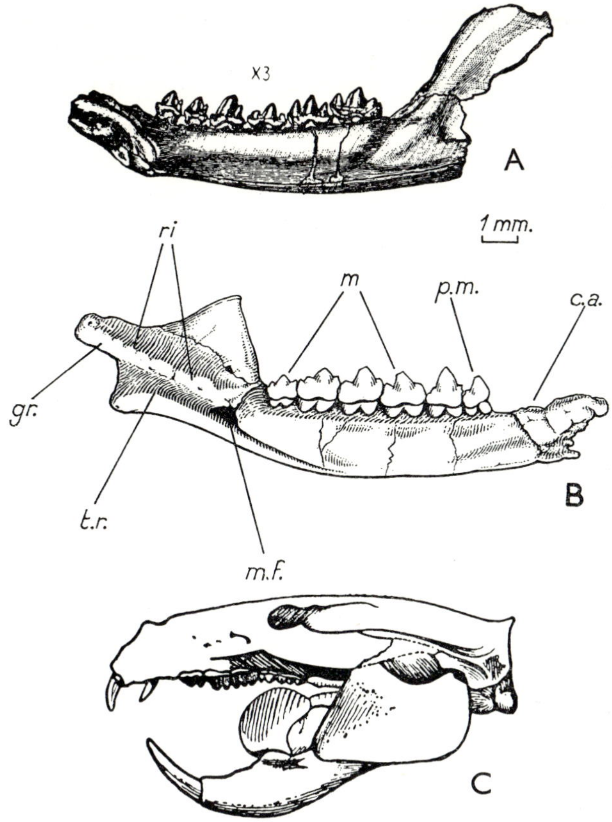

FIG. 320. A, triconodont (*Priacodon*) from the Jurassic; B, lingual view of lower jaw of a docodont; C, multituberculate (*Ptilodus*) from the Palaeocene. *c.a.* canine alveolus; *gr.* groove; *m.* molars; *m.f.* mandibular foramen; *p.m.* premolar; *ri.* ridge; *tr.* trough. (A and C after Lull, *Organic Evolution*, copyright 1917, 1929 by The Macmillan Company and used with their permission. B after Kermack and Mussett.)

without the groove have long been known from the Jurassic (**Amphilestes*). They may represent an offshoot from a stock like the docodonts.

The multituberculata were the most numerous and long-lasting Mesozoic mammals, surviving from the early Jurassic period to the lower Eocene. They were herbivorous and sometimes of large size (Fig. 320). Between the incisors and the molariform teeth there was a gap (diastema), as in other mammals that chew large amounts of vegetable matter. The cheek teeth carried longitudinally arranged rows of cusps, presumably used for grinding the food. The arrangement of the muscles can be deduced from the jaws. The temporal muscle was small and there was a small coronoid process on the mandible for its attachment. There was therefore no wide sweeping up and down movement of the jaw as in carnivores. On the other hand, there was a large masseter, pulling anteriorly, and pterygoid transversely and a shallow glenoid fossa. All these features will be found again in placental herbivores. The relationship of these animals to earlier and later types is quite uncertain. It has been suggested that the multituberculates gave rise to the monotremes or marsupials, but the features they have in common with these are mostly those of all primitive mammals.

The symmetrodonts and eupantotheres are imperfectly known but probably included various carnivorous and omnivorous types, not unlike modern opossums and some insectivores. The lower jaw was formed of a single dentary bone, the hinder jaw bones having presumably already formed ear ossicles. In **Amphitherium* there were 4 pairs of incisors on each side of the lower jaw, one canine and 11 cheek teeth, 4 of these being preceded by milk teeth and hence classed as premolars. Several different sorts of eupantotherian are known from the Jurassic and could have given rise to the earliest placental insectivores, which appear in the late Cretaceous (p. 583). The reason for believing in this relationship is the cusp-pattern of the teeth.

8. Original cusp-pattern of teeth of mammals

In the symmetrodonts and pantotheres the teeth were so arranged as to bite against each other. The lower cusps formed a triangle, with a surface behind, the talonid or heel, with which the main cusp of the upper molar made contact (Fig. 321). The upper cusps also formed approximately a triangle and this 'tribosphenic' condition of the molars is believed to have been the plan from which the modern mammalian condition is derived. The apical cusp, which lies on the inner side of the upper molars and the outer side of the lower molars, was at one time believed to represent the original reptilian cone and was therefore called the protocone in the upper and protoconid in the lower molars. The other two cusps are called the paracone (paraconid) in front and metacone (metaconid) behind (Fig. 321). The separation between these latter cones is not sharp in the pantotheres, especially in the upper jaw, and we shall find this condition again in the earliest placental mammals.

There have been various theories about how the triangular plan was reached. The original tritubercular theory supposed that there was a 'rotation' into a triangular position from the three cusps in line of a triconodont. Even if this could be shown to have happened in a series of fossil teeth, we should still require a knowledge of the change of morphogenetic process by which the 'rotation' was produced. There have also been attempts to explain the many-cusped mammalian tooth as due to the 'fusion' of a number of reptilian tooth germs, either those making one series on the gum or the teeth of successive series. It is plausible that changes in relative time and/or place of tooth development could occur in this way, leading to a partial 'fusion' and the production of many-cusped structures. At present there is too little information to decide how the reptilian became changed into the mammalian tooth, but there can be little doubt that the tritubercular theory shows us approximately the nature of the earliest mammalian cusp patterns. Indeed it was originally put forward because nearly all the Eocene representatives of the various mammalian orders showed signs of a triangular cusp pattern.

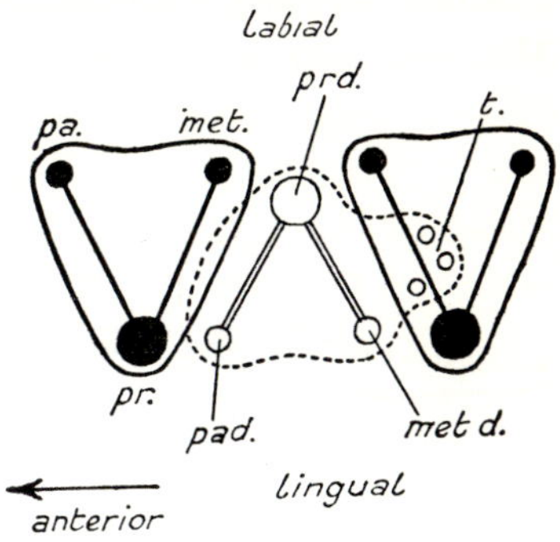

FIG. 321. Arrangement of occlusion of cusps of tritubercular teeth. Upper molars continuous outline, lower dotted.

met. metacone; *metd.* metaconid; *pa.* paracone; *pad.* paraconid; *pr.* protocone; *prd.* protoconid; *t.* talonid (heel). (After Osborn.)

FIG. 322. Duck-billed platypus (*Ornithorhynchus*). (From photographs.)

FIG. 323. Five-toed echidna (*Tachyglossus*). (From photographs.)

9. Egg-laying mammals. Subclass Prototheria (Monotremata)

The duck-billed platypus (*Ornithorhynchus*) and spiny ant-eater, usually called *Echidna* but strictly *Tachyglossus* (Figs. 322 and 323), are Australasian mammals, basically similar to each other, but so different from other mammals that it is certain that they left the main stock far back in the Mesozoic. Their organization possibly shows us many of the characteristics of mammalian populations at that time.

Since we are comparing the platypus and echidna chiefly with Mesozoic reptiles we shall deal first with their hard parts, examining

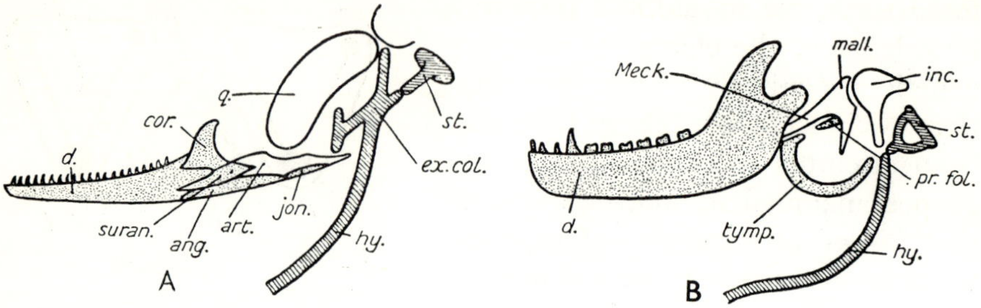

FIG. 324. Diagram of arrangement of jaw and auditory ossicles. A, in a reptile; B, in a mammal.

ang. angular; *art.* articular; *cor.* coronoid; *d.* dentary; *ex.col.* extracolumella; *hy.* hyoid; *inc.* incus (quadrate); *jon.* gonial; *mall.* malleus (articular); *Meck.* Meckel's cartilage; *pr. fol.* processus folianus (goniale); *q.* quadrate; *st.* stapes; *suran.* surangular; *tymp.* tympanic (angular). (After Ihle, from Gaupp.)

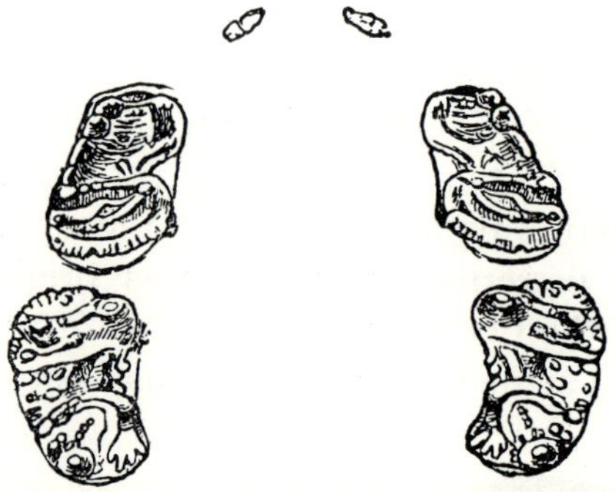

FIG. 325. The temporary upper teeth of the duck-billed platypus. (From British Museum Guide.)

the living animals as if they were fossils. The lower jaw consists of a single dentary bone. The quadrate, articular, and tympanic have entered the ear, but the malleus is large, the incus small, and the stapes elongated (Fig. 324). The tympanic bone forms a partial ring around the tympanum, and the whole apparatus is not enclosed in a bony 'bulla' as it is in modern mammals. Neither animal possesses true teeth in the adult, the platypus having a flattened bill covered with soft skin and used for 'paddling' for the small aquatic animals, especially mussels and snails, on which it lives. *Tachyglossus* and the related New Guinea form *Zaglossus* have long 'beaks' for eating ants. However, in the young platypus flattened, ridged teeth are present, unlike those of any other mammals (Fig. 325). These true teeth are replaced by horny structures, formed by an ingrowth of epidermis beneath them and apparently used for breaking the shells of the molluscs. It is particularly unfortunate, since most of our knowledge of early mammalian affinities comes from their teeth, that we can deduce so little from those of the monotremes.

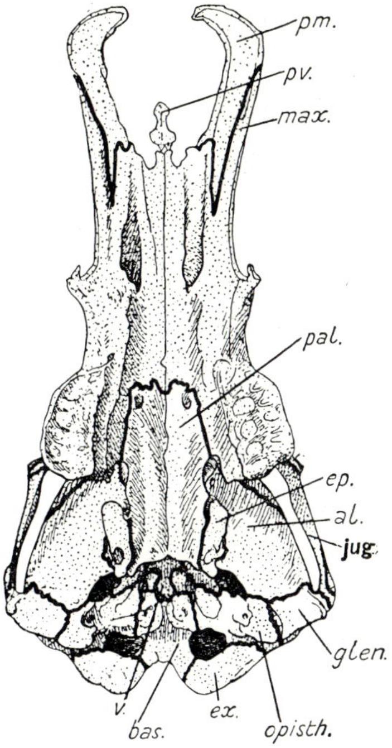

FIG. 326. Skull of platypus seen from below.

al. alisphenoid; *bas.* basioccipital; *ep.* epipterygoid; *ex.* exoccipital; *glen.* glenoid facet of squamosal; *jug.* jugal; *max.* maxilla; *opisth.* opisthotic; *pal.* palatine; *pm.* premaxilla; *pv.* prevomer; *sq.* squamosal; *v.* 'vomer'—basisphenoid. (From Ihle, after van Benneden.)

The skull (Fig. 326) is specialized in both genera, particularly at the front end, and many of the bones fuse early. There is a wide communication between the orbit and temporal fossa. There are many 'reptilian' features; for instance, separate pterygoid bones. The 'dumb-bell-shaped bones' are perhaps the remains of the prevomer but may be vestiges of the palatine processes of the premaxillae. Small prefrontal and postfrontal bones are present. In the temporal region there is a narrow canal that apparently represents the posterior temporal fossa of Therapsida. A curious feature is that the lateral wall of the brain-case is formed by an anterior extension of the petrosal and not by the alisphenoid.

The vertebrae are very reptile-like, especially the cervicals, which bear separate ribs, as in the synapsid reptiles. There are seven cervical vertebrae, but in the dorsal region differentiation has proceeded less far than in other mammals, there being 16 ribs in the spiny ant-eaters

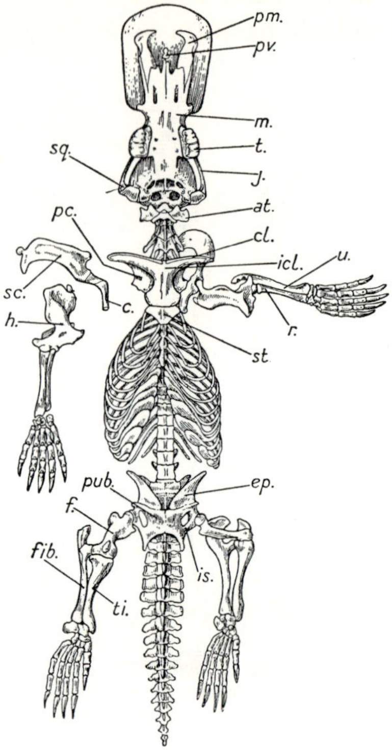

FIG. 327. Skull and skeleton of a female platypus.

at. atlas; *c.* coracoid; *cl.* clavicle; *ep.* epipubic bone; *f.* femur; *fib.* fibula; *h.* humerus; *icl.* interclavicle; *is.* ischium; *j.* jugal; *m.* maxilla; *pc.* precoracoid; *pm.* premaxilla; *pub.* pubis; *pv.* prevomer; *r.* radius; *sc.* scapula; *st.* sternum; *sq.* squamosal; *t.* tooth; *ti.* tibia; *u.* ulna. (Modified after Owen.)

and 17 in the platypus, with 3 or 4 lumbars in the former and only 2 in the latter. The ribs articulate only with the bodies of the vertebrae, not with the transverse processes. The tail is vestigial in *Zaglossus*, but forms a flattened swimming-organ in the platypus. The limbs and their muscles and girdles are remarkably reptilian (Fig. 327). They tend to be held laterally rather than beneath the body and in general the ventral parts are far better developed than in modern mammals, and this is sometimes spoken of as a 'plate-like' condition. In the pectoral girdle there are separate clavicles and a median interclavicle. The coracoid region includes two quite large and separate bones, the coracoid and 'precoracoid'. There is no spine on the scapula. The ventral portion of the pelvic girdle is enlarged by the development of epipubic bones, presumably partly for the support of the marsupial pouch. There is a large and broad humerus in both animals, held horizontally. In *Ornithorhynchus* the fibula is expanded at its upper end like the ulna of other mammals, for the attachment of the large muscles that produce the swimming action.

The condition of the skeleton is therefore quite sufficient to establish the early divergence of the monotremes from other mammals and we are justified when looking at the soft parts in supposing that many of the characters were possessed by the synapsid reptiles. However, it must always be borne in mind that many of the 'mammalian' characters could conceivably have been produced by parallel evolution, subsequent to the divergence of the two lines.

Perhaps the outstanding non-skeletal feature is the egg-laying habit. The large yolky eggs have a whitish shell and in the spiny ant-eater are transferred by the mother to a special marsupial pouch, which develops at the breeding season. The female platypus makes a nest in a burrow for her two or three eggs and remains with them continuously until after hatching. Monotremes are unique in possessing a caruncle on the head as well as the egg-tooth, suggesting that both were present in the ancestor of Amniotes as means of breaking out of the shell. After incubation and hatching the young enter the pouch and are fed by milk. The post-natal care of the young therefore developed before the egg-laying habit was lost. Both genera produce milk from specialized sweat glands on the ventral abdominal wall of the female, but the ducts of these are not united to open on nipples.

The presence of hair again gives us a valuable clue. Unless this feature has been separately evolved on several lines we may conclude that the Mesozoic mammals and perhaps even the synapsids had made some progress in temperature regulation. The mechanism is still imperfect in monotremes, whose temperature is lower and more variable than that of other mammals. The platypus has a fine short fur of dark brown colour; in the spiny ant-eater the back carries a mixture of spines and hairs, the belly carries hairs alone.

The rectum and urinogenital system open to a common cloaca, a 'reptilian' feature found also in marsupials (Fig. 328). The testes are undescended. The penis of the male is a simple groove in the cloacal floor and is used only for the passage of sperm, the urine entering the cloaca by a special urinary canal. A curious feature found in both monotremes but in no other mammals is a grooved poison spine on the tarsus of the male, served by a gland in the thigh. It is possible that this is used to immobilize the female during coition.

The brain is relatively large and arranged essentially on the mammalian plan (Fig. 329). The pallial portion of the cerebral hemispheres is well developed, not the striatal portion as in reptiles, and the surface is actually convoluted in spiny ant-eaters, though smooth in the

A

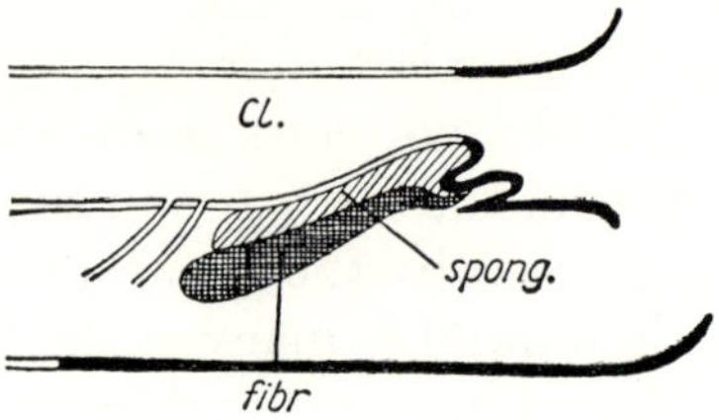

B

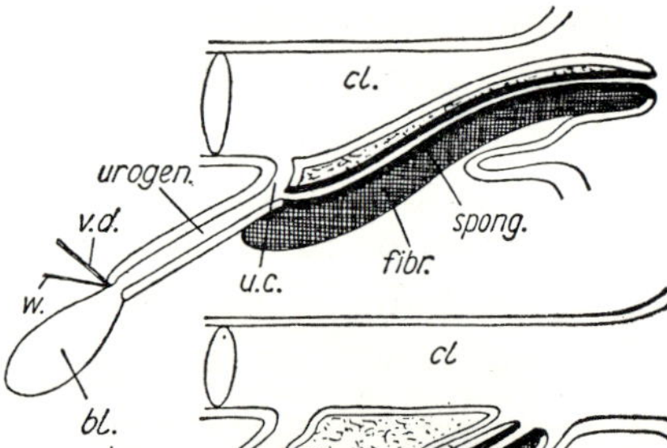

C

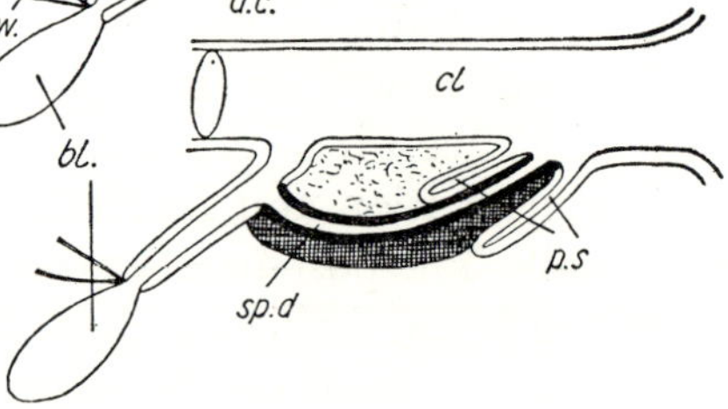

FIG. 328. Comparison of the cloaca and penis of tortoise (A) with monotreme (B and C). The penis is shown erect in B, withdrawn in C. (From Ihle.)

bl. bladder; *cl.* cloaca; *fibr.* corpus fibrosum; *p.s.* preputial sac; *spong.* corpus spongiosum; *sp.d.* sperm duct; *urogen.* urogenital canal; *u.c.* urinary canal; *v.d.* vas deferens; *w.* ureter.

platypus. Perhaps, therefore, the large brain and active, memorizing habits appeared early in the Mesozoic, but an interesting feature is that there is no corpus callosum joining the hemispheres.

The soft parts also show mammalian characteristics. The diaphragm is fully developed and the heart and single left aortic arch resemble those of other mammals. These animals have therefore advanced in their circulatory system beyond the anapsid condition, such as is probably shown today in Chelonia (p. 397).

With all their archaic features the monotremes also show many specializations. The platypus is highly modified for aquatic life. Apart from its bill there are the webbed feet, dorsal nostrils, long palate, short fur, thick tail, and perhaps absence of external ears. The animals burrow in the banks, making nests in which the young are reared. They are not uncommon in the rivers of southern and eastern

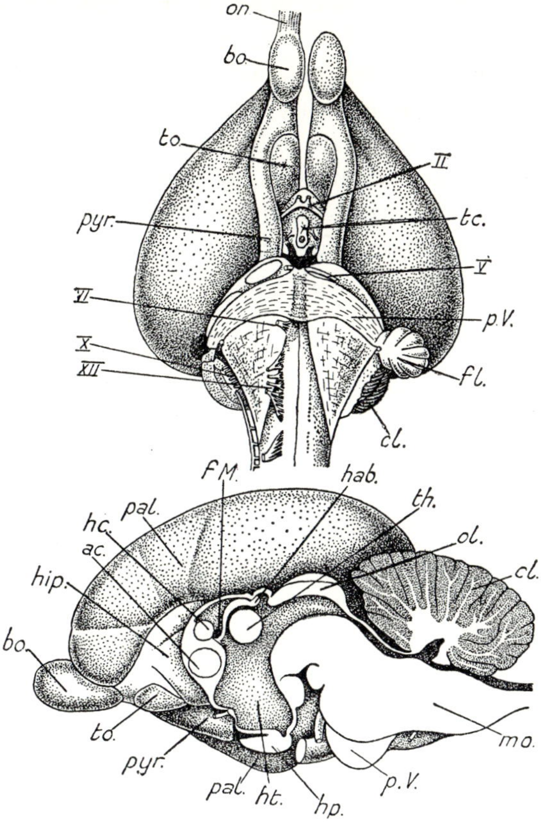

FIG. 329. Brain of the platypus.

ac. anterior commissure; *bo.* olfactory bulb; *cl.* cerebellum; *fl.* flocculus; *fM.* foramen of Munro; *hab.* habenula; *hc.* hippocampal commissure; *hip.* hippocampus; *hp.* hypophysis; *ht.* hypothalamus; *mo.* medulla oblongata; *ol.* optic lobe; *on.* olfactory nerve; *pal.* pallium; *pV.* pons Varolii; *pyr.* pyriform lobe; *tc.* tuber cinereum; *th.* thalamus; *to.* tuberculum olfactorium; *II–XII.* cranial nerves. (From Kingsley, *Comparative Anatomy of Vertebrates*, The Blakiston Company. After Elliot Smith.)

Australia and Tasmania and fortunately are difficult to catch and therefore in no danger of extinction, though their fur and flesh are both useful.

Several species of spiny ant-eater are known and they occur in New Guinea as well as Australia. They show specializations for ant-eating similar to those of *Myrmecophaga* (p. 397), namely, long snout, long tongue, and large salivary glands. The clawed feet are used to make burrows for the young, as well as for digging up the nests of ants. It is perhaps significant that both of these very early mammals burrow and make nests to assist in the protection of their young.

The Monotremata thus show a peculiar mixture of mammalian and reptilian characteristics. In their brain, hair, warm blood, heart, and diaphragm they are mammalian, but in skeleton and egg-laying habit they resemble reptiles. A large part of their interest is that they suggest an intermediate stage, in many features, between the two groups. Thus the pectoral girdle appears to be that of a reptile partly changed to a mammal. There are certainly many things to be discovered from these extraordinary creatures, which have remained with little change in fundamental organization for possibly nearly 150 million years. The characters they show literally provide us with a view of the past, yet the facts that these two alone have survived and that they show special features of their own remind us sharply that evolutionary change is almost universal: new types replace the old almost if not quite completely.

XX

MARSUPIALS

1. Marsupial characteristics

THE pouched mammals are essentially very similar to placentals, though they undoubtedly diverged from some early stage of the main mammalian stocks. They parallel, in the isolation of Australasia, the

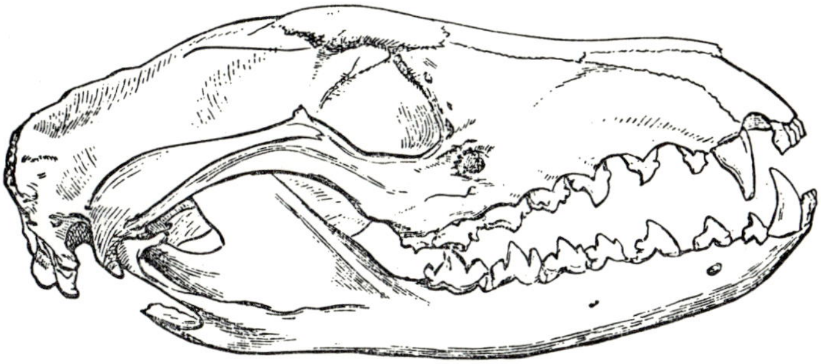

FIG. 330

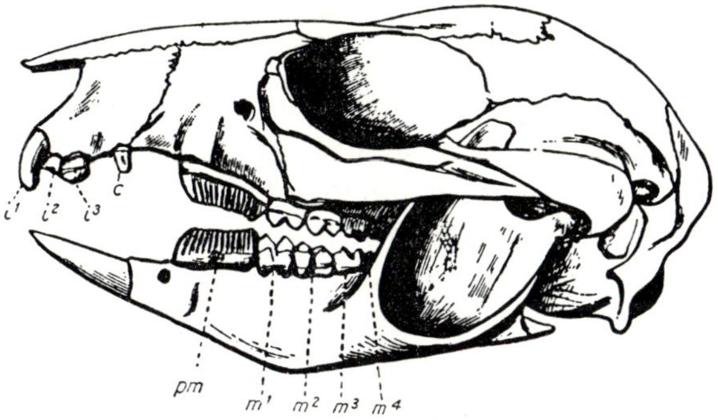

FIG. 331

FIGS. 330–1. Skulls of thylacine (330), and rat kangaroo (*Bettongia*) (331). (From Flower and Lyddeker, *Mammals, Living & Extinct*, A. & C. Black, Ltd.)

adaptive radiation accomplished in other parts of the world by the placentals. Many of their features are specialized, so that they represent not a stage on the way to placental evolution but a specialized side branch. Today some 230 species are found in the Australasian region, and there are successful representatives in North and South America. In Eocene times they occurred in Europe.

The skull shows many characters found also in Insectivora and other early mammalian groups (Figs. 330–1). The brain-case is small and the top of the skull therefore rather flat. The orbit and temporal fossa remain fully confluent and there is no post-orbital bar. The bony palate is incomplete posteriorly, there being large holes in the palatine portion of it. The jugal bone always reaches back to the glenoid articulation of the jaw. The lower jaw, consisting of course of a single dentary, has a characteristically inturned or 'inflected' inner 'angle'. Other special features of the skull, not usually found in placentals, are that the foramen for the optic nerve and that for the eye-muscle and trigeminal ophthalmic nerves are not separated from each other

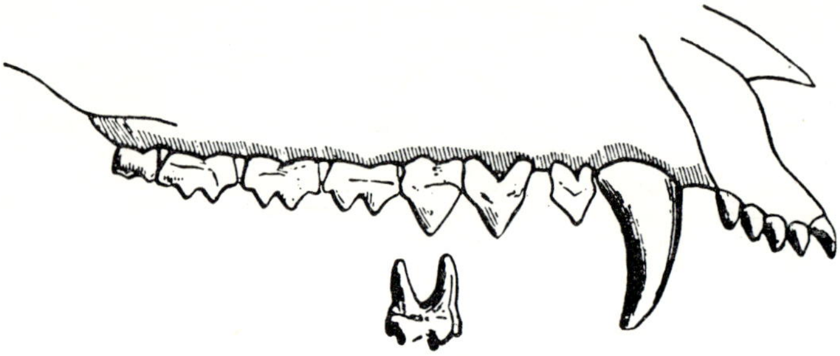

FIG. 332. Teeth of the upper jaw of the opossum (*Didelphys*) showing the last premolar, whose place is occupied in the young by a molariform tooth. (From Flower and Lyddeker.)

and that the lachrymal bone extends outside the orbit. More interesting than these apparently trivial and unconnected diagnostic features is the fact that the auditory region is not protected by the formation of a bulla of the petrous bone as it is in other mammals; instead the alisphenoid bone sends a wing over the middle ear.

The teeth are not easy to interpret (Fig. 332). The incisors are more numerous than in placentals, as many as 5 on each side in the upper and 3 in the lower jaw. Of the cheek teeth only one, the third of the series, is replaced in modern forms, and if this is regarded as the last premolar we have a dentition of 3 premolars and 4 molars, as against the 4 and 3 of a typical placental. However, fossil marsupials with three replacing teeth have been found (p. 566) and the significance of the peculiar condition of the modern forms remains obscure; it may be connected with the specialization of the mouth of the young for life in the pouch. The cusps of the teeth of many marsupials (e.g. opossum) show a close approach to the presumed primitive plan (p. 549), but in addition to the main triangle other cusps are present, especially on the outside. As in placentals the herbivorous types of marsupial develop grinding surfaces on the teeth.

The general plan of the muscles and backbone is essentially that found in placentals and has been greatly changed from the reptilian or monotreme condition. There are no cervical ribs. The thoracic region consists of about 13 rib-bearing vertebrae, as in placentals, and there are about the usual 7 lumbars. The pectoral girdle shows no interclavicle, but the clavicle remains large. The coracoid is reduced,

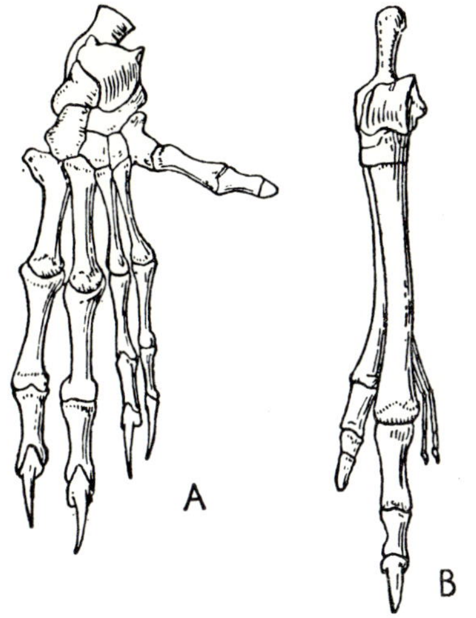

FIG. 333 FIG. 334

FIGS. 333–4 Hind foot of A, *Trichosurus*, and B, *Macropus*, showing elongation of the toes of the latter and reduction of the 1st, 2nd, and 3rd toes. (From Zittel, *Text-book of Palaeontology*, Macmillan, after Dollo.)

FIG. 335. Pouch young of *Dasyurus* about 10 days old. (After Hill and Osman Hill.)

as in placentals, and the scapula enlarged and provided with a spine. In fact all the developments of the dorsal region of the girdle that are typical of the mammalian method of locomotion have taken place. In the pelvic girdle there are epipubic bones, reminiscent of those of monotremes; they take the special stresses produced by the abdominal muscles and pouch, but are reduced in the fully terrestrial and quadrupedal Tasmanian wolf. The hands usually carry five digits, armed with claws, but the number of toes is often reduced and they may bear hoof-like structures (Figs. 333–4).

The Müllerian ducts are paired and differentiated into upper 'uterine' and lower 'vaginal' portions. In many species the latter are provided with median diverticula, the two meeting in the mid-line as a median vagina (Fig. 336). This ends blindly until the young are about to be born, when an opening is formed through the tissues—the pseudo-vagina or birth canal. This may then close until the next parturition. As in monotremes the rectum and urinogenital sinus open together at a common cloaca, though this is not very long, longer in the female than the male (Fig. 337). There is a well-developed penis, often bifid at the tip, in which case the clitoris is also double, the

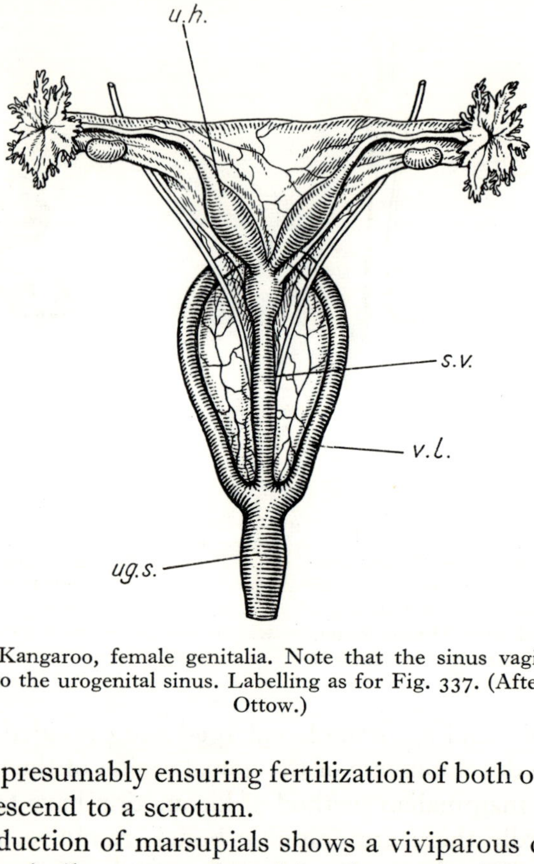

FIG. 336. Kangaroo, female genitalia. Note that the sinus vaginalis opens directly into the urogenital sinus. Labelling as for Fig. 337. (After Brass and Ottow.)

arrangement presumably ensuring fertilization of both oviducal tubes. The testes descend to a scrotum.

The reproduction of marsupials shows a viviparous condition that is not closely similar to any found in placentals. The egg is rather yolky and covered with albumen and a membrane; cleavage is very unequal. In some forms (*Dasyurus*) there is a contact of the vascular wall of the yolk-sac with the somewhat hypertrophied uterine wall (omphaloidean placenta). Only in the bandicoot, *Perameles*, does the allantois develop a nutritive function to some extent. In most marsupials no placental arrangement develops at all, instead, uterine milk may be taken up by the yolk-sac. The embryos are born very young, as little as 8 days from conception in the opossum (Fig. 335). They crawl along a track of saliva that is laid between the cloaca and the pouch by the mother with her tongue. They become attached to the teats and remain for a long time in the pouch.

This is not by any means a primitive plan of development, it involves many specializations. To make the journey to the pouch the fore-limbs and their nervous centres are precociously developed, being fully functional at birth, when the hind-limbs are mere buds. The method of suckling also involves special developments of both mother

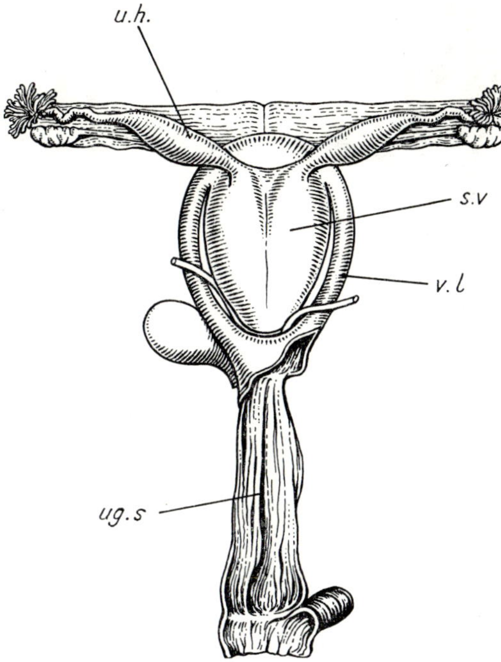

FIG. 337. *Phalanger* sp. Female genitalia. Note caudal blind ending of the sinus vaginalis.

u.h. uterine horn; *s.v.* sinus vaginalis; *v.l.* vagina lateralis; *ug.s.* urogenital sinus. (After Brass and Ottow.)

and foetus, so that milk is injected into the latter without choking it. The sides of the lips grow together round the teat, which is thrust far back in the pharynx, the larynx extending forwards into the nasal passage. Milk is pressed out by a special muscle (homologue of the male cremaster) attached to the epipubic bones.

The brain shows some reptilian features. The cerebral hemispheres are small for a mammal, and the olfactory bulbs large. The hemispheres are not prolonged backwards over the cerebellum, which is itself small and simple. There are dorsal (hippocampal) and ventral (anterior) commissures but no corpus callosum. The cochlea of the ear is spirally coiled.

2. Classification of marsupials

Marsupials are often divided into two suborders, the more primitive insectivorous or carnivorous polyprotodonts, found outside as well as within Australasia, and the diprotodonts, more specialized and restricted. The distinction is based on the presence of more than three

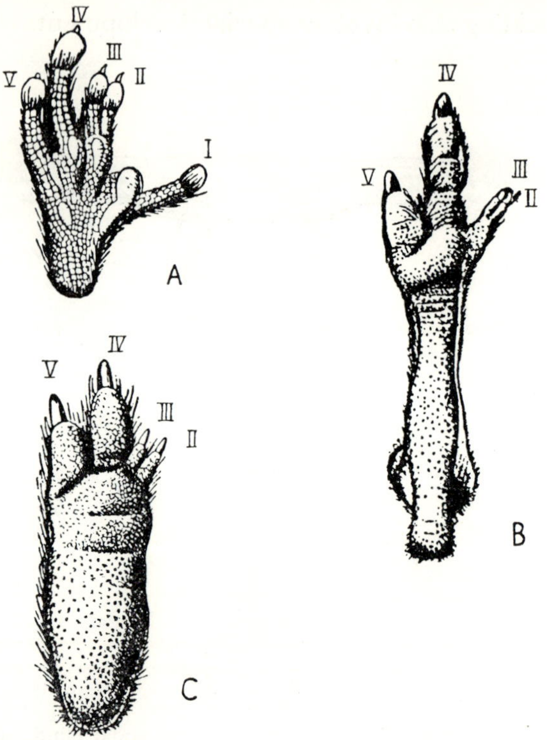

FIG. 338. Hind feet of marsupials. A, opossum, with grasping hallux, arboreal; B, kangaroo, without hallux, digits II and III syndactyl, cursorial; C, tree kangaroo, without hallux, arboreal (secondarily). (After Bensley, from Lull, *Organic Evolution*, copyright 1914, 1929 by The Macmillan Company, and used with their permission.)

pairs of incisors in each jaw in the first group while in the other only two remain in the lower jaw and protrude. A further interesting feature is that in most diprotodonts the second and third digits of the hindlimb are fused to make a comb for cleaning the hair (Fig. 338). This character is absent in nearly all polyprotodonts, which are hence said to be didactylous, and this is no doubt the primitive condition. However, the comb-like condition (syndactyly) is found not only in diprotodonts but also in the polyprotodont bandicoots; conversely the curious South American opossum-rat (*Caenolestes*), though didactylous, has diprotodont teeth.

The fact is that marsupials, having radiated perhaps over 60 million years ago, present us today with a number of distinct types of organization. Simpson groups them into six superfamilies and it is perhaps better not to attempt any higher grouping of these. The absence of Australian Tertiary deposits before the Pleistocene increases the difficulties of study of marsupial phylogenesis.

Infraclass 2. Metatheria

Order Marsupiala

Superfamily 1. Didelphoidea. Upper Cretaceous–Recent. Europe and America

**Eodelphis*. Upper Cretaceous, N. America; *Didelphis*, opossum, Pliocene–Recent, N. and S. America; *Chironectes*, water opossum, Central and S. America

*Superfamily 2. Borhyaenoidea. Palaeocene–Pliocene. S. America

**Thylacosmilus*, Miocene; **Borhyaena*, Oligocene–Miocene

Superfamily 3. Dasyuroidea. Pleistocene–Recent. Australasia

Dasyurus, native cat; *Sarcophilus*, Tasmanian devil; *Thylacinus*, Tasmanian wolf; *Myrmecobius*, banded ant-eater; *Notoryctes*, marsupial mole; *Sminthopsis*, pouched mouse

Superfamily 4. Perameloidea. Pleistocene–Recent. Australasia

Perameles, Bandicoot

Superfamily 5. Caenolestoidea. Eocene–Recent. S. America

**Palaeothentes* (= **Epanorthus*), Oligocene–Miocene; *Caenolestes*, opossum-rat, Ecuador, Colombia, Peru

Superfamily 6. Phalangeroidea. Pliocene–Recent. Australasia

Trichosurus, Australian opossum; *Petaurus*, flying opossum; *Phascolarctos*, koala bear; *Vombatus*, wombat; *Macropus*, kangaroo; *Bettongia*, rat kangaroo; **Diprotodon*, Pleistocene; **Thylacoleo*, Pleistocene.

3. Opossums

The opossums (Didelphoidea) were the earliest group to appear and the other families have probably evolved from them, with syndactyly appearing twice. They are arboreal, mainly nocturnal and omnivorous or insectivorous animals, with a prehensile tail, occurring over the southern United States and Central and South America (Fig. 339). The pouch is generally absent. Similar forms are found back to the Upper Cretaceous (**Eodelphis*) and the American opossums are certainly the closest of living marsupials to the ancestors of the group, perhaps they are the least modified of all therian mammals. *Chironectes* is a related South and Central American otter-like form, with webbed feet.

FIG. 339. American opossum (*Didelphis*). (From photographs.)

FIG. 341. Tasmanian devil (*Sarcophilus*). (From photographs.)

FIG. 342. Tasmanian tiger cat (*Dasyurus*). (From photographs.)

FIG. 340. Tasmanian wolf (*Thylacinus*). (From photographs.)

The bandicoots (*Perameles*) of Australia and New Guinea are burrowing animals, rather rabbit-like but mainly insectivorous. They have a polyprotodont dentition, quadritubercular grinding molars, syndactyly of the hind toes, and an allantoic placenta. Their affinities are uncertain.

4. Carnivorous marsupials

The Dasyuroidea include some nocturnal carnivorous polyprotodonts that show remarkable convergence with placental carnivores, the teeth being modified for cutting flesh in a way similar to the carnivore

FIG. 343. Eastern native cat (*Dasyurus*). (From photographs.)

FIG. 344. Banded ant-eater (*Myrmecobius*). (From photographs.)

FIG. 345. Marsupial mole (*Notoryctes*). (From photographs.)

FIG. 346. Pouched mouse (*Sminthopsis*). (From photographs.)

FIG. 347. Kangaroo (*Macropus*). (From photographs.)

carnassial. *Thylacinus* (Fig. 340), the Tasmanian wolf, is now nearly or quite extinct. *Sarcophilus*, the Tasmanian devil (Fig. 341), is rare, but (*Dasyurus* (native cat) (Figs. 342 and 343) includes several species of cat-like creatures common in Australia and New Guinea. Successful carnivorous marsupials formerly existed in South America. They may perhaps be related to the Dasyuridae; *_Borhyaena_ shows many similarities to *Thylacinus*. **Thylacosmilus* was a Miocene sabre-tooth, the size of a panther, whose huge upper canine and other features closely parallel the placental **Smilodon* (p. 689). It is said that in some of these borhyaenoids, two or more milk teeth were replaced; if true this suggests that the condition in modern marsupials is secondary.

5. Marsupial ant-eaters and other types

Myrmecobius is an ant-eating form, with elongated snout (Fig. 344). *Notoryctes*, the marsupial mole (Fig. 345), from South Australia, has reduced eyes, well-developed fore-limb, fused cervical vertebrae, and many features suiting it for burrowing and feeding upon ants. Its pouch opens backwards. *Sminthopsis*, the pouched mice (Fig. 346), are small marsupials occupying the niche taken in other parts of the world by the shrews.

Caenolestes, the opossum-rat of the forests of the Andes, is an interesting shrew-like creature, with the polyprotodont number of incisors but procumbent lower incisors, resembling those of diprotodonts. There is no syndactyly. It is the survivor of a group formerly abundant in South America, some with teeth similar to those of multituberculates.

6. Phalangers, wallabies, and kangaroos

The diprotodont marsupials form a compact group (leaving out *Caenolestes*) of Australian forms, here included as the superfamily Phalangeroidea. Their fossil history is little known, but they have become specialized for various modes of life, mainly as herbivores, in Australia and the neighbouring islands. The kangaroos and wallabies (Macropodidae) (Fig. 347) have become mostly terrestrial and developed a bipedal method of progression, involving modification of the ilia and thigh muscles, for whose attachment the tibia bears a marked anterior crest. The foot gains increased leverage by elongation of the metatarsal of digit 4. Digits 2 and 3 are very small and syndactylous. There are several modifications for a herbivorous diet; the single pair of lower incisors is directed forwards and their sharp inner edges can be moved in such a way as to cut grass like shears. This condition recalls that of Rodentia (p. 655) and, as in that group, a special transverse muscle is developed (m. orbicularis oris), but in this case it is part of the facial musculature and innervated from the seventh nerve, whereas the analogous muscle of the rodents is a part of the mylohyoid and innervated from the trigeminal. The molar teeth are modified for grinding, by the fusion of the cusps to make two transverse ridges, recalling those of ruminants. The stomach has a special sacculated non-glandular chamber, presumably allowing digestion by symbionts. *Bettongia* and other 'rat kangaroos' are terrestrial and bipedal jumpers, and also have a prehensile tail (Fig. 348).

FIG. 348. Rat kangaroo (*Bettongia*). (From photographs.)

FIG. 349. Cuscus (*Phalanger*). (From photographs.)

FIG. 351. Flying opossum (*Petaurus*). (From photographs.)

FIG. 350. Koala bear (*Phascolarctos*). (From photographs.)

The Australian opossums or phalangers are less modified than the kangaroos and are arboreal animals, with a prehensile tail and various special modifications, mostly for herbivorous diets; *Trichosurus* is the common phalanger, with four-cusped upper molars. *Phalanger*, the cuscus (Fig. 349), eats mainly leaves. The koala or native bear, *Phascolarctos* (Fig. 350), lives on the leaves of *Eucalyptus*; it has cheek-pouches, an enlarged caecum, and a reduced tail. Three distinct genera of phalangers have developed extensions of the skin for purposes of soaring; *Petaurus* (Fig. 351) is the best known of these flying phalangers. *Vombatus*, the wombat (Fig. 352), is a large, burrowing, tailless animal, with rodent-like grinding teeth; it eats roots. **Diprotodon* was a very large marsupial, of the size and form of a rhinoceros, which lived in Australia in the Pleistocene. **Thylacoleo* was a marsupial lion in which the incisors were developed as fangs.

FIG. 352. Wombat (*Vombatus*). (From photographs.)

7. Significance of marsupial isolation

The explanation of the curious distribution of the marsupials remains uncertain. The fossil evidence suggests a palaearctic distribution in the Eocene, followed by radiation in South America and Australia with later reduction in the former. An antarctic bridge has, however, been postulated by some authors.

The opossums show that marsupial life can continue effectively in competition with placentals. But it can hardly be an accident that the diversification of marsupials in Australia has been accomplished in isolation. It is true that there are 108 species of placentals in Australia, as against 119 marsupials, but the placentals are almost all bats (40 species) and murid rodents. The marsupials, on the other hand, have become differentiated into numerous types, arboreal, fruit-eating, grazing, gnawing, digging, burrowing, ant-eating, insectivorous or carnivorous, in each case with appropriate structure. It will be interesting to see how this assemblage stands up to competition with placentals in the future. Carnivores, ruminants, lagomorphs, rodents, and primates have recently become firmly established in Australia and it can hardly be an accident that some of the corresponding marsupial types are already becoming rare or extinct.

XXI

EVOLUTION OF PLACENTAL MAMMALS AND ITS RELATION TO THE CLIMATIC AND GEOGRAPHICAL HISTORY OF THE CENOZOIC

1. Eutherians at the end of the Mesozoic

SEVERAL different lines of evidence converge to show that all the eutherians (placentals) have been derived from small, perhaps nocturnal, insectivorous or omnivorous animals, living in the Cretaceous period about 100 million years ago. Many features of marsupials and placentals alike suggest origin from a small Cretaceous shrew-like form, perhaps itself descended from some animal like the Jurassic pantotheres. It is especially interesting, therefore, that fossil evidence is now available to show that both opossums (p. 563) and placental insectivores (p. 583) existed in the Cretaceous. We may be reasonably sure that the population from which those groups were derived resembled both of these animals, which are indeed basically similar. At this Cretaceous period the arrangements for nourishing the young were presumably not yet fully developed and in the marsupials (Metatheria) they have remained at a simple level, little above ovo-viviparity, though the condition in *Perameles* makes it doubtful whether an allantoic placenta has been lost by the other marsupials.

The stock that was to give rise to the eutherians was therefore already differentiated in the Cretaceous; as the revolution proceeded these animals began to flourish and to develop into several divergent populations. The only placental types known to have lived during Cretaceous times were insectivores; yet by the very beginning of the Cenozoic period, in the formations known as the Palaeocene, several different types of placental are found.

2. The end of the Mesozoic

It is not easy to discover any close connexion between the climatic changes and this early flowering of the placentals. Throughout the period of the Cretaceous earth movements the land gradually became higher and the climate probably colder, at least in some parts of the world. Indeed, this process had been going on intermittently throughout the Mesozoic. In the Permian there was a series of glaciations even more profound than those of the Pleistocene and it has been suggested that this may have been responsible for the production of mammal-like characteristics among the synapsids (p. 538). It is tempting to make similar suggestions about the development of placental mammals at the end of the Cretaceous. The presence of cold uplands may indeed have been responsible for such success as the multituberculates and other Mesozoic mammals achieved during the Cretaceous. But this slow revolution does not explain, by itself, the success of the placentals, because they only became differentiated into varied groups towards the end of the Cretaceous. There is some evidence that this latter time was not very cold, and it will be remembered that some dinosaurs persisted to the very close of the Mesozoic. On the other hand, pterodactyls had died out earlier and so had some groups of the dinosaurs.

Although there may well have been large climatic changes at the end of the Cretaceous period it is unwise to make simple statements about their relation to the evolution of the mammals. The whole Cretaceous period lasted for more than 60 million years and we cannot trace in detail the numerous changes of climate that must have taken place, probably in different directions in different parts of the world. Even the marine faunas were affected; for instance, the ichthyosaurs and plesiosaurs died out before the end of the Cretaceous and were followed by the mosasaurs (p. 409), which had a sudden period of success lasting for some millions of years.

The change of fauna at the end of the Cretaceous was as remarkable for the animals that appeared as for those that were lost. The remaining dinosaurs died out on land, as did the mosasaurs, and also ammonites and belemnites, in the sea. At the same time there appeared not only numerous placental mammals, but also a great variety of true birds, and in the seas teleostean fishes and cephalopods of modern type.

The apparent suddenness of the change may be deceptive. Our knowledge of the conditions at the beginning of the Cenozoic period is fragmentary. It was a time when the land stood high, at least in the regions we know best. A great part of the continental shelf was above water, and therefore producing few fossils. Most of our information about the fossils laid down at this time comes not from marine beds, but from the 'Palaeocene' continental deposits, known chiefly in America. These deposits lie on top of undoubted Cretaceous, dinosaur-containing beds, and they contain a variety of placental mammals. At the upper end these Palaeocene deposits are continuous with beds that can be correlated with the previously known Eocene marine beds in other parts of the world, at which level there is a still wider variety of placentals.

It is clear, therefore, that there was a period of time of unknown duration, following the disappearance of the dinosaurs, during which the placentals were becoming differentiated. The old belief in the sudden appearance of various types in the Eocene was, therefore, perhaps chiefly an artefact of the conditions of fossilization at the time.

3. Divisions and climates of the Tertiary Period

The whole Tertiary or Cenozoic period is now divided as follows:

Epoch	*Time from beginning of epoch to present (millions of years)*	*Per cent modern species*
Recent	..	100
Pleistocene ('most recent')	1	90
Pliocene ('more recent')	10	50
Miocene ('less recent')	25	20
Oligocene ('few recent')	40	10
Eocene ('dawn recent')	60	5
Palaeocene ('ancient recent')	70	0

The names were originally given by Lyell to indicate the percentages of modern species of shells; for curiosity these latter are given (approximately) in the third column. During the Palaeocene, Eocene, and Oligocene the mountains raised during the Laramide revolution were eroded. The climate was cold in the palaearctic region early in the period, but it later became warmer and damper and there were probably extensive forests during the later Eocene and Oligocene. Palms then grew over much of Europe and there were forests where are now immense areas of steppe. The climate probably showed marked seasonal changes, at least in many parts of the world, and deposits of the time often show a lamination ('varving') that indicates an alternation of wet and dry seasons. During this first part of the Cenozoic there was some invasion of the land by water, but this was on a much smaller scale than during the inundations of earlier periods; indeed, the main land-masses have remained approximately constant throughout the Tertiary.

During the early Miocene there was a time of intense crustal disturbance, the Cascadian revolution. In this many of the earth's main mountain chains, the Rockies, Andes, Alps, Himalayas, as well as humbler ranges such as the English chalkdowns, were raised into their present form. This revolution marks the subdivision of the Cenozoic into two periods. After it there was a further denudation during the Miocene and Pliocene, with some renewed sea invasion. The uplifting of the land during the Miocene probably produced arid conditions unfavourable to the growth of forests, and at this time there emerged several types of animal suitable for life on open prairies. Conditions probably became gradually colder throughout the Pliocene, and the subsequent Pleistocene epoch was characterized by the extensive glaciations of the 'Ice Ages', occurring in at least four peaks of maximum cold, during each of which the ice caps advanced over large parts of the continents (p. 648).

Study of the history of the Cenozoic in the palaearctic region therefore gives us some idea of the climatic changes that have taken place, and we can try to correlate these with the succession of types of animal life. On the other hand, it is important not to be over-impressed with any simple account of climatic changes. The periods of time involved are enormously long and it is unsafe to assume that conditions remained constant for any length of time that can be easily imagined, or even that conditions varied at a constant rate. For example, in Yellowstone Park, U.S.A., there are exposures of the remains of Eocene tree-trunks and these are arranged in layers, showing that at least twenty forests grew up and were covered by volcanic ash one after the other. Each of these eruptions presumably produced a major revolution for the animals and plants in the area concerned; we do not know how wide that area may have been. Obviously no broad generalization about the presence of 'humid conditions and forests' throughout the Eocene can give us any clear picture of the ecological conditions even in one area. The geological history shows us that conditions were continually changing, though perhaps often at a rate very slow in comparison with the duration of animal lives. We can well imagine that these slow changes were responsible for producing new conditions and hence new types of life, but the data of the rocks are too obscure to show us the detailed circumstances of the emergence of any particular type.

4. Geographical regions

Although the main land-masses have varied little during the Tertiary period, there have been considerable changes in the opportunities for communication between them, both by the making and breaking of narrow land-bridges and by the development of sharp temperature gradients and desert areas. Zoogeographers divide the land into six regions (Fig. 353), Palaearctic (Central and N. Asia, Europe), Nearctic (N. America), Neotropical (Central and S. America), Ethiopian (Africa), Oriental (S. Asia and E. Indies), and Australasian. The Palaearctic and Nearctic have similar climates and have been

FIG. 353. Polar projection, showing the zoogeographical realms. (From Lull, *Organic Evolution*, copyright 1917, 1929 by The Macmillan Company, and used with their permission.)

connected several times during the Tertiary by an Alaska–Siberian bridge; they are therefore often grouped together as Holarctic. The Ethiopian region is now separated from the Palaearctic by the sudden change of temperature and desert conditions of North Africa, but in earlier times the animals of the two regions mixed freely. The same is true of the Palaearctic and Oriental regions. South America was connected with the Nearctic region in the Eocene, but the bridge was then broken until the Pleistocene; the Neotropical land faunas therefore differ considerably from the others. Australasia east of Wallace's line has been separated since the late Cretaceous.

The land masses have therefore been connected, in the main, in the north and separated in the south. In other words, America, North and South Africa, and the Oriental regions are capes projecting into the sea from the central Palaearctic (Eurasian) land-mass. There is evidence that many forms of animal life evolved in this central area and migrated away towards the extremities. Several of the types that evolved earlier remain as vestiges at these 'ends of the world', long after newer types have replaced them nearer to the centre. The lung-fishes are a conspicuous example of this and at another extreme it is perhaps no accident that Australian Aborigines and South African Bushmen are among the more primitive of men. This is the most probable explanation of the similarity in the fauna between these southern continents, but many suggestions of Antarctic land bridges and continental drift have been put forward, in particular one connecting South Africa and South America in the early Tertiary. For example, we have to explain the appearance of porcupine-like animals (p. 660) in Africa and South America at the same time (Oligocene).

5. The earliest eutherians

Fossil placentals found in the Cretaceous period have all been insectivorans (p. 583). Those of the Palaeocene include also some that can be referred to the primates and to the carnivora, but they are very unlike modern members of those groups and could almost equally well be classed as Insectivora. Similarly, the ungulates and various other types that appeared in the Palaeocene could have had an insectivoran ancestry. Evidently, therefore, during the late Cretaceous and Palaeocene, the original placental stock was branching out into various habitats, and the branching was rapid. Simpson's careful classification recognizes twenty-six orders of placentals (p. 577), and nearly all of these had become distinct by the Eocene. Ten of them have since become extinct. Evidently the great period of mammalian expansion was in the earlier part of the Cenozoic and the group may be considered to have passed its peak for the present. Only the bats, rodents, lagomorphs, and perhaps the primates and carnivores can be considered really successful land animals at the present time; to these we may add the whales in the sea. The Artiodactyla are also abundant, but most of the remaining placental orders are today poorly represented in numbers.

In order to gain a comprehensive general understanding of the great placental history during its 80 million or so years duration, we will first give a technical definition of a placental, then the characteristics of the earlier types, and finally try to list some of the tendencies to change that are widely found in different groups. For simplification we can attempt some grouping of the great list of orders, before dealing with them individually.

6. Definition of a eutherian (placental) mammal

The distinguishing characteristics of placentals are often listed somewhat as follows: The young are retained for a considerable time in the uterus and nourished by means of an allantoic placenta; there is no pouch or epipubic bones. In the skull there is usually a separate optic foramen, no palatal vacuities, and no in-turned angle of the jaw. The tympanic bone is either ring-like or forms a bulla, there is never an alisphenoid bulla. The brain has large cerebral hemispheres connected by a corpus callosum. There is no cloaca. The dental formula is $\frac{3 \cdot 1 \cdot 4 \cdot 3}{3 \cdot 1 \cdot 4 \cdot 3}$, or some number reduced from this.

Many of these are obviously small points of formal definition, artificially abstracted for the purpose of classification. They are not really satisfactory as a definition of the life of a placental, such as we may hope to have in a more developed biology. The early population of Cretaceous insectivores presumably possessed most of these features, and showed characteristics that are common to all the earlier mammals. Among these we may list (1) small size; (2) short legs with a plantigrade type of foot, having five digits; (3) the full eutherian number of teeth $\frac{3 \cdot 1 \cdot 4 \cdot 3}{3 \cdot 1 \cdot 4 \cdot 3}$, the molariform teeth being based on the tuberculo-sectorial pattern (p. 549); (4) long face and tubular skull, enclosing a relatively small brain.

7. Evolutionary trends of eutherians

In the descendants of these early mammals we can recognize changes in each of these four sets of characters; changes occurring, independently, in some members at least of all the later lines. (1) Many of the mammals became larger. Increase in size seems to be advantageous to many animal types and may be connected with the advantages of a large brain storing much information during the life of the animal and so allowing slow reproduction. It is presumably especially so for warm-blooded animals in cold climates, since it reduces the relative area of heat loss, though also introducing new problems of obtaining adequate amounts of food. This may have to be finely ground by tooth-surfaces whose increase with size is less rapid than that of the weight of tissue they must support. (2) The limbs became longer and specialized in various ways for locomotion, often by raising the heel off the ground, so that the animals came to walk on the digits instead of the sole of the foot (Fig. 354) and the number of toes became reduced. (3) The number of teeth became reduced and their shape specialized, often

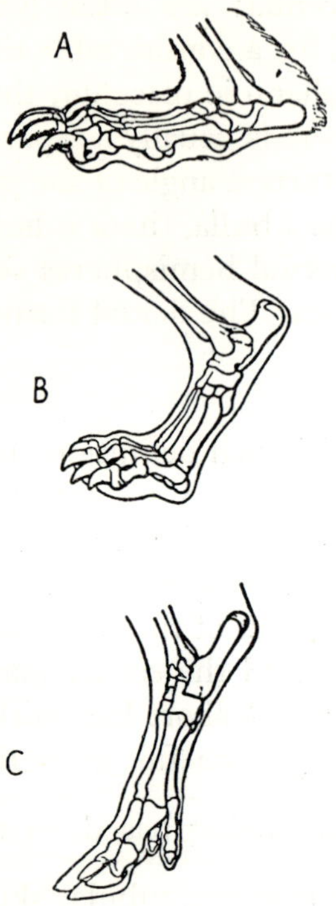

FIG. 354

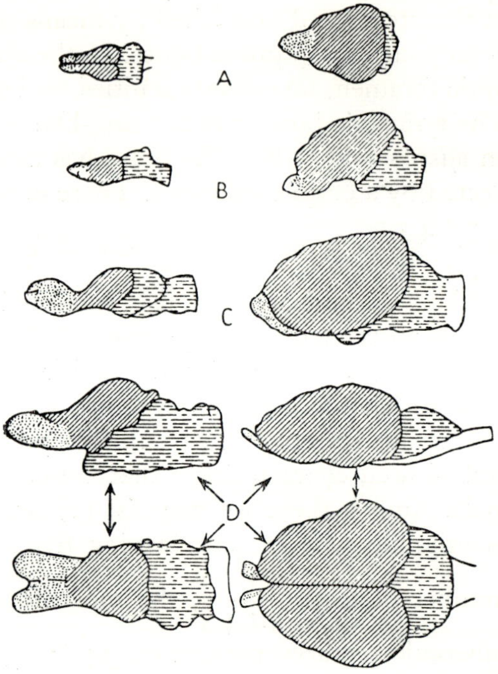

FIG. 355

FIG. 354. Postures of the foot in mammals. A, plantigrade (bear); B, digitigrade (hyaena); C, unguligrade (pig). (From Lull, after Pander and D'Alton.)

FIG. 355. Comparison of pairs of brains of archaic (left) and modern (right) mammals of similar size. Olfactory lobes dotted, cerebral hemispheres oblique lines, cerebellum and medulla, dashes. A, **Arctocyon* and *Canis*; B, **Phenacodus* and *Sus*; C, **Coryphodon* and *Rhinoceros*; D, *Uintatherium* and *Hippopotamus*. (After Osborn, from Lull, *Organic Evolution*, copyright 1917, 1929 by The Macmillan Co., and used with their permission.)

by the addition of cusps and their fusion to make transverse or longitudinal grinding ridges in herbivorous animals or cutting blades in carnivores. (4) The brain of the earlier mammals resembled that of reptiles (Fig. 355); later forms showed increasing development of the non-olfactory part of the cortex, increase of the frontal lobes, and other changes probably correlated with more complicated behaviour and better memory. It has often been supposed that the brain becomes relatively larger in later forms, but Edinger has shown by study of fossil horse brains that as the body size increases the brain-body ratio decreases.

8. Conservative eutherians

Much of this change was the result of adopting a fully terrestrial and often herbivorous mode of life, in place of the earlier arboreal and insectivorous or carnivorous one. Changes in these four directions have taken place in many separate mammalian lines, but the evidence contradicts the thesis that they are the result of some force of orthogenesis, driving the animals infallibly along. In nearly every group there are examples of some animals that have remained nearly unchanged for long periods, e.g. opossums and shrews since the Cretaceous (80 million years), lemurs and tarsiers since the Eocene (50 million years), pigs and tapirs since the Oligocene (35 million years), and deer since the Miocene (20 million years), to name only a few.

It is important to study such animals in which there has been little change; they form the 'controls', and may enable us to recognize the factors inducing change when it does occur. Moreover, in many further lines there has been change in some but not all of the above directions, for instance, many of the most successful mammals have remained small. There may even be changes in the directions opposite to those listed, for instance, some edentates and whales have more than the original number of teeth. Mammals do not commonly decrease in size during evolution (but they may do so), and they probably never reacquire lost digits, though in a few claws have reappeared after they had been lost.

9. Divisions and classification of Eutheria

Infraclass 3. Eutheria
- Cohort 1. Unguiculata
  - Order 1. Insectivora
  - Order 2. Chiroptera
  - Order 3. Dermoptera
  - *Order 4. Taeniodonta
  - *Order 5. Tillodontia
  - Order 6. Edentata
  - Order 7. Pholidota
  - Order 8. Primates
- Cohort 2. Glires
  - Order 1. Rodentia
  - Order 2. Lagomorpha

Classification (*cont.*)

Cohort 3. Mutica
  Order Cetacea
Cohort 4. Ferungulata
  Superorder 1. Ferae
    Order Carnivora
  Superorder 2. Protungulata
    *Order 1. Condylarthra
    *Order 2. Notoungulata
    *Order 3. Litopterna
    *Order 4. Astrapotheria
    Order 5. Tubulidentata
  Superorder 3. Paenungulata
    Order 1. Hyracoidea
    Order 2. Proboscidea
    *Order 3. Pantodonta
    *Order 4. Dinocerata
    *Order 5. Pyrotheria
    *Order 6. Embrithopoda
    Order 7. Sirenia
  Superorder 4. Mesaxonia
    Order Perissodactyla
  Superorder 5. Paraxonia
    Order Artiodactyla

For purposes of phylogenetic study as well as classificatory convenience it is desirable to attempt to discover how the original eutherian population became divided at its Cretaceous origin, and whether there were main trunks of the placental tree. By Eocene times most of the existing orders were already well established (Fig. 356) and it is often stated that the branching of the population occurred relatively rapidly, though it is doubtful if we may use this term for a process occupying perhaps 30 million years during the late Cretaceous and Palaeocene! This early expansion into varied branches, occurring at a time when few fossils were being formed, makes it difficult to discover the outlines of the main divisions. Nevertheless, careful piecing together of evidence suggests grouping of the twenty-six eutherian orders into four main cohorts, and the division corresponds in the main to the classification originally proposed by Linnaeus in 1766 on a basis of the foot structure. The Unguiculata include orders in which the original characteristics of the mammalian type have been

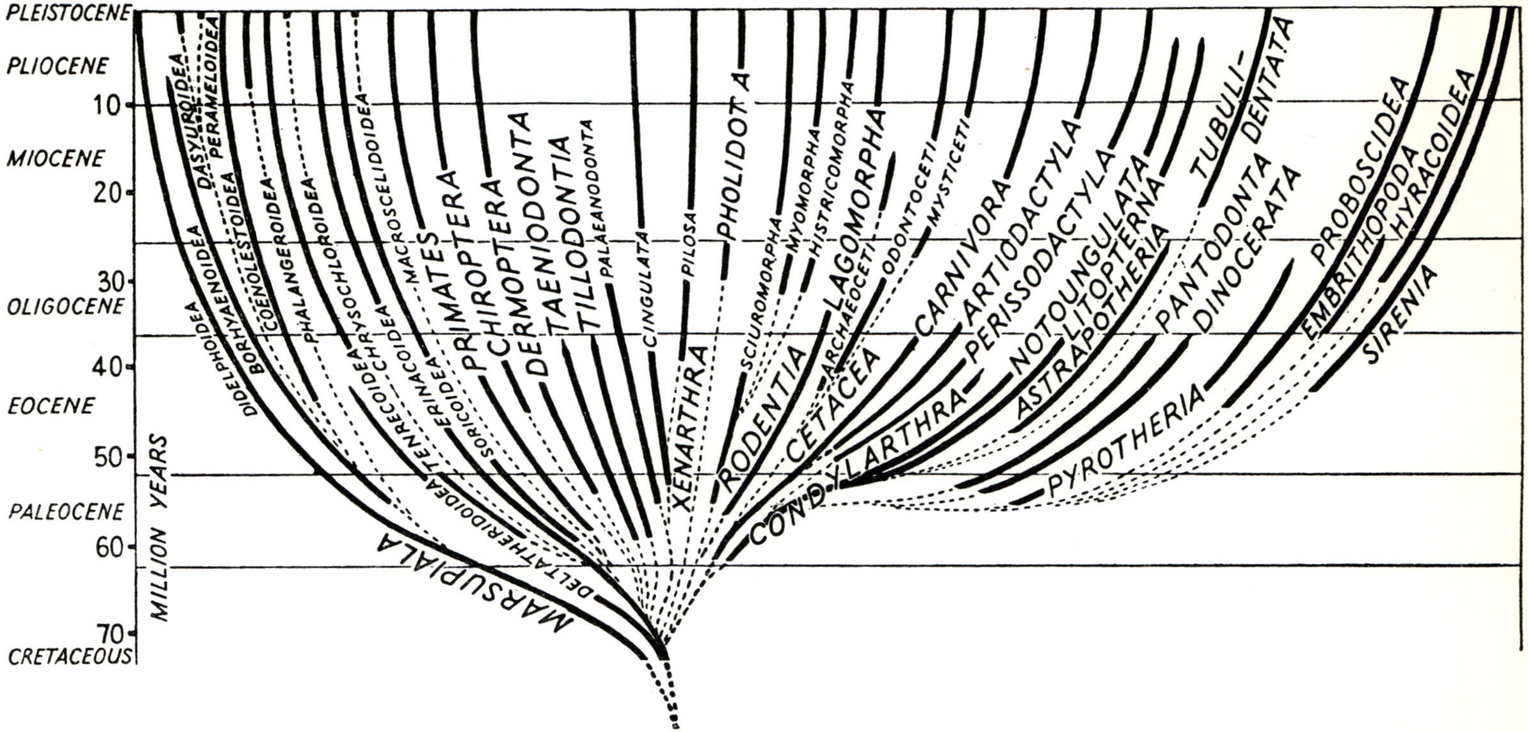

FIG. 356. Chart showing the probable affinities of the orders of Mammals (large capitals) and the lesser divisions of some of the more primitive orders (smaller capitals).

largely preserved, the Insectivora themselves and the Chiroptera (bats), the Primates and the Edentata (sloths, ant-eaters, and armadilloes), all of which can be easily and directly derived from the insectivores. Here also probably belong the scaly ant-eater, *Manis* (Pholidota), and the extinct *Taeniodontia and *Tillodonta, which survived only for a short time and of which little is known and they will not be mentioned further. The Rodentia and the rabbits and hares (Lagomorpha) are often also considered to belong here, but they appear fully developed in the Eocene and must have diverged very early and are therefore placed alone in a separate cohort Glires. Similarly the whales (Cetacea) have certainly been distinct since the Eocene. Their affinities are quite obscure, and they are classed as a separate cohort named by Linnaeus Mutica (most inappropriately since they communicate by sounds).

All the remaining mammals show signs of a common origin and Simpson suggests grouping them together as a cohort Ferungulata. It has long been realized that the hoofed animals include two distinct types, those with an uneven number of toes, Perissodactyla, and those with even toes, Artiodactyla. The former can be derived from the Palaeocene and Eocene animals known as *Condylarthra. The Carnivora seem at first sight to have no similarity to either of the ungulate types, but it has been suspected for some time that the ancestral Carnivora, the *Creodonta, resembled the earliest artiodactyls. Further study has shown that creodonts and condylarths are often so alike as to be hardly separable (they are also very like Insectivora). Simpson's suggested cohort Ferungulata recognizes the existence of this common creodont-condylarth stock, perhaps in early Palaeocene times. The cohort may then conveniently be subdivided into five superorders, Ferae for the Carnivora, Protungulata (='first ungulates') for the condylarths, the extinct South American ungulates (*Liptopterna and *Notoungulata), the obscure *Astrapotheria and Tubulidentata (*Orycteropus*, the Cape ant-eater). The third superorder Paenungulata (='near ungulates') includes the elephants (Proboscidea), hyraxes (Hyracoidea), and sea cows (Sirenia), as well as the extinct *Pantodonta, *Dinocerata, *Pyrotheria, and *Embrithopoda. The fourth superorder Perissodactyla is then made to include only the horses, tapirs, and rhinoceroses; and the fifth superorder Artiodactyla the pigs, camels, and ruminants. This system gives us a means of grouping that is phylogenetically reasonably accurate and also conveniently close to the usually accepted uses of familiar names.

XXII

INSECTIVORES, BATS, AND EDENTATES

1. Order 1. Insectivora

*Suborder 1. Deltatheridioidea. Upper Cretaceous–Eocene. Asia and N. America

**Deltatheridium*, Cretaceous, Asia

Suborder 2. Lipotyphla

Superfamily 1. Erinaceoidea. Paleocene–Recent. Holarctic, Oriental, Africa, N. America

Echinosorex, moonrat, Asia; *Erinaceus*, hedgehog, Miocene–Recent, Old World

Superfamily 2. Soricoidea. Paleocene–Recent

Family Soricidae. Oligocene–Recent. Holarctic, Africa

Sorex, shrew, Miocene–Recent; *Neomys*, water shrew, Eurasia

Family Talpidae. U. Eocene–Recent. Holarctic

Talpa, mole, Miocene–Recent; *Myogale*, desman, water mole

Family Solenodontidae. Recent. W. Indies

Solenodon, alamiqui

Family Tenrecidae. Miocene–Recent. Africa, Madagascar

Potamogale, otter shrew, W. Africa; *Tenrec* (= *Centetes*), tenrec, Madagascar

Family Chrysochloridae. Miocene–Recent. Africa

Chrysochloris, golden mole, S. Africa

Suborder 3. Menotyphla

*Superfamily 1. Leptictoidea. Upper Cretaceous–Oligocene

**Zalambdalestes*, Cretaceous, Asia; **Ictops*, Oligocene, N. America

Superfamily 2. Macroscelidoidea. Miocene–Recent. Africa

Macroscelides, elephant shrew

Superfamily 3. Tupaioidea. Oligocene–Recent. Asia

**Anagale*, Oligocene; *Tupaia*, tree shrew; *Ptilocercus*, pen-tailed tree shrew

Modern insectivores are mostly small, nocturnal animals, maintaining, possibly because of some special habitat, the earliest mammalian features. They have persisted with little change since the Cretaceous. Apart from the retention of primitive characters, they have little else in common. With them have been classified a number of early, primitive eutherians that cannot be placed in any other order.

The order Insectivora is thus a convenience rather than a natural group.

The full dentition of $\frac{3.1.4.3}{3.1.4.3}$ is usually preserved, and the cusps have diverged little from the tritubercular-tuberculo-sectorial pattern but two extra cusps are often added on the outer side of the tooth to make a W pattern ('dilambdodont'). The skull (Fig. 357) shows many primitive features. The orbit is broadly continuous with the temporal fossa (except in tree shrews). There is an incomplete bony palate and an open tympanic cavity, in which the tympanic bone (the old angular) forms a partial ring. Non-primitive characters include the incomplete bony palate of the hedgehog and the loss of the zygomatic arch in tenrec and the shrews. In the post-cranial skeleton there is usually found a clavicle, five digits with claws in both limbs, and the method of locomotion is plantigrade. A specialization of the Lipotyphla is reduction or elimination of the pubic symphysis. In the soft parts the primitive characters again predominate. The stomach is simple. The brain has large olfactory bulbs and small cerebral hemispheres, composed mainly of large pyriform lobes (rhinopallium), usually not covering the corpora quadrigemina or cerebellum, and with little convolution. The neopallium and corpus callosum are small. Besides the nasal receptors insectivores have a sensitive snout often drawn out into a short trunk. There are vibrissae and acute hearing (especially in moles). The eyes are large in Menotyphla but smaller in Lipotyphla, sometimes rudimentary (moles). In many species the retina contains only rods but there are cones in some species of *Sorex* and in the tree shrews, *Tupaia*. Some insectivores retain the cloaca. The uterus is bicornuate and the testes are never fully descended into a scrotum. The placenta is discoidal and haemochorial, that is to say of a type not obviously close to the presumed ancestral mammalian condition. Numerous young are produced (up to 32 in *Tenrec*). Many insectivores hibernate in winter and are provided with special reserves of fat for this purpose. Most insectivores are solitary but some have social habits, exchanging auditory and olfactory signals (*Solenodon*). They may make simple nests.

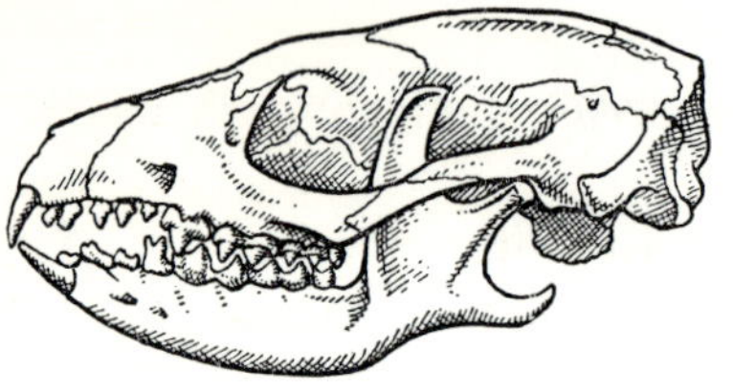

FIG. 357. Skull of hedgehog (*Erinaceus*).

**Deltatheridium* is a fossil insectivore found in the Upper Cretaceous of Mongolia and having characters very close to those of the ancestor not only of insectivores but of all placental mammals. The skull was tubular and elongated, but without the special snout of some modern forms. The teeth are of special interest in that in the upper molars the central cusp (amphicone) showed only partial division into paracone and metacone. There were large canines and other features suggesting creodonts.

Classification of the various lines of insectivores has been difficult, as might be expected in a group containing surviving ancient as well

FIG. 358. *Tenrec*. (From photographs.)

FIG. 359. *Solenodon*, alamiqui. (After *Cambridge Natural History*.)

as modified types. It is not feasible here to discuss the possible affinities of the various groups.

Tenrec (Fig. 358) from Madagascar, and *Solenodon*, the alamiqui (Fig. 359), from the West Indies are remarkably similar animals, showing in their dentition, brain, and other features characters more primitive even than those of other insectivores. The teeth have a tritubercular V pattern, by which they are sometimes distinguished as 'zalambdodont' from the remaining or 'dilambdodont' insectivores. The resemblance of the alamiqui and the tenrec has often been cited as evidence of a land bridge, but is probably a result of retention of primitive features. *Potamogale* (the otter shrew) is a related aquatic African form, feeding on fish.

The golden moles (*Chrysochloris*) of Africa are burrowing animals, with interesting features of similarity to the marsupial and true moles. Hedgehogs (*Erinaceus*) (Fig. 360) are mainly nocturnal creatures, feeding on a mixed animal diet of insects, slugs, small birds, and snakes or even fruit. They have a remarkable immunity to snake-bite and indeed to bacterial and other toxins. Related genera in South-east Asia, such as *Echinosorex*, are more primitive in that the hairs of the back are normal, and not converted into spines as in the hedgehogs. In the Oligocene and Miocene of Europe both types were equally common. The shrews (Soricidae) are mouse-like, insectivorous and omnivorous animals of various types, some terrestrial, others aquatic, found throughout the world. The incisors are specialized as pincers. *Sorex* (Fig. 361) is a very ancient genus, found from the Miocene onwards with little change. Moles (*Talpa*) (Fig. 362) are related to shrews and are found throughout the Holarctic region. They are highly specialized for burrowing, with rudimentary eyes (sometimes covered

FIG. 360. *Erinaceus*, hedgehog. (From photographs.)

FIG. 361. *Sorex*, common shrew. (From photographs.)

FIG. 362. *Talpa*, common mole.

FIG. 363. *Tupaia*, tree shrew. (From photographs.)

FIG. 364. *Macroscelides*, elephant shrew. (From a photograph.)

with opaque skin) and no external ears, smooth fur, fused cervical vertebrae, massive pectoral girdle, including a procoracoid, and broad digging claws on the hands. They feed mainly on earthworms. *Myogale*, the desman, of south Europe is an aquatic mole, with webbed feet.

Apart from these familiar forms the Insectivora includes the interesting oriental tree shrews (*Tupaia*) (Fig. 363) and African jumping shrews or elephant shrews (*Macroscelides*) (Fig. 364), sometimes placed together in a separate suborder ('Menotyphla'). This group can be traced back to the Oligocene and more primitive precursors (**Leptictoidea*) of the Cretaceous, where **Zalambdalestes* is found in the same deposits as **Deltatheridium*. The Menotyphla are interesting in that they show in several ways indications of primate organization. The brain is larger than in the insectivores that we have already considered

('Lipotyphla'). There are various genera of elephant shrews in Africa, all with long trunk-like snouts. They are partly diurnal and proceed by a series of jumps with their long back legs. There are five digits in hand and foot. The tree shrews, *Tupaia* and its allies, are diurnal, arboreal, squirrel-like creatures with a long tail. They feed on insects or fruit and show many lemuroid characters. For instance, there is a complete post-orbital bar, and the brain has larger hemispheres than in *Macroscelides* or other insectivores, and less development of the olfactory regions. The eyes, lateral geniculate body, and visual cortex are well developed. These animals, though they are like insectivores, have a life very like that of lemurs and they are often classified with the primates; they show how narrow is the gap between the two groups, but we need not worry unduly whether they 'really' belong in one group or the other.

2. Order Chiroptera. Bats

Cohort Unguiculata
- Order 2. Chiroptera
  - Suborder 1. Megachiroptera. Oligocene–Recent
    - Family Pteropidae. Asia, Australia, Africa
      - *Pteropus*, fruit bat
  - Suborder 2. Microchiroptera. Eocene–Recent
    - 18 families, including:
    - Family Rhinolophidae
      - *Rhinolophus*, horseshoe bats, Europe, Asia, Australasia
    - Family Phyllostomatidae
      - *Desmodus*, vampire bats, S. America
    - Family Vespertilionidae
      - *Vespertilio*, European bats, Palaearctic
- Order 3. Dermoptera. Palaeocene–Recent
  - *Cynocephalus* (= *Galeopithecus*), colugo or flying lemur, Asia
- *Order 4. Taeniodonta. Palaeocene–Eocene. N. America
  - **Psittacotherium*
- *Order 5. Tillodontia. Palaeocene–Eocene. Europe, N. America
  - **Tillotherium*

Except for their specializations for flight the bats stand very close to the insectivores. They diverged early, however, and their characteristics were already developed in the Eocene. They are the only mammals that truly fly, by flapping the wings, as distinct from the soaring of flying phalangers, colugos, and others. The wing is a patagium or fold of skin, involving all the digits of the hand except the first, and extending also along the sides of the body to include the legs (but not the feet) and, usually, the tail. The chief skeletal modification is therefore a great elongation of the arm, and especially of its more distal bones (Fig. 365). The sternum carries a keel for the attachment of the

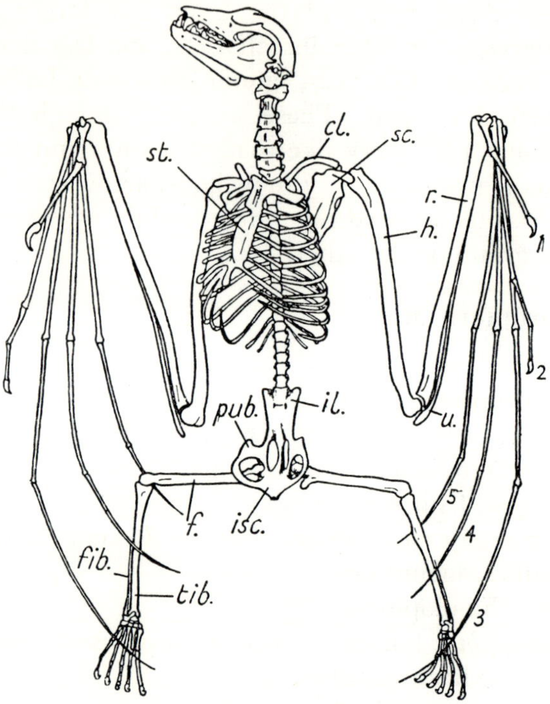

FIG. 365. Skeleton of fruit bat (*Pteropus*).

cl. clavicle; *f.* femur; *fib.* fibula; *h.* humerus; *il.* ilium; *isc.* ischium; *pub.* pubis; *r.* radius; *sc.* scapula; *st.* sternum; *tib.* tibia; *u.* ulna. (From Reynolds, *The Vertebrate Skeleton*, after Shipley and MacBride.)

large pectoral muscles and the clavicle is stout, often fused with the sternum and with the scapula. Since the thorax is used as a fixation point for the flight muscles the ribs move relatively little on each other and respiration is mainly by the diaphragm. The ribs are flattened and may indeed be fused together and with the vertebrae. The characteristics of the arms and thorax are rather similar in these flying animals to those found in brachiating arboreal creatures, such as gibbons, which also have long arms and large, fixed thoracic cages.

The humerus is very long and carries a large greater tuberosity, which may acquire a special articulation with the scapula. The flight movements occur mostly at the shoulder, with the rest of the limb held stiff. The radius is long and the ulna reduced and fused with the radius; the elbow joint allows only flexion and extension. The carpus is much specialized by fusion of bones, allowing flexion-extension and spreading of the digits. Of the five fingers the first is stout and free of the wing; it bears a claw in Microchiroptera, as

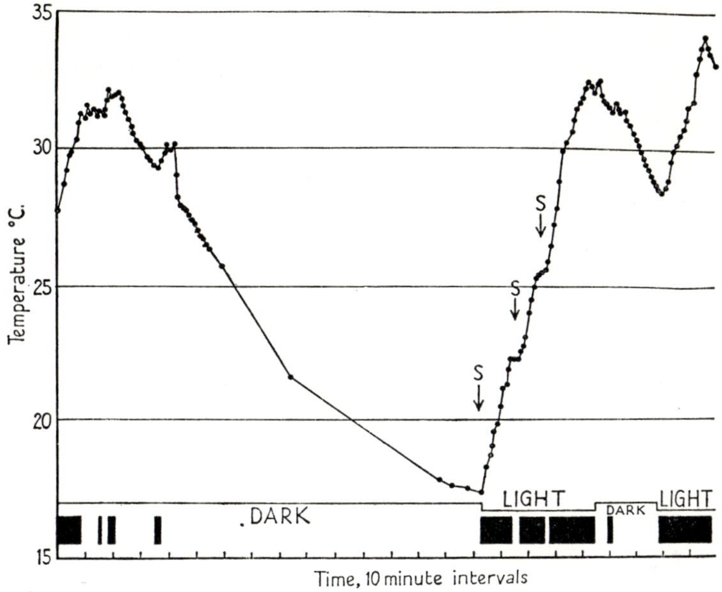

FIG. 366. Temperature chart of greater horseshoe bat. Manipulation during attachment of thermocouple to back has caused warming at beginning of experiment. Bat stimulated at points marked *s*. The black rectangles mark periods during which the animal was shivering or moving. Room temperature 15·5° C. (From Burbank and Young.)

does the second also in fruit bats. The remaining metacarpals and proximal phalanges are enormously elongated to support the wing, the distal phalanges being relatively short. As in birds the wings are short and broad in the slower fliers (horseshoe bats), long and narrow in those that fly faster, with long rapid beats (noctules). On landing horseshoe bats turn a somersault forwards and catch on with the hind legs. Others land on all fours and can walk reasonably well.

The pelvis is rotated so that the acetabulum lies dorsally and the limb is held outwards and upwards. The ventral portions of the girdle are thus drawn apart and are often not united in a symphysis. The hind legs are weak and carry five clawed digits, by which the animal is suspended upside down when at rest, the tendons providing a catch mechanism so that no muscular effort is needed. Even a large fruit bat remains suspended if shot while hanging.

The great development of the arm and patagium makes it impossible for bats to walk actively, but they can climb and crawl. When not flying they usually hang head downwards. When they excrete they turn up and hang by the claw of the pollex, so that the wing is not soiled. When a bat is hanging, there is no upward temperature regulation. The animal becomes cold every time it rests (Fig. 366). When stimulated it can walk, open the mouth and cry out, but can only fly after a period of some minutes of warming up by jerking the legs and shivering. Hibernation is probably only an accentuated form of this daily sleep, the animal living for months on a 'hibernation gland', the subcutaneous fat reserve.

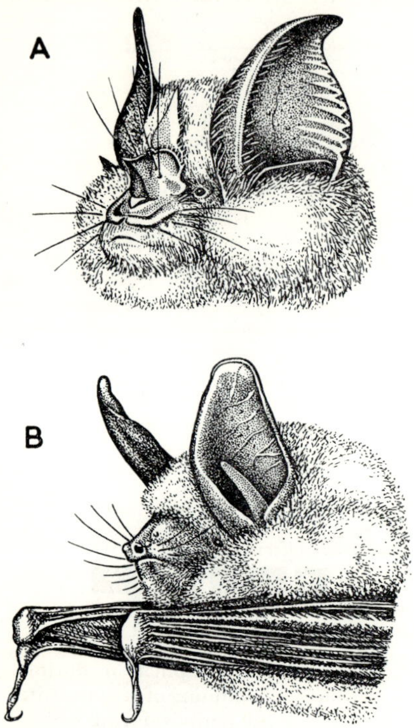

FIG. 367. Heads of two bats.
A. Lesser horseshoe bat, *Rhinolophus hipposideros*.
B. Whiskered bat, *Myotis mystacinus*. (After Grassé.)

The typical microchiropteran is an insectivore, often with molars of the ancestral tritubercular type arranged in a dilambdodont W pattern. Insects are caught on the wing with the large mouth and the wings bitten off neatly. Diets differ considerably: some lick nectar from flowers, which they thus pollinate. The vampires *Desmodus* and *Diphylla* of South America drink blood and have the upper incisors modified into cutting blades. Other Microchiroptera eat fruit, fish, or flesh, and the Megachiroptera (flying foxes) of the tropics are wholly fruit-eaters and have flattened grinding teeth. The skull of flying foxes retains many primitive features and resembles that of an insectivoran, in the Microchiroptera it is shorter. There is an annular tympanic bone and no post-orbital bar.

Bats obtain the major part of their information through the ears, and the reflection of the sound waves that they themselves emit. The cerebral hemispheres are small and the olfactory portions reduced, but the inferior corpora quadragemina and cerebellum are large. The eyes are often moderately large and presumably used in twilight; the retina contains mainly rods. The touch receptors are well developed, especially on the wings.

The echolocation is performed by discrete pulses of high intensity and up to 120 kc frequency. These are produced by the very large larynx, whose cartilages are ossified to make a rigid framework. The

FIG. 368. Big-eared bat (*Plecotus*).
(After *American Mammals*, by W. J. Hamilton, McGraw-Hill Book Company.)

strong cricothyroid muscles put great tension on the light vocal cords. In the horseshoe bats there are special resonating chambers and the face is elaborately modified to beam the sound forwards (Fig. 367).

The ears of bats are greatly specialized, often with very large pinnae. The cochlea is large and the basilar membrane narrow and tightly stretched. The tensor tympani and stapedius muscles are large.

The bat is almost entirely dependent on echolocation for avoiding obstacles and catching insects. If the larynx is damaged or the ears blocked it blunders against even large obstacles. The normal animal can avoid wires less than 0.5 mm thick in complete darkness and if blinded. The presence of loud noise at high frequency disturbs the bat, but lower frequencies do not. This is evidence that the bat hunts by echolocation and not (usually) by listening to the sounds made by the insects.

The mechanism adopted is not fully understood and certainly is not always the same. In Vespertilionidae the sound is emitted by the mouth in pulses of 1–4 m. sec. duration. The note falls through about an octave in each pulse. The pulse repetition rate varies from less than 10/sec at rest to over 100/sec when avoiding obstacles or hunting. The pinna of these bats is very large and folded below to form an anti-tragus (Figs. 367, 368). Vespertilionids commonly detect insects at 50 cm and may do so at 1 m.

In the horseshoe bats the pulse is much longer (40–100 m. sec), and of high and constant frequency (85–100 kc). It is emitted through the nose (Dijkgraaf) and beamed by interference at the nostrils, which

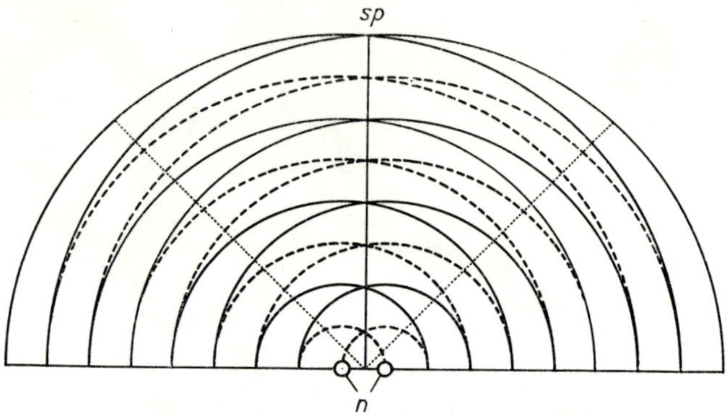

FIG. 369. Diagram showing interference pattern of the ultrasonic waves from the nares of a horseshoe bat. Peaks are indicated by a continuous line, valleys by dashes. The dotted lines limit the sector in which the waves are most intense, maximal in the sagittal plane.

n. nares; *s.p.* sagittal plane. (After Möhres.)

are set half a wavelength apart (Fig. 369). The pulse repetition rate is low (<10/sec). This mechanism is even more effective than the other and is said to detect insects even at 6 m (Möhres).

It was first suggested by Hartridge by analogy with early audio-location and radar devices that bats estimate distance by measuring the echo delay. The middle ear muscles and intra-aural reflexes do indeed allow a very rapid recovery of sensitivity after short loud sounds, as would be necessary, forming a sort of transmit-receive switch (Griffin, 1958). Yet it hardly seems possible that the reflex can work fast enough to allow accuracy at short distances. An alternative hypothesis is that the bat measures the loudness of the echoes, especially the horsehoe bats, with their long pulses (Möhres). The beam movements might give direction and searching movements the range by triangulation. This theory seems to require a very complete acoustic separation of ear and nasopharynx and special cerebral capacities for calculation.

A third suggestion is that the ear receives both the outgoing and reflected notes and constructs difference or summation tones by the introduction of a specific non-linear device (Pye, 1960). Since the sound is used only for location its absolute qualities are not important for a bat as they are for man. With this method any object within range will be located by the variation in the beat notes that are produced as its position changes.

The distortion required to produce the beat notes might perhaps be a function of the middle ear muscles, but is more probably a property of the cohlea. Distortion of cochlear microphonic potentials is known to occur in other animals at sound intensities lower than those emitted by bats. Such a mechanism could thus readily have been evolved from a more conventional hearing system without any fundamental change either in the ear or brain. Only the cochlear microphonics need to follow the high notes of the bat's voice. The auditory nerve carries information only about the difference notes of a few kc, which could be readily recognized by the brain.

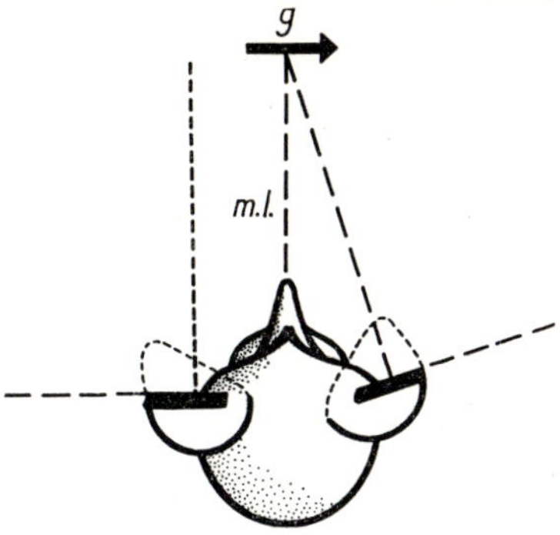

FIG. 370. Diagrammatic view from above the head of a horseshoe bat, showing a possible means of estimating distances using the principle of differences in intensities.

The left ear does not receive an echo, but the right ear because it is turned inwards, receives the reflected waves. The thick black line that limits the pinna anteriorly is the antitragus, which is used as the echo receptor.

m.l. the median line is the zone of maximum intensity of the ultrasonic waves; *g.* is an object in the wave field. (After Möhres.)

Horseshoe bats often seem to search around when they are stationary and there is some evidence that under these conditions they introduce an artificial velocity factor, and hence a Doppler shift, by movements of the ears, which may occur at as much as 50/sec. However, it is not clear exactly how this is achieved or how it is related to a second opening of the meatus that lies at the non-moving base of the pinna and leads by a groove to the nose-leaf, with its lance and shield (Fig. 367).

Some bats (Hipposideridae) may use both methods, the later part of each pulse being modulated. All bats also produce sounds of lower frequency, perhaps as specific recognition signals.

The placenta is of a discoidal form and haemochorial, at least in some types. A peculiar feature is that copulation occurs in the autumn and the sperms remain alive (but presumably not active) within the female until fertilization takes place in the spring. The young are well formed at birth and have then already cut the milk dentition of special teeth, with sharp, backwardly directed hooks, which, with the claws, enable the baby to remain attached to the mother in flight. Most bats live massed together in colonies during the day, with a considerable social organization. They spread out at night and home accurately. After artificial displacement marked bats return home from 100 km or more. Some species hibernate in large colonies where there are suitable caves and then migrate for 1,000 km or more and return to the same cave next winter.

The Megachiroptera, fruit bats or flying foxes, are quite large animals (with wing span up to 5 ft.) living in Asia, the Pacific, Australia, and Africa. In spite of their diet they are in some ways the less-specialized group, having a snout, head, and ears of normal mammalian form.

The Microchiroptera is one of the most successful groups among modern mammals, including a large number of families, genera, and species, with differing habitats. As would be expected, the families often have wide geographical ranges, vespertilionids, for instance, are found all over the world. It is interesting, however, that the genera mostly have a rather restricted range. For instance, *Vespertilio* is limited to the Palaearctic, though *Pipistrellus* is found also in North America. The vampire bats (Phyllostomatidae) are restricted to Central and South America. The fact that even flying mammals should be so restricted is good evidence that the simple problem of communication is one of the least of the difficulties standing in the way of the dispersal of an animal type.

3. Order Dermoptera

The colugo or flying lemur of the orient, correctly called *Cynocephalus* (= *Galeopithecus*), was probably an early offshoot from the insectivoran stock, with a patagium developed for parachuting. The wing differs from that of bats in that the fingers are not elongated and the wing is not flapped. The animals are nocturnal and feed on leaves and fruit. A peculiarity is the forwardly projecting lower incisors with tips divided to form a comb, as in lemurs. A related Palaeocene form shows that this line has been separate for more than 50 million years.

4. Order Edentata

Cohort Unguiculata
  Order 6. Edentata
    *Suborder 1. Palaeanodonta. Palaeocene–Oligocene. N. America
        **Metacheiromys*, Eocene

Suborder 2. Xenarthra. Palaeocene–Recent. Central and S. America

Infraorder 1. Cingulata. Palaeocene–Recent

Superfamily 1. Dasypodoidea. Armadillos

Dasypus, nine-banded armadillo

*Superfamily 2. Glyptodontoidea. Upper Eocene–Pleistocene

**Glyptodon*

Infraorder 2. Pilosa. Upper Eocene–Recent. Central and S. America

*Superfamily 1. Megalonychoidea. Ground sloths. Upper Eocene–Pleistocene

**Megatherium*; **Mylodon*

Superfamily 2. Myrmecophagoidea. Ant-eaters. Pliocene–Recent

Myrmecophaga, giant ant-eater; *Tamandua*, tamandua; *Cyclopes*, two-toed ant-eater

Superfamily 3. Bradypodoidea. Sloths. Recent

Bradypus, three-toed sloth; *Choloepus*, two-toed sloth, unau

Order 7. Pholidota

Family Manidae. Oligocene–Recent

Manis, scaly ant-eater (pangolin), Asia, Africa

The reduction or loss of the teeth with adoption of a diet of soft invertebrates and especially ants has occurred independently at least five times among mammals; this habit is indeed to be expected, since the whole mammalian stock was at first insectivorous. We have already noticed the occurrence of ant-eating characteristics in the echidnas and in *Myrmecobius*, the marsupial ant-eater. Among eutherians the habit is well developed in animals of three different types, (1) the ant-eaters of South America, *Myrmecophaga* and its allies, (2) the pangolins of Africa and Asia, *Manis*, and (3) the aardvark or Cape ant-eater (*Orycteropus*). These ant-eating animals have many features in common. They all possess a long snout and tongue, very large salivary glands, and reduced teeth; because of these similarities they were for a long time classed together as Edentata. It has gradually become apparent, however, that the three groups of placental ant-eaters have evolved separately. The aardvark was probably an early offshoot from the ferungulate stock (p. 704). The pangolins are placed in the Unguiculata, but they represent a separate line, diverging from the insectivoran stock very early (p. 601). The South American ant-eaters form a natural group with the armadillos and sloths, having, like the South American ungulates and other animals, proceeded along several courses of evolution of their own during the long isolation of their continent throughout the Cenozoic period. The term Edentata is now reserved for this South American group.

In many ways the Edentata remain close to the basic eutherian condition. The characteristic feature has been a simplification of the teeth, which are absent altogether in the ant-eaters themselves. In sloths and armadillos the front teeth are absent and the hinder ones are rows of similar pegs, with no covering of enamel. Except in sloths, there is considerable elongation of the snout and the whole cranium is of tubular form, with a low brain-case, containing a small brain with poorly developed hemispheres, having a large olfactory region. The jugal bar is often incomplete, but the hind end of the jugal carries a large downward extension in sloths and ground sloths. A characteristic common to all the Edentata is the presence of extra articulations between the lumbar vertebrae, a striking feature in view of many different modes of locomotion in the group. From these articulations the group gets its name, Xenarthra. Several other peculiar features of the skeleton are common to most or all these animals, such as a fusion of the coracoid with the acromion to enclose a coracoscapular foramen and a union between the ischium and the caudal vertebrae. The feet have well-developed claws, often used for digging, and the animals may walk on the outside of the claws, though some species are arboreal and use the claws for hanging.

Many of the characteristics of the group are obviously those of all generalized eutherians, the edentates having departed little from the original mammalian plan. For example, they all have rather low temperatures, fluctuating widely with the environment. Their features are mostly the result of special ways of life, often leading to bizarre external appearances, such as the long snout of the great ant-eater or the carapace of the armadillo.

The order Edentata is divided into two suborders, the first *Palaeanodonta for a few Palaeocene and Eocene types such as **Metacheiromys*, which had not yet acquired the structure of the vertebrae found in all the remaining edentates (suborder Xenarthra). The palaeanodonts are found in North America and are held by some to be survivors of the original stock, existing before the separation of the continents. The xenarthrous population itself split up early and we can recognize two main groups (infraorders), the Cingulata for the armadillos and glyptodons, and Pilosa for the ant-eaters, sloths, and ground sloths.

5. Armadillos

The armadillos (Dasypodidae) (Fig. 371) have departed least from the ancestral plan and are a very ancient group, already differentiated in Palaeocene times. They are nocturnal and fossorial and obtain protection by the development of bony plates in the skin, these being

FIG. 371. Hairy armadillo, *Dasypus*. (From photographs.)

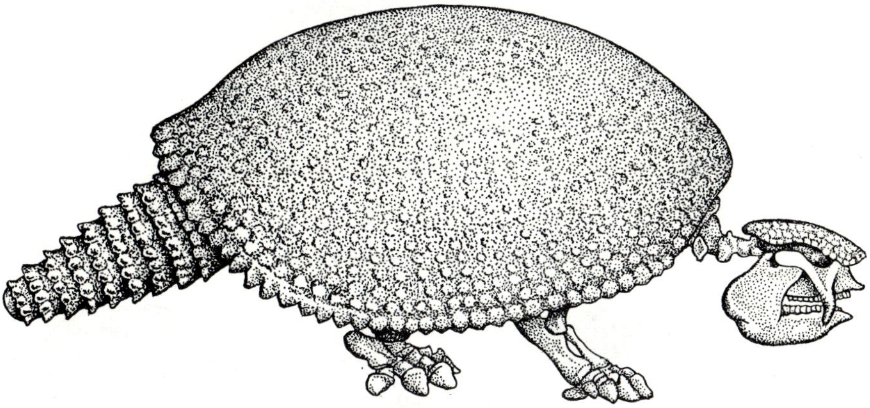

FIG. 372. *Glyptodon*. (From a reconstruction lent by the Trustees of the British Museum.)

covered by horny scutes. The plates are usually arranged in rings round the body and in some genera they allow the animal to roll up into a ball. The vertebrae tend to be fused to support the shield, and many vertebrae unite in the sacrum. The teeth are simple uniform pegs, without enamel, and with open roots and continuous growth. They are often more numerous than in other mammals (as many as twenty-five in each jaw); with simplification of the system of tooth morphogenesis, repetition becomes possible, as we see also in whales. The armadillos are insectivores and omnivorous scavengers in tropical Central and South America; there are many different genera and species. The nine-banded armadillo (*D. novemcinctatus*) is a very active burrower and is

FIG. 373. Giant South American edentates of the Pleistocene. *Megatherium,* the giant ground sloth, was the size of a modern elephant. The glyptodonts were related to the armadillos. (From a mural by C. R. Knight.)

FIG. 374. Great ant-eater, *Myrmecophaga,* from life.

FIG. 375. Lesser ant-eater, *Tamandua.* (From photographs.)

FIG. 376. Tree ant-eater, *Cyclopes,* showing one of the peculiar attitudes adopted, perhaps to startle an attacker (dymantic posture). (From a photograph.)

spreading northwards in the United States with the destruction of its carnivore enemies by man. The haemochorial placenta, at first diffuse then discoidal, is modified as a result of the process of polyembryony. After cleavage the embryo divides into eight or twelve secondary embryos, all developing within a single amnion.

During the Pleistocene and earlier periods, besides the modern armadillos, there were also giant armadillos. The glyptodonts (Fig. 372) were a related type, diverging as early as the Upper Eocene, with a skull and carapace composed of many fused small pieces and sometimes the well-known 'battle-axe' tail. They show a remarkable convergence with tortoises and some dinosaurs and probably lived in deserts.

6. Ant-eaters and sloths

The modern soft-skinned edentates (ant-eaters and sloths) are very specialized and not superficially like the armadillos. The enormous ground sloths (Fig. 373), of which there were several families living

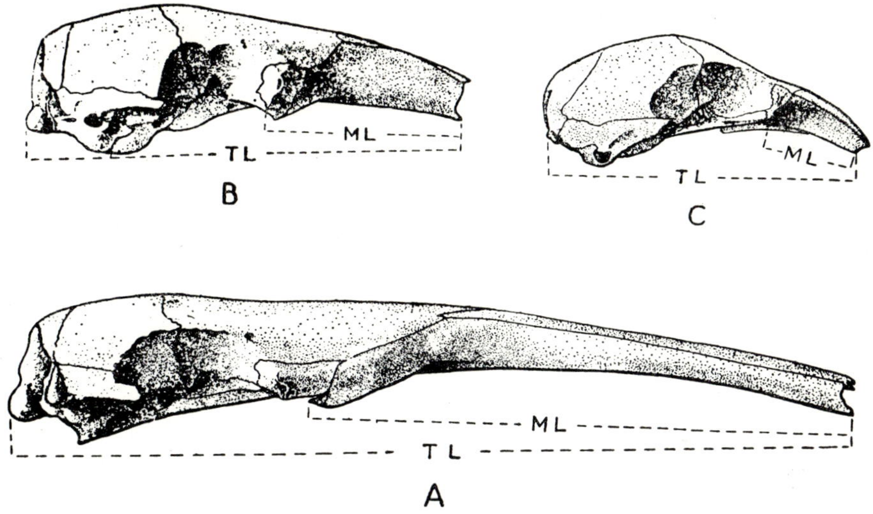

FIG. 377. Side views of adult skulls of A, *Myrmecophaga*; B, *Tamandua*; C, *Cyclopes*. *Tamandua* and *Cyclopes* are approximately 1½ and 3 times the scale of *Myrmecophaga*. *T.L.* measurement of total length; *M.L.* of maxilla length. (From Reeve.)

between the Oligocene and Pleistocene, were in some respects intermediate between the two types. They were quadrupedal animals, but the fore-limbs were shorter than the hind and provided with long claws. Probably the ground sloths were largely bipedal, perhaps crawling slowly about with their fore-limbs among the branches and bearing them down with their weight. The brain was small but the teeth large and hypsodont. Nearly fifty genera of ground sloths have been recognized and they were evidently successful in the South American forests. **Megatherium* was larger than an elephant and a similar form, **Neomylodon*, persisted nearly to the present day. Pieces of its skin have been found, and these, even in the favourable conditions, can hardly have maintained their appearance for more than a few hundreds or at the most thousands of years.

The ant-eaters (Myrmecophagidae) have a characteristic elongated snout, without teeth. There are three genera, differing in size, and

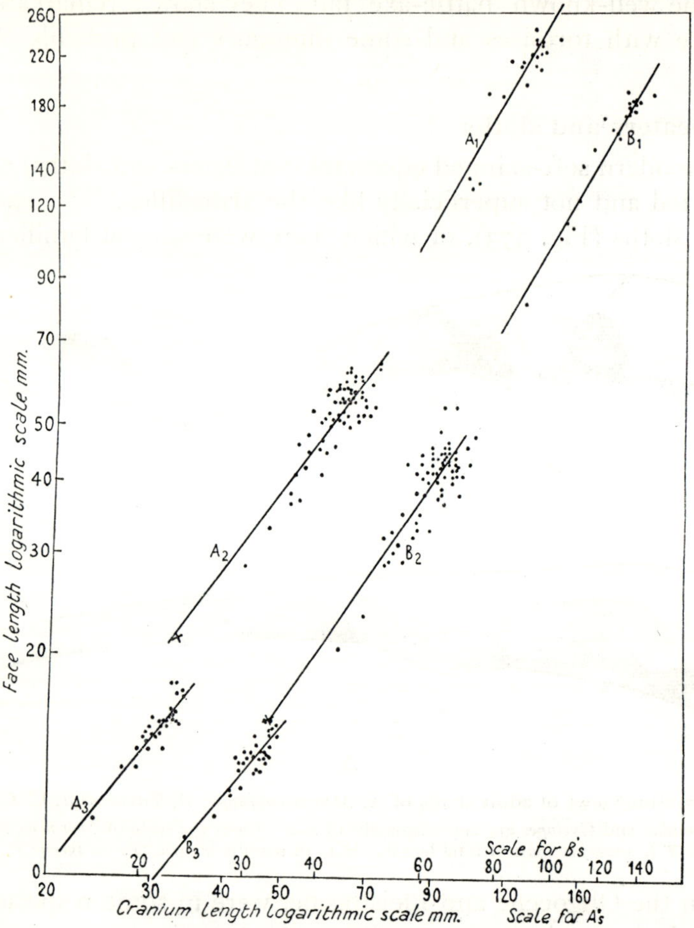

FIG. 378. Logarithmic plots of lengths of maxillae (A's) and nasals (B's) against cranium.

The suffixes represent: 1, *Myrmecophaga*; 2, *Tamandua*; 3, *Cyclopes*. Scale for B graphs is shifted to the right compared with A's. Crosses at bottom of lines for A_2 and B_2 represent a very young *Tamandua*. The lines were fitted to each sample by least squares. (From Reeve.)

the larger species have relatively much the longer snouts. The great ant-eater *Myrmecophaga* (Fig. 374) has an enormously elongated face, but this is much shorter in the smaller *Tamandua* (Fig. 375), and the very small tree-living *Cyclopes* (Fig. 376) has a head of normal mammalian shape. Analysis shows that there is little difference between the relative rate of growth of the face in these three genera, and the differing final forms result mainly, though not wholly, from the differences in absolute size (Figs. 377 and 378). In all ant-eaters the face becomes relatively longer as the animal increases in size, and the enormous snout of the great ant-eater is produced by a relative growth-rate only slightly higher than that found in *Tamandua* and *Cyclopes*. This is an excellent example of the way in which the proportions of an organ will vary in animals of different sizes if its growth is allometric,

FIG. 379. Two-toed sloth, *Choloepus*. (From a photograph in Scott, *Land Mammals*, copyright 1913, 1937 by the American Philosophical Society and used with the permission of the Macmillan Company.)

that is to say, relatively faster or slower than that of the body as a whole (p. 737).

The hard palate is prolonged backwards in *Myrmecophaga* by union of the pterygoids, a condition found also in some armadillos (*Dasypus*).

The great ant-eater is a fine animal, over 6 ft long, with a long hairy coat, including a very bushy tail and with a black stripe edged with brown at the shoulder. It has a long thin tongue for collecting ants, and enormous submaxillary salivary glands. The claws of the front legs are very large and used for defence as well as for digging. *Tamandua* and *Cyclopes* differ from *Myrmecophaga* in other features besides the length of snout. They are arboreal and the tail is prehensile.

The sloths (Bradypodidae) (Fig. 379) are fully adapted for arboreal life and can hardly walk on the ground. They show, as do the bats, how the mammalian skeleton can be used with surprisingly little change to support weight by hanging, the limbs being used as tension members rather than as pillars. In marked contrast to the ant-eaters the face is short and the head rounded, with large frontal air sinuses. The neck is peculiar for the presence of nine or ten cervical vertebrae in the three-toed sloth, *Bradypus*. This might be supposed to provide a flexible neck for an animal that must often face backward, were it not that in the two-toed sloth *Choloepus* there are but six cervical vertebrae.

The limbs are long, especially the fore-limbs, and the digits carry hooked claws for hanging (Fig. 380). In the pectoral girdle the clavicle articulates with the coracoid, a unique condition among mammals. As in ant-eaters the acromion is connected with the coracoid, enclosing a coraco-scapular foramen. The significance of these special features is not clear, but the habit of hanging upside down has produced some obvious modifications, for instance all the vertebral neural spines are

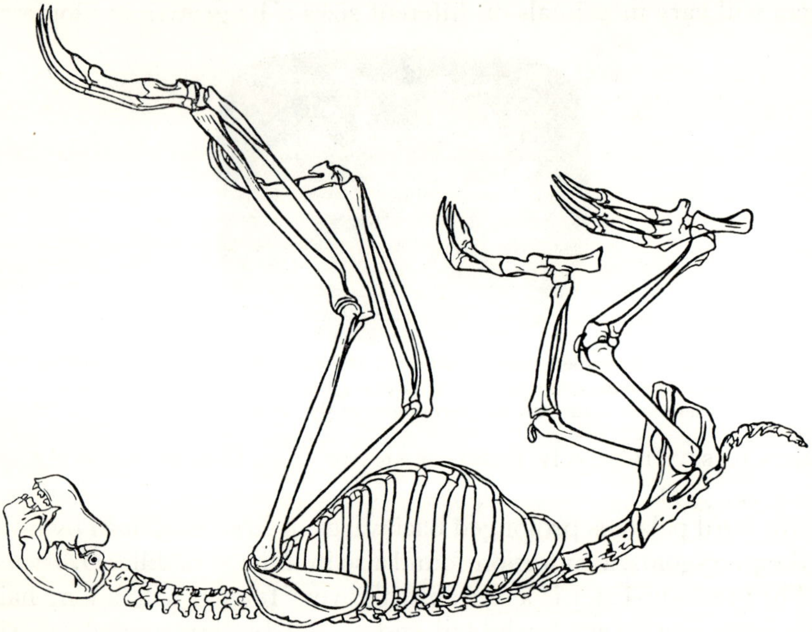

FIG. 380. Skeleton of three-toed sloth, *Bradypus*. (After Blainville.)

low and the pelvis is short. Even here, however, we find peculiar similarities to the ant-eaters, in the union of the ischium with caudal vertebrae, a feature whose adaptive significance is obscure.

The sloths live on foliage, but this herbivorous diet is perhaps secondary to a long period of insectivorous life, during which there was a reduction of the teeth and loss of enamel. On adoption of the new way of life the enamel could not be restored, but a grinding surface is provided by the presence of cement and continuous growth of the teeth. The stomach is large and divided into several chambers, recalling those of ruminants. The rectum is enormous and the masses of faeces are retained for several days, intestinal peristalsis being as slow as all the other movements of these creatures. Interesting features connected with this slow life are the small size of the thyroid and adrenals.

The sloths live in the rain forests of South and Central America, moving very slowly among the branches, which they come to resemble closely by the growth of blue-green algae in special grooves in the hairs. Their upside-down posture has led to many changes from the typical mammalian organization, including, it is said, a reversal of the usual mechanism for maintaining posture. When a normal mammal is decerebrated its legs assume a pillar-like extensor rigidity, because of the overaction of the reflexes of standing. A decerebrate sloth is said to show the opposite, flexor rigidity.

7. Order Pholidota: pangolins

The pangolins or scaly ant-eaters, *Manis* (Fig. 381) of the Old World (Africa and Asia) have many features superficially like those of the New World ant-eaters and the groups may be remotely related. Unfortunately nothing is known of the fossil history of *Manis* and its position among the unguiculates is therefore provisional. The body is up to 5 ft long, covered with horny epidermal scales, interspersed with hairs. The absence of teeth, the elongated snout, long thin tongue, simple stomach, reduced ears, and long claws are all features found in the other ant-eaters. Rods of cartilage extending backwards from the xiphisternum have been compared with the abdominal ribs of reptiles, but are probably a special development, connected with the protrusion of the enormous tongue, which is carried in a special sac and operated by muscles attached to the xiphisternal processes. The animals are macrosmatic, with small eyes. The brain is very small, but the hemispheres are folded. The placenta is diffuse and epithelio-chorial, with a large allantois and a yolk-sac persisting until birth. Evidently the pangolins preserve many very ancient mammalian features. There are various species of *Manis*; some live in open savannah, others are able to climb trees. All are nocturnal and eat ants and termites.

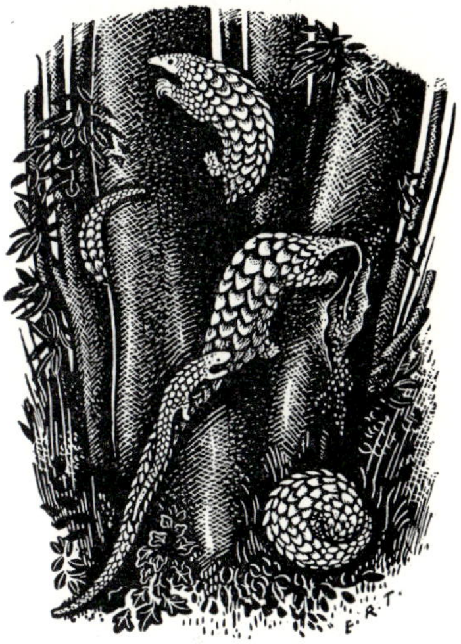

FIG. 381. Black-bellied tree pangolin, *Manis*. (From photographs.)

XXIII

PRIMATES

1. Classification

Order 8. Primates

Suborder 1. Prosimii. Palaeocene–Recent

Infraorder 1. Lemuriformes. Palaeocene–Recent

*Family 1. Plesiadapidae. Palaeocene–Eocene. Europe, N. America

**Plesiadapis*, Palaeocene

*Family 2. Adapidae. Eocene. Europe, N. America

**Notharctus*; **Adapis*

Family 3. Lemuridae. Pleistocene–Recent. Madagascar

**Megaladapis*, Pleistocene; *Lemur*, common lemur

Family 4. Indridae. Pleistocene–Recent. Madagascar

Indri, indris

Family 5. Daubentoniidae. Recent. Madagascar

Daubentonia (= *Cheiromys*), aye-aye

Infraorder 2. Lorisiformes. Pliocene–Recent. Asia and Africa

Family. Lorisidae.

Loris, slender loris, India; *Galago*, bush baby, Africa; *Perodicticus*, potto, Africa

Infraorder 3. Tarsiiformes. Palaeocene–Recent. Holarctic, Asia

*Family 1. Anaptomorphidae. Palaeocene–Oligocene

**Necrolemur*, Eocene, Europe; **Pseudoloris*, Eocene, Europe

Family 2. Tarsiidae. Recent. E. Indies

Tarsius, tarsier

Suborder 2. Anthropoidea. Oligocene–Recent

Superfamily 1. Ceboidea. New World monkeys. Miocene–Recent. S. America

Family 1. Callithricidae. Recent

Callithrix (= *Hapale*), marmoset

Family 2. Cebidae. Miocene–Recent

**Homunculus*, Miocene; *Cebus*, capuchin; *Ateles*, spider monkey; *Alouatta*, howler monkey

Superfamily 2. Cercopithecoidea. Oligocene–Recent

*Family 1. Parapithecidae. Oligocene. Africa

**Parapithecus*

Family 2. Cercopithecidae. Old World monkeys. Oligocene–Recent. Africa, Asia

**Mesopithecus*, Miocene; *Macaca*, rhesus monkey, macaque, Asia, N. Africa; *Papio*, baboon, Africa; *Mandrillus*, mandrill, Africa; *Cercopithecus*, guenon, Africa; *Presbytis*, langur, E. Asia; *Colobus*, guereza, Africa

Superfamily 3. Hominoidea

Family 1. Pongidae. Apes. Oligocene–Recent

**Propliopithecus*, Lower Oligocene, Egypt; **Pliopithecus*, Lower Miocene, Europe, Africa; **Dryopithecus*, Miocene, Africa, Asia; **Oreopithecus*, Pliocene, Europe; **Australopithecus*, Pleistocene, S. Africa; **Proconsul*, Miocene, Africa; *Hylobates*, gibbon, SE. Asia; *Pongo*, orang-utan, E. Indies; *Pan*, chimpanzee, Africa; *Gorilla*, gorilla, Africa

Family 2. Hominidae. Man. Pleistocene–Recent

**Pithecanthropus* (= **Sinanthropus*), Java and Pekin man, Pleistocene, E. Asia; *Homo*, man (all living races). Pleistocene–Recent

2. Characters of primates

Linnaeus reserved his order Primates for the monkeys, apes, and men, distinguishing them thus from the other mammals, Secundates, and all other animals, Tertiates. The term primate carries with it the implication that the animals in the group are not only the nearest to ourselves but are also in some sense the first or most completely developed members of the animal world. We shall try to examine this belief in accordance with the principles adopted earlier and to inquire whether we and our relatives can be said to be the highest animals in the sense that we possess a system of life able to survive under the most varied and unpromising conditions.

The earliest eutherians of the Cretaceous were probably arboreal; the primates have continued this habit and with it they retain many of the features present at the beginning of mammalian history, for instance the five fingers and toes and the clavicle. Primates already existed in the Palaeocene, 65 million years ago and have a longer geological history than any other placentals except the insectivores and carnivores. It is not surprising, therefore, that it is difficult to separate the primates from the insectivores; the tree shrews, for instance (p. 584), have several times been transferred from the one order to the other.

The primates have retained many primitive mammalian features, some of which have become strongly accentuated for arboreal life.

Their characters are those of animals raised up from the ground; the opportunities offered in the trees for the use of hand and brain have no doubt been important influences in the shaping of man.

The general plan of primate life has thus been to retain the original eutherian conditions, with emphasis on those features important for tree life. In such an existence continual quick reaction to circumstances is likely to be necessary, the environment is varied, and the mechanical supports it offers are often precarious. Under these conditions safety is achieved by quick reactions rather than by stability; thus primate more than any other life tends to be a matter of continual exploration and change. The information that ensures the life of the species is obtained by the individuals and stored in their brains, rather than by selection among large numbers of rapidly breeding individuals. The time taken for development thus increases in the primate series. Growth continues for about 3 years in prosimians, 7 in monkeys, 9 in gibbons, 12 in other apes, and 20 in man. To obtain this information receptors are obviously of first importance, but in the tree-tops one cannot hunt by smell; the eyes and ears therefore became developed, at the expense of the nose. Primates are microsmatic, with reduction of the number and length of the turbinal bones and hence of the long snout that houses them. Consequently the eyes come to face forwards, so that their fields overlap, binocular vision becomes possible, and central areas appear in the retinas. Monkeys are certainly more dependent on vision than are most animals and for this reason they approach the birds in the adoption of colour patterns for sexual recognition and excitation.

The changes in the receptors were accompanied by conspicuous changes in the brain, which becomes very large in later primates, with cerebral hemispheres reaching far backwards. The olfactory bulbs and rhinopallium become small and the neopallium very large, differentiated into areas and provided with a large corpus callosum. The occipital pole, concerned with vision, and the frontal areas, become especially well developed in the apes and man. Stereoscopic eyes with numerous cones would be of no value without a central analyser to allow the animal to discriminate shapes, retain the impression of past situations, and otherwise make use of the available information. The marked differences in the rate of growth of the brain of different primates are shown in Fig. 382, from Schultz's careful measurements. At early stages of development all the primates studied have the same (high) relative brain weight, but in the adults the brain is relatively and absolutely larger in man than in monkeys or apes.

The special developments of the receptors and brain have marked effects on the skull, whose facial portion becomes shorter and the brain-case relatively larger and rounder; the foramen magnum comes to face downwards, rather than backwards. As the eyes are directed forwards the orbits become closed off from the temporal fossae behind.

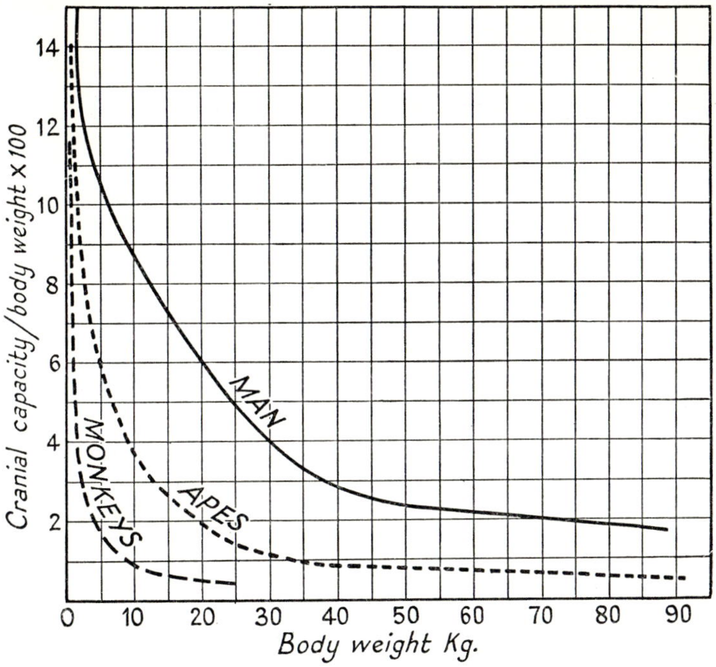

FIG. 382. The relative cranial capacity as a function of body-weight in various primates. The brain grows relatively faster in man than in either monkeys or apes. The curves are constructed by measuring cranial capacity and body-weight of individuals of differing ages. The monkeys included various Cercopithecidae, the apes only gorillas, chimpanzees, and orangs. (Modified after Schultz, *Am. J. phys. Anthropol.* **28**.)

The head is more clearly marked off from the body than is usual in mammals and the neck is very mobile, allowing the eyes to be turned in all directions.

The skeletal and muscular systems become arranged to allow jumping, swinging, and grasping. The pentadactyl plan is retained, without the loss of digits or fusion of bones that are found in most mammals, but the hand and foot are made suitable for grasping by development of adduction movements of pollex and hallux. The digits mostly have sensitive pads and the original claw is replaced by a flat nail. The clavicle remains large and indeed is specially developed in primates to allow mobility of the fore-limb, the muscles being arranged to allow rotation of the scapula, increasing the range of movement. Still further mobility is provided by improving the joints between radius and ulna and humerus at which movements of pronation–supination take place.

The primates early ceased to feed only on insects, and took to a mixed diet; the teeth have not become so specialized as in ungulate mammals. The hands are frequently and ingeniously used to obtain

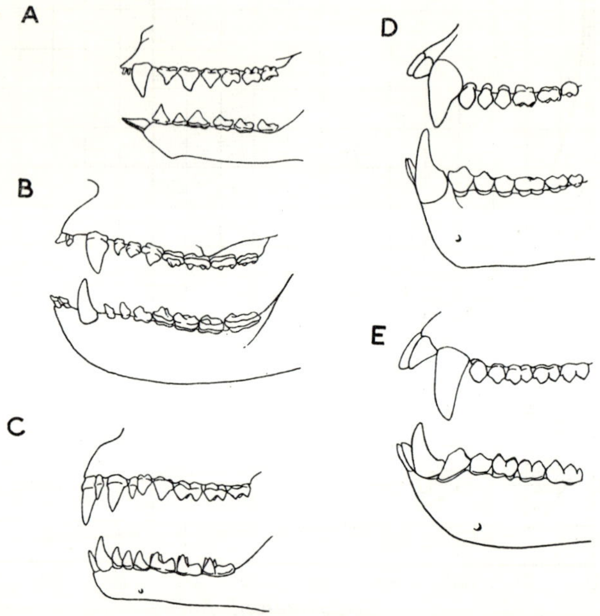

FIG. 383. Upper and lower dentition. A. Modern lemur, *Lemur varius*. B, Fossil adapid, *Notharctus osborni*. C, Tarsius, *Tarsius spectrum*. D, Platyrrhine monkey, *Cebus*. E. Catarrhine monkey, *Macaca*. (After Le Gros Clark.)

food. Omnivorous or frugivorous diets are common, and the molars have become quadritubercular (Fig. 383), the upper adding a hypocone and the lower losing the paraconid of the original pattern, leaving the metaconid and protoconid, while the hypoconid and entoconid become raised to make a posterior pair, sometimes with addition of a fifth cusp, the hypoconulid, posteriorly. The cusps are usually not of the sharp insectivorous type, but are low (bunodont) cones and extra ones may be added, or the cusps joined to make ridges. These changes are associated with the adoption by many primates of a diet of fruit or leaves, requiring treatment by biting and grinding.

The method of reproduction is one of the most characteristic of primate features. The uterus retains signs of its double nature in the earlier types but later becomes a single chamber. The number of young produced is small, as in other animals with large brains that learn well. There is often only a single pair of teats and in association with the arboreal habit these are pectoral. The arrangements for placentation involve elaborate changes in the uterine mucosa in preparation for reception of the embryo, followed by breakdown at regular intervals (menstruation). This special nature of the uterine mucosa makes possible the efficient haemochorial form of placentation. Besides these arrangements for nutrition of the young the primates also extend parental care for a long time after birth.

In many features of their life, therefore, the primates show to a high degree the adjustability and power to obtain sustenance from varying environments that is characteristic of all life. The receptors, brain, and hand provide means for doing this in more elaborate ways than are used by any other animals. The monkeys and apes have exploited these powers to a considerable extent and are successful animals, living, as we might say, by their wits, in a wide variety of circumstances. However, non-human primates are unable to adjust to conditions outside the tropical and subtropical regions. Man has made still better use of his talents and by creating his own environment manages to support a population of nearly 3,000 million large individuals, scattered all over the globe.

3. Divisions of the primates

Fortunately many of the changes of habit characteristic of primates involved changes in the skull and these can be followed in the fossils. Our knowledge of the evolutionary development of primate life, though far from complete, is less so than might be expected from the rarity of preservation of arboreal skeletons. During the 50 million years since the Eocene the various primate stocks have, of course, divided and subdivided many times, and invaded many special habitats. The forms at present known, living and as fossils, are placed by Simpson in 150 genera, two-thirds of them extinct, 70 of these being prosimians. Most of primate evolution has occurred in the Old World; there are no fossil primates known from North America between the Oligocene and modern times. Only ten fossil primates throughout the whole Tertiary are even moderately well known, probably because animals living in trees are seldom preserved as fossils.

Bitter controversy still rages around the question of the best means of classification of Primates. Earlier zoologists tended to postulate a series of stages successively closer to man, the latest product of evolution. There has been increasing awareness of the unwisdom of this procedure. Recognition that many of the surviving stocks have been separate for a long time has led systematists to emphasize the distinctions between the groups more sharply. Thus lemurs, far from being regarded, as they were formerly, as rather primitive monkeys, are now often placed in a distinct order, having in common with other primates only 'the retention of certain primitive characters and an adaption to arboreal life' (Wood Jones). There is no general agreement

FIG. 384. Ring-tailed lemur, *Lemur*. (From life.)

about the best means of classification; the more traditional schemes, such as that adopted here, probably give an over-simplified idea of a progression of forms. A classification on more 'natural' or phyletic lines could be devised, but would necessitate the postulation of a large number of distinct categories, unless these were simplified by admitting speculations about the affinities of the lines.

We shall, as usual, in the main follow Simpson. His arrangement retains the order Primates and recognizes two great suborders, Prosimii and Anthropoidea. The division is 'horizontal' rather than 'vertical'; the two groups are not separate and divergent lines, they contain respectively the ancestral and the 'developed' forms. Two primate stocks are indeed known to have existed in the Palaeocene and these are both included in the prosimians, whereas no anthropoids are known before the Oligocene. The Prosimii includes three sorts of primate, all 'primitive' in the sense of retaining insectivoran characters, such as long face, lateral eyes, and small brain; they are grouped here as three infraorders: Lemuriformes for the lemurs of

Madagascar and their fossil allies; Lorisiformes for the rather similar animals outside Madagascar; and Tarsiiformes for the living tarsier and its Eocene relatives. The suborder Anthropoidea includes two distinct types, first the New World monkeys, superfamily Ceboidea, secondly the Old World monkeys, apes and man, grouped together as Cercopithecoidea.

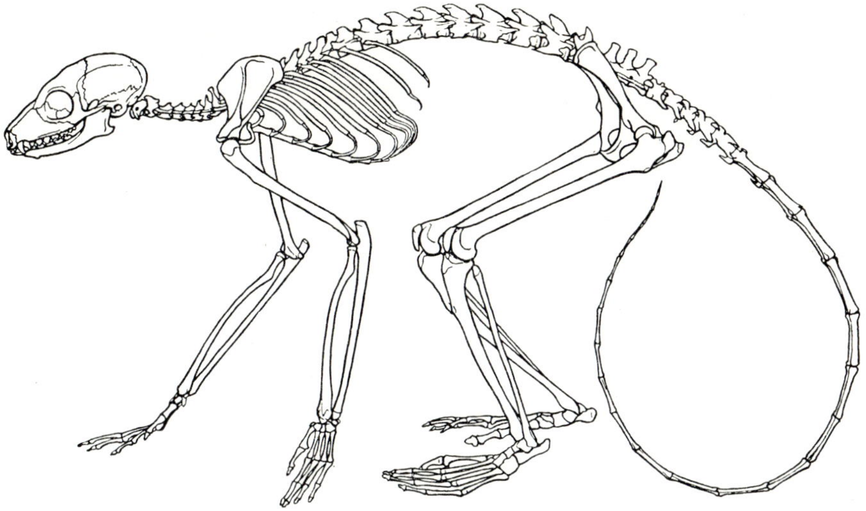

FIG. 385. Skeleton of ring-tailed lemur.

We propose, therefore, to arrange our examination of primates around the idea of three main stocks diverging in the Palaeocene, namely lemurs, lorises, and tarsiers, with an anthropoid stock arising from one of these, probably the tarsioid, in the Eocene, and itself early becoming separated into two lines, the New World monkeys on the one hand, and Old World monkeys and apes on the other (Fig. 416).

4. Lemurs and lorises

The lemurs (Fig. 384) living in Madagascar today resemble certain fossils, known as plesiadapids and adapids, that existed in various parts of the world in Palaeocene and Eocene times. We may, therefore, perhaps assume that they show us the characters of part at least of the primate stock more than 50 million years ago. Lemurs show their 'primitive' nature in their habits and appearance, as well as in the details of their structure. They are mostly nocturnal, arboreal, insectivorous, omnivorous, or fruit-eating animals; the name means 'ghost', but it is more interesting that they are often said to be rather like squirrels in behaviour. The brain (Fig. 386) has relatively small cerebral hemispheres, not overlapping the cerebellum, but with olfactory regions large for a primate, though smaller than in insectivores or other primitive mammals. The nose has numerous well-developed turbinal bones (Fig. 394). The cerebral sulci tend to run longitudinally,

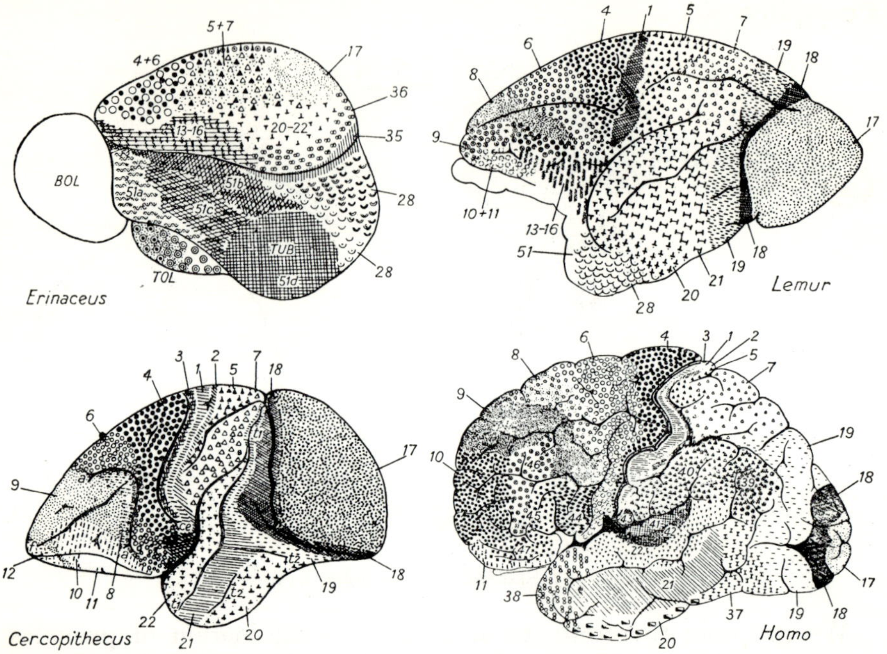

FIG. 386. Brains of hedgehog and various primates to show the relative development of various parts. The numbers refer to the areas recognized by Brodmann on a basis of their structure. 4 and 6 are the precentral motor areas, 8–12 the frontal and prefrontal areas, lacking in the earliest forms. 1–3 are the end station of skin sensations, and 5 and 7 are also concerned with these. 17 is the visual end station, and 18 and 19 are also concerned with this sense. 22 is the auditory end station. BOL, bulbus olfactorius; TOL, tuberculum olfactorium. (From Brodmann.)

rather than transversely as in anthropoids. The whole behaviour is not like that of a monkey; the animals move from branch to branch by sudden leaps, balancing with the long, bushy tail, which is not prehensile. Social habits are little developed.

The snout is long, with a cleft and moist upper lip, the eyes are directed sideways, and the retina contains only rods except in the genus *Lemur*, which is diurnal and possesses cones. There is no fovea and no binocular vision. The external ears may be large, as in other nocturnal animals. In the skull (Fig. 387) there is a post-orbital bar, but the temporal fossa opens widely to the orbit. The tympanic region shows several peculiar features. The tympanic bone forms a ring, lying

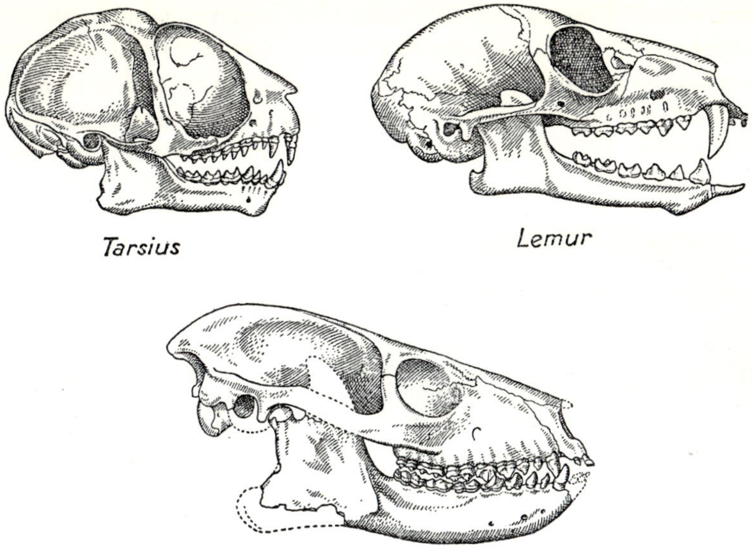

FIG. 387. Skulls of early primates. (After Gregory and Flower and Lydekker, *Mammals, Living and Extinct*, A. & C. Black, Ltd.)

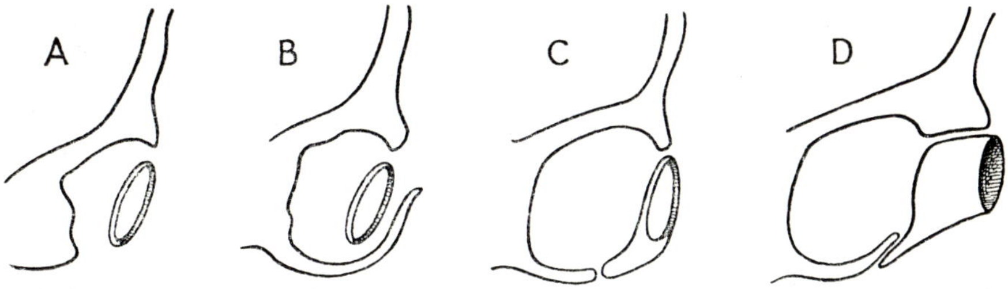

FIG. 388. Tympanic ring and tympanic bulla.

A, Primitive mammalian condition, floor of cavity unossified. B, Lemuriformes, ring enclosed within bulla. C, Lorisiformes and platyrrhines, ring part of bulla. D, *Tarsius* and catarrhines, bony meatus. (After Le Gros Clark, *Early Forerunners of Man.*)

within a petrosal bulla, but not fused with it (Fig. 388), a condition like that in *Tupaia*, but not found in higher primates. The pollex and hallux are used for grasping; most of the digits have nails, but the second digit of the foot has a toilet claw. The fourth digit is usually the longest, whereas in anthropoids the whole symmetry of the hand and foot is arranged about a long third digit. The teeth show the typical primate number $\frac{2.1.3.3}{2.1.3.3}$, but the upper incisors are very small and the lower incisors and canines are procumbent, that is to say directed forwards and are used for combing the fur. The first lower premolar is caniniform. The molars are triangular in some genera, in others a hypocone gives them a square shape. The lower molars are of typical tuberculosectorial type, with a heel. The reproduction shows several primitive features. There are marked breeding-seasons and the females are polyoestrous. The uterus is bicornuate and the placenta of a

FIG. 389. Aye-Aye, *Daubentonia*. (From a photograph.)

FIG. 390. Slender loris, *Loris*. (From life.)

FIG. 391. Bush-baby, *Galago*. (From photograph.)

remarkably simple type, epithelio-chorial and diffuse, with villi all over the surface of the chorion, which is vascularized directly by a large allantois, filled with fluid. The amnion arises as folds and not by cavitation as in higher primates.

Ten genera of lemurs occur today in Madagascar, where they have evidently flourished in isolation throughout the Tertiary. *Indri*, the largest, is an animal nearly 3 ft long, able both to jump and to walk on its hind legs. Earlier lemurs became larger still, the skull of the Pleistocene **Megaladapis* was nearly a foot long. *Daubentonia* (= *Cheiromys*), the aye-aye (Fig. 389) of Madagascar, is like the lemurs in many ways but has large, continually growing upper and lower incisors, like a rodent. It has a very long and thin third finger, which it uses, with its teeth, to find insects deep in the bark. It also eats the inside of bamboo and sugar-canes among which it lives.

The Lorisformes (Fig. 390) include the slow lorises and other lemur-like animals that are found outside Madagascar. They are known as fossils back to the Miocene. The slow lorises (*Nycticebus* and *Loris*) of India and Ceylon are arboreal and nocturnal, proceeding by remarkably slow and deliberate movements and often hanging upside down. They eat fruit or small animals. Lorises also show some features that recall the higher primates, for instance the tympanic ring is fused to the petrosal bulla. In some of them the face is shorter and the brain-case rounder than in true lemurs. It is therefore possible that they

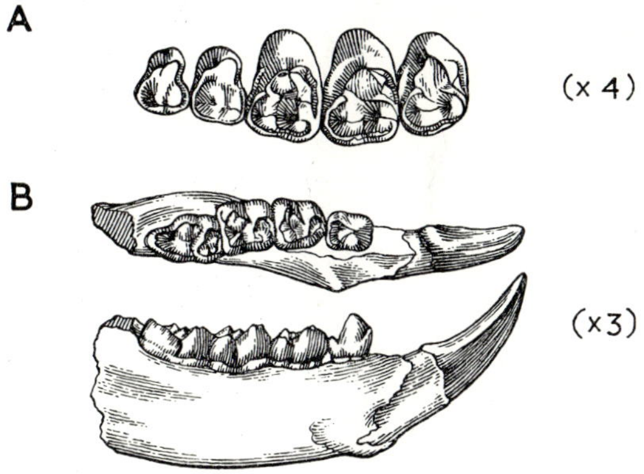

FIG. 392. Dentition of *Plesiadapis anceps*. A. Right upper teeth showing P^3P^4 and 3 molars. B. Lower dentition. (After G. G. Simpson.)

are survivors of an earlier stock, closer to our own than are the lemurs, and some of the features, such as procumbent incisors, may be developed independently in the two groups. However, traces of very early features remain, including a transverse skin fold on the abdomen of the female, which is considered by some to represent a marsupium. On the African mainland there are also two successful genera of this type, *Galago*, the bush baby (Fig. 391), and *Perodicticus*, the potto. The former are jumping animals and can leave the trees; their elongated tarsus somewhat recalls that of *Tarsius*.

5. Fossil Prosimians

The earliest primates of the Palaeocene and Eocene were insectivorous and fruit-eating animals. They may be distributed among five families, whose relationships are difficult to decide. The **Plesiadapidae*, from the Paleocene and Eocene of both Old and New Worlds, had large upper and lower incisors and have been considered as related both to tree-shrews and to the aye-aye (Fig. 392). They are probably too specialized to be directly ancestral to either the lemurs or higher primates, but may be close to both. **Plesiadapis* is the only primate genus except *Homo* that occurs in both Old and New Worlds.

The Adapidae were also Palaeocene and Eocene animals, like lemurs in many ways but without procumbent incisors. The Old World members of the family (**Adapis*) could have given rise to both the lemurs and lorises. The adapids were large creatures with heads a foot or more long. The brain case was small (**Notharctus*, Fig. 387) but carried temporal crests. There was a very full dentition $(\frac{2.1.4.3}{2.1.4.3})$. The incisors were not procumbent but the canines were incisiform. The tympanic ring was included in the bulla. These animals therefore showed many features common to other early mammals but with distinctly lemuroid tendencies.

FIG. 393. Spectral tarsier, *Tarsius*. (From life.)

6. Tarsiers

The third group of the Prosimii, the Tarsiiformes, includes one living form, *Tarsius*, and a number of early Tertiary fossils, placed in a separate family *Anaptomorphidae. The whole group could be described by saying that its members show many characteristics similar to those of Insectivora and lemurs, but also others suggestive of the anthropoid primates. Yet there are present specializations that rule out the possibility that these animals are in the direct line of descent of the higher forms, and we must therefore regard them as an early offshoot, showing us something of the characteristics that were possessed by the anthropoid stock in Palaeocene or early Eocene times.

Tarsius itself (Fig. 393) is an arboreal, nocturnal, insectivorous creature, the size of a small rat, living in the East Indian islands. The dental formula is $\frac{2.1.3.3}{1.1.3.3}$; the molars retain a very simple tritubercular pattern and the incisors and canines do not show the specializations found in lemurs (Fig. 383). The head is more like that of a monkey than of a lemur; it is set on a mobile neck, indeed the animal has the uncanny power of turning its head through 180° so that it faces backwards, while the eyes, like those of owls, are so large that they are little movable. The foramen magnum opens downwards. The eyes face more nearly forwards than in lemurs, the snout is shortened, and the turbinals of the nose reduced. The nose thus resembles that of a

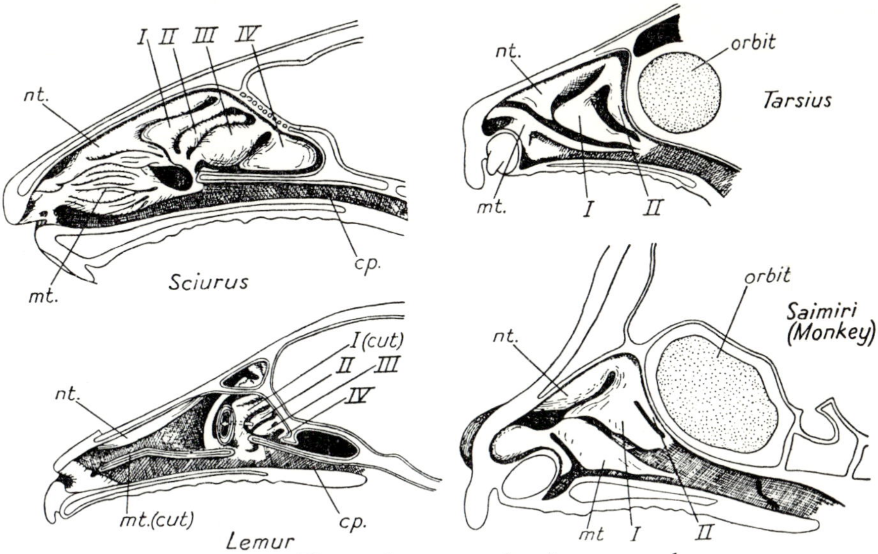

FIG. 394. The nasal passages of various mammals.
cp. choanal passage; *mt.* maxillo-turbinal; *nt.* naso-turbinal; *I–IV*, endo-turbinals.
(From Cave, *B.M.A. Ann. Proc.* 1948.)

monkey (Fig. 394), and there is neither a cleft in the upper lip nor a moist rhinarium, such as is present in most mammals and in lemurs, but absent in anthropoids. This reduction of the snout as a tactile organ perhaps goes with the development of the hand for that purpose.

The eyes are enormous, relatively larger than in any other Primate, but suited for night vision, with the retina containing only rods, though, nevertheless, possessing a yellow macula and a small fovea. The external ears are large and mobile and the sense of hearing is keen. The orbit is partly divided off from the temporal fossa (Fig. 387). The tympanic bone is not only fused to the very large petrosal bulla but also somewhat drawn out into a spout, as in anthropoids (Fig. 388). The brain is small and shows a curious mixture of early mammalian and advanced primate characters. The olfactory regions are small and the cerebral hemispheres large, though smooth. The visual (occipital)

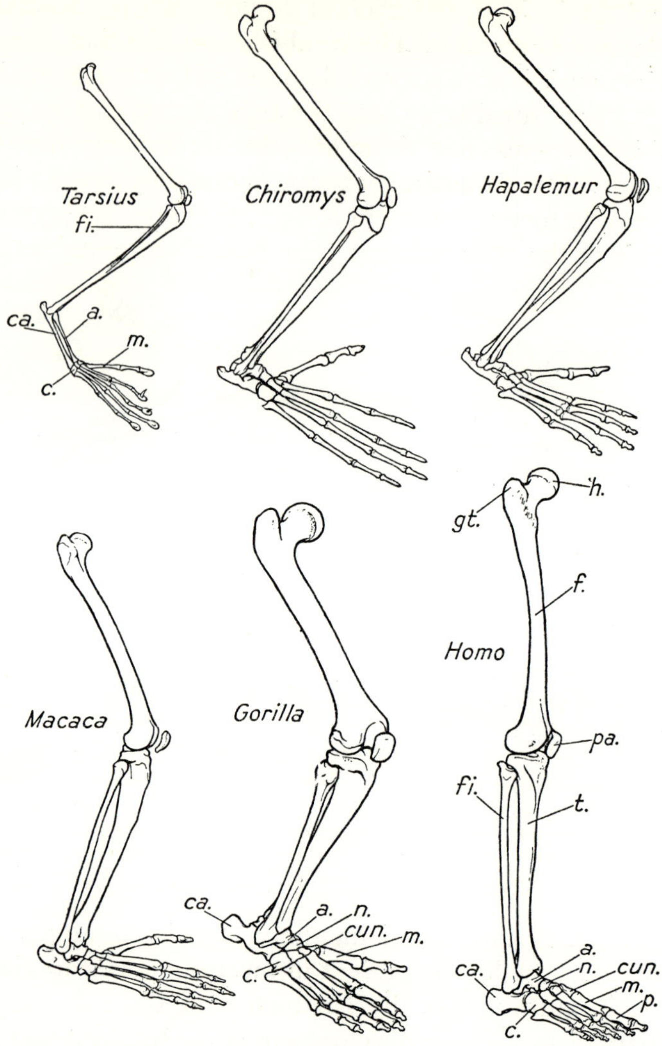

FIG. 395. Bones of hind leg and foot of various primates (not to the same scale).

a. astragalus (talus); *c.* cuboid; *ca.* calcaneum; *cun.* medial cuneiform; *f.* femur; *fi.* fibula; *gt.* greater trochanter; *h.* head; *m.* first metatarsal; *n.* navicular; *p.* proximal phalanx; *pa.* patella; *t.* tibia.

cortex shows remarkably well-differentiated layers. The corpus callosum is small and the anterior commissure large. The cerebellum is small and simple. The posterior corpora quadragemina are large.

Tarsiers are said to live in pairs with no social organization. The reproduction shows much similarity to that of Anthropoidea. There is a menstrual cycle. The uterus is double, as in lemurs, but the placenta is of discoidal shape and haemochorial organization, with a much reduced allantois, almost like that of apes and men. The amnion is formed by folding.

Some of these characters indicate a type of organization so similar to that of anthropoids that the resemblance can hardly be entirely due to convergence. Many of the monkey-like features of the head could, however, be due to the large size of the eyes. Moreover, the reduction of the turbinals has taken place differently in *Tarsius* and anthropoids.

In its legs (Fig. 395) *Tarsius* shows considerable specialization for its jumping method of progression, the fibula being fused with the tibia and the calcaneum and astragalus elongated so as to provide an extra leg segment while retaining the grasping foot. Both first digits can be used for grasping and the digits bear adhesive pads; all have nails except the second and third in the hind-limb, which carry claws used for cleaning the fur. As in other jumping animals the ilium is very long. There is also a long tail.

The *Anaptomorphidae are fossil tarsioids found from Palaeocene to Oligocene in Europe and America. At least 20 genera are known, mostly from skulls and teeth; where limb bones are found they already have the tarsioid specializations. In most the eyes were large and the face short. The brain was like that of *Tarsius* but with olfactory regions better developed. The teeth of some (**Pseudoloris*) were remarkably like those of *Tarsius* but of very generalized pattern $(\frac{3.1.3.3}{2.1.4.3})$ and tribosphenic (Fig. 383). Other lines were specialized, for instance by reduction of the lower incisors and procumbency of the canines (**Necrolemur*). Some of these animals may have been close to the primate ancestry, but it must be recognized that the tarsioids show more similarity to the Anthropoidea than to the lemurs. We are ignorant of tarsioid history from the Oligocene to recent times, but it seems likely that they have remained an isolated stock, their relationship to higher primates being one of common ancestry in early Tertiary times, when all these primates were so alike that it is best to class them together as Prosimii, some of which went on to develop into anthropoids (Fig. 416).

7. Characteristics of Anthropoidea

The monkeys, apes, and men form a natural group, almost certainly of common descent from some Eocene population. They first appear as fossils in the Oligocene and have flourished greatly since; Simpson lists 66 genera in the suborder, of which only 30 are extinct; evidently the type has been successful and is expanding. The outstanding characteristic of the anthropoids might be said to be their liveliness and exploratory activity, coming perhaps originally from life in the tree-tops, necessitating continual use of eye, brain, and limbs. With this is associated the development of an elaborate social life, based not, as in most mammals, on smell, but on sight. Monkeys show more bright colours than do other mammals, especially the curious reds and blues worn on the head and rear. The species may become subdivided into distinct races showing great differences of coat colour (Fig. 400). Communication between individuals is ensured by elaborate systems of vocal signals and the platysma muscle becomes differentiated into a set of facial muscles used to signal 'emotions'.

Many of the characters of the group are those of *Tarsius*, listed already, but the Anthropoidea are mostly diurnal and microsmatic, with a short snout, large forwardly-directed eyes, many cones, a well-marked central area in the retina and partial decussation in the optic tract, features that are associated with binocular vision and large powers of visual form discrimination. The external ears, no longer serving as tactile organs or for direction-finding, are small and the edge is usually rolled over. The orbit is closed off behind. The tympanic bone is fused to the petrosal and in later forms drawn out to a spout. The tactile sense is greatly developed on the fingers and toes, which carry characteristic ridges. The brain (Fig. 386) is relatively much larger than in lemurs or *Tarsius* and its cerebral hemispheres are especially well developed, overhanging the cerebellum and medulla. The olfactory parts of the brain are reduced and the pyriform lobe becomes displaced on to the medial surface by the extension of the neopallium. The surface of the neopallium is highly fissured, showing a characteristic form of Sylvian fissure, and a well-marked central sulcus, separating the motor and sensory areas. A parieto-occipital sulcus separates these lobes and a large lunate sulcus marks the visual area, especially in monkeys (simian fissure). The neocortex thus shows four distinct lobes, frontal, parietal, occipital, and temporal. The occipital (visual) and frontal regions are especially large.

The head is rounded to fit the brain, with the foramen magnum below, so that the head is carried up on a mobile neck. The gait of monkeys is typically quadrupedal and plantigrade when on the ground, with the fore-limbs somewhat longer than the hind. In the trees some may be described as low canopy runners (e.g. guenons) others are high canopy acrobatic types (spider monkeys). Only the great apes habitually swing along with the arms (brachiation). Some

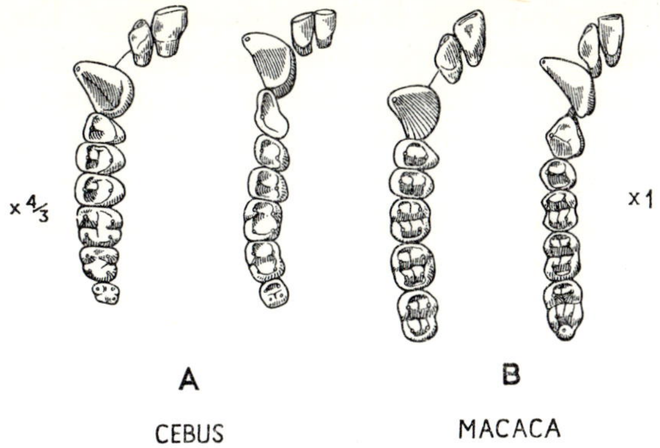

FIG. 396. Tooth rows of New World and Old World monkeys. (After Le Gros Clark.)

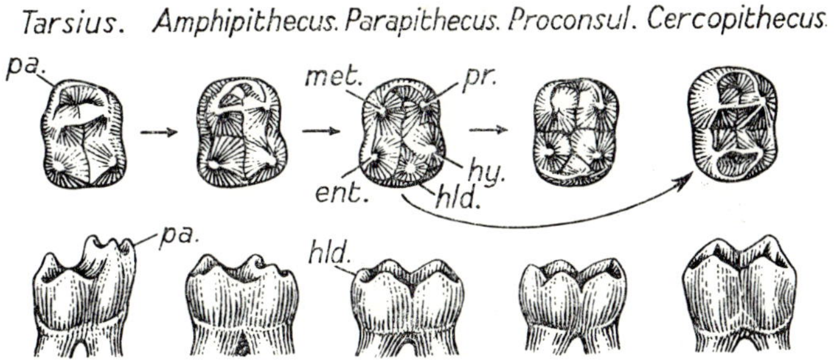

FIG. 397. Diagram of the right lower molar cusp pattern in some Primates, to show the presumed evolutionary stages in the development of the cusp pattern. *Tarsius* shows the primitive (tribosphenic) type. The paraconid is involved in the formation of the trigonid. **Amphipithecus* shows the paraconid undergoing reduction while the talonid and trigonid portions of the crown are at the same level. **Parapithecus*: the paraconid has completely disappeared, the talonid bears the hypoconid, the entoconid, and a relatively well-developed hypoconulid; the trigonid portion bears the metaconid and protoconid. **Proconsul*: the five cusps are more or less equally developed and separated by a characteristic pattern of intervening grooves. *Cercopithecus* shows the characteristic bilophodont pattern, with transverse ridges.

ent. entoconid; *hld.* hypoconulid; *hy.* hypoconid; *met.* metaconid; *pa.* paraconid; *pr.* protoconid. (After Le Gros Clark.)

monkeys have become terrestrial (baboons). The pollex and hallux are opposable and the digits all carry nails. The hands and feet are used for feeding as well as for locomotion.

The characteristic of the tooth row is a tendency to shortening, presumably connected with the shortening of the face. There are three premolars in the earlier Anthropoidea, later reduced to two (Fig. 383). In the line leading to man there is a tendency to still further reduction, with the last molar becoming smaller than the others. The cusp-pattern is tritubercular in earlier anthropoids, but later the molars become square and have four or more bunodont cusps in higher anthropoids (Fig. 383). The incisors become spatulate, rather than pointed, and the premolars bicuspid (Figs. 396, 397).

The reproduction is characterized by the presence of menstrual cycles, continuing throughout the year. Ovulation occurs once in each cycle, often accompanied by the development of sexual signals and behaviour patterns by the female. There is a discoidal, haemochorial placenta, with very early development of the extra-embryonic mesoderm and reduction of the yolk-sac, amniotic folds, and allantois. Usually a single offspring is produced and there is one pair of pectoral mammae. The young are looked after for a long period. The social life is often based upon families of one male and several females and young.

FIG. 398. Spider monkey, *Ateles*. (From life.)

8. New World monkeys, Ceboidea

The continent of South America houses a special type of monkeys, as of so many other mammalian groups. These platyrrhine (flat-nosed) monkeys have presumably been isolated since Eocene times. They could not have been a later immigration from North America because, so far as we know, no cercopithecoids or hominoids reached that continent until man came. The differences from the Old World monkeys are not very profound, however; therefore either the characteristic monkey organization had appeared in the Eocene or the platyrrhines and catarrhines have evolved on parallel lines.

In the teeth the second premolar is retained ($\frac{2.1.3.3}{2.1.3.3}$), whereas it is lost in all Old World forms; the molars are quadritubercular (Figs. 383 and 396). The brain is relatively larger in marmosets even than in man, but this results from the small size of the animal. The smaller species show little fissuring, but this develops in the larger ones (*Ateles*), showing a pattern similar to that of Old World monkeys. The nasal apparatus, though smaller than in lemurs, is larger than in Old World monkeys, producing the wide separation of the nostrils, from which the name platyrrhine derives. Facial vibrissae are present, but usually small. In the ear the tympanic bone is a ring, fused with the petrosal, but is not drawn out into a tube as it is in catarrhines, and there is a large bulla, which is absent in the latter. The coecum is relatively large. The reproductive system does not show the full 'anthropoid' pattern; for instance, there are at most only slight signs of menstrual bleeding at the end of the luteal phase of the oestrus cycle. Social life is well

FIG. 399. Common marmoset, *Callithrix*. (From life.)

developed but the sexual signalling system is probably less complicated than in catarrhines. Thus the colour is seldom brilliant, and the facial musculature around the mouth relatively simple. The loud voice of the howler monkeys (*Alouatta*), which have special laryngeal sacs, is used in the assertion of territorial rights by the clan. This, unlike the families of most monkeys, includes several mature males as well as females and young. Cooperation is ensured by a language of at least nine distinct sounds with separate 'meanings'. Spider monkeys (*Ateles*, Fig. 398) have a somewhat similar organization.

These New World monkeys are very well adapted for arboreal life, with long limbs, delicate hands, and tail for balancing or seizing. The tail pad has special tactile sensitivity, with ridges like those on the digits and a large representation in the cerebral cortex. The animals swing along freely among the branches and may make jumps in which they advance by as much as 15 ft while falling 50 ft.

The fourteen living genera of New World monkeys are divided into two families, the marmosets, *Callithrix* (= *Hapale*), being more primitive than the remainder (Cebidae). The marmosets (Fig. 399) are very small insect- and fruit-eating animals, of somewhat squirrel-like appearance and habits, living in tropical South America. They have a thick non-prehensile tail and there are claws on all the digits except the first, allowing the animals to run up trunks they cannot grasp. These are probably true claws and not secondarily modified nails. The pollex is not opposable. Three premolars are present, but the molars are reduced to two, a condition found in no other anthropoid. The cusp-pattern is tritubercular. Unlike other anthropoids the marmosets give birth to two or three young and there are signs of ancient conditions in the placenta, where the yolk-sac becomes larger than in most primates.

Unfortunately, little is known of the evolutionary history of the New World monkeys; fossils are known only back to the Miocene of South America (**Homunculus*). In view of the isolation and compactness of the group we may feel reasonably confident that it has evolved independently since its origin from Eocene tarsioids.

XXIV

MONKEYS, APES, AND MEN

1. Common origin of Old World monkeys, apes, and men

THE seventeen living genera of Old World monkeys, apes, and men are sometimes classified together as Catarrhina because of the common characteristics in which they contrast with the New World monkeys. Perhaps this union is justified and the Catarrhina is a monophyletic group, with a common ancestor in the late Eocene. However, these creatures are obviously much more diverse than the New World monkeys and have entered a wide variety of habitats. The Old World monkeys proper, the super-family Cercopithecoidea, diverged very early from the apes and men (Hominoidea). Some distinguish between the monkey and ape-human branches of the stock, by calling the former cynomorphs, the latter anthropomorphs or hominoids. Others regard all three groups as widely separate. However, the earliest definite catarrhine known, **Parapithecus* of the Oligocene, is close to the origin of all three groups, so we may reasonably keep them together.

2. Old World monkeys, Cercopithecoidea

The cercopithecid or Old World monkeys do not differ very strikingly in general habits and organization from the monkeys of the New World, though they are mostly larger. We must conclude either that the two groups have made many changes in parallel or that in the Eocene there were already animals with the good senses, active brains, and skilled movements of the monkeys. The distinguishing features of the Old World types are rather trivial, for instance they sit upon ischial callosities, surrounded by naked and often highly coloured skin, which becomes enlarged in the female before ovulation. There are often cheek pouches in which to store food, usually complicated laryngeal sacs, a bony tympanic tube, and never a prehensile tail. The great reduction of the olfactory turbinals leaves the nostrils close together and pointing downwards. The dentition is reduced to $\frac{2.1.2.3}{2.1.2.3}$, the upper molars carry four cusps, and the lower four except for the last, which has five (Figs. 383 and 396). The diet of the more generalized cercopithecids is omnivorous, including insects, lizards, eggs, and fruit, but many monkeys are specialized fruit-eaters, and the molar teeth are quadrangular, with the four cusps united to make two transverse ridges used for grinding, somewhat as in ungulates. These specialized molars make it unlikely that the modern cercopithecids could have been ancestral to the apes or man. The colon usually has a sigmoid flexure and a small caecum and appendix. The reproduction is similar to that of apes and man; there is menstrual bleeding and haemochorial placenta. The ano-genital region (sexual skin) of the female may show marked signals (swelling and coloration) at the time

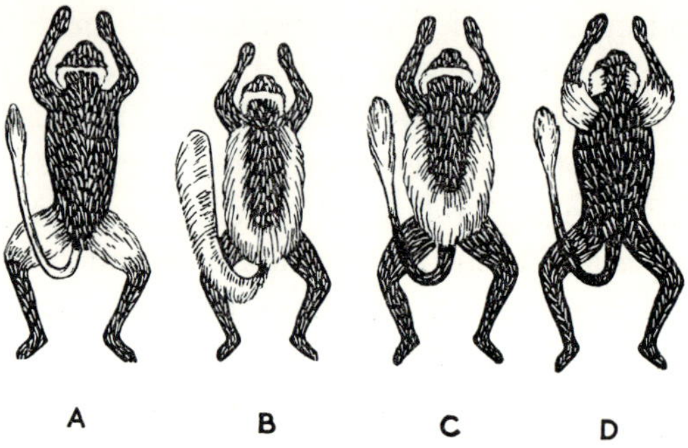

FIG. 400. Coloration types of various sub-species of *Colobus polykomos*. A. *C.p. vellerosus* Graff. B. *C.p. caudatus* Thomas. C. *C.p. abyssinicus* Oken. D. *C.p. angolensis* Sclater. (After P. Rodt.)

of ovulation. The male shows continuous spermatogenesis, without a breeding season. In some species there are, however, seasonal fluctuations in the number of births.

The cerebral cortex is always large and fissured and its frontal regions well developed. Behaviour is exploratory and manipulative and learning powers high. Social behaviour is elaborate and often based upon polygamous families. The coat colour is often ornate (Fig. 400) and there are elaborate communication systems by the facial musculature and vocal apparatus.

The Cercopithecoidea include a number of variant types, many common and well known. *Macaca*, the rhesus monkey (Fig. 401), has many species in Asia and North Africa, one reaching Gibraltar. *Cercopithecus*, the guenons, are similar animals in Africa, many highly coloured. The baboons, *Papio* (Fig. 402), of Africa and related forms in Arabia, and mandrills, *Mandrillus* (Fig. 403) of West Africa have become secondarily terrestrial and quadrupedal and the face has become elongated ('dog-face'), allowing for a long tooth-row of grinding molars. The Colobinae are fully arboreal, leaf-eating monkeys, without cheek pouches but with the stomach sacculated; the guerezas

(*Colobus*) live in tropical Africa and the langurs or leaf monkeys (*Presbytis* = *Semnopithecus*) in south Asia.

The monkey type thus shows us something of the condition of catarrhines in Oligocene and Miocene times. A fragment of a lower jaw from the Lower Oligocene of Egypt, **Apidium*, may have been a

FIG. 401. Rhesus monkey, *Macaca*. (From life.)

FIG. 402. Sacred baboon, *Papio*. (From life.)

FIG. 403. Mandrill, *Mandrillus*. (From life.)

very early catarrhine, if it belongs to a primate at all. The molars carry a quadrangle of four cusps and a hypoconulid behind. **Parapithecus* from the same deposits is known from a single mandible (Figs. 397 and 409) and was an animal the size of a squirrel, which might have been derived from the anaptomorphids and led on to the catarrhines. The two rami diverge posteriorly, as in tarsioids, rather than running parallel. However, the number of teeth is reduced to that of catarrhines and the canines are incisiform. The molars carry five cusps, not united by ridges. The cercopithecid type was well established by the Miocene and abundant remains are available of **Mesopithecus*, with the molar cusps united to ridges.

3. The great apes : Pongidae

The question of the exact degree of affinity between the existing apes and man remains unsettled. There were plenty of fossil apes in

FIG. 404. Gibbon, *Hylobates*. (From life.)

Miocene times and man-like creatures are found in the early Pleistocene, but we have no undoubted evidence of human remains from the Pliocene and it is therefore impossible to say whether the human stock was derived from apes after Miocene times or whether it separated much earlier, either from the ancestral catarrhine stock, say, in the Oligocene, or, as a few believe, even earlier still, from some *Tarsius*-like prosimian. Evidence of definitely man-like creatures that can be placed with certainty in a family Hominidae is found only back to about 1 million years ago, whereas the longest of the above estimates would say that our stock has been distinct and evolving separately for nearly 60 million years, without leaving any remains. Although the view that men have descended from apes is probably the more widely held, we shall first survey the structure of the great apes by treating them as members of a separate family Pongidae.

The living apes include the gibbon, *Hylobates* (Fig. 404), and orang-utan, *Pongo* = *Simia* (Fig. 405), from east Asia and the chimpanzee, *Pan* (Fig. 406), and *Gorilla* (Fig. 407) from Africa. They and related fossil forms are marked off from the cercopithecids by their teeth and methods of locomotion. Many apes are rather large animals and this has made it impossible for them to walk along the branches as monkeys do. They therefore swing by the arms, which are longer than the legs

FIG. 405. Orang-utan, *Pongo*. (From life.)

FIG. 406. Chimpanzee, *Pan*. (From life.)

FIG. 407. Gorilla, *Gorilla*. (From life.)

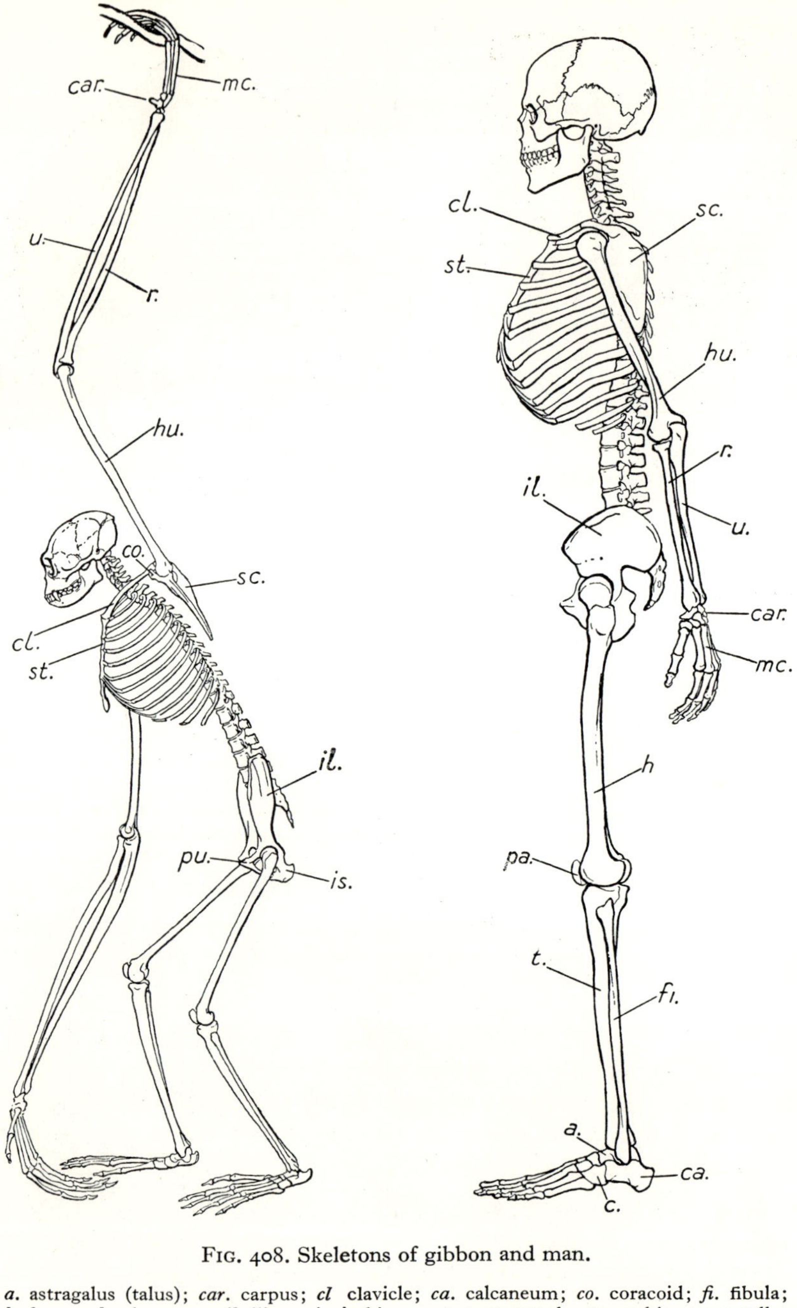

FIG. 408. Skeletons of gibbon and man.

a. astragalus (talus); *car.* carpus; *cl* clavicle; *ca.* calcaneum; *co.* coracoid; *fi.* fibula; *h.* femur; *hu.* humerus; *il.* ilium; *is.* ischium; *mc.* metacarpals; *pu.* pubis; *pa.* patella; *r.* radius; *sc.* scapula; *st.* sternum; *t.* tibia; *u.* ulna.

and are provided with very powerful muscles, the hands and feet being efficient grasping organs. There is no tail. These brachiating habits have affected the entire skeleton. All the apes and men differ from the cercopithecids in having wider chests, longer necks, longer limbs, and larger heads (Fig. 408). The cervical and sacral regions are longer in

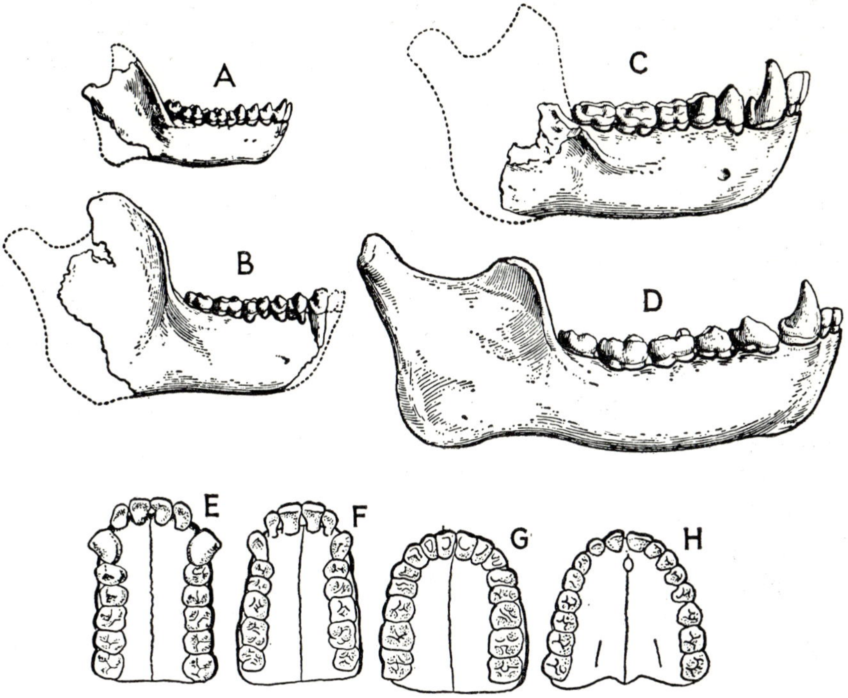

FIG. 409. Mandibles of A, **Parapithecus*; B, **Propliopithecus*; C, **Pliopithecus*, D, *Hylobates*. Dental arches of E, *Gorilla*; F, **Dryopithecus*; G, **Plesianthropus*; H, *Homo*.

(A–D after Le Gros Clark, *B.M. Guide*, and Gregory; E–H after Howells, *Mankind So Far*, Doubleday and Company, Inc.)

apes than in the monkeys and the lumbar region shorter. When proceeding on the ground the apes cannot balance on two legs for long; instead, the long forearms prop up the front of the body and produce a semi-erect position. The hands are specialized for brachiating, with a short thumb and long metacarpals and digits (Figs. 410 and 411). The foot is in general similar but in the chimpanzee and gorilla it is more suited for walking, with broader sole and shorter toes. In the terrestrial *Gorilla gorilla beringei* the hallux lies parallel to the other toes almost as in man.

The teeth (Fig. 409) are of a rather generalized type. The canines are often large, especially in males, and the lower front premolar forms a sectorial blade. The molars carry grinding tubercles and often show a crenation of the enamel, which is characteristic of apes, though found as an abnormality in monkeys and man. The trigon is still present in the upper molar, the hypocone being small. The lower molars have a hypoconulid, making five cusps, as contrasted with the four of monkeys. All the apes are mainly vegetarians, but they may eat meat occasionally. The use of the teeth for grinding is associated with powerful masticatory muscles and the development of temporal and occipital crests. The supraorbital ridges are also large. The canines are used for attack and defence. In the digestive system apes and man differ from other primates in the presence of a vermiform appendix.

The brain is much larger than in cercopithecids, and shows a pattern of convolutions similar to that of man, though more simple. The behaviour provides many signs of efficient memory, leading to the attitudes that we characterize as imitation and association of ideas. There are extensive powers of manipulation and for obtaining ends by indirect means.

The communication system is highly developed. The young chimpanzee is said to be able to make at least thirty-two distinct sounds. The facial musculature is more highly differentiated than in monkeys and produces a wide range of expressions such as of rage, surprise, pleasure, and laughter.

Social organization is always well developed. The gibbons are monogamous, the family consisting of a pair and the young of current and previous years. Chimpanzees, and so far as is known gorillas, live in bands, led by a dominant male. The individuals of a band cooperate in helping each other and the groups show differing social traditions. The apes are diurnal animals, eating in the day. They make platforms on which they rest at night (except the gibbons).

Reproduction is based on a menstrual cycle of thirty-five days in chimpanzees, with a great development of the sexual skin at mid-cycle. Gestation is long, as is the growth period, seven to nine years in the gibbon, ten to twelve in the chimpanzee. The life span is also long, reaching forty years in chimpanzees, perhaps fifty in gorillas.

The gibbons are fully arboreal, swinging rapidly with their extremely long arms (Fig. 404). This extreme brachiation may be a quite recent specialization. They eat mainly fruit and leaves, also insects or eggs. Gibbons are numerous throughout south Asia and the type is certainly the most successful among modern great apes. The characteristic cries are part of the defence of the territory that is occupied by each group. The orang-utan of Borneo and Sumatra is larger and also has very long arms. The chimpanzees (*Pan*) and gorillas (*Gorilla*), living in the forests of tropical Africa, are so alike that it is doubtful

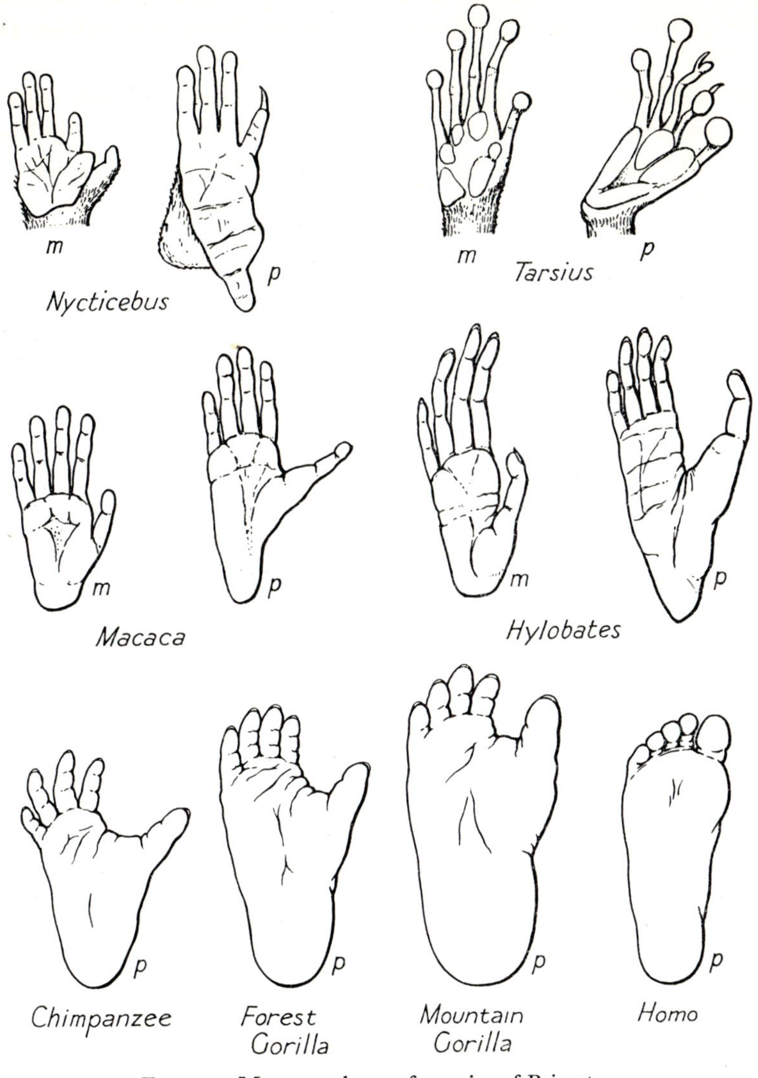

FIG. 410. Manus and pes of a series of Primates.
m. manus; *p.* pes. (After Le Gros Clark, Morton & Fuller.)

if the generic separation is justified. The chimpanzee is the smaller and less muscular animal, lacking, for example, the large parietal and occipital crests found in the male gorilla. There is a corresponding difference of temperament, the chimpanzee being lively and sometimes tameable, the gorilla gloomy, ferocious, and unafraid. Gorillas are mainly terrestrial, walking on all fours and sleeping either on the ground (males) or at a small height (females and young). Like the other apes they show much local variation but all may be referred to a single species, *G. gorilla*. *G. gorilla beringei* is a mountain form that is more fully terrestrial than the others (Fig. 410).

The similarities and differences between these animals and man will be discussed later. Their relationship with the other catarrhines is clearer. The lower jaw of **Parapithecus* of the Egyptian lower Oligocene contained teeth of a pattern that could have given rise to those of the Pongidae as well as the Cercopithecidae (Fig. 397). In the same beds was found another jaw, which is definitely that of an ape, **Propliopithecus*. Here the molars have a distinctly five-cusped pattern, with protoconid and metaconid in front and a large heel, carrying a hypoconid laterally and entoconid medially, and also a posterior hypoconulid. Some such animal could have given rise to **Limnopithecus* of the Miocene and **Pliopithecus* of the Pliocene, animals similar to the gibbons and living in the woods of Europe and Africa. Great apes were found quite widely in the Old World during the Miocene and Pliocene. The earliest of these, **Proconsul* from the lower Miocene of Kenya, showed a combination of characters of cercopithecids, great apes, and man. The skull was more lightly built than in apes, with no brow ridges. The tooth rows converged anteriorly as in **Parapithecus*. The incisors were small and like those of man but the canines were large and the first lower premolar was sectorial as in apes. The limb bones suggest that the gait was terrestrial and quadrupedal, and that the brachiating habit had not yet evolved.

**Dryopithecus* from the middle and upper Miocene of Africa, Europe, and India was closer to the apes, with U-shaped dental arcades. On the other hand, **Ramapithecus* from the Miocene and Pliocene of India showed human characteristics in the rounded upper arcade of the teeth, small canines, and other features.

Several other types are known and evidently the apes were widespread, varied, and successful animals in the Miocene and there are among them plenty of signs of the characteristics both of the modern apes and men. The remains of **Oreopithecus* from the Pliocene of Italy show a curious mixture of characters. It was not a brachiator, but its method of locomotion is not clear. The lower molars have four cusps arranged in pairs as in monkeys but not united by ridges. The upper molars resemble those of apes but the small canines, absence of diastema, and bicuspid first lower premolar have led some to place it close to man.

4. The ancestry of man

In order to discover the position of man in relationship to the living and fossil ape populations we may try to specify the characters distinctive of the family Hominidae and then discuss whether they could have been derived from those of monkeys or apes. Schultz, who has made careful measurement of many features of primates, lists the following as the chief specializations of man: (1) elaboration of the brain and behaviour, including communication by facial gestures and speech; (2) the erect posture; (3) prolongation of post-natal development; and (4) the great rise in population in recent years. Others might make up the list differently, but we may use it as a basis for discussion of the differences between men and other creatures.

5. Brain of apes and man

The brain is much larger absolutely and relatively in man than any living ape; Fig. 382 shows that man stands farther apart from the apes in this respect than they do from other anthropoids. The cranial capacity for males of modern (Caucasian) man may be taken as 1,500 c.c., whereas that of chimpanzees is given as 410, gorillas as 510, and orangs as 450. The general arrangement of function within the brains is similar in man and apes, but the parts especially well developed in man are the frontal and occipital lobes. The latter are concerned with the sense of sight and are related to our intensely visual life. The frontal lobes, so far as is known, serve to maintain the balance between caution or restraint and sustained active pursuit of distant ends, which, above all else, ensures human survival in such a variety of situations, and makes possible the social life by which so great a population is maintained. The difference of behaviour between men and apes exceeds all the structural differences; our lives are so widely different from theirs that any attempt to specify the divergences in detail is apt to seem ridiculous. Perhaps the more striking of them are related to the powers of communication by speech which, besides its obvious social advantage, gives to man the power of abstract thought. Whatever we may think about the consciousness of animals there is no doubt that our own awareness of life, being expressed in words, is widely different from that of all other creatures. The speech system depends upon a complex of features of the brain, larynx, tongue, mouth, and auditory apparatus. In addition, the facial musculature is more fully differentiated even than in apes, especially around the eyes and mouth.

6. The posture and gait of man

The gait of man differs from that of any ape in that the body can be fully and continuously balanced on the two legs. This involves considerable modifications throughout the skeleton and musculature (Fig. 408). The backbone, instead of the single thoracic curve of quadrupeds, has an S shape, being convex forward in the lumbar, backward in the thoracic, and again forward in the cervical region. The thoracic curve develops before birth, but the cervical only as the baby holds its head up and the lumbar as it begins to walk. The vertebral column, which in quadrupeds is a horizontal girder, in man becomes vertical, carrying bending and compression stresses along its length. This entirely alters the arrangement of its secondary struts and ties. The bodies of the vertebrae carry much of the weight and are massive, tapering in size upwards. They are separated by well-developed intervertebral disks, acting as elastic cushions. The weight of the head is balanced on the backbone through the neck, and the thorax acts as a bracket from which the viscera are suspended. The muscles of the back, the ties of the vertebral girder, though arranged on the same general morphological plan as in quadrupeds, now carry very different stresses and no long neural spines or large transverse processes develop, since the girder is not now of cantilever type. For the same reason there is no sharp change in the direction of the neural spines at the hind end of the thoracic region; the girder is now one unit, with bending stressing along its whole length.

The balancing of the body on the legs also involves many changes. The muscles around the hip joint achieve this balance, and the changes to allow this affect especially the gluteal muscles and the ilium and sacrum to which they are attached, these being the extensor and abductor muscles, which raise the body from the quadrupedal position and prevent it falling medially when the weight is on one leg. The buttocks are therefore a characteristic human structure. The adoption of a bipedal position imposes entirely new requirements on the musculature of the limbs. In quadrupedal progression the retractor muscles are the main means of locomotion, drawing the leg backward at the hips while straightening the knee. In man the propulsive thrust is obtained mainly from the calf muscles and in particular from the soleus, which runs from the tibia to the heel, the gastrocnemius, since it tends also to bend the knee, being reduced. The quadriceps femoris becomes very large, serving to keep the knee extended both while the calf muscles develop their thrust and, as a check to the forward momentum, when the foot touches the ground.

The ilium is very broad in man, increasing the surfaces for attachment of the glutei, iliacus (a flexor of the hip), and for the abdominal muscles, which are attached along its crest and have an important part to play in carrying the weight of the viscera.

7. The limbs of man

Many changes would be needed to convert an ape-like leg and foot to the human condition (Fig. 395). The femur of man is straight and the articular surface at its lower end set at an angle to the shaft. This allows the lower legs and feet to be as nearly as possible below the centre of gravity in standing, in other words, for the knees to be held together although the femoral heads are wide apart. At the ankle joint, on the other hand, the articular surface is at right angles to the tibia in man, at an oblique angle in apes, since in the latter the foot is turned outwards. In ourselves the weight is transferred from the tibia to the talus and then partly backwards to the calcaneum and partly forwards through the tarsus to the metatarsal heads (Fig. 412). The calcaneum is modified for this weight-bearing and the tarsus and digits even more so, the whole foot being converted into an arched system, no trace of which is found in apes. With this arrangement the hallux is not used for grasping and is very large. It is held in line with the other digits and the whole forms a compact wedge with a joint at the metatarsal heads. In walking, when the foot is raised by the calf muscles, the toes remain on the ground, to prevent slipping forwards. The condition in which the first toe is the longest is peculiar to man, but in some monkeys and apes the axis tends to shift from the third digit medially and the human condition is an accentuation of this change, with the metatarsal and first phalanx of the first digit becoming long and strong. Even in modern human populations the second toe as a whole is often longer than the first; this condition was perhaps commoner in historical antiquity (the 'Grecian toe'), and may be a cause of foot trouble, the long second digit being unsuited to the stresses it is made to bear.

The differences between apes and men in the arms and hands (Fig. 410) are marked, though perhaps less striking than in the feet. The human fore-limb is, of course, relatively much shorter than that of any ape and its muscles far less powerful. In order to carry the whole weight of the large body an ape needs enormous muscles all along the limb. Thus the serratus anterior, which pulls the body up on the scapula, is very large and the ribs to which it is attached have large flattened surfaces, are very long, and extend far caudally; the chest of man is much more lightly built. Similarly, the muscles of the shoulder and the flexor muscles of the elbow, wrist, and hand are all much larger in apes, as are the ridges to which they are attached, for instance on the palmar surfaces of the phalanges (Fig. 411). The human arm has specialized in mobility. The hand can be brought into almost any position in relation to the body by virtue of the wide range of

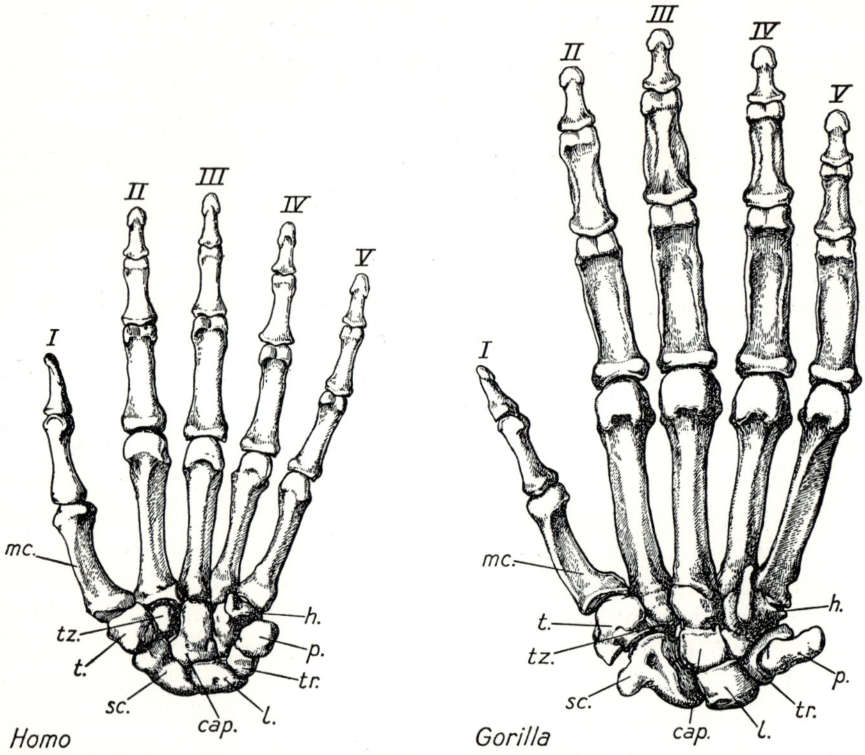

FIG. 411. Bones of the left hands of man and gorilla, palmar view.

cap. capitate; *h.* hamate; *l.* lunate; *mc.* metacarpal; *p.* pisiform; *sc.* scaphoid; *t.* trapezium; *tr.* triquetral; *tz.* trapezoid.

Notice the large bony points of attachment of the flexor muscles of the gorilla on the trapezium, scaphoid, hamate, and proximal phalanges.

movement at the shoulder, pronation and supination of the forearm and movements at the wrist.

In the hand itself the thumb is characteristically long in man and moved by powerful muscles. Man is the only animal in which the thumb can be in the fullest sense opposed to the other digits, so that the pads face each other. This is achieved by special development of the joint between the first carpal and metacarpal. The third digit is the longest in apes, as in men, but the second digit (index) of man is generally at least as long as the fourth, often longer (the 'Napoleonic finger'). In lower primates the digits of the ulnar side are relatively much longer. Apart from proportions and skeletal features the human hand also has a very well developed sensory supply, which is essential for its use as a handling organ.

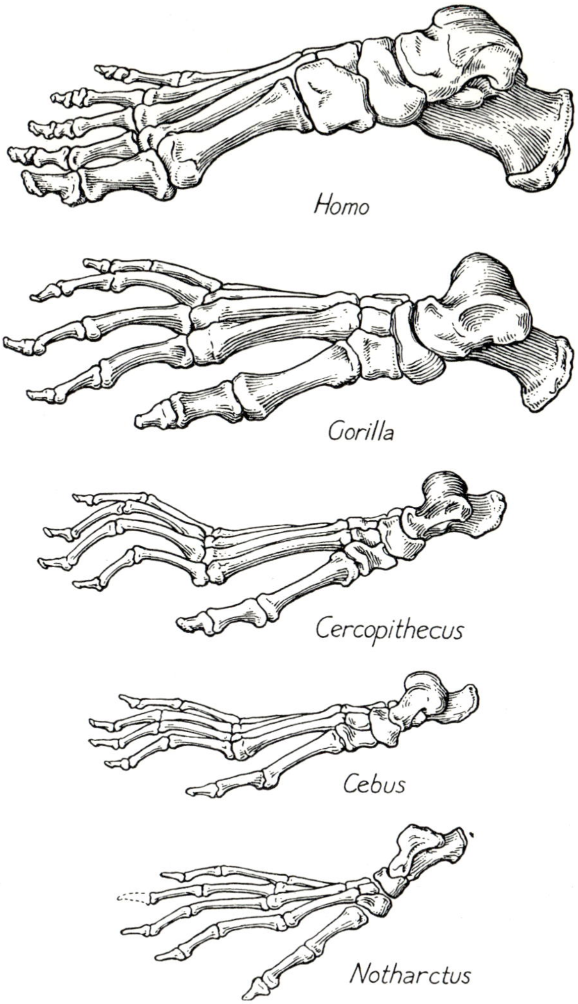

FIG. 412. Foot skeleton of a series of Primates. (After W. K. Gregory.)

8. The skull and jaws of man

Comparisons between the skulls of apes and men have attracted special attention because so many of the finds of early human types have been of skulls (Fig. 413). The differences are mainly referable to changes in the brain, dentition, and method of balancing the head upon the neck. The enlargement of the brain has been in the occipital and especially in the frontal region (p. 633), giving a high forehead and the characteristic upright face. At the same time the jaws have receded, so that the human tooth-row is unusually short. Moreover, the dental

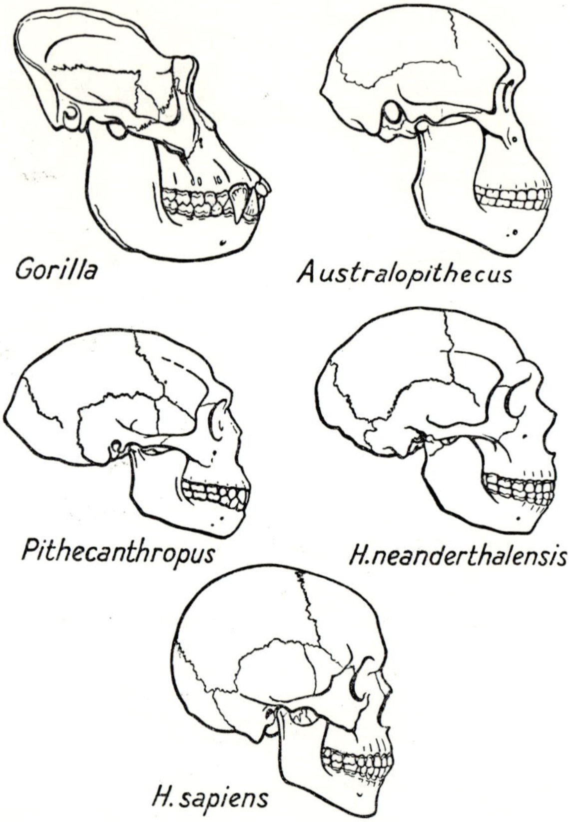

FIG. 413. Skulls of apes and man. The face and back of jaw of **Pithecanthropus* has been restored. (Partly after Romer, *Vertebrate Paleontology*, University of Chicago Press.)

arcade is characteristically rounded in front, that of apes is U-shaped, with large canines at the bend (Fig. 409). In man the canines are small and incisiform; the first lower premolar is bicuspid, like the rest and not sectorial as in other catarrhines. The molars show a characteristic pattern that may be regarded as based on four cusps above and five below. The cusps are arranged roughly as a rectangle, so that the grooves between them make a + as compared with the Y patterns typical of the dryopithecine molar (Fig. 414). Thus the protoconid meets the entoconid in the human but not in the earlier type. However, there is great variation in these patterns, both in man and apes. Little can be said therefore about a single tooth, but the proportion of molars with four cusps and a + pattern is higher in man than in apes (Fig. 415). The last molar (wisdom tooth) is smaller than the others in man, but not in modern apes. There are, however, many signs of possible ape-like ancestry in our teeth; for instance the canine has a long root and erupts late.

The lower jaw of man is less shortened than the upper; whereas in apes it is strengthened by a 'simian shelf' of bone on its inner side, in man this strengthening is on the outside, making the chin. The jaw is less massive in man than in apes, especially its posterior ramus;

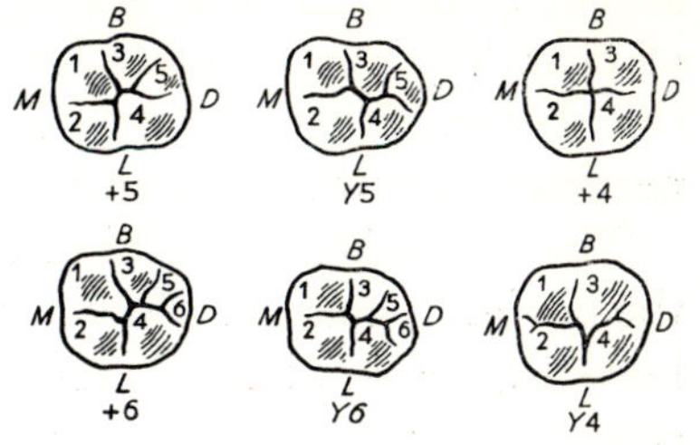

FIG. 414. Mandibular molar patterns in the Liberian chimpanzee and human dentition. The chief distinguishing feature between Y and + patterns is the relationship of cusps 2 and 3 to each other. In the Y pattern they are in contact, in the + pattern they are separated by cusps 1 and 4.

B, buccal; *D*, distal; *L*, lingual; *M*, mesial; 1 protoconid; 2, metaconid; 3, hypoconid; 4, entoconid; 5, hypoconulid. (After Schuman and Brace.)

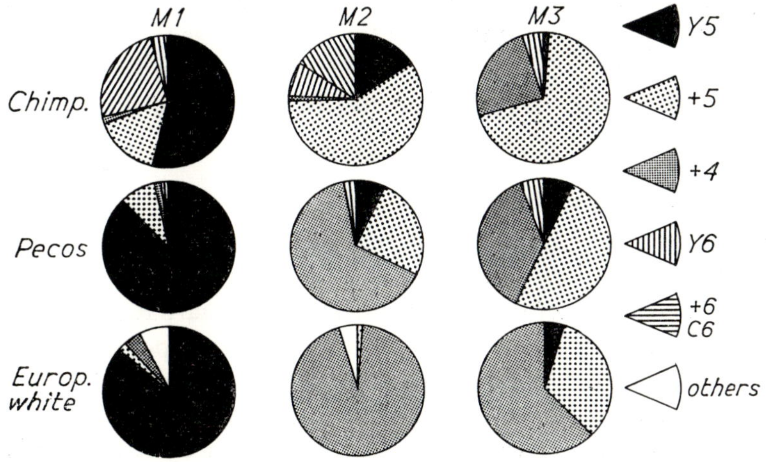

FIG. 415. Proportions of the various mandibular molar patterns found in Chimpanzees, Pecos Indians, and European Whites. (After Schuman and Brace.)

the muscles for moving it are less powerful. Correlated with this weakening of the jaw has been a rounding of the surface of the skull. Occipital and temporal crests for the attachment of the neck and jaw muscles are well developed in the male gorilla, suggested in other apes, but absent in man. The brow ridges, also characteristic of the apes, are large masses of bone above the eyes, probably produced to meet the compression stresses set up by the powerful action of the jaw-muscles. Their absence, together with the large forehead, produces the human type of face. The large external nose is presumably another corollary of the shortened face; it provides some extension of the nasal cavity, necessary for warming and filtering the air.

The balancing of the head on the neck is a result of the adoption of the upright position. Movement of the foramen magnum to a position beneath the skull has been noted as a primate characteristic and it reaches its extreme in man, allowing considerable reduction of the musculature at the back of the neck; the splenius and semispinalis capitis muscles are much smaller in man than in apes. The small size of the trapezius is partly a consequence of the good balance of the head, partly of the absence of brachiating habits. Reduction of these muscles leads to simplification of the bones at both ends of them. The area of their attachment to the occipital surface of the skull becomes much reduced and remains smooth, instead of being roughened and even raised into ridges as in apes. At the same time the spines of the cervical vertebrae, very long in the gorilla, are short and almost vestigial in man. When the head is properly balanced on the backbone it can be freely turned around, and for this purpose the sternomastoid muscles are well developed and the large mastoid ('breast-like') swellings where they are attached to the base of the skull provide a characteristic human feature.

9. Rate of development of man

One of the most striking differences between man and apes is the slow rate of our own growth and development; there is a strong suspicion that many of our features are due to retardation of the time of onset of maturity. Schultz has shown that in the apes growth ceases between the ages of 10 and 12 and that the epiphyses finally close between 12 and 14. Many of the features of man, such as the reduction of hair and the large head, presence of a prepuce on the penis and hymen in the vagina, are those to be found in foetal apes, and it is therefore suggested that one of the main changes leading to our development has been delay in the rate of differentiation and onset of maturity. This might well depend on the endocrine balance, perhaps particularly on the action of the anterior lobe of the pituitary. It is only possible to guess at the process of habit change and selection by which the appropriate genetic change has occurred. It may well be that those family organizations were more efficient in which individuals developed late and were therefore better behaved, in early years because of immaturity, and later by the great development of the 'inhibitory' or balancing functions made possible by growth of the frontal lobes (p. 633). Families composed of such slow-developing and restrained individuals would therefore survive and the genetic factors involving delay of maturity be selected.

10. Growth of human populations

This increase of the post-natal developmental period may well be connected with the appearance of the fourth outstanding feature of man noted by Schultz (p. 633), the great population increase in recent times. No exact figures are available, but it is probable that a first increase occurred when the Neolithic agricultural civilization developed, perhaps 10,000 years ago. This development presumably depended on factors making for orderly and restrained behaviour, such as we have been discussing; it is no accident that family customs are closely linked with those of tribes and nations in all stages of society. A further great increase of human population, probably at least a doubling, has occurred during the past 200 years, and we may associate this with the further extension of habits of thought and restraint in the conduct of affairs, making possible the development of logic and science and their application to human productivity.

11. Time of development of the Hominidae

Thus there are seen to be profound differences between man and the existing apes, and it must be remembered that we have considered mainly skeletal features and hardly touched on the details of the inner life of the animals, or their powers of communication or social organization. The most significant difference between man and all other animals is in the size of the brain (Fig. 382) and the difference of life and behaviour that goes with this. In studying the documents of our history, however, we can discover only little of the brains and less of the behaviour of our ancestors; we must rely mainly on study of the skeleton.

No undoubted human remains are found before the beginning of the Pleistocene, less than 1·5 million years ago. They are not common until 500,000 years later, but their total absence from the Pliocene, Miocene, and Oligocene epochs must certainly be considered suspicious. During those periods we have admittedly only few remains of apes, but they do occur and men do not; there is therefore a prima facie case for considering that men have evolved from the same

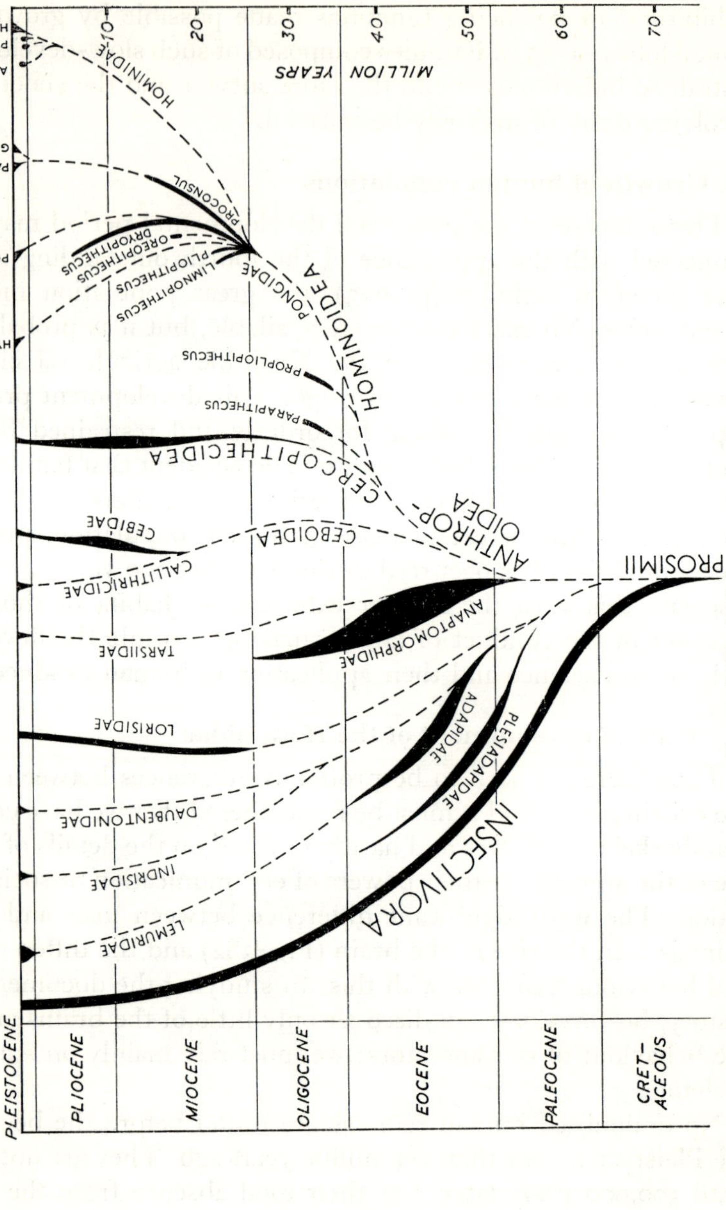

FIG. 416. Evolutionary tree of primates.

Miocene stock as the apes. Some hold, however, that the human line has been distinct for a much longer period, perhaps even back to a separate tarsioid ancestor in the Eocene.

A survey of the evidence about the affinities of man and apes will probably lead the unprejudiced to the conclusion that although we do not know enough to be certain, the human stock probably diverged from that of the apes in early Miocene times, perhaps from a form like **Proconsul*, before the brachiating habit had become fully developed. Fig. 416 shows the possible relationship based on this hypothesis, but

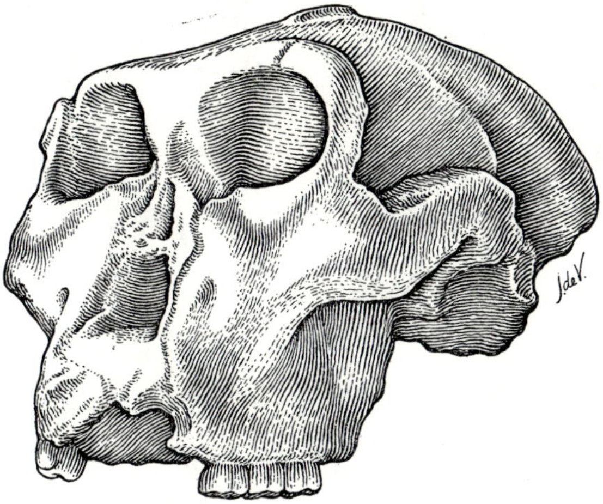

FIG. 417. Skull of *Paranthropus*. (From a cast.)

we shall remain uncertain of the exact course of our descent until Pliocene and Miocene fossils that could have been our ancestors are found.

12. The Australopithecinae

A series of fossils found in Africa shows a curious combination of the characters of men and apes. The specimens occur in lime deposits probably of early Pleistocene date. Tools occur associated with them but the cranial capacity was between 500 and 750 c.c., hardly greater than in living apes. Several types occur and it is not agreed whether all should be included in the single genus **Australopithecus*. The first skull found, in 1924, is that of a young individual, whose rather prognathous jaws and low cranium have an ape-like appearance, though the brow ridges and crests are slight (Fig. 413). Later finds include a different type (**Paranthropus*) with marked brow ridges and sagittal and occipital crests (Fig. 417). Evidently the muscles of mastication were powerful and the jaws are massive. However, the dental arcades are smoothly rounded, as in man, with little development of the canines, small incisors, and bicuspid lower first premolar. The molars, however, are heavily built and the third is the largest. They carry cusps with a 'dryopithecus' pattern (Fig. 418).

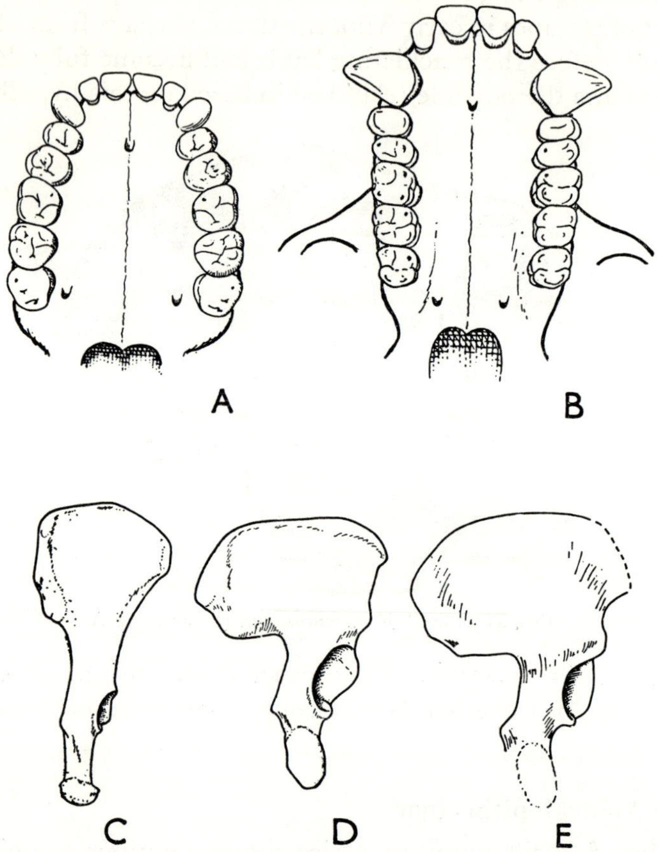

FIG. 418. A, diagrammatic reconstructions of palate of an australopithecine for comparison with that of gorilla (B); right side of pelvis of chimpanzee (C), modern man (D), and an australopithecine (E) (not to the same scale). The missing part of the front of the ilium of E extended farther forward than the dots show. (After Le Gros Clark, *B.M. Guide*.)

There is much evidence that these creatures walked upright. The foramen magnum lies far forward and the area for attachment of nuchal muscles is reduced (Fig. 413). Several specimens of the pelvis have been found and they show a broad ilium very different from that of apes (Fig. 418). The lower end of the femur and the astragalus also show human characteristics. It may be that these creatures were bipeds, though it has not yet been established that the ilium was placed as in man or that the gluteus medius and minimus acted as abductors.†

This complicated mixture of human and ape features makes it difficult to assess the correct position of these fossils. They suggest that the upright gait was not necessarily associated with a large brain. Although the skull and dentition have undoubtedly many similarities with those of man there are also great differences, especially in the **Paranthropus* type. These points make it doubtful whether the Australopithecines lie close to the direct ancestry of man, which would perhaps be unlikely in any case because of their late date. They may well be descendants of the Pliocene population that gave rise to the true hominids.

13. Early Hominids, *Pithecanthropus

Fossil remains that are beyond doubt those of creatures close to man have been found from the first interglacial period of the Pleistocene onwards. The most primitive type, **Pithecanthropus*, was first named from fossils found in Java, but similar bones have since been found in China; although named **Sinanthropus* by some investigators Pekin man is often included in the same genus as the Java man. These creatures had a long, low brain-case, with low forehead, large brow ridge, and a very thick skull (Fig. 413). The cranial capacity was about 900–1,000 c.c., much less than in modern men, but more than in any known ape. The face was rather prognathous, the lower jaw long and strong, but with a receding chin. The teeth were in general of human type, but with large canines in the males. The head must have been quite well balanced on the neck, for the mastoid processes were large. Further evidence that the creature walked erect is found in the straight and very modern femur, but the post-cranial skeleton is not well known.

The differences between these people and modern man are sufficient to make the use of a separate generic name a convenience, yet they must have been in many ways very like ourselves. There is evidence from the caves of the use of fire and pottery, and also the bones there show clear signs of being broken open for their marrow. There is,

† The most recent Australopithecine finds have been referred to a separate genus **Zinjanthropus*, somewhat similar to **Paranthropus*. Its canine teeth are even more human than those of previous finds. Further interesting features are that the fossil was found in association with primitive stone tools (Chellean or pre-Chellean) and finally that the potassium-argon method gives a date of one and three-quarter million years for the stratum. (Leakey 1961.)

however, a slight possibility that the fire, the instruments, and the cannibalism were all the work of later invaders, of a still more human type! The remains undoubtedly belonging to **Pithecanthropus* all come from the Far East, but some large jaws from the middle Pleistocene of Algeria, named **Atlantropus*, may be related.

14. Man

All the remaining finds are usually referred to the genus *Homo*. All living races of man are included in a single species *H. sapiens*, but some of the fossils are placed in other species, *H. heidelbergensis* and *H. neandertalensis*. The first of these is the name given to a jaw of the first interglacial period. It is of a very heavy type, rather like that of **Pithecanthropus*, but with a chin and without a simian shelf. The teeth are not far from the modern type. The true position of this Heidelberg man is uncertain and so therefore the age of the genus *Homo*. Much more abundant remains are found in the third interglacial and last glacial period and are referred to as Neanderthal man (Figs. 419 and 420). The brain-case was larger than that of many modern men (nearly 1,600 c.c.), but was long and low and especially prominent behind. The brow ridges are large, the face rather prognathous, and the chin present but receding. The whole structure of the skull was stouter than our own, with a thick jugal bar, thick roof bones, and large mandible. The teeth were larger than in modern man and the Y5 pattern is said to be frequent. The third molars were smaller than the second, however. The foramen magnum pointed rather backwards and it is possible that Neanderthal man did not walk fully upright. Long cervical spines have been described, but these are very variable and do not justify the reconstruction of these people as having a slouching gait.

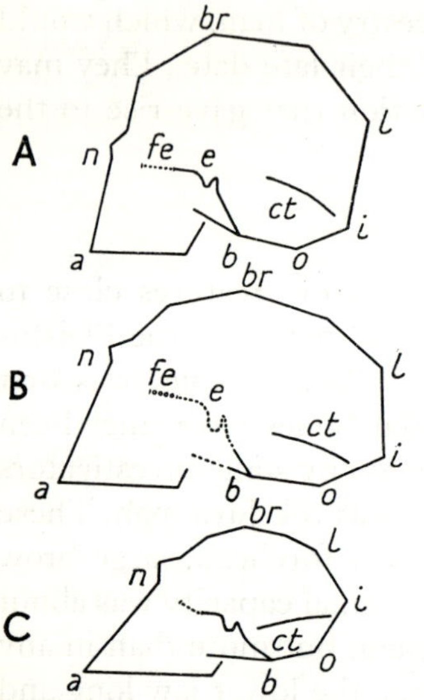

FIG. 419. Comparisons between the geometric outlines of the skulls of: A. Modern European; B. **Homo neandertalensis* (La Chapelle); C. Chimpanzee.

a. prosthion; *b.* basion; *br.* bregma; *ct.* upper border of the cerebellum; *e.* ephippion; *fe.* fossa ethmoidalis; *i.* inion; *l.* lambda; *n.* nasion; *o.* opisthion. (After Boule.)

Casts of the brain show that it differed in several ways from our own, especially in the relatively small frontal lobes and backwardly extending occipital region.

Typical Neanderthal skeletons are common in Europe but the type was probably widespread since similar fossils have been found in

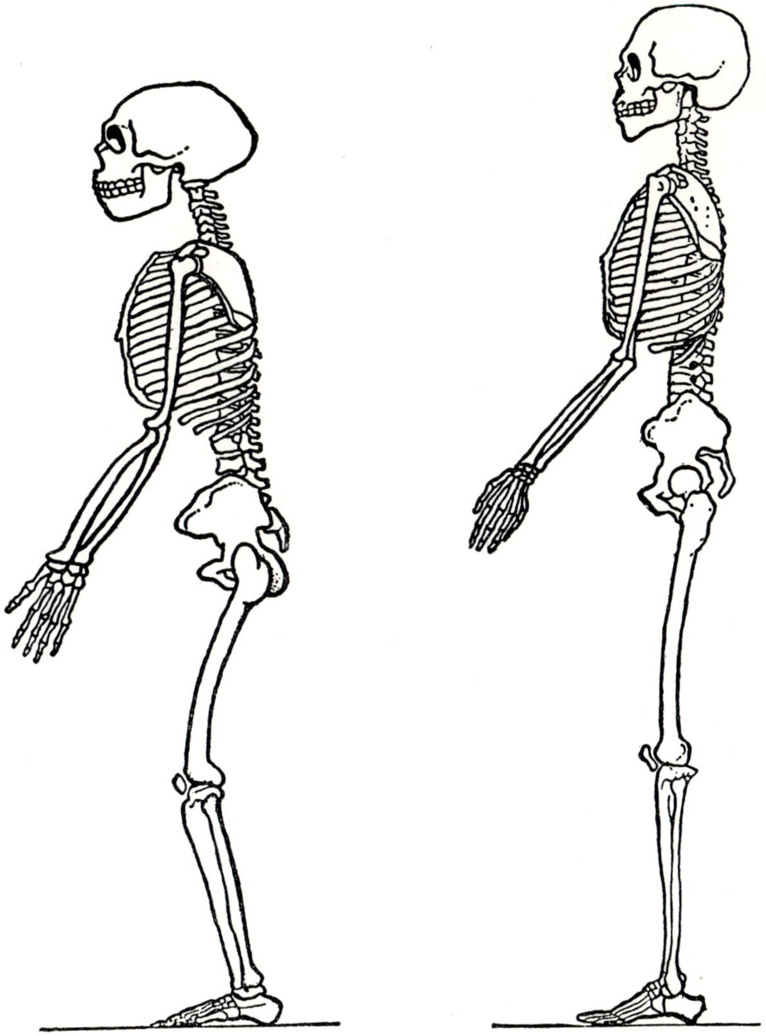

FIG. 420. Skeletons of Neanderthal and modern man. (After Howells, *Mankind So Far*, Doubleday and Co. Inc.)

South Africa (**H.n. rhodesiensis*) and Java (**H.n. soloensis*), the latter showing primitive features that perhaps connect it directly with **Pithecanthropus*.

15. Human cultures

All of these early men were hunters, living apparently in small families, in caves. They used fire, but their only known instruments were of wood or stone. Chipped flints found from the time of the first

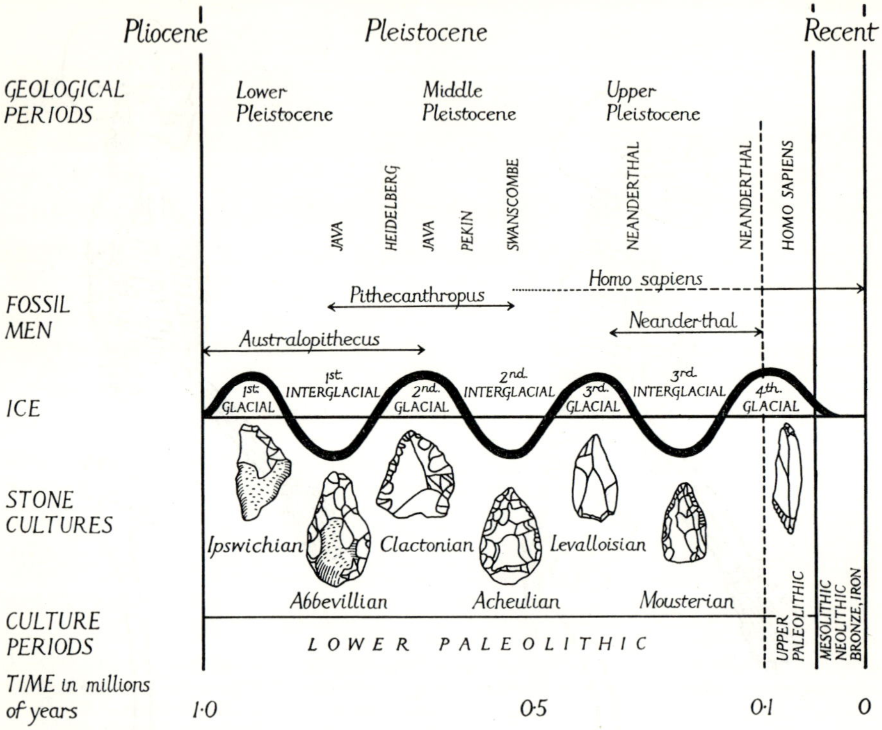

FIG. 421. Diagram of Pleistocene time and some stages of human cultural evolution. (After Howells, *Mankind So Far,* Doubleday and Co. Inc.)

glaciation and following warm period are so crude as to be hardly recognizable as artefacts and these 'eoliths' are said to be signs of a pre-palaeolithic culture. More definitely shaped flints, used as axes, are found during the second ice age, second interglacial, and third ice advance and are referred to the lower Palaeolithic (Chellean and Acheulian) stages. The third interglacial and last glacial period constitute the middle Palaeolithic (Mousterian), with well-made flints, probably produced by Neanderthal man. The making of tools is therefore probably characteristic of the creatures we refer to the genus *Homo*. This power may be connected with the capacity for communication by symbols that indicate abstractions, which marks off men from animals more clearly than any other feature. The recession of the last main glaciation began less than 100,000 years ago and at that time there were a series of upper Palaeolithic cultures (Aurignacian, Solutrean, and Magdalenian) in which besides wonderfully chipped flints there were also fine bone and ivory needles, blades, and other instruments. These were probably made by men of *H. sapiens* type, whose skulls first became abundant in caves and deposits of the last glacial period.

The question of the first appearance of *H. sapiens* is, however, complicated by the skulls found at Swanscombe and Fontechevade, from Middle and Upper Pleistocene levels. These 'presapiens' fossils lack the Neanderthal characters (brow ridges, occiput, &c.) but are thick and more heavily built than modern skulls. Evidently from the Middle Pleistocene onwards there has been a considerable variety in the human populations, but all may be included in one species.

The details of the replacement of these species and races by each other remain obscure. Probably invading races often took over parts of the cultures of their victims, as well as bringing in their own culture, so that the whole story becomes very confused. It is usually considered that *H. neanderthalensis* did not evolve into *H. sapiens*, at least in Europe, but was replaced by a wave of invaders coming from central Asia. Similarly, at the end of the Palaeolithic period, about 14,000 years ago, the hunters of that time gave place first to a series of little-known Mesolithic invaders, builders of the lake-dwellings found in various parts, and then to the Neolithic people, who were farmers and city-builders and from about 9000 B.C. onwards dominated the Middle East and south Mediterranean region and thence spread outwards, developing the series of cultures known as the Bronze and Iron Ages.

The history of man, like that of so many other mammals, has therefore possibly been of a series of invasions from the central Asiatic land-mass, and it is not surprising that remnants of earlier types should be found surviving today on the tips of the continents, towards the south. The most conspicuous of these are the Australian Aborigines, who show several primitive features, both of structure and of culture. The brow ridges are better developed than in any other men and the forehead and chin recede. These Blackfellows are nomadic hunters, without fixed abode, and their social organization and instruments are those of a Palaeolithic stage of life.

In something the same way the Bushmen of South Africa, though not physically of a primitive type, preserve traces of Palaeolithic culture, including art conventions similar to those used by the Aurignacian people who painted bison and mammoths in the caves of Altamira in Spain 50,000 years ago.

There is much evidence that change of the physical characteristics of human populations has continued during the last 10,000 years. Thus *H. sapiens* before the Mesolithic period was long-headed (dolichocephalic). Short, broad skulls (brachycephalic) first appeared at that time and are now found in more than half the population of the world. Changes in the shape of the jaw and face and reduction of the teeth, which have been going on for a long time, probably continue today.

Study of the gradual mixing and changing of human populations, besides its personal interest, is of value to a zoologist in calling the process of evolution to our imagination and showing us its complexity and slowness. Without undue difficulty we can have in mind a picture of great populations of human beings, composed of individuals differing slightly in structure and habits, warring and competing with each other, so that one group comes to dominate another, the invader taking over some of the genes and the gods of its victims. We can at least guess how such a process would lead the population of any area to change, either if a set of individuals arises that is more active or in some way more efficient than another, or following a change in conditions that gives preference to a particular set of structures or habits. Though evolution is a slow process it is always going on before us, and the best way to see it is to look at our own species, vastly more familiar and revealing than any other. All sheep or shrews look to us much alike, but we can more readily tell what sort of man or woman we are meeting and what they are likely to do in the world. Moreover we can reconstitute their capacities by the remains of their cultures and recognize clearly how different our ancestors were from ourselves 50,000 years ago and still more so at 500,000 years. In this way we may get some faint picture of the slow and confused changes that constitute evolution. Even with such a slow breeding species as our own the effort of thought is very great. Allowing only so few as four generations per century we may conclude that we are separated from **Pithecanthropus* populations by 40,000 generations. If we ever had an ape-like hairy ancestor it was perhaps our great-grandfather 500,000 times removed, counting back only to the beginning of the Pliocene. Yet men and apes are zoologically much alike.

Many factors have been suggested to account for the appearance of man's particular characteristics. Disappearance of forests may have favoured a terrestrial bipedal life and use of the hands. The great size of the brain might have followed from this. It has been claimed that this might have resulted from single mutations that increased the number of divisions of the neuroblasts from 31 in apes, giving about

$2{\cdot}5 \times 10^9$ cortical cells, to 32 in **Pithecanthropus* ($4{\cdot}3 \times 10^9$) and 33 in *Homo* (10×10^9).

In development of powers of obtaining information with the nervous system man is only showing an accentuation of the characteristics of mammals in general and especially primates. With this goes the low reproduction rate, slow development, and great post-natal care. The retardation of development of somatic characters seems to have been the basis of many human features. As Bolk pointed out man resembles in many ways a foetal ape, rather than an adult one. Thus the position of the head, at right angles to the vertebral column, is that found in foetal apes. The late fusion of ossification centres, lack of hair, external genitalia, structure of hand and foot and many other features point the same way. In addition to this general retardation there have no doubt been many special developments, such as formation of the nose, lengthening of the pollex and elaboration of the organs of speech and the muscles of expression.

This paedomorphosis in man, producing teachable, cooperative individuals may well be the factor that has made possible the development of complex societies and their tools. By efficient communication man is able to produce a cumulative store of information outside his mortal body and passed on not only to few individuals, as is the genetic store, or that passed by word of mouth, but to many. Each individual thus 'inherits' not from two or few parents but from the accumulated memory store of a large population. It is perhaps this 'multi-parental' inheritance of information that has changed man so rapidly in the past and is likely to do so even faster in the future. Success will clearly be for those populations that are able so to cooperate as to discover and transmit more and more information.

XXV

RODENTS AND RABBITS

1. Characteristics of rodent life

THE animals loosely known as rodents are the most successful of modern mammals other than man. They live in all parts of the world, from the tropics nearly to the poles. Three thousand species are known, as many as are found in all other mammalian orders put together. They inhabit a considerable variety of ecological niches, mostly on the land, often in burrows, but many in the trees and some in the water. This most successful type of mammalian life is, however, in several ways untypical of the rest and indeed has been isolated since the early Tertiary. One striking point is that the animals have never become large in size, although such increase is a tendency found in almost all other mammalian groups. The South American capybara, the largest living rodent, is the size of a small pig, and few fossil forms were much larger. Rodent life has specialized in rapid breeding and this system of production of large numbers of small animals has been very successful. The total rodent biomass today may well be greater than that of the whales, which are at the other extreme, and have all the advantages of aquatic life. The rapid reproduction presumably brings considerable evolutionary advantages, enabling the population to make the adjustments necessary to meet changing circumstances. One of the characteristics of rodent populations today is their great fluctuations (p. 663), notorious in the case of the voles, mice, and lemmings, but marked also in rats and other forms. The pressure of rodent life is such that no stable equilibrium is reached with the environment and extreme oscillations occur, often with results of great importance to man and to his crops.

In spite of the similarities of all these animals with gnawing teeth, zoologists consider that the rabbits and hares are not closely related to the others and are therefore to be placed in a distinct order Lagomorpha, the order Rodentia being retained for all other 'rodents'. It is not even certain that the two orders are in any way related. Fossil rodents (in the strict sense) certainly occurred in the late Palaeocene period and a probable lagomorph is reported from the same time. Both groups retain many primitive mammalian characters, for instance, a long, low skull, with small brain and small cerebral hemispheres, temporal fossa widely open to the orbit, pentadactyle limbs, separate radius and ulna, and so on. These features, being found in all early mammals, indicate no closer affinity of the two orders than depends on evolution from a common stock. It is not even clear exactly how the two groups are related to the ancestral eutherians, and we must be content to say that it is probable that animals with rodent specializations diverged from the insectivoran eutherian stock in the late Cretaceous or Palaeocene and then rapidly became differentiated into lagomorphan and rodent types. The two orders are therefore placed together in an isolated cohort Glires.

2. Classification

Cohort 2. Glires
  Order 1. Rodentia (= Simplicidentata)
    Suborder 1. Sciuromorpha
      Thirteen families, including
      *Family Ischyromyidae. Palaeocene–Miocene. Eurasia, N. America
        **Paramys*, Palaeocene–Eocene
      Family Aplodontidae. Eocene–Recent
        Aplodontia, mountain beaver, N. America
      Family Sciuridae. Squirrels. Miocene–Recent
        Sciurus, squirrel, Holarctic; *Marmota*, marmot, woodchuck, Holarctic; *Tamias*, chipmunk, N. America; *Petaurista*, flying squirrel, Eurasia
      Family Geomidae. Gophers. Oligocene–Recent. N. America
        Geomys, pocket gopher
      Family Castoridae. Beavers. Oligocene–Recent. Holarctic
        Castor, beaver
    Suborder 2. Myomorpha
      Nine families, including
      Family Dipodidae. Jerboas. Pliocene–Recent. Palaearctic
        Dipus, jerboa, Eurasia
      Family Cricetidae. Voles. Oligocene–Recent. World-wide (except Australasia)
        Peromyscus, deer mouse, N. America; *Lemmus*, lemming, Holarctic; *Microtus*, vole, Holarctic
      Family Muridae. Rats and mice. Pliocene–Recent. Native to Old World
        Apodemus, field mouse; *Rattus*, rat; *Mus*, house mouse; *Glis*, dormouse; *Notomys*, jerboa-rat

2. Classification (*cont.*)

Family Zapodidae. Jumping mice. Oligocene–Recent. Holarctic
Zapus, jumping mouse
Suborder 3. Hystricomorpha
Nineteen families, including
Family Hystricidae. Oligocene–Recent. Palaearctic, Africa
Hystrix, porcupine, Asia, Africa; *Erethizon*, N. American porcupine, N. America
Family Caviidae. Pliocene–Recent. S. America
Cavia, guinea-pig
Family Hydrochoeridae. Pliocene–Recent. S. America
Hydrochoerus, capybara
Family Dasyproctidae. Recent. S. America
Cuniculus, agouti (pacas)
Family Chinchillidae. Oligocene–Recent. S. America
Lagostomus, vizcacha
Family Bathyergidae. Pleistocene–Recent. Africa
Bathyergus, mole-rat
Order 2. Lagomorpha (= Duplicidentata)
Family 1. *Eurymylidae. Palaeocene
**Eurymylus*, Asia
Family 2. Ochotonidae. Upper Oligocene–Recent
Ochotona (= *Lagomys*), pika (cony), N. America
Family 3. Leporidae. Upper Eocene–Recent
Lepus, hare, Pleistocene–Recent, Palaearctic, N. Africa; *Oryctolagus*, rabbit, Pleistocene–Recent, Europe, N. Africa; *Sylvilagus*, cotton-tail, Pleistocene–Recent, N. and S. America

3. Order Rodentia

Rodents are mostly herbivorous and their most characteristic features are, of course, in their teeth, especially the incisors, one pair only of which persists in each jaw; hence they were the suborder 'Simplicidentata' of the older order Rodentia, which included also the rabbits and hares. These latter have a second pair of upper incisors, hence 'Duplicidentata'. The incisor has enamel only on its labial surface and thus maintains a cutting edge. It is worn away at the rate of perhaps several millimetres a week and is replaced by continual growth, for which it has a very wide open pulp cavity, or in the conventional term is said to be a 'rootless' tooth. The incisors are often very large and curved and their gnawing action against each other gives them chisel edges. If one incisor is lost the other continues to grow round in a spiral, until it enters the skull.

The remaining incisors, canines, and anterior premolars are missing, leaving a large diastema in front of the cheek teeth. Folds of skin can be inserted into this gap to close off the front part of the mouth, so that material bitten off during gnawing is not necessarily swallowed. A distinct anterior chamber of the mouth is thus formed, and may be prolonged into deep pockets in which food is stored for transport to

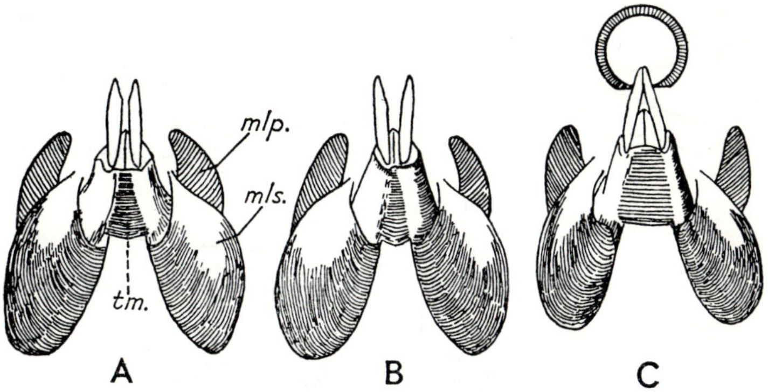

FIG. 422. Diagrams of lower jaws of *Sciurus*.

A, at rest; B, opened for seizing; C, closed for gnawing or to prise open a nut. *mlp.* deep part of lateral masseter; *mls.* superficial part of the same; *tm.* transversus mandibulae. (From Weber after Krumbach.)

the hoards that are collected by many species. The teeth are not used only for obtaining food; rats will gnaw their way through a lead pipe.

The premolars are reduced to two above and one below in the more primitive squirrels; even fewer are present in other rodents. The molars and premolars are usually alike in pattern and show modifications similar to those found in the grinders of other herbivorous mammals. The cusps of the original eutherian molar can still be recognized in the squirrels, but they are arranged in transverse rows, paracone and protocone in front, metacone and hypocone behind. The cusps become joined in pairs to form ridges, giving a bilophodont grinding tooth. In most of the rodents further ridges are then added in front, behind, and between the original ones, and these are also joined by cross-ridges, giving a multi-lophodont molar. Similar changes occur in the lower teeth. The teeth also become very high-crowned or 'hypsodont' and the enamel, dentine, and bone ('cement') wear at differing rates. The roots remain wide open and the teeth grow continually. All of these features are like those arrived at in ungulates by a convergent process of evolution. A further feature of the rodents is that the upper tooth-rows are set closer together than the lower and bite inside the latter, often giving an oblique grinding surface. The milk teeth are shed very early and are not functional.

The lower jaw and its muscles show many modifications. The articulation is very long, and the lower jaw moves backwards and forwards on the upper; indeed the lower incisors are thrust so far forward that while gnawing the molar surfaces no longer occlude. A

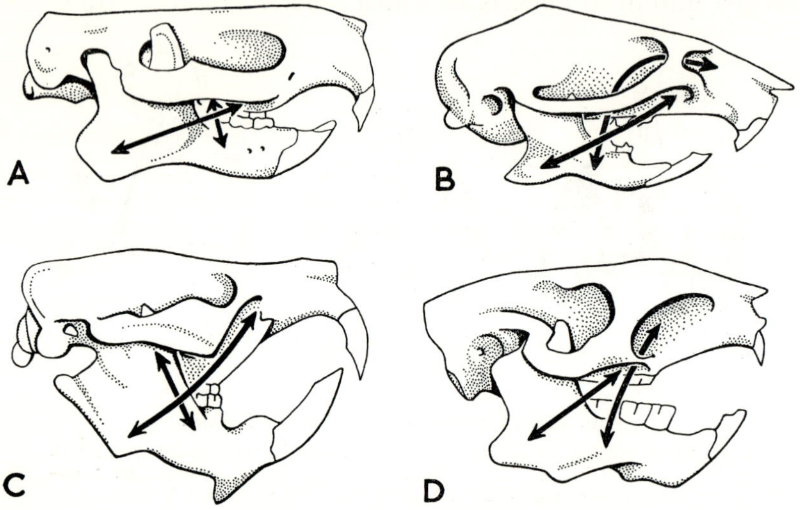

FIG. 423. The masseter of rodents.

A, *Aplodontia*, primitive sciuromorph: the muscle takes origin from the zygomatic arch; B, myomorph: the deep masseter passing through the orbit is also attached to the face; C, advanced sciuromorph: the superficial masseter is attached to the skull in front of the orbit; D, hystricomorph: superficial masseter unspecialized, large foramen for deep portion. (After Romer, *Vertebrate Paleontology*, University of Chicago Press.)

curious feature is that the mandibles are not united in a symphysis but are freely movable, with a joint cavity between them. A special portion of the mylohyoid muscle, known as the m. transversus mandibulae, draws the two mandibles together, causing the lower incisors to separate (Fig. 422). The action of the lateral portions of the masseters then brings the two teeth together again with a scissor action.

The jaw-muscles are very large and modified to produce the backward and forward motion of the jaw (Fig. 423). In the more primitive condition, found in squirrels (Sciuromorpha), the masseter is attached to the zygomatic arch as it is in other mammals, but is divided into a more lateral portion with simple up and down action, and a medial part that pulls the jaw forward. In more advanced rodents both of these parts of the muscle obtain extra insertions. In rats and mice (Myomorpha) the lateral part extends forward on to the face and the medial passes out of the orbit through the much-enlarged infra-orbital canal. In the porcupines and their allies (Hystricomorpha) the lateral portion remains simple and the medial proceeds to a large insertion on the face below (but not through) the infra-orbital canal. The lower jaw often carries a large flange, for the attachment of the masseter muscle. The pterygoid muscles and their attachments are also often large, but the temporal muscle is usually small.

Apart from their gnawing mechanism the rodents remain rather unspecialized mammals. The gait is plantigrade, the fore-limbs often shorter than the hind and used for handling the food. In some this tendency is carried to the extent of producing a hopping, bipedal gait. The gut shows a large coecum and food is passed twice through the gut (p. 662). A division of the stomach is found only in mice, which have a horny cardiac region. Rodents are macrosmatic, with large olfactory bulbs and relatively small, smooth neopallium. The eyes are often well developed, especially in arboreal rodents and those living in open country or steppe. Hearing is often good, and in some desert species the tympanic bulla is greatly dilated, perhaps to detect sounds made by the widely separated individuals. Many rodents have well-developed social habits, with olfactory and visual, as well as auditory and tactile signalling.

Rodents are mostly polyoestrous, often breeding throughout the year, at least in captivity. Numerous young are produced and are often cared for in a nest. The uterus is usually double and the placentation usually of the discoidal and haemochorial type.

No thoroughly satisfactory scheme for grouping the various types of rodent has been devised; the best that we can do is to keep to the classical division of the order into three suborders, Sciuromorpha, Myomorpha, and Histricomorpha. The first includes besides the more primitive surviving forms, *Aplodontia*, the mountain beaver of North America, also the most ancient fossil rodents (**Ischyromyidae*) and some families of uncertain affinities. The characteristics of the sub-order are that there are two upper premolars and one lower and a masseter muscle that does not pass through the infra-orbital canal. The suborder includes the squirrels (Sciuridae) found in all major regions except Australasia. They are diurnal, with large eyes and often bright colouring. The flying squirrels, *Petaurista*, glide for long distances by means of the patagium, whose muscles enable them to change direction in the air (Fig. 424).

The marmots (*Marmota*, Fig. 425), ground squirrels (*Citellus*), and prairie dogs (*Cynomys*) are burrowers, with elaborate underground

FIG. 424. Flying squirrel, *Petaurista*. (From a photograph.)

FIG. 425. Marmot, *Marmota*. (From photograph.)

FIG. 426. Beaver, *Castor*. (From photographs.)

FIG. 427. Lodge and food store of the beaver. (From *American Mammals*, by W. J. Hamilton, McGraw Hill Book Company.)

societies. The gophers (*Geomys*) and kangaroo rats (*Dipodomys*) of the United States are also sometimes placed here. They are jumpers, paralleling the jerboas, gerbils, and jerboa-rats.

Possibly also in this group are the beavers, Castoridae, found throughout the Holarctic. These are aquatic rodents (Fig. 426) that show remarkable habits in preparing a house and store of food for the winter (Fig. 427). The house ('lodge') is built on a mass of debris so as to be surrounded by water. Sticks are built up to make a wall round the platform and the whole finally closed by a dome of sticks

FIG. 428. Jerboa, *Dipus*. (From life.)

and mud, which is carried by the beavers with their fore-paws. When this damp structure freezes it makes a strong protection against bears and other enemies, and the beavers keep warm inside it. Food is obtained from the bark of branches kept in a store under water and brought up to the lodge through a plunge hole. The beaver dams are made by the beavers during the summer in order to deepen the streams; they may reach a height of 12 ft and a length of several hundred. The lodges and dams are the result of cooperative work by successive generations. The beavers work compulsively, repairing the structures whether they need it or not. Like other rodents they mark their territory by smell, there being large anal oil-glands. When an animal smells a deposit it visits it and adds its own 'castoreum'.

The Myomorpha is a very large group, including the rats, mice, voles, jerboas, and other types, all having the medial portion of the masseter running through the infra-orbital canal, but probably not really closely related. They include many families and genera, with specializations for individual ecological niches. The jerboas, Dipodidae (Fig. 428), are members of this group, with limbs specialized for hopping, but in other respects with somewhat primitive characteristics. There has been a great elongation of the metatarsals of the three central digits, which alone are well developed. The gerbils (*Gerbillus*) are yet another family of jumping animals, living in deserts. The Muridae (Fig. 429) are among the most successful of all mammals and are an ancient family, recognizable back to the Pliocene and invading all parts of the world, including in relatively recent times South America and Australasia, reaching the latter before man. They are the only mammals indigenous to New Zealand. The family includes an enormous variety of mice, rats, dormice, field mice, hamsters and so on, and also some special forms such as the jerboa-rat (*Notomys*) of Australia, which has paralleled the true jerboas in its jumping habits. The rats and mice may be considered as parasites of man and are still changing their distribution. *Rattus* has been widely spread in recent centuries. The black rat (*R. rattus*) prefers warmer and drier conditions than the brown rat (*R. norvegicus*) but the two often compete.

The voles (Cricetidae) are a related family, including partly aquatic as well as terrestrial forms. The field vole (*Microtus agrestis*) is apt to increase greatly in numbers, producing notoriously destructive plagues. *Lemmus*, the lemming (Fig. 430), is a mouse-like form living on grasses and roots on the Norwegian mountains. At irregular intervals of 3–5 years the population grows greatly by increase in the numbers in the litters and in the number of litters in the season. The population becomes too great for the area to support and large numbers emigrate to the lowlands and die of starvation or from predators. Others reaching the sea-shore swim out and are drowned.

The histricoid rodents are also a large group, with the infra-orbital canal enlarged for the medial part of the masseter, but with the lateral part attached to the zygoma. They include the porcupines (*Hystrix*) of Africa and Asia (Fig. 431) and the somewhat different porcupines of North America, but all the rest of the group occurs in South America, an area invaded by hardly any other rodents. Fossil histricomorphs are found in South America from the Oligocene. The Hystricidae all have long spines, used for attack by a rapid backward movement as well as for defence. The cavies and capybaras (Fig. 432) are closely related South American forms, the latter being a large semi-aquatic animal with a greatly enlarged and folded last upper molar. The agoutis (Fig. 434) are also rather large, for rodents, and are terrestrial, often burrowing forms. The vizcachas, *Lagostomus* (Fig. 433), also burrow underground, often making large colonies.

Possibly related to the hystricomorphs are still more modified digging animals, the Bathyergidae or African mole-rats, which have lost most of the hair and developed long claws on the forelimbs.

4. Order Lagomorpha

The rabbits and hares are nowadays considered to be a very isolated offshoot of the early eutherian stock, whose similarities to the Rodentia

Fig. 429. Common brown rat, *Rattus*. (From photographs.)

Fig. 430. Lemming, *Lemmus*. (From a photograph.)

Fig. 431. Porcupine, *Hystrix*. (From a photograph.)

Fig. 432. Capybara, *Hydrochoerus*. (From a photograph.)

Fig. 433. Vizcacha, *Lagostomus*. (From a photograph.)

Fig. 434. Agouti, *Dasyprocta*. (From life.)

may be only superficial. The two orders are kept together in one cohort Glires, more as a convenience than because of characters they have in common. The arrangements for gnawing found in the lagomorphs have indeed a superficial similarity to those of rodents, but on inspection the differences appear profound. Continually growing incisors

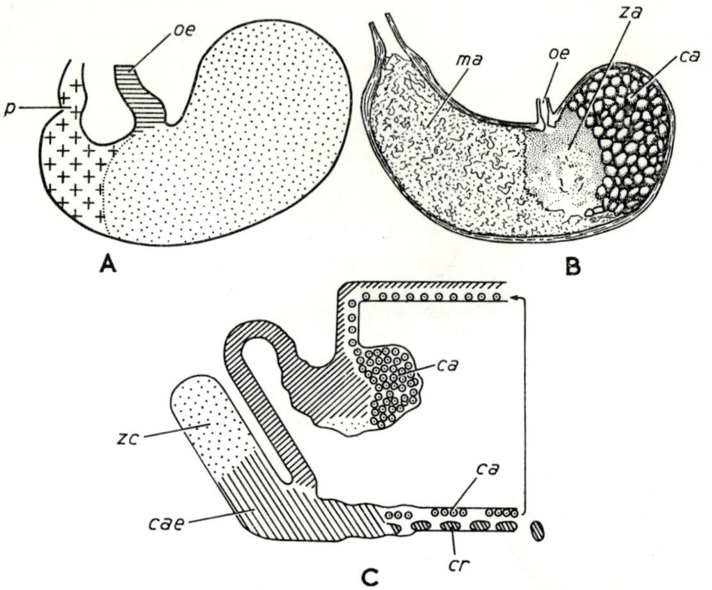

FIG. 435. A. Diagram of stomach of *Lepus*: dashes, cardiac zone; dots, fundus; crosses, pyloric glands. B. L.S. of rabbit stomach. C. Diagram showing the passage of food and caecotrophs traversing the digestive tracts. In the stomach, under the mechanical action of peristalsis the caecotrophs are mixed with the food. In the lowest part of the caecum, the chyme under the action of bacteria is made into caecotrophs, which the animal eats soon after they are expelled from the anus.

ca. caecotrophs; *cae.* caecum; *cr.* waste pellets; *ma.* alimentary mass; *oe.* oesophagus; *p.* pylorus; *za.* zone where the caecotrophs are mixed with the alimentary mass; *zc.* zone of formation of caecotrophs. (B. and C. after W. Harder.)

are present in both groups, but in lagomorphs the upper pair is accompanied by a small second pair (hence 'Duplicidentata'). The diastema is common to the two groups, and both have cheek pouches with similar functions. The similarity of the molariform teeth is only superficial, however; in the lagomorphs three premolars remain in the upper jaw and two in the lower. The premolars and the molars all acquire sharp transverse ridges, usually two each, used for cutting rather than grinding, the upper teeth biting outside the lower ones, not inside as in rodents. The masseter is powerful, but simpler than in rodents, and it does not extend into the infra-orbital canal. The temporalis muscle is reduced in both groups, but the lagomorphs lack the power of movement between the two halves of the lower jaw.

Taken all together, therefore, the differences are as great as the similarities, even in the gnawing mechanism, which the rabbits share with the other rodents. In the remaining parts there are few points of close similarity, other than those due to the fact that neither set of animals has departed far from the original eutherian condition. Moreover, serological studies do not show any signs of closer affinity of the lagomorphs with the rodents than with other mammalian orders. If anything they are more like artiodactyls. However, the lagomorphs share with rodents the habit of passing food twice through the alimentary canal (caecotrophy). Dried faecal pellets are produced only during the day. At night soft pellets covered with mucus are formed in the caecum and are immediately taken from the anus by the lips. They are stored in the stomach and later mixed with further food taken (Fig. 435). The double passage of the food is necessary for the life of mice and guinea pigs as well as rabbits. The animals die in two to three weeks if prevented from reaching the anus. The moist pellets probably contain the metabolites that have been produced by breakdown of cellulose by the bacteria of the caecum, which cannot be absorbed by the organ itself.

Rabbits and hares have characteristically developed the hind legs for a jumping method of locomotion and there has been a reduction of the tail. The rabbits show a number of specializations for burrowing life. They are among the most successful of all mammals, especially in the Holarctic region, but have made relatively little progress in Africa or South America. Their enormous spread in Australia since introduction by man in the eighteenth century shows how accidental limitations of access, and their alteration, affect the distribution of animal life.

Fossil lagomorphs, quite like modern hares and rabbits, are found back to the Oligocene. Few remains are known from the Eocene, but there is evidence that the type was already distinct in the Palaeocene and has persisted with relatively little change ever since.

5. Fluctuations in numbers of mammals

Fluctuation in numbers is characteristic of many rodents and other small mammals. The phenomenon is usually first recorded as a 'plague' of the rats, mice, voles, or rabbits, and these may be cases of local and sporadic abundance. Study of some species has shown, however, that there are rather regular cyclical fluctuations in numbers, extending over many years. Thus figures for the furs collected by trappers for the Hudson Bay Company enable the variations in numbers of the varying hare to be followed back to 1850 (Fig. 436). The cycle is surprisingly regular, with a period of 9·7 years. During the periods of abundance of the hares the mammal and bird predators of these also increase and fall off a little later than the herbivore populations, changes in food taken by the predators producing all sorts of secondary effects on the animals of the area.

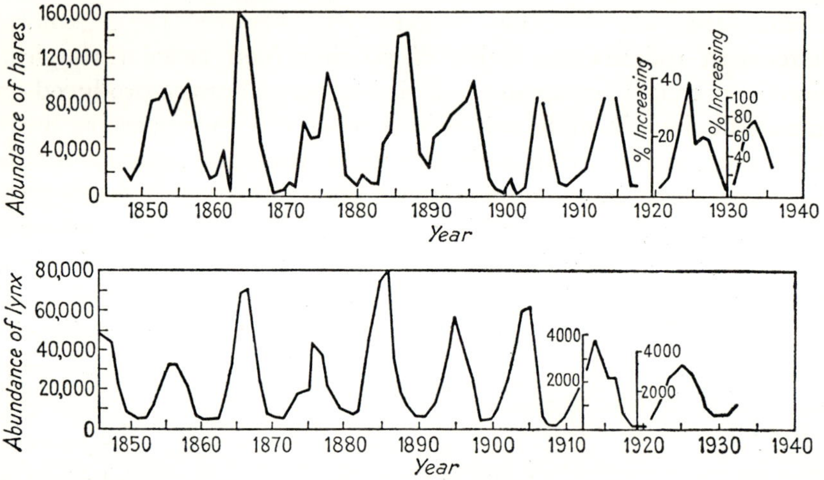

FIG. 436. Fluctuations in numbers of varying hares (above), and lynx (below). Estimated from the Hudson Bay Company's fur returns. (From W. J. Hamilton, *American Mammals*, McGraw Hill Book Company. After MacLulich.)

The existence of these fluctuations is striking evidence that animal populations are not maintained in any stable equilibrium. It is often suggested that the cycles depend on changes in the physical conditions, for instance, on the sunspot cycles, but no close correspondences have been discovered. The sunspot cycle has a period of 11·1 years, whereas rodent and lagomorph cycles varying from the 9·7 years for the hares to 5 years for lemmings and 3 for voles have been recorded. Changes in the amount of solar radiation no doubt produce considerable effects on the animals. There is some evidence for a direct effect of diet increasing the fertility. In South America plagues of rats have been observed to coincide with abundance of bamboos. But it seems likely that the cycles of numbers depend on the particular balances set up within the animal communities. These presumably depend on the interactions of the reproductive pressures of the plants, herbivores, carnivores, and parasites, and it is easy to believe that in some conditions these factors, either alone or in association with cycles of solar radiation or other climatic factors, would produce an oscillatory system.

When the numbers are at a maximum the animals show unusual behaviour patterns, including migration, and may enter into a pathological state, becoming cold and torpid and with a very low blood-sugar content. Probably the pressure of competition and lack of food operates on the hypophyso-adrenal system to produce a state of shock. It is this, rather than disease or predators, that finally reduces the numbers.

Such phenomena are probably not peculiar to small mammals, but appear markedly in them because the rapid breeding and short life provide cycles of relatively short period. The facts are all consistent with the interpretation already reached that the control of animal populations depends very largely on the interaction of biotic factors. This continual pressure of the animals on each other is probably the main factor responsible for the great variations in the characteristics of animal populations in different parts of their range, which is found wherever sufficiently careful study is undertaken. This variation in turn leads to the phenomenon of organic evolution and the continual replacement of types by their descendants, which appears so clearly as one follows the history of the vertebrates.

XXVI

WHALES

Cohort 3. Mutica
Order Cetacea
*Suborder 1. Archaeoceti. Eocene–Miocene
Protocetus, Eocene; *Basilosaurus* (= **Zeuglodon*), Eocene
Suborder 2. Odontoceti. Toothed whales. Eocene–Recent
*Superfamily 1. Squalodontoidea. Eocene–Miocene
**Squalodon*, Miocene
Superfamily 2. Platanistoidea. River dolphins. Miocene–Recent
Platanista, Ganges dolphin, Recent
Superfamily 3. Physeteroidea. Miocene–Recent
Hyperoodon, bottle-nosed whale, N. Atlantic, Antarctic; *Physeter*, sperm whale, all oceans
Superfamily 4. Delphinoidea. Miocene–Recent
Monodon, narwhal, Arctic; *Delphinus*, dolphin, all oceans; *Phocaena*, porpoise, all oceans
Suborder 3. Mysticeti. Baleen whales. Oligocene–Recent
Balaenoptera, rorqual, all oceans; *Balaena*, right whale, temperate and polar seas

The whales are a successful set of populations of aquatic, carnivorous mammals, which diverged from the eutherian stock, possibly from the creodont branch, at a very remote date. Their specializations for aquatic life involve divergences from the typical mammalian plan greater than those found in any other order, and their organization shows in a remarkable way the effects of habit of life and environment. In many respects the whales have reverted to the characteristics of a fish form of life, most noticeably in the body shape, with elongated head, no neck, and tapering 'streamlined' body. In terrestrial mammals the air resistance is not a serious factor and the body shape does not conform to it, but in water it becomes of first importance.

The swimming mechanism depends on up-and-down oscillations of the flukes and tail stock, and the efficiency of the system is shown by Gray's demonstration that the power necessary to drag a wooden model of a dolphin through the water with the speed of the animal would necessitate a horse-power in the muscles several times greater than that of any known mammal. Probably in life the accommodation of the skin by deformation absorbs part of the turbulence energy and reduces drag. The surface is completely smoothed off by the loss of all hair, except for a few sensory bristles round the snout in some species. There is a thick layer of dermal fat (blubber), which, besides acting as a heat insulator, may also provide a reservoir of food and perhaps, when metabolized, of water. The fat also reduces the specific gravity of the animal and perhaps provides an elastic covering to allow for changes in volume during deep diving. There are no glands in the skin.

The propulsive thrust comes largely from the horizontally placed tail flukes, which constitute a cambered aerofoil that is moved up and down by the tail muscles. These consist of upper and lower sets of longitudinal fibres inserted in the tail-stock vertebrae by means of long tendons from muscles originating from more forwardly situated vertebrae. More caudally placed muscles inserted on the hindmost vertebrae in the tail flukes allow movement of the flukes relative to the tail stock to produce the forward thrust. The arrangement allows the whole tail to be bent up and down on the body, while the fluke is bent relative to the tail and produces the thrust. Stability is provided by the paddle-like fore-limbs, and there is often a large dorsal fin, especially in fast swimmers such as the killer-whales (*Orcinus*). The plasticity of animal form is shown by the fact that the flipper is a modification of the ambulatory fore-limb, whereas the dorsal fins and tail flukes are 'neomorphs', folds of skin with no skeletal support, radical innovations indeed: it is not easy to imagine by what alterations of habit an early eutherian could come to develop fins out of such folds of skin. Equally remarkable has been the disappearance of the hind-limb, leaving no external trace and internally only paired pelvic vestiges with additional bony nodules in some whales representing limb bones. These rods serve as attachments for the corpora cavernosa of the penis and may therefore be regarded as ischia.

The vertebral column (Fig. 437) carries no weight except when the animal jumps out of the water. In the vertebral column the zygapophyses are reduced, and the centra are well developed to make a compression strut, as in fishes. The vertebral epiphyses remain separate to a late age. The neural spines and transverse processes are well developed for the attachment of muscles in a typically mammalian manner giving a dorso-ventral movement of the body, this being in contrast with the lateral movement produced by the segmental myotome arrangement of fishes. The neck is very short and the cervical vertebrae partly or wholly fused together. The ribs, as in other aquatic vertebrates, are rounded and mobile; they are the chief agents of respiration, the diaphragm containing little muscle.

In the fore-limb the humerus is short, the elbow-joint hardly mobile, and the hand increased in length and sometimes expanded (right whales, killers, river dolphins). The number of fingers is often reduced to four; the phalanges of some of the digits may be considerably increased in number (hyperphalangy). The scapula is flattened and there is no clavicle.

Some striking modifications are seen in the head. The skull shows a curious telescoping of the bones over each other. The maxilla of toothed cetaceans extends above the frontal, combining with the latter to make a roof to the temporal fossa. Thus the maxilla almost reaches the supra-occipital. In baleen whales the backward prolongation of the maxilla is mainly *below* the supraorbital process of the frontal, although there is a medial process extending dorsally towards the supra-occipital. This telescoping occurs only towards the end of foetal life. A further curious feature is that the skull in the toothed whales is asymmetrical, the vertex being shifted over to the left side; no satisfactory explanation for this phenomenon has yet been offered. The jaws are always greatly elongated, in the Greenland right whales they make up one-third of the total body length. The masticatory muscles and the coronoid process are reduced, the latter most extremely in *Balaena* in which it is distinguished only as an inconspicuous ridge.

Whales are microsmatic or even in some species anosmatic (with no olfactory nerve). The brain-case is therefore short and rounded while the nostrils have moved backwards and open upwards. The nasal bones have become reduced in length and no longer roof over the nasal cavity. The process has proceeded somewhat differently in the toothed and in the whalebone whales. The auditory region is much modified and the whole petrosal bone is free from the rest of the skull; there is a large tympanic bulla, fused with the petrosal. Extensions of the middle ear cavity form pneumatic sacs, below the base of the skull, which serve to insulate sound and equilibrate the varying pressures experienced under water.

The feeding arrangements provide further special features. Many of the toothed whales, such as the porpoises and dolphins, eat mainly fish and the teeth form a row of numerous (65/58), similar peg-like structures, usually in both jaws. With elongation of the jaws the masticatory function of the teeth has been reduced and they probably serve to hold the prey. *Orcinus*, the killer-whale, has large powerful jaws and teeth and its diet includes dolphins, birds, seals, and the flesh of large whales. Squids are also eaten, but in species that feed predominantly on these there is some tendency to a reduction in the number of teeth. For example, in the sperm whale functional teeth are confined to the lower jaw and in the bottle-nosed or beaked whales only one or two pairs of teeth are visible. In whalebone whales (Mysticeti), however, there are teeth only in the foetus; the food is plankton. This is collected by the fringed baleen, which consists of rows of transverse plates of keratin. The tongue of the right whales is powerfully muscular and forces the water from the mouth. In the

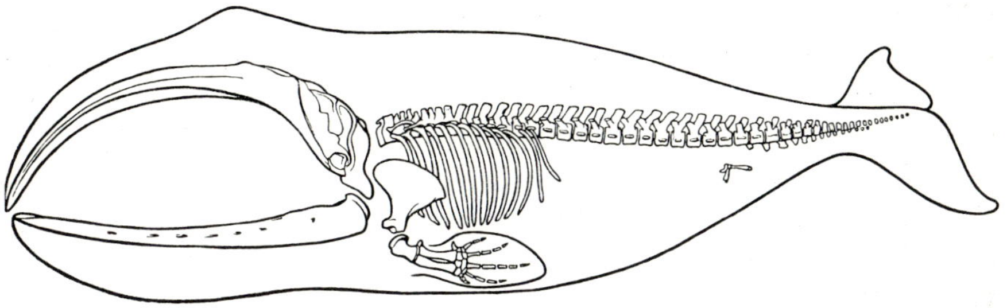

FIG. 437. Skeleton and outline of the right whale (*Balaena*). (After *British Museum Guide*.)

rorquals the same purpose is achieved by the contraction of subcutaneous muscles associated with the external throat grooving. The shrimp-like 'krill' (*Euphausia*) and other organisms are then swallowed through the narrow oesophagus to a special stomach with several chambers. The first stomach is non-glandular and is a crop-like oesophageal specialization, the hindmost stomach into which the bile and pancreatic ducts open is a specialized intestinal cavity. The intestine may be as much as sixteen times as long as the body.

Both types of nutrition are evidently efficient and the whales are abundant and of course very large. It is more easy to see the advantage of large size for aquatic than for land animals. There are no problems of support of weight, and on the other hand a great premium is placed on large size by the fact that skin friction is thereby relatively reduced, and this, which forms but a small element in the work to be done by a land animal, must be important in the water. Further, the heat loss is greatly reduced by the size, and this may be a large factor in cold water, with its high thermal conductivity (Parry, 1949). However, it has also been claimed that downward temperature regulation may be a problem and that the flippers and fins act as 'radiators'.

The respiratory system shows many special developments in the air passages, lungs, and nostrils. There are valves for closing the nostrils during diving, and special cartilaginous rings and muscles in the bronchioles. The epiglottis is extended as a tube inserted into the posterior narial cavity so that an uninterrupted air passage is provided from the blow-hole to the lungs. The very elastic and extensible lungs can thus be quickly filled with large volumes of air. In spite of the enlarged tracheo-bronchial tree the respiratory surface is small, but there are special arrangements of valves and venous plexuses to ensure economical distribution of the air and blood. Some whales can remain submerged for half an hour and reach depths of 500 metres or more. There is relatively little experimental evidence about the means by which they obtain oxygen and resist compression. The whole arrangement ensures the taking down of a maximum of air. There is rapid ventilation while on the surface, followed by slower heart-beat and presumably reduced tissue respiration while below, so that the whale can take down enough oxygen to last throughout its dives. However, the heart rate slows only to one-half in the only cetacean fully investigated (*Tursiops*) and there is no evidence about retention in venous sinuses or other means of reducing circulation such as are found in seals (p. 692). The respiratory centre in the medulla has a lesser sensitivity to CO_2 than in land animals.

Besides the air in the lungs there may also be some provision for storage of extra oxygen in the large blood-volume of the retia mirabilia, networks of blood-vessels, which abound throughout the body, especially in the thorax. However, the function of the retia mirabilia is probably connected with the accommodation of the animal to varying hydrostatic pressures. They expand and contract to occupy the space in the thorax as the air in the lung is diminished or increased as the animal rises or descends while swimming. There is much myoglobin in the muscles. The brain is supplied with blood entirely from meningeal arteries, which draw on the thoracic retia. The basilar artery and intracranial carotid close early.

No doubt the metabolism is also arranged to allow accumulation of a high oxygen debt, but the special metabolic peculiarities that allow for this are not known. It is perhaps not necessary for whales to have a special defence against caisson sickness if they are using oxygen reserves. There is no continuous addition to the nitrogen dissolved in the blood such as would lead to the formation of the bubbles that occurs when miners or divers rise suddenly after breathing air at great depths. The sudden expiration on surfacing produces a cloud of foetid vapour, the blow. This is generally supposed to be due to condensation, but it occurs even if the air is hot.

The brain is absolutely larger in whales than in any other animals (up to 7,000 g), and the hemispheres are elaborately folded. The cerebellum is very large. Little is known in detail about the sensory equipment or powers of the animals. The eyes are small (vestigial in *Platanista*) and in all whales much modified for aquatic life and diving. The cornea is more flattened than in subaerial mammals, and the lens rounded; in these respects the whales have returned to fish-like conditions. The whale eye is enclosed in a thickened sclera and further has special lid muscles. The tear glands and their duct are absent; instead, the surface of the eye is protected by a special fatty secretion of the Harderian glands.

The ear provides the major receptor system. The auditory nerve, lateral lemniscus, superior olive, inferior colliculus, and medial geniculate are all very large. Presumably much of the cortex serves the sense of hearing. The apparatus concerned with reception of air-borne vibrations is reduced. The external opening is very small and the long meatus is often filled with secretion. The tympanum is thick and ligamentous. It is to a normal ear drum as a closed umbrella is to an open one. The tip of the tympanic 'ligament' is attached only to the tip of manubrium mallei. The distal end of the processus gracilis of the malleus is fused to the adjoining bone of the tympanic bulla. The ossicles have articulations with one another as in terrestrial mammals and the tip of the stapes is movable in the foramen ovale.

The petro-tympanic bone is free from the skull, rests on a thick fibrous pad and is otherwise almost completely enveloped in a system of foam-filled air sinuses. The whole arrangement is believed to be designed to isolate the essential organ of hearing from vibrations extraneous to those reaching it by means of the meatus and auditory ossicles, and so to provide the means for directional hearing.

In the ears of terrestrial mammals the ossicles provide an arrangement for converting the relatively large displacement at low amplitudes of air-borne waves at the tympanum to waves with a sixtyfold greater pressure amplitude at the fenestra ovalis. The physical properties of water-borne vibrations, however, differ markedly from those that are air-borne. The pressure amplitude for the same intensity and frequency of water-borne and air-borne sound is in the ratio 61 : 1, and the displacement amplitude 1 : 61. It can be shown that adjustments of amplitude and pressure to values normally experienced in the cochlea by terrestrial mammals are achieved in cetaceans by the modifications of the middle ear mechanism (Fraser and Purves, 1959).

Whales emit a variety of sounds but little is known of their method of production or use. *Tursiops* in a tank recognize the sex of a new arrival in another tank, out of sight. Several such social reactions have been reported, for example, between mother and young. *Tursiops* can also avoid obstacles in the dusk and since they react to frequencies up to 120 kc/sec it is possible that they emit these for echolocation.

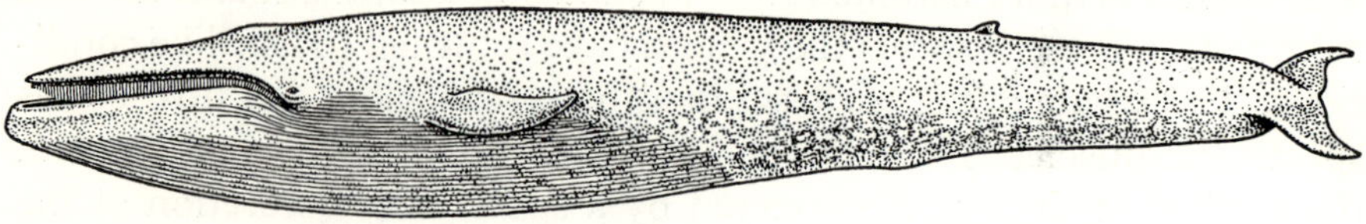

FIG. 438. Blue whale, *Balaenoptera*. (After Mackintosh and Wheeler.)

FIG. 439. Killer-whale, *Orcinus*. (After *British Museum Guide*.)

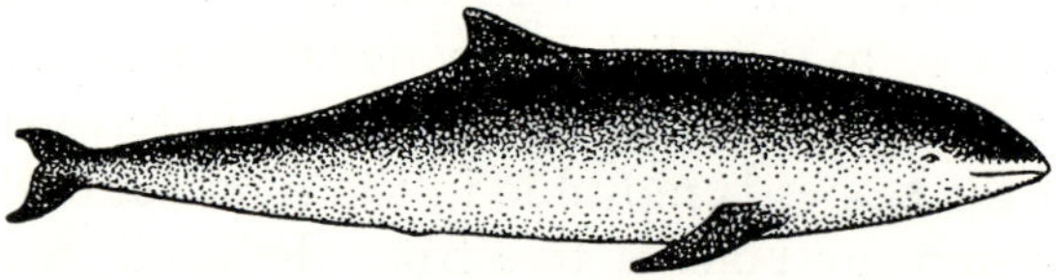

FIG. 440. Porpoise, *Phocaena*. (After *British Museum Guide*.)

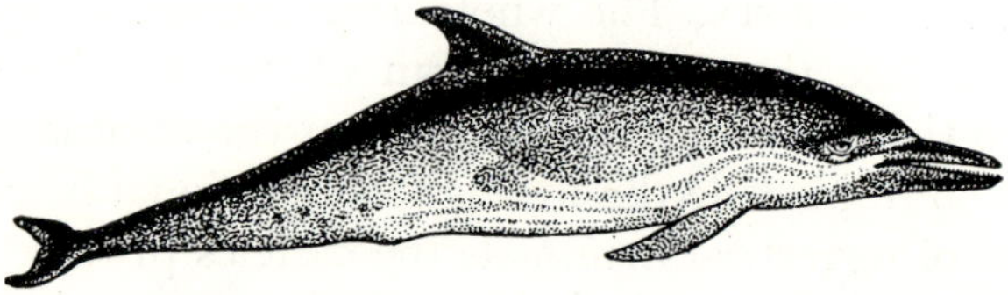

FIG. 441. Dolphin, *Delphinus*. (After *British Museum Guide*.)

The head of odontocetes carries an organ known as the 'melon', which is possibly a receptor. It is a mass of fat in front of the nostril, traversed by muscle-fibres and richly innervated by the trigeminal. It may serve to detect pressure changes in the water, substituting for the vibrissae of seals. In the sperm whale it is enormous (p. 674).

The behaviour of whales is undoubtedly elaborate, involving social life, communication by sound, and probably much learning. There is cooperation between individuals in helping to keep a wounded companion or a new-born at the surface. Play is common and rhythmical 'dancing' has been observed and also homosexual behaviour, in captivity. Many species migrate, for example, the humpbacks (*Megaptera*) and others spend the summer in the Antarctic feeding on krill and then come north to tropical waters to breed.

The reproduction shows various modifications for aquatic life. The testes do not descend into sacs, but to a position just below the body surface. The penis is very long, and curled when not erect. The uterus is bicornuate, but only one young is carried and is retained for a long time (more than a year in large whales), so that it is as much as a third the length of the mother at birth. The placenta is diffuse but with a few villosities like cotyledons. Its structure is epitheliochorial and there is a large allantois. There is a pair of teats in the inguinal region and the mammary glands are provided with a special receptacle and muscle so that milk is pumped into the mouth of the young. Some species of dolphins migrate to shallow protected water at the time of parturition.

It is clear that many factors have collaborated to concentrate the biomass of whale life into large units. Indeed, whales include the largest known animals, either fossil or recent. The blue whale, *Balaenoptera musculus* (Fig. 438), reaches nearly 150 tons, with a length of 100 ft. This is one of the whalebone whales, which are in general larger than the odontocetes, perhaps because of the immense sources of food directly available in the plankton; they have grown fat by eliminating the 'middle-men' upon which all toothed whales must feed. These mysticetes appeared in the Oligocene and radiated in the Miocene and since into a relatively small number of types, all of large size. *Balaena*, the right whale of the Arctic, is now very rare. The chief modern prey of the whalers are the fin whales (*Balaenoptera physalus*) of the Antarctic.

The odontocete whales are a more varied group; their history can be traced back to the late Eocene. The squalodonts of the Oligocene and Miocene were like the porpoises, but with triangular teeth. Most of them disappeared in the Miocene, but the river-porpoise *Platanista* of the Ganges, and related forms in the Amazon and in China may be descendants. The modern porpoises (Delphinoidea) are a very successful and numerous group of relatively small animals, with a dorsal fin and teeth in both jaws. They are all active predators, but the habits vary from those of the killer-whale *Orcinus* (Fig. 439), which is a fierce and cunning hunter, attacking even the largest whales, to the omnivorous porpoise *Phocaena* (Fig. 440), whose food includes crustacea as well as fishes and cephalopods. This is the commonest and smallest British cetacean, the largest individuals reaching 6 ft; the jaws are rather short, especially the upper one, and the teeth are spade-shaped. The true dolphins, *Delphinus* (Fig. 441), are larger animals (8 ft), living mostly on fish and having long, many-toothed jaws, the upper forming a beak. They are very fast swimmers, reaching 20 knots. The narwhal, *Monodon* (Fig. 442), is a large delphinoid (15 ft long) with a single tooth, usually the left upper incisor, remaining in the male and growing continually

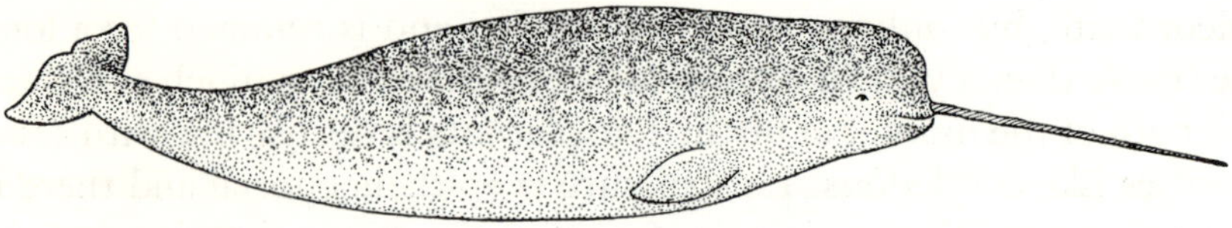

FIG. 442. Narwhal, *Monodon*. (After Norman and Fraser.)

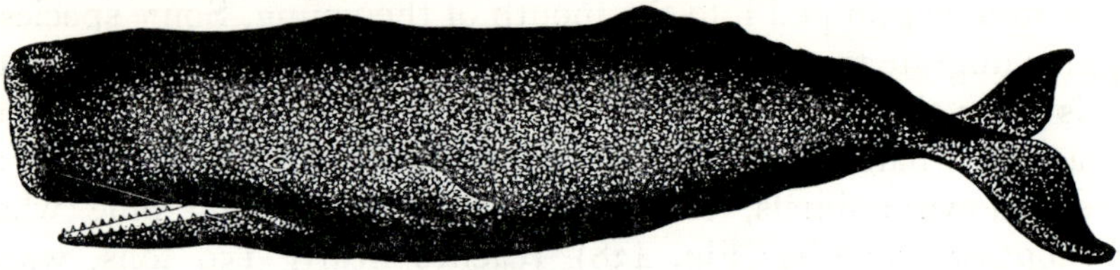

FIG. 443. Sperm whale, *Physeter*. (After Flower and Lydekker, *Mammals, Living and Extinct*, A. & C. Black, Ltd.)

FIG. 444. Bottle-nosed whale, *Hyperoodon*. (After Norman and Fraser.)

to make the spirally twisted horn, up to 9 feet long, whose use is not known. The female retains a pair of small incisors buried in the premaxillae.

In the sperm whales, Physeteroidea (Fig. 443), the rostrum is overlaid by an enormous reservoir containing spermaceti, a pure white waxy liquid solidifying in air. Either this, or some other feature, enables them to reach a larger size than other toothed whales (60 ft). Sperm whales have functional teeth only in the lower jaw and vestigial ones in the upper. They feed on cephalopods, the remains of whose jaws, combined with solid secretions to make stones in the intestine, form the substance ambergris used as an absorbent in the manufacture of scent. The bottle-nosed whales, such as *Hyperoodon* (Fig. 444) of polar seas, are also cephalopod feeders and reach 30 ft in length.

The whales known from the Eocene were different from either of the modern groups and are placed in a separate suborder *Archaeoceti (Fig. 445). The body was very long (up to 70 ft) and apparently thinner than in modern whales, suggesting a sea-serpent form, probably of low swimming efficiency. The hind legs had already disappeared, though vestiges of their skeleton remained. The skull was

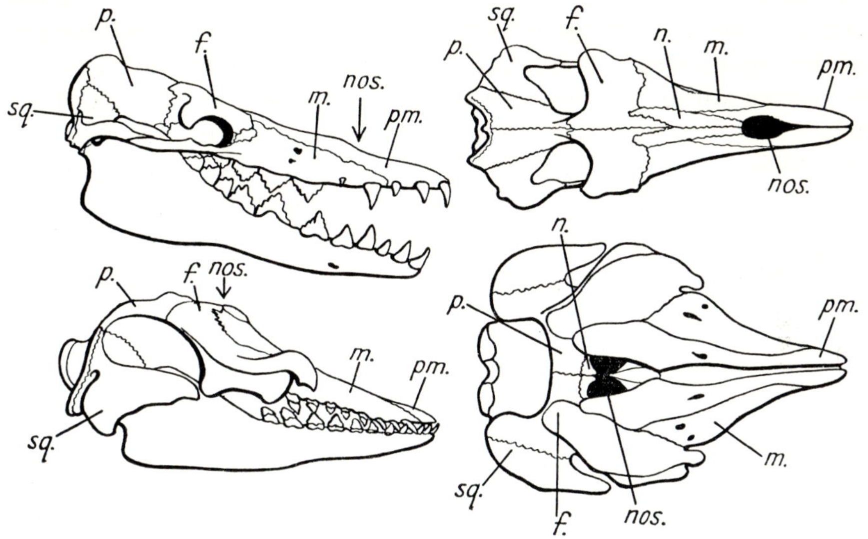

FIG. 445. Change in position of the blow-hole (*nos.*) during the evolution of whales. The two upper figures show the condition in the Eocene **Basilosaurus*, the lower figures a Miocene squalodont.

f. frontal; *m.* maxilla; *n.* nasal; *p.* parietal; *pm.* premaxilla; *sq.* squamosal. (Modified after Romer, *Man and the Vertebrates*, University of Chicago Press.)

long, the nostril had moved some way back. The teeth were of the normal mammalian number, 44, and were heterodont. The molars had sharp crenated edges, as in other fish-eaters, and the animals were obviously carnivorous, suggesting a possible creodont ancestry. Casts of the brain show large olfactory centres but a small, little folded, cortex. The cerebellum was enormous (if it has been correctly reconstructed).

These animals, such as **Basilosaurus* (= **Zeuglodon*), had already developed a long way from the main eutherian stock by middle Eocene times. This is an example of relatively rapid evolutionary change; it may be presumed that their ancestors had been in the condition of small insectivores not much later than the end of the Cretaceous, at the most 20 million years earlier. The basilosaur type persisted to the Miocene, but the exact connexions with the two modern sorts of whale are not clear. Whales have been abundant throughout the Tertiary, but there is not yet sufficient evidence available to reconstruct their full phylogenetic history. In broad outlines, however, it is clear that there has been a progressive adoption of features suitable for aquatic life, the long-bodied, heterodont basilosaurs giving place to the shorter, stream-lined modern whales, provided with suitable stabilizing fins and with the mouth highly specialized for eating fish, cephalopods, or plankton. It is impossible to say whether the changing of the populations is due to indirect influences of climatic changes or to factors within the animal populations themselves. The course of whale evolution, like that of teleosts, appears to have produced increasing efficiency within a single habitat, rather than a progressive colonization of new fields; but this appearance may be only a reflection of our ignorance and lack of knowledge of the varied and changing condition of the sea.

XXVII

CARNIVORES

1. Affinities of carnivores and ungulates: Cohort Ferungulata

THE union of the modern carnivores and hoofed animals in a single cohort Ferungulata is based on palaeontological work that has shown how both groups, together with some isolated surviving types such as the elephants and sea-cows, and many other forms now extinct, all arose from a common population in Palaeocene times. The ungulates have been grouped together for a long time because of their obvious common characteristics of herbivorous diet and hoofed feet, but it is clear that 'ungulates' include two very different sorts of creature, the even-toed artiodactyls, such as the pigs, sheep, and cows, and the uneven-toed perissodactyls, the rhinoceroses and horses. The latter can be traced backwards to a very ancient group, the Condylarthra of the Palaeocene and Eocene, and they were therefore for some time placed rather widely apart from the Artiodactyla, whose origin was mysterious. It has now been shown, however, that there is a resemblance between the Eocene artiodactyls and some of the creodonts, animals that were also the ancestors of the modern Carnivora. The creodonts and condylarths are in many ways alike, and it now seems probable that the whole group makes a single unit, diverging first from the insectivorous eutherian ancestral population in the late Cretaceous or Palaeocene, probably as a carnivorous stock. Some members then diverged almost at once to make the condylarths, perhaps from a stock that already possessed hoofs and then proceeded later to produce the Artiodactyla (Fig. 446).

It is a convenience to use these relationships as a basis for classification, but it must be recognized that modern carnivores have little more in common with ruminants than with, say, monkeys or rats. The three great groups that make up the Ferungulata diverged from each other only a short time after their common stock had diverged from that of the other mammals; at that time all eutherians were so alike that we should probably place them in a single order if they had left no descendants.

The Ferungulata have become much more diversified than the other cohorts into which we have divided the Eutheria and we have to recognize no fewer than fifteen orders in the group. It is therefore convenient to make further subdivision into five superorders. The first of these, Ferae, makes the central group, including the Carnivora. The second superorder, Protoungulata, includes the earliest ungulates, the condylarths, and it is convenient to place here also certain early offshoots, such as the South American ungulates and one living survivor, the aardvark. The third superorder is known as Paenungulata

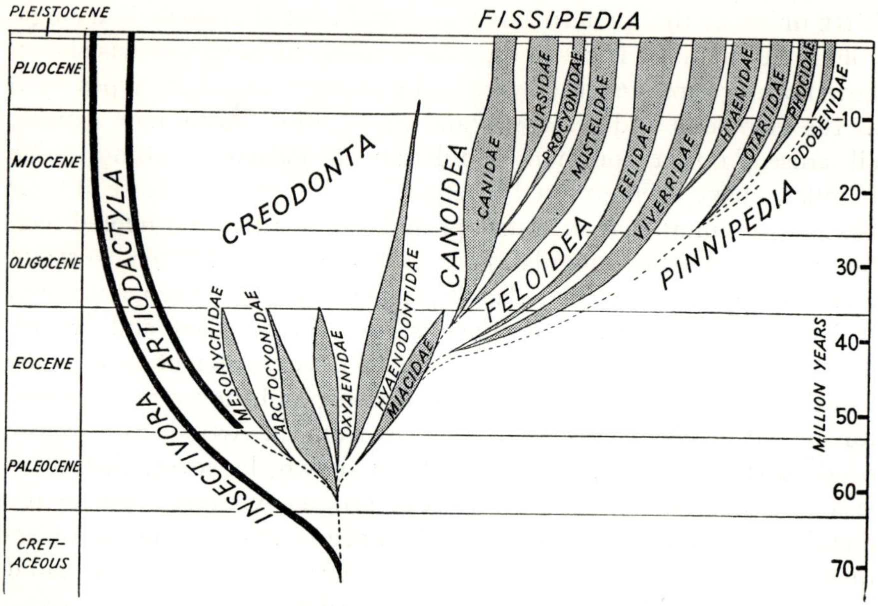

FIG. 446. Chart of the evolution of Carnivora.

('near ungulates') and includes a group of orders with rather primitive organization, most of them extinct. The elephants, hyraxes, and sea-cows remain as isolated vestiges of this great group, which included the huge pantodonts and dinocerates, formerly classed together as amblypods, also the pyrotheres, elephant-like animals from South America, and the embrithopods, large, horned animals from Africa. The fourth superorder, Mesaxonia, includes only the Perissodactyla, descended from the condylarths, while the fifth superorder, Paraxonia, contains the artiodactyls, derived a little later from the condylarths.

Of all this assembly of varied types only the carnivores and ruminant artiodactyls remain successful types at the present day, abundant in species and individuals, the remaining orders are either extinct or are represented only by few species.

2. Classification

Cohort 4. Ferungulata

Superorder 1. Ferae

Order Carnivora

*Suborder 1. Creodonta. Palaeocene–Pliocene. Holarctic

*Family 1. Arctocyonidae. Palaeocene–Eocene

**Arctocyon*, Paleocene; **Tricentes*, Palaeocene

*Family 2. Mesonychidae. Palaeocene–Eocene

**Mesonyx*, Eocene

*Family 3. Oxyaenidae. Palaeocene–Eocene

**Oxyaena*, Palaeocene–Eocene

*Family 4. Hyaenodontidae. Eocene–Pliocene

**Hyaenodon*, Eocene–Oligocene; **Apataelurus*, Eocene

Suborder 2. Fissipeda. Palaeocene–Recent

*Superfamily 1. Miacoidea. Palaeocene–Eocene, Holarctic

*Family Miacidae

**Miacis*, Eocene; **Vulpavus*, Eocene

Superfamily 2. Canoidea

Family 1. Canidae. Dogs. Eocene–Recent

Canis, wolves, dogs, jackals, Pliocene–Recent, world wide, but not originally wild in S. America or Australia; *Vulpes*, fox, Miocene–Recent, Holarctic, N. Africa

Family 2. Ursidae. Bears. Miocene–Recent

Ursus, bear, Pliocene–Recent, Holarctic

Family 3. Procyonidae. Raccoons. Miocene–Recent

Procyon, raccoon, Pliocene–Recent; N. and S. America; *Ailurus*, panda, Recent, Asia, *Ailuropoda*, giant panda, Recent, Asia

Family 4. Mustelidae. Weasels. Oligocene–Recent

Mustela, weasel, ferret, stoat, Miocene–Recent, Holarctic, S. America, N. Africa; *Meles*, badger, Pliocene–Recent, Eurasia; *Taxidea*, American badger, Pliocene–Recent, N. America; *Mephitis*, skunk, Recent, N. America; *Lutra*, otter, Pliocene–Recent, Holarctic, S. America, Africa; *Martes*, marten, Pliocene–Recent, Holarctic

2. Classification (*cont.*)

Superfamily 3. Feloidea

Family 1. Viverridae. Civets and mongooses. Oligocene–Recent

Viverra, civet, Miocene–Recent, Asia; *Herpestes*, mongoose, Oligocene–Recent, Eurasia, Africa, and introduced to W. Indies

Family 2. Hyaenidae. Miocene–Recent. Eurasia, Africa

Hyaena, hyena, Pliocene–Recent, Asia, Africa

Family 3. Felidae. Cats. Upper Eocene–Recent

Felis, cats, pumas, ocelots, leopards, lions, tigers, jaguars, Pliocene–Recent, world-wide; **Hoplophoneus*, sabre-tooth, Oligocene–Miocene; **Smilodon*, sabre-tooth, Pleistocene

Suborder 3. Pinnipedia

Family 1. Otariidae. Eared seals. Miocene–Recent

Eumetopias, sea lion, Atlantic and Pacific

Family 2. Odobenidae. Walruses. Miocene–Recent

Odobenus, walrus, Arctic

Family 3. Phocidae. Seals. Miocene–Recent

Phoca, seal, Atlantic and Pacific; *Halichoerus*, grey seal, N. Atlantic.

3. Order Carnivora

The earliest Cretaceous mammals were probably insectivorous and it is not therefore surprising that some of their descendants became flesh-eating; indeed it is curious that a single stock has provided nearly all the hunters found among the mammals ever since, though the marsupials have produced carnivorous types in South America and Australasia. It is difficult to see why carnivores have not developed more often from the insectivoran or some other stock; that they have not done so may remind us that special circumstances are necessary for the origin even of a type for which a means of life would seem to be readily available.

4. The Cats

The changes that convert a mammal into an effective hunter occur in many parts of the body, without, as it were, radically distorting any. We may illustrate this by considering the most specialized members of the group, the cats (Fig. 447). The head is large, with long ears, long whiskers, and nose with many turbinals. The brain is large, the cerebral hemispheres overlap the cerebellum; the olfactory centres are large. As is usual with carnivores, behaviour is complicated; in order to continue pursuit of prey that cannot be seen, or perhaps even smelt, the animals learn to associate the presence of food with obscure clues such as footmarks, and to make use of these clues they must lie in

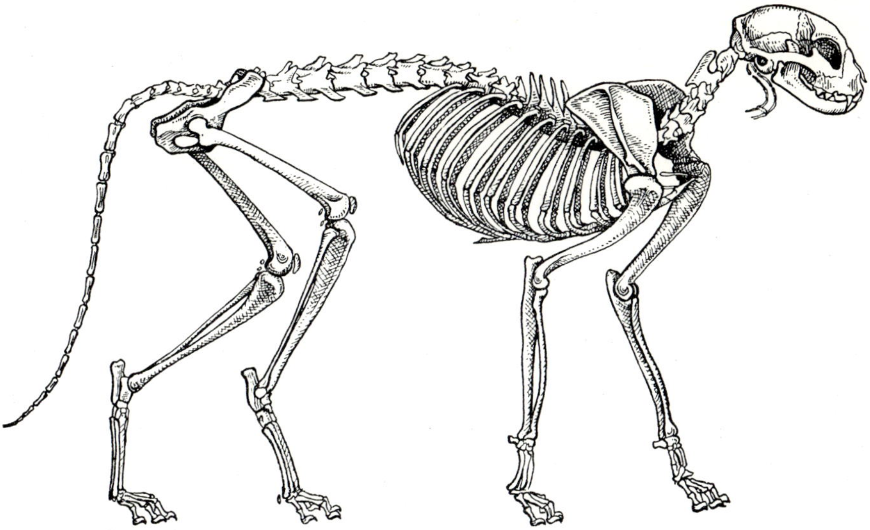

FIG. 447. Skeleton of the cat (*Felis*).

wait for the prey. All of this involves an elaborate balance of internal motivation with activity and restraint. This power of 'abstraction' of ultimate satisfaction from the immediate situation may perhaps be associated with the familiar play of the kitten or the less edifying treatment of a captured mouse by an adult cat.

Social or family groups are commonly well marked in carnivores and there are usually characteristic odours for recognition, often associated with large anal glands, especially well known as producers of civet and the 'poison' of the skunks. The back of the head is enlarged to take the brain and there is a well-developed snout for the nose, but the face is nevertheless short, and it is characteristic of the specialized carnivores that the tooth-row is shortened, developed especially at the front end, producing the incisors for piercing, canines for tearing, and premolars and anterior molars for cutting. In contrast to the ungulate type of dentition the hinder molars, not being needed for grinding, are reduced. In the cats, as in all modern carnivores except seals, the teeth most favourably placed for biting by their position relative to the jaw muscles, namely, the last upper premolar and first lower molar are specially developed into cutting-blades, the carnassials. This is done by formation of a ridge along the outer side of the upper molar, the paracone and metacone making a single cutting-edge. The protocone remains as an inwardly projecting ridge at the front of the tooth, which otherwise makes a single blade, shearing outside a similar blade formed by the paraconid and protoconid of the lower molar. This restriction to long sharp ridges also affects the teeth in front of the carnassials, but behind them the molars are so reduced that in true cats they are represented only by a single vestige in each jaw. More of the posterior molars remain in some of the primitive carnivores (dogs), and in some, such as the bears, they may acquire a bunodont surface and hence the power of grinding.

The jaws are, of course, powerful in carnivores, the articulation being a tight, transverse hinge, allowing none of the rotatory movements found in other mammals. The jaw-muscles include especially powerful temporals, for whose attachment there is a large coronoid process on the jaw and often large sagittal crests on the top of the skull. The temporal fossa is very wide and never closed off from the orbit, since there is no need for specially increased surfaces for the masseter, which is only moderately strong. The pterygoid muscles (and their fossa) are reduced, since the jaw has no rotary action.

The post-cranial skeleton shows a generalized mammalian build, with specializations for sudden leaping movements. There are five digits in the hand and four in the foot in cats; in other carnivores the number is never less than four. The toes are armed with the characteristic claws, which are held drawn back by elastic ligaments and pulled out when needed by the action of the flexor digitorum profundus muscles on the terminal phalanx, to which the claw is attached. The weight of the body is carried on special pads on the second interphalangeal joints and the metatarsal heads. A curious feature of the carpus of all modern carnivores is the fusion of the scaphoid and lunate bones. The arrangement of the limbs and backbone is that of a quadruped able to proceed over uneven surfaces and also steeply upwards and downwards, especially in the carnivores that are arboreal. This involves a long body, with much of the weight carried on the fore-limbs; the thoracic neural spines are therefore high. On the other hand, the vertebral girder has to take the strain of powerful sacrospinalis muscles for leaping; the transverse processes are broad in the lumbar region, and again in the neck for the muscles that move the head. The clavicle is reduced. The tail, well developed for the maintenance of balance in wild cats, tends to become reduced under domestication, the extreme being the Manx variety, with only three caudal vertebrae.

As in so many carnivores, the alimentary canal is short and the stomach never complex or the coecum large. The uterus retains the primitive mammalian bicornuate form. The chorio-vitelline placenta is important early in pregnancy. The chorio-allantoic placenta is of a type known as vasochorial with the villous portion of the chorion restricted to a characteristic band around the embryo (hence 'zonary' placenta).

5. *Suborder Creodonta

Besides animals such as the cats, adapted in this detailed way for hunting live prey, the order Carnivora contains a variety of less-modified forms, such as the dogs and their offshoot the bears, which are partly scavengers. Others are suited for special types of carnivorous life, such as the weasels, mongooses, and other small animals. The seals and walruses are carnivores much changed by aquatic life.

Great numbers of fossil carnivores are known and we can trace much of the history of the order. The *Creodonta (= 'flesh-tooth') of the Palaeocene and Eocene included four distinct families, one, the *Arctocyonidae, is perhaps close to the ancestry of the whole Ferungulate stock, the other three are more specialized. The earliest creodonts, such as **Tricentes* of the North American Palaeocene, were small semi-plantigrade creatures, very like the contemporary small insectivores, which were the prototype of all eutherian mammals. The skull was long and low, with a small macrosmatic brain, but already having sagittal crests for the temporalis muscle. The dentition included the full number of teeth and these carried sharp cusps, arranged in the tritubercular pattern, with the beginning of the development of a hypocone in the upper molar. There was no carnassial, however; animals of this type could therefore well have given rise to non-carnivorous forms, such as the ungulates. Other differences from modern carnivores were that the scaphoid and lunate were not fused and there was no ossified auditory bulla.

It is not surprising that these early creodonts were for a long time classed as Insectivora. Their descendants soon began to show various specializations. Thus **Arctocyon* of the Upper Palaeocene of Europe was a large animal, with tuberculated molars, probably omnivorous like a bear. Some Eocene descendants of this type, such as **Mesonyx*

(Fig. 448), retained the triangular molars and became very large, perhaps they were scavenging creatures. Their toes carried hoofs, and this, together with their simple dentition and other features, has suggested to some that smaller members of the group may have been ancestral to the Artiodactyla. Other Eocene creodonts, however, became more typical carnivores, with shearing carnassials, developed from various teeth, often the upper first and lower second molar.

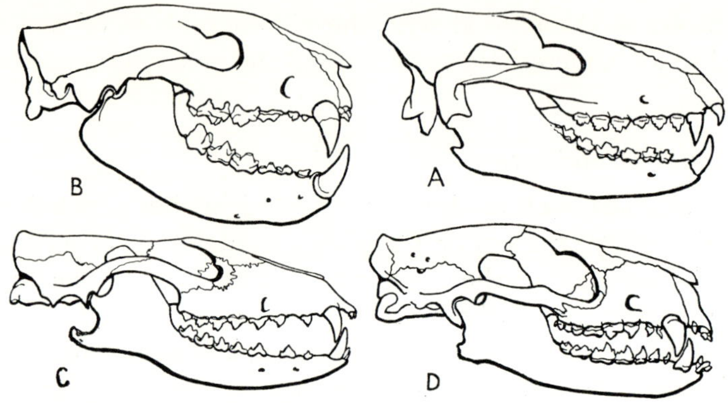

FIG. 448. The skulls of some early carnivores.

A, **Mesonyx*; B, **Oxyaena*; C, **Sinopa*; D, **Vulpavus*. A–C are creodonts, D is a miacid fissipede. (After Romer, A after Scott, B and C after Wartmann, and D after Matthew.)

These early carnivores radiated into lines that parallel those found today. Thus **Dromocyon* resembled a dog, **Oxyaena* an otter, **Dissacus* a cat, and **Sinopa* a weasel. **Hyaenodon* and its allies were abundant hyena-like animals of various sizes; they survived until the Eocene. **Apataelurus* was an Eocene sabre-tooth. By their similarities to later carnivores these arrays of early carnivore types constitute a remarkable example of convergent evolution.

After their abundance in the Eocene the creodonts declined, although a few genera survived until the Pliocene. Perhaps the type, though able to catch the cumbrous early herbivores, was unable to make a living from the later, faster-moving ungulates.

6. Suborder Fissipeda

Towards the end of the Eocene a new type of carnivore became abundant. The earliest of these, the *Miacidae, were so like creodonts that they are sometimes classified in that group, sometimes with the Fissipeda. Thus miacids of the Eocene (Fig. 448) were small animals, perhaps arboreal, very close to the *Arctocyonidae and still with triangular molars, unossified bulla, and separate carpal bones. There was, however, a sign of the beginning of the fissipede carnassials, the fourth upper premolar and first molar being elongated, the cusps partly united to form a ridge. At the end of the Eocene and beginning of the Oligocene early representatives of the various fissipede families appeared and were presumably derived from these miacids. The dogs (Canidae) (Fig. 449) appeared very early and have since changed relatively little, the modern *Canis* being practically a survivor showing us the Eocene stage of carnivore evolution. Numerous fossil dogs

FIG. 449. Wolf, *Canis*. (From a photograph.)

are known and the type has been a very successful one. The diet is varied and partly herbivorous. The European red fox, for instance, feeds on mice, hares, rabbits, and chickens, but also on snails, insects, and berries. The type is not suited for climbing but for running in open country, for which purpose long legs and a digitigrade habit have been developed, the pollex and hallux being reduced (Fig. 354). The teeth have remained unspecialized, with at least two post-carnassial grinding molars and still distinct signs of the triangular form in the carnassial (Fig. 450). Dogs, wolves, and foxes are found throughout the world, including South America, but not in Madagascar or New Zealand. So many types are known that the exact ancestry of the various wolves and foxes has not been fully disentangled.

The bears (Ursidae) (Fig. 451) were a Miocene offshoot from this dog stock and here the tendency to non-carnivorous diet became accentuated; there are no carnassials and the molars acquire bunodont grinding surfaces. The gait is plantigrade. Various types are found through the Holarctic region and South America. The raccoons (Procyonidae) (Fig. 452) are a rather similar though smaller type of animal, with dentition suited to an omnivorous diet, the upper carnassial having developed (? redeveloped) a hypocone. They are all American, except *Ailurus* and *Ailuropoda*, the panda and giant panda (Fig. 453), large, herbivorous creatures living in Asia. The latter has a special bone near the pollex, making a grasping organ for holding bamboo shoots.

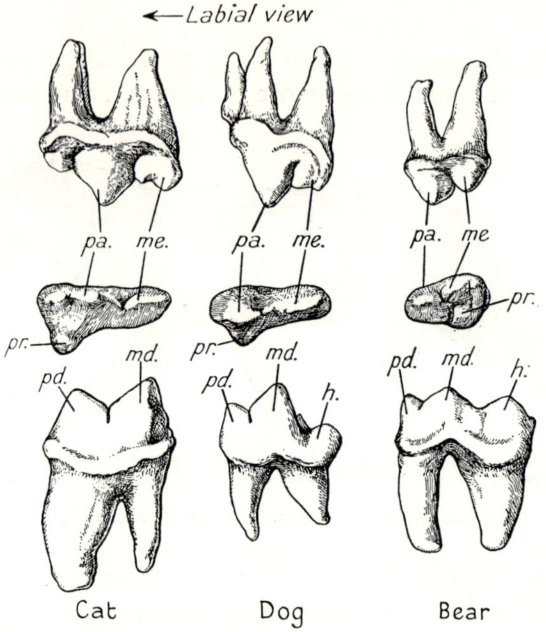

FIG. 450. Carnassial teeth of carnivores.

The top row shows the last left upper premolar from the labial side; middle row the same from below; bottom row the first left lower molar from the labial side. *h.* heel (talonid); *me.* metacone; *md.* metaconid; *pa.* paracone; *pd.* paraconid; *pr.* protocone. (Partly after Flower and Lydekker.)

These families of rather primitive carnivores are united in a superfamily Canoidea, and with them we may associate the weasels (Mustelidae), which are the typical small carnivores found throughout the world and retain many primitive characters. They have well-developed carnassials and never more than one post-carnassial molar. They can be recognized back to the Eocene and now include the stoats and weasels, *Mustela* (Fig. 455), which live on rats and mice, rabbits, and other small herbivores; *Meles* the badger (Fig. 454), which is omnivorous and has a hypocone; *Mephitis* the skunk (Fig. 456), a burrowing animal that ejects a stream of foul-smelling liquid from special anal glands; and the otters, *Lutra* (Fig. 457), with webbed feet, short fur, small ears, and other features suited for life in the water. In many

FIG. 451. Brown bear, *Ursus*. (From life.)

FIG. 452. Raccoon, *Procyon*.
(From a photograph.)

FIG. 453. Giant panda, *Ailuropoda*.
(From a photograph.)

FIG. 454. Badger, *Meles*.
(From photographs.)

FIG. 455. Stoat, *Mustela*.
(From a photograph.)

FIG. 456. Common skunk, *Mephitis*.
(From photographs.)

FIG. 457. Otter, *Lutra*. (From photographs.)

FIG. 458. Indian mongoose, *Herpestes*.
(After a photograph by F. W. Bond.)

FIG. 459. Hyena, *Hyaena*. (From photographs.)

FIG. 460. Sabre-tooth, **Smilodon*. (After Scott.)

mustelids there is delayed implantation of the blastocyst and other interesting reproductive phenomena.

The most-modified Carnivora are placed in the superfamily Feloidea. The civets and mongooses (Viverridae) (Fig. 458) are survivors that show us many of the characters possessed by this type in the Oligocene. They are the small carnivores that occupy in the Old World tropics the position taken farther north by the weasels. In general they are like the ancestral miacids, with long skull, small brain, and short legs. *Herpestes*, the mongoose, is abundant throughout Africa and Asia.

The hyenas (Fig. 459) have become large running creatures, with massive teeth specialized for crushing, and hence allowing a scavenging life. The true cats (Felidae) were already differentiated from the miacid ancestry in the Eocene, but at that early period they all had the very great development of the upper canines as cutting and piercing sabreteeth (Figs. 460, 461). There were numerous genera with this characteristic from the Eocene onwards until the Pleistocene, when they disappeared simultaneously from Europe, Asia, and America. Probably they attacked large, thick-skinned herbivores. The jaw could be opened to a right angle to allow the fang to strike, and there were such associated developments as large mastoid processes for the sterno-mastoid muscles that pulled the head downwards and forwards in the strike. The closing muscles of the jaw and the coronoid process were, however, small in the sabre-tooths.

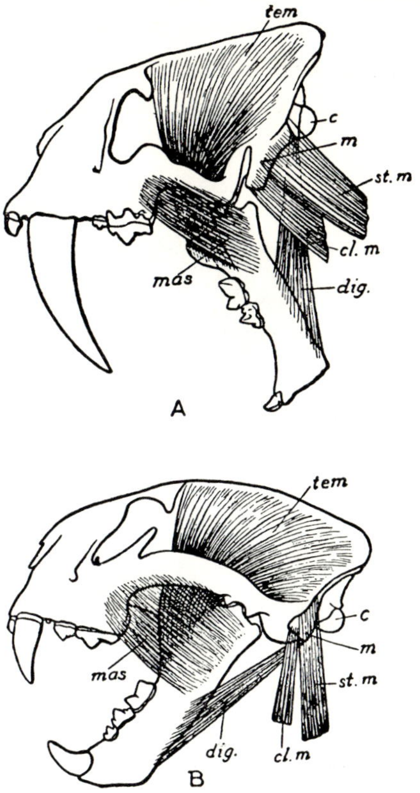

FIG. 461. Neck and jaw muscles of A, sabretooth, **Smilodon*, compared with those of a modern cat B, to show modifications for striking with the whole head and biting, respectively.

c. condyle; *cl. m.* cleidomastoid; *dig.* digastric; *m.* mastoid process; *mas.* masseter; *st. m.* sternomastoid; *tem.* temporalis. (From Lull, *Organic Evolution*, copyright 1917, 1929 by The Macmillan Company and used with their permission, after Matthew.)

The cats themselves, with smaller canines, appeared in the Oligocene. They are very successful carnivores, partly arboreal and hence most common in tropical regions, where there are large forests. There is much difference in detailed habits between the many species of the family, but all are alike in bony structure. Numerous attempts have been made to divide the group into genera and subgenera, but all may reasonably be retained in a single genus *Felis*. Lions (*F. leo*) have many distinct races in Africa and Asia and are mainly terrestrial, hunting

FIG. 462. Tiger, *Felis*. (From photographs.)

in open country. Tigers (*F. tigris*) (Fig. 462) occur throughout Asia to Siberia, are usually solitary, and often frequent damp places. They also have many races, and in captivity lions and tigers can be crossed. Other cats are mostly smaller and more fully arboreal than the lion and tiger. The leopard (*F. pardus*) of Africa and Asia reaches 5 ft in body length. *Felis catus*, the wild cat, formerly common, still exists in Britain and has a Palearctic distribution. The domestic cat probably arose in Egypt from the caffre cat, *F. ocreata*. The jaguar (*F. onca*), puma (*F. cougar*), and ocelot (*F. pardalis*) are large Central and North American cats, which have recently invaded South America.

Felis is thus one of the most widespread of all mammalian genera, occurring in all main parts except Australasia, Madagascar, and oceanic islands. Because of their striking and familiar characteristics we can form a vivid picture of all these slightly different sorts of cat, pursuing varying prey in the different regions. In order to visualize past evolution properly we should need to have an equally detailed knowledge of past populations. It is difficult enough to classify and describe a modern population of this sort and we need not be surprised that the palaeontologist, who has to consider also variations with time, finds that the growth of his material introduces intolerable problems of classification.

7. Suborder Pinnepedia

The seals, sea-lions, and walruses are marine carnivores that have existed since the Miocene. Their exact affinities are doubtful; they show some similarities to otters, perhaps due to common ancestry but possibly to convergence. They also have a remote likeness to the bears.

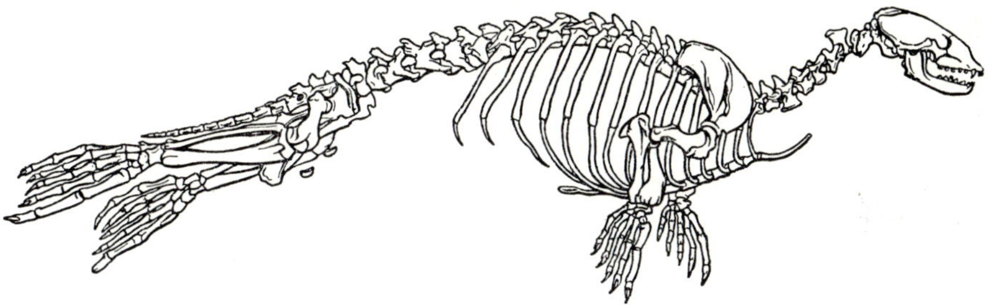

FIG. 463. Skeleton of the seal, *Phoca*. (After Blainville.)

In the seals (*Phocidae*) the body is streamlined, covered with thick fur, below which is a thin epidermis and a thick layer of blubber, making a quarter of the weight of the animal. Swimming is by means of the paddle-like limbs and flexion of the whole body, there is webbing between the digits, and the tail is reduced to a short rudiment. The basal segments of the limbs are shortened and some of the digits lengthened, without any increase in their number, though there are some extra phalanges. The speed of swimming may reach 15 knots if the seal is frightened. The cervical vertebrae are massive, with complex articulations, but the hinder ones are simplified and the column very flexible, so that it can be bent dorsally or laterally, allowing sudden turns in the water and complicated balancing feats on land (Fig. 463). All the seals leave the water to breed and therefore need some support for their large bodies.

The teeth show a reduction in number and are rows of nearly similar, laterally compressed spines. They may carry three cusps in a row, a reversion to 'reptilian' conditions, which serves to prevent escape of the slippery prey. There are large canines. The milk dentition is lost very early, sometimes *in utero*. The intestine is long. The water-supply is obtained from metabolic water.

The pinnipedes resemble the whales in being microsmatic but have good eyes, with flat cornea, round lens, and a muscular palpebral sphincter. The eyes are directed upwards and prey is often caught from below. The external ears are reduced but hearing is probably acute; the auditory ossicles are massive. There are numerous large vibrissae on the muzzle. The brain is large and rounded, with convoluted hemispheres and large midbrain and cerebellum. There is an elaborate vocal communication system, the calls varying to human ears from booming to chirping.

Young seals can remain submerged for up to 25 minutes and have been shown to be able to stand a pressure equivalent to a dive to 95 m. Larger seals can remain submerged even longer and at greater depths. The nostrils are closed by special muscles. The lungs are large and the bronchi contain myoelastic valves. During a dive the heart slows from 120 to 4 beats a minute. There is no drop in blood-pressure, because of a widespread reflex vaso-constriction, which prevents blood reaching the tissues, except the brain and heart muscle. Blood from the brain returns to the abdomen, by a large vessel above the spinal cord, and then accumulates there in extensive sinuses, including a huge dilatation of the vena cava above the liver, which is occluded by a sphincter of striated muscle above the diaphragm. There are few true retia mirabilia but abundant venous plexuses. The blood can carry as much as 35 c.c. of oxygen for 100 c.c. of blood (20 c.c. in man under the same conditions), and there is much myoglobin in the muscles. The respiratory centre tolerates a high CO_2 level. Lactic acid accumulates in the muscles, reducing metabolic levels. By these means the animal is provided with sufficient oxygen for the dive, without absorbing nitrogen and risking 'bends'.

Copulation takes place in the water in most seals and the penis bone is very large, especially in the walrus. The external genitalia, like the nipples, are withdrawn into folds of the surface. The eggs are fertilized shortly after parturition, the two horns of the uterus carrying alternate pregnancies. Implantation of the blastocyst is delayed for two months or more. As in other carnivores the placenta is zonary, with coloured margins due to the presence of bilirubin.

In the sea-lions (Otariidae) (Fig. 464) the legs can still be turned forward for use on land, and there are other primitive features, including external ears. They are more mobile on land than are the seals and can even climb cliffs. The family dates back to the Miocene. The walrus, *Odobenus*, is a related form, highly specialized for eating bottom-living molluscs, which it digs up with its enormous canines.

The Phocidae are the modern seals, found in all seas and fully aquatic, the hind limbs being attached to the tail. They come ashore only for short periods in isolated places to breed, and can only just drag themselves along the beaches. Many seals migrate for long distances to

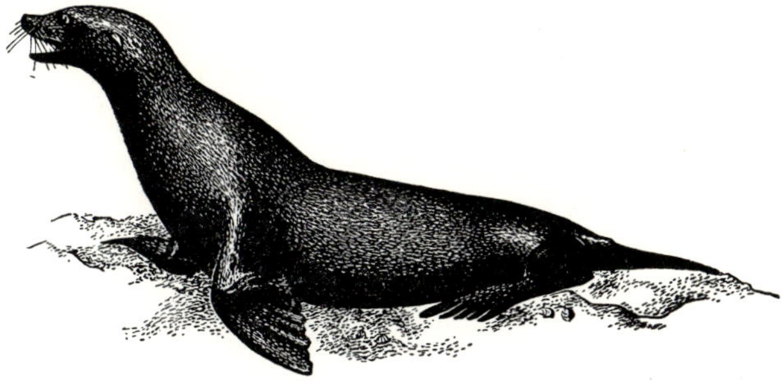

FIG. 464. Sea-lion, *Eumetopias*. (From a drawing belonging to the Zoological Society.)

particular breeding-places, such as the Pribilof Islands in the case of the fur seals (*Callorhinus*). The males, arriving first, fight furiously with each other, the victors then collecting harems of twenty or more females, who give birth to their young and are soon afterwards impregnated again. The bulls remain on shore without feeding, guarding the family, while the females return to suckle the young at each tide for a period of about three weeks.

XXVIII

PROTOUNGULATES

1. Origin of the ungulates

No herbivorous eutherians are known from the Cretaceous period, but during the Palaeocene epoch a number of animals abandoned the insectivorous habit and began to eat plants. These condylarths rapidly radiated into numerous types, so that by the end of the Palaeocene several distinct orders descended from this stock can be recognized. In North America and elsewhere there appeared large, clumsy animals, the *Pantodonta (Amblypoda) and *Dinocerata, while in South America a special fauna, the *Notoungulata and *Litopterna, developed. Further types then arose in the Eocene, including the early elephants and Perissodactyla. The Artiodactyla first appeared in the lower Eocene, as rather pig-like creatures; their origin is uncertain but they may have come from some form not very distinct from the *Condylarthra.

During the Eocene and Oligocene there were, therefore, numerous large, heavy-bodied herbivorous mammals, perhaps mainly suited to forest life and living upon relatively soft green food, since their teeth were mostly not highly developed for grinding. They were, however, gradually replaced during the Miocene by swifter, grazing animals suited to the plains of that period.

Following Simpson we shall classify the numerous orders of herbivorous (ungulate) mammals into four superorders. The Protoungulata, including *Condylarthra, *Litopterna, *Notoungulata, *Astrapotheria, and Tubulidentata, include the oldest forms, together with various early offshoots and one living creature the aardvark or Cape ant-eater, which is difficult to place elsewhere. A second superorder, Paenungulata, includes a number of descendants of the condylarths that early achieved success, the *Pantodonta and *Dinocerata of the Holarctic region, *Pyrotheria of South America, and *Embrithopoda in Africa. With these are placed the Proboscidea, which succeeded them as large herbivores in the Miocene. The conies (Hyracoidea) are an isolated group that still shows some of the Eocene characteristics of this Paenungulate group, and the sea-cows (Sirenia) are an early offshoot that took to aquatic life. Finally, the orders Perissodactyla and Artiodactyla occupy two separate superorders, Mesaxonia and Paraxonia.

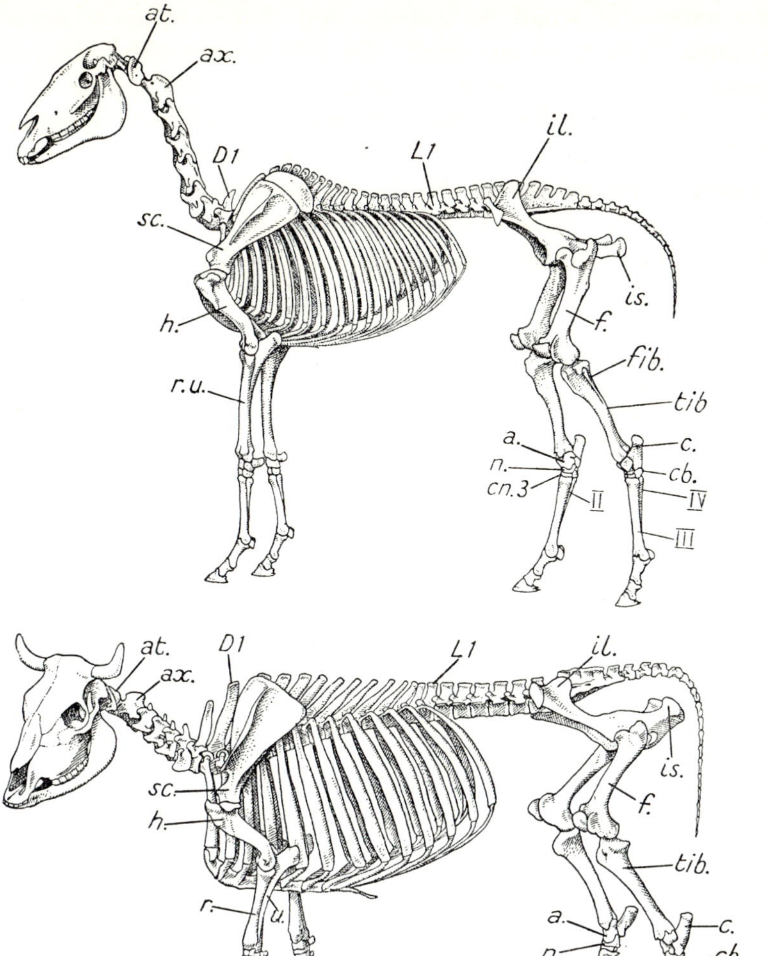

FIG. 465. Skeletons of horse and cow. (After Ellenberger and Sisson.)

a. astragalus; *at.* atlas; *ax.* axis; *c.* calcaneum, *c2–4*, distal carpals; *cb.* cuboid; *cn3.* 3rd cuneiform; *D1.* 1st thoracic vertebra; *f.* femur; *fib.* fibula; *h.* humerus; *i.* intermedium; *il.* ilium; *is.* ischium; *L1.* 1st lumbar vertebra; *n.* navicular; *r.* radius, *rad.* radials; *sc.* scapula; *t1–3.* distal tarsals; *tib.* tibia; *u.* ulna; *ul.* ulnae, *II–IV*, metatarsals.

2. Ungulate characters

When mammals adopt a herbivorous diet they assume certain characteristics, which it is convenient to recognize before dealing with the individual groups (Fig. 465). The animals often become large, but it must be remembered that, outside the ungulates, the rodents include many successful small herbivores, and that conversely among the ungulates the hyraxes are small. The skin is often thick and a variety of protective coloration schemes of spots and stripes appear, the under side usually being paler to eliminate shadows. Defensive weapons such as tusks or horns are common, but the problem of security often leads an unaggressive animal to the development of a swift gait. For this the limbs are lengthened by raising up on the toes (p. 576), producing first digitigrade and then unguligrade locomotion. When this happens

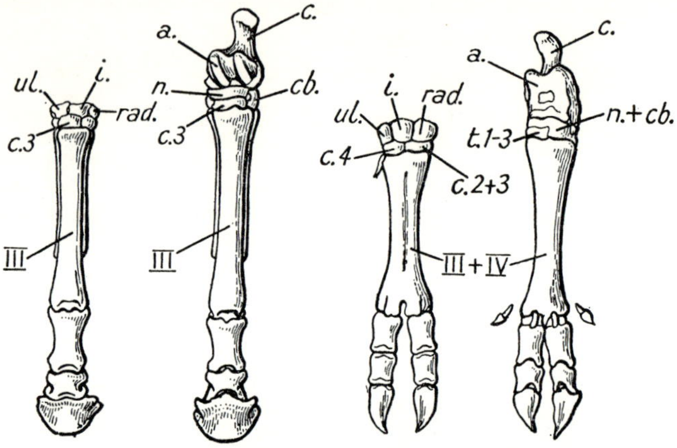

FIG. 466. Skeletons of feet of horse and cow. Lettering as Fig. 465.

the more lateral digits, failing to reach the ground, become reduced and may disappear, leaving finally the characteristic one or two. The movement of the limb becomes restricted to a fore-and-aft direction, and the joints assume a pulley-like form, especially characteristic in the trochlea of the talus, deeply grooved in artiodactyls but markedly so also in perissodactyls (Fig. 466). The carpal and tarsal bones of these swift-moving animals become arranged on the so-called interlocking plan, by which each elongated metapodial thrusts up against two carpals or tarsals. No movements of pronation occur, and the ulna and fibula tend to be reduced and fused with the radius and tibia. The hoofs themselves are a characteristic development, the terminal phalanx is broadened, and the claw becomes modified to surround it, while a pad forms below. The elongation of the limbs is mainly in the lower sections, the humerus and femur being short. Locomotion is by movement of the whole limb by the action of its proximal muscles, the hind-limbs being the main propellents and the fore-limbs weight-bearers, with corresponding modification of the vertebral girder (p. 728). The neural spines are very high above the fore-legs and the ribs are numerous, so that the girder has large compression struts above and below and it balances largely on the fore-legs and is pushed from behind. The ilium is broad and raised vertically, providing large attachments for the glutei, which are the important locomotor muscles, and for the abdominal muscles, which carry the weight of the belly. This arrangement of the column is essentially preserved even in the very large animals, such as elephants, rhinoceroses, and many extinct types, which are said to be 'graviportal'. In these latter, however, the legs are arranged on a different plan, since the great weight can only be carried by very massive struts of large cross-section. The proximal parts are therefore enlarged and several digits are retained to make broad supports for the pillars, as is well seen in the elephant's foot.

In many ungulates the neck becomes considerably lengthened, probably both in order to reach up or down for food and also to give a good look-out for the head. The ears are long and hearing acute, so that the direction of sounds may be easily detected. Sight is not especially developed, but the pupil is often horizontal in animals that live on the plains, giving a wide visual angle. The sense of smell is well developed and the animals often graze advancing up-wind, using for this purpose the receptors of the moist muzzle. The tongue is large and taste receptors sensitive.

The brain is large and the life of these herbivores is conducted with the use of much information learned during each lifetime. This enables them to range over large territories and to vary these with the seasons, in search of food and water. They show a remarkable alertness to changes of sound or scent.

Many herbivores are social animals, and information is shared among a large group. They have elaborate means of communication by scent glands, which are used to mark trails and territory, as well as for exchange of signals between individuals. Their sexual organization is complicated, involving elaborate interchange of visual, auditory, and olfactory signals and often combat between males. The establishment of a leader is apparently often needed to allow the advantages of social organization for protection and finding food. Gestation is long and relatively few young are produced (as in other large animals with efficient brains). The new-born is well developed and soon able to run with the herd.

Perhaps the most significant modifications are in the means of obtaining and digesting the food. The triangular molar pattern gives place to a square one, by development of a hypocone on the posterior interior side of the upper molar. The lower molar also becomes square, by loss of the paraconid and raising the heel, whose outer hypoconid and inner entoconid make a pair, behind the metaconid and protoconid (Fig. 467). Even more characteristic is the change in the cusps themselves. Instead of the original sharp points they develop first low cones, giving so-called bunodont grinding surfaces. Then, in later evolutionary stages, ridges or lophs appear between the cusps; an ectoloph between paracone and metacone, transverse protolph at the front of the tooth (between protocone and protoconule), and metaloph behind between hypocone and metaconule. All sorts of

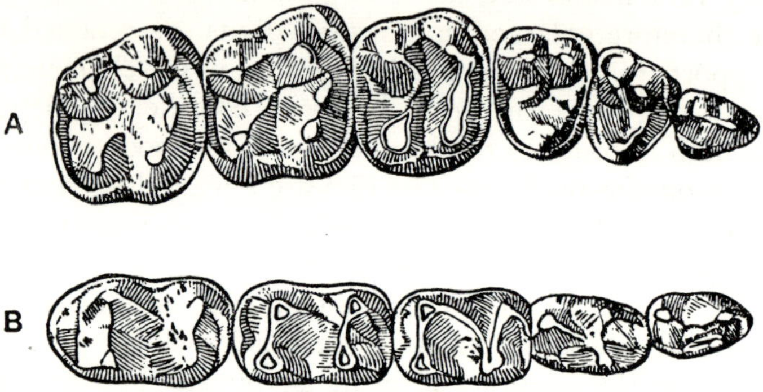

FIG. 467. *Hyracotherium.* Upper (A) and lower (B) premolars and molars. (After Wortmann from *Outlines of Vertebrate Palaeontology*, Cambridge University Press.)

further developments and cross-connexions may then take place in such lophodont molars; moreover, the whole tooth becomes surrounded by 'cement' (bone), so that the ridges are supported as they wear away and continually maintain a rough surface for grinding. Short (brachydont) molars, which would wear away too quickly, are replaced by deep (hyposodont) ones, which grow continually from open roots in extreme instances.

In these animals that need to increase the grinding surfaces the whole set of teeth is usually retained and the molar structure extends forwards to the premolars. This molarization may be said to be the opposite of the condition in carnivores, where the tooth row is shortened and the hinder teeth come to have cutting edges like the front ones. The incisors of ungulates become specialized for cropping the food; in artiodactyls the upper ones are lost and the lower work against a horny upper lip. The canine is often absent, leaving a diastema. The cropping and grinding mechanisms involve various modifications of the lips, palate, tongue, and, of course, the jaws and their muscles. The articulation of the jaw with the skull is usually made by a flattened facet, allowing rotatory action of the lower jaw. The pterygoid and masseter muscles are well developed, the temporal less so and the skull is flat and without a sagittal crest, in contrast with carnivores. To provide lateral attachment for these muscles there is a tendency for a redevelopment of the post-orbital bar.

In the digestive system of ungulates there is usually some chamber in which bacterial action upon cellulose can take place, but this has evidently evolved independently in the different groups, being in the stomach of artiodactyls but in the caecum of perissodactyls.

This set of 'ungulate' characteristics has developed independently many times in descendants of the insectivoran eutherian ancestor, and shows strikingly how the adoption of a particular method of life leads to selection of variations of structure tending in similar directions. There is no special difficulty in understanding how this has happened if we imagine that each part varies genetically in its dimensions. A herbivorous diet will be easier for animals with ridged teeth and long legs, whereas those with sharper teeth can become carnivores. Types are selected that combine a nervous organization leading to certain habits with other features that make these habits successful. In the evolution of any population there is evidently an elaborate interplay between variation in different directions in various organ systems and changes in the environmental circumstances.

3. Classification

Superorder 2. Protoungulata

*Order 1. Condylarthra. Palaeocene–Eocene

*Family 1. Hyopsodontidae. Palaeocene–Eocene. N. America

**Mioclaenus*, Palaeocene; **Hyopsodus*, Eocene

*Family 2. Phenacodontidae. Palaeocene–Eocene. Holarctic

**Tetraclaenodon*, Palaeocene; **Phenacodus*, Palaeocene–Eocene

*Family 3. Didolodontidae. Palaeocene–Miocene. S. America

**Didolodus*

*Family 4. Periptychidae. Palaeocene. N. America

**Periptychus*

*Family 5. Meniscotheriidae. Palaeocene–Eocene

**Meniscotherium*, Holarctic

*Order 2. Notoungulata. Palaeocene–Pleistocene

**Palaeostylops*, Palaeocene, Asia; **Notostylops*, Eocene, S. America; **Toxodon*, Pleistocene, S. America; **Homalodotherium*, Miocene, S. America; **Hegetotherium*, Oligocene–Miocene, S. America

3. Classification (*cont.*)

*Order 3. Litopterna. Palaeocene–Pleistocene. S. America
**Thoatherium*, Miocene; **Macrauchenia*, Pleistocene

*Order 4. Astrapotheria. Eocene–Miocene. S. America
**Astrapotherium*, Oligocene–Miocene

Order 5. Tubulidentata. Pliocene–Recent
Orycteropus, aardvark, Cape ant-eater, Africa

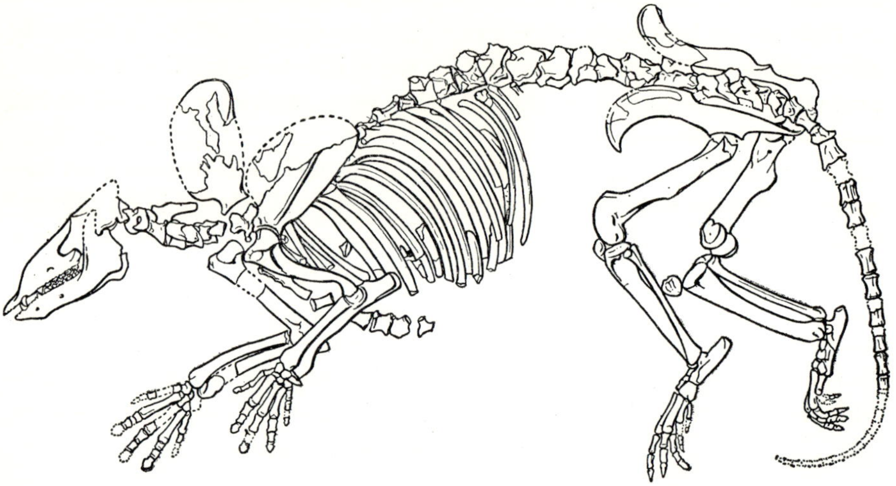

FIG. 468. Skeleton of *Phenacodus* as found in the rock. (Simplified after S. Woodward and Cope.)

4. Superorder Protoungulata

*Order Condylarthra

This group includes animals so close to the central eutherian stock that it is still disputed whether some of them should be classified as insectivores, primates, or creodonts. Five families are recognized, all from the Palaeocene and Eocene periods. The more 'primitive' in structure, such as the Eocene *Hyopsodus*, had a complete row of bunodont, quadritubercular teeth, and also short legs with clawed digits. They were small (1 ft long) and perhaps arboreal, and therefore could be classified with lemurs or insectivores. *Mioclaenus* is an even older type, which possessed sharp-cusped teeth. *Phenacodus* (Fig. 468) is perhaps the best known condylarth. It was an Eocene form with ungulate characters already present, including hoofs and square bunodont molars. The build was, however, still that of a generalized carnivorous or insectivorous mammal, with a markedly curved spine, small brain-case, sagittal crests, rather short limbs with slightly elongated metapodials, the central digit the longest, complete ulna and fibula, carpus and tarsus not interlocking, and a long tail. The animal was, however, rather large (4 ft long). Smaller Palaeocene phenacodonts, such as **Tetraclaenodon*, still possessed claws and may have been very close to the ancestry of all protoungulate types. Evidently some 10 million years of herbivorous life in the Palaeocene had produced only suggestions of the 'ungulate' facies. Other condylarths became more specialized in the Eocene. Thus **Meniscotherium* had lophodont grinders, though retaining the clawed digits. **Didolodus* and similar forms from South America are condylarths that may perhaps have given rise to some of the characteristic South American ungulates though they themselves survived to the Miocene. The *Periptychidae were Eocene condylarths probably ancestral to the pantodonts. They also show similarities to the surviving *Orycteropus*, which may be descended from them.

5. South American ungulates

*Order Notoungulata

The ungulate fauna of South America provides a case of geographic isolation as striking as that of the marsupials in Australia. In Palaeocene times there were ungulates common to South America and the rest of the world. Besides the condylarths, considered above, there was the **Palaeostylops*, in the Palaeocene of Asia, and the similar **Notostylops* found in North and South America. These earliest notoungulates showed only a slight advance in size and other features from the basal condylarth condition. The teeth possessed simple ridges. From some such beginnings there quickly developed, after the isolation of South America in the Eocene, a very rich fauna, including many large animals. Specimens of these peculiar animals were first collected by Darwin during the voyage of the *Beagle* and were later described by Owen. Darwin records that their characteristics were among the earliest stimuli that turned his thoughts to evolution (see p. 524). A characteristic of the group was the very large tympanic bulla. The brain was small and especially the cerebral hemispheres.

As many as nine families of notoungulate can be recognized in the Oligocene; after this period they became less numerous. Some of them persisted throughout the Tertiary, but all became extinct in the Pleistocene, after the connexion with North America was re-made and competition was felt from more modern types, both ungulates and carnivores. The notoungulates known as toxodonts were very large graviportal animals, the tooth row being curved to form a bow, from which the group takes its name. The limbs were massive, with as few as three digits, the middle the longest, bearing hoofs. **Toxodon* itself

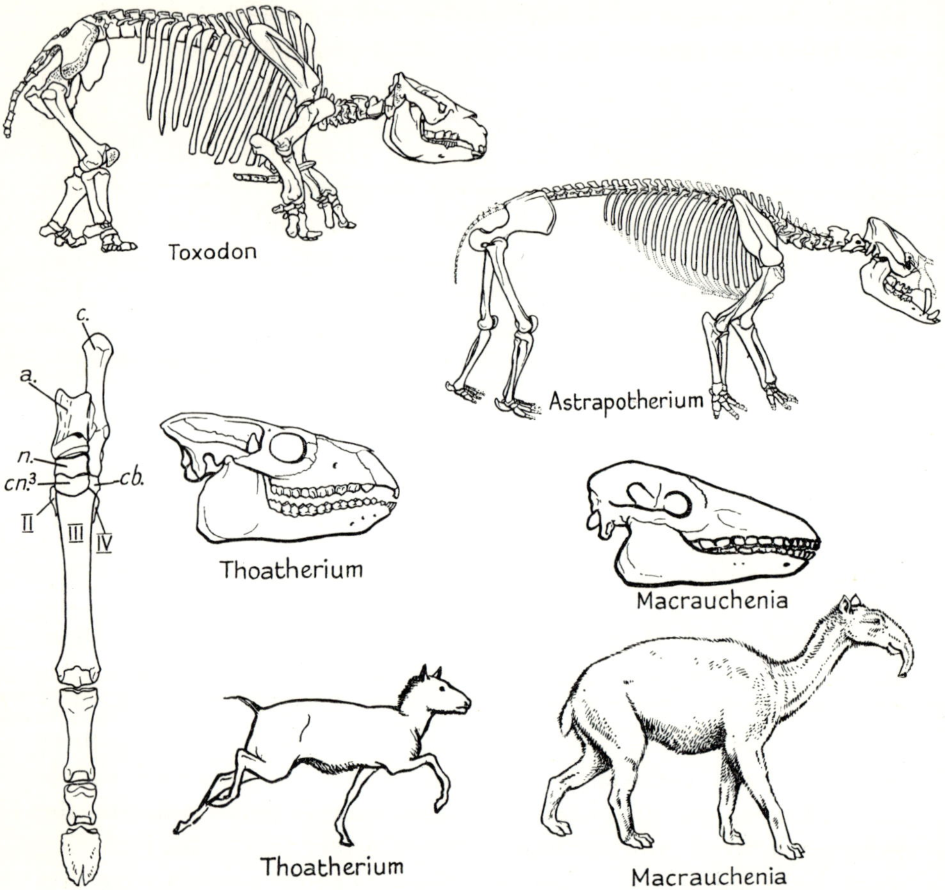

FIG. 469. Various ungulates. (After Zittel, K. A., *Text-book of Palaeontology* (revised A. S. Woodward), Macmillan & Co., Romer, *Vertebrate Paleontology*, Chicago University Press, and Scott, *Land Mammals*, The Macmillan Co.) Lettering as Fig. 465.

(Fig. 469) survived to the Pleistocene and was a creature nearly 10 ft long, with enormous head, short front and long hind legs. **Homalodotherium*, on the other hand, had front legs longer than the hind, and provided with claws. The incisors were small but formed rows suitable for cropping. Probably the animals reared up on their hind legs to reach branches and the large ischia show that the muscles for this were strong (p. 634). The typotheres and hegetotheres were small rabbit-like creatures with gnawing incisors. The notoungulates thus radiated to form various types and for many millions of years they were the dominant herbivores of the South American forests.

6. *Order Litopterna

Some descendants of the condylarths in South America developed along lines astonishingly similar to the horses. The didolodonts already show tendencies in this direction and are, indeed, sometimes removed from the condylarths and placed with the South American horse-like forms in the *order Litopterna. A series of fossils shows that members of this order became first digitigrade and then unguligrade, the central metapodials elongating and the lateral ones reducing, first to three and then, in **Thoatherium* (Fig. 469), to a single one, with splint bones even smaller than remain in our horses. The general appearance of the limbs was very horse-like, for instance in the grooved talus, but the carpus never became interlocking. Other respects in which these litopterns developed less far than the horses were that the tooth row remained nearly complete and the molars low-crowned, though provided with ridges. A post-orbital bar was developed. These differences from our horses are as interesting as the similarities, and they show that the features of the ungulate facies do not necessarily all evolve together. It is impossible to say what difference in conditions was responsible for forming horse feet on an animal whose head was only partly horse-like.

These ungulates were presumably evolved to meet conditions on the South American plains in the Miocene. It is interesting that the three-toed types outlived the one-toed **Thoatherium*, lasting into the Pliocene. **Macrauchenia* (Fig. 469), a creature looking like a camel but perhaps living in swamps, was the only Pleistocene survivor.

7. *Order Astrapotheria

This order includes some Oligocene and Miocene South American ungulates, with a short skull but long lower jaw, long canines, and massive molar teeth (Fig. 469). There was probably a proboscis. The feet were small and perhaps rested on pads. The weak vertebral spines and transverse processes suggest that the animals may have been aquatic and the large lower canines, diverging in older animals, resemble those of a hippopotamus.

8. Order Tubulidentata

The aardvark ('earth-pig') or Cape ant-eater, *Orycteropus* (Fig. 470), is a zoologically very isolated form, of unknown affinities, placed by Simpson with the Protoungulata, because it is possibly not very remote from the condylarths. It is the size of a small pig, with a highly curved back, and is much given to digging, both for protection and to obtain termites, which are its main food. It occurs from South Africa to the Sudan. There is an elongated snout, round mouth, and long tongue, as in other ant-eaters. The peglike teeth are unlike those in any other mammal. They consist of numerous hexagonal columns

FIG. 470. *Orycteropus*, the aardvark. (From life.)

of dentine, separated by tubes of pulp. There is no enamel, though enamel organs are present in the tooth germs. In the adult there are about five teeth in each jaw, but there is a full series of rudimentary milk teeth.

There are special arrangements in the mouth and throat to allow the animal to bury its snout in a mass of termites and then to swallow them while continuing to breathe. There are large salivary glands. The digits (4 in hand and 5 in foot) are covered by structures sometimes referred to as compressed nails, sometimes as hoofs. There is a strong clavicle and complete radius and ulna and tibia and fibula. The limbs are thus specialized for digging, but retain the characters of the earliest mammals. The head is long and the brain small and of an extremely primitive type, with extensive olfactory regions and very small neopallium. The olfactory turbinals are better developed than in any other mammal; the aardvarks find termites by their scent. The animals are nocturnal and the retina has only rods, and a tapetum. The ears are long and hearing acute and there are bristles on the long mobile snout. The uterus is paired and the placenta of a zonary type, somewhat like that of carnivores. There is a very large allantois.

Orycteropus occurs as fossils back to the Miocene. Its earlier history is unknown, but similar teeth have been reported from the Eocene, and many features of the skeleton are strikingly like those of condylarths. The animal obviously retains many characters that were present in the earliest eutherians, the fact that it is placed by some as an edentate or insectivore and by others close to the base of the ungulate stock suggests that it has diverged relatively little from the ancestor of all eutherians.

XXIX

ELEPHANTS AND RELATED FORMS

1. 'Near-ungulates', superorder Paenungulata

FROM the Palaeocene ungulate stock, when it was yet hardly differentiated from that of other mammals, there diverged several lines of herbivorous animals and these rapidly increased and diversified in the Eocene, many of them becoming very large. Most of these lines declined in the Oligocene and only the huge elephants and tiny hyraxes remain today to show approximately the structure of this range of Eocene pantodonts, dinocerates, and other forms. The highly specialized Sirenia (sea-cows) were also an early offshoot of this type of animal. The various lines diverged so very long ago that we should hardly expect to find that they have much in common that they do not share with other ungulates, or indeed with all mammals, but it has long been recognized that there is a loose grouping of orders around the elephants and hyraxes. Simpson suggests the name Paenungulata ('near ungulates') for these forms that are all slightly, but not much, beyond the protoungulate level. The legs of all of them remain rather primitive, with long upper segments, complete ulna and fibula, and several digits, and without well-marked hoofs. The incisors and canines often become reduced to single pairs of large tusks in each jaw and the molars are specialized for grinding, with the development of cross-ridges.

2. Classification

Superorder 3. Paenungulata

Order 1. Hyracoidea. Oligocene–Recent. Palearctic, Africa
Procavia (= *Hyrax*), hyrax, Africa, Asia

Order 2. Proboscidea. Eocene–Recent

*Family 1. Moeritheriidae. Eocene–Oligocene. Africa
**Moeritherium*

*Family 2. Deinotheriidae. Miocene–Pleistocene. Eurasia, Africa
**Deinotherium*

*Family 3. Gomphotheriidae. Oligocene–Pleistocene. Holarctic, S. America, Africa

**Palaeomastodon*, Lower Oligocene, Africa; **Phiomia*, Oligocene, Africa; **Gomphotherium* (= **Trilophodon*), Miocene–

Pliocene, Holarctic, Africa; **Serridentinus*, Miocene–Pliocene, Holarctic; **Anancus* (= **Pentalophodon*), Pliocene–Pleistocene, Eurasia; **Stegomastodon*, Pliocene–Pleistocene, N. and S. America

*Family 4. Mammutidae. Miocene–Pleistocene. Holarctic
**Mammut* (= **Mastodon* = **Zygolophodon* = **Turicius*)

Family 5. Elephantidae. Pliocene–Recent. Holarctic, Africa
**Stegolophodon*, Miocene–Pleistocene, Eurasia; **Stegodon*, Pliocene–Pleistocene, Asia; **Mammuthus* (= **Mammonteus* = **Archidiskodon*), mammoth, Pleistocene, Holarctic, Africa, S. America; *Loxodonta*, African elephant, Pleistocene–Recent; *Elephas*, Indian elephant, Pleistocene–Recent.

*Order 3. Pantodonta. Palaeocene–Eocene. Holarctic
**Pantolambda*, Palaeocene; **Coryphodon*, Palaeocene–Eocene

*Order 4. Dinocerata. Palaeocene–Eocene. Holarctic
**Uintatherium*, Eocene

*Order 5. Pyrotheria. Palaeocene–Oligocene. S. America
**Pyrotherium*, Oligocene

*Order 6. Embrithopoda. Oligocene. Africa
**Arsinoitherium*

Order 7. Sirenia. Sea-cows. Eocene–Recent
**Protosiren*, Eocene; *Dugong* (= *Halicore*), sea-cow, Indian Ocean and Pacific; *Manatus* (= *Trichechus*), manatee, Atlantic

FIG. 471. *Procavia*, hyrax. (From life.)

3. Order Hyracoidea

The hyraxes or conies (Fig. 471) are animals that live in Africa and neighbouring regions and have persisted throughout the Tertiary as small herbivorous creatures, occupying similar niches to rabbits, which they resemble superficially in some ways. Fossils are known in Africa back to the Oligocene and probably the group existed before that time and therefore shows us something of the appearance of smaller Eocene and Oligocene ungulates. The gait is plantigrade, with four anterior and three posterior digits, carrying somewhat hoof-like nails, except for a sharp bifid claw, used for toilet purposes, on the inner hind toe (Fig. 472). There is a single pair of continually growing incisors in the upper and two in the comb-like lower jaw. There is a diastema and seven grinding molariform teeth of bunoselenodont type, with transverse ridges, recalling those of brontotheres. The lower jaw is very deep, for the attachment of the masseter muscle, and, as is usual in

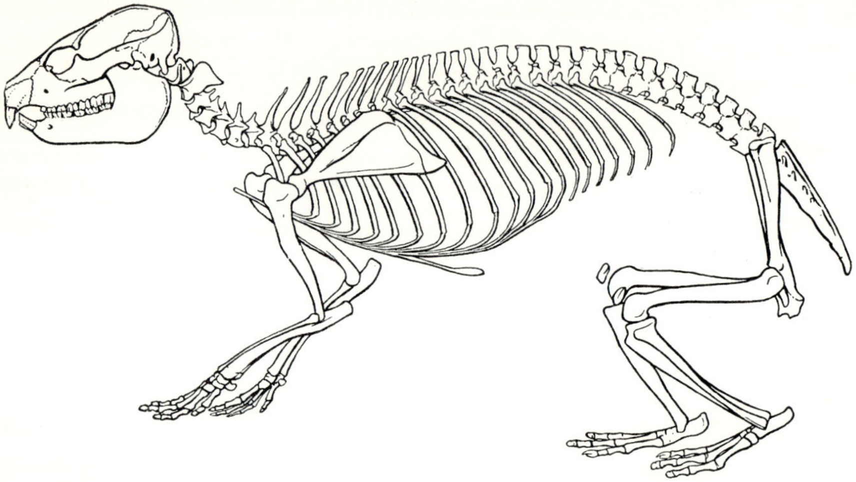

FIG. 472. Skeleton of *Procavia*.

ungulates, the post-orbital bar is nearly or quite complete. There is a serial carpus with a centrale, an unusually primitive feature for an ungulate. The intestine provides chambers for digestion by symbionts. In the large median caecum are found enormous ciliates (*Pycnothrix*), up to 5 mm long, and a fauna of cellulose-splitting bacteria. Beyond this lies a further pair of caeca.

The brain is of macrosmatic type. As in the elephants, the testis fails to descend and remains close to the kidney, there being no sign of a scrotum or inguinal canal. The uterus is paired, and the placenta at first covered with foetal villi, later restricted to a zonary arrangement, superficially similar to that of carnivores, but with a haemochorial structure with a resemblance to that of *Tarsius*. There is a single pair of pectoral mammae.

Various species of *Procavia* are common throughout Africa (not Madagascar), Arabia, Palestine, and Syria, living in desert regions in colonies under rocks. They do not dig burrows, the feet being flattened and well suited for moving over smooth rock surfaces. The related *Dendrohyrax* live in the trees. Earlier hyraxes were sometimes as large as small horses (**Megalohyrax* from the Oligocene).

4. Elephants. Order Proboscidea

The late Henry Fairfield Osborn, one of the greatest zoologists of his time, devoted a great part of a long working life and the large resources available to him at the American Museum of Natural History to the study of the evolution of elephants. It cannot be said that he was able as a result of all this study to draw conclusions that have revolutionized biology, and this failure is in a sense a measure of the immaturity of our science. The elephants have shown great and relatively rapid changes in recent geological times and have left abundant remains, especially of their large and hard teeth. We may therefore take knowledge about evolution of elephants as a fair example of the most that can be known of the evolutionary processes in large mammals; if the study of this great mass of material leaves us in a state of confusion rather than of certainty we shall be warned to suspect the apparent clarity of other alleged evolutionary sequences, and to distrust dogmatic statements about the 'causes' of evolutionary change.

The two existing types of elephant, referred to distinct genera, live still in considerable numbers in Africa (*Loxodonta*) and Asia (*Elephas*). They are survivors of a much larger population, reaching its greatest variety in Pliocene times. The essential feature of the type is the great size ($11\frac{1}{2}$ ft high in *Loxodonta*) and the presence of a special food-collecting system able to gather enough raw material to support such a large living mass. Of the various factors influencing the optimum size for a given animal type, all those favouring increase must be present in the ingredients of elephant life. Elephants are larger than any other land animals, living or extinct, except perhaps the huge Oligocene rhinoceros **Baluchitherium* and some of the largest dinosaurs (if these were indeed terrestrial).

The basal metabolism of a mammal increases only with about the two-thirds power of its weight, so that larger animals need, on this account, relatively less food. But, as Watson (1946) has pointed out, the output of energy by the muscles is proportional to the weight of the muscle; the total intake needed is therefore proportional to some power between two-thirds and the first power of the weight of the whole, being 'larger the greater and more continuous the activity of the animal'. Elephants, as he further comments, are very active when wild, their playfulness and strength are proverbial, and often a nuisance to man. They manage to collect their food with sufficient economy of energy, though to do this they must eat throughout a large part of the day, perhaps for as much as 18 hours. Here another factor has to be considered, namely, the area of tooth available for grinding the food. This will vary approximately as the two-thirds power of the total weight; as the animals become larger the tooth surface needs therefore to be relatively increased.

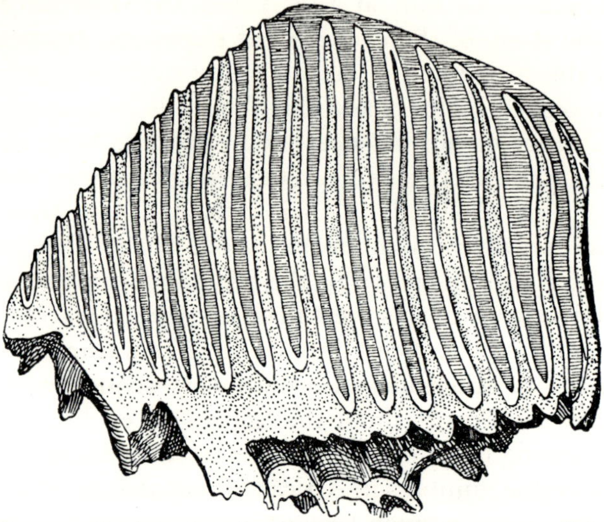

FIG. 473. Section of tooth of elephant. The front part of the crown (on the left) is already worn away. Notice the upstanding enamel lamellae, which reach to the base of the tooth. Dentine is shown dotted, cement by lines (From Weber.)

With these factors in mind we shall recognize that the significant features in the organization of elephants are that they are very large animals, with an efficient nervous organization for finding the food, efficient means of collecting it, and large surfaces for grinding it.

The trunk is the main means of collection—an enormously elongated nose and upper lip, with appropriate muscles and sensitive grasping tip. The muscles have been developed chiefly from the parts of the facial musculature that are responsible for moving the sides of the nose. The trunk probably developed rather quickly, in late Miocene times, perhaps 10–15 million years ago; the earlier elephants of the Miocene possessed very long lower jaws, which became shortened as the trunk developed. Any tall animal must have means of reaching the ground and the trunk is probably superior even to a very long neck for this purpose, because it can reach upwards and sideways as well as downwards.

Only one pair of continually growing incisors remains in modern elephants, forming the two enormous upcurved tusks, up to 11½ ft long in *Loxodonta*, composed of solid dentine except for a temporary cap of enamel at the tip. This mass of ivory is no doubt useful for defence and perhaps in food collection, but it seems to be a considerable waste of calcium and phosphorus, not to mention of the energy necessary to carry the 350 lb weight. This weight is balanced against that of the body, upon the pillar-like front legs, and it is perhaps not fantastic to suggest that the tusks serve partly as counterweights, for

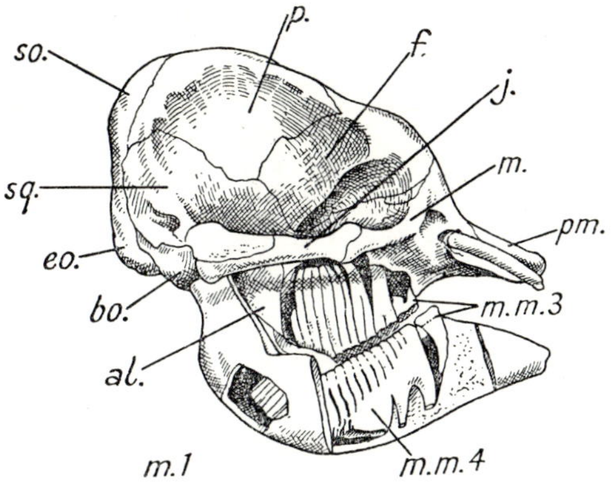

FIG. 474. Skull of young Indian elephant. The roots of the teeth have been exposed.

al. alisphenoid; *bo.* basioccipital; *eo.* exoccipital; *f.* frontal; *j.* jugal; *m.* maxillary; *m.1.* first molar; *m.m. 3* and 4. third and fourth milk molars; *p.* parietal; *pm.* premaxillary; *so.* supraoccipital; *sq.* squamosal. (After Reynolds, *The Vertebrate Skeleton*, Cambridge University Press.)

purposes of balance (see p. 697), extravagant though such an arrangement may be. However, the weight of the head is reduced by extensive development of air sinuses between the inner and outer tables of the bones of the skull. The tusks are smaller in females and in the Indian than in the African elephant; in the female Indian elephant they do not project beyond the lips. This difference may be connected with the relatively small size of these females.

The essential features of the grinding apparatus are the immensely large molars (Fig. 473), with numerous sharp transverse ridges, by which all sorts of plants, including hard grasses, are chopped into small fragments. There are three small premolars, which are soon shed, and the three molars are then developed in a series and used one after the other. By a special arrangement of the palate (Fig. 474) the teeth are allowed to form high up in the skull, so that each tooth has a very great area, made up by the fusion of as many as twenty-seven separate 'plates', which develop as separate cones of dentine and enamel, each with its own pulp cavity, the cones being finally joined together by cement. The teeth are placed just above the ascending ramus of the mandible, so that the large jaw-muscles work at maximum advantage. For their attachment the skull becomes extremely short and high, with the development of large air spaces

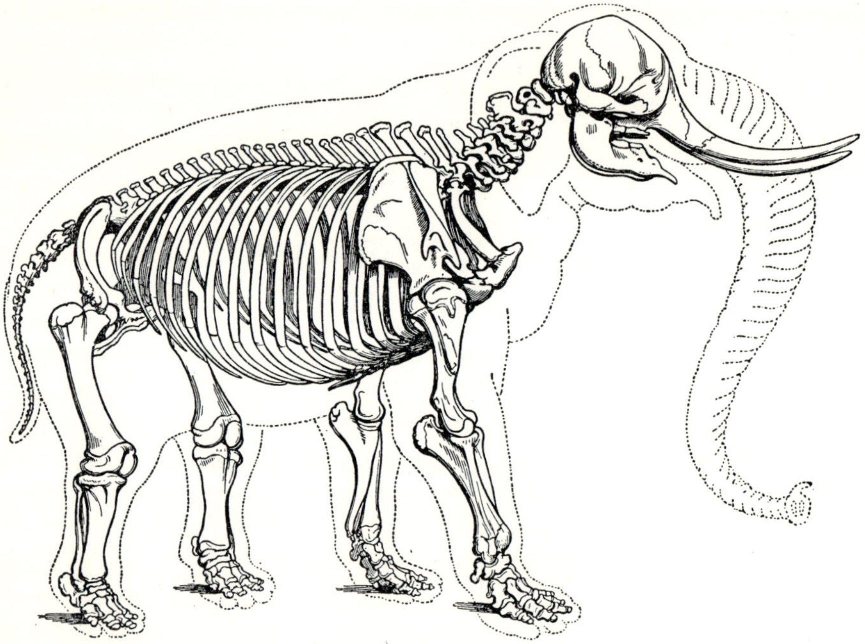

FIG. 475. Skeleton of an Indian elephant. (From Owen, *The Anatomy of Vertebrates*, Longmans, Green & Co.)

between its tables. This shape also allows a large occipial region for the muscles that hold up the head.

With this head structure the elephants have been able to grow to a size that must approach the limit possible for a fully terrestrial animal. The backbone (Fig. 475) is based on a 'single girder' plan, with as many as twenty ribs, and high thoracic neural spines, forming together a huge beam that carries the weight of the abdomen and balances it on the fore-legs against the weight of the head, the hind-legs acting as propellents. The ilium is nearly vertical and expanded transversely for the attachment of the large gluteal, iliacus, abdominal and sacrospinalis muscles. The acetabulum faces downwards.

As in other heavy animals the legs are enormous pillars, with long upper segments and no great extension of the lower. The ulna and fibula are complete and bear part of the weight, the ulna and radius being held permanently crossed in a fixed position of pronation. Walking is of a modified digitigrade type; all three of the short phalanges of each digit reach the ground, but the greater part of the weight is taken by a pad of elastic tissue at the back of the foot. There are five digits in each foot, united by a web to make a firm basis, and having small, flat, somewhat hoof-like nails at the tips. The ribs carry so much weight that respiration is almost wholly diaphragmatic and the lungs are fused to the walls of the thoracic cavity by elastic tissue.

The soft parts of elephants show some features retained unmodified from their early ungulate ancestry. Thus the cerebral hemispheres are relatively rather small and leave the cerebellum uncovered. In other respects the brain is well developed, it has a greater absolute size than that of any other land mammal (6,700 cm^3). The proverbial intelligence and memory capacity have been verified by experiment. Smell, hearing, and the tactile organs of the trunk provide the main receptors, vision being less developed. Like many other animals with large brains there is a long period of post-natal growth and life is social. Much information is no doubt learned from other individuals, and it has been shown that elephants can learn to discriminate between upwards of 100 pairs of visual situations.

In spite of the specialization of the head for a herbivorous diet, the stomach and intestine remain simple and there is no special large fermentative chamber, though the caecum is long and sacculated and there is an ileocaecal sphincter.

The testes are remarkable in that, as in other paenungulates, they lie close to the kidneys, and have made no movement of descent into a scrotum. The two horns of the uterus remain separate, though united externally. Only one young is born at a time, after a gestation of twenty-two months. The placenta has a superficial similarity to the zonary arrangement of carnivores, but in structure resembles that of hyraxes and sirenians. At the poles are areas of diffuse, non-deciduate placenta while in an annular zone round the middle there is much invasion of the trophoblast, the details of which are not known. Development is slow and Asian elephants reach puberty at about 13–14 years, African elephants rather earlier.

The earliest-known member of the elephant line, **Moeritherium*, from the Upper Eocene of Egypt, was only 2 ft high and was probably partly aquatic, with eyes and ears placed high, as in the hippopotamus.

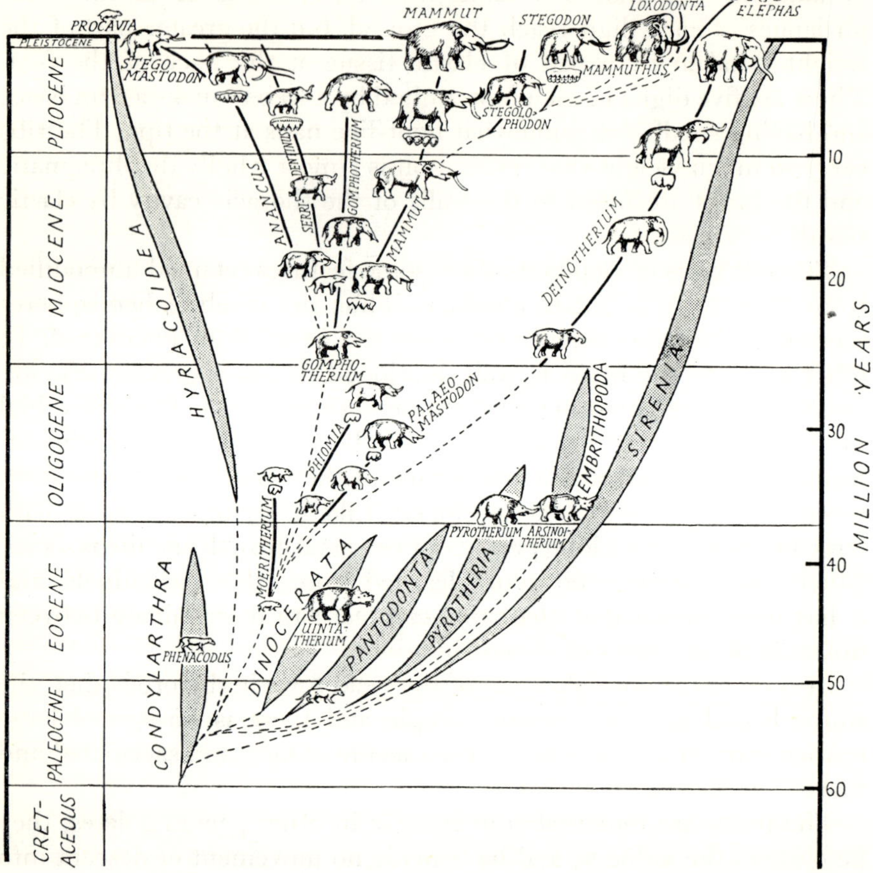

FIG. 476. Chart of the evolution of the paenungulate orders. The animals are shown reconstructed following Andrews, Osborn, and others and their size is indicated approximately. Some stages in development of the molar teeth are also shown.

The skull was elongated and small tusks were present, but the dentition was nearly complete, $\frac{3.1.3.3}{2.0.3.3}$. The molars were bunodont and carried only two cross lophs, a condition easily derived from that of a condylarth quadritubercular tooth. From some such animal arose ultimately so great a collection of types that Osborn in his study of the group recognized 350 species, only two living at the present day. **Moeritherium* survived into the Lower Oligocene, where there is found also **Phiomia*, about twice as large. Both upper and lower jaws of this animal carried tusks and the whole front of the head was greatly elongated, with formation of a long diastema. The wear of the lower incisors shows that they were used for digging. The molars carried three low ridges and were all used together, not successively. The similar **Palaeomastodon* lived at the same time and was about 6 ft high.

After this period there is a gap, probably covering 10 million years or more, in our knowledge of elephant evolution, but it is clear that throughout this long period the stock must have continued with little change, as a race of animals with long digging tusks and rather mobile face and lips, becoming gradually larger and elongating the face more and more to enable the ground to be reached. In the Miocene is found a considerable variety of these long-jawed animals, the various species of the genus **Gomphotherium* (= **Trilophodon*). The teeth of the earliest of these long-faced elephants were used all at once, not in series; later the premolar teeth tended to be reduced and the molars became covered with an increasing number of cusps, arranged to make a number of cross-ridges, seldom, however, more than five. From the low cusps they are known loosely as 'mastodonts', but the colloquial terms for description of elephants are used in senses almost as varied as the 'scientific' names; as in other branches of knowledge, great abundance of information has led to confusion of terminology.

From this Miocene stage onwards the study of proboscidean evolution becomes a desperate attempt to sort out huge numbers of fossil specimens (often, however, only molar teeth) into truly phylogenetic lines. Osborn made an heroic effort to recognize only fully documented sequences, but even with the wealth of material available it is rarely possible to say with complete certainty that one type has evolved into another. The interpretations of the sequences and their expression in classificatory terms vary considerably even in the hands of the most careful interpreters of Osborn's work. However, it is probable that in this mass of material there can be seen several distinct lines, in which evolution proceeded in a parallel manner. From the end of Miocene times onwards the very elongated jaws began to shorten, and it was presumably at this time that the trunk became long and the typical elephant-like habit was adopted. Certainly not all animals having elephant-like appearance belong to the 'main' elephant line, and the shortening of the lower jaw took place at different times in the various stocks. Thus the members of the Pliocene and Pleistocene genus **Anancus*, though extremely like elephants in general shape, had bunomastodont molars. This line retained little cusps between the main ridges, such as were present in the Miocene gomphotheres. **Stegomastodon* was a related animal that lived on in South America until as late as A.D. 200.

The species of **Serridentinus* represent another line. Here there were no little cusps on the teeth, but the grinding area was increased by extra 'serrate' ridges. This type retained the long lower tusks into the Pliocene, but then tended to shorten them, though never to the full elephant condition. The line that produced the modern elephants can first be recognized in the Lower Miocene, by the fact that the cusps are united into sharp ridges. This condition presumably marks the transition to feeding on hard grasses, which have to be cut, rather than on softer stalks, which can be ground. Some members of this 'zygolophodont' stock retained few ridges, and presumably a browsing diet, even into the Pleistocene, although they acquired short lower jaws like those of the elephants. These animals are therefore 'mastodons', and it is unfortunate that the rules of priority require that they shall be called **Mammut* (= **Zygolophodon*). In **Stegolophodon* and **Stegodon* of the Upper Pliocene and Pleistocene, however, the lower jaw was already short and the arrangement of the skull elephant-like, with huge curved upper tusks. From some such form the two modern elephants (*Loxodonta* and *Elephas*) and the mammoths (**Mammuthus*) have been derived, but the details of the evolutionary sequence become here even more involved. We possess such great numbers of teeth, showing all sorts of detailed differences, that to arrange them in evolutionary sequences is mostly a matter of guesswork. Osborn preferred not to deal with individual teeth but only with the fifty or so complete skulls that have been found, and these fall into three groups. The mammoths, **Mammuthus* (= **Archidiskodon* = **Mammonteus*) had very long curved tusks, turned upwards at the tips. There were several 'species', including such forms as *M. primigenius*, the woolly mammoth, common in Europe during the Pleistocene and surviving until recently in Alaska and Siberia. Many carcasses of these animals have been found in frozen soil and glaciers, allowing study of the soft parts and contents of the stomach.

Loxodonta, the African elephant, has straighter tusks and the surface of the molars wears to a diamond-shaped pattern. In *Elephas* the tusks are also nearly straight and the molar ridges are parallel. There are, of course, numerous other differences between the two modern elephants and it is not possible to trace out in detail the ancestry of the two types and of the mammoths. In fact one lesson to be learned from the study of elephant evolution is that mammalian fossil remains are seldom sufficiently abundant to allow study of the details of evolutionary history. There is usually some doubt about the exact relationship of the bones and teeth that are found. Phyletic lines are constructed by careful comparison of the characters of the fossils, there is seldom direct evidence of the genetic relationship of any two types. In the present instance it is probable that from the gomphotheres of Miocene times onwards the lower jaw began to shorten and the skull to achieve an elephant-like form in at least four separate stocks (perhaps far more). **Anancus* and the true elephantines evolved faster in this direction than **Serridentinus* and **Mammut*, both of which retained the 'mastodont' characters associated with browsing even into the Pleistocene. The existence of parallel evolution may be regarded as established beyond reasonable doubt in this case; evidently there was some feature either in environmental change or internal 'tendency', or in both, leading all these stocks to change in similar ways, though at different times and rates.

**Deinotherium* was a distinct type of elephant, separate from all others from Miocene times or earlier. There were down-turned lower tusks and probably also a trunk. There were several molars in the tooth row and the full elephant specializations did not develop. The animals remained similar in structure for a long period, but became very large before they disappeared in the Pleistocene. This is a good example of the development of different variants of a type; deinotheres had the trunks but not the molars of elephants.

5. *Order Pantodonta (Amblypoda)

During the Palaeocene and Eocene the ungulate stock produced various large herbivores and these may be referred to the paenungulate group. The relationships of the numerous types discovered are still obscure and classification is probably not yet final. The animals here placed (following Simpson) in the order *Pantodonta were formerly, with others, known as amblypods ('blunt feet'). The Palaeocene

**Pantolambda* was about 3–4 ft long, with a long face and tricuspid molars. The limbs were short and broad, and the pelvis very like that of *Phenacodus*.

Later members of the group, such as **Coryphodon* (Fig. 477), were over 8 ft long and heavily built, with some formation of ridges on the teeth, and feet with five digits; some had simple hoofs, others claws. The brain was small and evidently these were clumsy creatures, successful for a time in Europe, Asia, and America, but unable to compete with later herbivores.

6. *Order Dinocerata

These were even larger animals, of the same general graviportal build as the Pantodonta, and were previously classed with the latter as Amblypoda. The two pairs of horns as well as nasal protuberances and very large dagger-like canines provided weapons of defence. The molars showed folds and ridges and provided a reasonably efficient grinding battery. **Uintatherium* (Fig. 477) was a typical Eocene form; even though the brain was small and the gait clumsy, the animals were evidently successful at the time, and reached a size as great as that achieved by any other land mammals except the elephants.

7. *Order Pyrotheria

**Pyrotherium* and its allies (Fig. 477) were Eocene and Oligocene South American ungulates and they are usually classed with Notoungulata, but more for geographical than phylogenetic reasons. They were remarkably similar to elephants, for instance in their large size and in the dorsal nostril, suggesting the presence of a trunk. The incisors were developed into tusks and the molar teeth carried two transverse rows of cusps, as in bilophodont early proboscidians. The similarities of the two groups are striking, but they probably indicate only common early ungulate derivation and provide another instance of convergence.

8. *Order Embrithopoda

**Arsinoitherium* from the Lower Oligocene of Egypt was another large creature that may be placed here. Its limbs resembled those of elephants, with five semi-plantigrade digits. There was a pair of enormous nasal horns, with a keratinous covering like that of ruminants, also smaller frontal horns. There was a regular tooth row, with no enlargement of the incisors or canines and hypsodont molars.

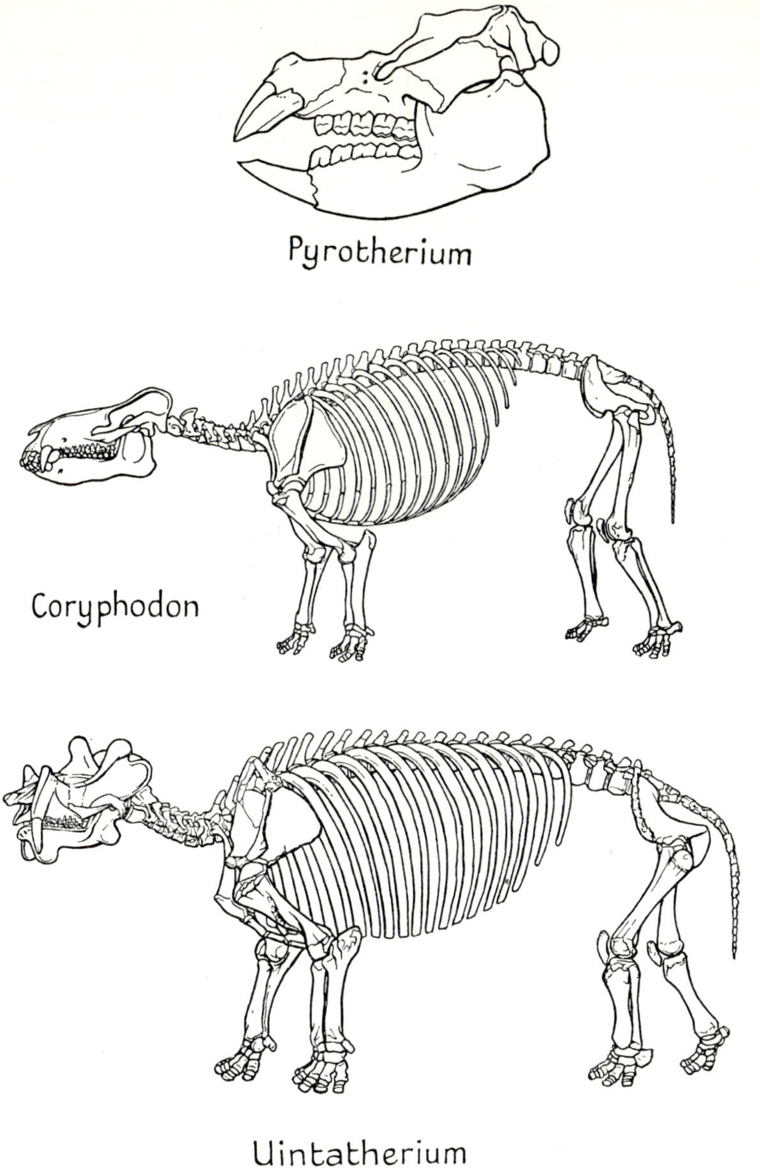

FIG. 477. Skeletons of a pantodont, a dinocerate, and a pyrothere. (After Woodward, *Outlines of Vertebrate Palaeontology*, Cambridge University Press, Flower and Lydekker, *Mammals*, A. & C. Black, Ltd., and Romer, *Vertebrate Paleontology*, University of Chicago Press.)

9. Order Sirenia

The sea-cows are herbivorous creatures, living along the coasts and in rivers, and highly adapted to aquatic life. There is little doubt, however, that they have reached this condition by modification of a basic ungulate type of organization, probably not very different from that of the early proboscidians. The two modern forms, the manatee of the Atlantic (Fig. 478) and dugong of the Pacific and Indian oceans, are different in many respects and represent lines that have been separate for a long time, probably since the Eocene. *Manatus* (= *Trichechus*) has three species on the Atlantic coasts and in the rivers of

FIG. 478. Manatee, *Manatus*. (From photographs.)

Africa and America. *Dugong* (= *Halicore*) is a purely marine animal extending from the Red Sea throughout the Indian Ocean to Formosa and Australia. *Rhytina* (Steller's sea-cow) was an Arctic form that became extinct in the eighteenth century.

Sea-cows have a 'streamlined' body-form, with few hairs and thick 'blubber'. There are no hind-limbs and the pelvic girdle remains only as small rods to which the corpus cavernosum is attached in the male. The fore-limbs are large, the digits joined to form paddles, with a full pentadactyl structure and no hyperphalangy or hyperdactyly. The caudal vertebrae are well developed and swimming is effected by the body and tail, the latter carrying a terminal horizontal fin. The vertebrae articulate with each other by flat surfaces, as in other aquatic forms, but there are zygapophyses, and the whole column is not quite reduced to the condition of a simple compression strut. The bones have a characteristic structure (pachyostosis), probably produced by lack of stressing. The manatee has only six cervical vertebrae. The ribs are round and the diaphragm is oblique, as in elephants and whales, allowing the lungs to reach far back. Respiration is probably mainly by means of the barrel-like ribs. The lungs contain large air-sacs. Sea-cows remain submerged only for relatively short periods (10 minutes). The blood system shows retia mirabilia in the brain and elsewhere, as in other aquatic mammals (p. 692). The brain is small and the ventricles exceptionally large. The forebrain is rounded but the rhinencephalon less reduced than might be expected by comparison with whales. The neopallium is smaller and less folded than in almost any other mammal of comparable size. The eyes are small and protected by muscular lids; the animals do not see well. The external auditory meatus is reduced to a channel a few millimetres wide, as in whales. Little is known of the hearing but reports are that it is acute.

In the manatees, the upper lip is greatly developed to form a strong yet sensitive pad, used for cropping. The front parts of the jaws carry horny pads for chewing. The teeth form a series of pegs, with two transverse ridges; there may be up to twenty of them and those in front drop out when worn. It has been supposed that there is a continual replacement from behind, as in elephants, but this is doubtful. In the dugong the teeth are much reduced and the lower jaw carries a horny pad; the upper carries a pair of tusks in the male and the premaxillae are very large. The stomach is complex but not like that of either the whales or other ungulates. The intestine is very long.

The reproductive system shows such primitive features as abdominal testes (with no signs that there was a descent in the ancestors) and a bicornuate uterus. The placenta shows a zonary arrangement and haemochorial structure, resembling that of elephants and conies. The young are born in the water and nursed at pectoral teats, which habit, with other features, may have produced some of the legends of mermaids.

Eocene fossils are known (**Protosiren*) which, while definitely sirenians, show distinct similarity to the ungulates of those times. The nostrils were directed dorsally as in modern forms, but the tooth row was complete and a small hind-limb was present.

XXX

PERISSODACTYLS

1. Perissodactyl characteristics

THE protungulate and paenungulate herbivorous types achieved their chief radiation and greatest numbers early in the Tertiary period. Their organization was not profoundly different from that of the original eutherians and although a few of them, such as the elephants, have persisted to the present day, most have been supplanted by ungulates that appeared by later modification of the original type. Very roughly we may say that the protungulate is the chief Palaeocene mammalian herbivorous type and the paenungulate that of the Eocene. The Perissodactyla, including horses, rhinoceroses, tapirs, and certain early extinct types, then represent a third or Oligocene–Miocene development, supplanting the paenungulates and itself then largely replaced in Pliocene, Pleistocene, and Recent times by the Artiodactyla. This analysis must of course be taken only as a very rough approximation, especially as it is given unsupported by the quantitative data that it evidently requires. It is subject to many exceptions, for example the large development of the elephants in post-Miocene times.

The early perissodactyls were much like all other early ungulates and it is not easy to characterize the group as a whole. The limb structure developed the mesaxonic condition, with the digits arranged around the third as the main weight-bearing member, the others being reduced. With the power of fast movement the lower part of the limbs became elongated and the upper segments shortened, with reduction of the ulna and fibula, but these are characters found also in artiodactyls. A distinctive feature of the perissodactyls was the plan of the carpus and tarsus (Fig. 466). One distal carpal, the capitate (magnum), became enlarged and interlocked with the proximal carpals, while in the foot the ectocuneiform developed into a large flat bone, transmitting the thrust through a flat navicular to the talus, which has a flat undersurface, not a pulley-like one as in Artiodactyla. Modifications of the backbone for carrying great weight or for running were similar to those of other orders (elephants, Dinocerata), including increase in the number of ribs and the vertical position of the ilium (p. 697).

The feeding mechanism, though it has been the basis of the success of the perissodactyls, is in several ways less specialized than that of artiodactyls. The incisors are preserved and used for cropping, having a pit on the free surface, so that sharp edges are presented as the tooth wears away (incidentally allowing the age of the animal to be determined). The canine may be reduced or absent, and there is often a diastema. The molars of many of the earlier types remained bunodont and low-crowned, but those of the later rhinoceroses and horses developed an elaborate grinding surface. This was achieved by formation of a longitudinal ectoloph along the outer edge of the upper molar and parallel transverse ridges, the protoloph and metaloph (Fig. 465). Even with the secondary complications of the latest forms these teeth remain recognizably of quadritubercular pattern, and the same might be said of the lower molars. The premolars come to resemble the molars, giving a long battery of teeth. The gut shows less specialization than in artiodactyls, the stomach being undivided, but in horses there is a large cardiac area of non-glandular, oesophageal structure. Digestion of cellulose takes place in the caecum and large intestine, which may be greatly developed. The brain of Perissodactyla is relatively small, especially in the earlier forms, such as the tapirs. It is of macrosmatic type and the sensory portion of the nose is highly developed.

The reproductive system also shows primitive features. The uterus is bicornuate and the placenta of the diffuse epitheliochorial type, with a large allantoic sac. The yolk sac grows to a large size and forms a yolk-sac placenta during the early part of the development.

2. Classification

Superorder 4. Mesaxonia
  Order Perissodactyla
    Suborder 1. Hippomorpha
      *Family 1. Palaeotheriidae. Eocene–Oligocene. Eurasia
        **Palaeotherium*
      Family 2. Equidae. Horses. Eocene–Recent
        **Hyracotherium* (= **Eohippus*), Lower Eocene, Holarctic; **Orohippus*, Eocene, N. America; **Epihippus*, Upper Eocene, N. America; **Mesohippus*, Oligocene, N. America; **Miohippus*, Oligocene–Miocene, N. America; **Anchitherium*, Miocene, Holarctic; **Parahippus*, Miocene, N. America; **Merychippus*, Miocene, N. America; **Hipparion*, Pliocene, Holarctic, Africa; **Pliohippus*, Pliocene, N.

Superorder 4. Mesaxonia. Family 2 (*cont.*)

America; **Hippidion*, Pleistocene, S. America; *Equus*, horses, asses, zebras, Pliocene–Recent, world-wide

*Family 3. Brontotheriidae (= *Titanotheridae). Eocene–Oligocene. Holarctic

**Lambdotherium*, Eocene; **Eotitanops*, Eocene; **Brontops*, Oligocene

*Family 4. Chalicotheriidae. Eocene–Pliocene. Holarctic

**Eomoropus*, Eocene; **Chalicotherium*, Oligocene–Pliocene, Eurasia, Africa; **Moropus*, Miocene, N. America

Suborder 2. Ceratomorpha

Superfamily 1. Tapiroidea. Tapirs. Eocene–Recent

**Homogalax*, Eocene; *Tapirus*, tapir, Miocene–Recent, Asia, S. America

Superfamily 2. Rhinocerotoidea. Rhinoceroses. Eocene–Recent

*Family Hyrachyidae. Eocene. Holarctic

**Hyrachyus*. Eocene

*Family Hyracodontidae. Eocene–Oligocene. Holarctic

**Hyracodon*, Oligocene

**Family* Amynodontidae. Eocene–Miocene. Holarctic

**Amynodon*, Eocene

Family Rhinocerotidae. Oligocene–Recent

**Aceratherium*, Oligocene–Pliocene; **Baluchitherium*, Oligocene–Miocene, Asia; *Rhinoceros*, Indian and Javan rhinoceros, Pliocene–Recent, Asia; *Diceros*, black African rhinoceros, Pleistocene–Recent, Africa

3. Perissodactyl radiation

The fossil history of animals with the perissodactyl structure is perhaps better known than that of any other mammals; the type reached its peak during a period from which many fossils have been preserved and we have therefore a better opportunity to study the development, flowering, and decay of the group than in the case of forms whose maximum development occurred either earlier or later. Here if anywhere we should be able to learn lessons about the nature of the evolutionary process and to study the forces that produce change in animal form. Because of the very abundance of the fossils it is necessary, however, to be cautious in interpretation and to recognize exactly what can be proved from the evidence.

The known types of horses are divided into 350 species, but only a small proportion of these can be confidently placed close to the

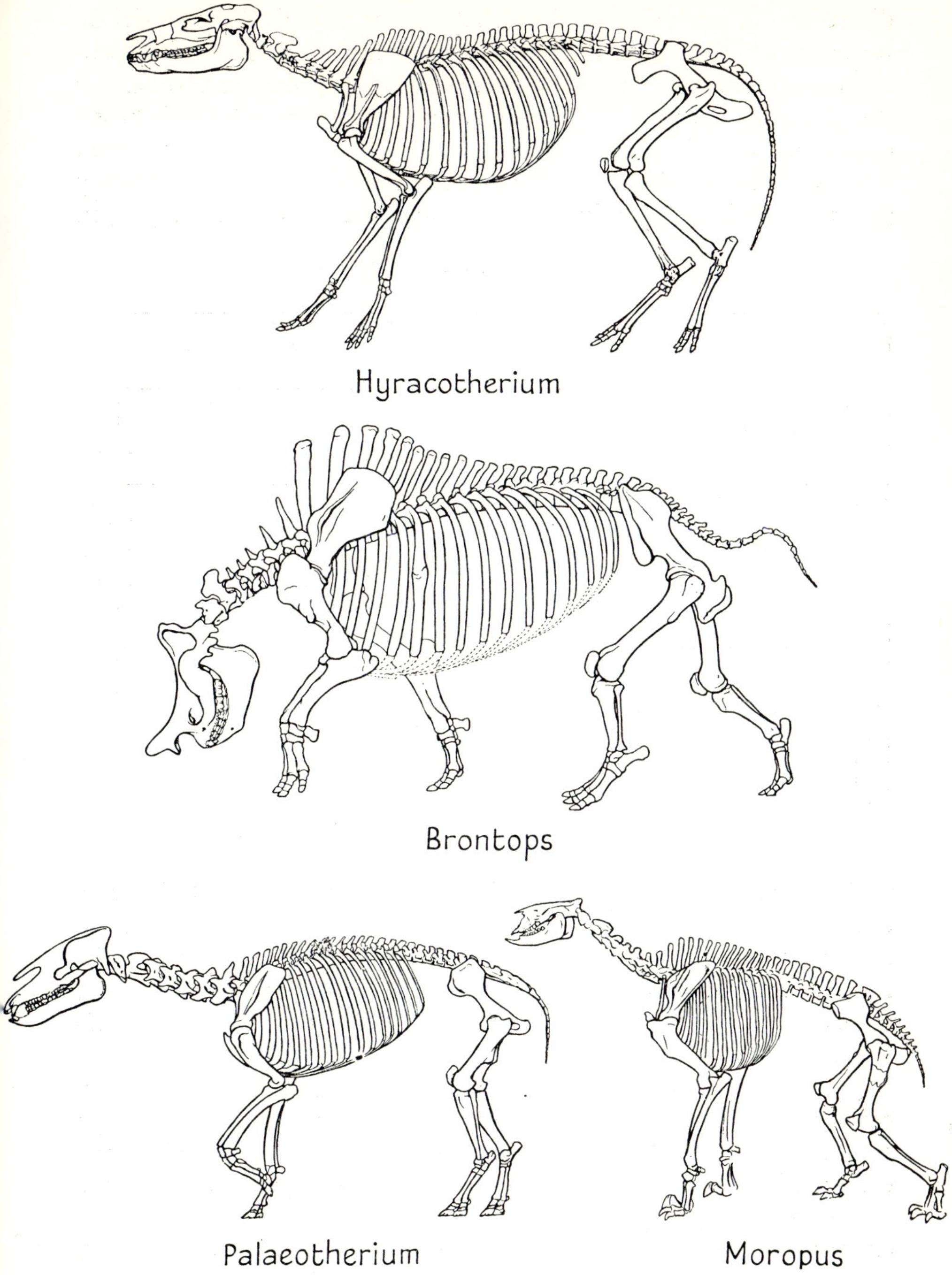

FIG. 479. Skeletons of some early perissodactyls. (After Woodward, *Palaeontology*, Cambridge, Osborn, Abel, and Scott, *Land Mammals*, The Macmillan Company.)

direct line of evolution to *Equus*. Abundant though the material is, we have not, therefore, anything like a complete series of fossils to show every shade and grade of change of the populations throughout the 50 million years or so of their evolution. Our knowledge is based on a small sample of individuals, preserved at random at scattered intervals. The remains often suggest evolutionary sequences and many accounts speak confidently of changes and trends. We shall try, even

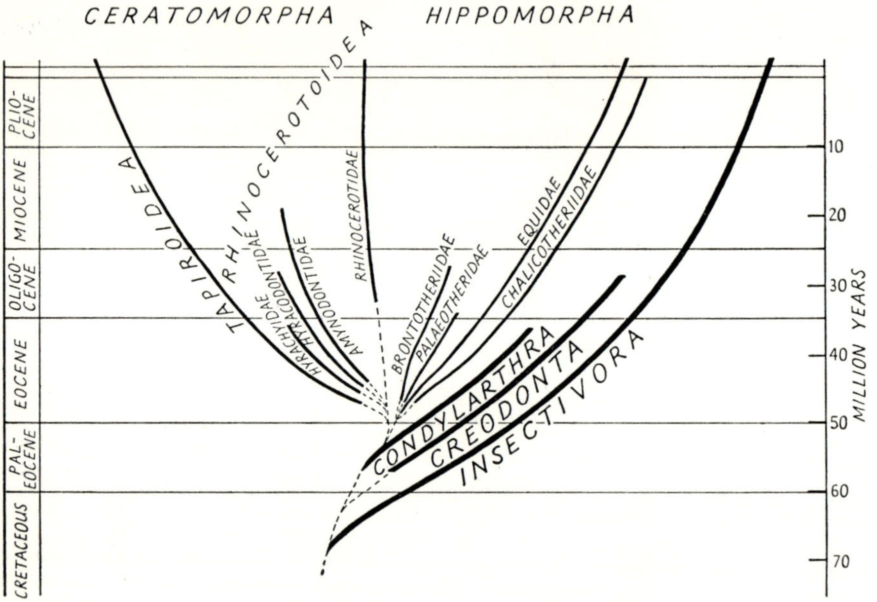

FIG. 480. Chart of the evolution of the perissodactyls.

in this brief account, to describe the actual discoveries and to indicate clearly what evidence is available for evolutionary speculation. With all the mass of information we possess it must yet be realized that the study of the details of perissodactyl evolution has hardly been begun, for example we have little quantitative information about the variability of the characters concerned.

The earliest perissodactyls had departed but little from condylarth conditions. **Hyracotherium* (= **Eohippus*) from the Eocene of Europe and North America (Fig. 479) was the size of a dog and resembled the condylarth **Phenacodus*. The tooth row was complete, with square bunodont molars (Fig. 467). In addition to the four main upper molar cusps an anterior protoconule and posterior metaconule are suggested, and between these and their neighbours the dentine is partly built up to form two transverse ridges. The premolars were tritubercular. The gait was digitigrade, with rather long metapodials and with hoofs, the front leg having four and the hind-leg three toes, the central ones the longest. These animals were already distinctly horse-like, in spite of their small size, and they probably lived in forests, browsing on the leaves.

From a population of animals of this type there evolved a varied

FIG. 481. Malayan tapir, *Tapirus*. (From photographs.)

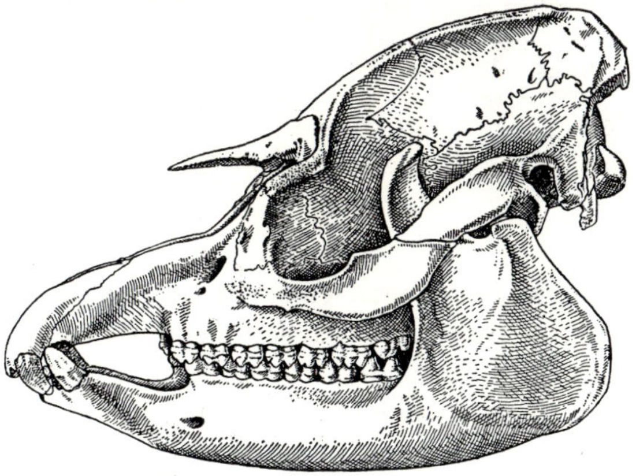

FIG. 482. Skull of the tapir. (After Reynolds, *The Vertebrate Skeleton*, Cambridge.)

host of herbivores (Fig. 480), which may be divided into six main types: the tapirs, remaining with little change; the rhinoceroses, becoming large and heavy-bodied; brontotheres (titanotheres), also becoming large; palaeotheres, an early horse-like line; chalicotheres, which secondarily acquired claws, and finally the horses themselves. Of course each of these lines had many subdivisions and branches, producing a most complex evolutionary bush.

4. Suborder Ceratomorpha, tapirs and rhinoceroses

The modern tapirs of Central and South America, Malaya, and the East Indies (Fig. 481) are nocturnal creatures, mostly living in forests on damp, soft ground; they have retained many of the conditions of the ancestors of all Perissodactyla. There are four digits in the fore-foot and three in the hind; the ulna and fibula are complete and distinct. The tooth row is complete (42 teeth in tapirs), but the premolars are molariform, though of a simple square pattern, with low crowns (Fig. 482). The nose has developed into a short trunk, with characteristic shortening of the nasal bones. The stomach is like that of horses and there is a large caecum. The placenta is diffuse. The brain is relatively smaller than in horses.

Fossil tapirs, very similar to modern forms, are found back to the Oligocene, and somewhat more primitive Eocene related forms

FIG. 483. Indian rhinoceros (*Rhinoceros*).

(**Homogalax*) might have been close to the ancestors of **Hyracotherium*. We may say therefore with some confidence that the modern tapirs show us with little change the condition of the perissodactyl stock in late Eocene or Oligocene times, perhaps 40 million years ago. Fossils almost exactly similar to the existing genus occur back to the Miocene, say, 20 million years ago, and the tapirs were then of widespread distribution. We may notice once again the important fact that, given suitable environments, types persist with little change even when their relatives, moving into other conditions, become greatly changed.

5. Rhinoceroses

The rhinoceroses (Fig. 483) are the only surviving large perissodactyls; they show the graviportal type of body form that was adopted by many of the extinct forms (brontotheres, &c.) and also by many other large eutherians. Dinocerates, elephants, hippopotamuses, and rhinoceroses all show the same type of skeleton. The vertebral column (Fig. 484) has long neural spines above the fore-legs, there are many ribs, reaching back nearly to the pelvis. The whole column thus makes a girder balanced on the fore-legs, and the head, being very heavy, counterbalances the body weight. The hind-legs provide the main locomotor thrust. It is characteristic of this 'single-girder' type of backbone that the ilia are wide and vertically placed. The feet are basically similar in all these groups in that several digits (usually three in the

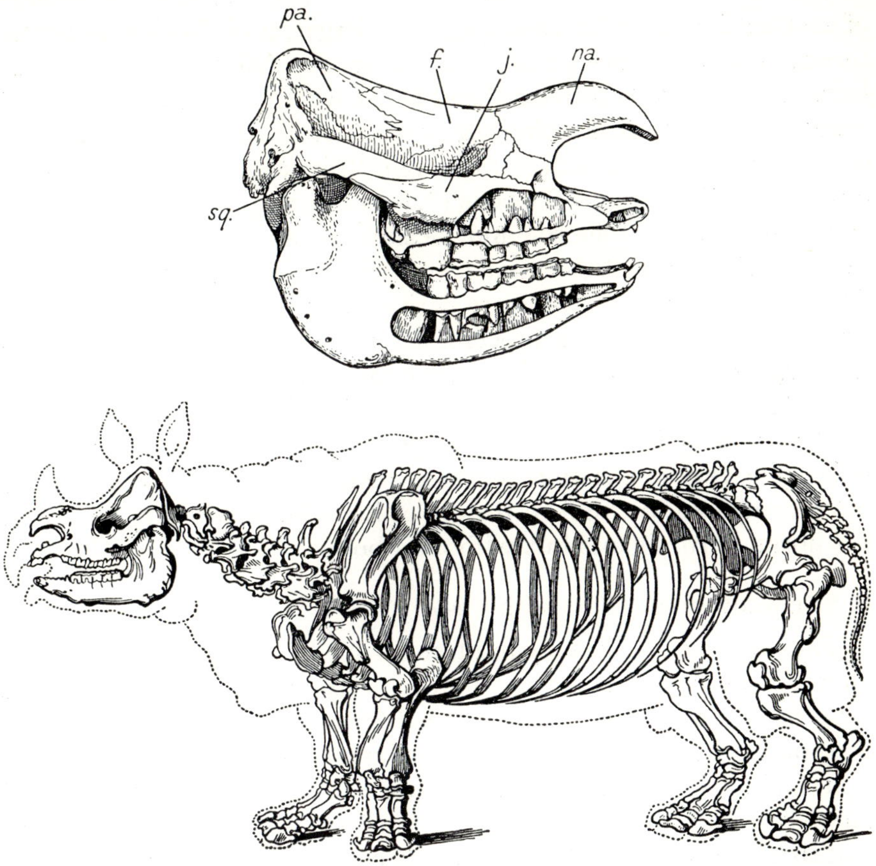

FIG. 484. *Below*: Skeleton of the Indian rhinoceros. (From Owen.) *Above*: Skull and teeth of a young Indian rhinoceros. The grinding surface is made up of four milk premolars and one adult molar on each jaw. The remaining permanent teeth have not erupted. *f.* frontal; *j.* jugal; *na.* nasal; *pa.* parietal; *sq.* squamosal. (After Reynolds, *The Vertebrate Skeleton*, Cambridge.)

rhinoceros) are preserved, making supports of large area. The brain of the rhinoceros is small and the chief receptors are those of smell and hearing; the eyes are mainly used in weak light. Like the tapirs they are essentially timid animals, mainly nocturnal, though defending themselves with a charge if attacked. They live singly or in pairs.

The earliest members of the rhinoceros group, such as **Hyrachyus* of the Eocene, were very like other primitive perissodactyls, mostly small and with a complete tooth row, in which the molars already show an ectoloph and the parallel transverse lophs characteristic of the group. The *Hyracodonts were an Eocene and Oligocene line specialized for swifter movement ('running rhinoceroses') with long legs and three toes in each foot, much as in the earlier horses. The members of another line, the *Amynodonts, were larger, probably semi-aquatic forms. The true rhinoceroses appear in the Oligocene, already as large creatures, fully terrestrial and hence with stout limbs and a good grinding battery, with molarized premolars. **Baluchitherium* became of enormous size, as much as 18 ft high. Rhinoceroses became numerous in the Miocene and Pliocene (**Aceratherium*). Various types persisted through the Pliocene and Pleistocene, and the modern single-horned *Rhinoceros* of Asia and two-horned *Dicerorhinus* of the East Indies and *Diceros* of Africa are derived from some of these. Extinct types such as the woolly rhinoceros are known from Palaeolithic drawings and from partly embalmed specimens. Modern rhinoceroses all have a very thick, almost hairless skin, with characteristic folds. The tendency to active keratin development also produces the horns, either one, two, or occasionally three median outgrowths on the head, often compared to clumped masses of hairs but essentially similar to the horns of ruminants but without a bony core.

6. *Brontotheres (*Titanotheres)

These were early, heavily built ungulates, reaching large size in the later Eocene and Oligocene, in fact preceding the rhinoceroses as large herbivores. The fully developed forms, such as **Brontops* (Fig. 479) of the Lower Oligocene, were of typical graviportal type, up to 8 ft high, with high thoracic spines, numerous ribs, vertical and laterally expanded ilia, and rather short legs, with four digits in front and three behind. The tooth row was complete and the molars large but low-crowned, with a ridge along the outer side, but isolated cusps on the inner (hence 'bunolophodont'). A single pair of large horns was carried on the front of the skull. The brain was even smaller than that of rhinoceroses.

The earliest fossils that can be referred to this type, **Lambdotherium* of the Lower Eocene, were much smaller, and without horns; they could well have been derived from **Hyracotherium*. **Eotitanops* from the Middle Eocene was larger, but still hornless. From some such stage numerous lines probably diverged, each becoming larger and independently acquiring horns. For obvious reasons it is difficult to obtain a proper idea of the evolution of such giants, as Simpson points out, genera and perhaps even subfamilies have probably been created on a basis of differences that may be only sexual or individual.

7. *Chalicotheres (= *Ancylopoda)

One remarkable side-line of perissodactyl evolution, while becoming large and horse-like in some ways, acquired structures resembling claws instead of hoofs (**Moropus*, Fig. 485). These chalicotheres, all rather alike, existed from the Eocene to the Pleistocene and were therefore a successful group. The terminal phalanges of the three toes of each foot were cleft and undoubtedly carried a nail or claw of some

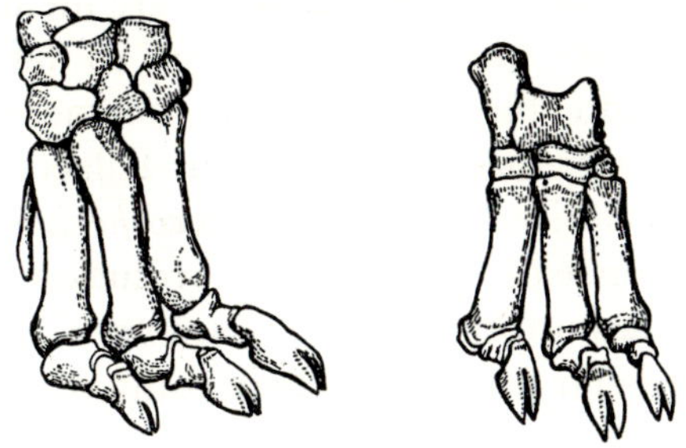

FIG. 485. Feet of a chalicothere. (After Romer, *Vertebrate Paleontology*, University of Chicago Press.)

sort, though not necessarily one like that of true unguiculates. There is no doubt that in a sense this is a case of reversal of evolution, but we cannot assert much about its possible genetic implications unless we can find details of the nails.

When chalicothere digits were first discovered in 1823 Cuvier applied his 'law of correlation' and suggested that this was the remains of an ant-eater, 'un Pangolin gigantesque', while teeth and other bones found near by he referred to an ungulate. It was only when skeletons were found in such a position that the association of the bones could not be denied that the danger of this attempt to apply deductive principles in biology was exposed. The other parts of the skeleton are unambiguously perissodactyl (Fig. 479), the teeth rather like those of brontotheres. There may have been a short proboscis. The neck vertebrae of some forms show very strong zygapophyses, and it has been suggested that the snout and claws were used for digging for roots or water. It is more probable that the chalicotheres reared up on their hind-legs and used the claws to cling to tree-trunks while reaching for leaves with their flexible necks, or perhaps to drag down branches. Like the toxodont **Homalodotherium* they had long front legs and large ischia. Moreover, their remains are found in association with those of forest dwellers. The attraction of speculating about these creatures has not diminished with the demonstration of its dangers.

8. Palaeotheres

These animals, from the later Eocene and early Oligocene of Europe, were an early offshoot that paralleled in many ways the evolution in North America. For example, they developed three-toed feet, and the premolars became molarized. The teeth developed ridges on a similar plan to the horses, but differing in details. Some forms became hypsodont. Several lines of descent are included in the group. **Palaeotherium* became large, though not so large as the graviportal brontotheres and rhinoceroses. The shortness of its nasal bones suggests that it had a proboscis like a tapir. Palaeotheres, like horses, have probably been derived from a **Hyracotherium*-like stock. They illustrate the importance of parallelism in evolution, and serve to warn us against the easy assumption that a character that is shown by two animals must have been present in their common ancestor.

9. Horses

The horse, besides its special interest as one of our oldest and most useful commensals, has provided a rather complete and convincing record of its origin. We shall therefore first describe its present structure and then analyse the fossil record to discover exactly what can be demonstrated about the evolution. Existing horses, asses, and zebras, all referred to the genus *Equus* (Figs. 486–88), are highly specialized for swift movement and eating grasses (p. 697). Only the third digits are developed and covered with hoofs. These are elaborately organized pads, including several sorts of keratin, harder in front, more elastic behind. The metapodials of digits II and IV are present as small splint-bones. There is a horny callosity on the inner side of the fore-limb in all species (also on the hind-limbs in *E. caballus*, the domestic horse), representing the vestigial hoofs of lateral digits.

There are three incisors in each jaw, usually one small canine (the 'tush', absent in females). The first premolar is vestigial in each jaw ('wolf-tooth'); the remaining three resemble the three molars. All the cheek teeth are hypsodont, square in cross-section, with ectoloph and transverse protoloph and metaloph, joined by longitudinal ridges that give the tooth a certain resemblance to the selenodont molars of artiodactyls, hence 'selenolophodont'. The skull is modified to allow space

FIG. 486. Zebra (*Equus*). (From photographs.)

FIG. 487. Racehorse (*Equus*). (From a photograph.)

FIG. 488. Shire horse (*Equus*). (From a photograph.)

for the deep, continually growing teeth and for the large jaw-muscles, and there is a complete post-orbital bar.

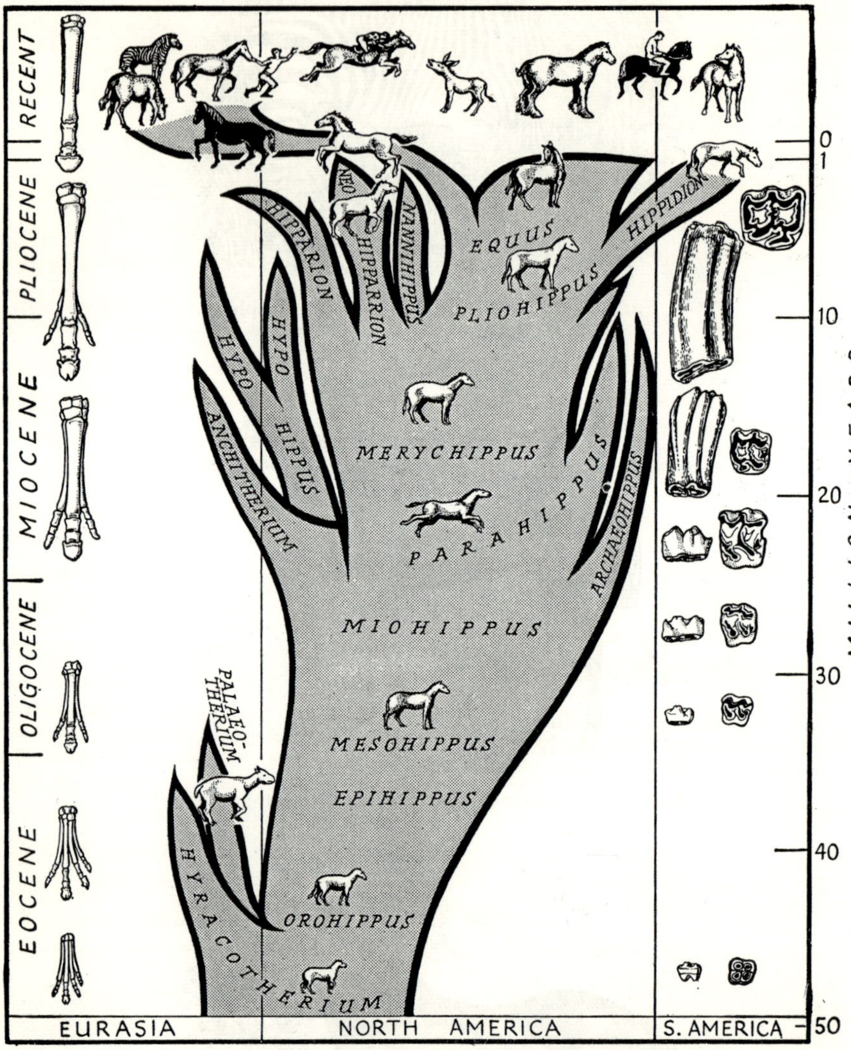

FIG. 489. Table to show the evolution of horses. (Based on Stirton.) The approximate condition of the limbs and teeth at each epoch are shown to the left and right.

The hair is long all over the body and tends to show the pattern of vertical stripes that is so marked in zebras, often clear in asses, and occasionally present in horses. The tail is long, its hairs beginning close to the base in horses, half-way along in the others.

All horses in the native state live in herds, as do so many herbivores that dwell on plains. The brain is large, and although the organs of smell are well developed the eyes are also large and the neopallium is extensive. Receptors for touch are well developed in the muzzle, in the skin beneath the hoofs, and elsewhere. Hearing is exceptionally acute. Besides the keen senses common to many herbivores the horse, with its large brain, also has considerable powers of learning and ability to vary and restrain its behaviour. There is an elaborate communication system, involving not only sounds but movements of the ears, tail, and lips. In these respects horses and elephants, and perhaps also modern artiodactyls, are probably very different from the small-brained herbivores of the Eocene, though, of course, we can only guess at the behaviour of these.

Modern horses show considerable genetical diversity (Figs. 487 and 488), but none of the 'species' are mutually sterile, though the F 1 resulting from the cross may be nearly so, as in the case of the mule, produced from the horse–ass cross. Evidently the population is in process of divergence. The domestic horse *E. caballus* is not found truly wild, but *E. przewalskii* of central Asia may be. There are several species of wild asses, such as *E. onager* of Asia and *E. asinus* of Africa. Several species of zebra live in Africa, one of them being *E. zebra*.

Between the modern *Equus* and the lower Eocene **Hyracotherium* a great number of fossil stages can be recognized (Fig. 489). The chief changes that can be followed may be listed as (1) increase of size, (2) lengthening of the distal portion of the legs, (3) reduction of lateral digits, (4) increase in the relative length of the front part of the skull, (5) increase of depth (hypsodonty) and of the grinding lophs of the molars, (6) approximation of premolars to molar structure, (7) completion of post-orbital bar. No doubt there has been change also in many other characteristics, for instance the brain and behaviour; these are difficult to follow in a fossil series, but study of cranial casts suggests that a rapid increase in size and folding of the cerebrum occurred relatively early in the evolution.

The fossil remains are not usually available in long series of layered beds, such that we can be sure that one population has evolved into the next. However, the dating of the fossils can often be done with considerable accuracy by means of the associated animals, and a series can thus be produced such as would be expected in the progress from **Hyracotherium* to *Equus*. There are, however, many fossils that show special developments, and cannot be fitted into the direct series. These are presumed to be divergent lines: it must be emphasized that this is an arbitrary though probably justified procedure. These 'side-lines' are so numerous that they immediately throw doubt on the idea that there has been any single uniform 'trend' in horse evolution. At least twelve types sufficiently marked to be classified as genera are known, in addition to those directly on the line leading to *Equus*; of course there is a much larger number of shorter, independent, evolutionary lines within these genera. We have enough evidence to glimpse the extraordinary complexity that would be revealed by the complete evolutionary 'bush', even in this single family. A further complication is produced by migrations. It is at present believed that the main course of horse evolution went on in North America, with migration at various times to the Old World and South America. Certainly a more continuous series of forms has been revealed in North America than elsewhere, but it must be remembered that they have been looked for intensively, and brilliantly studied. It is not impossible that further study of Old World horses will produce still greater complications by revealing sequences of evolution within that area.

Throughout the Eocene epoch the horses all possessed four toes in each limb. The fossils classed as **Orohippus* and **Epihippus* from the Middle and Upper North American Eocene are little different from **Hyracotherium*, except for molarization of the hinder premolars. The size remains small.

The Oligocene horses, **Mesohippus* and **Miohippus*, walked with three toes on the ground, and all the premolars were molarized. The ectoloph was well formed but the inner cusps were still separate, and the teeth low-crowned. Some horses of this type (**Anchitherium* and its descendant **Hypohippus*) persisted into the Miocene, presumably surviving as browsers in forests, while other descendants took to the plains. These browsing horses migrated to the Old World in the Miocene, then died out there, as they did also in North America.

**Parahippus* of the American Lower Miocene shows the beginning of the adaptation for life on the plains. The lateral digits II and IV still carried hoofs, but since the central proximal phalanx was much the longest and strongest, it is probable that the lateral ones touched the ground only to maintain balance over uneven surfaces, or in soft conditions. The teeth were still rather low, but were beginning to be elongated and to show cement on the crowns. The protoloph and metaloph were connected by a narrow bridge. There was a partial post-orbital bar.

**Merychippus* comes from later Miocene beds and could have been directly derived from **Parahippus* by increase in the depth of the teeth and reduction of the lateral digits to short stumps, still three-jointed and carrying hoofs, but vestigial in the sense of never touching the ground. The presumption is that this type of structure was found advantageous for life on large grassy plains produced by arid Miocene conditions, the high-crowned teeth being needed to grind the tough siliceous grasses.

Apparently the type was very successful and in the Pliocene it produced various populations. **Hipparion*, with the two lateral toes remaining as vestiges, spread through Eurasia in the Pliocene. **Nannippus* was a small form that remained in America. **Pliohippus* was another American descendant from **Merychippus*, and here the lateral digits were lost altogether in the Pliocene, the metapodials remaining as long thin vestiges. When the land connexion with South America became open this type of horse migrated there and produced a special development, **Hippidion* of the Pleistocene, with rather short legs, perhaps correlated with a mountain habit.

Meanwhile in the late Pliocene or early Pleistocene the **Pliohippus* stock of North America finally reduced the lateral metapodials to short splint bones and produced the *Equus*-type, which spread thence over all the available land-masses, becoming then extinct in North and South America until reintroduced by man.

10. Allometry in the evolution of horses

Although *Equus* is certainly a very different creature from **Hyracotherium*, we are fortunate in that many of the differences are due to measurable changes in proportions. A beginning has been made with attempts to estimate the rate of evolutionary change, as a preliminary to study of the factors that influence it. Some of the changes in proportion seen during horse evolution are a consequence of the increase in size. If an organ grows relatively faster or slower than the body as a whole it is obvious that its proportions will differ in animals of differing adult size. The size of an organ, y, in relation to that of the body, x, is often expressed as $y = bx^k$, where the constant k describes the relative growth rate. If $k > 1$ the organ becomes larger in larger animals and is said to be positively allometric (J. S. Huxley). The demonstration that growth actually follows this law in particular cases is not easy, and the underlying assumptions have been questioned. It is probably true, however, that organs do sometimes differ in relative growth-rates, and the method provides a means of investigation of the proportions of an organ not only at one stage but throughout the growth period, and indeed also between adults throughout an evolutionary sequence. Thus Robb (Fig. 490) shows that the length of a horse's face increases between embryo and adult along a line similar to that found in the series **Hyracotherium* to *Equus*, and that adult horses of different sizes vary similarly in face proportion. A nearly fitting line gives constants $b = 0{\cdot}25$ and $k = 1{\cdot}23$. Other methods of plotting, for instance face length against cranium length, give somewhat different results and it cannot be considered certain that no new genetic factors have been involved in the increase of face-length throughout the whole evolutionary sequence.

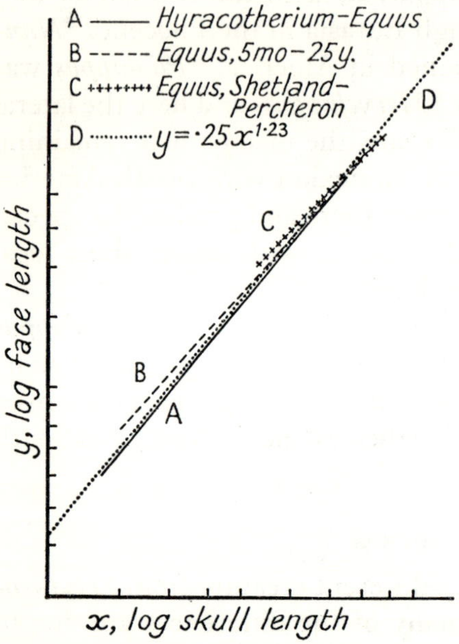

FIG. 490. Relative rates of growth in horses. The lines show regression of log skull length on log face length; in A, the line of horse phylogeny; B, the ontogeny of *Equus*; C, various races of *Equus* of different sizes; D, $y = 0{\cdot}25\ x^{1\cdot 23}$. (After Simpson from the data of Robb.)

Again, reduction of the lateral digits of the toes is perhaps not based on any steady genetic change except that related to size. However, in this case there was probably one relatively sudden change in the constant b (which may be said to express the body size at which the toe begins to form). Thus in the three-toed horses the length (y) of the side toes is related to that of the cannon bone of the central digit (x) by the equation

$$y = 1{\cdot}5\ x^{0\cdot 97}.$$

In the one-toed horses from **Pliohippus* onwards

$$y = 0{\cdot}8\ x^{0\cdot 99}.$$

There has therefore been little change in the relative growth-rate, which is negatively allometric in all horses, so that the lateral toes are relatively smaller in the larger animals.

11. Rate of evolution of horses

Study in the same way of other horse characters, for instance those involved in hypsodonty, shows that special genetic changes may be involved and that genetic change does not go on at a constant rate.

Estimates of rate of evolution for the whole animal have also been made by Matthew, Simpson, and others. Assuming that the genus is assessed as a comparable entity throughout (a large assumption, this!) and dividing the eight genera (excluding *Equus*) on the direct line into the time involved between **Hyracotherium* and **Pliohippus* (50 million years), we should have 6·3 million years per genus. However, there is reason to suppose that individual genera lasted for very different times, **Miohippus*, for instance, less than a third of the time of **Merychippus*. Therefore if the criterion of a genus is constant, the rate of evolution must vary.

Sufficient fossil horse material is available to allow consideration whether known rates of mutation are likely to be adequate to account for the observed evolutionary changes. Simpson calculates that from **Hyracotherium* to *Equus* there must have been at least 15 million generations, which, with a population in North America of 100,000 (a low estimate), gives a total of $1{\cdot}5 \times 10^{12}$ individuals in the 'real and potential ancestry of the modern horse'. One in a million is a moderate rate for large mutations at any locus in *Drosophila*, and this would give 1·5 million such mutations for a single locus in the horse ancestry. It would be safe to assume that one-fifth of these (300,000) were in the direction favoured by selection and that one-tenth of all such genes affect a structural change, such as ectoloph length. The actual increase in this length between **Hyracotherium* and *Equus* was from 8 to 40 mm which, divided into 300 steps, gives an increase per mutation of only 0·1 mm. This is a reasonable figure, and such calculations suggest that observed mutation rates are quite adequate to account for the evolutionary changes, even neglecting possible multiple actions and interactions of genes, by which the speed of evolution could be further increased.

12. Conclusions from the study of the evolution of horses

Careful consideration of the fossil horse material therefore shows reason to suppose that evolution has proceeded by gradual change. As more and more evidence becomes available the series becomes more and more complete, and incidentally the nomenclature increasingly confusing. Incompleteness of material may give an impression of evolution by jumps and saltation, especially when, as in the Old World horses, there have been successive migrations into one region from another. The European palaeontologists, finding **Hyracotherium*, **Anchitherium*, **Hipparion*, and *Equus*, without intermediate forms, interpreted the evidence as showing evolution by saltation. This was indeed a reasonable deduction from the facts, but was not the only possible one, as has since been shown by the discovery of the much fuller sequence in North America. It is probable that there are many equally unjustified conclusions in our current beliefs about evolution.

The outstanding conclusion from a study of horse evolution is that it is very difficult to describe the change as occurring in a single direction, as supposed by believers in 'orthogenesis'. Apart from the fact that many 'side-lines' can be detected beside the line that happens to have survived, it is important to remember that not every line evolves in the same direction. Thus in at least two genera of horses size became progressively *smaller* (**Nannippus* and **Archaeohippus*). However, it is certainly true that in some lines evolution may proceed for long periods in one direction. We have no clear evidence why this should be so, but it is reasonable to suppose that it is due to 'ortho-selection', that is to say, the survival of animals that adopt a particular method of life for which they are suited by a particular make-up. The effect of this would be gradually to select all those genetic factors that make for success in one environment (say, grazing on grassy plains) and hence to produce evolutionary change in one direction.

XXXI

ARTIODACTYLS

1. Characteristics of artiodactyls

THE even-toed ungulates, though they can be traced as a distinct line back to the Eocene, may be considered as the latest mammalian herbivores, having radiated out chiefly in the Miocene and attained then a dominance that has persisted to the present day. Except for man and the horse all the large mammals really well established and successful at the present time are artiodactyls. Any attempt to be dogmatic about the reasons for the success of a group of animals is apt to be superficial, but it is not unreasonable to suggest that in this case the result is due to swiftness of foot, combined with keenness of sense and brain, efficient cropping and grinding mechanisms, and especially a complex stomach, allowing the digestion of cellulose by symbionts. Two families of artiodactyls survive without these special features, the pigs and hippopotamuses, water- and forest-living remnants that show us approximately the condition of the group in the Eocene.

The characters of artiodactyls show a fascinating 'similarity with a difference' to those of perissodactyls. The common origin of the two groups (p. 694) was little above the insectivore stage and nearly every feature has been evolved independently; the general structural similarities and detailed differences therefore show the effect produced by similar ways of life on slightly differing populations. For example, a postorbital bar developed in both groups, for attachment of the large masseter muscle, but whereas in the horses it is formed wholly of a process of the frontal bone, in ruminants there is a union of processes of the jugal and frontal. Many such similarities and differences are seen throughout the body, and especially in the limbs.

The skull of later artiodactyls shows changes of shape to accommodate the very deep molars and to support the horns that are commonly found. It becomes very high (as in horses), and there is a sharp kink between the basisphenoid and presphenoid, so that the face slopes steeply downward. The facial bones become large and the parietals restricted to the vertical posterior face of the skull, to which the powerful neck-muscles are attached. In many ruminants there is a scent-gland, lying in a pre-orbital fossa of the skull and opening on the side of the head. The pre-lacrymal fossa is a gap in the skull, where the nasal cavity is separated from the outside only by the skin.

The vertebral column shows the characteristics of other large mammals in the development of high thoracic spines. Some of the heavier types have a long rib series and graviportal 'single girder' structure, but the tendency has been to retain and develop the break in structure of the vertebral column behind the thoracic region, giving a long lumbar region with forwardly directed transverse processes. In rabbits and other mammals this division of the column is associated with the jumping habit, and this is also found, though in a different form, in ruminants (Young, 1955, p. 139). Associated with this method of progress is a fore-and-aft elongation of the pelvic girdle, the ischium being well developed for the attachment of the retractor muscles of the thigh. In making the jumping movements, which are common in all artiodactyls and are especially used by the mountain-loving types, the extensor muscles of the back (sacrospinalis and multifidus) work with the retractors of the two hind limbs to give a powerful thrust.

The characteristic of the limbs is, of course, the equal development of digits III and IV, with reduction of the rest. The gait was at first plantigrade, then digitigrade; hoofs, differing from those of perissodactyls, have developed on the toes. The elongation of the lower segments of the limbs and shortening of the upper has been similar to that of perissodactyls, but the long metapodials have become united in later forms to make the 'cannon bone'. The ulna and fibula become reduced, as in horses. The presence of two digits has led to the retention of two bones in the distal row of carpals, the hamate and fused magnum-trapezoid, and these articulate in interlocking fashion with the three proximal carpals (Fig. 466). Similarly in the foot the two lateral cuneiforms are fused to thrust upon the third digit, while the fourth sends its thrusts to the cuboid and the latter is fused with the navicular. Between this compound bone and the talus there is a very characteristic joint, the under surface of the talus being grooved like its upper surface. These joints of the carpus and tarsus are evidently an important part of the apparatus of locomotion; probably in both limbs they serve to take strain when the animal is moving over uneven ground, and in the leg they are also the seat of a considerable propulsive thrust from the calf-muscles. In walking, the limb of artiodactyls is moved as a whole at the shoulder and hip, by action of the upper muscles. The wrist and ankle joints bend just enough to raise the feet off the ground, and the elbow and knee joints, lying so high as to be hardly visible externally, also bend little. The essence of artiodactyl locomotion is the use of the upper limb muscles; indeed the hinder part of the vertebral column has almost become part of the limb!

The dentition of artiodactyls is highly specialized. The upper

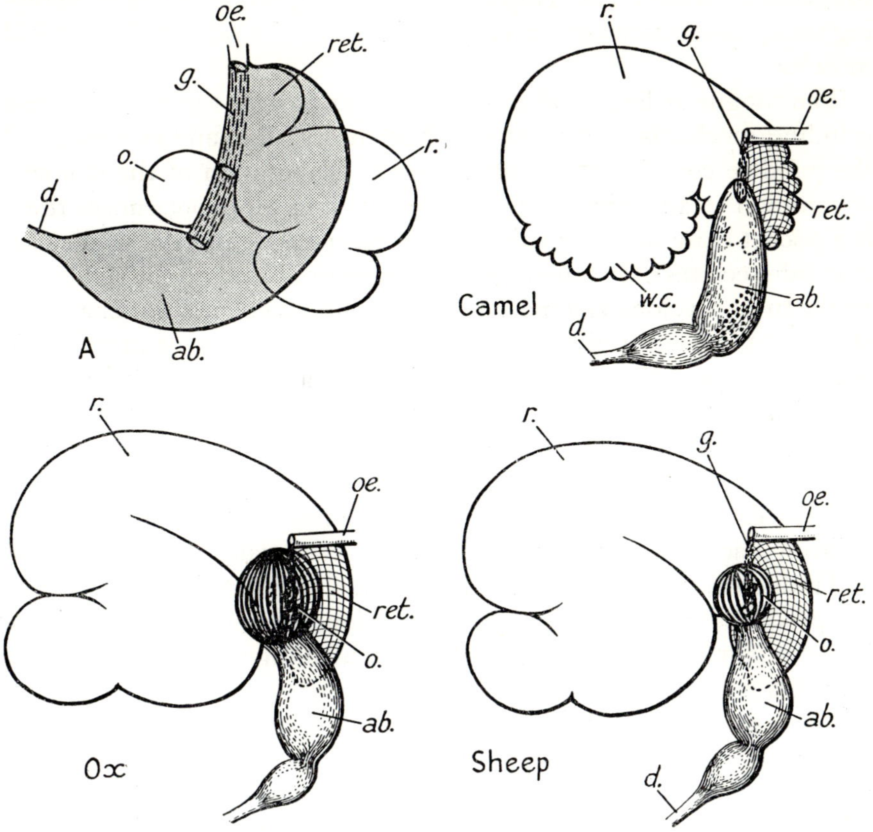

FIG. 491. Stomach of camels and ruminants.

A shows the relationship of the normal mammalian stomach (stippled) to that of ruminants. The rumen (*r.*) represents the cardiac region, the reticulum (*ret.*) the body. The oesophageal groove (*g.*) and omasum (*o.*) are derived from the lesser curvature as far as the incisura angularis and the abomasum (*ab.*) represents the pyloric antrum; *d.* duodenum; *oe.* oesophagus.

The omasum and abomasum are shown as if pulled downwards. In the camel *w.c.* are the water cells. The abomasum is mostly lined with stratified squamous epithelium; fundic glands are found only in the dotted area. (Material for figure kindly supplied by Dr. A. T. Phillipson, partly after Pemkopf.)

incisors are lost in later types, which crop by means of their gums. The canines may form tusks. Premolars are not molarized, but an efficient grinding battery is often provided by the very elongated, hypsodont molars. These acquire a grinding surface by the development of each of the four original cusps into a longitudinal ridge—

the selenodont ('moon-tooth') condition. The effect is similar to that arrived at, by very different means, in horses, and the enamel, dentine, and cement, wearing at differing rates, provide a continually roughened surface. The temporo-mandibular joint is flattened, allowing rotary movements of the jaw, produced by the powerful pterygoid muscles.

The tongue is large and is an important part of the cropping and grinding mechanism; it is very mobile, protrusible and pointed, and the papillae covering it are often horny. Elaboration of the stomach is common to all artiodactyls. In the pigs and hippopotamuses there is a pocket close to the opening of the oesophagus and the whole cardiac side secretes only mucus, pepsin being produced on the right side. In the fully developed stomach of Ruminantia there are four chambers (Fig. 491), rumen, reticulum, omasum (= psalterium or manyplies), and abomasum. The first three are lined by a stratified epithelium of oesophageal type, folded into muscular ridges. These are low in the rumen, form a network in the reticulum, and are overlapping leaves in the omasum. Food is first swallowed into the rumen, where it is mixed with mucus and acted upon by a fauna of anaerobic cellulose-splitting bacteria, whose enzymes break up the walls of the plant food and reduce the whole to pulp. Organic acids, from acetic acid upwards, are produced, absorbed into the circulation, and metabolized. There is also a fauna of ciliates in the rumen, which digest cellulose and are themselves later digested.

The process of rumination depends upon an oesophageal groove running from the cardia to the opening of the omasum. When the lips of this are brought together food does not enter the reticulum and is returned from the rumen to the mouth. After chewing, the bolus is again swallowed, the groove opens, and the food passes to the reticulum and omasum. Here water is pressed out and absorbed and the remainder proceeds to the abomasum, the 'true' stomach, with peptic glands. This elaborate digestive mechanism has no doubt contributed largely to the success of the artiodactyls, allowing them to eat their food rapidly and then retire to digest it in security. The efficient cellulose-splitting system also enables them to make use of hard grasses and other unpromising sources of nutriment.

The brain is moderately well developed in later artiodactyls, but even here the cerebral hemispheres only partly cover the cerebellum, and in the earlier forms the brain was relatively small, as it is today in hippopotamuses and pigs. The olfactory organ and related parts of the brain are well developed and most artiodactyls also have large eyes, with a horizontal pupil, and long ears and an acute sense of hearing.

Artiodactyls have an elaborate system of scent-glands, on the head, between the digits, in the inguinal region, and elsewhere, though not usually around the anus. These glands are used for marking territory and in the sexual and social life, which is often elaborately organized. The colour of the coat and especially the form of the head and horns also play an important part in the communication system between individuals.

The reproductive system remains rather close to the presumed original eutherian condition. The uterus is bicornuate and in pigs the placenta is of the diffuse epitheliochorial type. In ruminants there is a cotyledonary placenta, but the contact between maternal and foetal tissues is never very close (syndesmo-chorial) and the allantois is usually large.

2. Classification

Superorder 5. Paraxonia

Order Artiodactyla

Suborder 1. Suiformes

Infraorder 1. Palaeodonta. Pigs and peccaries. Eocene–Recent

**Diacodexis*, Lower Eocene, N. America; **Homacodon*, Middle Eocene, N. America; **Entelodon*, Lower Oligocene, Holarctic; *Sus*, pigs, Lower Pliocene–Recent, Eurasia (then world-wide); *Phacochoerus*, wart-hog, Pleistocene–Recent, Africa; *Dicotyles*, peccary, Pleistocene–Recent, Central and S. America; *Potamochoerus*, water-hog, Pleistocene–Recent, Africa

Infraorder 2. Ancodonta. Hippopotamuses. Oligocene–Recent

**Anthracotherium*, Oligocene–Pliocene; *Hippopotamus*, Pliocene–Recent, Eurasia, Africa

*Infraorder 3. Oreodonta. Eocene–Pliocene. N. America

**Merycoidodon* (= **Oreodon*), Oligocene; **Agriochoerus*, Oligocene–Miocene

Suborder 2. Tylopoda. Camels. Eocene–Recent

**Protylopus*, Eocene, N. America; **Poebrotherium*, Oligocene, N. America; **Procamelus*, Miocene–Pliocene, N. America; **Alticamelus*, Miocene–Pliocene, N. America; *Lama*, alpaca, Pleistocene–Recent, S. America; *Camelus*, camel, dromedary, Pleistocene–Recent, Asia

Order Artiodactyla (*cont.*)

Suborder 3. Ruminantia. Eocene–Recent

Infraorder 1. Tragulina. Eocene–Recent. Holarctic, Africa

**Archaeomeryx*, Eocene. Asia; *Tragulus*, chevrotain, Pliocene–Recent, Asia; *Hyemoschus*, water chevrotain, Pleistocene–Recent, Africa

Infraorder 2. Pecora. Oligocene–Recent. Holarctic, Africa, S. America

Family 1. Cervidae. Oligocene–Recent. Holarctic, S. America

**Blastomeryx*, Miocene–Pliocene, N. America; **Palaeomeryx*, Miocene, Europe; *Moschus*, musk-deer, Pliocene–Recent, Asia; *Cervus*, red deer, American elk, &c., Pliocene–Recent, Holarctic; *Dama*, fallow deer, Pleistocene–Recent, Eurasia; *Rangifer*, reindeer, Pleistocene–Recent, Holarctic; *Capreolus*, roe deer, Pliocene–Recent, Eurasia; *Alce*, moose, European elk, Pleistocene–Recent, Holarctic

Family 2. Giraffidae. Miocene–Recent. Eurasia, Africa

Giraffa, giraffe, Pliocene–Pleistocene, Asia; Recent, Africa; *Okapia*, okapi, Recent, Africa; **Palaeotragus*, Miocene–Pliocene, Eurasia; **Sivatherium*, Pleistocene, Asia

Family 3. Antilocapridae. Miocene–Recent. N. America

**Merycodus*, Miocene–Pliocene; *Antilocapra*, prong-buck, Pleistocene–Recent

Family 4. Bovidae. Miocene–Recent

**Eotragus*, Miocene, Europe, Africa; *Gazella*, gazelles, Pliocene–Recent, Eurasia, Africa; *Taurotragus*, eland, Pleistocene–Recent, Africa; *Aepyceros*, impala, Recent, Africa; *Bos*, cattle, yak, Pleistocene–Recent, Eurasia and N. America, now world-wide; *Bison*, buffalo, Pleistocene–Recent, Holarctic; *Capra*, goat, Pleistocene–Recent, Eurasia, Africa (now world-wide); *Ovis*, sheep, Pliocene–Recent, Holarctic, Africa (now world-wide).

3. The evolution of artiodactyls

Although abundant fossil material is available, the lines of evolution within the artiodactyls are not altogether clear, and numerous classificatory arrangements have been suggested. We shall, as usual, follow that of Simpson, who recognizes three suborders. The suborder Suiformes contains the ancestral Eocene forms and some of their little-modified descendants; it is represented today by the pigs and hippopotamuses. Probably no members of this group developed the ruminating habit, and they are sometimes called 'non ruminantia'. The suborder Tylopoda is for the camels, and the third suborder, Ruminantia, includes all the other modern forms of artiodactyl.

The earliest artiodactyls (Fig. 492), included in the Suiformes, were

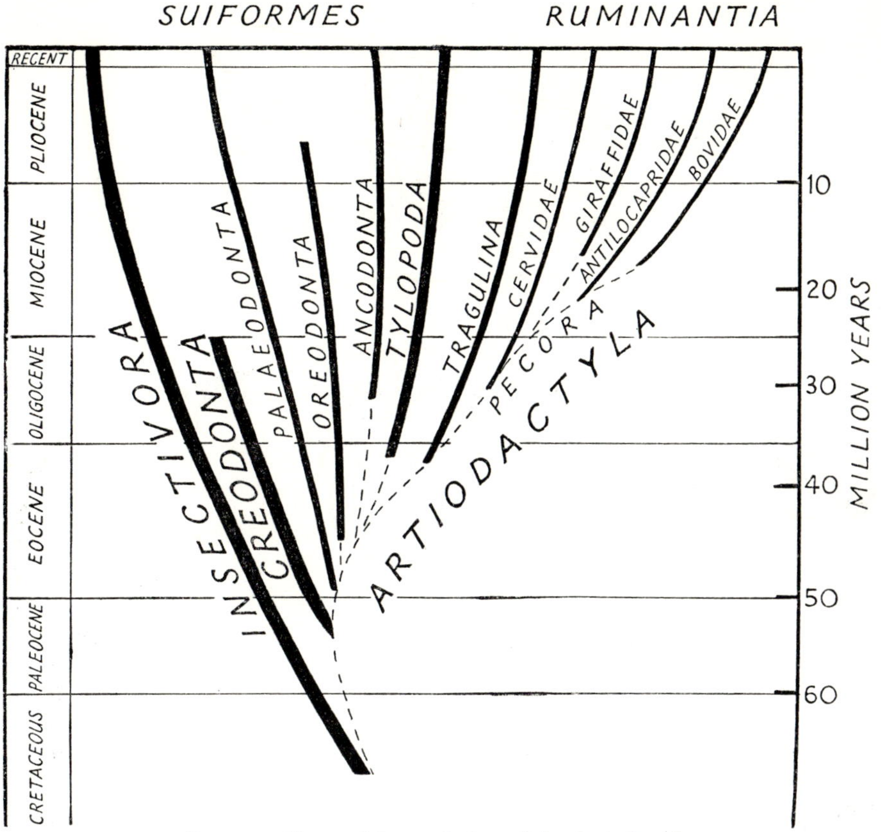

FIG. 492. Chart of the evolution of the Artiodactyla.

close to the ancestral stock of all placentals. In **Diacodexis* from the North American Lower Eocene there were tritubercular molars and it was probably a small, running, omnivorous form, with four toes on each foot. These animals could indeed almost equally well be classified as insectivores or creodonts, and the only reason for placing them as artiodactyls is that the talus had the typical pulley-like lower surface. Later Eocene and early Oligocene forms developed a bunodont condition, with sometimes six cusps in the upper molars, protocone, paracone, metacone, hypocone, protoconule, and metaconule. In some later forms these cusps show a selenodont condition.

4. Pigs and hippopotamuses

The pigs have remained essentially in this Eocene condition; Simpson recognizes this by classifying them with the Eocene forms in one infraorder Palaeodonta, distinct from the amphibious Ancodonta (hippopotamuses and anthracotheres) and the *Oreodonta.

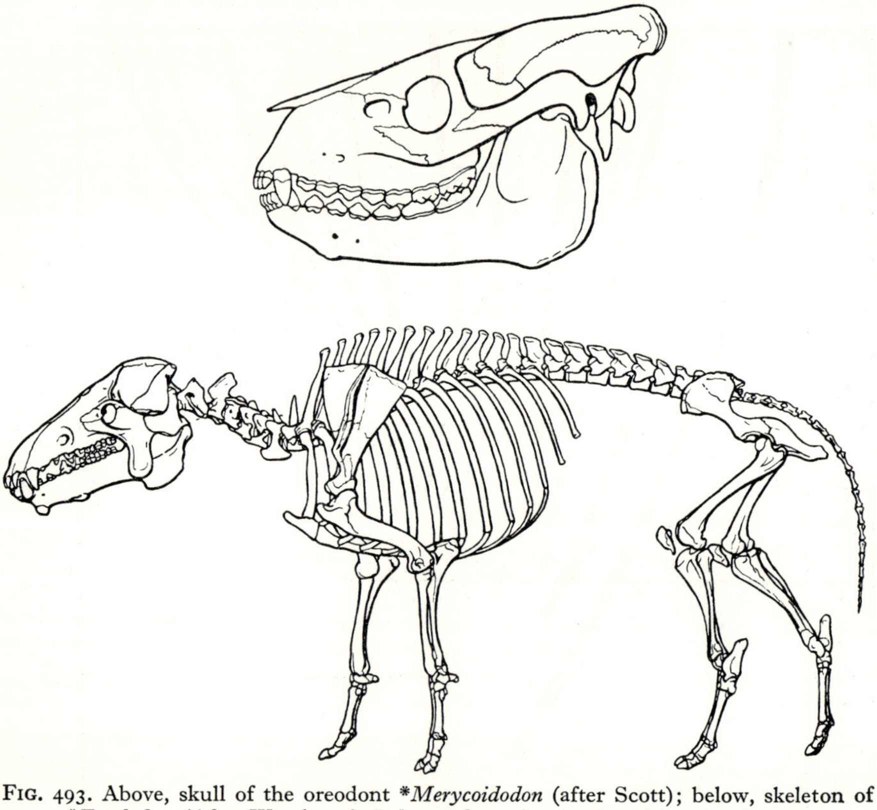

FIG. 493. Above, skull of the oreodont **Merycoidodon* (after Scott); below, skeleton of **Entelodon*. (After Woodward, *Palaeontology*, Cambridge University Press.)

Close relatives of the pigs are found from the Eocene, with bunodont molars. Several lines can be recognized, including the entelodonts, giant pigs of the Oligocene, over 5 ft high and 12 ft long and of graviportal structure (Fig. 493). The modern pigs (Fig. 494) show a nearly complete dentition, with persistently growing canine tusks, used for defence and for digging roots. The orbit is continuous with the temporal fossa. There are no horns. There are four toes, but only two reach the ground. The brain is small. They mostly live in marshy, forest conditions and are omnivorous, digging with the long snout for food detected by smell. The neck muscles are very large. Pigs live in families or small troops. A large number of young are produced and the male produces a great volume of semen. *Sus* is mainly an Old World genus, found from the Miocene onwards. It was represented in Great Britain by the wild boar, common until the sixteenth cen-

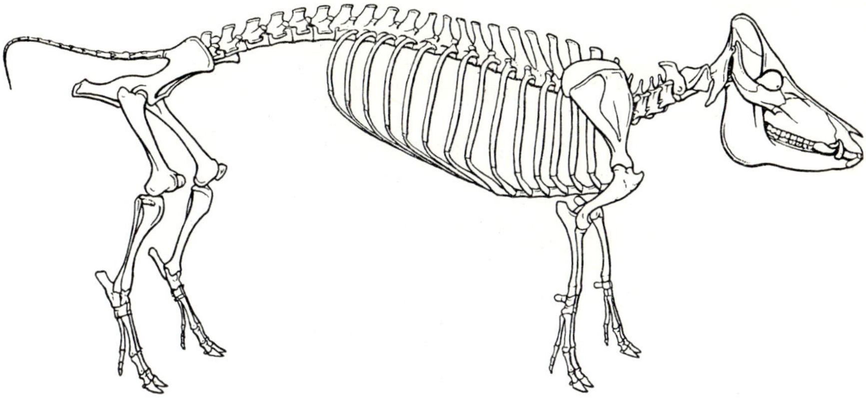

FIG. 494. Skeleton of the pig. (Modified after Ellenberger from Sisson and Grossman, *Anatomy of Domestic Animals*, 3rd edition 1938, published by W. B. Saunders Company, Philadelphia.)

FIG. 495. The wart-hog, *Phacochoerus*. (From photographs.)

tury. The African wart-hog (*Phacochoerus*) (Fig. 495) is not very dissimilar. *Potamochoerus* is the red river-hog of Africa. The peccaries of Central and South America are similar to the pigs, but have been distinct since the Oligocene. There are two small lateral digits in the fore-foot and one in the hind. A large scent-gland opening on the back resembles a second navel, hence the name, *Dicotyles*.

An offshoot of the palaeodont line in the Eocene led to the development of a race of large amphibious animals, the anthracotheres and hippopotamuses (Figs. 496–7), classed together as an infraorder Ancodonta of the Suiformes. They have an enormous barrel-like thorax, with large lungs, short, relatively thin legs with four digits, and a complete dentition with low-crowned bunodont molars, wearing to a foliage pattern. The stomach is enormous and partly divided.

FIG. 496. Hippopotamus, *Hippopotamus*. (From photographs.)

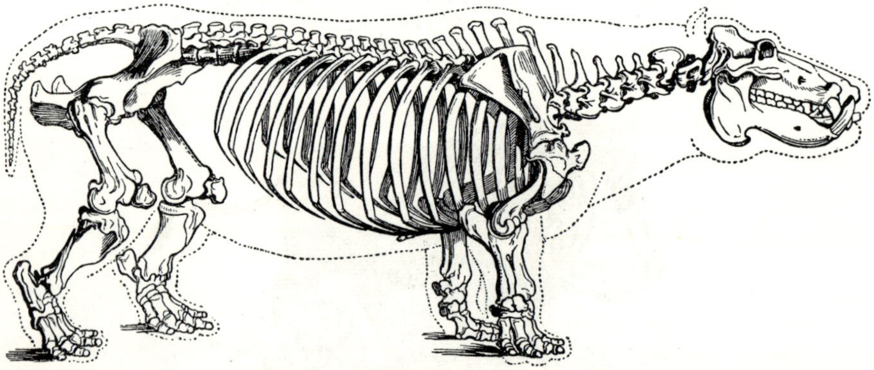

FIG. 497. Skeleton of the hippopotamus. (From Owen, *The Anatomy of Vertebrates*, Longmans, Green & Co.)

There are many specializations for life in the water, including eyes, ears, and nose on the top of the head, muscles for closing the nostrils, and a broad muzzle. They can remain submerged for five minutes. Modern hippopotamuses are found only in Africa, but they were widespread throughout the Old World until recent times.

5. *Oreodonts

The oreodonts were abundant and successful herbivores, living in North America from the Eocene to the early Pliocene (Fig. 493). They had long bodies and short legs and perhaps somewhat resembled pigs. There were four functional digits in each foot, and a complete tooth row, including molars whose cusps were selenodont and in some later forms quite high-crowned. Although Eocene intermediate forms have not been found, it may be presumed that the oreodonts arose from a basal palaeodont ancestor. They pursued an independent evolution in North America parallel in some ways to that of the ruminants in the Old World. Unlike the latter, they were at a disadvantage in the changed conditions of the Pliocene and then died out.

FIG. 498. Llama, *Lama*. (From photographs.)

**Agriochoerus* and its relatives were oreodonts that acquired claws and therefore represent a parallel to ancylopod perissodactyls, with which they were for a long time confused. It has been guessed by some that the claws were used for digging roots, by others that they were for climbing.

6. Camels

All other artiodactyls chew the cud and are often included in a single group Ruminantia. However, the camels have been a separate stock since the Eocene and are so distinct from the remainder that it is convenient to keep them in a separate suborder Tylopoda. They have been common animals since the late Eocene, flourishing especially in North America, although, like the horses, they died out there very recently and survive today only as remnants, which migrated from North America in the Pleistocene, the camels to the Old World and the llamas to South America. The bactrian camel of the Gobi Desert in central Asia is perhaps a wild form, all others being commensals of man. The llamas (Fig. 498), though similar in basic structure to the camels, differ in the smaller size, long hair, and lack of hump. When wild they are mainly mountain-dwellers. The spitting of the llama is a protective device, the whole contents of the stomach being thrown at the attacker!

The Tylopoda show some features retained from the Eocene condition, some developments parallel to those found in the Ruminantia, and various special features of their own, the latter mainly in characters that suit them for life in sandy desert conditions. In the limbs

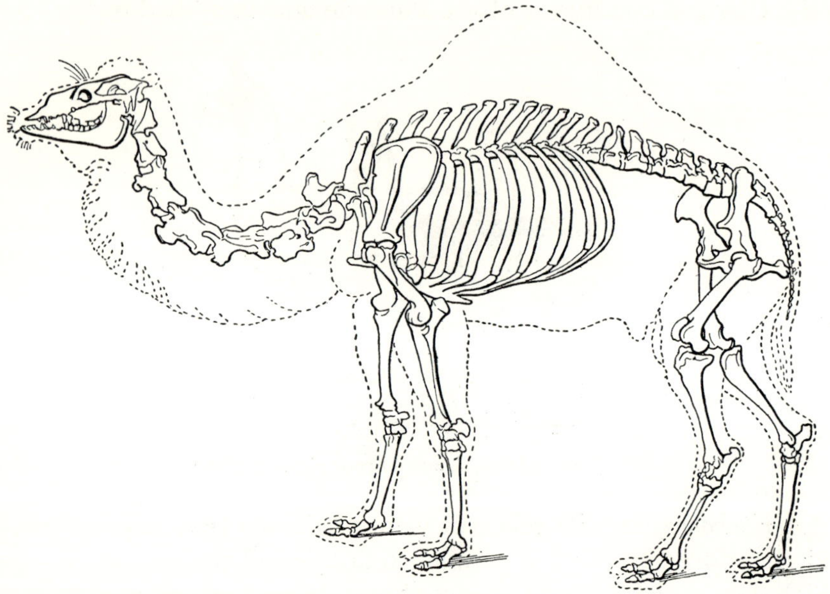

FIG. 499. Skeleton of dromedary. (From Owen, *The Anatomy of Vertebrates*, Longmans, Green & Co.)

(Fig. 499) there has been complete loss of the lateral digits and of some carpals and tarsals, but not the fusion of navicular and cuboid that is so characteristic of the Pecora. A specialized feature is the loss of the hoofs. They were present in early camels but are replaced in modern forms by a nail and large pad. The toes thus spread sideways and enable the animals to walk on soft or sandy ground. The large hump of fat on the back provides water as well as calories when metabolized. The ruminating mechanism is different from that of the Pecora and simpler (Fig. 491).

The wall of the rumen contains a number of pockets separated by muscular walls. These are usually called water pockets and have been supposed to have a storage function, with sphincters. However their walls are glandular and their function may be digestive. There is no external separation of omasum and abomasum, which form a single tubular organ with glandular lining. These differences suggest that the ruminating habit may have been evolved separately in camels and Ruminantia (Bohlken, 1960).

In the head there are many features of similarity to the Pecora. Cropping is by means of procumbent lower incisors, working against specialized premaxillary gums; but an upper incisor and canine are still present (three incisors in the young). The molars have a typical selenodont pattern, and the structure of the skull shows the developments so commonly seen as a result of herbivorous life, such as closure of the post-orbital bar. The lips and tongue are tough and able to chew spiny desert plants. A peculiar feature of camels is that the red blood corpuscles are oval, as in no other mammal. The placenta is of a diffuse (non-cotyledonary) syndesmochorial type.

There is no doubt that many of these features have developed independently in camels and Pecora; the Eocene camel-ancestors did not show them. At the stage of **Protylopus* the camels were small and had short limbs, with separate radius and ulna and four digits in the manus. Throughout the Oligocene many primitive features still remained. **Poebrotherium* was about 3 ft high, with a complete dentition, and orbit only partly closed behind. However, the lateral toes had been lost and the digits began to diverge distally, though probably still carrying hoofs. The remaining increase of size, and the development of other special camel features can be traced slowly through such types as **Procamelus* of the Miocene and Pliocene. During the dry Miocene times the type was very successful in North America and developed various lines, such as **Alticamelus*, the giraffe-camels, with long necks.

7. Ruminants

The most successful modern artiodactyls, the deer (Cervidae) and cattle, sheep, and antelopes (Bovidae), have flourished only since the Miocene and are thus the most recent ungulate group, largely replacing the tylopods, oreodonts, perissodactyls, and still earlier protungulate and paenungulate types. They have always been mainly an Old World group and this remains their headquarters, though some have reached other parts of the world. The early ancestors of the ruminants can be traced to Oligocene and late Eocene forms very like the early camels, oreodonts, and other primitive artiodactyls; the modern chevrotains (*Tragulus*) retain some of the features of this early stage.

The characteristic features of the ruminants are the full development of the feeding system described on p. 743, with loss of the upper incisors and often also of the canines, development of selenodont molars, and of a four-chambered stomach. In the legs only two digits are functional, though traces of others may be found. Besides loss of the extra carpal and tarsal bones there is fusion of those that remain and in particular of the navicular and cuboid. Protection is afforded mainly by swift running and keen senses, but in nearly all ruminants also by antlers or horns on the head.

This definition will not quite cover all members of the groups. The chevrotains keep so many primitive features that although they are

FIG. 500. Chevrotain, *Tragulus*. (After Beddard, *Cambridge Nat. Hist.*, Macmillan & Co.)

certainly related to the ancestors of the other ruminants it would be almost as easy to class them with the camels. This is another case where it is difficult to decide whether to make horizontal or vertical classificatory divisions; any system is bound to be arbitrary.

8. Chevrotains

The deerlets or mouse-deer of Africa and Asia (Fig. 500) are peculiar little creatures, only a foot high, with more external resemblance to a rodent than to modern deer. In some features they show suggestive similarity to the pigs. There are no horns, but the upper canines are large and tusk-like. The upper incisors have been lost and the molars are selenodont, but the stomach has only three chambers. The feet have four, hoofed digits in each limb, and although the two main metatarsals are fused to form a cannon bone the metacarpals are still partly separate. The navicular, cuboid, and external cuneiform make a single bone, this being a 'diagnostic' ruminant character. The placenta is diffuse, as in camels. The individuals live alone in the forests, pairing only for breeding.

In many of these characteristics the chevrotains show signs of retention of an ancient organization, and fossil forms similar to them are found in the Pliocene. Animals not very different (**Archaeomeryx*) are found back to the Eocene and were then like the early camels or palaeodonts; they may well be close to the ancestry of all ruminants. Several similar types and lines can be recognized. Evidently the group represents a population persisting with rather little change since the Eocene, and this status is represented by recognizing an infraorder Tragulina of the Ruminantia, contrasting with the Pecora, which includes all the higher ruminants.

9. Pecora

The true ruminants show the full development of artiodactyl characteristics. They have been an actively expanding group since the Miocene and are now the most successful of the ungulates, existing in vast herds in Africa and to a lesser extent in Asia and North America, though strangely enough hardly penetrating to South America. The modern and fossil forms are of three types. The deer (Cervidae), browsing creatures with bony deciduous antlers, are closest to the central stock, from which have been derived on the one hand the giraffes and on the other the great groups of grazing ruminants or Bovidae, including the primitive prong-buck and the host of sheep, goats, cattle, and antelopes.

FIG. 501. Musk-deer, *Moschus*. (After Beddard, *Cambridge Natural History*, Macmillan & Co.)

10. Cervidae

The ancestral population from which these forms arose must have been quite similar to the Eocene traguline **Archaeomeryx*, but the group does not become distinctly recognizable until the Oligocene. The early members either had large canines and no antlers, as in **Blastomeryx* of the Miocene of North America, or, as in **Palaeomeryx* of the Miocene of Europe, they possessed a bony outgrowth covered with skin and not shed. *Moschus*, the musk-deer (Fig. 501) of central Asia, also has large permanently growing canines and no antlers and is probably a survivor of this Miocene stage of evolution. It is intermediate in many respects between *Tragulus* and the Cervidae and some classify it with the former. Musk-deer are about 2 ft high and the individuals live alone in mountain forests. The much-

FIG. 502. Growth of the antlers of a mule deer. The bony outgrowth is covered with very vascular skin (velvet) which is shed when growth is complete, shortly before the rut. The antlers are then shed. (From *American Mammals*, by W. J. Hamilton, McGraw Hill Book Company.)

valued musk of the male is a pre-putial gland, in the form of a sac. The true deer, Cervidae, have developed antlers in the males (Fig. 502), bony growths shed each year and forming progressively more branches as the animal grows older (Fig. 503). In the reindeer the females also have antlers. A sign of the rather primitive nature of the deer is the retention of definite rudiments of the first two phalanges of the lateral digits. The molars are brachydont, but the placenta cotyledonary as in bovidae. The deer (Fig. 504) have been common since the Pliocene, as browsing animals of the forests of the Holarctic region and South America, but not Africa. They live in herds with an elaborate social organization, based on the supremacy of a leading male, maintained by a succession of 'fights' with his rivals. These fights are very fierce, but do not necessarily result in death, and indeed the complicated horns interlock in such a way as to mitigate their danger to the challenger. Red deer (*Cervus elaphus*) are still wild in Britain in Scotland, the Lake District, Exmoor, and the New Forest. The antlers have six or more points. Roe deer (*Caproelus*) have smaller antlers (three points). They are also indigenous in Great Britain; fallow deer (*Dama*) have been introduced, and are usually spotted, with palmate antlers.

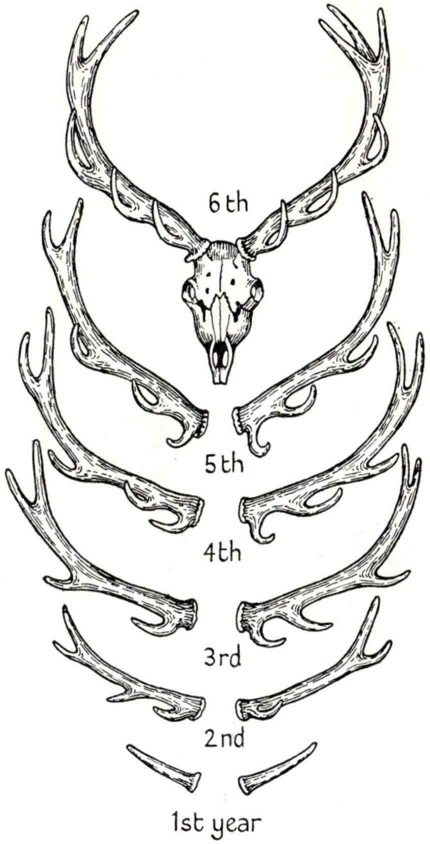

FIG. 503. Series of antlers in the British Museum (Natural History) showing the increasing number of tines in successive years. (After Romanes.)

11. Giraffidae

The giraffes (Fig. 505), like the Cervidae, from which they diverged in the Miocene, are browsing animals, now restricted to tropical Africa. The teeth are low-crowned and the head bears up to five simple skin-covered bony prongs in both sexes. This has been held to be a condition similar to that of Miocene cervids, whereas others believe that the bony core is of dermal origin, as in bovids, fusing later with the frontal. In the okapi there is a rudimentary keratinous horn at the tip. The lateral digits are completely absent and the legs are very long; the whole structure is specialized to carry the great bulk on the fore-legs, the head and neck balancing the weight of the body and the hind-legs being used mainly for propulsion. In walking the fore- and hind-legs of one side move together; since the weight is balanced on the fore-legs there is no use of the tripodal method of

FIG. 504. Deer, *Cervus*. (From life.)

movement that is usual in quadrupeds. This is an extreme development of the type of vertebral organization in which the weight-carrying beam ends in the middle of the back, there being a small number of ribs, so that the hinder part of the column functions as an upper segment of the hind limbs and the extensor muscles of the back aid in propulsion. The long neck makes it possible to balance the great weight on the fore-legs, and is of use not only for reaching high branches but also as a look-out among the long grasses. The rare *Okapia* (Fig. 506), discovered in 1900 in the Belgian Congo, is a form with shorter legs and neck, very similar to **Palaeotragus* and other Pliocene animals, all possessing small horns. Other lines (**Sivatherium*) acquired a pair of large non-deciduous horns, a course of evolution

FIG. 505. Giraffe (*Giraffa*). (From photographs.)

FIG. 506. Okapi, *Okapia*. (From photographs.)

parallel to that found in Cervidae. The exact origin and affinities of the family remain uncertain.

12. Antilocapridae and Bovidae

The remaining Pecora are all rather alike and are often placed in a single family, Bovidae. However, the prong-buck, *Antilocapra* (Fig. 507) of the North American west, and its numerous fossil allies, all New World forms, have been distinct since before the Miocene from the true bovids, evolving in the Old World. The origin of the two groups is obscure and we have no Eocene or Oligocene fossils that are certainly on the bovid line of evolution. As already mentioned **Archaeomeryx* and other Eocene tragulines show us a type of population from which the Pecora could all have been evolved, but the stages of the transformation have not been found.

FIG. 507. Prong-buck, *Antilocapra*. (From a photograph.)

Antilocaprids and bovids are alike in living in herds and in their grazing habits, with which are associated deeply hypsodont molars. The side toes have been almost or completely lost, a development occurring parallel to that of the cervids, since the common ancestry almost certainly possessed lateral toes, which are indeed present in rudimentary form in some bovids. In *Antilocapra* (Fig. 507) the horns, present only in the males, are two-branched and have a bony core and rather soft keratinous covering, the latter but not the former being shed. This therefore suggests how a skin-covered antler, such as that of the Cervidae, may have become converted into the bovid horn. In earlier antilocaprids, such as the Miocene **Merycodus*, the horns were more elaborately branched; evidently the group has developed a horn structure parallel to that of the Cervidae. In all true bovids the horns are permanent coverings for the bony core. They are unbranched, though curved and twisted in various geometrically interesting ways. Moreover, they are usually borne in both sexes (though often larger in the male) and their function is definitely defensive, as well as social and sexual. Correspondingly the social organization is often into large herds, rather than into the small family groups under a dominant stag, such as are found among Cervidae. Grazing on open plains and mountains has presumably led to the formation of the larger herds, smaller and more closely knit groups being more suitable for forest life. The placenta is cotyledonary.

The Bovidae, with more than 100 genera, is much the largest ungulate family. The original centre of evolution of the family was in Eurasia, where they are now less common, whereas in Africa they are particularly successful at present. A few types, such as the bison, reached North America, but none entered South America until man showed that they can flourish there, and indeed also in Australia. The fact that we possess numerous fossil remains and that the group is still at the height of its development makes classification very difficult. This is the situation that we should expect, remembering that evolution consists in the slow change of the characteristics of populations. At first thought it may seem paradoxical that in a group so recently evolved and of which we know so much it should be exceptionally difficult to trace affinities and lines of descent. The fact is that the numerous remains of fossil bovids from the Pliocene and Miocene are still quite insufficient to enable us to reconstruct the changes in the populations. It is not really to be expected that the relatively few specimens of these large animals that can be collected and studied should show us the detailed changes, extending over 20 million years or more, by which a population of perhaps a million small creatures such as **Eotragus* of the Miocene, developed into the present bovid population of, say, a thousand million animals, divisible into hundreds of non-interbreeding populations that range in structure and habits from the gazelle to the bison. An imaginative look at the details of evolutionary change reveals a terrifyingly complicated system, which we can hardly hope to follow in detail. The geological information can surely never be sufficient to show us the necessary facts about the variation of such great populations, and their gradual changes, at least in the case of animals as large and rarely preserved as Bovidae. We know hardly anything about variation and heredity in our own cattle, so how can we hope to follow the genetics of their ancestors? Yet nothing less than a full view of the gradual population changes will show us how the evolution of a group has proceeded.

Following types of organization over long geological periods gives a deceptively simplified idea of the stages traversed. We recognize the 'stages' because, fortunately for us, only tiny remnants of the populations have been preserved, and perhaps some 'primitive' types remain to the present day. Thus in the long history of the perissodactyls we can refer all our modern and fossil forms to some 160 genera; the tapirs are there to show us a very ancient condition, and we know just enough fossil horses to arrange them in a number of series with side branches, so that we feel that we can imagine the whole evolution of the group. It is interesting that the sequence of evolution used as a type-specimen for students is so often that of the horses and not of the bovids, although we have very much more material for the latter, at least in the later stages. It would be wise to study the two together and to learn from the difficulty of recognizing clear-cut boundaries among the millions in the herds of intergrading sheep, goats, oxen, and antelopes that the most important result of the discovery of evolutionary change was the realization that our logic and use of words can no longer depend, as the ancients thought and many backward-lookers still wish today, on the recognition of a certain number of 'species', to one of which every individual can be referred.

FIG. 508. Impala antelope, *Aepyceros*. (From life.)

In trying to classify the Bovidae we may perhaps recognize a central group of 'antelopes', but the term is vague and certainly includes several diverse lines; it is not even possible to find criteria for saying 'this is an antelope, that a cow, and this other a sheep'. 'Typical' antelopes (Fig. 508) live in Eurasia and Africa, especially the latter. They are rather tall and slender, with smooth hair and backward-curving horns, living mostly on warm or tropical plains. The gazelles may be taken as an example among many. The oxen are heavier animals, often almost of graviportal structure, but with very high thoracic spines; they live in cooler conditions on more northern plains and have more and shaggier hair. Their horns curve forwards and are not twisted. They originated in Eurasia. The domestic cattle and yaks, *Bos* (Fig. 509), are good examples, and the *Bison* (Fig. 510) are related creatures, now almost restricted to North America. There are animals, however, that, with the criteria used, cannot be classed as either 'antelopes' or 'oxen'; for instance, the elands (*Taurotragus*) (Fig. 511) of Africa are large and cow-like, but have backwardly directed and twisted horns. Similarly the ovine (sheep and goat) section of the

FIG. 509. Central Asian Yak, *Bos*. (From photographs.)

FIG. 510. American bison, *Bison*. (From photographs.)

FIG. 511. Eland, *Taurotragus*. (From photographs.)

family is not clearly marked off from the antelopine. The goats (*Capra*) are characteristically mountain-living animals of the Holarctic region, with backwardly curved but not twisted horns, and the sheep, *Ovis* (Fig. 512), are closely related, but addicted to less mountainous country and with a slight spiral on the horns. The beard and scent glands of the male are signs that there are great differences in social and sexual organization between sheep and goat life, in spite of the similarity in structure.

Thus there is very great variety of life and structure in the modern bovids, and one of the most striking conclusions about the group is that the specialists have not yet succeeded in finding an agreed system of classification. More important perhaps than these problems is the fact that with their fine cropping mechanism, grinding battery and stomach the bovids provide the only satisfactory intermediary by which grass can be used as a contributor to human life. This, with their peaceful and gregarious disposition, has made them our most important commensal. If there were no Bovidae there would be fewer human beings in the world, and our social organization would be very different. The obverse is also true; it is probable that their lives and ours will continue to evolve together and mutually to modify each other.

FIG. 512. Barbary sheep, *Ovis*. (From life.)

XXXII

CONCLUSION. EVOLUTIONARY CHANGES OF THE LIFE OF VERTEBRATES

1. The life of the earliest chordates

WE set out to try to define the features that are characteristic of vertebrate life, hoping then to show how these features have changed during evolutionary history. We may now summarize the evidence collected and see how far it is possible to make general statements about vertebrate life and the factors that change it.

The vertebrate type of organization has proved capable of supporting life under a wide variety of circumstances; most of its modern forms operate under conditions very different from those in which the type first appeared. According to the most probable theory (p. 47) chordate life began at the sea surface, as the ciliated larvae of some creatures rather like sessile echinoderms. The first fish-like animals, with the characteristic chordate organization, appeared when such larvae acquired powers of rhythmic metachronal muscular movement, in order to allow support of a large weight. We can still see approximately this stage today in amphioxus. It is not possible to summarize the nature of this organization in any brief general statement. The science of morphology is still young and ill equipped with general principles; it does not allow us to define the varieties of living organization with precision and completeness. At present we cannot give a full description such as we might wish for, specifying the composition and activities of an organism or the inherited code of instructions under which it operates. We can only describe some of the methods by which the system maintains itself, for example its means of nutrition, respiration, and reproduction.

The earliest chordates showed a rather low level of metazoan organization, with a relatively small number of distinct cell types and few special organs. Nitrogen and other raw materials were obtained in the form of minute plants, collected by ciliary action of the pharynx and gill-slits. The food was broken down by a system of enzymes working in alkaline solution, and absorbed through the walls of a simple intestine. There was probably no specialization of cells of the walls of the gut to produce enzymes or to perform particular operations of conversion or storage; at least no special liver, pancreas, or other organs were present for these purposes. There were no special respiratory surfaces and the oxygen was carried to the tissues in simple solution in colourless blood. The circulatory system perhaps at first involved little more than an irregular set of spaces among the cells, but quite early there must have appeared the distinct contractile vessels, containing a blood with composition distinct from that of the surrounding lymphatic or tissue spaces. The method of excretion of the earliest chordates is not clearly known; it perhaps involved no highly specialized cells, but occurred all over the body surface. Since the animals were marine there were no serious osmotic problems. Movement was by the metachronal contraction of a series of blocks of longitudinally arranged muscle-fibres and this serial repetition of the muscles and their attendant nerves and blood-vessels has left a large mark on the chordate plan of life.

The nervous organization was at first based on a system of nerve-cells and fibres lying spread out below the epidermis, but then became concentrated dorsally in the walls of a neural tube. The special receptor organs were probably simple and lay either in the skin or within the tube, perhaps along its whole length, with little concentration at the front end and no definite anterior enlargement or brain. The system functioned as a series of more or less separate reflex arcs, activation coming from the stimulation of receptor organs by changes in the world around. There were no large masses of nervous tissue and little possibility of sustained independent action by the creatures, which probably showed little flexibility of behaviour or variation of action with experience. The only endocrine influences were the effects of cell by-products on neighbouring tissues; there were no specialized glands of internal secretion.

Reproduction was presumably sexual (perhaps also by budding) and development followed the pattern of radial (indeterminate) cleavage and gastrulation by invagination, with chordo-mesoderm separating from the endoderm. The young were provided with yolk for their development, but were probably not otherwise cared for by their parents.

This gives a rough picture of chordate organization in the Cambrian period when it probably first arose, 500 million years ago, after the paedomorphic change by which a previously larval creature became sexually mature. It was an organization that had already proceeded far from the aggregation of similar cells that presumably characterized the first metazoans. Its embryological processes were already sufficiently elaborate to produce a creature with well-marked organ systems, though these did not have the numerous cell types and anatomically separate parts that are found later.

2. Comparison of the life of early chordates with that of mammals

Such an early chordate is immensely complicated when considered as a chemical system, yet it lacks the specializations that later became so characteristic of vertebrates. The difference appears very clearly if we contrast the organization and life of some such simple, amphioxus-like chordate with those of a mammal. In almost every part of the body of the latter we find cell types and organs that are not yet differentiated in the former. For illustration of this difference we can look at almost any tissue, say the skin, the blood, the gut, or the brain. In a mammal the skin contains far more types of cell than are present in amphioxus; there are hairs and these are different in various parts of the body; there are several types of gland and of receptor organ. The blood, again, circulates with great rapidity and in two circuits; it is delicately adjusted in composition so as to allow rapid flow of materials to the tissues; it contains haemoglobin in special corpuscles. In addition there are numerous types of cell able to be used for defence, and a system of antibodies for the same purpose that is almost certainly also far beyond anything found in the earlier creature. Digestion in a mammal involves an elaborate arrangement of mouth, oesophagus, stomach, and intestine, each with a controlled pH and special masses of cells aggregated into groups, such as the salivary glands, pancreas, and liver, the latter a most elaborate chemical workshop. Finally the nervous system possesses an enormous number of cells and elaborate receptor organs. It gives the power to react to many aspects of environmental change that cannot be discriminated by the simpler organism. Nervous conduction is rapid, allowing these large creatures to be well coordinated. The nervous system, working through the many contractile parts that are provided by the muscular system, enables the performance of numerous elaborate acts, helpful in obtaining food, escaping enemies, and perhaps particularly in providing for the care of young, which is another characteristic mammalian feature. The pattern of behaviour does not always follow one single course, but is adaptable and suited to the conditions that are likely in view of past experience to be encountered. There is an elaborate system of chemical signalling by many endocrine glands.

No doubt this greater complexity found in mammalian organs reflects a similar complication of the metabolic processes throughout the body, though as yet we have little information about this. Moreover, an organism with so many diverse parts presumably depends for its propagation on a genetical system that is very elaborate. There is evidence that the genetic mechanism is more complicated in the more elaborately organized later animals. In reptiles, birds, and mammals each individual has a genetic constitution so specific that a piece of tissue grafted from one individual to another of the same species (a homograft) nearly always sets up an immunity reaction and is ultimately destroyed. However, in urodeles such homografts are successful, presumably because the genetic mechanism is less specific.

3. The increasing complexity and variety of vertebrates

The above comparison is not intended to be a complete analysis of the organization of early and late chordates, but only an indication that the difference between the two is in the greater number of diverse parts and actions found in the later type. At every stage of the life-cycle there are more alternative possible actions available and better methods for selecting the appropriate one. In other words in higher organisms more information passes through the system. It was suggested in the first chapter that this greater complexity of the higher animals enables their life to be carried on under conditions that would have been impossible for the simpler ancestors. The survey of the evolution of chordates has certainly shown that since the Cambrian the chordate organization has invaded situations very different from the sea surface in which it probably arose. It would not be profitable now to recapitulate all the stages of this change—they have already been described throughout the book. If we consider ecological niches in detail the number of fresh situations invaded by vertebrate life is almost as great as that of the species in the group. Among the earlier changes were the transfer from the sea surface to other waters and to the sea bottom. The entrance into fresh water must have called into play many special mechanisms of adjustment. Development of jaws, perhaps 350 million years ago, probably from the anterior branchial arches, gave the possibility not only of eating new types of food (including fellow fishes) but also of performing simple acts of 'handling' of the environment. The heavy armour of the early types was given up and the body form was then greatly improved from a hydrodynamic point of view and with development of the air-bladder into a hydrostatic organ the fishes achieved their full mastery of the water.

Meanwhile other fishes left the water, probably in the Devonian period, rather less than 300 million years ago. At first they operated with little modification of the method of life they had used in the water, but they later developed all sorts of devices to meet the new conditions, the earlier types dying out as the later developed. This process has been going on ever since, to produce the modern amphibia, reptiles, birds, and mammals, inhabiting a great variety of situations.

In the water tetrapods are found at all levels, including great depths and in perpetually dark caverns. They live in the most varied situations on the land and also by burrowing beneath its surface. Not a few are able to move in the air, some even to feed there. It is hardly possible to overestimate the great variety of vertebrate life; at each new examination of any phase one is amazed at the extraordinary number of special modes of life that are adopted by variants of each type.

There are no sure means of telling the number of types or of individuals constituting the biomass that was present in past times, but it is probable that by means of the above special devices the vertebrate stock has increased and colonized new regions, though not perhaps continuously or at a uniform rate. It is not unlikely that today there are more and more varied vertebrates than at any previous period. It has been pointed out that the number of species described from deposits tends to increase geometrically with time (Caillaux, 1950). This is not an artefact due to poor preservation.

Moreover, as Lotka has pointed out, the total energy flux through the system has probably also been enlarged. It would not be easy to demonstrate these conclusions rigorously with our present knowledge: it is difficult to believe that they are true of all populations. We should return from these speculations to reconsider the nature of the evidence about evolutionary change, to discover the changes that we are sure have taken place since the vertebrate organization first appeared.

4. The variety of evidence of evolutionary change

At various points throughout the book attention has been called to the conclusions that the evidence allows us to draw, and it is important to notice that they vary considerably from group to group within the chordates. For example, we can draw from morphology some conclusions about the changes that produced the original fish-like vertebrate, but these conclusions are unsupported by fossil evidence. The fossils available for study of evolution of the earliest gnathostomes are too few to allow us a detailed view of the change from jawless ostracoderm to placoderms with jaws, and from these to more modern fish. At the other end of the scale, there are so many fossil elephants' teeth to be studied that only an obscure picture of parallel lines of evolution has emerged. Again, in some groups, for instance birds, palaeontology is only of limited help in studying evolutionary change, but nevertheless we have a considerable knowledge about the course of evolution from study of the existing forms, because the birds are conspicuous, well known, and varied (p. 522).

A proper appraisal of the nature of evolutionary change demands an understanding of the fact that evidence about it comes from very different sources and varies in different animal groups. We may therefore profitably extract from the results of various parts of our study such simple propositions as are strictly justified by the evidence and provide us with a sure foundation of knowledge about the subject.

5. Rate of evolutionary change

It has recently become possible to consider several ways of measuring the rate of evolutionary change. Simpson (1953) distinguishes between measurement of (1) genetic rates, (2) morphological rates, (3) taxonomic rates, and we may add as a possibility (4) rates of change of information flow. Although the first and last of these are biologically the most instructive, they are impossible to measure on a large scale or in extinct populations. Rates of change of linear or other dimensions can be estimated in suitable cases such as horses' teeth (p. 738). Haldane suggests that changes should be considered on a percentage rather than an absolute basis, for instance by considering the time needed for a unit increase in the natural logarithm of a variate or one standard deviation. Change of one s.d. per million years might be called a 'darwin'; the horse tooth change (p. 737) being then at a rate of 40 millidarwins.

The only easily available quantitative data about large groups of animals are the number of taxonomic units (species, genera, &c.) into which they are divided. If it were true that differences between, say, genera, meant the same when used by different authors and in different animal groups then we could measure rates of evolutionary change by the numbers appearing at each taxonomic level. The condition is unfortunately not strictly fulfilled and there are inevitably examples of what has been called 'monographic evolution'. Nevertheless, the definition of a difference as of, say, 'generic' or 'ordinal' rank by a competent worker is in effect a kind of measure of general morphological difference. Indeed in view of the subtle efficiency of the human receptors and brain for this sort of comparison, the measure is perhaps as accurate as could be expected to result from any artificial 'morphometer' that can at present be imagined.

Using taxonomic criteria it is clear that rates of evolution vary in different groups. Thus the living prosimians have changed little since the Eocene while their descendants have gone on to produce the whole range of modern primates. The rate of evolution of horses and chalicotheres has been about the same (0·13 genera per million years) and much faster than that of ammonites (0·05), assuming that 'genus' has a similar meaning in the two cases.

Using taxonomic rates it has been shown that many vertebrate major groups seem to evolve fast at their 'first appearance'. This rapid evolution (tachytely) is presumably the result of moving into a new adaptive zone, which we notice *ex post facto* as the beginning of a higher taxonomic group. The chances of finding these transition types as fossils may be unduly small if evolution is rapid and especially if it occurs in a small population (or a large one divided into small units; S. Wright). Claims to have found the 'centre of origin' of a major group must therefore be looked upon with suspicion. In any case, parallel evolution may carry several lines over the arbitrary line we use to mark a higher taxonomic order; at least five lines of therapsid reptiles crossed to become mammals (p. 545).

This discussion may make it seem unreal to speak of 'origins' of higher taxonomic groups. Indeed, it is probably misleading to look for major 'branches' in what must be a multiple evolutionary 'bush'. Nevertheless there is evidence that rate of change is not constant. Apart from such examples as those already discussed, there are many others. Bats, as Simpson points out, have certainly evolved more slowly since they first got wings and 'broke through' to a new environment than in the period of that change itself.

6. Vertebrates that have evolved slowly

We may accept then the concept of bursts of rapid evolution, followed by slower change. In many lines after the rapid change there is a period over which many genera become extinct. However, a few linger on for times longer than would be expected (bradytely). This general pattern can be seen for fishes in Fig. 513*a*. This phenomenon of bradytely produces phylogenetic relicts, of which there seem to be so many that some general explanation of them is desirable. *Neoceratodus* has a good claim to be considered the 'oldest' living vertebrate; it is very similar to fossils found in the Triassic, nearly 200 million years ago. Even in this case, however, there have been slight changes and the Triassic form is placed in a distinct genus **Ceratodus*. *Latimeria* provides us with an example of survival with little change for nearly 100 million years, as well as the humbling thought that no fossil relatives are known throughout that time. *Heterodontus*, the Port Jackson shark, is another very ancient fish; it is closely similar to fossils found in the Triassic; indeed, all sharks are quite like their Palaeozoic ancestors.

Sphenodon has changed little since Permian times and hardly at all since the Jurassic, perhaps 140 million years ago. Among mammals,

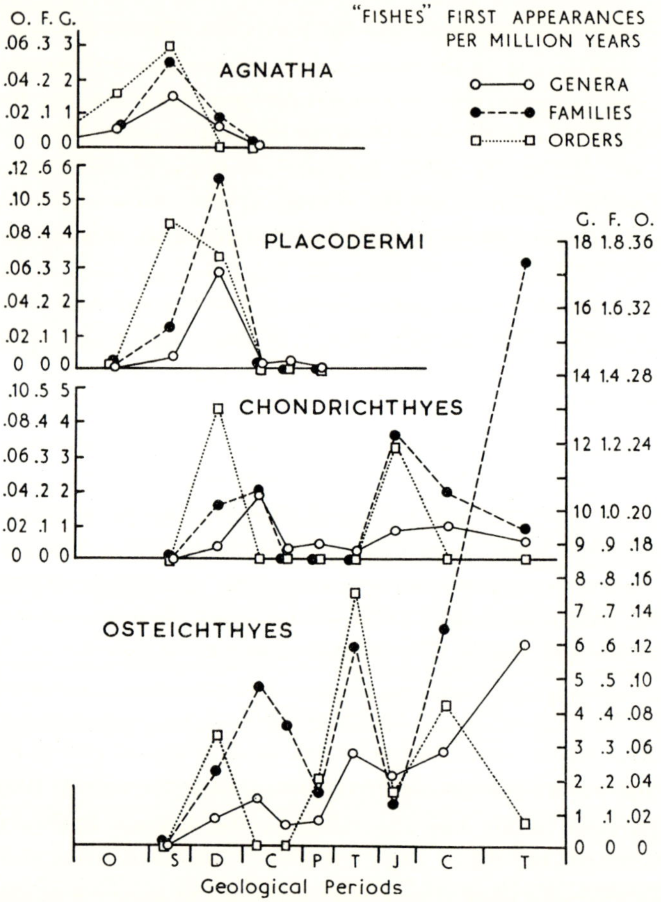

FIG. 513*a*. Graphs showing the first appearances per million years of known orders, families, and genera of the four classes of 'fishes'. *O.* orders; *F.* families; and *G.* genera per million years. Time scale is in geological periods. (After Simpson.)

the opossums and hedgehogs are quite similar to those of Cretaceous times, nearly 100 million years ago, and there are several mammals that have survived with little change for the 50 million years since the Eocene, for instance dogs, pigs, and lemurs. In none of these, however, has a form survived absolutely without change; they are examples of the persistence of a type of organization rather than of superficial details.

It is most interesting to consider possible reasons for these very slowly evolving (bradytelic) populations (Simpson, 1953). (1) It might be low mutation rate or low variability, but there is no evidence that opossums (say) are less variable than other mammals and indeed they have undergone much speciation. (2) It is often implied that survival is assisted by some special habit, such as being nocturnal or abyssal, but others with the same habits evolve fast. (3) Survival sometimes seems to be assisted by isolation (e.g. lemurs) but there is no evidence that this is necessarily a factor (*Latimeria*). (4) Low rate of evolutionary change is not a function of 'primitive' organization as such, indeed as we have seen the reverse is true. In any case *Sphenodon* and *Crocodylus* were not especially 'primitive' when they stopped evolving in the Triassic and Cretaceous. (5) Long survival must depend upon some special relationship between the genetical and information-carrying powers of the species, the risks imposed by the environment, and the stability of the latter. (6) If the adaptive zone is a narrow one it must be stable and persist. This would seem to be unlikely in very 'difficult' habitats such as deserts or impermanent ones (salt lakes) or variable ones, such as Alps. (7) Long survival is perhaps more to be expected in a broad adaptive zone such as the ocean or shore, lowland rivers or forest belts, especially in the tropics. Such environments present, however, many niches that can be considered as 'corridors', leading to diversification, and it is surprising that forms nevertheless remain stable in them. Thus opossum-like creatures gave rise to various offshoots in South America but themselves changed little. (8) Bradytelic populations must be genetically so integrated that any deviation is subject to counter-selection (though in that case it is hard to see how the offshoots have arisen). (9) Simpson concludes that these bradytelic organisms 'have run the whole repertory of baffles and . . . persist indefinitely'. Most organisms are turned off into one or other of the corridors presented by the environment; when a group has met and passed them all it persists.

The discussion of organisms that have evolved only very slowly is thus a stimulus to considering the whole balance of factors by which a population of organisms maintains its homeostasis. Evidently there are some circumstances in which it can do this with little genetic change. In the great majority, however, change of the genes and hence of the structure and physiology is a part of the very mechanism by which the living system continues to survive in spite of changes around it.

7. Varying rates of evolutionary changes

Although gradual modification of living organization is almost universal it is clear that the change is often extremely slow. The transition from an osteolepid fish, say **Sauripterus*, through stages like **Eogyrinus* to **Seymouria* took nearly 90 million years. The change of the horses from **Hyracotherium* to *Equus* is not very profound, considering that it took at least 50 million years. Probably few populations stay the same for long periods, but there may be marked differences in rate of change in groups not otherwise dissimilar. Thus some lines of

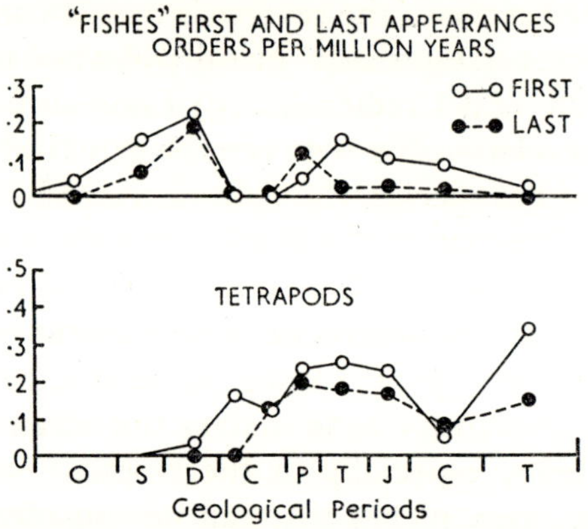

FIG. 513*b*. Graphs showing first and last appearances of orders per million years in 'fishes' and in tetrapods. (After Simpson.)

elephant shortened the lower jaw earlier and faster than others (p. 716). Whereas the majority of elephant populations changed greatly between the Oligocene and the Pliocene, the deinotheres remained almost the same throughout this long period.

Taxonomic methods show that for each class of vertebrates the rate of diversification increases rapidly shortly after the origin of a class and thereafter falls though sometimes showing a second rise (Fig. 513*a*). The maximum rate of formation of new orders precedes that of formation of new genera by 25–50 million years in each case. Rates of first and last appearances follow each other rather closely (Fig. 513*b*) suggesting continuous replacement in the populations.

8. Vertebrates that have disappeared

Set against the few examples of relative constancy of organization there are the wholesale extinctions that we deduce from study of the rocks. The ostracoderms, osteolepids, stegocephalians, dinosaurs, and pterodactyls (to mention only a few) became completely extinct or changed into some very different type of animal. Looking at any of the evolutionary trees in this book, which represent, as it were, the summary of the evidence about the populations, one notices at once how frequently types common in earlier periods become extinct or are replaced by their own descendants. Occasionally some line is recorded as continuing over a long period, but the common picture is of change, each type disappearing after a span of years.

9. Successive replacement among aquatic vertebrates

Does the examination of the sequences of types enable us to say anything about the nature of these evolutionary changes? Can we record any sense in which it represents a 'progress' or 'advance'? One striking feature that we have noticed is that often one type of organism seems to replace another. There is always a difficulty in establishing that this has occurred, because the fossil record does not leave us sufficient information to show for certain that the two types occupied identical 'niches'. However, if we take broad 'habitats', and particularly those that change relatively little, such as the waters, we cannot but be impressed with the succession of tenants that appears, each replacing the one before (p. 237). Thus, among fish-like vertebrates we can recognize ostracoderms, placoderms, crossopterygians, palaeoniscids, holosteans, and teleosteans, each almost completely replacing the one before.

Again, there has been an astonishing series of tetrapods returning from the land to water, developing characters suitable for aquatic life and then becoming extinct, apparently displaced by later migrants, also returning from the land. To name only a few of these returners we have among amphibians the phyllospondyls, lepospondyls, branchiosaurs, and some urodeles; among reptiles the phytosaurs, crocodiles, plesiosaurs, ichthyosaurs, mesosaurs, mosasaurs, aigialosaurs, dolichosaurs, and snakes. Finally of the mammals there are the basilosaurs, modern whales, seals, and sea-cows, as well as some less completely aquatic types.

However much we make allowance for the fact that the sea itself may be changing, it is difficult not to find in these facts a suggestion that the later types are replacing the earlier by their greater 'efficiency'. These returned aquatics are especially interesting because each type when it first re-enters the water seems to be not very well suited to that medium—because of its shape for instance—and would therefore be expected to be at a disadvantage in relation to the 'streamlined' creatures that were already there.

10. Successive replacement among land vertebrates

It is equally easy to trace out successions of types occupying habitats on land, though here it is even more difficult to be sure that the successive animals are occupying identical niches. There has been a long series of large land herbivores, including the labyrinthodonts, pareiasaurs, herbivorous synapsids, various dinosaurs, multituberculates, condylarths, dinocerates, pantodonts, brontotheres, horses, pigs, rhinoceroses, elephants, and artiodactyls. Clearly not all of these lived in similar surroundings (and there were, of course, other herbivores), but the succession is impressive. As with the aquatic animals we have the curious phenomenon that the earlier members of each group seem to be clumsy creatures, no better fitted for their life than those they are replacing. The early mammalian herbivores, with their large limbs and small brains, do not seem greatly superior to the stegosaurs and ceratopsians of the Cretaceous. It is, of course, exceedingly difficult to know enough to settle such questions, for instance to assess the value of warm-bloodedness.

If we look at other ecological niches we see the same picture of continued replacement. Thus there has been a succession of land carnivores, first synapsid reptiles, then archosaurian reptiles, followed in the Tertiary period by the creodonts, which were replaced by modern carnivores and carnivorous birds.

11. Is successive replacement due to climatic change?

A very careful analysis is needed before we can venture to say much about the nature of this successive replacement of types. We have several times noticed how easy and dangerous it is to find superficial 'causes' for evolutionary change. The periods of time involved are so long that a stern discipline is needed to prevent oneself from using analogies that are really only applicable to much shorter periods (p. 572). We have to try to imagine vast communities of animals of various sorts, interacting with each other to produce fluctuating populations, all living in climatic conditions that vary from year to year and also, very slowly, through the centuries. Only if we hold such a picture in mind can we begin to answer questions about whether the stimulus to evolutionary change comes from the changing environment.

It is now very doubtful whether there have been periods of 'revolutionary' geological change (p. 16). Local rises and falls and foldings of the crust have certainly occurred and must have influenced the fauna. Even so it is not certain that the new types appearing after such events, originated during them. They may well have evolved elsewhere and migrated into the area to meet the new conditions left after the 'revolution'.

In the past it has been usual to try to find rather simple correlations of this sort. We are told that emergence of land vertebrates was due to drying up of large areas of sea during the Devonian, that the mammals emerged because of the colder conditions at the end of the Cretaceous, or the horses because of the appearance of wide plains in the Miocene. For the reasons already given we must regard such suggestions with suspicion, especially when they relate to conditions extending over a long period of time. There are, however, certainly some valid correlations of climatic and faunistic changes, especially in relatively recent periods. Thus the finding in Britain of woolly mammoths, cave bears, and other animals to be expected in a cold climate may reasonably be associated with advances of the ice cap (p. 572). Indeed we have evidence of the climatic change independent of the animal remains. No doubt change of climate has been one of the variable factors that has led to the continual change of vertebrate life, which we are seeking to understand. It is probable, however, that animal populations change their character independently of any climatic change. It is not easy to find critical situations to test this belief, but examples such as the faunas of the Galapagos and other islands (p. 524) suggest that diversity can arise as animals explore the possibilities of their environment, especially if there are factors that divide up a population into a number of nearly isolated units.

Often a population undergoes an 'adaptive radiation', branching out to form a number of types, each suited for a particular environment or niche. The phenomenon is so widespread that it suggests a type of evolution common at least to many populations. The conception of adaptive radiation originally put forward by Osborn was that each 'stem form' (of mammals) diverged in five directions, giving cursorial, fossorial, scansorial, volant, and aquatic types. These are, of course, only particular aspects of the radiation. When we examine, say, the Galapagos finches, or the marsupials, we obtain the strong impression that members of a particular animal population seek out a variety of new habitats, and gradually become suited for them, until a range of new types is thus produced. This is the history of each of the groups of vertebrates, they radiate into many different types and then disappear.

12. Convergent and parallel evolution

A remarkable fact that has appeared many times in our survey is that during these radiations similar features repeatedly appear in distinct lines. It is as if the vertebrate organization produced time after time slightly different variations on a series of themes. Thus animals that feed on fishes acquire long jaws and numerous teeth. We have examples of these characteristics among the fishes themselves (garpike, *Belone*)

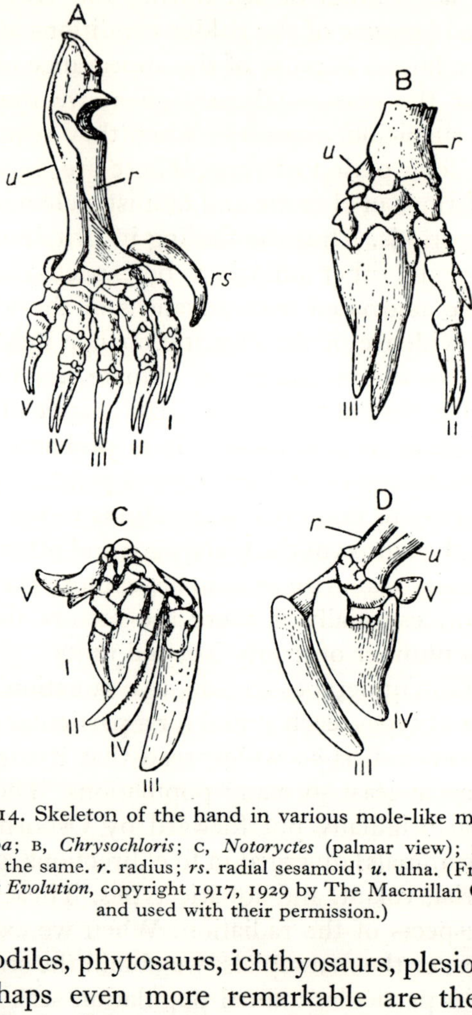

FIG. 514. Skeleton of the hand in various mole-like mammals. A, *Talpa*; B, *Chrysochloris*; C, *Notoryctes* (palmar view); D, dorsal view of the same. *r.* radius; *rs.* radial sesamoid; *u.* ulna. (From Lull, *Organic Evolution*, copyright 1917, 1929 by The Macmillan Company and used with their permission.)

and in the crocodiles, phytosaurs, ichthyosaurs, plesiosaurs, birds, and mammals. Perhaps even more remarkable are the 'duck-bills' of animals that sift small invertebrates from mud. There is a distinct similarity in the structures used for this purpose by *Polyodon*, the ducks themselves, and the platypus. One could continue with endless examples of the same sort, for instance, the large mouth of insect-eating animals—the frogs, swallows, swifts, and bats. The five sorts of ant-eater found among mammals provide another remarkable example of this 'convergence'; all have an elongated snout, long sticky tongue, large salivary glands, and other features. The hands of the moles show how a similar result may be arrived at by various slightly differing means (Fig. 514).

There can be no doubt that vertebrates adopting a given mode of life tend to acquire a particular structure. There is also evidence of what we might call the converse, namely that animals with a particular form tend to develop in certain directions. This is one of the forms of the situation described by some as 'pre-adaptation'. The elasmobranch fishes maintain equilibrium in the horizontal plane by a heterocercal tail, driving the head downwards, and horizontal pectoral fins and flattened front end of the body, having the opposite effect (p. 140). From this type of organization creatures of ray-like type have several times developed, flattened dorso-ventrally and obtaining their propulsive thrust from the pectoral fins. Conversely among teleosts, where the compression is in a transverse plane, the bottom-living types are flattened laterally, as in the sole or plaice. Where the swim-bladder has been lost, however, there may be dorso-ventral flattening.

The more closely we examine the evolution of populations the more signs of these similar tendencies in evolution appear. Parallel evolution is so common that it is almost a rule that detailed study of any group produces a confused taxonomy. Investigators are unable to distinguish populations that are parallel new developments from those truly descended from each other. Examples we have noticed from study of modern populations are the various types of tree-living and burrowing anurans (p. 366). The 'tree-frog' and 'burrowing' conditions have been evolved both from true frogs and from toads, probably several times in each case. Again, the habit of burrowing underground, with loss of the limbs, has appeared in a number of squamate reptiles; the slow-worms, amphisbaenas, and subterranean skinks are certainly distinct lines, possibly each contains more than one, and the snakes are probably derived from another group that went underground.

13. Some tendencies in vertebrate evolution

These are only a few scattered examples, but they already suggest that the evolutionary changes of vertebrate populations follow certain patterns. Although in the history of each type there is no doubt that much is unique, yet most types tend to follow along certain lines, according to the situations they have reached. It is not possible yet for the systematic morphologist to make any complete classification or analysis of these tendencies. He can only point to a few of them, already so familiar as to be almost banal. Thus vertebrates that move in the water tend to have the fish form, with streamlined body, vertebrae with flat articulations and paddle-like limbs. We are so used to this that perhaps its interest is often overlooked, especially the fact that the pectoral limb may revert from an elongated pentadactyl structure to a paddle, sometimes with increased number of digits and of phalanges. Evidently this form of limb is suitable for the uses demanded in the water and tends to be developed again when needed.

Similar examples that can be called in a sense a reversal of evolution are interesting not only from genetic and embryological points of view, but especially because they show strikingly that under given conditions vertebrate populations tend to react alike, giving us the possibility of developing a reasonable science of morphology. Thus vertebrates that take to the air develop light and thin bones, those that burrow underground often lose their limbs (but among mammals these are often the main digging agents). Similar general evolutionary principles could probably be developed for each organ system. Thus the eyes develop cone-like structures in diurnal, rod-like in nocturnal species, and they become buried under the skin or lost altogether in populations that live in the dark.

14. Evolution of the whole organization

For an analysis of the nature of the changes occurring during evolution to be satisfactory it would have to deal not only with the evolution of isolated parts but with that of the whole organization of the activities of the population. In our present ignorance of the morphogenetic processes it is very difficult for us to provide descriptions of the nature of the whole organization that will be satisfactory. We know enough of the underlying hereditary material and the way it produces the chemical action systems and organs of the body to be able to say that there are some genetic factors that affect almost the entire organization, but also that sometimes individual parts change separately. Changes in the activities inherited through the egg and sperm may affect a single process, such as the deposition of pigment in the skin, but usually they influence a wide range of activities and the form of many organs. Conversely each gene is usually influenced in its action by several, perhaps many, others, and the consequent wide range of expression of each character provides the basis of variability on which natural selection can act.

There are indications that during evolution sets of characters tend to evolve together. For instance, the various features of the mammal-like organization appeared several times during the Mesozoic—the teeth, jaw and skull bones, and limbs apparently evolving together. On the other hand differing combinations are also undoubtedly possible; the notoungulate **Thoatherium* developed the limbs but not the teeth of a horse; **Australopithecus* had the legs but not the brain of a man.

Unfortunately we have no very satisfactory techniques for describing the organization and its changes. All populations are homeostatic self-reproducing systems. They are able to remain alive by selecting from their repertoire of possible actions those that ensure survival. This selection may be done either between the genes ('natural selection') or between possible courses of morphogenesis ('functional adaptation') or between possible actions of the neuromuscular system ('behaviour'). We need means for measuring the information flow involved in these selections and the amount that is stored in the memory system of the species.

Change in the genetic system (evolution) is one of the means adopted to ensure homeostasis, and evolution is a feature that is essential for prolonged maintenance of the organization. If storage powers are effective and adequate such a system, as it receives information, must presumably develop a widening repertoire of responses, both morphogenetic and behavioural. The appearance of increasing complexity as organisms become as we say 'more highly evolved' is a measure of the extent to which they have found ways of encoding new information about the environment and so channeling it as to produce responses that keep the species alive in the face of new risks. The higher organisms are thus those that pass the greater amounts of information, using more complex codes. Unfortunately we have little quantitative information about genetic, morphogenetic, or neural codes to support such an assertion.

15. Summary of evidence about evolution of vertebrates

The evidence about the history of vertebrates may be said, then, to show us the following facts: (1) In all or nearly all populations the organization of life-processes changes, though often only slowly. (2) The later types of organization usually replace the earlier in any one environment. (3) Evolutionary change is not always obviously associated with environmental change, though it may be so. (4) When different populations adopt the same habit of life they often develop similar but not identical organizations. Probably no population is stable either in numbers or in genetic or phenotypic characteristics. The evidence suggests that in most species evolution is going on now and all the time. The number of animals in a population is seldom if ever constant. Probably the genetic and phenotypic characters of the members also vary with time, perhaps actually in correlation with the fluctuations of number.

We have, therefore, a picture of an animal species as a set of individuals that are similar but not all identical, interbreeding with each other, though perhaps with degrees of difficulty that vary with genetical and geographical differences (themselves probably correlated). The characteristic features of such a population are given, as we have seen in Chapter I, by its power and ability to produce later populations, both like and unlike itself. Unfortunately, we know little about these powers in vertebrates, or indeed other organisms; the present study has not dealt with them fully. They are presumably influenced by such variables as the frequency of reproduction, number of offspring produced, and the viability of these in the face of various climatic and biotic factors, availability of food, persistence of predators, and so on. It has been pointed out by Haldane that these factors may lead to the development of various distinct sorts of organization. If the death-rate is low the productivity will also be low and those members of the population will be selected whose characters allow for a long life—for example, hypsodont molars will develop. In populations with a high death-rate to predators or disease, however, selection will choose those individuals with high productivity and rapid development, incidentally perhaps also allowing greater variability, by which the predators and pathogens may be avoided.

There must be a complicated relationship between such factors as the frequency of reproduction and numbers of young, rate of growth, time of maturity, size, likelihood of death from predators and pathogens, capacity to 'adapt' during the lifetime and especially to store information in the nervous system. Members of species that breed fast usually also grow fast, learn little, and are killed before they grow old. Where there is a brain with a large memory it usually directs a massive individual and keeps it alive for a long time.

Evidently the particular characteristics of a population (including its productivity) will depend on the influences to which it is subjected. Many of these variables act with an intensity that depends on the activity of the adult organisms and the energy and ingenuity with which these find situations suitable for the life of themselves and their offspring. Since this adult activity is itself influenced by hereditary and by environmental factors it is clear that the productivity and increase of a population depend on a very complicated system of inter-connected variables. J. B. S. Haldane, R. A. Fisher, Sewall Wright, and others have made some progress towards analysing this situation into its elements by mathematical reasoning, but we have at present estimates for only few of the variables involved and no approach to an exact solution is possible.

16. Conservative and radical influences in evolution

In such a system we have factors that tend to keep the population the same and others that tend to make it vary; according to the preponderance of one or the other the characters of the populations will either stay steady about a mean or tend continually to change, that is to say, to evolve. We cannot make any very precise list of the factors in the two classes. Those tending to reduce evolutionary change presumably include (1) The copying or reproductive tendency that makes like produce like. Here we may include the restraints imposed by the fact that development is a highly integrated process, so that any wide departures from normality are liable to interfere with it. (2) Anything that reduces the productivity in terms of number of offspring. (3) Prevalence of predators, if these act so as to prohibit minor deviations from the previous mode of life. (4) Absence of geographical barriers. (5) Existence of stable external climatic and other physical factors. (6) Characteristics within the population tending to keep the animals in their existing conditions, rather than to seek new ones; 'adventurousness' is essential if the individuals are to enter new conditions; it corresponds to the adaptations for dispersal that are found among plants (Salisbury, 1942).

The contrary circumstances, those that encourage evolutionary development, are presumably (1) High rate of mutation, that is to say, failure of the tendency to copy. Spurway has pointed out that the extent to which change in the hereditary and developmental pattern can be varied probably differs in different populations. In some almost any change is lethal and the population appears immutable, whereas other organizations can stand considerable variations, allowing the possibility of rapid evolution. (2) High productivity, if it leads to high intra-specific competition and hence pressure to find new conditions for life. (3) Absence of predators, if this makes new adventures possible. (4) Presence of geographical barriers, which allow separation into various races that do not interbreed (or only seldom do so) and hence become more and more distinct. (5) Change in external conditions. (6) Presence of a high power of adaptability and 'adventurousness' in the various action systems of the animals. (7) Haldane has suggested that the influence of pathogens also puts a premium on variation. The disease organism, being small, can usually evolve more rapidly than the host, and it is therefore an advantage for a member of the host population to be different from the majority, to which the pathogen is adapted.

During the present study there has been little progress towards demonstrating that these are the factors influencing evolution, though there are clear signs that some of them are at work, for instance, geographical isolation. They are treated in detail by such works as those of Huxley (1942) and Simpson (1953) and there properly discussed. These questions are raised here only as a reminder that they will have to be further considered if any satisfactory general evolutionary theory is to emerge and to be applicable to the vertebrates.

17. The direction of evolutionary change

Since we cannot closely specify the factors influencing evolution we can hardly expect to go far towards the solution of the still more difficult problem of the direction of evolutionary change. The facts of importance that have emerged from the evidence about vertebrates are (1) That populations tend to be replaced by others of different form, often themselves descended from the first. (2) That the later populations often have more complicated organizations than the earlier ones; we can hardly be said to have established this last clearly as a rule in vertebrates, but it has repeatedly been suggested that the evidence supports some such thesis. (3) Some later populations invade habitats not previously occupied by vertebrates (e.g. the land).

Facts of this sort have led us repeatedly to the suspicion that the later types are often in some way better able to carry on the self-maintenance of life than their predecessors, and that vertebrate life is continually invading new situations. With this goes the suspicion that the total biomass of vertebrate life (perhaps of all life) has been increasing and the energy flux becoming more rapid, though we have no exact estimates to verify this.

For organizing our knowledge about the evolution of vertebrates or other animals the above three facts about the progress or direction of change are of great value. The invasion of new habitats is of particular interest, because it is usually made possible by the increase of complexity, this provides a system of elaborate adjustments that maintains the internal conditions nearly constant in face of fluctuations in the environment. This internal constancy or homeostasis is, of course, only a development of the general tendency to self-maintenance, which characterizes all living organisms; but it is carried to its extreme in the higher vertebrates, giving them the power to maintain life under varied and unpropitious conditions.

18. The influences controlling evolutionary progress

Our analysis of the factors affecting evolution gives us clues about the influences that have produced this increase of complexity. Excess productivity and intra-specific competition force sections of the population to try new habits. Those with the ability to do so may, if they can tolerate new external conditions, find situations in which they can flourish. So a new type becomes developed, only to meet later the competition of its own descendants or other invaders and hence be driven to extinction or to the colonization of still other fields.

If we have interpreted the situation correctly there can be said to be an evolutionary path that is progressive, in the sense of enabling life to be lived effectively under wider and wider conditions. Lotka has suggested that it is possible to recognize a 'basic principle defining the direction of organic evolution', namely, that the collective effects of organisms 'tend to maximize, on the one hand, the energy intake of organic nature from the sun, and on the other, the outgo of free energy by dissipative processes in living and in decaying dead organisms'. This increasing turn-over of energy is presumably a sign of the development of more and more complicated mechanisms for ensuring homeostasis in spite of external changes. These mechanisms in turn depend upon an increasing store of instructions in the genotype. If this view is correct there is a tendency for organisms to come to represent more and more features of the environment, which is another way of saying that they have more information about it. More simply still we may say that they come to be able to live under ever more 'difficult' conditions, gathering and expending more energy in order to keep alive (Young, 1938). It seems reasonable that this increase of complexity should be progressive; as the organisms acquire more information about the environment they also gain in possibility of acquiring still more. In particular those that develop mechanisms for learning directly with the nervous system will be successful and will evolve fast. It is possible in this way to see the basis for the direction of evolutionary change. Such a formulation is far from exact, however, and the process clearly depends upon many conditions not here specified, for example as to the nature of environmental changes and the effects of interaction between organisms.

During the present study we have not been able to show rigorously that all evolution follows such principles, but the data are not inconsistent with this. It may be that vertebrates have proceeded farther along the path suggested than any other animals. The birds and mammals probably turn over energy faster than other vertebrates. The point that must not be overlooked is that they do this in order to provide the means by which they can remain constant, in spite of fluctuations in the external conditions of an environment that is different in composition from the living system.

We cannot then at present discern with certainty the principles that determine the change of animal form, but we can see something of the influences that have modified the original vertebrate type to produce the great variety of creatures that has existed and remains today. We cannot give tables of numbers to show how the variables have operated to keep *Ceratodus* nearly constant for 200 million years while related descendants have gone on to produce the whole variety of tetrapod life. But we can suggest that it is worth while pursuing the study as a means of building a truly general science of zoology. We begin to see signs of the principles according to which the vast populations of animals interact with each other, with the plants, and with the inorganic world. The changes that these interactions produce seem to be, if not constantly in one direction, at least often such as to allow the appearance of more varied and complicated forms of life, ever more able to maintain themselves constant and apart from the environment, and hence to exist in a wider range of conditions. The effect of this evolutionary change has probably been to increase the amount of material organized into living things, while the total energy flux through the system has also been enlarged.

This finding is not a mere rehabilitation of the complacent anthropocentric prejudices of the nineteenth century. The conclusion that there is a sense in which the mammals and man are among the highest animals should be based on objective analysis of the behaviour of the populations of vertebrates and the flow of information and energy through them. However, it has been repeatedly emphasized that we still have a long way to go before we have the knowledge necessary to understand these very slow changes. It has only been possible in this volume to suggest certain principles that may one day make possible a satisfactory systematic study of the life of vertebrates.

REFERENCES

GENERAL WORKS ON EVOLUTION

BRITISH MUSEUM (Natural History). 1959. *A Handbook on Evolution.* 2nd edition. London.

BURTON, M. 1949. *The Story of Animal Life.* 2 vols. London.

CAIN, A. J. 1954. *Animal Species and their Evolution.* London.

DE BEER, G. R. 1958. *Embryos and Ancestors.* 3rd edition. Oxford.

DOBZHANSKY, T. 1951. *Genetics and the Origin of Species.* 3rd edition. New York.

FISHER, R. A. 1930. *The Genetical Theory of Natural Selection.* Oxford. (Also Re-issue London. Penguin Books.)

FORD, E. B. 1960. *Mendelism and Evolution.* 7th edition. London.

GREGORY, W. K. 1951. *Evolution Emerging.* 2 vols. New York.

HALDANE, J. B. S. 1933. *The Causes of Evolution.* London.

HUXLEY, J. S., HARDY, A. C., and FORD, E. B. 1954. *Evolution as a Process.* London.

HUXLEY, J. S. 1942. *Evolution, the Modern Synthesis.* London.

KIMURA, M. 1961. 'Natural Selection as the Process of Accumulating Genetic Information in Adaptive Evolution.' *Genet. Res.* **2**, 127–40.

LAMARCK, J. B. 1809. *Zoological Philosophy,* translated by H. Elliot, 1914. London.

LOTKA, A. J. 1945. 'The Law of Evolution as a Maximal Principle.' *Hum. Biol.* **17**, 167.

LULL, R. S. 1947. *Organic Evolution.* Revised edition. New York.

MAYR, E., and others. 1953. *Methods and Principles of Systematic Zoology.* New York.

MULLER, H. J., and others. 1947. *Genetics, Medicine, and Man.* Ithaca.

—— 1939. 'Reversibility in Evolution from the Standpoint of Genetics.' *Biol. Rev.* **14**, 261.

OPARIN, A. I. 1959. *The Origin of Life on the Earth.* Proceedings of the first international symposium, Moscow, 1957. London.

RENSCH, B. 1959. *Evolution above the Species Level.* (Translation of 2nd German ed.) London.

SALISBURY, E. J. 1942. *The Reproductive Capacity of Plants.* London.

SCHOENHEIMER, R. 1946. *The Dynamic State of Body Constituents.* 2nd edition. Cambridge, Mass.

SIMPSON, G. G. 1953. *The Major Features of Evolution.* New York.

—— 1961. *Principles of Animal Taxonomy.* New York.

SMITH, J. M. 1958. *The Theory of Evolution.* London. Penguin Books.

THOMPSON, D'ARCY W. 1961. *On Growth and Form.* Abridged edition by J. T. Bonner. Cambridge.

WADDINGTON, C. H. 1956. *Principles of Embryology.* London.

—— 1957. *The Strategy of the Genes.* London.

YOCKEY, H. P., and others. 1958. *Symposium on Information Theory in Biology.* London.

GENERAL WORKS ON VERTEBRATE ZOOLOGY

BALDWIN, E. H. F. 1948. *Introduction to Comparative Biochemistry.* 3rd edition. Cambridge.

BARCROFT, J. 1936. *Features in the Architecture of Physiological Function.* Cambridge.

BLAIR, W. F., and others. 1957. *Vertebrates of the United States.* New York.

BOLK, L., and others. 1931–9. *Handbuch der vergleichenden Anatomie der Wirbeltiere.* 7 vols. Berlin.

BRACHET, A. 1935. *Traité d'embryologie des vertébrés.* 2nd edition. Paris.

COTT, H. B. 1940. *Adaptive Coloration in Animals.* London.

DE BEER, G. R. 1951. *Vertebrate Zoology.* 2nd edition. London.

—— 1937. *The Development of the Vertebrate Skull.* Oxford.

FLOWER, S. S., and others. 1929. *List of Vertebrated Animals Exhibited in the Gardens of the Zoological Society of London. 1828–1927.* 3 vols. London.

GEGENBAUR, C. 1878. *Elements of Comparative Anatomy.* Translated by F. J. Bell and E. R. Lankester. London.

GOODRICH, E. S. 1930. *Studies on the Structure and Development of Vertebrates.* London. (Re-issue Dover Books.)

GRASSÉ, P. P., and others. 1948 onwards. *Traité de Zoologie.* Paris.

HUXLEY, T. H. 1871. *Manual of the Anatomy of Vertebrated Animals.* London.

HYMAN, L. H. 1942. *Comparative Vertebrate Anatomy.* Philadelphia.

IHLE, J. F. W., and others. 1927. *Vergleichende Anatomie der Wirbeltiere.* Berlin.

JOHNSTON, J. B. 1906. *The Nervous System of Vertebrates.* Philadelphia.

KAPPERS, C. U. ARIËNS. 1929. *The Evolution of the Nervous System in Invertebrates, Vertebrates and Man.* Haarlem.

—— HUBER, G. C., and CROSBY, E. 1936. *The Comparative Anatomy of the Nervous System of Vertebrates.* 2 vols. New York.

KÜKENTHAL, W. 1923 onwards. *Handbuch der Zoologie.* Leipzig.

NEAL, H. V., and RAND, H. W. 1936. *Comparative Anatomy.* Philadelphia.

—— —— 1939. *Chordate Anatomy.* London.

OWEN, R. 1866–8. *On the Anatomy of Vertebrates.* 3 vols. London.

PARKER, T. J., and HASWELL, W. A. 1940. *A Text-book of Zoology.* 6th edition, revised by C. Foster Cooper. London.

RANVIER, L. 1878. *Leçons sur l'histologie du système nerveux.* 2 vols. Paris.

REYNOLDS, S. H. 1913. *The Vertebrate Skeleton.* 2nd edition. Cambridge.

ROMER, A. S. 1955. *The Vertebrate Body.* 2nd edition. London.

—— 1959. *The Vertebrate Story.* London.

SAUNDERS, J. T., and MANTON, S. M. 1949. *A Manual of Practical Vertebrate Morphology.* 2nd edition. Oxford.

SHERBORN, C. D. 1902–32. *Index Animalium.* 10 vols. London.

VERTEBRATE LOCOMOTION. 1961. *Symposium Zool. Soc. London.* No. 5. London.

WALLS, G. L. 1942. *The Vertebrate Eye.* Michigan.

WIEDERSHEIM, R. 1886. *Lehrbuch der vergleichenden Anatomie der Wirbelthiere.* 2nd edition. Jena.

—— 1907. *Elements of the Comparative Anatomy of Vertebrates.* Translated by W. N. Parker, 3rd edition. London.

WILLIER, B. H., WEISS, P. A., and HAMBURGER, V. 1955. *Analysis of Development.* Philadelphia.

WINTERSTEIN, H. 1910–25. *Handbuch der vergleichenden Physiologie.* 4 vols. Jena.

YOUNG, J. Z. 1938. 'The Evolution of the Nervous System and of the Relationship of Organism and Environment.' In *Evolution,* essays presented to E. S. Goodrich, Ed. G. R. de Beer. Oxford.

—— 1957. *The Life of Mammals.* Oxford.

GENERAL WORKS ON GEOLOGY AND VERTEBRATE PALAEONTOLOGY

ABEL, O. 1919. *Die Stämme der Wirbeltiere.* Berlin.

EDINGER, T. 1929. 'Die fossilen Gehirne.' *Ergebn. Anat. EntwGesch.* **28,** 1.

—— 1948. 'Evolution of the Horse Brain.' *Mem. geol. Soc. Amer.* **25.**

HENBEST, L. G., and others. 1952. 'Significance of Evolutionary Explosions for Diastrophic Division of Earth History.' *J. Palaeont.* **26**, 299.
HOLMES, A. 1944. *Principles of Physical Geology*. London.
—— 1959. 'A Revised Geological Time Scale.' *Trans. Edinb. geol. Soc.* **17**, 183.
KNOPF, J. 1949. 'Time in Earth History.' In *Genetics, Palaeontology and Evolution*. Ed. by G. L. Jepson and others. Princeton.
ROBERTSON, J. D. 1959. 'The Origin of Vertebrates Marine and Freshwater.' *Rep. Brit. Ass.* **61**, 516.
ROMER, A. S. 1945. *Vertebrate Paleontology*. 2nd edition. Chicago.
SCHUCHERT, C., and DUNBAR, C. O. 1933. *Text-book of Geology*. 3rd edition. New York.
STIRTON, R. A. 1959. *Time, Life and Man*. New York, London.
WESTOLL, T. S. (editor). 1958. *Studies on Fossil Vertebrates*. London.
ZEUNER, F. E. 1958. *Dating the Past*. 4th edition. London.
ZITTEL, K. A. VON. 1925–32. *Text-book of Palaeontology*. Revised by A. Smith Woodward. 3 vols. London.

CHAPTERS II AND III. EARLY CHORDATES

BARRINGTON, E. J. W. 1937. 'The Digestive System of *Amphioxus*.' *Phil. Trans.* B, **228**, 269.
—— 1940. 'Feeding and Digestion in *Glossobalanus*.' *Quart. J. micr. Sci.* **82**, 227.
—— 1958. 'The Localization of Organically Bound Iodine in the Endostyle of *Amphioxus*.' *J. Mar. biol. Ass. U.K.* **37**, 117.
BERRILL, N. J. 1936. 'The Evolution and Classification of Ascidians.' *Phil. Trans.* B, **226**, 43.
—— 1950. 'The Tunicata. With an Account of the British Species.' *Ray Society*. London.
—— 1955. *The Origin of Vertebrates*. Oxford.
BONE, Q. 1959. 'The Central Nervous System in Larval Acraniates.' *Quart. J. micr. Sci.* **100**, 509.
—— 1959. 'Observations upon the Nervous Systems of Pelagic Tunicates.' *Quart. J. micr. Sci.* **100**, 167.
—— 1960. 'A Note on the Innervation of the Integument in Amphioxus and its Bearing on the Mechanism of Cutaneous Sensibility.' *Quart. J. micr. Sci.* **101**, 371.
—— 1961. 'The Organization of the Atrial Nervous System of *Amphioxus* (*Branchiostoma lanceolatum* (Pallas)). *Phil. Trans.* B, **243**, 241.
GARSTANG, S., and GARSTANG, W. 1928. 'On the Development of Botrylloides and its bearing on Some Morphological Problems.' *Quart. J. micr. Sci.* **72**, 1.
GARSTANG, W. 1928. 'The Morphology of Tunicata.' *Quart. J. micr. Sci.* **72**, 51.
GRASSÉ, P. P. 1948. *Traité de zoologie*. Tome XI. Paris.
HARMER, S. F. 1910. 'Hemichordata.' *Cambridge Natural History*. London.
HERDMAN, W. A. 1910. 'Ascidians and Amphioxus.' *Cambridge Natural History*. London.
HORST, C. J. VAN DER. 1927–36. 'Hemichordata.' Bronn's *Klassen und Ordnungen des Tierreichs*, **4**, Abt. 4, Buch 2, Teil 2.
—— 1932. 'Enteropneusta.' In Kükenthal, *Handbuch der Zoologie*, **3**, Abt. 2.
LELE, P. P., PALMER, E., and WEDDELL, G. 1958. 'Innervation of the Integument of Amphioxus.' *Quart. J. micr. Sci.* **99**, 421.
PIETSCHMANN, V. 1929. 'Acrania.' In Kükenthal, *Handbuch der Zoologie*, **6**, Berlin.
WEICHERT, C. K. 1951. *The Anatomy of the Chordates*. New York.
WILLEY, A. 1894. *Amphioxus and the Ancestry of Vertebrates*. London.

CHAPTER IV. AGNATHA

BARRINGTON, E. J. W. 1936. 'Proteolytic Digestion and the Problem of the Pancreas in *Lampetra*.' *Proc. roy. Soc.* B, **121**, 221, also 1942, *J. exp. Biol.* **19**, 45.

GAGE, S. H. 1929. 'Lampreys and Their Ways.' *Sci. Mon. N.Y.* **27**, 401.

HUBBS, C. 1925. 'The Life Cycle and Growth of Lampreys.' *Pap. Mich. Acad. Sci.* **4**, 587.

JOHNELS, A. G. 1956. 'On the Peripheral Autonomic Nervous System of the Trunk Region of *Lampetra planeri*.' *Acta zool. Stockh.* **37**, 251.

JONES, F. R. HARDEN. 1955. 'Photo-kinesis in the Ammocoete Larva of the Brook Lamprey.' *J. exp. Biol.* **32**, 492.

KNOWLES, F. G. W. 1941. 'The Duration of Larval Life in Ammocoetes.' *Proc. zool. Soc. Lond.* A, **111**, 101.

PIETSCHMANN, V. In Kükenthal, *Handbuch der Zoologie*, **6**. Leipzig.

RITCHIE, A. 1960. 'A New Interpretation of *Jamoytius kerwoodi* White.' *Nature, Lond.* **188**, 647.

SCHULTZ, L. P. 1930. 'The Life History of *Lampetra planeri*.' *Occ. Pap. Mus. Zool. Univ. Mich.* **221**, 1.

STENSIÖ, E. 1925. *Downtonian and Devonian Vertebrates of Spitzbergen.* Oslo.

—— 1958. In Grassé, P. P. *Traité de Zoologie.* Tome XIII. Paris.

WHITE, E. I. 1935. 'On the Ostracoderm *Pteraspis*, and the Relationships of the Agnathous Vertebrates.' *Phil. Trans.* B, **225**, 381.

—— 1946. '*Jamoytius kerwoodi*, a New Chordate from the Silurian of Lanarkshire.' *Geol. Mag.* **83**, 89.

CHAPTERS V–X. FISHES

BERG, L. S. 1940. *Classification of Fishes, Both Recent and Fossil.* Moscow (with English translation). Also American edition, 1947.

BRIDGE, T. W., and BOULANGER, G. A. 1904. 'Fishes.' *Cambridge Natural History*, **7**. London.

BROWN, M. E. 1957. *The Physiology of Fishes.* 2 vols. New York.

BURGESS, G. H. O. 1956. 'Absence of Keratin in Teleost Epidermis.' *Nature, Lond.* **178**, 93.

BURNSTOCK, G. 1958. 'The Effect of Drugs on the Spontaneous Motility and on Response to Stimulation of the Extrinsic Nerves of the Gut of a Teleostean Fish.' *Brit. J. Pharmacol.* **13**, 216.

DEAN, B. 1895. *Fishes, Living and Fossil.* New York.

GARDINER, B. G. 1960. 'A Revision of Certain Actinopterygian and Coelacanth Fishes, chiefly from the Lower Lias.' *Bull. Brit. Mus.* (*nat. Hist.*), *Geol.* **4**, 7.

GRASSÉ, P. P. 1958. *Traité de zoologie.* Tome XIII. Paris.

GRAY, J. 1933–36. 'Studies in Animal Locomotion.' *Proc. roy. Soc.* B, **113**, and *J. exp. Biol.* **10**, 386 and **13**, 170.

—— 1953. *How Animals Move.* Cambridge.

HARRIS, J. E. 1936–8. 'The Role of the Fins in the Equilibrium of the Swimming Fish.' *J. exp. Biol.* **13**, 476 and **15**, 32.

—— 1938. 'The Dorsal Spine of *Cladoselache*.' *Sci. Publ. Cleveland Mus. nat. Hist.* **8**, 1.

—— 1950. '*Diademodus hydei* a New Fossil Shark from the Cleveland Shale.' *Proc. zool. Soc.* **120**, 683.

JARVIK, E. 1950. 'On Some Osteolepiform Crossopterygians from the Upper Old Red Sandstone of Scotland.' *K. svenska VetenskAkad. Handl.* (4), **2**, 1.

Jones, I. C. 1960. 'Hormones in Fishes.' *Symp. zool. Soc. Lond.*, **1**, 1.
Jones, J. W. 1959. *The Salmon.* New Nat. Series. Special volume **16**. London.
Jordan, D. S. 1905. *A Guide to the Study of Fishes.* London.
Marshall, N. B. 1960. 'Swimbladder Structure of Deep Sea Fishes in Relation to their Systematics and Biology.' *'Discovery' Rep.* **31**, 1.
Millot, J., and Anthony J. 1958. *Anatomie de Latimeria chalumnae.* Part I. Paris.
Moy-Thomas, J. A. 1939. *Palaeozoic Fishes.* London.
Norman, J. R. 1931. *A History of Fishes.* London.
Olsson, R. 1958. 'A Bucco-hypophyseal Canal in *Elops saurus*.' *Nature, Lond.* **182**, 1745.
Regan, C. T. 1932. *Guide to the British Freshwater Fishes.* 2nd edition. Brit. Mus. (Nat. Hist.). London.
Romer, A. S. 1946. 'The Early Evolution of Fishes.' *Quart. Rev. Biol.* **21**, 33.
Schaeffer, B. 1952. 'The Triassic Coelacanth Fish *Diplurus*, with Observations on the Evolution of the Coelacanthini.' *Bull. Amer. Mus. nat. Hist.* **99**, 31.
Smith, J. L. B. 1940. 'A Living Coelacanthid Fish from South Africa.' *Trans. Roy. Soc. South Africa*, **28**, 1.
Watson, D. M. S. 1959. 'The Myotomes of Acanthodians.' *Proc. roy. Soc.* B, **151**, 23.
Westoll, T. S. 1944. 'The Haplolepidae—A study in Taxonomy and Evolution.' *Bull. Amer. Mus. nat. Hist.* **83**, 1.
—— 1945. 'The Paired Fins of Placoderms.' *Trans. roy. Soc. Edinb.* **61**, 381.
—— 1949. 'On the Evolution of the Dipnoi.' In *Genetics, Palaeontology and Evolution.* Edited by G. L. Jepson and others. Princeton.

CHAPTER XI. FISHERIES

Davis, F. M. 1937. 'An Account of the Fishing Gear of England and Wales.' *Fish. Invest., Lond.* (Ser. 2), **15**.
Graham, M. 1943. *The Fish Gate.* London.
—— and others. 1956. *Sea Fisheries.* London.
Gross, F. 1949. 'Further Observations on Fish Growth in a Fertilized Sea Loch.' *J. Mar. biol. Assoc. U.K.* **28**, 1.
Hardy, A. 1956–9. *The Open Sea.* 2 vols. London.
Hickling, C. F. 1935. *The Hake.* London.
Ommanney, F. D. 1949. *The Ocean.* Oxford.
Russell, E. S. 1942. *The Overfishing Problem.* Cambridge.
Thompson, W. F. 1935. 'Conservation of the Pacific Halibut.' *Ann. Rep. Smithsonian Inst.* 361.

CHAPTERS XII, XIII. AMPHIBIA

Bellairs, A. d'A., and Boyd, J. D. 1950. 'Jacobson's Organ.' *Proc. zool. Soc. Lond.* **120**, 269.
Case, E. C. 1946. 'A Census of Determinable Genera of Stegocephalia.' *Trans. Amer. phil. Soc.* **35**, 325.
Ecker, A., and Wiedersheim, R. 1896–1904. *Anatomie des Frosches.* Based on Gaupp. 3 vols. (in 2). Brunswick.
Evans, F. G. 1946. 'Anatomy and Function of the Foreleg in Salamander Locomotion.' *Anat. Rec.* **95**, 257.
Foxon, G. E. H. 1955. 'Problems of the Double Circulation in Vertebrates.' *Biol. Rev.* **30**, 196.

FRANCIS, E. T. N. 1934. *The Anatomy of the Salamander.* London.
GADOW, H. 1909. 'Amphibia and Reptiles.' *Cambridge Natural History.* London.
GRAY, J., and LISSMAN, H. W. 1946. 'The Co-ordination of Limb Movements in the Amphibia.' *J. exp. Biol.* **23**, 133.
GREGORY, W. K., and RAVEN, H. C. 1941. 'The Origin and Early Evolution of Paired Fins and Limbs.' *Ann. N.Y. Acad. Sci.* **42**, 273.
JARVIK, E. 1955. 'The Oldest Tetrapods and their Forerunners.' *Sci. Mon., N.Y.* **80**, 141.
MARSHALL, A. M. 1920. *The Frog.* 11th edition. Edited by F. W. Gamble. London.
NOBLE, G. K. 1931. *Biology of the Amphibia.* Paperback edition. 1959. London.
ROMER, A. S. 1947. 'A Review of Labyrinthodonts.' *Bull. Mus. comp. Zool. Harv.* **99**, 368.
—— 1955. 'Herpetichthyes, Amphibioidei, Choanichthyes or Sarcopterygii?' *Nature, Lond.* **176**, 126.
—— 1956. 'The Early Evolution of Land Vertebrates.' *Proc. Amer. phil. Soc.* **100**, 157.
SMITH, M. 1954. *The British Amphibians and Reptiles.* Collins. London.
SWINTON, W. E. 1958. *Fossil Amphibians and Reptiles.* Brit. Mus. (Nat. Hist.). 2nd edition. London.
TUMARKIN, A. 1955. 'On the Evolution of the Auditory Conducting Apparatus.' *Evolution,* **9**, 221.
WATSON, D. M. S. 1919 and 1925. 'Evolution and Origin of Amphibia.' *Phil. Trans.* B, **209**, 1 and **214**, 189.
—— 1939. 'The Origin of Frogs.' *Trans. roy. Soc. Edinb.* **60**, 195.
WESTOLL, T. S. 1943. 'The Origin of Tetrapods.' *Biol. Rev.* **18**, 78 and *Proc. roy. Soc.* B, **131**, 373.
WHITING, H. P. 1961. 'Pelvic Girdle in Amphibian Locomotion.' *Symp. zool. Soc. Lond.* **5**, 43.

CHAPTERS XIV, XV. REPTILES

BELLAIRS, A. D'A. 1957. *Reptiles.* London.
DITMARS, R. L. 1936. *The Reptiles of North America.* New York.
FOXON, G. E. H., GRIFFITH, J., and PRICE, MYFANWY. 1956. 'The Mode of Action of the Heart of the Green Lizard *Lacerta viridis.*' *Proc. zool. Soc. Lond.* **126**, 145.
GADOW, H. 1909. 'Amphibia and Reptiles.' *Cambridge Natural History.* London.
OLIVER, J. A. 1955. *The Natural History of North American Amphibians and Reptiles.* Princeton.
OWEN, R. 1849–84. *A History of British Fossil Reptiles.* 4 vols. London.
POPE, C. H. 1956. *The Reptile World.* London.
ROMER, A. S. 1956. *Osteology of Reptiles.* Chicago.
SWINTON, W. E. 1934. *The Dinosaurs.* London.
WATSON, D. M. S. 1918. 'Seymouria.' *Proc. zool. Soc. Lond.,* **1918**, 267.
WHITE, T. E. 1939. 'The Osteology of *Seymouria.*' *Bull. Mus. comp. Zool. Harv.* **85**, 323.
WILLISTON, S. W. 1925. *The Osteology of the Reptiles.* Cambridge, Mass.

CHAPTERS XVI–XVIII. BIRDS

AYMAR, G. C. 1936. *Bird Flight.* London.
BEDDARD, F. E. 1898. *The Structure and Classification of Birds.* London.
BERGER, A. J. 1961. *Bird Study.* New York.

BRADLEY, O. C. 1950. *The Structure of the Fowl.* 3rd edition. London.
BROWN, R. H. J. 1948. 'The Flight of Birds.' *J. exp. biol.* **25**, 322.
DE BEER, G. R. 1954. *Archaeopteryx lithographica.* Brit. Mus. (Nat. Hist.). London.
—— 1956. 'Evolution of Ratites.' *Bull. Brit. Mus. (nat. Hist.), Zool.* **4**, 2.
EVANS, A.H. 1900. 'Birds.' *Cambridge Natural History,* **9**. London.
FISHER, J. 1939. *Birds as Animals.* London.
FISHER, J., and LOCKLEY, R. M. 1954. *Sea Birds.* London.
HEILMANN, G. 1926. *The Origin of Birds.* London.
HINDE, R. A., and TINBERGEN, N. 1958. 'The Comparative Study of Species-specific Behaviour.' In *Behaviour and Evolution.* Ed. A. Roe and G. G. Simpson. Yale.
HORTON-SMITH, C. 1926. *The Flight of Birds.* London.
HUXLEY, J. S. 1914. 'Courtship of the Great Crested Grebe.' *Proc. zool. Soc. Lond.* **1914**, 491.
LACK, D. 1947. *Darwin's Finches.* Cambridge.
LAMBRECHT, K. 1933. *Handbuch der Palaeornithologie.* Berlin.
LILLIE, F. R., and JUHN, M. 1932 and 1938. 'The Physiology of the Development of Feathers.' *Physiol. Zoöl.* **5**, 124 and **11**, 434.
MARSHALL, A. J. 1960–61. *Biology and Comparative Physiology of Birds.* 2 vols. London.
PETERS, J. L. 1931 (onwards). *Check List of Birds of the World.* Harvard.
STORER, J. H. 1948. *The Flight of Birds.* Michigan.
STRESEMAN, E. 1934. In Kükenthal, *Handbuch der Zoologie,* **7**, 2. Leipzig.
SWINTON, W. E. 1958. *Fossil Birds.* Brit. Mus. (Nat. Hist.). London.
THOMSON, J. A. 1923. *The Biology of Birds.* London.
TINBERGEN, N. 1948. 'Social Releasers.' *Wilson Bull.* **60**, 6.
TYNE, J. VAN, and BERGER, A. J. 1959. *Fundamentals of Ornithology.* New York.
WETMORE, A. 1930. 'A Systematic Classification for the Birds of the World.' *Proc. U.S. Nat. Mus.* **76**, 1.
WOLFSON, A. Ed. 1955. *Recent Studies in Avian Biology.* Urbana.

CHAPTERS XIX–XXXI. MAMMALS

BEDDARD, F. E. 1909. 'Mammalia.' *Cambridge Natural History.* London.
—— 1902. *A Text-book of Zoogeography.* Cambridge.
BENSLEY, R. A. 1948. *Anatomy of the Rabbit.* 8th edition. Toronto.
BOHLKEN, H. 1960. 'Remarks on the Stomach and the Systematic Position of Tylopoda.' *Proc. zool. Soc. Lond.* **134**, 207.
BRADLEY, O. C. 1946–7. *Topographical Anatomy of the Horse.* 2nd edition. 3 vols. Edinburgh.
—— 1959. *Topographical Anatomy of the Dog.* 6th edition. Edinburgh.
BRONN, H. G. 1859 (onwards). *Die Klassen und Ordnungen des Thier-Reichs.* vols. i–v. *Mammalia.* Leipzig.
BROOM, R. 1932. *The Mammal-like Reptiles of South Africa.* London.
BURRELL, H. 1927. *The Platypus.* Sydney.
CLARK, W. E. LE GROS. 1934. *Early Forerunners of Man.* London.
—— 1954. *The Fossil Evidence for Human Evolution.* Chicago.
—— 1955. 'The Os Innominatum of the Recent Ponginae with Special Reference to that of the Australopithecinae.' *Amer. J. phys. Anthrop.* N.S. **13**, 19.
—— 1959. *The Antecedents of Man.* Edinburgh.
—— 1960. *History of the Primates.* 6th edition. Brit. Mus. (Nat. Hist.). London.

CROMPTON, A. W. 1955. 'A Possible Explanation for the Origin of the Mammalian Brain and Skull.' *S. Afr. J. Sci.* **52,** 130.

CUNNINGHAM, D. J. 1951. *Text-book of Anatomy.* 9th edition. London.

DAVISON, A. 1937. *Mammalian Anatomy, with Special Reference to the Cat.* Philadelphia.

FLOWER, W. H. 1885. *Introduction to the Osteology of the Mammalia.* London.

—— and LYDEKKER, R. 1891. *An Introduction to the Study of Mammals, Living and Extinct.* London.

FRASER, F. C., and PURVES, P. E. 1959. 'Hearing in Whales.' *Endeavour,* **18,** 93.

GAVAN, J. A. (editor). 1955. *The Non-human Primates and Human Evolution.* Detroit.

GERHARDT, U. 1909. *Das Kaninchen.* Leipzig.

GRAY, H. 1958. *Gray's Anatomy.* 32nd edition. London.

GREENE, E. C. 1935. 'Anatomy of the Rat.' *Trans. Amer. phil. Soc.* **27.** (Also re-issued, 1959, New York.)

GREGORY, W. K. 1922. *The Origin and Evolution of the Human Dentition.* Baltimore.

—— 1934. 'A Half Century of Trituberculy.' *Proc. Amer. phil. Soc.* **73,** 169.

GRIFFIN, D. R. 1958. *Listening in the Dark.* Newhaven, Conn.

—— 1960. 'Bats Feeding.' *Anim. Behav.* **8.**

HAMILTON, W. J. 1939. *American Mammals.* New York.

HARTMAN, C. G., and STRAUS, W. L. 1933. *The Anatomy of the Rhesus Monkey.* London.

HENDERSON, J., and CRAIG, E. C. 1932. *Economic Mammalogy.* London.

HILL, W. C. O. 1953 onwards. *The Primates.* Edinburgh.

HOTTON, N. 1959. 'The Pelycosaur Tympanum.' *Evolution,* **13,** 99.

HOWELL, A. B. 1930. *Aquatic Mammals.* Springfield.

KERMACK, K. S., and MUSSETT, F. 1958. 'The Jaw Articulation of the Docodonta and the Classification of Mesozoic Mammals.' *Proc. roy. Soc.* B, **148,** 204.

LEAKEY, L. S. B. 1959. 'A New Fossil Skull from Olduvai.' *Nature, Lond.* **184,** 491.

LYDEKKER, R. 1896. *A Geographical History of Mammals.* Cambridge.

OSBORN, H. F. 1929. *Titanotheres.* Washington.

PARRINGTON, F. R. 'Cranial Anatomy of some Gorgonopsids and the Synapsid Middle Ear.' *Proc. zool. Soc. Lond.* **125.**

PYE, J. D. 1960. 'A Theory of Echolocation by Bats.' *J. Laryng.* **74,** 718.

REEVE, E. C. R. 1940. 'Relative Growth in the Snout of Anteaters.' *Proc. zool. Soc. Lond.* **110,** 47.

REIGHARD, J. E., and JENNINGS, H. S. 1944. *Anatomy of the Cat.* 3rd edition. New York.

SCOTT, W. B. 1913. *A History of Land Mammals in the Western Hemisphere.* New York.

SIMPSON, G. G. 1928. *A Catalogue of Mesozoic Mammalia in the British Museum (Natural History).* Brit. Mus. (Nat. Hist.). London.

—— 1929. 'American Mesozoic Mammalia.' *Mem. Peabody Mus. Yale,* Pt. 1. New Haven and London.

—— 1945. 'The Principles of Classification and a Classification of the Mammals.' *Bull. Amer. Mus. Nat. Hist.* **85.**

—— 1959. 'Mesozoic Mammals and the Polyphyletic Origin of Mammals.' *Evolution,* **13,** 405.

SISSON, S., and GROSSMAN, J. D. 1938. *The Anatomy of the Domestic Animals.* 3rd edition. Philadelphia and London.

WATSON, D. M. S., and ROMER, A. S. 1956. 'A Classification of Therapsid Reptiles.' *Bull. Mus. comp. Zool. Harv.* **114**, 38.

WEBER, M. 1927–8. *Die Säugetiere.* 2 Auflage. Unter Mitwirkung von O. Abel und H. M. de Burlet. 2 vols. Jena.

WOOD JONES, F. 1916. *Arboreal Man.* London.

—— 1929. *Man's Place among the Mammals.* London.

—— 1941. *The Principles of Anatomy as seen in the Hand.* 2nd edition. London.

—— 1949. *Structure and Function as seen in the Foot.* 2nd edition. London.

—— 1948. *Hallmarks of Mankind.* London.

INDEX

Major topics are shown in **Bold Type**, genera in *italic*, and authors' names in CAPITALS and SMALL CAPITALS